QUANTUM MECHANICS

D. I. BLOKHINTSEV

QUANTUM MECHANICS

D. REIDEL PUBLISHING COMPANY
A MEMBER OF THE KLUWER ACADEMIC PUBLISHERS GROUP
DORDRECHT / BOSTON / LANCASTER / TOKYO

OSNOVY KVANTOVOĬ MEKHANIKI
Gosudarstvennoe izdatel'stvo tekhniko-teoreticheskoĭ literatury
Moskva-Leningrad, 1944
4. izd., Gosudarstvennoe izdatel'stvo Vysshaya Shkola, Moskva, 1963
Translated from the third and fourth Russian editons by J. B. Sykes and M. J. Kearsley

2-0487-300 ts

CONTENTS

PREFACE TO THE SECOND RUSSIAN EDITION

The second edition of *Osnovy kvantovoĭ mekhaniki*, like the first (published in 1944 under the title *Vvedenie v kvantovuyu mekhaniku* [Introduction to quantum mechanics]), is essentially a series of lectures on quantum mechanics given by the author for a number of years in the Department of Physics at the Lomonosov Moscow State University.

The inevitable changes in these lectures have led me to make a number of corrections and additions in the second edition. The chapter concerning the concept of states in quantum mechanics and the uncertainty relation has been considerably altered and clarified. The new edition includes also a treatment of methodological problems in quantum mechanics, and a criticism of idealistic views on quantum theory which are now widely held in other countries. Some additions have also been necessitated by the further development of applications of quantum mechanics in recent years.

In this book, as in the first edition, I have striven to provide the student beginning quantum mechanics with a correct understanding of its physical basis and mathematical formalism, and to indicate the value of the subject by means of some important applications.

The improvement of this book has been greatly assisted by many useful comments from my colleagues; I am very grateful to them, and especially to S. I. Drabkina, M. A. Markov, A. A. Sokolov, S. G. Suvorov and E. L. Feĭnberg. The writing of the last section of the book was considerably helped by discussions at the philosophy seminar of Moscow State University and with theoreticians at the USSR Academy of Sciences' Institute of Physics.

I am also obliged to physics students at Moscow State University who have helped to remove misprints and other errors in the first edition.

PREFACE TO THE ENGLISH EDITION

The English translation of *Osnovy kvantovoĭ mekhaniki* has been made from the third and fourth Russian editions. These contained a number of important additions and changes as compared with the first two editions. The main additions concern collision theory, and applications of quantum mechanics to the theory of the atomic nucleus and to the theory of elementary particles. The development of these branches in recent years, resulting from the very rapid progress made in nuclear physics, has been so great that such additions need scarcely be defended. Some additions relating to methods have also been made, for example concerning the quasiclassical approximation, the theory of the Clebsch-Gordan coefficients and several other matters with which the modern physicist needs to be acquainted.

The alterations that have been made involve not only the elimination of obviously out-of-date material but also the refinement of various formulations and statements. For these refinements I am indebted to many persons who at different times have expressed to me their critical comments and suggestions.

Particularly important changes have been made regarding the definition of a quantum ensemble in Section 14.

The underlying idea and spirit of the book remain as in the first two editions: to provide the student beginning quantum mechanics with a correct understanding of its physical basis and mathematical formalism, and to indicate by simple examples the ways in which it can be applied in various branches of atomic physics: the theory of the solid state, atomic and molecular physics, optics, magnetism, the theory of the atomic nucleus, and so on.

I have also attached great importance to the use of correct methods; without a mastery of methods, even the loftiest intellect betrays some touch of the labourer. In consequence, the materialistic methodology, explicitly or implicitly, pervades the whole of the book.

In recent years this book has been published in many countries, and I am glad that it has helped in the diffusion of knowledge of and interest in modern atomic physics among many nations.

I am now deeply indebted to Mr. A. Reidel, the publisher of the English translation, and to Dr. J. B. Sykes and Dr. M. J. Kearsley, the translators, for making this book accessible to a much wider public.

Finally, I am grateful also to those who helped me improve this book, and to my colleagues and students, in particular M. A. Markov, who read the revised manuscript and made a number of useful suggestions and comments, and S. I. Drabkina for her enthusiastic help in preparing the corrections and additions. I am also obliged to the staff of the 'Vysshaya shkola' publishing house, who gave much assistance in connection with the publication of the book in the original Russian.

D. I. BLOKHINTSEV

INTRODUCTION

In recent decades the science of atomic phenomena has not only formed one of the most important branches of modern physics but also found many practical applications. Even the most superficial examination of the field of atomic phenomena reveals features considerably differing from those of the macrouniverse.

The first novel aspect of the microuniverse is its atomism. The elementary particles have entirely definite properties of charge, mass, etc., which are the same for all particles of a given kind. No such atomism occurs in the macrouniverse. Macroscopic objects are assemblies of large numbers of elementary particles, and the laws of macroscopic phenomena are those appropriate to such assemblies.

This shows that it would be incorrect to regard microparticles as being analogous to macroscopic bodies. Even the point mass of classical mechanics is an abstract idealised picture not of a microparticle but of a macroscopic body whose dimensions are small compared with the distances occurring in a given problem.

The atomism of the microuniverse is not restricted to the definiteness of the properties of the microparticles; it also leads to the existence of an absolute measure of mechanical motion, namely Planck's constant $\hbar = 1.05 \times 10^{-27}$ erg sec. This is of prime importance in the mechanics of microparticles. Physicists were for long unaware that quantitative changes can become qualitative ones and attempted to understand atomic phenomena on the basis of classical macroscopic theories. The discovery of Planck's constant was the first real indication of the invalidity of mechanically applying large-scale laws to small-scale objects.

In the 1920's further experimental facts were discovered which finally forced the abandonment of this approach. It was shown that electrons possess wave properties: if a beam of electrons is passed through a crystal they are distributed on a screen in the same manner as the intensity of waves of an appropriate wavelength. This is the diffraction of microparticles, a phenomenon unknown to classical mechanics. Later it was shown that not only electrons but all microparticles exhibit this behaviour. In this way a fundamentally new and completely general law was revealed.

The motion of microparticles was found to be in many respects more akin to the motion of waves than to that of point masses along paths. The phenomenon of diffraction is incompatible with the supposition that the particles move in paths. Hence the principles of classical mechanics, where the concept of the path is fundamental, cannot be used to examine the motion of microparticles.

The word 'particle' itself, when applied to individual entities of the microuniverse, creates the idea of an analogy with the point masses of classical mechanics much closer than that which actually exists. This should be borne in mind whenever the word 'particle' is used in this book, for brevity, in place of 'microparticle'.

Classical mechanics is only a certain approximation suitable for the discussion of the motion of bodies of large mass moving in fields which vary sufficiently smoothly (macroscopic fields). Under these conditions Planck's constant is not significant, and may be regarded as negligibly small. Diffraction phenomena also are unimportant. In the small-scale microuniverse classical mechanics is replaced by quantum mechanics. Thus the object of study in quantum mechanics is the motion of microparticles.

Quantum mechanics is a statistical theory. For example, it can be used to predict the mean distribution over a photographic plate of electrons reflected from a crystal, but only a probability can be derived regarding the point of incidence of each individual electron, in the form of a statement that it will appear in a given place with a given probability.

A similar situation occurs in statistical mechanics, but there is a profound difference between quantum mechanics and classical statistical mechanics. The latter is based on Newtonian mechanics, which in principle allows the history of each particle to be traced. Modern quantum mechanics, by contrast, is not based on any theory of individual microprocesses. It deals with the individual properties of microparticles and individual microprocesses by working with statistical ensembles. These are defined by properties taken over from classical macroscopic physics, such as momentum, energy and co-ordinate. When, therefore, the reproducibility of a microphenomenon is discussed in quantum mechanics (e.g. the repetition of a given experiment), this refers to the reproducing of the *macroscopic* conditions for the *microscopic* phenomenon, i.e. the establishment of the same statistical ensemble.

Thus quantum mechanics considers statistical ensembles of microparticles in their relation to macroscopic measuring apparatus with which the 'state of the particles' can be determined, i.e. the statistical ensemble can be specified.

Within the scope defined by the foregoing formulation, quantum mechanics is a great advance in the development of twentieth-century atomic physics – which has, indeed, outstepped the bounds of physics and entered the realm of the industrial arts.

CHAPTER I

FOUNDATIONS OF QUANTUM THEORY

1. Energy and momentum of light quanta

The development of quantum mechanics was preceded by that of the quantum theory of light. At the end of the last century it seemed that in the contest between the corpuscular and wave theories of the nature of light the wave theory had finally triumphed in the form due to Maxwell. The experiments of Hertz with electromagnetic waves, the proof by Lebedev of the existence of radiation pressure, and other facts revealed by the experimenter's skill seemed to prove irrefutably the correctness of Maxwell's theory.

Yet the victory of the electromagnetic theory of light was incomplete. Although all problems relating to the propagation of light were successfully solved by the wave theory, a number of important phenomena relating to the emission and absorption of light refused to be accommodated by the wave treatment. For instance, despite all attempts by theoreticians, the energy spectrum of a black body derived on the basis of the wave theory not only was in flat contradiction with experiment but also involved internal inconsistencies.

In 1901 Planck formulated for radiation from a black body in thermal equilibrium an energy spectrum which was in agreement with experiment. This was the starting point of the development of the quantum theory. It was based on the assumption that the emission and absorption of light by matter is discontinuous, and that light is emitted and absorbed in finite amounts called *light quanta*.

The energy ε of such a quantum is proportional to the frequency ω of the light oscillations, and is given by the formula

$$\varepsilon = \hbar\omega. \tag{1.1}$$

Here $\hbar = 1.05 \times 10^{-27}$ erg sec is the well-known Planck's constant.[1]

The concept of light quanta received its final form after Einstein had shown the necessity of assigning to the quantum, besides the energy ε, a momentum $p = \varepsilon/c$, whose direction is that in which the light is propagated. In terms of the wave vector $\mathbf{k}$, whose components are

$$k_x = (2\pi/\lambda)\cos\alpha\,, \qquad k_y = (2\pi/\lambda)\cos\beta\,, \qquad k_z = (2\pi/\lambda)\cos\gamma\,,$$

[1] In the older literature 'Planck's constant' was usually denoted by h, which signified a quantity 2π times greater, i.e. 6.62×10^{-27} erg sec, and the frequency $\nu = 1/T$ was used instead of the angular frequency $\omega = 2\pi/T$ (T being the period of oscillation).

where λ is the wavelength and $\cos\alpha$, $\cos\beta$, $\cos\gamma$ the direction cosines of the normal to the light wave, the formula for the momentum of the light quantum may be written in the vector form

$$\mathbf{p} = \hbar\mathbf{k}. \tag{1.2}$$

Formulae (1.1) and (1.2) are the fundamental equations of the quantum theory of light, and relate the energy ε and momentum $\mathbf{p}$ of the light quantum to the frequency ω and wavelength λ of a monochromatic plane wave whose direction of propagation is given by the vector $\mathbf{k}$.[2]

The deeper significance of the quantum theory of light is not that we imagine light as a gas consisting of particles with energy $\hbar\omega$ and momentum $\hbar\mathbf{k}$ (which is a useful concept for visualisation, but incomplete); it is that the exchange of energy and momentum between microsystems (electrons, atoms, molecules, etc.) and light occurs by the creation and annihilation of light quanta.

This view is more exactly expressed by the application of the law of conservation of energy and momentum to any system interacting with light (or, more precisely, with any electromagnetic radiation). For clarity we shall use instead of *interaction* the more vivid term '*collision*'.

Let E and $\mathbf{P}$ be the energy and momentum of the system before the 'collision' with a light quantum, and E' and $\mathbf{P}'$ its energy and momentum after the 'collision'; $\hbar\omega$ and $\hbar\mathbf{k}$ the energy and momentum of the light quantum before the 'collision', and $\hbar\omega'$ and $\hbar\mathbf{k}'$ its energy and momentum after the 'collision'. With this notation the law of conservation of energy and momentum is

$$\hbar\omega + E = \hbar\omega' + E', \tag{1.3}$$

$$\hbar\mathbf{k} + \mathbf{P} = \hbar\mathbf{k}' + \mathbf{P}'. \tag{1.4}$$

These equations include all three of the fundamental processes: absorption, emission and scattering of light.

If $\omega' = 0$ (and so $\mathbf{k}' = 0$), Equations (1.3) and (1.4) relate to the absorption of a light quantum $\hbar\omega$; if $\omega = 0$ (and $\mathbf{k} = 0$), these equations define the emission of a quantum $\hbar\omega$.

If ω and ω' are not zero, the equations describe the scattering of light, when a quantum $\hbar\omega$, $\hbar\mathbf{k}$ is transformed into a quantum with a different energy $\hbar\omega'$ and a different momentum $\hbar\mathbf{k}'$.

The law of conservation of energy and momentum in the form (1.3) and (1.4) contradicts both the wave and the corpuscular concepts of light and cannot be interpreted within the framework of classical physics.

According to the wave theory, the energy of a wave field is determined not by the frequency ω of the waves but by the amplitudes of the waves which form the field. There is no general relation between the wave amplitude and the oscillation frequency

[2] Formulae (1.1) and (1.2) are assumed valid for any frequency ω; they are equally valid for visible light and for γ radiation. For this reason, instead of 'quantum of light', 'quantum of γ radiation', etc., the expression *photon* is used for brevity.

which would allow the energy of a single quantum to be related to the wave amplitude. Let us suppose that a beam of light encounters a transparent plate. Some light is reflected and some transmitted by the plate. The wave theory indicates that the amplitudes of the incident, transmitted and reflected waves should be different. If we now relate the quantum energies ε to the wave amplitudes in any way, we must conclude that the quantum energies in the three beams are different. But Equation (1.1) shows that the quantum energy cannot be changed without changing the frequency: some of the quanta are always 'coloured' differently from the initial quanta.

Our assumption that the quantum energy can be determined by the amplitude therefore leads to the result that the colour of the incident, reflected and transmitted beams would have to be different, which of course does not occur when light passes through a transparent body.

The assumption that a light quantum is a particle located somewhere in space, as it were 'floating' on the wave, is also invalid. A light quantum, by definition (Equations (1.1) and (1.2)) is associated with a monochromatic plane wave. Such a wave is a purely periodic process, infinite in both space and time. The assumption that the quantum is localised is in contradiction with the complete periodicity of the wave: a sinusoidal wave deformed in any way is no longer a single sinusoidal wave but an assembly of different sinusoidal waves.

Thus, if we accept the conservation laws (1.3) and (1.4), we must agree that classical ideas are inadequate to express phenomena which occur on the atomic scale. Light is of a twofold nature and has both wave and corpuscular properties.

The modern quantum theory of the electromagnetic field allows both these aspects to be taken into account, but a discussion of it lies outside the scope of this book, which is concerned with the non-relativistic mechanics of microparticles.

2. Experimental test of the laws of conservation of energy and momentum for light quanta

It was shown by Einstein that the conservation law (1.3) makes it possible to interpret certain features of the photoelectric effect which are puzzling from the classical point of view. This effect consists essentially in the emission of electrons by metals under the action of light incident on their surfaces.[3]

The observed properties have no classical interpretation. Experiment shows that the velocity of the photoelectrons depends solely on the frequency ω of the light (for a given metal) and not at all on the intensity of the incident light. The latter determines only the number of electrons emitted by the metal in unit time.

However ingenious a model is devised for this phenomenon, according to Newton's law the increase in the electron velocity must be proportional to the force acting. The latter is equal to the product of the charge e on the electron and the field $\mathbf{E}$ of the light wave (the effect of the magnetic field of the wave can be neglected). Thus the velocity acquired by the electron must be proportional to E, and the energy must be proportional to E^2, i.e. to the intensity of the light, which is not what is observed. Ioffe and

[3] The laws of the photoelectric effect were originally investigated by Stoletov, Hallwachs, Righi and others.

Dobronravov (see [83]) have shown that at low intensities a photoelectric effect can still be observed, and it is found that electrons are emitted by the metal in accordance with the laws of statistics, so that only the mean number of electrons is proportional to the intensity of the incident beam. Particularly important results were obtained in experiments by Millikan, who showed rigorously that the energy of electrons emitted in the photoelectric effect is entirely determined by the frequency of the light and not by its intensity.

The reason for this result becomes evident if we apply the law of conservation of energy (1.3) to the photoelectric effect. Let us assume that monochromatic light of frequency ω is incident on the surface of a metal. Since a certain amount of work has to be done to extract electrons from the metal (this work is called the work function and will be denoted by χ), the initial energy of the electron in the metal must be taken as $-\chi$. In the photoelectric effect the light quantum is completely absorbed, i.e. $\hbar\omega' = 0$. The energy E of the electron after absorption of the light quantum is $\frac{1}{2}m_0v^2$, where m_0 is the mass of the electron and v its velocity on leaving the metal. Equation (1.3) in this case therefore becomes[4]

$$\hbar\omega - \chi = \tfrac{1}{2}m_0v^2. \tag{2.1}$$

This is Einstein's well-known equation for the photoelectric effect.

According to this equation the energy $\frac{1}{2}m_0v^2$ of the photoelectron increases linearly with the frequency ω of the light. If we measure the energy of the electron by means of a retarding potential V, so that $eV = \frac{1}{2}m_0v^2$ (as Millikan did), the slope of the straight line on the graph of V against ω will be determined by the quantity $\hbar/e$. If we know the charge e and determine the slope experimentally, we can find $\hbar$. Millikan showed that the value of $\hbar$ obtained is the same as from the theory of black-body radiation. This proves the validity of Equation (1.3) for the photoelectric effect.

Einstein's equation is now one of the fundamental equations underlying the theory of applied electronics.[5]

Equations (1.3) and (1.4) have been experimentally confirmed by Compton, who studied the frequency of scattered X-rays as a function of the angle of scattering. The scattering substances which he used were those in which electrons are only weakly bound to the atoms, such as paraffin wax and graphite. Since the quantum energy of X-rays is high, the energy of the electron in the atom can be ignored in the calculation (at least for electrons in the outer shells) and we can regard the electrons as free particles at rest. Accordingly the initial energy E and momentum P of the electron will be taken to be zero.

After collision with an X-ray quantum the energy of the electron may be very large, and so we shall use the formulae of relativity theory, taking into account the dependence of the mass of a particle on its velocity. According to relativity theory, the kinetic

[4] Equation (1.4) is of no significance here, as it simply states that the momentum of the light quantum is transferred to the block of metal as a whole.

[5] Details of Millikan's experiments are given in [66, 77, 83].

energy of an electron moving with velocity v is

$$E' = \frac{m_0 c^2}{\sqrt{(1 - v^2/c^2)}} - m_0 c^2 , \tag{2.2}$$

where m_0 is the rest mass and c the velocity of light; its momentum is

$$\mathbf{P}' = \frac{m_0 \mathbf{v}}{\sqrt{(1 - v^2/c^2)}} . \tag{2.3}$$

Substituting these values in (1.3) and (1.4) we have, since $E = 0$ and $P = 0$,

$$\hbar\omega = \hbar\omega' + m_0 c^2 \left[\frac{1}{\sqrt{(1 - \beta^2)}} - 1\right], \tag{2.4}$$

$$\hbar\mathbf{k} = \hbar\mathbf{k}' + \frac{m_0 \mathbf{v}}{\sqrt{(1 - \beta^2)}}, \qquad \beta = v/c . \tag{2.4'}$$

Here ω and $\mathbf{k}$ are the frequency and wave vector of the incident radiation, and ω' and $\mathbf{k}'$ those of the scattered radiation.

The first equation shows immediately that $\omega > \omega'$. Thus the scattered radiation must have a longer wavelength than the incident radiation. This result is confirmed by Compton's experiments, whereas according to the classical theory the frequency of the scattered light should be equal to that of the incident light (Rayleigh scattering).

One important conclusion follows from Equations (2.4) and (2.4′): a free electron cannot absorb light, but can only scatter it, since complete absorption would mean that $\omega' = 0$ (and $\mathbf{k}' = 0$). Then (2.4′) shows that $\mathbf{k}$ and $\mathbf{v}$ are in the same direction, and this formula can be written in the scalar form

$$\hbar k = m_0 v / \sqrt{(1 - \beta^2)} .$$

This together with Equation (2.4) gives for absorption

$$\frac{1}{\sqrt{(1 - \beta^2)}} - 1 = \frac{\beta}{\sqrt{(1 - \beta^2)}},$$

whence $\beta = 0$, and hence $k = 0$. This proves that absorption is impossible.

The photoelectric effect discussed above, in which the quantum is totally absorbed, is possible only because the electron is bound to the metal, so that work χ is needed to extract it, and momentum can be transferred to the metal.

In order to be able to test the equations (2.4), (2.4′), Compton had to determine from them how the frequency of scattered light ω' depends on the scattering angle θ. In Figure 1 the line OA represents the direction of propagation of the beam of primary X-rays. The direction OC is that in which the rays scattered by electrons are observed. The parallelogram in Figure 1 represents the momentum $\hbar\mathbf{k}$ of the incident quantum as the sum of the momentum of the scattered quantum $\hbar\mathbf{k}'$ and that of the electron $\mathbf{P}'$. The angle θ is the angle of scattering, and α is the angle between the momentum of the primary quantum and that of the electron after the impact, called the 'recoil electron'. To find the relation between the angle θ and the value of $\hbar\omega'$ for the scattered quantum we take components of Equation (2.4′) along two perpendicular axes, OA and OB. Since $|\mathbf{k}| = \omega/c$ and $|\mathbf{k}'| = \omega'/c$, we have

$$\frac{\hbar\omega}{c} = \frac{\hbar\omega'}{c}\cos\theta + \frac{m_0 v}{\sqrt{(1-\beta^2)}}\cos\alpha\,,$$

$$0 = \frac{\hbar\omega'}{c}\sin\theta - \frac{m_0 v}{\sqrt{(1-\beta^2)}}\sin\alpha\,.$$

A simple elimination of β and the angle α shows that

$$\omega - \omega' = (2\hbar/m_0 c^2)\,\omega\omega' \sin^2 \tfrac{1}{2}\theta\,.$$

If we now put $\omega = 2\pi c/\lambda$, $\omega' = 2\pi c/\lambda'$, we easily find the wavelength change

$$\Delta\lambda = (4\pi\hbar/m_0 c)\sin^2 \tfrac{1}{2}\theta\,. \tag{2.5}$$

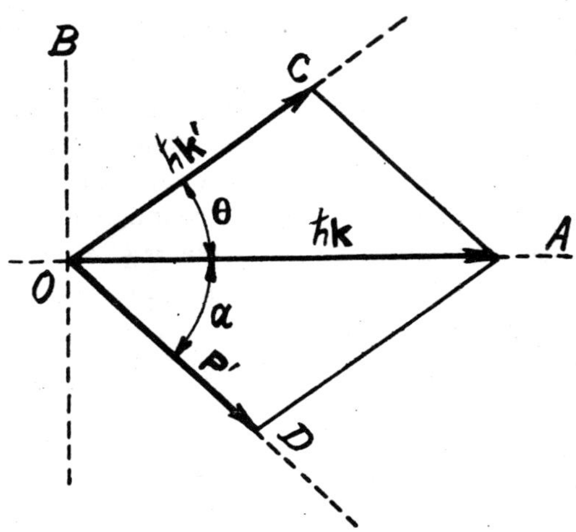

Fig. 1. The Compton parallelogram.

This formula was first derived by Compton. By changing the angle at which the scattered radiation was observed and measuring the experimental wavelength change $\Delta\lambda$, Compton and Woo compared their results with the theoretical predictions from Formula (2.5) and found complete agreement.[6]

Thus Compton's experiments demonstrate that the hypothesis of the existence of the momentum of a light quantum determined by Formula (1.2) is entirely correct.

It may be noted that in some cases cloud-chamber pictures make it possible to establish the direction of motion of the quantum scattered in the Compton effect, and also the path and energy of the recoil electron, and thus to 'see' the composition of the momenta of the electron and of the light quantum, as shown in Figure 1.

The length $\Lambda = \hbar/m_0 c = 3.9 \times 10^{-11}$ cm which occurs in Formula (2.5) is called

[6] Details of the experiments are given in [83]; [96], p. 155.

the *Compton wavelength*. This quantity is of fundamental importance in the relativistic theory of the electron, being one of the scales of the micro-universe. If we know $\Delta\lambda$ (2.5) we can find $\hbar$, so that the Compton effect gives another method of determining $\hbar$.

Phenomena in which the constant $\hbar$ plays a significant part are called *quantum phenomena*. Any such phenomenon can serve to determine $\hbar$.

As we should expect, a quantum phenomenon cannot be interpreted classically. According to the classical theory, which assumes continuity of the exchange of energy between the field and the microsystems, $\hbar = 0$ and no frequency shift should occur when light is scattered by a free electron, $\Delta\lambda$ being proportional to $\hbar$ by (2.5). A direct calculation by the classical theory leads to the same result. Under the action of the variable field of frequency ω, the electron executes a forced oscillation of the same frequency. Thus oscillations of the charge e with frequency ω occur. Such oscillations produce a variable field of the same frequency (since the field equations are linear), and so the scattered radiation has the same frequency as the incident radiation.

3. Atomism

In the microuniverse we encounter a number of what are usually called elementary particles. In recent years, as a result of the study of cosmic rays and experiments with

TABLE

ELEMENTARY PARTICLES*

Name	*Symbol*	*Mass* m	*Charge* e	*Spin* σ	*Isotopic spin* T	T_3	*Strangeness* S	*Lifetime* (*sec*)
Photon	γ	0	0	1	—	—	—	∞
Neutrino	$\nu(\tilde{\nu})$	0	0	$\frac{1}{2}$	—	—	—	∞
Electron	$e(\tilde{e})$	1	-1	$\frac{1}{2}$	—	—	—	∞
μ meson	$\mu(\tilde{\mu})$	206.7	-1	$\frac{1}{2}$	—	—	—	2.22×10^{-6}
π mesons	π^+	273.3	$+1$	0	1	$+1$	—	2.56×10^{-8}
	π^0	264.3	0	0	1	0	—	$\sim 1 \times 10^{-16}$
	π^-	272.8	-1	0	1	-1	—	2.56×10^{-8}
K mesons	$K^-(\tilde{K}^-)$	966	-1	0	$\frac{1}{2}$	$\frac{1}{2}$	1	$\sim 10^{-8}$
	$K^0(\tilde{K}^0)$	~ 961	0	0	$\frac{1}{2}$	$-\frac{1}{2}$	1	$\sim 10^{-10}$
Nucleons	$p(\tilde{p})$	1836.1	$+1$	$\frac{1}{2}$	$\frac{1}{2}$	$\frac{1}{2}$	0	∞
	$n(\tilde{n})$	1838.6	0	$\frac{1}{2}$	$\frac{1}{2}$	$-\frac{1}{2}$	0	1.04×10^3
Λ hyperon	$\Lambda_0(\tilde{\Lambda}_0)$	2181	0	$\frac{1}{2}$	0	0	-1	2.77×10^{-10}
Σ hyperons	Σ^+	2327	$+1$	$\frac{1}{2}$	1	1	-1	0.78×10^{-10}
	Σ^0	2340	0	$\frac{1}{2}$	1	0	-1	$< 10^{-10}$
	$\Sigma^-(\tilde{\Sigma}^-)$	2340	-1	$\frac{1}{2}$	1	-1	-1	1.58×10^{-10}
Cascade hyperon	Ξ^-	2585	-1	$\frac{1}{2}$	$\frac{1}{2}$	$-\frac{1}{2}$	-2	$\sim 10^{-10}$
	Ξ^0	2585	0	$\frac{1}{2}$	$\frac{1}{2}$	$\frac{1}{2}$	-2	

* Only particles are given in this table; antiparticles are denoted by the tilde ($\sim$). The $\tilde{\Sigma}^-$ was discovered at Dubna.

artificially accelerated particles, the list of elementary particles has grown considerably longer. The table gives the main properties of these particles.[7]

The mass, charge and other properties of all the elementary particles of a given kind are completely identical and invariable. The only changes of elementary particles which are known with certainty in present-day physics consist in the transformation of one type of particle into another, wherein particles are either annihilated or produced in their entirety.

This does not mean that the 'elementary' particles are without structure.[8] It means only that for a very wide range of phenomena the elementary particles may be regarded as structureless objects having certain overall properties of mass, charge, spin, etc.

In the present book, which deals with non-relativistic quantum mechanics, we shall be concerned only with processes in which the change in the energy of particles is much less than their rest energy $E_0 = m_0c^2$. The processes which involve transformations of elementary particles lie outside the scope of non-relativistic mechanics.[9]

The existence of elementary particles is not the only aspect of the atomism which is the most important distinguishing feature of the microuniverse. The complex particles formed from the elementary particles (for instance, molecules, atoms and atomic nuclei) also possess atomistic properties.

These properties arise from two facts. Firstly, each type of complex particle is formed from completely definite elementary particles (for example, a hydrogen atom consists of one proton and one electron; a nucleus of uranium 238 consists of 92 protons and 146 neutrons, and so on). Secondly, the internal states of complex particles are discrete: for each complex particle there is a series of completely definite possible states, each separated from the next by discontinuous changes. For this reason, by no means every interaction can bring a complex system from the state of lowest energy (called the *ground state*) to the neighbouring *excited state*.

If the energy of the external interaction is insufficient to cause a transition of the system from the ground state to the excited state, then after the external interaction ceases the system remains in its original state (the ground state). In consequence, atomic systems subjected to any external interaction remain largely as they were before, or enter new and definite states. This discontinuity in the changes of state of complex atomic systems was the physical reason (though not explicitly understood) which led chemists to the idea of the indivisibility of the atom and allowed physicists to regard atoms in the kinetic theory as unchangeable material points. This unchangeability and indivisibility persist only until the external interactions become so strong that transitions of the complex particle to neighbouring states become possible.

[7] The table shows that the properties of microparticles include some (the spin σ, the isotopic spin T and the strangeness S) which do not occur for macroparticles. These properties are discussed in Sections 58, 59, 131 and 136. The new particles are described in [64]; [102].

[8] For example, the structure of the proton and the neutron has now been shown to exist; see the review article [15].

[9] Photons and neutrinos have a rest mass $m_0 = 0$. They are therefore relativistic particles at all energies, and cannot be studied by the methods of non-relativistic quantum mechanics. The limits of quantum mechanics are further discussed in Section 138.

Owing to the identity of properties of elementary particles and the discreteness of states of complex particles, the particles of the microuniverse have no individual 'character'. Events involving an electron or a hydrogen atom do not affect their characteristic properties. A macroscopic system, on the other hand, usually reveals its history to some extent, and more so for more complex systems.

The discreteness of states of microsystems can be demonstrated experimentally. Franck and Hertz passed a beam of electrons, i.e. an electric current, through mercury vapour. The transmitted current was found to have maxima and minima as a function of the electron energy, as shown in Figure 2.

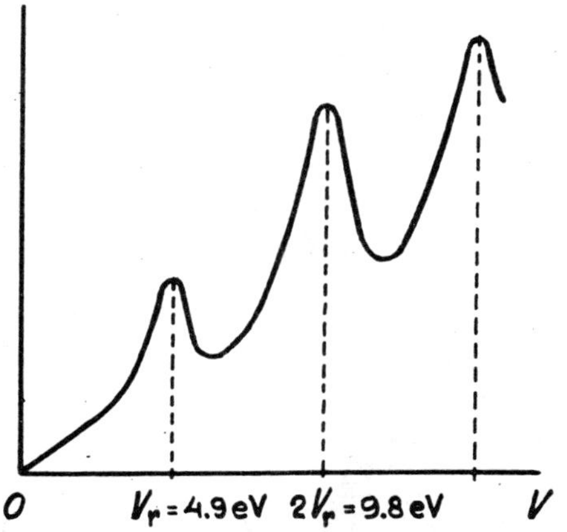

Fig. 2. Results of Franck and Hertz's experiment.

Initially, when the energy of the electrons does not exceed 4.9 eV, the electron beam passes through the mercury vapour without loss of energy, and so the current increases with the voltage. (In actual fact there is some exchange of energy when an electron collides with a mercury atom as a whole, but since the mass of the mercury atom is much greater than that of the electron and the collision is elastic, this energy exchange may be neglected.) As soon as the energy reaches 4.9 eV, the current drops, since the electrons begin to lose energy by changing the internal state of mercury atoms with which they collide.

This proves the discreteness of the possible values of the internal energy of the mercury atom. The energy of the state of this atom nearest to the ground state is 4.9 eV greater than the energy of the ground state.[10]

Stern and Gerlach were able to show that the angular momentum of atoms has only certain discrete values, like the energy. These authors measured the magnetic moments of atoms, which are due to currents within the atoms; since these currents are caused by the motion of electrons, there is a relation (which we shall discuss in Sections 53 and 64) between the magnetic moment of the atom and its angular momentum. Stern and Gerlach's experiments consisted essentially of passing a narrow beam of atoms

[10] Details of Franck and Hertz's experiments are given in [77]; [94], p. 48.

through an inhomogeneous magnetic field. If the magnetic moment of the atom is $\mathfrak{M}$, its potential energy in a magnetic field $\mathbf{H}$ is

$$U = -\mathfrak{M}\cdot\mathbf{H} = -\mathfrak{M}H\cos\alpha,$$

where α is the angle between the direction of the magnetic field and that of the magnetic moment of the atom. The force exerted on the atom by the inhomogeneous field (if it varies in the direction of the axis OZ) is

$$F = -\partial U/\partial z = \mathfrak{M}(\partial H/\partial z)\cos\alpha.$$

The gradient of the field was perpendicular to the beam of atoms, and so the force F caused a deflection of the atoms from their original direction of motion. If all orientations of the magnetic moment of the atom (i.e. all values of α) were possible, as given by classical theory, the force F would take all values from $-\mathfrak{M}\partial H/\partial z$ to $+\mathfrak{M}\partial H/\partial z$. Different atoms would undergo different deflections, and when the beam is incident on a screen we should obtain a blurred image of the slit which defines the beam. In reality, two sharp images of the slit are obtained (Figure 3). This experimental result shows

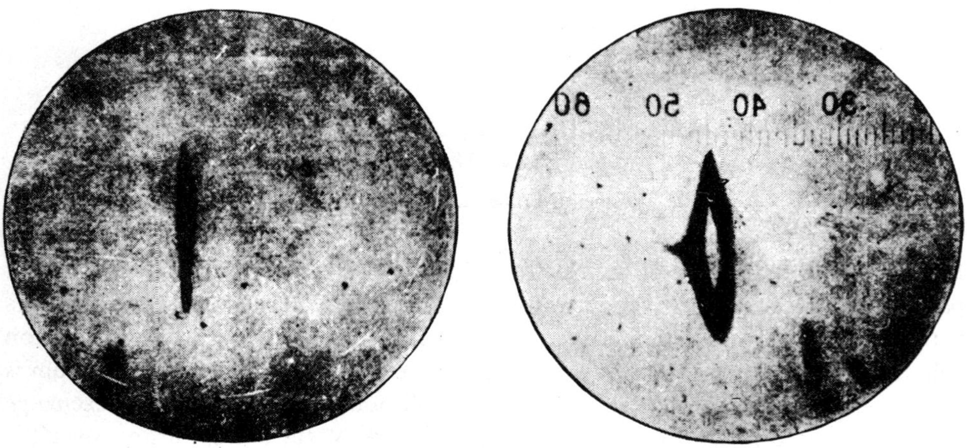

Fig. 3. Splitting of a beam of silver vapour in a magnetic field.

that only two discrete orientations of the magnetic moment of the atom are possible: $\cos\alpha = \pm 1$.

Calculation shows that the amount of the deflection of the beams corresponds to a value of the magnetic moment $\mathfrak{M}$ of the atom

$$\mathfrak{M}_B = e\hbar/2\mu c = 9\times 10^{-21}\ \text{CGS unit},$$

where e is the charge on the electron, μ its mass, and c the velocity of light. This value was first derived theoretically by Bohr from elementary quantum theory, and is called the *Bohr magneton*. It is a kind of quantum of magnetic moment.

The phenomenon discovered by Stern and Gerlach is called *spatial quantisation*, since what is involved is the discreteness of orientations of the magnetic moment relative to the magnetic field. From the above-mentioned relation between the angular momentum and the magnetic field we can say that Stern and Gerlach's experiments also prove the discreteness of the possible values of the angular momentum.

We shall show later (Chapter X) that the magnetic moment of the atom observed by Stern and Gerlach is due not to the orbital motion of the electron (as was originally thought) but to an intrinsic magnetic moment of the electron itself.[11]

From the point of view considered here we can say that Stern and Gerlach's experiments show that the magnetic moment of the atom as a whole has discrete quantum values, and so they afford a further proof of the discreteness of the possible states of the states of the atom.[12]

The reader's attention may also be drawn to the fact that the discreteness of atomic states is of importance in a quite different group of phenomena. According to very general principles of classical statistical mechanics, the average energy per degree of freedom of a system in equilibrium at temperature T is $\frac{1}{2}kT$, where $k = 1.38 \times 10^{-16}$ erg/degree is *Boltzmann's constant*. According to this, for example, monatomic gases have a mean energy per atom of $\frac{3}{2}kT$ and a specific heat $\frac{3}{2}k$. This theoretical conclusion is well confirmed by experiment, but it contains the tacit assumption that the atom resembles a point mass with three degrees of freedom (corresponding to the three coordinates of the centre of gravity). It is known, however, that the helium atom, for example, consists of three particles: the nucleus and two electrons. It is assumed that these electrons are not able to supply or receive energy and so do not participate in the establishment of thermal equilibrium in the gas. This assumption cannot be justified by classical mechanics, which states that, if there is a stable motion with energy E, there is also a motion with energy almost equal to E; this means that atomic electrons must transmit and acquire energy in atomic collisions, i.e. must participate in the establishment of the equilibrium distribution of energy. From the point of view of the quantum theory, on the other hand, an atom may indeed be regarded, to a considerable extent, as an object having only three degrees of freedom. According to quantum theory, a finite amount of energy ΔE is necessary in order to transfer an atom from the ground state to the neighbouring excited state. Hence, if $\Delta E \gg \frac{3}{2}kT$, the electrons will not be excited in collisions between atoms, and the atoms will behave as 'rigid' point masses. The internal degrees of freedom are 'frozen'.

Since the time of the experiments described above, the number of experimental proofs of the discreteness of states of atomic systems has become very large.

The study of the atomic nucleus has furnished a particularly large number of new facts. It has been shown that atomic nuclei also have a discrete system of levels. Figure 4 shows the cross-section for the interaction of a neutron with an oxygen atom as a function of the neutron energy. This curve has sharp resonance peaks at certain

11 This applies to Stern and Gerlach's first experiments with hydrogen and silver in the ground state. In general the magnetic moment of the atom is due both to the orbital motion of the electrons and to the intrinsic magnetic moment of these electrons.

12 Details of Stern and Gerlach's experiments are given in [77]; [95], p. 7–28.

energies, which indicate the existence of discrete energy levels in the nucleus (see [28]).

Resonance phenomena in elementary particles also are now known. These are found in the scattering of π mesons by nucleons and of γ rays by nucleons (Figure 5) (see the review article [31]).[13]

Hyperons decay with emission of π mesons or γ quanta and are transformed into nucleons. Thus hyperons may be regarded as discrete excited states of nucleons.

4. Bohr's theory

In order to describe the discontinuous properties of atomic systems discussed in the previous sections, Bohr proposed a modification of classical mechanics by including Planck's constant $\hbar$ in the laws of motion. The modification suggested was that not all motions permitted by classical mechanics actually occur in atomic systems, but only

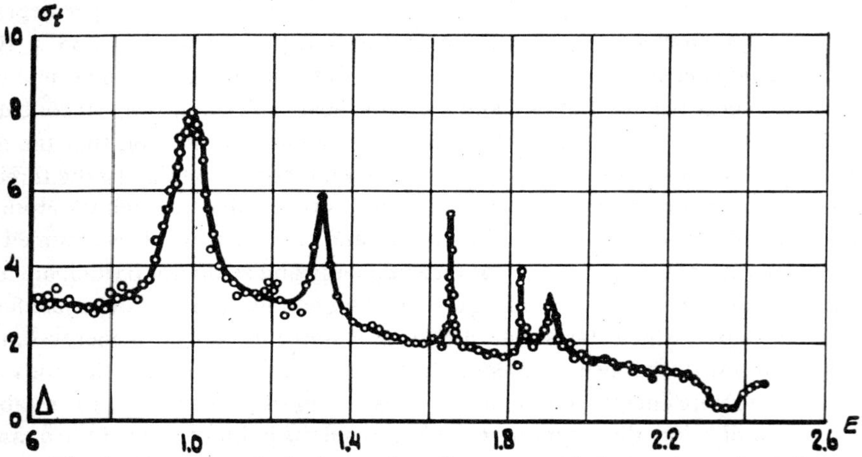

Fig. 4. Resonances in the interaction of neutrons with the oxygen nucleus $_8$O. The ordinate is the interaction cross-section in barns (10^{-24} cm^2); the abscissa is the neutron energy in MeV.

certain selected ones. Bohr formulated a special selection rule which will not be considered here.[14] The use of this rule provided a means of finding the possible values of the energy of the hydrogen atom, but Bohr's procedure was not valid for more complex atomic systems (for example, the helium atom). As applied to the energy E of the atom, Bohr's hypothesis implied that this energy could take only discrete quantum values:

$$E = E_1, \quad E_2, \dots, \quad E_n, \dots, \quad E_m, \dots. \tag{4.1}$$

The modern theory, as we shall see, requires no such postulate, and does not regard discreteness of states as a necessary property of a quantum system. Nevertheless, Bohr's postulate is still correct for a certain range of phenomena, since it may be regarded in such cases as a direct expression of experimental facts.

[13] Figure 5 is taken from [74].

[14] Details of Bohr's theory are given in [77]; [100], Ch. II.

Bohr's postulate is in contradiction with the classical theory of radiation, since according to the latter an excited atom emits continuously, and therefore its energy can lie between the permitted energy levels. Bohr therefore took the quantum view (Section 1), according to which the energy is emitted in discrete light quanta. Then, on combining the law of conservation of energy with Bohr's postulate regarding the discreteness of atomic states, we have the relation originally due to Bohr between the frequencies ω_{mn} which an atom can emit and absorb (its *spectrum*) and the quantum levels E_n belonging to that atom, viz.[15]

$$\hbar\,\omega_{mn} = E_m - E_n. \tag{4.2}$$

This equation is just the law of conservation of energy in light emission and absorption, and in the older Bohr theory was a postulate (Bohr's *frequency rule*). Dividing Equation (4.2) by Planck's constant, we find that the frequencies absorbed or emitted

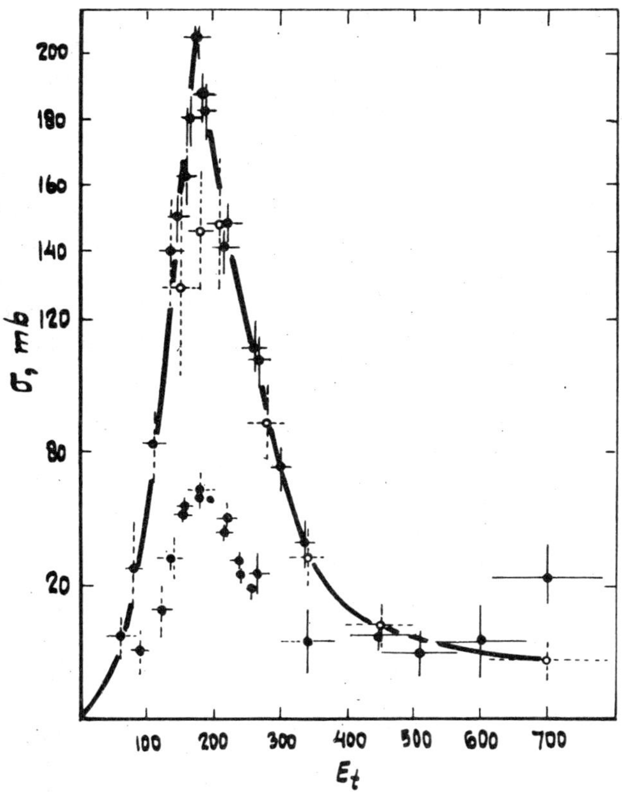

Fig. 5. Total cross-section for interaction of π mesons with nucleons. Resonance for π^+ mesons in the π meson energy range of about 200 MeV.

[15] For absorption we put in (1.3) $\omega' = 0$, $E' = E_m$, $E = E_n < E_m$, $\omega = \omega_{mn}$; for emission $\omega' = \omega_{mn}$, $E' = E_n$, $E = E_m$, $\omega = 0$.

by quantum systems can always be represented as a difference of two frequencies:

$$\omega_{mn} = \omega_m - \omega_n, \qquad \omega_m = E_m/\hbar, \qquad \omega_n = E_n/\hbar. \tag{4.3}$$

These are called *spectral terms.*

Long before Bohr's theory, it had been established purely empirically by Ritz that the observed frequencies of atoms can be represented as differences of terms (the Ritz *combination principle*). Hence (4.3) can be regarded as expressing Ritz's empirical rule.

The Ritz combination principle represents a fundamental contradiction between classical theory and experiment. If an electron is in an atom, it executes a periodic or quasi-periodic motion. In the simplest case, that of one-dimensional motion, its coordinate $x(t)$ can be expanded in Fourier series:

$$x(t) = \sum_{n=-\infty}^{\infty} x_n \cdot e^{i\omega_n t}, \tag{4.4}$$

where $\omega_n = n\omega_1$, ω_1 being the fundamental frequency and ω_n that of the $(n-1)$th harmonic. The intensity I_n of the radiation of frequency ω_n is given by the amplitude of the $(n-1)$th harmonic, i.e. the quantity x_n (see Section 87). The frequencies, according to the classical theory, can be arranged in a line:

$$\omega = \omega_1, \quad \omega_2, \ldots, \quad \omega_n, \ldots. \tag{4.5}$$

The corresponding intensities I_n or amplitudes x_n can be arranged in the same way. This very general consequence of the classical theory contradicts the empirical Ritz principle, according to which the experimentally observed frequencies are always determined by the two numbers n and m (the term numbers)[16], so that the line contains not the frequencies but the terms ($\omega_n = E_n/\hbar$), while the frequencies from a two-dimensional matrix:

$$\omega = \begin{matrix} 0 & \omega_{12} & \omega_{13} & \cdots & \omega_{1n} & \cdots \\ \omega_{21} & 0 & \omega_{23} & \cdots & \omega_{2n} & \cdots \\ \cdots & \cdots & \cdots & \cdots & \cdots & \cdots \\ \omega_{m1} & \omega_{m2} & \omega_{m3} & \cdots & \omega_{mn} & \cdots \\ \cdots & \cdots & \cdots & \cdots & \cdots & \cdots \end{matrix} \tag{4.6}$$

The corresponding intensities I_{mn} or oscillation amplitudes x_{mn} can be arranged in a similar table.

This contradiction could be overcome by assuming that each of the frequencies ω_{mn}

[16] If the system has f degrees of freedom, it can have f fundamental frequencies ω_a ($a = 1, 2, \ldots, f$), and then the general expression for the frequency ω_n, according to the classical theory, is

$$\omega_n = \sum_{a=1}^{f} n_a \omega_a,$$

where n_a are integers. The existence of several fundamental frequencies does not essentially affect our statement that there is a contradiction between the classical theory and Ritz's principle, since in this case each term $\omega_n = E_n/\hbar$ will be described by a group of numbers $n_1, n_2, \ldots, n_f$, and the emitted frequencies again by two groups of numbers: $n_1, n_2, \ldots, n_f$ and $m_1, m_2, \ldots, m_f$.

is one of the fundamentals and corresponds to a different degree of freedom. The atom would thus resemble a piano, with each degree of freedom represented by a key. But we should then have to assume the existence of a very large, essentially infinite, number of degrees of freedom, and should thus render even deeper the contradictions between the predictions of classical mechanics regarding the specific heat of atoms and the experimental facts.

In conclusion we may mention also that Bohr's theory, though it allows a determination of the frequencies ω_{mn}, i.e. of the spectrum, at least in the simple case of the hydrogen atom, gives no information regarding the intensities I_{mn} of emission of these frequencies, and the corresponding absorption coefficients. The calculation of these intensities presented a fundamental and insuperable obstacle to Bohr's theory. Only qualitative arguments were possible. The calculation by Bohr's theory for atoms more complex than that of hydrogen also led to difficulties of principle, which were resolved by quantum mechanics.[17]

In 1927 Heisenberg suggested that all quantities describing intra-atomic motions should be regarded as being matrices like (4.6). From this new point of view the coordinate and momentum of the electron must be represented as matrices x_{mn} and p_{mn}. In this way Heisenberg arrived at the famous 'uncertainty principle' and obtained correct values for the terms of the simplest quantum systems. His method was named 'matrix mechanics' and soon merged with another treatment, the wave theory of de Broglie and Schrödinger.

5. The elementary quantum theory of radiation

The elementary theory of radiation on the basis of quantum ideas is due to Einstein. It is to some extent phenomenological[18] but nevertheless, with the aid of modern quantum mechanics, it affords a solution of the problem of intensities of emission and absorption of light.

From the quantum viewpoint the intensity of emission or absorption of electromagnetic radiation is determined by the probability of the transition of an atom from one state to another. The solution of the problem of intensities amounts to calculating these probabilities.

Let us consider two states of any system, for example an atom, denoting one by the letter m and the other by n. Let the energy of the first state be E_m, and of the second be E_n. For definiteness we assume that $E_m > E_n$, so that the state m belongs to a higher quantum level E_m than the state n (quantum level E_n).

It is found from experiment that a system can spontaneously jump from a higher state m to a lower state n, emitting a light quantum $\hbar\omega = E_m - E_n$ with frequency $\omega = (E_m - E_n)/\hbar$; the quantum also has a definite polarisation and is propagated within a solid angle $d\Omega$ (Figure 6). Any polarisation for a given direction of propagation of light can be represented as a superposition of two independent polarisations l_1 and l_2 in perpendicular directions. In the transition $E_m \rightarrow E_n$ a light quantum can be

[17] These difficulties are discussed in [49]; [94], p. 47.

[18] Einstein's hypotheses are entirely justified by modern quantum electrodynamics; see [1, 47].

emitted with polarisation either l_1 or l_2. We shall denote the polarisation by the suffix α (= 1, 2). The probability per second of the transition $m \to n$, with emission of a quantum of frequency $\omega = (E_m - E_n)/\hbar$ into the solid angle $d\Omega$ and with polarisation α, is denoted by

$$dW_r' = a_{m\alpha}^n \, d\Omega \,. \tag{5.1}$$

This is called the *spontaneous* transition probability. In the classical theory such transitions correspond to emission by an excited oscillator.

Radiation in the neighbourhood of an atom affects it in two ways. Firstly, the radiation can be absorbed, the atom going from a lower state n to a higher state m. The probability per second of such a transition is denoted by dW_a. Secondly, if the atom is in the excited state m, the external radiation may cause a transition to the lower state n, increasing the emission probability by an amount dW_r''. This additional probability will be called the *induced* or *stimulated* transition probability.[19] Both types of transition have analogues in the classical theory: an oscillator under the action of external radiation can either absorb or emit energy, depending on the relation between the phase of its oscillations and the phase of the light wave.

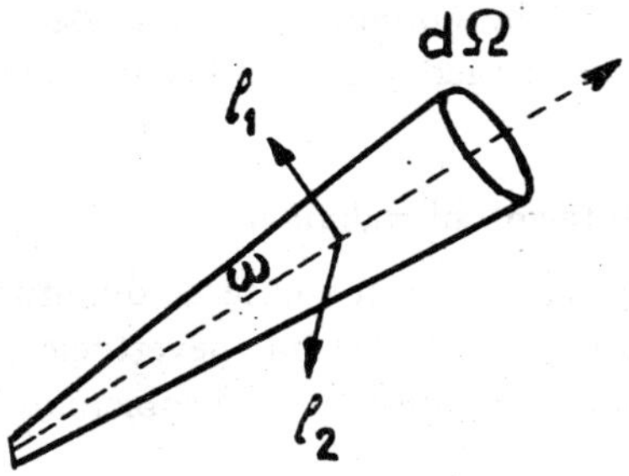

Fig. 6. Radiation characteristics. The direction of propagation (angle $d\Omega$), the frequency ω, and the two independent directions of polarisation l_1 and l_2.

Thus the total probability of emission is

$$dW_r = dW_r' + dW_r'' \,.$$

The absorption probability dW_a and the stimulated emission probability dW_r'' are, according to Einstein's hypothesis, proportional to the number of light quanta whose absorption and emission are under consideration. Let us now determine this number.

Radiation is not in general monochromatic, and has various directions of propagation and various polarisations. To determine the nature of the radiation, we introduce the quantity $\rho_\alpha(\omega, \Omega)\, d\omega\, d\Omega$, which gives the energy density of radiation propagated in directions within the solid angle $d\Omega$, with polarisation α and frequency between ω and $\omega + d\omega$. Since the quantum energy is $\hbar\omega$, the number of light quanta per cm^3 with frequency between ω and $\omega + d\omega$ which are propagated in

[19] Molecular amplifiers (masers) are based on the utilisation of induced emission.

the solid angle $d\Omega$ and have polarisation α is $\rho_\alpha(\omega, \Omega)\, d\omega\, d\Omega/\hbar\omega$. Owing to the above-mentioned proportionality between the number of quanta and the probabilities of absorption and stimulated emission, we can put

$$dW_a = b^m_{n\alpha}\, \rho_\alpha(\omega, \Omega)\, d\Omega\,, \tag{5.2}$$

$$dW_r'' = b^n_{m\alpha}\, \rho_\alpha(\omega, \Omega)\, d\Omega\,. \tag{5.3}$$

We call $a^n_{m\alpha}$, $b^m_{n\alpha}$ and $b^n_{m\alpha}$ the *Einstein differential coefficients*. They depend only on the nature of the systems which emit and absorb light, and can be calculated by the methods of quantum mechanics (see Section 88). Here we may draw some general conclusions regarding the properties of these coefficients, without actually calculating them.

Let us consider conditions in which there is equilibrium between emission and absorption. Let the number of atoms in the excited state m be n_m, and the number in the lower state be n_n. Then the number of light quanta emitted per second in transitions $m \to n$ will be $n_m\,(dW_r' + dW_r'')$, and the number of quanta absorbed per second in transitions $n \to m$ will be $n_n\, dW_a$. In equilibrium the number of absorptions must be equal to the number of emissions:

$$n_n\, dW_a = n_m (dW_r' + dW_r'')\,.$$

Substituting dW_r' from (5.1) and dW_r'', dW_a from (5.2) and (5.3), and cancelling $d\Omega$, we find

$$n_n\, b^m_{n\alpha}\, \rho_\alpha(\omega, \Omega) = n_m \left[b^n_{m\alpha}\, \rho_\alpha(\omega, \Omega) + a^n_{m\alpha} \right] \tag{5.4}$$

(where $\omega = \omega_{mn}$).

Let us assume that there is thermal equilibrium. Then the numbers of atoms in the various states will be functions of the temperature T. The radiation density $\rho(\omega, \Omega)$ will also be a function of temperature. This will be the density of radiation in equilibrium with matter at temperature T, i.e. the density of *black-body radiation.*

It is known that the properties of black-body radiation are independent of the particular properties of the matter with which it is in equilibrium. Hence all conclusions derived by consideration of black-body radiation are of general validity. This fact was used by Einstein to establish relations between the coefficients $a^n_{m\alpha}$, $b^m_{n\alpha}$ and $b^n_{m\alpha}$ in a general form.

The ratio of the numbers of atoms in various states can be determined from statistics. Usually (see, for example, Section 50) several different states of a quantum system correspond to any given energy level E_n. The number f_n of such states is called the *statistical weight* or *degree of degeneracy.*

According to the canonical distribution, which is valid for both classical and quantum systems, the number of atoms N_n in a state with energy E_n is

$$N_n = \text{constant} \times f_n\, e^{-E_n/kT}\,, \tag{5.5}$$

where k is Boltzmann's constant. If we are concerned with the number of atoms in any one state with energy E_n, the same distribution gives

$$n_n = N_n/f_n = \text{constant} \times e^{-E_n/kT}\,. \tag{5.5'}$$

Substituting n_n and n_m from (5.5′) in (5.4) and cancelling the constant, we obtain

$$e^{-E_n/kT}\, b_{n\alpha}^{m} \rho_\alpha(\omega, \Omega, T) = e^{-E_m/kT} \left[b_{m\alpha}^{n} \rho_\alpha(\omega, \Omega, T) + a_{m\alpha}^{n} \right], \tag{5.6}$$

where the temperature has been included as an additional argument of ρ, since in thermal equilibrium, as stated above, the density of equilibrium radiation depends on the temperature. As $T \to \infty$, the radiation density ρ must increase without limit, i.e. $\rho \to \infty$. From (5.6) with $T \to \infty$ we obtain one important relation:

$$b_{n\alpha}^{m} = b_{m\alpha}^{n}. \tag{5.7}$$

On the basis of this relation, and since $E_m - E_n = \hbar\omega$, we have from (5.6)

$$\rho_\alpha(\omega, \Omega, T) = \frac{a_{m\alpha}^{n}}{b_{m\alpha}^{n}} \frac{1}{e^{\hbar\omega/kT} - 1}. \tag{5.8}$$

In order to find the ratio $a_{m\alpha}^{n}/b_{m\alpha}^{n}$, Einstein made ingenious use of the fact that at high temperatures, when $kT \gg \hbar\omega$, the quantum formula (5.8) obtained for the density of equilibrium radiation must become the classical Rayleigh-Jeans formula. The classical formula for the density of equilibrium radiation is derived on the asumption that radiation of frequency ω may have energy as small as we please. According to the quantum theory, on the other hand, the smallest energy of such radiation is $\hbar\omega$. If $kT \gg \hbar\omega$, the quantity $\hbar\omega$ is small, and the basic assumption of the classical theory is fulfilled. From (5.8), expanding $e^{\hbar\omega/kT}$ in series, we obtain for $\hbar\omega/kT \ll 1$

$$\rho_\alpha(\omega, \Omega, T) = \frac{a_{m\alpha}^{n}}{b_{m\alpha}^{n}} \frac{kT}{\hbar\omega}. \tag{5.9}$$

The classical Rayleigh-Jeans formula gives for the density of equilibrium radiation the expression

$$\rho_\alpha(\omega, \Omega, T) = \omega^2 kT/8\pi^3 c^3. \tag{5.10}$$

For $kT \gg \hbar\omega$ the two formulae (5.8) and (5.10) must be the same. Hence, comparing (5.9) and (5.10), we find

$$a_{m\alpha}^{n}/b_{m\alpha}^{n} = \hbar\omega^3/8\pi^3 c^3, \qquad \hbar\omega = E_m - E_n. \tag{5.11}$$

This important formula enables us to calculate one coefficient from the other, since the relation does not depend on the nature of the substance (which is as it should be), and depends only on the frequency of the radiation.

Substitution in (5.8) finally gives the formula for the density of equilibrium radiation:

$$\rho_\alpha(\omega, \Omega, T) = \frac{\hbar\omega^3}{8\pi^3 c^3} \frac{1}{e^{\hbar\omega/kT} - 1}. \tag{5.12}$$

6. Black-body radiation

Integrating $\rho_\alpha(\omega, \Omega, T)$ over the whole solid angle ($\Omega = 4\pi$) and summing over the two polarisations ($\alpha = 1, 2$), we obtain the density $\rho(\omega, T)$ of radiation in the frequency range ω to $\omega + d\omega$ irrespective of polarisation and direction of propagation.

According to (5.12) the equilibrium radiation is isotropic, i.e. is independent of the direction of propagation, and it is the same for both polarisations. Hence we have

$$\rho(\omega, T) = 8\pi\rho_{\alpha}(\omega, \Omega, T), \tag{6.1}$$

and so the density of equilibrium radiation of frequency ω and temperature T is

$$\rho(\omega, T) = \frac{\hbar\,\omega^3}{\pi^2 c^3} \cdot \frac{1}{e^{\hbar\omega/kT} - 1}. \tag{6.2}$$

This formula gives the distribution of energy in the spectrum of black-body radiation, and was first derived by Planck.[20] Figure 7 shows graphs of this distribution for

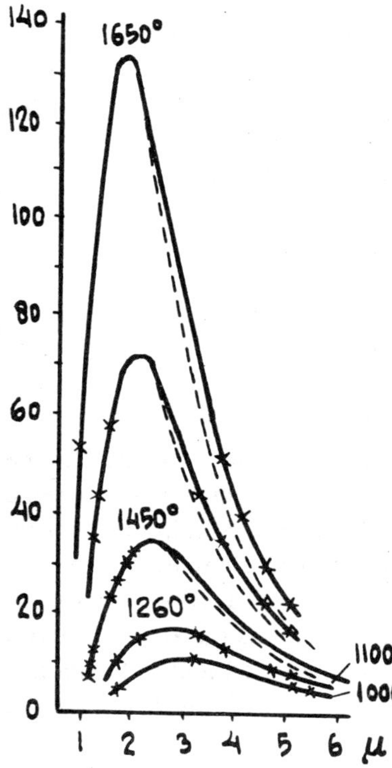

Fig. 7. Energy distribution in the spectrum of black-body radiation at various temperatures. The abscissa is the wavelength in microns.

various temperatures T. In the range $\hbar\omega \ll kT$, Planck's law agrees with the classical Rayleigh-Jeans law

$$\rho_{cl}(\omega, T) = \omega^2 kT/\pi^2 c^3. \tag{6.3}$$

[20] It may be noted that in the older literature Planck's formula is written in a different manner: (1) instead of our constant $\hbar$, the constant 2π times greater (used by Planck) appears; (2) instead of the angular frequency ω the ordinary frequency $\nu = \omega/2\pi$ is used, and $\rho(\omega, T)\mathrm{d}\omega = \rho(\nu, T)\mathrm{d}\nu$, i.e. $\rho(\nu, T) = 2\pi\rho(\omega, T)$.

For large quanta, $\hbar\omega \gg kT$, since $e^{\hbar\omega/kT} \gg 1$, (6.2) gives

$$\rho(\omega, T) = (\hbar\omega^3/\pi^2 c^3)\, e^{-\hbar\omega/kT}. \tag{6.4}$$

The Rayleigh-Jeans formula is derived from a consideration of light as continuous waves. Formula (6.4) can be obtained if light is regarded as a gas consisting of particles of energy $\varepsilon = \hbar\omega$. The former is the wave picture of light, and the latter is the corpuscular picture. Both are inadequate: Planck's formula corresponds to neither. It is easy to see that the wave picture is valid when the light quanta are small and numerous; the corpuscular picture is valid where the quanta are large and few in number. For, the number of quanta per cm^3 in the Rayleigh range ($\hbar\omega_1 \ll kT$) in the frequency interval ω_1 to $\omega_1 + d\omega$ is

$$dN_1 = \frac{\rho(\omega_1, T)\, d\omega}{\hbar\omega_1} = \frac{kT}{\pi^2 c^3}\frac{\omega_1}{\hbar}\, d\omega, \tag{6.5}$$

and in the Wien range ($\hbar\omega \gg kT$) and the frequency interval $\omega_2 + d\omega$ it is

$$dN_2 = \frac{\omega_2^2}{\pi^2 c^3}\, e^{-\hbar\omega_2/kT}\, d\omega. \tag{6.5'}$$

The ratio is

$$\frac{dN_2}{dN_1} = e^{-\hbar\omega_2/kT} \cdot \frac{\hbar\omega_2^2}{kT\omega_1} \tag{6.6}$$

and since $\omega_2 \gg kT/\hbar$, $dN_2/dN_1 \ll 1$.

7. De Broglie waves. The group velocity

We shall not follow the historical sequence of development of quantum mechanics, and in particular shall not describe the analogies between mechanics and optics which led de Broglie and later Schrödinger to establish the starting points of wave mechanics (now more often called quantum mechanics), interesting as this history is.[21] Disregarding those features of the original theory which are now of purely historical significance, de Broglie's basic idea was the application of the fundamental Equations (1.1) and (1.2) of the quantum theory of light to the motion of particles.

De Broglie associated the motion of any freely moving particle of energy E and momentum $\mathbf{p}$ with the plane wave

$$\psi(\mathbf{r}, t) = C e^{i(\omega t - \mathbf{k}\cdot\mathbf{r})}, \tag{7.1}$$

where $\mathbf{r}$ is the radius vector of an arbitrary point in space and t is the time. The frequency ω of this wave and its wave vector $\mathbf{k}$ are related to the energy and momentum of the particle by the same equations as are valid for light quanta, i.e.

$$E = \hbar\omega, \tag{7.2}$$

$$\mathbf{p} = \hbar\mathbf{k}. \tag{7.3}$$

[21] The reader who wishes to pursue this aspect will find an excellent account of de Broglie's ideas in [24].

These are de Broglie's fundamental equations. The historical development here is the converse of that which led to the quantum theory of light. For light we originally had a wave picture, which in quantum mechanics was supplemented by a corpuscular picture and the concepts of the momentum and energy of a light quantum. For particles (electrons, atoms, etc.), on the other hand, we have as the starting point the classical conception of the motion of particles, and following de Broglie we supplement this classical corpuscular picture in the quantum theory by the ideas of wave theory, using the frequency ω and the wavelength $\lambda = 2\pi/|\mathbf{k}|$ of waves associated with the motion of the particle.

Substituting in (7.1) ω and $\mathbf{k}$ from (7.2) and (7.3), we obtain a new expression for the wave (7.1), showing explicitly the relation between the frequency and wavelength and the corpuscular quantities, the energy E and momentum $\mathbf{p}$ of the particle:

$$\psi(\mathbf{r}, t) = Ce^{i(Et - \mathbf{p}\cdot\mathbf{r})/\hbar}. \tag{7.1'}$$

We shall call this a *de Broglie wave*. The problem of the nature of these waves and of the interpretation of their amplitude C is not a simple one, and will be deferred to the next chapter.

At first sight it may appear that the motion of the wave (7.1) cannot be related to the mechanical laws of motion of the particles. This is not so, however. In order to see the relation, let us consider the fundamental properties of a de Broglie wave. To simplify the calculations, we take the direction of the axis OX to coincide with the direction of propagation of the wave. Then (7.1) becomes

$$\psi(x, t) = Ce^{i(\omega t - kx)}. \tag{7.4}$$

The quantity $\omega t - kx$ is the phase of the wave. Let us consider some point x where the phase has a given value α. The co-ordinate of this point is defined by $\alpha = \omega t - kx$, whence, by differentiating this equation with respect to t, we see that such a point will move through space in the course of time at a velocity u, where

$$u = \omega/k. \tag{7.5}$$

This is called the *wave velocity* or *phase velocity*. If it depends on k, and therefore on the wavelength λ (since $\lambda = 2\pi/k$), there is dispersion of the wave. Unlike electromagnetic waves, de Broglie waves exhibit dispersion in empty space. This follows from de Broglie's Equations (7.2) and (7.3), since there is a certain relation between the energy E and the momentum p. According to relativity theory, for a particle velocity $v \ll c$ (the velocity of light), i.e. where Newtonian mechanics is valid, the energy of a freely moving particle is

$$E = +\sqrt{(m_0^2c^4 + p^2c^2)} = m_0c^2 + p^2/2m_0 + \cdots, \tag{7.6}$$

where m_0 is the rest mass of the particle.[22] Substituting this value of E in (7.2) and

[22] In non-relativistic theory the energy is always indeterminate as regards an additive constant, and so the rest energy m_0c^2 of the particle is usually omitted in defining the kinetic energy.

expressing p^2 in terms of k^2, we obtain

$$\omega = \frac{m_0 c^2}{\hbar} + \frac{\hbar k^2}{2m_0} + \cdots, \tag{7.7}$$

and therefore $u = \omega/k$ is a function of k.

Let us now establish the relation between the motion of the wave and that of the particle. To do this, we consider not the strictly monochromatic wave (7.4), which has a definite frequency and wavelength $\lambda = 2\pi/k$, but an almost monochromatic wave, which we shall call a *wave packet* or *wave group*. This term will more generally signify a superposition of waves differing only slightly in wavelength and direction of propagation. For simplicity, let us consider a *wave packet* (7.4) propagated in the direction OX. According to the above definition of a wave packet, the oscillations $\psi(x, t)$ are given by

$$\psi(x, t) = \int_{k_0 - \Delta k}^{k_0 + \Delta k} c(k)\, e^{i(\omega t - kx)}\, \mathrm{d}k, \tag{7.8}$$

where $k_0 = 2\pi/\lambda_0$ is the wave number around which lie the wave numbers of the waves forming the wave packet (Δk being assumed small).

Expanding the frequency ω as a function of k (see Formula (7.7)) in powers of $k - k_0$, we have

$$\omega = \omega_0 + (\mathrm{d}\omega/\mathrm{d}k)_0 (k - k_0) + \ldots, \qquad k = k_0 + (k - k_0).$$

Taking $k - k_0$ as a new variable of integration ξ and assuming the amplitude $c(k)$ to be a slowly varying function of k, we find that $\psi(x, t)$ can be put in the form

$$\psi(x, t) = c(k_0)\, e^{i(\omega_0 t - k_0 x)} \int_{-\Delta k}^{\Delta k} e^{i[(\mathrm{d}\omega/\mathrm{d}k)_0 t - x]\xi}\, \mathrm{d}\xi .$$

The integration with respect to ξ is straightforward, and gives

$$\begin{aligned} \psi(x, t) &= 2c(k_0) \frac{\sin\{[(\mathrm{d}\omega/\mathrm{d}k)_0 t - x]\,\Delta k\}}{(\mathrm{d}\omega/\mathrm{d}k)_0 t - x}\, e^{i(\omega_0 t - k_0 x)} \\ &= c(x, t)\, e^{i(\omega_0 t - k_0 x)} . \end{aligned} \tag{7.9}$$

Since the argument of the sine includes the small quantity Δk, the quantity $c(x, t)$ will vary only slowly as a function of the time t and the co-ordinate x, and so $c(x, t)$ may be regarded as the amplitude of the almost monochromatic wave, and $\omega_0 t - k_0 x$ as its phase. Let us find the co-ordinate x for which the amplitude $c(x, t)$ has a maximum. We call this the *centroid* of the wave packet. The maximum will evidently be at the point $x = (\mathrm{d}\omega/\mathrm{d}k)_0 t$. Hence it follows that the centroid of the wave packet will move with a velocity V obtained by differentiating this equation with respect to t:

$$V = (\mathrm{d}\omega/\mathrm{d}k)_0 . \tag{7.10}$$

We shall call this quantity the *group velocity* (as distinct from the wave velocity ω_0/k_0). If the waves under consideration were dispersionless, we should have $V = u$, but for de Broglie waves, on account of the dispersion, $V \neq u$. Using (7.7), we can calculate the group velocity V:

$$V = \mathrm{d}\omega/\mathrm{d}k = \hbar k/m_0 .$$

According to (7.3) $\hbar k = p$, and also $p = m_0 v$, where v is the velocity of the particle. Thus we have the important result

$$V = v ; \tag{7.11}$$

the group velocity of the de Broglie waves is equal to the mechanical velocity v of the particle.

The relations (7.10) and (7.11) which we have obtained can easily be extended to cover the propagation of waves in any direction relative to the axes OX, OY, OZ. The derivation is left to the reader; the final result is

$$V_x = \partial\omega/\partial k_x = \partial E/\partial p_x ,$$
$$V_y = \partial\omega/\partial k_y = \partial E/\partial p_y ,$$
$$V_z = \partial\omega/\partial k_z = \partial E/\partial p_z ,$$

or, in vector form,

$$\mathbf{V} = \mathbf{grad}_{\mathbf{k}}\omega = \mathbf{grad}_{\mathbf{p}} E = \mathbf{v} . \tag{7.11'}$$

The wavelength λ of the de Broglie waves may be calculated for two cases. From (7.3) it follows that

$$\lambda = 2\pi/k = 2\pi\hbar/p . \tag{7.12}$$

Taking only the case of low velocities $v \ll c$ and using the equation $E = p^2/2m_0$, we have

$$\lambda = 2\pi\hbar/\sqrt{(2m_0 E)} . \tag{7.12'}$$

This formula allows the wavelength λ to be calculated from a knowledge of the mass m_0 and energy E of the particle.

Let us apply this formula to the electron, for which $m_0 = 9 \times 10^{-28}$ g. Expressing the electron energy in eV by putting $E = eV$, where e is the electron charge and V the potential difference which accelerates the electron, in volts, we find

$$\lambda = \sqrt{(150/V)}\,\text{Å} . \tag{7.13}$$

For $V = 1$ V this gives $\lambda = 12.2$ Å; for $V = 10{,}000$ V, $\lambda = 0.122$ Å. For a hydrogen molecule of energy 6×10^{-14} erg (the mean energy of a hydrogen molecule at 300° K) and mass $2 \times 1.66 \times 10^{-24}$ g, substitution of these values in (7.12′) gives $\lambda = 1$ Å.

It is seen that the wavelength λ of de Broglie waves is very small, and decreases as the energy and mass of the particle increase. In practice, for example, it is impossible to obtain a wavelength λ equal to that of visible light, since even for electrons of energy 1 eV experiments are very difficult, and with $\lambda = 10^{-5}$ cm we have electrons of energy only 1.2×10^{-4} eV.

Particles of very high energy are obtained in present-day accelerators. Such machines can therefore be regarded as sources of very short waves. If the particle energy is much greater than the rest energy ($E \gg m_0c^2$), (7.6) gives $E \cong pc$, and so the wavelength in this case is

$$\lambda = 2\pi\hbar c/E. \tag{7.14}$$

For protons or mesons at energy $E = 10$ to 20 GeV, $\lambda = 1.26 \times 10^{-14}$ to 6.3×10^{-15} cm. Such short waves can be used to study the internal structure of the elementary particles.

The idea of associating the motion of a particle with that of a wave was so much at variance with the established concepts of mechanics that it appeared merely fanciful, and only experiment was able to impose its acceptance as a useful contribution to science. In what phenomena should one seek the confirmation (or refutation) of the idea of wave phenomena in the motion of particles? There are a number of effects pertaining only to waves, which exist regardless of the nature of the waves. These are the phenomena of diffraction and interference. Both are due to the combination of waves with certain phases and amplitudes, and their existence follows from the nature of wave motion. To test de Broglie's idea it was therefore necessary to consider experiments using particles in which these phenomena might be detected. It is known from optics that diffraction is appreciable only when the distance between the lines of the diffraction grating is comparable with the wavelength of the diffracted waves. If experiments are carried out with electrons, the above calculation shows that the de Broglie wavelengths are of the order of 1 Å, and for atoms even less. Thus the conditions for observing diffraction of electrons are approximately the same as those for observing diffraction of X-rays, so that the only suitable diffraction gratings are provided by crystals, where the distances between the 'lines' – the atoms of the crystal – are of the order of 1 Å. Experiments which confirm the correctness of de Broglie's view are briefly described in the following section.

8. Diffraction of microparticles

We shall now give a description of the experiments which demonstrate the correctness of de Broglie's idea, and begin with the classical experiments of Davisson and Germer. These authors examined the scattering of a beam of electrons at the surface of a crystal. On observing the beam intensity as a function of the angle of scattering, it is found that the angular distribution of electrons is very similar to the wave intensity distribution in diffraction. Figure 8 is a diagrammatic representation of Davisson and Germer's experiment. An electron gun served as the source of the electron beam. The Faraday cylinder was connected to a galvanometer, and the current strength indicated the number of electrons scattered by the surface of the single crystal at an angle θ to the initial beam, which was incident normally to the surface. The reader is referred to specialised monographs (e.g. [84]; [94], p. 71) for details of Davisson and Germer's experiment. The electron scattering picture obtained is very complex.

Electrons of low energy do not penetrate deeply into the crystal, and so a considerable fraction of the scattered electrons are scattered by a surface layer of the crystal. Thus diffraction occurs mainly from a plane diffraction grating formed by the atoms on the surface of the crystal. According to elementary diffraction theory, the position of the diffraction maxima is given by the formula

$$n\lambda = d \sin \theta , \tag{8.1}$$

where n is the order of the maximum, λ the wavelength of the diffracted radiation, d the constant of the plane surface lattice of the crystal, and θ the angle between the normal to the lattice and the direction of the scattered beam. Knowing the energy of the primary electrons incident on the crystal (which in their experiments could be varied between about 30 and 400 eV), Davisson and Germer were able to calculate for each energy the wavelength λ from de Broglie's formula (7.13) and to find from Formula (8.1) the position of the maximum for scattered ('diffracted') electrons.[23] Another method of testing de Broglie's formula was to test the validity of (8.1) for elec-

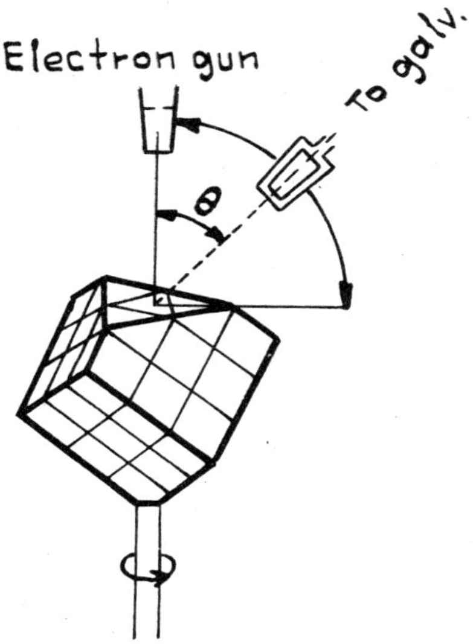

Fig. 8. Diagram of electron diffraction experiment (Davisson and Germer).

trons of various energies. Substituting in (8.1) λ from (7.13), we find that if de Broglie's formula is correct we must have

$$\sqrt{V} \sin \theta = \text{constant} , \tag{8.2}$$

if the angle θ corresponds to the position of maximum intensity of the scattered electrons. Both methods led Davisson and Germer to the conclusion that de Broglie's formula (7.12) relating the wavelength λ and the momentum p of the electrons is entirely valid.

Diffraction of X-rays can be observed not only from single crystals but also from

[23] It was afterwards found that for electrons of low energy, in contrast to fast electrons and X-rays, the refractive index of the crystal differs from unity. This fact has to be taken into account in calculating the diffraction of slow electrons.

polycrystalline substances, for example crystal powders (Debye and Scherrer's method). Tartakovskiĭ and G. P. Thomson first used this method to observe diffraction of electrons. Here a primary beam of electrons is passed through a polycrystalline film; to avoid strong absorption of electrons, the film is made very thin, its thickness being about 10^{-5} cm. In such a film the individual single crystallites are arranged at random. The beam in this method passes through the crystal, which forms a three-dimensional diffraction grating. Bragg and Vul'f's condition for such a system is

$$n\lambda = 2d \sin \phi , \tag{8.3}$$

where d is the lattice constant, ϕ the angle between the beam and the plane of the grating, and n and λ are as before. If any crystallite in the film satisfies this condition (see Figure 9), a spot Q is formed on the photographic plate P at the point of incidence

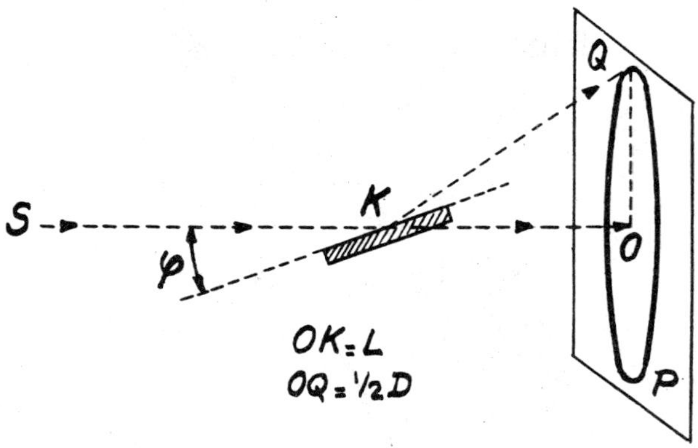

Fig. 9. Diagram of electron diffraction experiments (Tartakovskiĭ and Thomson).

on the plate of the diffracted ray KQ. Since the crystallites are arranged randomly, they include some whose position differs from that of the crystallite K only by a rotation about the axis SO, which is in the direction of the incident beam. Thus we obtain on the plate, instead of the spot Q, a ring of radius OQ. In the Debye-Scherrer method a diffraction ring corresponds to each spot in diffraction from a single crystal. The diameter D of these rings is easily found. If the distance from the plate to the film is L, then $\tan 2\phi = D/2L$. Combining this with (8.3), we obtain for small angles ϕ

$$n\lambda = d \cdot D/2L .$$

Substituting for λ its expression in terms of the energy of the electrons from de Broglie's formula (7.13), we find

$$D\sqrt{V} = \text{constant} . \tag{8.4}$$

The validity of this relation has been entirely confirmed by the observations of Tartakovskiĭ and Thomson [84]; [104], Ch. IV.

At the present time these experimental methods have been considerably improved,

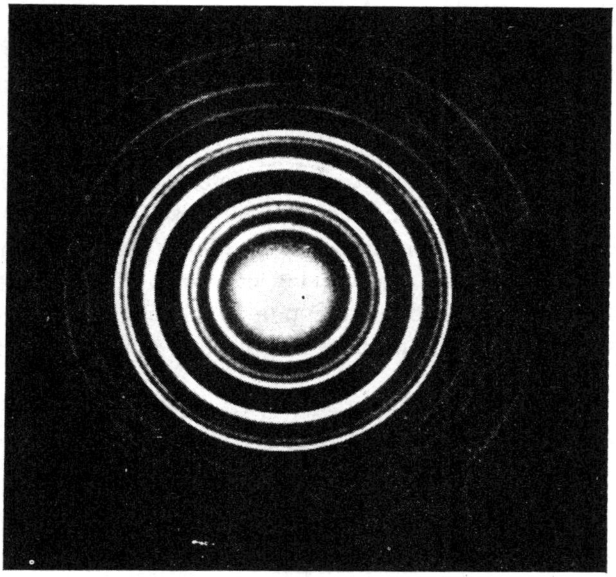

Fig. 10. Diffraction of electrons transmitted through a polycrystalline film of SnO_2.

Fig. 11. Neutron diffraction (Laue pattern).

and electron diffraction is used as successfully as that of X-rays for the analysis of crystal structure (particularly surface structure). Figure 10 shows the pattern of electron diffraction from a film of SnO_2. Thus the reality of electron diffraction is no longer doubted.

The problem of the applicability of de Broglie's formula (7.12) to particles more complex than the electron, such as atoms and molecules, is a very fundamental one. The possibility of such application to complex systems means that wave phenomena are not the result of peculiarities of the structure of any particular particle, but are of general significance, and express a general law of motion of microparticles.

Stern and Estermann examined the reflection of helium and hydrogen from lithium fluoride crystals in order to test de Broglie's formula for atoms and molecules. By

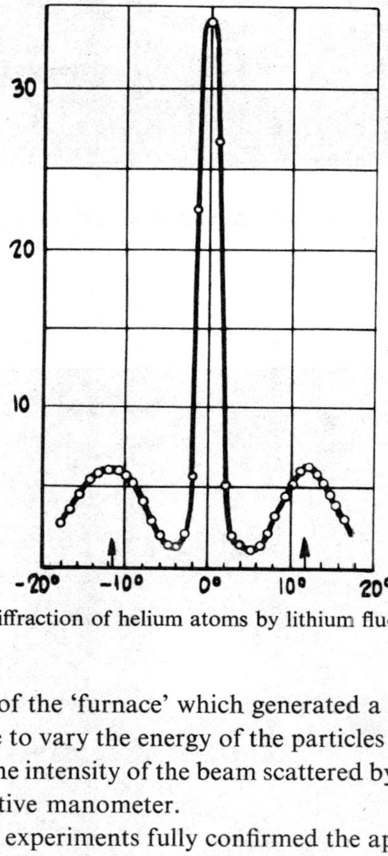

Fig. 12. Diffraction of helium atoms by lithium fluoride crystal.

varying the temperature of the 'furnace' which generated a narrow beam of atoms or molecules, they were able to vary the energy of the particles investigated, and thus the de Broglie wavelength. The intensity of the beam scattered by the crystal was measured by means of a very sensitive manometer.

Stern and Estermann's experiments fully confirmed the applicability of de Broglie's formula to these complex particles. Figure 12 shows the distribution of intensity in a scattered beam of helium atoms reflected from LiF crystals at $T = 295°$ K. The angle 0° corresponds to regular reflection of the helium beam from the crystal, and for this angle there is a sharp maximum. If we take account of the simple fact that the dimensions of an atom are of the same order as the distance between ions in the LiF lattice, then the presence of regular reflection is itself impossible to explain on the basis of corpuscular mechanics.

In addition to the maximum corresponding to regular reflection there are two diffraction maxima (first-order spectra). Their position agrees well with that calculated from de Broglie's formula. A similar result is obtained with hydrogen molecules.

Diffraction phenomena also occur with a beam of neutrons. Substituting in Formula (7.12) the neutron mass $m_0 = 1.66 \times 10^{-24}$ g, and expressing the neutron energy in electron-volts ($E = eV$), we find the neutron wavelength

$$\lambda = 0.285/\sqrt{V}\,\text{Å}. \tag{8.5}$$

Hence it is seen that, for a neutron energy of hundredths of an electron-volt (what are called *thermal neutrons*), λ is comparable with the lattice constants of crystals. Under this condition, diffraction is easily achieved. Since neutrons, unlike electrons but like X-rays, are not greatly absorbed by matter, they can be used to effect diffraction from the volume of a crystal (three-dimensional Laue diffraction). Figure 11 shows such diffraction of neutrons by a crystal of sodium chloride.

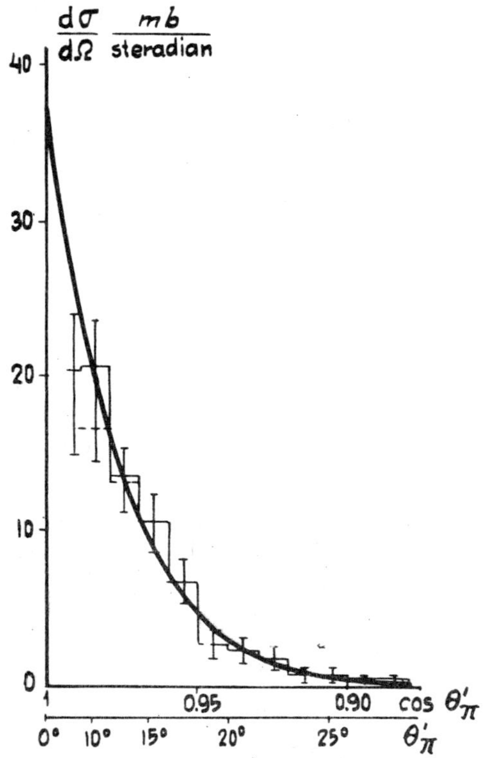

Fig. 13. Angular distribution of π mesons of momentum 6.8 GeV/c elastically scattered by protons. The scattering is strongly peaked forward and can be regarded as diffraction scattering by a highly absorbing sphere.

Finally, Figure 13 shows the diffraction pattern of π mesons of energy 7 GeV diffracted by protons [88]. This corresponds to diffraction of waves with $\lambda \sim 10^{-14}$ cm by a highly absorbing sphere of radius $\sim 10^{-13}$ cm.

The results given in this section show quite clearly that all particles, whatever their nature and structure, exhibit wave phenomena, and that de Broglie's formula relating the particle momentum and wavelength has universal validity.

CHAPTER II

FOUNDATIONS OF QUANTUM MECHANICS

9. Statistical interpretation of de Broglie waves

The physical significance of the waves which, according to de Broglie's concept, are related to the motion of particles was not discovered immediately. Attempts were made at first to regard the particles themselves as consisting of waves and distributed in some region of space. The intensity of the de Broglie wave was regarded, on this view, as a quantity representing the density of the medium of which the particle was formed. This conception of de Broglie waves was an entirely classical one. It was based on the fact that in some very special cases it is theoretically possible to construct from waves structures whose motion coincides with that of a particle moving in accordance with the laws of classical mechanics. An example is the wave packet discussed above. As shown in Section 7, the centroid of a wave packet moves as a particle. However, the motion of such a wave packet does not entirely agree with that of a particle. The reason is that the shape of the wave packet changes in the course of time. As will be shown in Section 34, the size of the wave packet increases, and it is broadened out. The necessity of this is easily understood from the existence of dispersion of de Broglie waves *in vacuo*. The individual waves of which the wave packet is composed are propagated at different velocities, and in consequence the wave packet spreads out.

Thus a particle composed of de Broglie waves will be unstable: even in motion in empty space, its size will increase continually and without limit. This instability is particularly striking if we consider the case where the particle moves from one medium into another. An example of this is furnished by the classical experiments on diffraction of particles. When, for instance, the beam of particles in Tartakovskiĭ and Thomson's experiment passes through a thin foil, it is separated into a series of conical diffracted beams. If we consider a particle, in this case an electron, as consisting of waves, we must originally identify the electron with the incident wave, whose dimensions are determined by the diaphragm in the apparatus, while after passage through the foil it must be identified with the whole system of diffracted waves. Each diffracted beam would have to represent some part of the electron. Let us now imagine that we have two apparatuses (e.g. photographic plates) recording the incidence of electrons, with only the first diffracted beam directed into the first apparatus, and only the second directed into the second apparatus. Then, if the particle is identified with the whole series of diffracted waves, we must conclude that each apparatus will receive only a part of the particle. This is a downright negation of the atomism of the particle, and brings the above conception of de Broglie waves into complete contradiction with experiment, since the particle always acts as a whole, and an apparatus detects the whole particle, not a part of it. In the example considered, the electron would enter either the first or the second apparatus, but not partly enter each.

The atomism observed in phenomena of the microuniverse consists in the fact that the elementary particles always act as a whole. Thus the conception of particles as consisting of de Broglie waves is in conflict with this atomism and must be rejected.

Neither can it be assumed that the waves themselves consist of particles, or more precisely originate in a medium composed of particles. Experiment shows that the diffraction pattern produced on a photographic plate is independent of the intensity of the incident beam of particles, and therefore of the number of particles per unit volume. In order to obtain the same diffraction pattern we can reduce the intensity but increase the exposure; only the total number of particles transmitted is of importance. This result shows clearly that each electron is diffracted independently of the others.[1] Hence the existence of wave phenomena cannot be related to the simultaneous presence of a large number of particles.

In order to emphasise this conclusion, we may note that wave phenomena appear in the motion of electrons in atoms, where a medium consisting of a large number of particles need not be present: such properties are found in electrons moving in atoms where there are very few electrons (one in hydrogen, two in helium, etc.).

The correct interpretation of de Broglie waves was found by Born in a quite different manner. In order to understand Born's basic idea, let us imagine that electrons are diffracted and the incidence of 'diffracted' electrons on a photographic plate is recorded. Let a small number of electrons be initially transmitted. Each electron which passes through the diffracting system (e.g. a foil) is found at some point on the plate and has a photochemical effect there. The passage of a small number of electrons will produce on the plate a pattern resembling a target used by an inexpert marksman. A regular distribution of electrons on the plate is obtained only when a large number of electrons is transmitted, and finally a distribution is obtained which corresponds exactly to the intensity distribution in the diffraction of waves (for example, the series of diffraction rings shown in Figure 10).

This behaviour of particles led Born to the statistical interpretation of de Broglie waves, whereby the atomism of the particles can be combined with wave phenomena. According to this interpretation, the intensity of de Broglie waves at any point in space is proportional to the probability of finding the particle at that point. For instance, if two diffracted beams are directed on to two different photographic plates, when the number of electrons passing through the apparatus is large, the number striking each

[1] At very high densities in the incident beam, additional scattering may occur owing to Coulomb interaction, but this is of secondary significance in the present problem; the important point is that at low intensities the wave-type interference phenomena still take place. This is shown in direct experiments by Biberman, Sushkin and Fabrikant [6] for electrons and in experiments by Jánossy and Náray [51] for photons.

plate is proportional to the intensity of de Broglie waves propagated in the direction of that plate.

If the photographic plate is so placed that the direction from the diffraction system to the plate is that of a diffraction minimum (the waves cancelling one another in this direction), the particles will not reach the plate. If, however, only one electron is considered, rather than a large number of them, then the intensity of the de Broglie waves indicates only the probability of incidence of the electron, but does not commit the electron to any definite behaviour.

On this view, de Broglie waves have nothing in common with the waves considered in classical physics. In all 'classical' waves the absolute value of the wave amplitude determines the physical state. If, for example, the amplitude of vibrations of air in one case is everywhere twice its value in another case, this means a fourfold energy of the vibrations and a different physical state of the medium.

In de Broglie waves, the intensities determine the position probability of the particle. Thus only the ratio of intensities in various regions of space is important, and not the absolute values of the intensities. The ratio shows the factor by which the probability of finding the particle at one point in space exceeds that of finding it at another. Hence, if in one case the intensity of de Broglie waves is *everywhere* twice that in another case, the physical state of the particle is the same in each case, since this increase in the amplitude of the waves causes no change in the ratio of intensities in various regions of space.

Thus de Broglie waves give a statistical description of the motion of microparticles: they determine the probability of discovering the particle at a given point in space at a given time.[2]

10. The position probability of a microparticle

Let x, y, z denote the co-ordinates of a particle. According to the discussion in Section 9, the exact significance of x, y, z consists in the following measuring operation: the quantities x, y, z are defined as the co-ordinates of the point in space where the particle is located. For example, these will be the co-ordinates of the dot on a photographic plate resulting from the incidence of a particle on the plate, or the co-ordinates which define the position of a slit through which the particle has passed.

The co-ordinates of the dot or slit can be determined by applying a rigid scale. We call this a '*direct*' measurement of a co-ordinate, since it is the measurement on which is based the macroscopic definition of the concept of the co-ordinate of a particle. Where such a determination of the co-ordinate of the particle is impossible (for instance, if the particle is inside an atom), we determine its co-ordinates by '*indirect*' experiment[3], i.e. by measuring in the way described above the co-ordinates of some other particle, which has undergone a collision with the particle under consideration,

[2] We shall see later that, if the de Broglie wave describing the state of a particle is known, we can find not only the position probability of the particle but also the probability of any result of measuring any mechanical quantity pertaining to the particle considered.

[3] The division of experiments into 'direct' and 'indirect' is due to L. I. Mandel'shtam.

and deducing from this measurement information about the co-ordinates of a particle within an atom which is not accessible to direct measurement. An example of such 'indirect' measurement will be given in Section 16.

Let us now give a mathematical formulation of the statistical interpretation of de Broglie waves. We may note first of all that the word 'waves' is here used in a specialised sense. Only in very particular cases will the state of particles be described by simple plane waves. In general, what we are calling de Broglie waves can be regarded as a very complex function of the co-ordinates x, y, z of the particle and the time t. Nevertheless, for these complex cases also we shall use the term *wave function* and denote the latter by ψ:[4]

$$\psi = \psi(x, y, z, t). \tag{10.1}$$

As explained in Section 9, we assume on the basis of the facts given that the position probability of a particle is determined by the intensity of the waves, i.e. by the squared amplitude of ψ. Since ψ may be a complex quantity, and the probability must always be real and positive, we take as the measure of intensity not ψ^2 but the squared modulus of ψ, i.e. the quantity $|\psi|^2 = \psi^*\psi$, where ψ^* is the complex conjugate[5] of ψ.

It should next be noted that the probability of finding the particle in a neighbourhood of the point x, y, z depends, of course, on the size of the region considered. If we take an infinitesimal region defined by $x, x + dx; y, y + dy; z, z + dz$, then ψ may be regarded as constant within that region, and so the probability of finding the particle may be taken as proportional to the volume of the region. We denote this volume element by $dv = dx\, dy\, dz$.

If the (infinitesimal) probability of finding the particle in the volume element dv near the point x, y, z at time t is denoted by $dW(x, y, z, t)$, we can express the statistical treatment of de Broglie waves by the equation

$$dW(x, y, z, t) = |\psi(x, y, z, t)|^2 \, dv. \tag{10.2}$$

By means of this equation a known wave function $\psi(x, y, z, t)$ can be used to calculate the position probability $dW(x, y, z, t)$ of the particle. The quantity

$$w(x, y, z, t) = dW/dv = |\psi(x, y, z, t)|^2 \tag{10.3}$$

will be called the *probability density*.

The probability of finding the particle in a volume V at time t is, according to the law of composition of probabilities,

$$W(V, t) = \int_V dW = \int_V w\, dv = \int_V |\psi(x, y, z, t)|^2 \, dv. \tag{10.4}$$

[4] Here it may be mentioned that we already know the wave function for two simple cases. For particles moving with given momentum p, the wave function ψ_p is the monochromatic plane wave (7.1). We also know the function for an almost monochromatic wave, i.e. for the wave packet (7.8). In the following discussion we shall use arbitrary wave functions, ignoring for the present the question whether such functions can be defined for given physical conditions (cf. Section 28). We shall assume that such a definition is possible, and say that the ψ function describes (statistically) the state of the particle.

[5] Henceforward the asterisk will always denote the complex conjugate quantity.

If we integrate over all space, we obtain the probability that at time t the particle is somewhere. This is the probability of an event which is certain to occur. In probability theory the probability of such an event is customarily taken to be unity. Then the integral of $|\psi|^2$ over all space must be unity:

$$\int_{\infty} |\psi(x, y, z, t)|^2 \, \mathrm{d}v = 1 . \tag{10.5}$$

This is called the *normalisation condition*, and a function ψ which satisfies this condition is said to be *normalised.*

Normalisation may be impossible if the integral of $|\psi|^2$ over the entire volume diverges, i.e. if the function ψ is not of integrable square. In physically real conditions the motion of a particle always takes place in a bounded space. This restriction is due to the geometrical dimensions of the apparatus and the finite velocity of particles. Hence the probability of finding the particle is non-zero only in a finite region of space, so that the function ψ must be integrable. In some cases, however, a certain degree of idealisation is necessary, and leads to functions which cannot be integrated. A simple example is the plane wave (7.1). Whereas in reality a parallel beam is always limited sideways by diaphragms and forwards by its front, when the beam is sufficiently large and edge effects are unimportant, we may regard the beam as a plane wave. The latter is assumed to occupy all space.

It follows from (7.1) that $|\psi|^2 = |c|^2 =$ constant. This means that the particle may with equal probability be found at any point. In this case normalisation to unity is not possible. We shall later, however, give a reasonable normalisation for this case also.

A further remark relates to the dependence on time. Normalisation is meaningful only if it is preserved in the course of time, i.e. the equation (10.5) must be valid at all times; otherwise, the probabilities at different times could not be compared. When discussing the laws of variation of the wave function with time it will be shown (Section 28) that in fact the normalisation does not change, i.e. the integral (10.5) is independent of time.

11. The principle of superposition of states

In given physical conditions a particle may be in various states, depending on the way in which it enters these conditions. If we consider the simplest case of free motion of a particle under no external forces and not interacting with other particles, we may find states of motion differing in both magnitude and direction of momenta. Each such state can occur. There are more complex cases, however; for example, in Davisson and Germer's diffraction experiment, the beam incident on the crystal is separated into a series of diffracted beams. After the interaction with the crystal the motion is again in empty space, but consists of an assembly of de Broglie waves with different directions of propagation.

On causing a beam of definite direction and definite wavelength λ to strike the surface of the crystal, we cannot obtain any one of the diffracted waves, but must obtain all of them (together with the incident wave); they will be found with certain definite phase relations to one another and are therefore capable of interference. This assembly of waves forms a single wave field and is represented by a single wave function ψ. Such a wave field is, however, a combination of simple de Broglie waves ψ_p, each of which by itself can describe a possible state of motion of a particle in empty space. This may be seen by selecting with a diaphragm one diffracted beam from the whole wave field ψ, and again diffracting it.

We say that the state arising in the diffraction of particles at the surface of a crystal is a *superposition* of states of free motion described by simple de Broglie waves. This case of superposition is a particular expression of the general principle of superposition of states, which forms one of the foundations of quantum mechanics.

This principle may be formulated as follows. If any system (a particle or assembly of particles) is able to be in a state represented by the wave function ψ_1, and in another state ψ_2, then it can also be in a state represented by a wave function ψ such that $\psi = c_1\psi_1 + c_2\psi_2$, where c_1 and c_2 are arbitrary, in general complex, numbers which determine the amplitudes and phases of the particular states ψ_1 and ψ_2. Hence it follows that, if there is a series of possible states of the system differing in the value of some quantity (momentum, energy, angular momentum, etc.), the states being represented by wave functions $\psi_1, \psi_2, \ldots$, then according to the principle of superposition there is a composite state

$$\psi = c_1\psi_1 + c_2\psi_2 + \ldots + c_n\psi_n + \ldots, \tag{11.1}$$

where $c_1, c_2, \ldots, c_n, \ldots$ are arbitrary complex amplitudes. If the superposed states differ by infinitesimal amounts, the sum in (11.1) will be replaced by an integral.

An important example of the latter type of superposition is the representation of an arbitrary wave field $\psi(x, y, z, t)$ as a superposition of de Broglie waves[6]:

$$\psi_p(x, y, z, t) = \frac{1}{(2\pi\hbar)^{\frac{3}{2}}} e^{-i(Et - \mathbf{p}\cdot\mathbf{r})/\hbar}. \tag{11.2}$$

The wave function of any state can be written in the form

$$\psi(x, y, z, t) = \iiint_{-\infty}^{\infty} c(p_x, p_y, p_z, t)\, \psi_p(x, y, z, t)\, \mathrm{d}p_x\, \mathrm{d}p_y\, \mathrm{d}p_z, \tag{11.3}$$

where $c(p_x, p_y, p_z, t)$ is the amplitude of the de Broglie wave having momentum $\mathbf{p}(p_x, p_y, p_z)$. This is evident, since (11.3) is just a representation of $\psi(x, y, z, t)$ as a triple Fourier integral. In order to see this, we put

$$\phi(p_x, p_y, p_z, t) = c(p_x, p_y, p_z, t)\, e^{-iEt/\hbar}. \tag{11.4}$$

Then, from (11.2), Formula (11.3) can be written

$$\psi(x, y, z, t) = \iiint_{-\infty}^{\infty} \phi(p_x, p_y, p_z, t)\, e^{i(p_x x + p_y y + p_z z)/\hbar} \frac{\mathrm{d}p_x\, \mathrm{d}p_y\, \mathrm{d}p_z}{(2\pi\hbar)^{\frac{3}{2}}}. \tag{11.5}$$

[6] The factor $1/(2\pi\hbar)^{\frac{3}{2}}$ is included for reasons of normalisation, the convenience of which will be seen shortly (cf. (12.6)).

Hence, using the familiar Fourier theorem to invert the integral (11.5), we find for each function ψ the amplitudes ϕ, and therefore c:

$$\phi(p_x, p_y, p_z, t) = \iiint_{-\infty}^{\infty} \psi(x, y, z, t)\, e^{-i(p_x x + p_y y + p_z z)/\hbar} \frac{\mathrm{d}x\,\mathrm{d}y\,\mathrm{d}z}{(2\pi\hbar)^{\frac{3}{2}}}. \qquad (11.6)$$

Thus we see that any state may be regarded as a superposition of de Broglie waves, i.e. of states with a given momentum $\mathbf{p}(p_x, p_y, p_z)$ of the particle.

12. Momentum probability distribution of a microparticle

We have shown how the position probability of a particle can be determined from the statistical interpretation of de Broglie waves. We shall now see that the superposition principle makes it possible to extend the statistical interpretation, so that we can determine not only the probability of various values of the co-ordinates of the particle, but also that of various values of its momentum $\mathbf{p}$.

We shall regard de Broglie's formula $\mathbf{p} = \hbar\mathbf{k}$ ($|\mathbf{k}| = 2\pi/\lambda$) as a definition of a quantity $\mathbf{p}$, which in quantum mechanics is called the *momentum* of a particle.[7] Consequently, the measurement operations which determine $\mathbf{p}$ are similar to those which are needed to determine the direction of propagation of the wave and its wavelength λ. Hence a diffraction grating can be used as an apparatus for measuring the momenta of particles. The diffraction grating resolves the spectrum and separates waves with different $\mathbf{k}$, and therefore also 'sorts' particles with different momenta $\mathbf{p} = \hbar\mathbf{k}$.

We shall regard a diffraction experiment which provides a determination of $\mathbf{k}$ as a 'direct' experiment determining also the momentum $\mathbf{p}$ of the particle.

In order now to discuss the problem of determining the probability of some value of the momentum of the particle, let us consider the experiment of diffraction of particles (for instance, electrons) at the surface of a crystal. The superposition of de Broglie waves which form the wave field $\psi(x, y, z, t)$ in diffraction at a crystal surface is shown diagrammatically in Figure 14, including the incident (i), reflected (r) and one diffracted (d) wave. In accordance with the actual conditions it is assumed that the primary wave is a beam bounded by a diaphragm. The secondary waves are similar beams.

Each beam can be represented as a de Broglie wave $\psi_p(x, y, z, t)$ with amplitude $c(\mathbf{p})$ which varies only slowly in a direction perpendicular to the beam.[8] The total wave field ψ is represented as a superposition of fields belonging to the individual beams:

[7] There may arise, in connection with our definition of the momentum of a microparticle, the question why the quantity $\mathbf{p} = \hbar\mathbf{k}$ should be called the momentum. The answer is that the quantity thus defined in fact has properties which are entirely analogous to those of the momentum $\mathbf{p}_{cl}$ in classical mechanics (cf. Sections 32, 33, 103). It is shown in Section 34 that the classical momentum $\mathbf{p}_{cl}$ (which satisfies Newton's equation) is the mean quantum momentum: $\mathbf{p}_{cl} = \bar{\mathbf{p}}$. In particular, for a state with a definite value of $\mathbf{p}$ we have $\mathbf{p}_{cl} = \mathbf{p}$. Consequently, $\mathbf{p}$ can be measured, for instance by recoil in collision, in the same way as $\mathbf{p}_{cl}$ in classical mechanics.

[8] Outside the beam $c(\mathbf{p}) = 0$. Thus, unlike (11.3), the amplitudes considered here are functions of the co-ordinates. But, in view of the slowness of the variation, they are close to the true Fourier amplitudes given in (11.3).

$$\psi = \sum_p c(\mathbf{p})\,\psi_p, \tag{12.1}$$

where the sum is taken over all the beams.

The state ψ as a whole is one with an indeterminate particle momentum, since it is a superposition of states ψ_p with different momenta. Hence, if we carry out a measurement of the particle momentum, we can obtain in each separate measurement one of the values $\mathbf{p}$ contained in the superposition (12.1).

What is the probability that we shall obtain the value $\mathbf{p}$ of the momentum? The diffraction grating resolves the wave field into monochromatic (actually, almost mono-

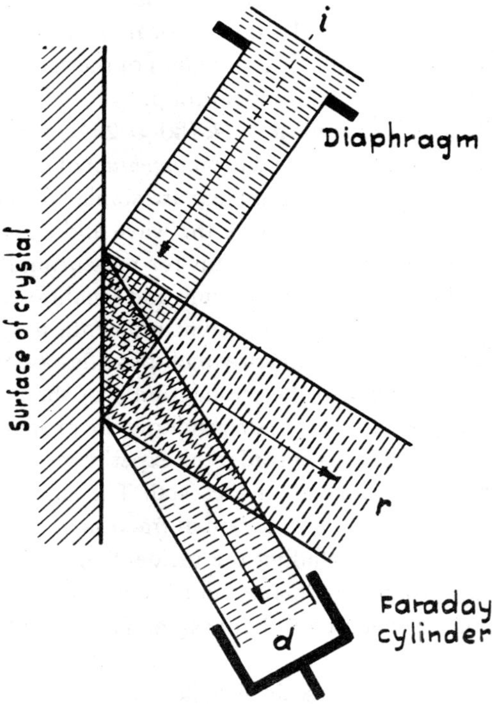

Fig. 14. With a bounded primary beam i the individual partial waves r, d, etc. are separated in space.

chromatic) beams, just as it resolves white light into the various pure spectral components. In order to compute the number of particles having momentum $\mathbf{p}$, we place a Faraday cylinder in various positions and determine the number of particles which enter it. Near the surface of the crystal we have a complex wave field resulting from the interference of all the beams, but far from the crystal the beams are separated. The probability that a particle is found in the cylinder will, according to the statistical interpretation of the wave function, be proportional to $|\psi(x, y, z, t)|^2$. If the Faraday cylinder is placed sufficiently far from the crystal, the individual beams will have separated, and

$$|\psi(x, y, z, t)|^2 = |c(\mathbf{p})|^2 \, |\psi_p(x, y, z, t)|^2 \tag{12.2}$$

(for one beam). Using the value of ψ_p (11.2), we obtain

$$|\psi(x, y, z, t)|^2 = |c(\mathbf{p})|^2/(2\pi \hbar)^3 \,. \tag{12.3}$$

Consequently $|c(\mathbf{p})|^2$ is proportional to the probability of finding an electron in the Faraday cylinder when the latter is placed so as to receive the wave ψ_p. This wave belongs to electrons having momentum $\mathbf{p}$. Hence $|c(\mathbf{p})|^2$ is proportional to the probability of finding in the state ψ an electron with momentum $\mathbf{p}$.

Bearing in mind (10.2) and the fact that the probability of finding a particle momentum in the range $p_x, p_x + \mathrm{d}p_x; p_y, p_y + \mathrm{d}p_y; p_z, p_z + \mathrm{d}p_z$ must be proportional to $\mathrm{d}p_x \, \mathrm{d}p_y \, \mathrm{d}p_z$, we arrive at the expression

$$\mathrm{d}W(p_x, p_y, p_z, t) = |c(p_x, p_y, p_z, t)|^2 \, \mathrm{d}p_x \mathrm{d}p_y \mathrm{d}p_z \tag{12.4}$$

and for the probability density

$$w(p_x, p_y, p_z, t) = |c(p_x, p_y, p_z, t)|^2 \,. \tag{12.5}$$

The above formulae involve a certain choice of normalisation of probabilities for the momentum.

Since $\phi(p_x, p_y, p_z, t)$ is, according to (11.6), the component of the Fourier series expansion of the function $\psi(x, y, z, t)$, it is easy to show that

$$\iiint_{-\infty}^{\infty} |c(p_x, p_y, p_z, t)|^2 \, \mathrm{d}p_x \mathrm{d}p_y \mathrm{d}p_z = \iiint_{-\infty}^{\infty} |\psi(x, y, z, t)|^2 \, \mathrm{d}x \, \mathrm{d}y \, \mathrm{d}z \,. \tag{12.6}$$

The left-hand side is the probability of finding some value of the momentum of the particle (an event certain to occur); the right-hand side is the probability of finding the particle at some point in space (also an event certain to occur). Hence the choice of normalisation made above is appropriate, since the probabilities of events certain to occur are equal. In particular, if the probability of finding the particle at some point is taken as unity, then the probability of finding some momentum is also unity.

13. Mean values of functions of co-ordinates and functions of momenta

In the preceding sections we have determined the position probability of a particle (10.3) in a state ψ and the momentum probability distribution of a particle (12.5) in that state. This enables us to write down immediately the mean values of any function $F(x, y, z)$ of the co-ordinates of a particle and any function $F(p_x, p_y, p_z)$ of the momentum of a particle for a state represented by the wave function ψ. From (10.3) and (12.5) we have, using the definition of the mean value of a random quantity,

$$\begin{aligned} \overline{F(x, y, z)} &= \int F(x, y, z) \, |\psi(x, y, z)|^2 \, \mathrm{d}x \, \mathrm{d}y \, \mathrm{d}z \\ &= \int \psi^*(x, y, z) \, F(x, y, z) \, \psi(x, y, z) \, \mathrm{d}x \, \mathrm{d}y \, \mathrm{d}z \end{aligned} \tag{13.1}$$

with the condition

$$\int |\psi(x, y, z)|^2 \,\mathrm{d}x\,\mathrm{d}y\,\mathrm{d}z = 1 \tag{13.2}$$

and

$$\begin{aligned}\overline{F(p_x, p_y, p_z)} &= \int F(p_x, p_y, p_z)\, |c(p_x, p_y, p_z)|^2 \,\mathrm{d}p_x\,\mathrm{d}p_y\,\mathrm{d}p_z \\ &= \int c^*(p_x, p_y, p_z)\, F(p_x, p_y, p_z)\, c(p_x, p_y, p_z)\,\mathrm{d}p_x\,\mathrm{d}p_y\,\mathrm{d}p_z\,,\end{aligned} \tag{13.3}$$

with the condition

$$\int |c(p_x, p_y, p_z)|^2 \,\mathrm{d}p_x\,\mathrm{d}p_y\,\mathrm{d}p_z = 1\,; \tag{13.4}$$

here the integrals are taken over the whole range of variation of the variables x, y, z and p_x, p_y, p_z respectively.

Formulae (13.1) and (13.3) can undergo a very important transformation based on the properties of Fourier integrals. Let $F(x, y, z)$ be a rational integral function of x, y, z, and $F(p_x, p_y, p_z)$ a similar function of p_x, p_y, p_z. Then Formulae (13.1) and (13.3) can be written in the forms[9]

$$\begin{aligned}\overline{F(x, y, z)} = \int c^*(p_x, p_y, p_z)\, F\left(i\hbar\frac{\partial}{\partial p_x}, i\hbar\frac{\partial}{\partial p_y}, i\hbar\frac{\partial}{\partial p_z}\right) \times \\ \times\, c(p_x, p_y, p_z)\,\mathrm{d}p_x\,\mathrm{d}p_y\,\mathrm{d}p_z\,,\end{aligned} \tag{13.5}$$

$$\begin{aligned}\overline{F(p_x, p_y, p_z)} = \int \psi^*(x, y, z)\, F\left(-i\hbar\frac{\partial}{\partial x}, -i\hbar\frac{\partial}{\partial y}, -i\hbar\frac{\partial}{\partial z}\right) \times \\ \times\, \psi(x, y, z)\,\mathrm{d}x\,\mathrm{d}y\,\mathrm{d}z\,.\end{aligned} \tag{13.6}$$

These formulae signify that the arguments of the function F are to be replaced by differential operators with respect to the arguments shown, multiplied by $\pm\, i\hbar$, acting on the c or ψ function which follows. For example, to calculate the mean value of the momentum component p_x we proceed thus: $F(p_x, p_y, p_z) = p_x$ and so

$$\overline{p_x} = \int c^*(p_x, p_y, p_z)\, p_x\, c(p_x, p_y, p_z)\,\mathrm{d}p_x\,\mathrm{d}p_y\,\mathrm{d}p_z\,, \tag{13.7}$$

or, from Formula (13.6), replacing p_x by $-\, i\hbar\, \partial/\partial x$,

$$\overline{p_x} = -\int \psi^*(x, y, z)\, i\hbar \frac{\partial \psi(x, y, z)}{\partial x}\,\mathrm{d}x\,\mathrm{d}y\,\mathrm{d}z\,. \tag{13.8}$$

Similarly, the mean value of p_x^2 may be calculated either from Formula (13.3):

$$\overline{p_x^2} = \int c^*(p_x, p_y, p_z)\, p_x^2\, c(p_x, p_y, p_z)\,\mathrm{d}p_x\,\mathrm{d}p_y\,\mathrm{d}p_z\,, \tag{13.9}$$

or from Formula (13.6), replacing $F(p_x) = p_x^2$ by

$$F\left(-i\hbar\frac{\partial}{\partial x}\right) = \left(-i\hbar\frac{\partial}{\partial x}\right)^2 = -\hbar^2\frac{\partial^2}{\partial x^2}\,. \tag{13.10}$$

[9] The proof that (13.1) and (13.3) are respectively equivalent to (13.5) and (13.6) is given in Appendix I.

This gives

$$\overline{p_x^2} = -\hbar^2 \int \psi^*(x, y, z) \frac{\partial^2 \psi(x, y, z)}{\partial x^2} \mathrm{d}x \, \mathrm{d}y \, \mathrm{d}z \, . \tag{13.11}$$

14. Statistical ensembles in quantum mechanics

In practical physics and engineering there are two important types of problem to be answered by means of quantum mechanics.

The first problem is to predict from the wave function the possible results of measurements on a microparticle (the 'direct' problem). The second type of problem is to determine from experimental results the wave function of the particle (the 'inverse' problem).

The predictions which result from a knowledge of the wave function are in general statistical ones. If any single measurement is made, the result of it will therefore show only to what extent our expectations were justified – whether a probable or improbable event has occurred.

The only objective results are the distributions of results of measurement, obtained when measurements are repeatedly carried out in a large number of identical experiments.

It is important to note that in the quantum region we cannot repeat an experiment on the same particle, since measurement may in general change the state of microparticles (Section 16).

In carrying out a large number ($N \gg 1$) of identical experiments it is therefore necessary to imagine a large number of particles or systems which are independently in the same macroscopic conditions. We call such a group of microparticles (or systems) a *quantum ensemble of particles*, or simply an *ensemble*.

If these macroscopic conditions are such that they entirely determine the state of the microparticles (see Section 28, where the concept of a complete set of quantities needed to define this state is introduced), then the state of these particles may be described by a single wave function ψ. In this case the ensemble itself is called a *pure ensemble*.

All probabilities and all mean values calculated from the wave function pertain to measurements in such an ensemble. For example, the statement that the probability of finding the particle co-ordinate x to lie near x' is $|\psi(x')|^2 \, \mathrm{d}x'$ means that if a large number of measurements of the co-ordinate are made in a series of identical experiments (with the same ψ), we shall find x near x' in N' cases, where

$$N'/N = |\psi(x')|^2 \, \mathrm{d}x' \, . \tag{14.1}$$

Similarly, on measuring the particle momentum p_x in the same ensemble, out of a total of M measurements ($M \gg 1$), p_x will be found near p_x' in M' cases, where

$$M'/M = |c(p_x')|^2 \, \mathrm{d}p_x' \tag{14.2}$$

and $c(p_x')$ is the amplitude in the expansion of $\psi(x)$ in de Broglie waves (cf. Section 12).

Knowing the distribution of the results of measurements for x (14.1) and for p_x (14.2), we can calculate the mean values of any functions $F(x)$ and $\Phi(p)$, for instance the mean values $\bar{x}$ and $\overline{p_x}$, the mean square deviations

$$\overline{\Delta x^2} = \overline{(x - \bar{x})^2}\,, \tag{14.3}$$

$$\overline{\Delta p_x^2} = \overline{(p_x - \overline{p_x})^2}\,, \tag{14.4}$$

etc.

We shall subsequently show that if the wave function ψ is known the probabilities may be calculated not only for x and p_x but also of any result of measuring any mechanical quantity pertaining to the particle or system considered.

It is quite clear that the wave function of a single microparticle cannot be determined by one measurement on it. If the distributions of results of measurements in an ensemble are known, however, the inverse problem may be solved, i.e. the particle wave function may be derived from the results of measurement (apart from a common normalising factor, of course, which always remains indeterminate) (Section 79).

Thus not only do the predictions of quantum mechanics apply to measurements in a quantum ensemble, but conversely the nature of this ensemble may be determined from measurements. Hence the state of a particle (or system) described by a wave function must be regarded as expressing the fact that the particle (or system) belongs to a definite pure quantum ensemble. It is in this sense that we shall use the words 'state of a particle', 'state of a quantum system', etc.

Let us now give a specific example of a pure ensemble, considering first of all the scattering of one electron by an isolated atom. Let the momentum of the electron be **p**. Then the wave function $\Psi_p(x)$ of the electron is a superposition of a de Broglie wave $\psi_p(x)$ representing the initial state of the electron with momentum **p**, and a wave $u(x)$ representing the wave scattered by the atom:

$$\Psi_p(x) = \psi_p(x) + u(x)\,. \tag{14.5}$$

If the scattered wave is known, then the fate of the scattered electron can be statistically predicted (see collision theory, Chapter XIII). But how can this experiment be repeated many times?

Suppose that the electrons are emitted from a heated filament, that a beam in a certain direction is selected by diaphragms, and that the electrons are given a definite velocity by applying an accelerating potential. Let this beam be directed into a gas, and the intensity of electron scattering at various angles be observed. If the gas density is low and the layer in which electron scattering occurs is not very thick, multiple scattering of electrons may be neglected. If, moreover, the density of electrons in the primary beam is so low that their interaction with one another may be neglected, this immediately provides a large number of independent experiments on the scattering of one electron by one atom. Finally, if the velocity acquired by the electrons in the accelerating field is much greater than their thermal velocity, and if the diaphragms select the beam sufficiently well, we can say that we have electrons with a definite momentum **p**, and therefore assign to them a wave function ψ_p which, together with the scattered wave, gives Ψ_p.

Thus in practice we can bring about an assembly of identical phenomena described by the same wave function $\Psi_p(x)$, i.e. a pure quantum ensemble.

From the viewpoint of quantum mechanics, the specification of the state of a particle by means of the wave function is the most complete and exhaustive. In reality we often meet with cases where the ensemble contains from the start particles in different states, described by different wave functions, $\psi_1, \psi_2, \ldots, \psi_n$. The probabilities $P_1, P_2, \ldots, P_n$ of each of these states are given. This is called a *mixed ensemble*.

It is evident that the quantities $P_1, P_2, \ldots, P_n$ give the probability of finding in the mixed ensemble the corresponding pure ensembles with wave functions $\psi_1, \psi_2, \ldots, \psi_n$.

An example of a mixed ensemble is the case where no accelerating potential is applied to electrons leaving a heated filament. Here the momentum of the electrons is determined only by the temperature T of the filament. The primary electrons will then have a Maxwellian distribution. The probability that the electron momentum lies in the range $p_x, p_x + \mathrm{d}p_x; p_y, p_y + \mathrm{d}p_y; p_z, p_z + \mathrm{d}p_z$ is

$$\mathrm{d}P_p = Ce^{-p^2/2\mu kT}\,\mathrm{d}p_x\,\mathrm{d}p_y\,\mathrm{d}p_z\,, \tag{14.6}$$

where μ is the mass of the electron, k Boltzmann's constant, and C a normalising factor such that $\int \mathrm{d}P = 1$. The electrons with momentum $\mathbf{p}$ will be described by a de Broglie wave function $\psi_p(x)$; hence $\mathrm{d}P_p$ (14.6) is just the probability that the electron will have the wave function $\psi_p(x)$, i.e. will belong to the *pure* ensemble $\psi_p(x)$ which forms part of the entire *mixed* ensemble under consideration.

A similar mixed ensemble occurs in Stern and Estermann's experiments on the diffraction of helium by lithium fluoride, where the distribution of momenta of helium atoms in the primary beam is determined by the temperature of the furnace. On the other hand, in Davisson and Germer's experiments we can completely ignore the thermal velocities of the electrons in comparison with the velocity which they acquire in the accelerating field. It can be assumed without great error that all the electrons have the same momentum $\mathbf{p}$. In these experiments, therefore, we have practically a pure ensemble described by the wave function ψ_p.

It may be noted that frequently no measurements are made in determining the initial state of particles; it is simply assumed that some particular pure or mixed ensemble exists. The validity of this assumption is afterwards tested from its observable or measurable consequences. The wave function or (for a mixed ensemble) the set of wave functions must therefore be regarded as a purely objective characteristic of a quantum ensemble, independent of the observer.

In conclusion, we may point out one important difference between a pure and a mixed ensemble, which might escape notice. Either a pure or a mixed ensemble can be formed from given wave functions. If the particular states $\psi_1, \psi_2, \ldots, \psi_n, \ldots$ are given, we can form from them a wave function Ψ which is a superposition of these states:

$$\Psi = \sum c_n \psi_n\,, \tag{14.7}$$

which describes a pure ensemble. The particular states appear in this superposition with definite phases and amplitudes ($c_n = |c_n|e^{i\alpha_n}$, where α_n is the phase).

If, however, it is known that the system can be in the state ψ_1 with probability P_1, in the state ψ_2 with probability P_2, etc., then we have a mixed ensemble, which must be described by two series of quantities[10]:

$$\begin{cases} \psi_1, \psi_2, \ldots, \psi_n, \ldots, \\ P_1, P_2, \ldots, P_n, \ldots . \end{cases} \tag{14.8}$$

Let us now calculate the probability that a particle is at the point x. For a pure ensemble the probability density is

$$w(x) = |\psi(x)|^2 = \sum |c_n \psi_n(x)|^2 + \sum_{n \neq m} \sum_m c_n^* c_m \psi_n^*(x) \psi_m(x). \tag{14.9}$$

In a mixed ensemble this probability must be calculated as follows. The probability that a particle is at the point x, if in the state $\psi_n(x)$, is $|\psi_n(x)|^2$. The probability of being in the state $\psi_n(x)$ is P_n. Hence the probability of the composite event is $P_n|\psi_n(x)|^2$, and the total probability $w(x)$ is

$$w(x) = \sum_n P_n |\psi_n(x)|^2 . \tag{14.10}$$

A comparison of (14.9) and (14.10) shows that in a pure ensemble there is interference between the various particular states, represented by terms such as $c_n^* c_m \psi_n^*(x) \psi_m(x)$; in a mixed ensemble no such interference occurs.

Thus the difference between pure and mixed ensembles as regards particular states is analogous to the composition of coherent and non-coherent light; in calculating probabilities, we combine the amplitudes in a pure ensemble and the intensities in a mixed ensemble.

15. The uncertainty relation

Let us now consider a very important property of quantum ensembles, known as the *uncertainty relation.*

In classical mechanics, it will be recalled, we are interested in the paths of particles and in their motion along these paths. It might be thought that quantum mechanics would give a statistical description of such a classical motion, as happens in classical statistical mechanics, but simple considerations show that this is not so. In the micro-universe, mechanical quantities are related differently from those found in the macro-universe of classical mechanics.

The concept of the motion of a particle along a path necessarily involves the assumption that at every instant the particle has a definite co-ordinate x and a definite momentum p_x. The former indicates the position of the particle, and the latter indicates how this position changes during an infinitesimal interval of time:

$$x + \mathrm{d}x = x + (p_x/m)\,\mathrm{d}t = x + v_x\,\mathrm{d}t, \tag{15.1}$$

where m is the mass of the particle and v_x its velocity.

[10] In Section 45 another method of describing a mixed ensemble is given, using the 'density matrix', a quantity analogous to the distribution function in classical statistical mechanics.

In a statistical ensemble the particles can have various momenta and co-ordinates, but if the ensemble is classical, we can always choose from it sub-ensembles with completely definite momenta and with completely definite co-ordinates. Such a procedure is, on the other hand, impossible for a quantum ensemble, and this indicates a relation between the position and momentum of the particle which is quite different from the classical relation. In order to discuss this very important property of microphenomena, we shall take as a basis experiments on the diffraction of microparticles. The fundamental conclusion from these experiments is expressed by de Broglie's formula relating the momentum and the wavelength:

$$p = 2\pi \hbar / \lambda . \tag{15.2}$$

If λ is taken as the wavelength, it cannot be a function of the co-ordinate x, whatever the nature of the waves. The expression 'the wavelength at the point x is λ' has no meaning, since by definition the wavelength is a characteristic of a sinusoidal wave of infinite extent in space (from $x = -\infty$ to $x = +\infty$). λ is a 'function' of the shape of the wave, but not of the co-ordinate of any point. Hence the right-hand side of (15.2) cannot be a function of the co-ordinate x, and so the left-hand side also, i.e. the momentum p, cannot be a function of the co-ordinate x.

Similarly, there is no answer to the question of the frequency with which a pendulum oscillates at a given instant, since the definition of the concept of frequency assumes that a large number of oscillations of the pendulum must be considered.[11]

Thus we conclude that, if de Broglie's relation (15.2) is accepted as valid, the particle momentum p cannot be a function of the particle co-ordinate x. In the microuniverse the statement 'the momentum of the particle at the point x is p' has no meaning. Accordingly, in the quantum region there are no ensembles in which both the momentum and the co-ordinates of particles simultaneously have completely definite values.

This important result may be proved first for the ensemble represented by the wave packet discussed in Section 7. As shown there, the wave packet

$$\psi(x, t) = \int_{k_0 - \Delta k}^{k_0 + \Delta k} c(k) e^{-i(\omega t - kx)} \mathrm{d}k \tag{15.3}$$

can be put in the form (see (7.9))

$$\psi(x, t) = 2c(k_0) \frac{\sin\{[(\mathrm{d}\omega/\mathrm{d}k)t - x]\Delta k\}}{(\mathrm{d}\omega/\mathrm{d}k)t - x} e^{-i(\omega_0 t - k_0 x)} . \tag{15.4}$$

The intensity $|\psi|^2$ in such a wave packet at some instant t is shown in Figure 15. As a measure of the size of the wave packet we can take twice the distance from the maximum of $|\psi|^2$ to the first minimum. Let this be $2\Delta x$. It follows from (15.4) that $\Delta x = \pi/\Delta k$. That is,

$$\Delta x \cdot \Delta k = \pi . \tag{15.5}$$

This is purely a wave relation valid for any waves, and shows that the product of the

[11] This analogy is due to L. I. Mandel'shtam.

linear dimensions Δx of the wave packet and the range Δk of wave numbers of the waves which form the wave packet is a constant and equal to π.

In particular, if a very short radio signal is to be transmitted (Δx small), then the various monochromatic waves represented in it with appreciable intensity will necessarily differ considerably in wavelength. Thus such a signal will be received by receivers tuned to various wavelengths. If, on the other hand, only receivers tuned in a certain way are to receive the signal, monochromatic signals must be used, and according to (15.5) they will be fairly long signals.

Returning now to quantum mechanics, we see from de Broglie's equation that $p_x = \hbar k$, and so, if k varies in the range Δk, the momentum p_x varies in the range

$$\Delta p_x = \hbar \Delta k . \tag{15.6}$$

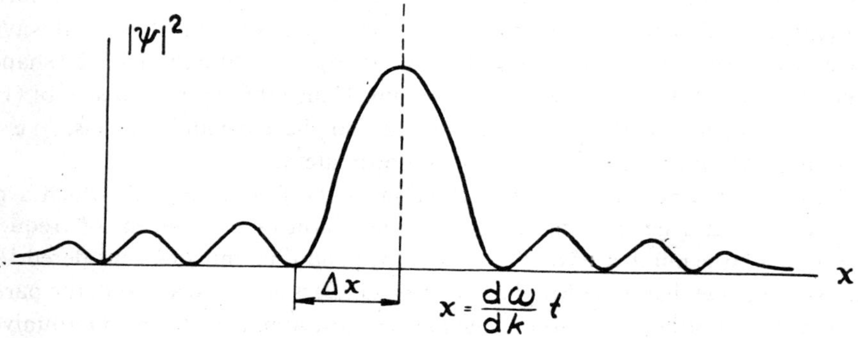

Fig. 15. The intensity $|\psi|^2$ in a wave packet as a function of x at some instant t.

Taking the wave packet (15.3) to be a de Broglie wave packet, we multiply Equation (15.5) by Planck's constant $\hbar$; then (15.6) gives

$$\Delta p_x \cdot \Delta x = \pi \hbar . \tag{15.7}$$

The meaning of Δp_x and Δx in Formula (15.7) may be seen as follows. If we make a measurement of the co-ordinates of particles in a state described by the de Broglie wave packet (15.3), then at time t the mean value of the results of measuring the co-ordinates will be $\bar{x} = (\mathrm{d}\omega/\mathrm{d}k)t$. The values of the results of individual measurements will be spread around $\bar{x}$, mainly in a range $\pm \Delta x$. The quantity Δx is the uncertainty in the co-ordinate x. If we measure the momentum p_x of the particles in the same state, the mean value will be $\overline{p_x} = p_0 = \hbar k_0$, and the individual values will be concentrated around p_0 in the range $\Delta p_x = \pm \hbar \Delta k_x$. The quantity Δp_x is the uncertainty in the momentum p_x.

The relation (15.7) is therefore called the *uncertainty relation* for the momentum p_x and the corresponding co-ordinate x. This relation was first established by Heisenberg. It is one of the most fundamental results of modern quantum mechanics and shows that the narrower the wave packet (i.e. the more definite the value of the particle co-ordinates; Δx small) the less definite the particle momentum (Δp_x large), and *vice versa.*

Let us now consider the proof of the uncertainty relation for any state of the particle, described by an arbitrary wave function ψ. For simplicity we shall take only one spatial dimension; the generalisation to a larger number of dimensions is entirely trivial. Suppose that some state of the particle is given, represented by the wave function $\psi(x)$.[12] We assume that the wave function is normalised to unity in the range from $-\infty$ to $+\infty$.

In order to establish the uncertainty relation in a rigorous form, we must first choose a measure of the deviation of individual results of measurements of the momentum p and the co-ordinate x from their mean values $\bar{p}$ and $\bar{x}$; in other words, we must define more precisely what we mean by the 'uncertainties' Δp_x and Δx.

As such a measure we take the mean square deviations $\overline{\Delta p_x^2}$ and $\overline{\Delta x^2}$ used in statistics.[13] These quantities are defined as follows. Let $\bar{x}$ be the mean value of the quantity x. If in some individual measurement we obtain a value x, then $\Delta x = x - \bar{x}$ is the deviation of this result from the mean value $\bar{x}$. The mean value of this deviation is evidently always zero: $\overline{\Delta x} = \bar{x} - \bar{\bar{x}} = \bar{x} - \bar{x} = 0$. For this reason the measure of deviation of individual measurements from the mean is taken not as $\overline{\Delta x}$ but as $\overline{\Delta x^2}$, the mean of the squares of the individual deviations.

Thus we can write

$$\overline{\Delta x^2} = \overline{(x - \bar{x})^2} = \overline{x^2} - \bar{x}^2, \tag{15.8}$$

$$\overline{\Delta p_x^2} = \overline{(p_x - \overline{p_x})^2} = \overline{p_x^2} - \overline{p_x}^2. \tag{15.9}$$

Without loss of generality in the proof we can use a suitable system of co-ordinates, taking the origin at the point $\bar{x}$. Then $\bar{x} = 0$. Also, let this system move with the centroid of the distribution $\bar{x}$. Then $\overline{p_x} = 0$. In this system of co-ordinates, (15.8) and (15.9) become

$$\overline{\Delta x^2} = \overline{x^2}, \tag{15.10}$$

$$\overline{\Delta p_x^2} = \overline{p_x^2}. \tag{15.10'}$$

From (13.1) and (13.11) we have

$$\overline{\Delta x^2} = \overline{x^2} = \int_{-\infty}^{\infty} \psi^*(x)\, x^2\, \psi(x)\, \mathrm{d}x, \tag{15.11}$$

$$\overline{\Delta p_x^2} = \overline{p_x^2} = -\hbar^2 \int_{-\infty}^{\infty} \psi^*(x) \frac{\mathrm{d}^2\psi(x)}{\mathrm{d}x^2}\, \mathrm{d}x. \tag{15.11'}$$

Our problem is to establish a relation between $\overline{\Delta p_x^2}$ and $\overline{\Delta x^2}$. To do so, we use the auxiliary integral

$$I(\xi) = \int_{-\infty}^{\infty} |\xi x \psi + \mathrm{d}\psi(x)/\mathrm{d}x|^2\, \mathrm{d}x, \tag{15.12}$$

[12] We need not include the time t explicitly, since the subsequent discussion is valid at any instant.
[13] The quantities $\sqrt{(\overline{\Delta p_x^2})}$ and $\sqrt{(\overline{\Delta x^2})}$ are called *standard deviations* or *dispersions*.

where ξ is a real auxiliary variable. Expanding the squared modulus gives

$$I(\xi) = \xi^2 \int_{-\infty}^{\infty} x^2 |\psi|^2 \, dx + \\ + \xi \int_{-\infty}^{\infty} x \left(\frac{d\psi^*}{dx} \psi + \psi^* \frac{d\psi}{dx} \right) dx + \int_{-\infty}^{\infty} \frac{d\psi^*}{dx} \frac{d\psi}{dx} \, dx . \tag{15.13}$$

Putting

$$A = \int_{-\infty}^{\infty} x^2 |\psi|^2 \, dx = \overline{\Delta x^2} , \tag{15.14}$$

$$B = - \int_{-\infty}^{\infty} x \frac{d}{dx} (\psi^* \psi) \, dx = \int \psi^* \psi \, dx = 1 , \tag{15.14'}$$

$$C = \int_{-\infty}^{\infty} \frac{d\psi^*}{dx} \frac{d\psi}{dx} \, dx = - \int \psi^* \frac{d^2 \psi}{dx^2} \, dx = \frac{\overline{\Delta p_x^2}}{\hbar^2} ; \tag{15.14''}$$

where integration by parts has been used[14], we find

$$I(\xi) = A\xi^2 - B\xi + C \geqslant 0 . \tag{15.15}$$

Since $I(\xi)$ is non-negative (for real ξ), this means that the roots of the equation

$$I(\xi) = 0 \tag{15.16}$$

are complex. A theorem concerning the roots of a quadratic equation shows that this can be true only if

$$4AC \geqslant B^2 . \tag{15.17}$$

Substituting in this inequality the values of A, B, C from (15.14), (15.14′) and (15.14″), we arrive at the desired relation between $\overline{\Delta p_x^2}$ and $\overline{\Delta x^2}$:

$$\overline{\Delta p_x^2} \cdot \overline{\Delta x^2} \geqslant \tfrac{1}{4} \hbar^2 . \tag{15.18}$$

This is the uncertainty relation in its most general and rigorous form. We have also shown that there are no quantum ensembles such that the mean square deviations for the momentum ($\overline{\Delta p_x^2}$) and the corresponding co-ordinate ($\overline{\Delta x^2}$) are simultaneously zero.

It is seen, on the contrary, that, as the mean square deviation of one quantity decreases, that of the other quantity increases. Hence it follows that no experiment can be imagined which should give a physical determination of both x and p_x, since the possibility of carrying out such an experiment assumes the existence of states in which $\overline{\Delta p_x^2} = 0$ an d$\overline{\Delta x^2} = 0$ simultaneously, and this contradicts the uncertainty relation, which is ultimately based on de Broglie's equation $p = 2\pi\hbar/\lambda$. The operations which are used where de Broglie's relation is valid (the microuniverse) to measure the co-

[14] Together with the fact that since $\psi^*\psi$ is integrable, ψ and its derivatives must vanish for $x \to \pm \infty$.

ordinate x of a particle and its momentum p_x must be mutually exclusive: particles can be classified either as regards momenta or as regards co-ordinates.[15] This is expressed by the fact that any determination of position leads to a change in momentum, predicted statistically by quantum mechanics.

The alteration of momentum by a determination of position makes it impossible to apply to the motion of microparticles the concept of a path. Thus quantum mechanics is concerned with fundamentally new types of object, which are not subject to the classical laws of motion of point masses.

The name 'uncertainty principle' emphasises this: the uncertainty arises only when classical quantities are invalidly employed for objects of a different nature. This will be illustrated in the next section.

16. Illustrations of the uncertainty relation

Let us first consider the measurement of the co-ordinate of a particle by means of a slit. The initial state will be described by a plane de Broglie wave ψ_p. Let the wave be propagated in the direction of the axis OX. This state has the property that the momentum of the particle has a completely definite value, viz.

$$p_x = p, \qquad p_y = p_z = 0. \tag{16.1}$$

Thus we have an ensemble of particles with given momentum.

The position (co-ordinates) of particles in this ensemble is, on the other hand, completely indeterminate; $|\psi_p|^2 =$ constant, and so all positions of the particles are equally probable. Let us suppose that we attempt to fix the value of just one co-ordinate, for instance y. To do so we place a screen with its plane perpendicular to the direction of propagation of the waves, the screen having a slit, as shown in Figure 16. Let the half-width of the slit be d. If a particle passes through the slit, then at the instant when it

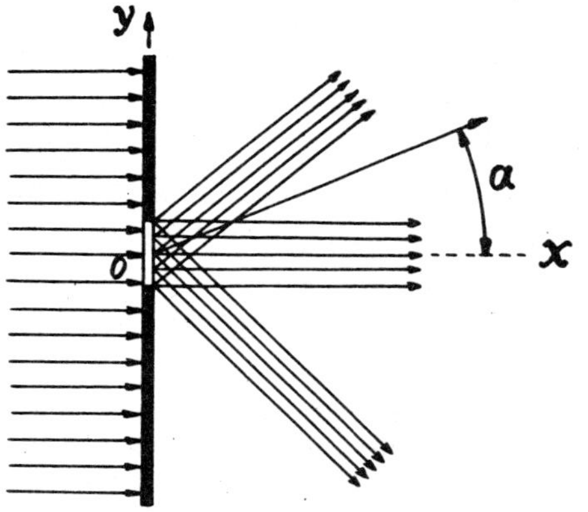

Fig. 16. Illustration of the measurement of p_y and y: diffraction from a slit in a screen.

[15] In the author's paper [8] it is shown that there is no distribution function depending on p and x which could represent a quantum ensemble. Cf. also Section 45 below.

does so its co-ordinate is fixed by the position of the slit to within the half-width d of the slit. Since the momentum in the y direction is known ($p_y = 0$), it appears at first sight that we have determined both the momentum p_y and the co-ordinate y. This is not so, however. In the above analysis the fact has been overlooked that near the slit diffraction will occur; the waves will be deflected from their original direction of propagation. The momentum of the particles is also changed when the screen is put in, and will not be the same as before.

The mean value of the momentum p_y along the axis OY remains unchanged: $\overline{p_y} = 0$, since diffraction around the slit is symmetrical. Let us estimate the order of magnitude of the possible deviation Δp_y of the momentum from its mean value. If we deflect a ray from the axis OX, it soon takes a position corresponding to the first diffraction minimum, then the first diffraction maximum, and so on. Let the angle made by this ray with the axis OX be α. Then the greatest wave intensity will occur in the range from $-\alpha$ to $+\alpha$. The angle α is given by the condition that for this direction the rays from the two halves of the slit cancel one another (phase difference π). If the wavelength is denoted by λ, the desired angle is given by the well-known relation

$$\sin\alpha = \lambda/2d\,. \tag{16.2}$$

The half-width d of the slit is just the uncertainty Δy in the measurement of the co-ordinate y.

The component of momentum along the axis OY is $p \sin\alpha$. Since the intensity of the de Broglie waves falls off mainly in the range from $-\alpha$ to $+\alpha$, when the momentum is measured most of the results will lie in the range from $-p\sin\alpha$ to $+p\sin\alpha$, i.e. the spread of the measured values around the mean value $\overline{p_y} = 0$ is $\Delta p_y = p\sin\alpha$. Since from de Broglie's relation $p = 2\pi\hbar/\lambda$, substitution in Equation (16.2) of Δp_y instead of $(2\pi\hbar/\lambda)\sin\alpha$ and Δy instead of d gives

$$\Delta p_y \cdot \Delta y = \pi\hbar\,. \tag{16.3}$$

This relation shows that the more accurate the determination of the position of the particles (Δy small, i.e. a narrow slit), the greater the uncertainty in their momentum (Δp_y large) and *vice versa*.[16]

Owing to diffraction at the slit, the measurement of the co-ordinate makes the momentum p_y indeterminate, i.e. after passing the slit a particle belongs to a new ensemble in which Δp_y is not zero.

Another example is provided by the photographic plate. Let us consider an idealised

[16] It may be noted that in our derivation of this relation we have used the fact that the wavelength λ, and therefore the total momentum p of the particle, are not altered by diffraction. The greatest value of Δp_y is therefore p, corresponding to a particle moving along the screen. It may therefore seem that by using no greater accuracy than $\Delta p_y = p$ we can achieve arbitrarily high accuracy in the determination of the co-ordinate y by reducing the width of the slit. This would, of course, contradict (15.6), and is in fact not so. Our treatment is approximate, and is valid provided that the wavelength λ is of the order of the width of the slit. As the width of the slit decreases, the wave field behind the screen becomes more complex. This field cannot be assigned a definite wavelength λ as we have done here. An analysis of this case shows that the relation (15.6) remains valid.

situation.[17] The idealisation is essentially that we identify the plate with a system of fixed atoms, and the formation of an image on the plate with the ionisation of such an atom. In reality the ionisation of an active atom is only the beginning of processes which ultimately lead to the formation on the plate of a developed grain which is experimentally observed.

An atom may be regarded as fixed or moving slowly near some point only if it is sufficiently heavy.[18] An 'ideal' plate must consist of infinitely heavy atoms which also have small dimensions a, since these determine the region in which ionisation has occurred.

It will be shown later (Section 51) that the wave function of an electron in an atom is non-zero in a region of order $a = \hbar/\sqrt{(2\mu I)}$, where I is the ionisation energy of the atom and μ the mass of the electron. The quantity a is equal, in order of magnitude, to the uncertainty in the position of the electron in the atom. This electron will therefore have an uncertainty in momentum $\Delta p \cong \hbar/a$. In this experiment we cannot establish the point at which ionisation of the atom occurred; we know only that the region in which the collision occurred has dimensions approximately equal to a. The co-ordinate x of an atom incident on the plate is therefore determined with, at best, an accuracy $\Delta x \cong a$. On the other hand, since the collision occurs with an electron of an atom having an uncertainty in momentum Δp of the order of $\hbar/a$, the electron whose co-ordinate is being measured will have a similar uncertainty in momentum Δp_x after the collision. Multiplication of $\Delta x \cong a$ by $\Delta p_x \cong \hbar/a$ gives

$$\Delta p_x \cdot \Delta x \cong \hbar . \tag{16.4}$$

Measurement of the co-ordinates of particles always involves an important interaction of the measuring apparatus with the particles. In the case of the photographic process for measuring the position of a particle, the condition for an observation of the co-ordinate to be possible is that an atom should be ionised. This requires energy I, which is taken from the energy of the particle itself. If the original momentum of the particles is p_0, we must have

$$p_0^2/2\mu > I = \hbar^2/2\mu a^2 = \hbar^2/2\mu \overline{\Delta x^2} . \tag{16.5}$$

If this does not hold, the photographic method is not possible.

Observation of the track of a particle in a bubble chamber is similar to this photographic method, since the track is formed by successive ionisations of the atoms of the liquid in the chamber, i.e. is a sequence of successive 'photographs' in the above sense (Figure 17).[19]

[17] The image formation process described here occurs, for instance, in the experiments of Zhdanov [92], where the track of a cosmic-ray particle through a photosensitive emulsion is found. The formation of tracks in a bubble chamber is of the same type (see below).

[18] For, putting in the uncertainty relation $\Delta p_x = M\Delta v_x$, where M is the mass of the atom and v_x its velocity, we find $\Delta v_x = \hbar/M\Delta_x$. Hence it follows that Δx and Δv_x can simultaneously be small only if M is large. An infinitely heavy particle could therefore have a definite position and also a definite velocity (in particular, be at rest).

[19] In a bubble chamber we observe the track of a particle not by means of ions but by means of bubbles

From (16.5) we can conclude that, to obtain a track in a bubble chamber, it is necessary that the momentum p_0 of the particle being photographed should satisfy the inequality $p_0 > \sqrt{(2\mu I)}$.

Let us now consider the indirect determination of the co-ordinates of micro-particles. We shall show that in this case also ensembles which satisfy the uncertainty relation will be produced. An example of an indirect experiment is the determination

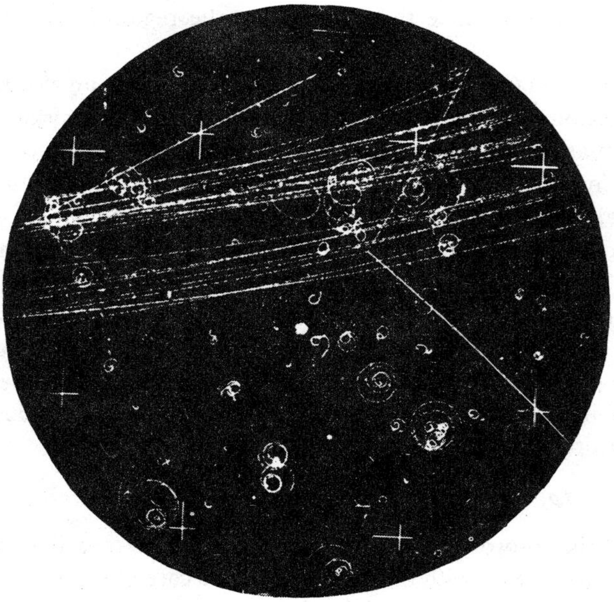

Fig. 17. Tracks of π^- mesons of energy 340 MeV in a liquid-hydrogen bubble chamber. In the centre of the picture is an elastic collision between a π^- meson and a proton. The thick track leading upwards is that of a recoil proton. (Nuclear Problems Laboratory, Dubna.)

of the position of particles by means of a microscope (Figure 18). A particle near $x = 0$ is illuminated with light of wavelength λ. The beam of light is parallel to the axis OX. Scattered light will enter the microscope objective. It is known from the theory of the microscope that the position of the particle is determined with accuracy $\Delta x \cong \lambda/\sin \varepsilon$, where 2ε is the angle subtended by the objective at the position of the object.[20] Thus an ensemble of particles with $\overline{\Delta x} \cong \lambda/\sin \varepsilon$ can be selected. When λ is sufficiently small, the quantity $\overline{\Delta x}$ may in principle be arbitrarily small. However, the momentum of

formed as a result of ionisation of the liquid. Before the track is photographed, the bubbles are able to grow considerably. Hence the practical accuracy of the determination of the position of a particle by the bubble-chamber method is very much less than the theoretical accuracy determined by the dimensions of an atom; it is in fact determined by the size of the bubbles and the optical properties of the photographic system.

[20] The uncertainty $\Delta x \cong \lambda/\sin \varepsilon$ is due to diffraction at the microscope objective; see, for instance, [59;45].

the photon is changed in each scattering process and, as is seen from the diagram, the component of the change of momentum along the axis OX will lie in the range $\pm (2\pi\hbar/\lambda) \sin \varepsilon$ (where $2\pi\hbar/\lambda = \hbar\omega/c$ is the momentum of the photon). This momentum will be transferred to the particles, giving them momenta in the range $\Delta p_x \cong (2\pi\hbar/\lambda) \sin \varepsilon$. Hence we see, firstly, that in creating an ensemble of particles localised in a small region Δx we must apply an interaction of very high energy (small λ, large quanta); secondly, that an ensemble with small Δx will have a large Δp_x. Multiplying Δp_x by Δx, we find $\Delta p_x \cdot \Delta x \cong 2\pi\hbar$.[21]

With particles which are not free, only indirect measurement is possible. For example, the co-ordinate of an electron within an atom is determined from the scattering of a beam of free particles (electrons or X-rays). In such cases, however, we always obtain information concerning not the position of an individual electron in an

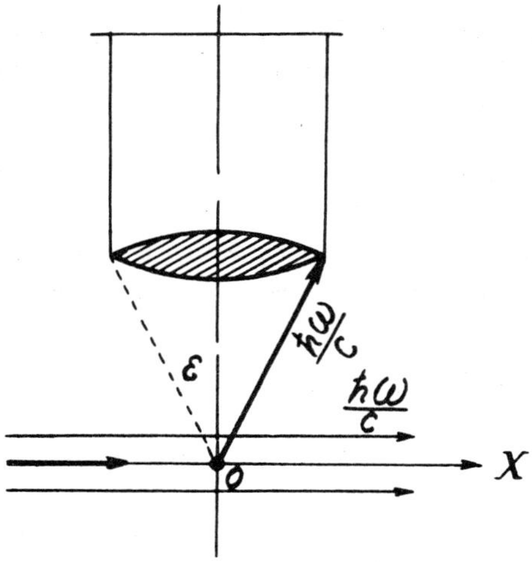

Fig. 18. Determination of particle co-ordinates by means of a microscope.

individual atom but the distribution of such positions in a large number of atoms in the same state, i.e. what we find directly is $|\psi(x)|^2$ (see collision theory, Section 79).

In conclusion, we may give one further example of the determination of the co-ordinates of a particle. Let us suppose that a particle is enclosed in a box with walls which it cannot penetrate. Let the dimension of the box be l. We now move the walls of the box together ($l \to 0$). Then the position x of the centre of the box determines the position of the particle. By hypothesis, the particle cannot escape from the box, and so the wave function of the particle is zero except inside the box; hence $\overline{\Delta x^2} \cong l^2$. As the

[21] In the non-Russian literature (e.g. [45]) this experiment is customarily considered as an experiment on one particle. However, one particle can give only one scattering (after which it belongs to a different ensemble), and the position of the particle cannot be judged from one scattered quantum (there being no image in the focal plane). The correct mathematical theory of this experiment, starting from the statistical interpretation of the ψ function, has been given by L. I. Mandel'shtam in his lectures on quantum mechanics.

volume of the box decreases, the momentum spread increases: $\overline{\Delta p^2} \geqslant \hbar^2/4l^2$. In this case $\bar{p} = 0$, and so the mean energy of the particle is $E = \overline{\Delta p^2}/2\mu \geqslant \hbar^2/8\mu l^2$. The contraction of the box therefore requires work to be done which increases without limit as the degree of localisation of the particle increases ($\Delta x = l \to 0$). Hence it follows that, the smaller the region of space in which particles are localised, the greater must be their energy. This remarkable deduction of the quantum theory is confirmed by experiment. For example, electrons in atoms (where the dimensions of the shells are 10^{-9} to 10^{-8} cm) have energy 10 to 100 eV, while nucleons in nuclei (dimensions 10^{-13} cm) have energy of the order of 1 MeV.

Let us now consider the measurement of momentum, and first of all discuss the diffraction experiment which was the basis of the determination of momentum. Figure 14 (Section 12) shows the grating, the primary beam (i) and the diffracted beams (r, d, ...).

Let the width of the primary beam be l and the grating constant be d. The number of grating lines effective in the experiment will be $N = l/d$. It is known from the theory of diffraction that such a grating makes it possible to separate two wavelengths λ and $\lambda + \Delta\lambda$, where

$$\Delta\lambda = \lambda/N = \lambda d/l\,. \tag{16.6}$$

This is the resolving power of a diffraction grating [59; 45]. Thus the grating divides the original ensemble into two ensembles, for instance (r) and (d), having two different momenta, if these momenta differ by more than

$$\Delta p = 2\pi\hbar\,\Delta\lambda/\lambda^2 = 2\pi\hbar\,d/\lambda l\,. \tag{16.7}$$

In order that the beams should be separated (so that the measurement is possible), the Faraday cylinder must be moved away a distance Δx (along the beam (r) or (d)) greater than l/α, where α is the angle between the beams (r) and (d). Hence $\Delta p \cdot \Delta x > 2\pi\hbar(d/\lambda) \cdot (1/\alpha)$. Since d and λ are of the same order[22], and the angle α is assumed small, this gives

$$\Delta p \cdot \Delta x > 2\pi\hbar\,, \tag{16.8}$$

i.e. the product of the beam dimension Δx (region of localisation of the particle) and the uncertainty in momentum Δp due to the finite resolving power of the grating must be greater than $2\pi\hbar$.

Another example is the determination of momentum of particles from the frequency of scattered light. For simplicity we shall take only one dimension. Let p_x be the momentum of the particles before collision with a light quantum, and p'_x the value after the collision. Let the frequency of the incident light be ω, and that of the scattered light be ω'. Then the law of conservation of energy gives

$$\hbar\omega - \hbar\omega' = \frac{1}{2\mu}(p'^2_x - p^2_x), \tag{16.9}$$

[22] If $\lambda \gg d$, no diffraction is observed.

and the law of conservation of momentum gives

$$\frac{\hbar\omega}{c} + \frac{\hbar\omega'}{c} = p_x' - p_x. \tag{16.10}$$

Hence

$$p_x = \mu c\frac{\omega - \omega'}{\omega + \omega'} - \frac{\hbar(\omega + \omega')}{2c}, \tag{16.11}$$

$$p_x' = \mu c\frac{\omega - \omega'}{\omega + \omega'} + \frac{\hbar(\omega + \omega')}{2c}. \tag{16.11'}$$

Thus, if we know ω and ω', we can determine the particle momentum p_x. However, this experiment gives no information on the position of the particle: the point of scattering is entirely indeterminate. We could determine this point with accuracy Δx if the monochromatic wave were replaced by a signal of width Δx, but then, as we know, there is a range of frequencies with $\Delta k_x = \Delta\omega/c \cong \pi/\Delta x$. As a result the particle momentum would be determined with accuracy $\Delta p_x = \hbar\Delta k_x = \hbar\Delta\omega/c$, so that $\Delta p \cdot \Delta x > \pi\hbar$.

We may conclude by considering another experiment which is widely used in practice. Suppose that we wish to determine the momentum p of a neutron by means of a collision with a proton, the momentum of the latter in the initial state being taken as zero. After the collision (assuming a central impact) the momentum of the neutron is zero, and that of the proton is equal to the original momentum p of the neutron (the masses of the two being supposed equal). This momentum can be measured, for example, by measuring the curvature of the track of the proton in a cloud chamber owing to the action of a magnetic field. This gives a measurement of the original momentum of the neutron. In this experiment, however, we know nothing regarding the place of the collision. By using a cloud chamber we can, of course, determine this place as the beginning of the track of the proton after the collision. But, as shown previously, the cloud chamber allows a determination of the position of a particle (and therefore of the beginning of a track) with accuracy at best $\Delta x \cong a$ (where a is the dimension of an atom).[23] The momentum of the particle is then determined with accuracy $\Delta p \cong \hbar/a$, i.e. we do not know the momentum of the proton any more accurately than this. This involves a similar uncertainty in the determination of the momentum of the neutron. The product of the uncertainties again gives $\Delta p \cdot \Delta x \geqslant h$.

These examples serve to illustrate the fact there is no contradiction between the statement that the uncertainty relation follows from the general principles of quantum mechanics and the possibilities afforded by measuring apparatus.

17. The significance of the measuring apparatus

In studying any phenomena by statistical methods, the measuring apparatus which serves both to define the statistical ensembles and to analyse the distribution within

[23] This is the 'ideal' accuracy, which is never achieved in practice; see footnote 19.

those ensembles must itself be outside the ensembles. In other words, it must not possess the random element which pertains to the statistical assemblages which it is used to investigate.[24] Nevertheless every apparatus, like any other body, consists of atoms, molecules and similar micro-objects executing certain motions, and from the viewpoint of quantum mechanics certainly belongs to some quantum ensemble. Thus there is at first sight a certain difficulty. From this difficulty quantum mechanics provides an escape of remarkable ingenuity and effectiveness: the measuring apparatus must be so constructed that only its classical properties (that is, those in which Planck's constant $\hbar$ is not involved) are ultimately used in bringing about its operation. We call such an apparatus *classical* or *macroscopic*. Its essential feature is that it has the maximum freedom from quantum statistical properties.

Any of the examples of determining p_x and x considered in Section 16 may serve as an illustration of a classical apparatus: fixed screens with slits, the heavy atom of an ideal photographic plate, the box with fixed impenetrable walls, the diffraction grating with rigidly fixed lines, or any spectroscope for determining the wavelength of scattered light. We have regarded all such apparatus as objects of classical physics; that is, in considering their operation we have neglected Planck's constant $\hbar$ at some essential point. Thus the apparatus measures classical corpuscular quantities.

A set of quantities which is sufficient to determine the wave function will be called a *complete set*, and the measurement itself a *complete measurement*.

In classical mechanics a complete measurement consists in measuring the momenta and co-ordinates of particles. Since in classical mechanics all quantities can be measured simultaneously, at least in principle, we can say that there is only *one* complete measurement.

If, for example, we have measured the Cartesian momenta and co-ordinates (p, x) of particles, we can calculate all other quantities, including the generalised momenta and co-ordinates (P, Q), which also form a complete set of quantities and define the motions just as well as (p, x). Moreover, there is nothing to prevent us from carrying out a more complicated measurement and measuring (p, x) and $(P. Q)$ simultaneously. Since classical mechanics is consistent, the calculated values of (P, Q) will agree with the measured values. Hence the change from one complete set of quantities to another, within classical mechanics, is unimportant.

In the quantum region the complete set of quantities which determines ψ, and therefore the quantum ensemble, is not unique. But the fundamental difference between quantum mechanics and classical mechanics is that in the former the various sets are, in general, mutually exclusive. Accordingly, in quantum mechanics there are many different complete measurements which are mutually incompatible. For example, there is a quantum ensemble defined with respect to an apparatus which determines the co-ordinates x, y, z of particles. This is one possible complete set.

Such an ensemble is described by a wave function $\psi_{x'y'z'}(x, y, z)$ (see Section 14), which expresses the fact that all the particles in the ensemble have co-ordinates $x = x', y = y', z = z'$. Another example is an ensemble with precisely determined

[24] The concept of randomness in dialectical materialism is discussed in [35].

momentum $p_x = p'_x, p_y = p'_y, p_z = p'_z$. The wave function of such an ensemble is $\psi_{p'_x p'_y p'_z}(x, y, z)$. This ensemble again is determined, from the viewpoint of quantum mechanics, with the maximum completeness, but it is entirely different from the previous one.

Thus the nature of a quantum ensemble is entirely different depending on the properties by which it is defined (i.e. depending on the complete set of quantities used) and will be considerably changed if measurements of a new complete set incompatible with the original one are made. Hence the state of a quantum ensemble cannot be considered independently of the complete set of quantities by which it is defined. In this respect, measuring apparatus which determines different complete sets must be regarded as different 'frames of reference' with which the state of a quantum ensemble is defined.[25]

The essence of this profound difference between the definitions of a state in classical and quantum treatments lies in the fact that according to classical ideas there was no absolute scale of smallness. The study of the microuniverse has revealed the existence of a number of atomic constants which provide such a scale: the elementary charge e; the elementary mass μ of the electron and positron; the masses of the simplest heavy particles, the proton (m_p) and the neutron (m_n); Planck's constant $\hbar$, etc.

We do not yet know precisely what restrictions of classical ideas and what new concepts must follow from the existence of the elementary charge and mass, but we know the results which follow from the existence of the quantum of action $\hbar$. The existence of this quantum of action leads to the phenomenon of diffraction of particles, which makes it impossible to apply simultaneously to the measurement of microparticles such quantities as, for example, p and x.

Let us now consider in more detail how measurement affects a quantum ensemble. We shall suppose that the ensemble is specified by a wave function $\psi(x)$ (a pure ensemble).[26] Let us first consider the measurement of momentum. To do so, we expand $\psi(x)$ as a spectrum of de Broglie waves

$$\psi_p(x) = \frac{e^{ipx/\hbar}}{\sqrt{(2\pi\hbar)}}:$$

$$\psi(x) = \int c(p)\,\psi_p(x)\,dp\,. \tag{17.1}$$

Let N measurements in all be made, of which N' give a value of p lying near p', N'' a value near p'', N''' a value near p''', etc. ($N = N' + N'' + N''' + \ldots$).

Then we have (cf. Section 14)

$$\frac{N'}{N} = |c(p')|^2\,dp', \qquad \frac{N''}{N} = |c(p'')|^2\,dp'',$$
$$\frac{N'''}{N} = |c(p''')|^2\,dp''', \ldots. \tag{17.2}$$

[25] This does not mean, of course, that without a measuring apparatus there is no quantum ensemble: situations spontaneously arise in Nature which define an ensemble, i.e. which correspond to a measurement.

[26] It is only for simplicity that we take the pure case and only one spatial dimension x; this is not

As a result of the measurements of N' particles, a new pure ensemble is found, with $p = p'$, described by a new wave function $\psi_{p'}(x)$. Thus the measurement selects from the original ensemble, with undetermined momentum, sub-ensembles with definite values of the momentum p', p'', p''', ..., described by new functions $\psi_{p'}(x)$, $\psi_{p''}(x)$, $\psi_{p'''}(x)$, ... respectively.

The original state $\psi(x)$ becomes one of the states of the form $\psi_p(x)$. This change in the wave function is called *reduction* of the wave packet. Its physical significance is that after the measurement the particle belongs to a new pure ensemble.

The whole ensemble resulting from the measurements is described by a series of wave functions $\psi_{p'}(x), \psi_{p''}(x), \psi_{p'''}(x), \ldots$, with corresponding probabilities $|c(p')|^2\,dp'$, $|c(p'')|^2\,dp''$, $|c(p''')|^2\,dp'''$, ..., i.e. it is a mixed ensemble.

A similar situation occurs in other cases also. Two further examples may be given. In a measurement of the co-ordinate x, let us expand $\psi(x)$ in a spectrum of wave functions describing a state with a definite value of x. Such a function is of the form $\psi_{x'}(x) = \delta(x' - x)$. The expansion is therefore

$$\psi(x) = \int c(x')\,\delta(x' - x)\,dx'. \tag{17.3}$$

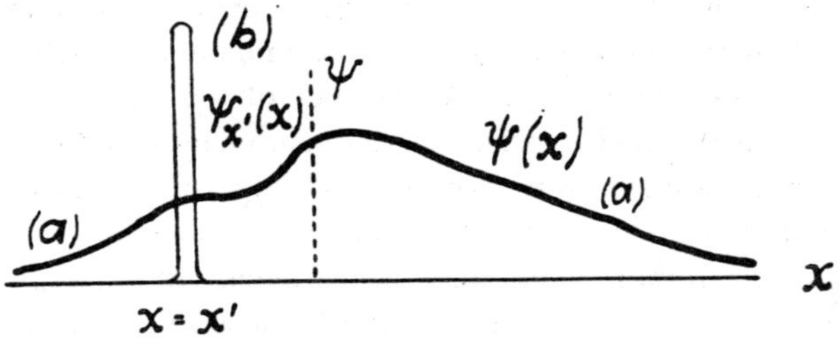

Fig. 19. Reduction of a wave packet $\psi(x)$ (curve a) to $\psi_{x'}(x)$ (curve b) after measurement of the co-ordinate x as x'.

The properties of the δ function show immediately that $c(x') = \psi(x')$. If we obtain x near x' in N' cases, near x'' in N'' cases, etc., then

$$\left.\begin{aligned} N'/N &= |c(x')|^2\,dx' = |\psi(x')|^2\,dx', \\ N''/N &= |c(x'')|^2\,dx'' = |\psi(x'')|^2\,dx'', \\ N'''/N &= |c(x''')|^2\,dx''' = |\psi(x''')|^2\,dx''', \ldots \end{aligned}\right\} \tag{17.4}$$

In each measurement the original function $\psi(x)$ is reduced to one of the functions of the form $\psi_{x'}(x) = \delta(x - x')$. This reduction is shown in Figure 19.[27]

We see that in the measurement of the co-ordinate a mixed ensemble is again formed, in which new pure sub-ensembles of the form $\psi_{x'}(x)$, $\psi_{x''}(x)$, ... are represented

essential in elucidating the nature of the problem. The effect of measurement on a mixed ensemble is described in Section 45.

[27] It may be recalled (see Section 16) that a measurement of the co-ordinate requires energy which is taken either from the apparatus or from the particle itself.

with probabilities $|\psi(x')|^2$, $|\psi(x'')|^2$, ..., i.e. this probability, as in the measurement of momentum, is determined by the intensity $|c(x')|^2$ with which the pure state $\psi_{x'}(x)$ is represented in the initial pure state $\psi(x)$ (in this special case, $c(x') = \psi(x')$).

It will be shown later (Section 22) that, if any mechanical quantity L is measured which can take values $L_1, L_2, L_3, \ldots, L_{n'}$, then in order to find the probability that $L = L_n$ it is necessary to expand $\psi(x)$ as a spectrum of states $\psi_n(x)$. Each of these states is characterised by the fact that in it the quantity L has a unique value $L = L_n$.[28]

Such an expansion may be written

$$\psi(x) = \sum_n c_n \psi_n(x). \tag{17.5}$$

Then the number of cases N_n where $L = L_n$ will be proportional to $|c_n|^2$, i.e.

$$\left.\begin{aligned} N'/N &= |c_{n'}|^2, \\ N''/N &= |c_{n''}|^2, \\ N'''/N &= |c_{n'''}|^2, \ldots; \end{aligned}\right\} \tag{17.6}$$

we again have a reduction of the original wave packet $\psi(x)$ to one of the states $\psi_n(x)$, and the whole set of measurements again forms a mixed ensemble.

Thus this behaviour of quantum ensembles in measurement is entirely general and may be formulated thus: measurement changes a pure ensemble into a mixed one.[29]

This changing of a pure ensemble into a mixed one is simply the resolution of the original ensemble into a spectrum of pure ensembles which are selected by the apparatus. The original ensemble, on 'passing through' the apparatus, is resolved into its constituent 'sub-ensembles', defined with respect to that apparatus. In quantum mechanics, the frame of reference, or classical measuring apparatus, is therefore just a spectral analyser of quantum ensembles which is used to examine their nature. The most important property of such analysers is that different analysers give (owing to the nature of the microuniverse) mutually exclusive spectral resolutions, since the simultaneous application to microparticles of additional characteristics is not consistent with reality.

These spectral analysers should not be thought of as necessarily having the form of laboratory apparatus. On the contrary, an experimenter or technician who chooses a particular apparatus merely makes a certain combination of what already exists in Nature, and it would be absurd to think that if there were no 'observer' quantum ensembles would no longer be meaningful. As soon as a situation arises in Nature where spectral resolution of the original ensemble occurs, there is a formation of new ensembles defined by new characteristics, i.e. what is usually called 'the intervention of measurement'. This process may or may not be observed by an observer; the objective phenomenon is unaffected.

[28] To vary the examples we here suppose that the quantity L has discrete values $L_1, L_2, \ldots$, unlike the quantities p and x previously considered, which have continuous values.

[29] Except for the case where the measurement is simply a repetition of that which defined the original ensemble. The ensemble then remains unchanged.

CHAPTER III

REPRESENTATION OF MECHANICAL QUANTITIES BY OPERATORS

18. Linear self-adjoint operators

As we have seen, in the quantum region there are no states in which the momentum and co-ordinate of particles have definite values simultaneously. This fact is reflected in the formalism of the theory: the mathematical treatment of quantum mechanics is quite different from that of classical mechanics, in which the specification of the pair of quantities p, x is entirely meaningful.

We shall begin the discussion of the mathematical formalism from the expressions given in Section 13 for the mean value of functions of co-ordinates or momenta in the state $\psi(x, y, z)$. Formula (13.1) was obtained for the mean value of a function of the co-ordinates of a particle:

$$\overline{F(x, y, z)} = \int \psi^*(x, y, z) F(x, y, z) \psi(x, y, z) \,\mathrm{d}x \,\mathrm{d}y \,\mathrm{d}z, \tag{18.1}$$

and Formula (13.6) for the mean value of a function of the momenta:

$$\begin{aligned}\overline{F(p_x, p_y, p_z)} = \int \psi^*(x, y, z) F\left(-i\hbar\frac{\partial}{\partial x}, -i\hbar\frac{\partial}{\partial y}, -i\hbar\frac{\partial}{\partial z}\right) \times \\ \times\, \psi(x, y, z)\,\mathrm{d}x\,\mathrm{d}y\,\mathrm{d}z.\end{aligned} \tag{18.2}$$

These formulae become identical in form if the momentum components p_x, p_y, p_z are represented by operators:

$$\boldsymbol{P}_x = -i\hbar\,\partial/\partial x, \qquad \boldsymbol{P}_y = -i\hbar\,\partial/\partial y, \qquad \boldsymbol{P}_z = -i\hbar\,\partial/\partial z, \tag{18.3}$$

and (18.2) is correspondingly written

$$\overline{F(p_x, p_y, p_z)} = \int \psi^*(x, y, z) F(\boldsymbol{P}_x, \boldsymbol{P}_y, \boldsymbol{P}_z) \psi(x, y, z)\,\mathrm{d}x\,\mathrm{d}y\,\mathrm{d}z. \tag{18.4}$$

Thus we have a representation of functions of momentum $F(p_x, p_y, p_z)$ by the operator $F(\boldsymbol{P}_x, \boldsymbol{P}_y, \boldsymbol{P}_z)$.

This result suggests that other more complex mechanical quantities $L(p_x, p_y, p_z, x, y, z)$, depending both on co-ordinates and on momenta, should also be represented by operators; and in fact it is found that all relations between mechanical quantities in the quantum region can be expressed in terms of operators of a certain class. This constitutes the fundamental significance of the use of operators in quantum mechanics.

In order to distinguish the class of operators encountered in quantum mechanics, let us first consider the general definition of an operator. Whatever the specific form, we shall understand an *operator* $\boldsymbol{L}$ to be a symbol showing how a function $v(x)$ is correlated with each of a given class of functions $u(x)$. This is symbolically written as a multiplication of u by $\boldsymbol{L}$:

$$\boldsymbol{L}u(x) = v(x). \tag{18.5}$$

In this equation $\boldsymbol{L}$ may signify, for example, multiplication by x ($\boldsymbol{L} = x$), differentiation with respect to $x(\boldsymbol{L} = \partial/\partial x)$, taking the square root ($\boldsymbol{L} = \sqrt{\ }$) and so on.

Among the variety of conceivable operators, only one particular class are used to represent mechanical quantities in the quantum region; these are called *linear self-adjoint* (or *Hermitian*) operators.

An operator $\boldsymbol{L}$ is said to be *linear* if it is such that

$$\boldsymbol{L}(c_1 u_1 + c_2 u_2) = c_1 \boldsymbol{L}u_1 + c_2 \boldsymbol{L}u_2, \tag{18.6}$$

where u_1 and u_2 are arbitrary functions, and c_1 and c_2 arbitrary constants. It is evident that $\partial/\partial x$ is a linear operator, while $\sqrt{\ }$ is not.

The restriction to linear operators follows from the principle of superposition of states. The property of linearity of an operator expressed by (18.6) means that the application of the operator to a superposition of two functions u_1 and u_2 gives the same result as the superposition of the functions obtained by applying that operator to u_1 and u_2 separately:

$$\boldsymbol{L}(c_1 u_1 + c_2 u_2) = c_1 v_1 + c_2 v_2,$$

where $v_1 = \boldsymbol{L}u_1$, $v_2 = \boldsymbol{L}u_2$; that is, we require that the application of operators should not violate the principle of superposition.

A linear operator is said to be *self-adjoint* (or *Hermitian*) if the equation

$$\int u_1^*(x)\cdot \boldsymbol{L}u_2(x)\cdot \mathrm{d}x = \int u_2(x)\cdot \boldsymbol{L}^* u_1^*(x)\cdot \mathrm{d}x \tag{18.7}$$

holds, where the integral is taken over the whole range of variation of the variable x, and u_1^* and u_2 are arbitrary functions within a very wide class.[1] If there are several variables, $\mathrm{d}x$ becomes $\mathrm{d}x\,\mathrm{d}y\,\mathrm{d}z\,\ldots$.

We shall see later that the significance of the self-adjointness condition is that only operators which satisfy this condition can represent real physical quantities.

Let us illustrate the property (18.7) by means of the momentum operator $\boldsymbol{P}_x = -i\hbar\,\partial/\partial x$. We have

$$\int_{-\infty}^{\infty} u_1^*\cdot \boldsymbol{P}_x u_2\cdot \mathrm{d}x = -i\hbar \int_{-\infty}^{\infty} u_1^* \frac{\partial u_2}{\partial x}\,\mathrm{d}x$$

$$= \left[-i\hbar\, u_1^* u_2\right]_{-\infty}^{\infty} + i\hbar \int u_2 \frac{\partial u_1^*}{\partial x}\,\mathrm{d}x = \int_{-\infty}^{\infty} u_2\cdot \boldsymbol{P}_x^* u_1^*\,\mathrm{d}x$$

[1] They must be integrable and their derivatives must be zero at the boundaries of the region of integration.

(since $u_1^*(\pm\infty) = u_2(\pm\infty) = 0$). Thus $\boldsymbol{P}_x$ is a linear self-adjoint operator. It is easy to see that the operator $\partial/\partial x$ is linear but not self-adjoint for

$$\int_{-\infty}^{\infty} u_1^* \frac{\partial u_2}{\partial x} \mathrm{d}x = - \int_{-\infty}^{\infty} u_2 \frac{\partial u_1^*}{\partial x} \mathrm{d}x \neq + \int_{-\infty}^{\infty} u_2 \frac{\partial u_1^*}{\partial x} \mathrm{d}x. \tag{18.8}$$

From various operators we can derive others. The methods of constructing more complicated operators from simple ones follow from the definition of an operator and can be formulated as simple algebraic rules. Let us consider two linear self-adjoint operators $\boldsymbol{A}$ and $\boldsymbol{B}$. We call an operator $\boldsymbol{C}$ such that

$$\boldsymbol{C}\psi = \boldsymbol{A}\psi + \boldsymbol{B}\psi \tag{18.9}$$

the *sum* of these two operators, and write symbolically

$$\boldsymbol{C} = \boldsymbol{A} + \boldsymbol{B}. \tag{18.10}$$

For example, if $\boldsymbol{A} = i\partial/\partial x$ and $\boldsymbol{B} = x$, (18.9) gives

$$\boldsymbol{C} = i\,\partial/\partial x + x.$$

The definition of multiplication is a little more complicated. The *product* of two operators $\boldsymbol{A}$ and $\boldsymbol{B}$ is defined to be an operator $\boldsymbol{C}$ such that

$$\boldsymbol{C}\psi = \boldsymbol{A}(\boldsymbol{B}\psi), \tag{18.11}$$

that is the operator $\boldsymbol{B}$ must first be applied to ψ, and then the operator $\boldsymbol{A}$ applied to the result. If the same final result can be obtained by means of an operator $\boldsymbol{C}$, then this is the product of $\boldsymbol{A}$ and $\boldsymbol{B}$. Symbolically,

$$\boldsymbol{C} = \boldsymbol{A}\boldsymbol{B}. \tag{18.12}$$

For example, if $\boldsymbol{A} = i\partial/\partial x$, $\boldsymbol{B} = x$, then

$$\boldsymbol{C}\psi = \boldsymbol{A}(\boldsymbol{B}\psi) = i\frac{\partial}{\partial x}(x\psi) = i\psi + ix\frac{\partial\psi}{\partial x},$$

so that

$$\boldsymbol{C} = i + ix\frac{\partial}{\partial x} = i\left(1 + x\frac{\partial}{\partial x}\right).$$

It is important to note that the product of operators depends on the order of the factors. In the above example,

$$\boldsymbol{C}'\psi = \boldsymbol{B}(\boldsymbol{A}\psi) = ix\,\partial\psi/\partial x, \qquad \text{i.e.} \qquad \boldsymbol{C}' = ix\,\partial/\partial x.$$

Hence, if we have two operators $\boldsymbol{A}$ and $\boldsymbol{B}$, in addition to the product $\boldsymbol{C}$ we can obtain another product:

$$\boldsymbol{C}' = \boldsymbol{B}\boldsymbol{A}. \tag{18.12$'$}$$

The foregoing rules allow operators to be added, subtracted and multiplied as in ordinary algebra, except that in general the order of factors can not be changed. For example,

$$\boldsymbol{C} = (\boldsymbol{A} - \boldsymbol{B})(\boldsymbol{A} + \boldsymbol{B}) = \boldsymbol{A}^2 - \boldsymbol{BA} + \boldsymbol{AB} - \boldsymbol{B}^2,$$

but this does not necessarily equal $\boldsymbol{A}^2 - \boldsymbol{B}^2$.

Such an algebra, in which factors cannot be interchanged, is called a *non-commutative algebra*, and the quantities involved are said to be *non-commutative*.

If the two products $\boldsymbol{C}$ and $\boldsymbol{C}'$ are equal:

$$\boldsymbol{AB} - \boldsymbol{BA} = 0, \tag{18.13}$$

then the operators $\boldsymbol{A}$ and $\boldsymbol{B}$ are said to *commute* or to be *commutative*. If not, they are said to be *non-commutative*. The operator $\boldsymbol{F} = \boldsymbol{AB} - \boldsymbol{BA}$ is called the *commutator* of the operators $\boldsymbol{A}$ and $\boldsymbol{B}$.

When multiplying linear self-adjoint operators it must be borne in mind that their product will not in general be self-adjoint. We have

$$\boldsymbol{AB} = \tfrac{1}{2}(\boldsymbol{AB} + \boldsymbol{BA}) + \tfrac{1}{2}(\boldsymbol{AB} - \boldsymbol{BA}); \tag{18.14}$$

using the self-adjointness of the operators $\boldsymbol{A}$ and $\boldsymbol{B}$, we can show from (18.7) that the operator

$$\boldsymbol{F} = \tfrac{1}{2}(\boldsymbol{AB} + \boldsymbol{BA}) \tag{18.15}$$

is self-adjoint, but the operator

$$\boldsymbol{G} = \tfrac{1}{2}(\boldsymbol{AB} - \boldsymbol{BA}) \tag{18.16}$$

is not, except when the operators commute and $\boldsymbol{G} = 0$. Since every operator commutes with itself, it follows that any (positive integral) power of a linear self-adjoint operator $\boldsymbol{A}$:

$$\boldsymbol{A}^n = \boldsymbol{A} \cdot \boldsymbol{A} \cdot \ldots \cdot \boldsymbol{A} \qquad (n \text{ factors}) \tag{18.17}$$

is a linear self-adjoint operator.

Using the above rules and the known operators of momentum components $\boldsymbol{P}_x, \boldsymbol{P}_y, \boldsymbol{P}_z$ (18.3) and co-ordinates x, y, z of a particle, we can derive more complicated linear self-adjoint operators $\boldsymbol{L}$.

19. The general formula for the mean value of a quantity and the mean square deviation

The fundamental idea in the use of operators in quantum mechanics is that each mechanical quantity L in quantum mechanics is correlated with a linear self-adjoint operator $\boldsymbol{L}$ which represents it. Symbolically,

$$L \to \boldsymbol{L}.$$

The problem of which physical quantity is represented by a given operator is determined by the properties of the quantity and the methods of observing it. Where the quantum quantity represented by an operator $\boldsymbol{L}$ has properties analogous to those of some classical quantity L, the same name is used for both.

For example, if there is a classical quantity $L(p_x, p_y, p_z, x, y, z)$ which is a function of momenta and co-ordinates, the linear self-adjoint operator $\boldsymbol{L}$ constructed according to the rules of Section 18 from the operators of the momentum components $\boldsymbol{P}_x, \boldsymbol{P}_y, \boldsymbol{P}_z$ and co-ordinates x, y, z is

$$\boldsymbol{L} = L(\boldsymbol{P}_x, \boldsymbol{P}_y, \boldsymbol{P}_z, x, y, z).$$

The self-adjoint operator $\boldsymbol{L}$ will represent a quantum quantity with properties anallogous to the classical quantity $L(p_x, p_y, p_z, x, y, z)$.[2]

Naturally, not all linear self-adjoint operators formed from $\boldsymbol{P}_x, \boldsymbol{P}_y, \boldsymbol{P}_z$ and x, y, z will represent quantities which have a simple physical significance and obey simple laws. The same is true in the classical theory. For example, the quantity $p^2/2m$ has the significance of kinetic energy and obeys a conservation law (in the absence of external forces), but the quantity px^2 has no general rule of behaviour and hence plays no part in mechanics.

The relation between operators and measured quantities is established by means of the formula for the mean value of the quantity L in an ensemble described by the wave function ψ: in quantum mechanics it is assumed that the mean value $\bar{L}$ of a quantity L represented by a linear self-adjoint operator $\boldsymbol{L}$, in a pure ensemble described by the function ψ, is given by

$$\bar{L} = \int \psi^* \cdot \boldsymbol{L} \psi \cdot \mathrm{d}x, \tag{19.1}$$

where $\mathrm{d}x$ signifies a volume element in the space of the independent variables and the integral is taken over the whole range of the independent variables. It is clear that our previous definitions (18.1) and (18.2) are particular cases of (19.1). To obtain (18.1) from (19.1), we must put $\boldsymbol{L} = F(x, y, z)$ and take $\mathrm{d}x$ as $\mathrm{d}x\,\mathrm{d}y\,\mathrm{d}z$. To obtain (18.2), we must put

$$\boldsymbol{L} = F\left(-i\hbar\frac{\partial}{\partial x}, -i\hbar\frac{\partial}{\partial y}, -i\hbar\frac{\partial}{\partial z}\right).$$

Using the self-adjointness of the operator $\boldsymbol{L}$, we can write (19.1) in the equivalent form

$$\bar{L} = \int \psi \cdot \boldsymbol{L}^* \psi^* \cdot \mathrm{d}x \tag{19.1'}$$

(with $u_1^* = \psi^*$, $u_2 = \psi$ in (18.7)). A comparison of (19.1) and (19.1′) shows that

$$\bar{L} = \bar{L}^*, \tag{19.2}$$

i.e. the mean value of a quantity represented by a self-adjoint operator is real.

We can obtain more information about the quantity L if we calculate, as well as its mean value $\bar{L}$, the mean square deviation $\overline{\Delta L^2}$, which shows by how much, on average, the results of individual measurements in the ensemble differ from the mean value. To find $\overline{\Delta L^2}$ we must construct an operator which represents ΔL^2. The deviation from the

[2] Since the wave function is regarded as a function of the co-ordinates x, y, z of the particle, the effect of the 'operators' x, y, z amounts simply to multiplication by x, y or z, and the effect of the operator $F(x, y, z)$ amounts to multiplication by $F(x, y, z)$.

mean is defined as $\Delta L = L - \bar{L}$. Thus the operator representing it is

$$\Delta \mathbf{L} = \mathbf{L} - \bar{L}. \tag{19.3}$$

Since the squared deviation $\Delta L^2 = (L - \bar{L})^2$, the operator for ΔL^2 will be

$$(\Delta \mathbf{L})^2 = (\mathbf{L} - \bar{L})^2. \tag{19.4}$$

Using the general definition of the mean value (19.1), we find

$$\overline{\Delta L^2} = \int \psi^* \cdot (\Delta \mathbf{L})^2 \psi \cdot \mathrm{d}x. \tag{19.5}$$

Thus, if we know the operator $\mathbf{L}$, we can calculate $\overline{\Delta L^2}$.

This quantity cannot be negative, as is easily shown by using the self-adjointness of the operator $\mathbf{L}$. Since $\bar{L}$ is a number, the operator $\Delta \mathbf{L}$ is also self-adjoint, and so, using (18.7) and putting $\psi^* = u_1^*$ and $\Delta \mathbf{L}\psi = u_2$ in (19.5), we get

$$\overline{\Delta L^2} = \int (\Delta \mathbf{L} \psi)(\Delta \mathbf{L}^* \psi^*) \, \mathrm{d}x = \int |\Delta \mathbf{L} \psi|^2 \, \mathrm{d}x\,; \tag{19.6}$$

since $|\Delta \mathbf{L}\psi|^2 \geqslant 0$, (19.6) shows that

$$\overline{\Delta L^2} \geqslant 0,$$

i.e. the mean square deviation is always positive or zero, as it should be.

20. Eigenvalues and eigenfunctions of operators and their physical significance. 'Quantisation'

The formulae of Section 19 give expressions for the mean value $\bar{L}$ and the mean square deviation $\overline{\Delta L^2}$. These formulae do not make any statement as to the values of L in individual measurements.

In order to find the possible values of L, let us consider the states ψ_L in which this quantity has only one value L. In such states the mean square deviation $\overline{\Delta L^2} = 0$. Thus for these states (19.6) gives

$$\int |\Delta \mathbf{L} \psi_L|^2 \, \mathrm{d}x = 0. \tag{20.1}$$

Since the integrand is essentially positive, it follows from (20.1) that $|\Delta \mathbf{L}\psi_L|^2 = 0$. The modulus of a complex number is zero only when that number is itself zero. Thus we have

$$\Delta \mathbf{L} \psi_L = 0$$

or, using the value of the operator $\Delta \mathbf{L}$ and the fact that in the state considered $\bar{L} = L$,

$$\mathbf{L} \psi_L = L \psi_L. \tag{20.2}$$

Since $\mathbf{L}$ is an operator, this equation is a linear equation for the wave function ψ_L of a state in which the quantity represented by the operator $\mathbf{L}$ has the unique value L. In most cases the operator $\mathbf{L}$ will be a differential operator, and Equation (20.2) is then a linear homogeneous differential equation.

It is known that the solution of a differential equation is uniquely defined only when the boundary conditions are specified.[3]

When the boundary conditions are given, however, the linear differential equation $\boldsymbol{L}\psi = L\psi$ has a non-trivial (i.e. non-zero) solution, in general, only for certain particular values $L_1, L_2, L_3, \ldots, L_n, \ldots$ of the parameter L. The corresponding solutions $\psi_1, \psi_2, \psi_3, \ldots, \psi_n, \ldots$ are called *eigenfunctions*, and the values $L_1, L_2, L_3, \ldots, L_n, \ldots$ of the parameter for which solutions exist are called *eigenvalues* (or *characteristic values*) *of the parameter* in Equation (20.2).

The best-known example of such a problem is that of vibrations of a string fixed at both ends. The equation of motion in this case is

$$d^2u/dx^2 + k^2u = 0, \tag{20.3}$$

so that $\boldsymbol{L} = -\, d^2/dx^2$ and $L = k^2$. The region in which a solution is sought is $0 \leqslant x \leqslant l$, where l is the length of the string. The boundary conditions are that $u = 0$ for $x = 0$ and $x = l$. The eigenfunctions for this problem are $u_n(x) = \sin(\pi n x/l)$, and the eigenvalues are $L_n = k_n^2 = \pi^2 n^2/l^2$, $n = 1, 2, 3, \ldots$.[4]

In quantum mechanics the wave function is always defined over the whole range of variation of its arguments; for instance, $\psi(x, y, z)$ is defined in the region $-\infty < x < \infty$, $-\infty < y < \infty$, $-\infty < z < \infty$, and so on. We therefore cannot formulate the boundary conditions for the wave function so directly as in classical problems of the vibration of bodies.

It can be shown, however[5], that the requirement of the conservation of the total number of particles leads to certain natural restrictions on the wave functions which are equivalent to boundary conditions. The requirement of the conservation of the total number of particles is equivalent to the condition that the probability of finding a particle somewhere in space must be independent of time, i.e. that

$$\frac{d}{dt}\int \psi^* \psi \, dv = 0. \tag{20.4}$$

Here the integral is taken over the whole range of variation of the arguments of the function ψ, and so it gives the probability that the particle is necessarily somewhere. The point is that the condition (20.4) can be fulfilled only if the wave functions behave in an appropriate manner, viz. if they (1) are finite throughout the range of the variables, except possibly at certain (singular) points where they may become infinite, though not too rapidly[6], (2) have a sufficient number of continuous derivatives (which

[3] The equations concerned do not involve time derivatives, so that no initial conditions are needed.
[4] The problem of finding eigenvalues is discussed in detail in [27], and [78], Vols. 3 and 4.
[5] See Appendix VIII.
[6] If the wave function does not vanish at infinity (e.g. a plane de Broglie wave), then for the integral in (20.4) to converge ψ must be replaced by the *eigendifferentials* (see Appendix III, (12) and (12′), where the normalisation rule is given for wave functions which do not vanish at infinity).

may also tend to infinity at certain points, though not too rapidly); (3) are single-valued. These conditions may be formulated more stringently, but in a form adequate for non-relativistic quantum mechanics, as follows: (1) finiteness, (2) continuity and (3) single-valuedness of the wave function throughout the range of its arguments.

These very mild conditions imposed on solutions of Equation (20.2) have the result that in many cases solutions having the properties (1), (2) and (3) exist not for all values of L but only for certain values $L = L_1, L_2, L_3, \ldots, L_n, \ldots$, i.e. we have a problem of finding eigenfunctions and eigenvalues of Equation (20.2) on the basis of natural conditions arising from the conservation of number of particles (20.4).

Instead of 'eigenfunctions of the equation' and 'eigenvalues of the parameter of the equation', we shall usually speak of eigenfunctions and eigenvalues of the operator $\boldsymbol{L}$, which determines the form of Equation (20.2).

We shall suppose that no values of the quantity L can be observed experimentally except those which are eigenvalues of the operator $\boldsymbol{L}$. In quantum mechanics, therefore, it is postulated that the set of eigenvalues of the operator $\boldsymbol{L}$: $L_1, L_2, L_3, \ldots, L_n, \ldots$ is identical with the set of all possible results of measuring the mechanical quantity L represented by the operator $\boldsymbol{L}$. This is precisely the postulate which establishes the connection between the operator representation of quantities and experiment: the set of eigenvalues is deduced mathematically, and by experiment it can be tested whether this set is in fact as predicted by the theory.

The states corresponding to the eigenvalues $L_1, L_2, \ldots, L_n$ are defined by the eigenfunctions $\psi_1, \psi_2, \ldots, \psi_n, \ldots$. In each of these states $\overline{\Delta L^2} = 0$ and the quantity L has only one of the values $L_1, L_2, \ldots, L_n$. The set of possible values of any quantity will be called its *spectrum*.

The spectrum may be *discrete*, when only individual values $L_1, L_2, \ldots, L_n, \ldots$ are possible, or consist of separate *bands*, so that the possible values of L lie in the ranges $L_1 \leqslant L \leqslant L_2, L_3 \leqslant L \leqslant L_4, \ldots, L_n \leqslant L \leqslant L_{n+1}$, or finally be *continuous*, when any value of L is possible. When the possible values of a quantity are discrete, it is said to have *quantised* values.

In Bohr's original theory there was no general method of solving the problem of the possible values of a quantity or, in particular, of finding the quantum values. Modern quantum mechanics fully resolves this problem by reducing it to the purely mathematical one of finding the eigenfunctions and eigenvalues of operators which represent mechanical quantities.

It follows from the self-adjointness of the operator $\boldsymbol{L}$ that the observed values of L are real:

$$L_n = L_n^* \qquad \text{or} \qquad L = L^* . \tag{20.5}$$

For the eigenvalue L_n (or L) may be regarded as the mean value of the quantity $\bar{L}$ in the eigenstate ψ_n (or ψ_L), and the mean value of a quantity represented by a self-adjoint operator is real (see (19.2)).

This makes quite clear the significance of self-adjointness of operators: self-adjoint operators represent real quantities.

21. Fundamental properties of eigenfunctions

Let us now consider the most important properties of eigenfunctions of self-adjoint operators, and to begin with take only the case of a discrete spectrum.

Two functions u_1 and u_2 are said to be *orthogonal* if

$$\int u_1^* u_2 \, dx = 0, \tag{21.1}$$

where the integral is taken over the whole range of the variables. For simplicity all the variables are denoted by a single letter x.

We shall show that eigenfunctions ψ_n and ψ_m of a self-adjoint operator $\boldsymbol{L}$ which belong to different eigenvalues L_n and L_m are orthogonal, i.e.

$$\int \psi_m^* \psi_n \, dx = 0. \tag{21.2}$$

Since ψ_n and ψ_m are eigenfunctions, we have

$$\boldsymbol{L}\psi_m = L_m \psi_m ; \qquad \boldsymbol{L}\psi_n = L_n \psi_n . \tag{21.3}$$

The complex conjugate of the first equation is

$$\boldsymbol{L}^* \psi_m^* = L_m \psi_m^* , \tag{21.3'}$$

since from (20.5) $L_m = L_m^*$. Multiplying the second Equation (21.3) by ψ_m^* and (21.3′) by ψ_n and subtracting, we have

$$\psi_m^* \, \boldsymbol{L}\psi_n - \psi_n \cdot \boldsymbol{L}^* \psi_m^* = (L_n - L_m) \psi_m^* \psi_n .$$

Integration of this equation over the whole range of the variables gives

$$\int \psi_m^* \cdot \boldsymbol{L}\psi_n \, dx - \int \psi_n \cdot \boldsymbol{L}^* \psi_m^* \, dx = (L_n - L_m) \int \psi_m^* \psi_n \, dx .$$

Since $\boldsymbol{L}$ is self-adjoint, the left-hand side is zero (in Equation (18.7), which defines self-adjointness, we put $\psi_m = u_1$, $\psi_n = u_2$), and so

$$(L_n - L_m) \int \psi_m^* \psi_n \, dx = 0 . \tag{21.4}$$

Since $L_n \neq L_m$, it follows that (21.2) is valid.

The functions of the discrete spectrum are always of integrable square, and so we can normalise them to unity:

$$\int \psi_m^* \psi_n \, dx = 1 . \tag{21.5}$$

This equation can be combined with (21.2) as

$$\int \psi_m^* \psi_n \, dx = \delta_{mn} , \tag{21.6}$$

where the symbol δ_{mn} is defined as follows:

$$\left.\begin{aligned} \delta_{mn} &= 1 \quad \text{if} \quad n = m , \\ \delta_{mn} &= 0 \quad \text{if} \quad n \neq m . \end{aligned}\right\} \tag{21.7}$$

Sets of functions which satisfy (21.6) are said to be *orthogonal and normalised.*

In the majority of cases encountered in quantum mechanics, an eigenvalue L_n of the operator $\boldsymbol{L}$ corresponds not to one function ψ_n but to several eigenfunctions ψ_{n1}, $\psi_{n2}, \ldots, \psi_{nk}, \ldots, \psi_{nf}$. This is called *degeneracy*. If the value $L = L_n$ corresponds to f eigenfunctions ($f > 1$), we say that there is *f-fold degeneracy*. The physical significance of degeneracy is that some particular value L_n of L can occur for more than one state.

The theorem proved above concerning the orthogonality of eigenfunctions applies only to functions belonging to different eigenvalues. Where there is degeneracy, the functions $\psi_{nk}(k = 1, 2, \ldots, f)$ belong to the same eigenvalue L_n:

$$\boldsymbol{L}\psi_{nk} = L_n \psi_{nk}, \qquad k = 1, 2, \ldots, f. \tag{21.8}$$

They will therefore not, in general, be orthogonal, but it can be shown[7] that these functions can always be so chosen that they are mutually orthogonal also:

$$\int \psi^*_{nk'} \psi_{nk} \,\mathrm{d}x = \delta_{k'k}. \tag{21.9}$$

Hence the condition (21.6) may be regarded as always satisfied if m and n are taken in general to be not one suffix but the whole set of suffixes describing an eigenfunction (e.g. two suffixes m and k' instead of m, and two suffixes n and k instead of n).

If the operator $\boldsymbol{L}$ has continuous eigenvalues, the above theorems are not directly applicable. In this case also, however, the eigenfunctions have properties analogous to those of the functions of a discrete spectrum.

The eigenfunctions of a continuous spectrum cannot be numbered. Here the functions depend on the eigenvalue L as a parameter, and we can put

$$\psi_L(x) = \psi(x, L), \tag{21.10}$$

where x denotes the variables in terms of which the operator $\boldsymbol{L}$ is expressed.

The orthogonality properties of eigenfunctions of the continuous spectrum can be most simply expressed in terms of the symbol $\delta(L' - L)$, called the *Dirac function* or the *δ function*. This function has the following properties:

$$\left.\begin{aligned} \int_a^b f(L')\,\delta(L' - L)\,\mathrm{d}L' &= 0 \text{ if the point } L' = L \text{ lies} \\ &\quad \text{outside the range } (a, b), \\ \int_a^b f(L')\,\delta(L' - L)\,\mathrm{d}L' &= f(L) \text{ if the point } L' = L \text{ lies} \\ &\quad \text{inside the range } (a, b), \end{aligned}\right\} \tag{21.11}$$

where $f(L')$ is any (sufficiently smooth) function. It can be shown[8] that functions of the continuous spectrum can be normalised so that

$$\int \psi^*(x, L')\,\psi(x, L)\,\mathrm{d}x = \delta(L' - L). \tag{21.12}$$

This equation is analogous to (21.6), since it follows from (21.11), putting $f(L') = 1$, that $\delta(L' - L) = 0$ everywhere except at the point $L' = L$, where δ becomes infinite.

[7] See Appendix II.
[8] See Appendix III.

Thus the symbol $\delta(L' - L)$ plays the same part as the symbol δ_{mn} for the discrete spectrum.

We know from mathematics that the set of eigenfunctions corresponding to any of a very wide class of operators is not only a set of orthogonal functions but is also *complete*. This means that any function $\psi(x)$ defined in the same range of the variables and subject to the same class of boundary conditions as the eigenfunctions $\psi_n(x)$ can be represented as a series of these eigenfunctions:

$$\psi(x) = \sum_n c_n \psi_n(x). \tag{21.13}$$

Using the orthogonality of the functions ψ_n, we can determine the coefficients c_n and so find the series which represents $\psi(x)$.[9] To do this we multiply (21.13) by $\psi_m^*(x)$ and integrate over all space:

$$\int \psi_m^*(x)\psi(x)\,\mathrm{d}x = \sum_n c_n \int \psi_m^*(x)\psi_n(x)\,\mathrm{d}x.$$

Since the functions ψ_n are orthogonal and normalised, the integrals in the sum are equal to δ_{mn} (see (21.6)); thus

$$\int \psi_m^*(x)\psi(x)\,\mathrm{d}x = \sum_n c_n \delta_{mn} = c_m.$$

Hence, interchanging m and n,

$$c_n = \int \psi_n^*(x)\psi(x)\,\mathrm{d}x. \tag{21.14}$$

Thus, if we know ψ and the set of orthogonal functions ψ_n, we can find all the amplitudes c_n in the series (21.13). Fourier series are a particular case of such expansions in orthogonal functions.

For a continuous spectrum we have an expansion as an integral resembling a Fourier integral. In this case

$$\psi(x) = \int c(L)\psi(x, L)\,\mathrm{d}L. \tag{21.15}$$

To determine the coefficients $c(L)$, we multiply (21.15) by $\psi^*(x, L')$ and integrate with respect to x:

$$\begin{aligned}\int \psi^*(x, L')\psi(x)\,\mathrm{d}x &= \int c(L)\,\mathrm{d}L \cdot \int \psi^*(x, L')\psi(x, L)\,\mathrm{d}x \\ &= \int c(L)\,\mathrm{d}L \cdot \delta(L' - L) = c(L').\end{aligned}$$

Interchanging L' and L we have finally

$$c(L) = \int \psi^*(x, L)\psi(x)\,\mathrm{d}x. \tag{21.16}$$

[9] Here we shall merely give the method of finding the coefficients c_n in the series (21.13). The concept of completeness of the set of orthogonal functions and the convergence of this series to the function $\psi(x)$ are discussed in [27], Chapter II, Section 1. The term 'completeness' of a set of functions (sometimes the set is said to be 'closed') can be easily explained as follows. Suppose that we have a sequence of orthogonal functions which is the same as the sequence ψ_n but does not include, say, the function ψ_1 ($n = 1$). This set of functions ($\psi_2, \psi_3, \ldots, \psi_n, \ldots$), like the set ($\psi_1, \psi_2, \psi_3, \ldots, \psi_n, \ldots$), is orthogonal, but not complete, since it cannot be used to express an arbitrary function as a series (to wit, the function ψ_1 cannot be so expressed). The proof of completeness consists in demonstrating that there are no such 'gaps'.

These representations of any function as an expansion (21.13) or (21.15) in terms of the eigenfunctions of operators lead to a very important result: any state represented by a wave function $\psi(x)$ can be expressed as a superposition (21.13) or (21.15) of states pertaining to definite values of some mechanical quantity. The states ψ_n or $\psi(x, L)$ are by definition states in which some mechanical quantity L has a definite value L_n or L respectively. The expressions (21.13) and (21.15) represent $\psi(x)$ as a sum or integral of these particular states.

22. General method of calculating the probabilities of the results of measurement

It has been shown above how the mean value $\bar{L}$ of any quantity represented by an operator L may be found, as well as the possible values $L_1, L_2, \ldots, L_n$ of such a quantity. We shall now calculate the probability that in some state $\psi(x)$ the measurement of the mechanical quantity L will give the value L_n. The basis of the calculation is the principle of superposition of states. Let the eigenfunctions of the operator $\boldsymbol{L}$ be $\psi_n(x)$. Using the completeness and orthogonality of these functions, we can write the wave function ψ as a superposition

$$\psi(x) = \sum_n c_n \psi_n(x). \tag{22.1}$$

The conjugate function is

$$\psi^*(x) = \sum_m c_m^* \psi_m^*(x), \tag{22.1'}$$

where m takes the same values as n.

Substituting these expressions for ψ and ψ^* in the formula for the mean value of the quantity L in the state ψ, we find

$$\bar{L} = \int \psi^* \cdot \boldsymbol{L} \psi \cdot \mathrm{d}x = \sum_n \sum_m c_m^* c_n \int \psi_m^* \cdot \boldsymbol{L} \psi_n \cdot \mathrm{d}x. \tag{22.2}$$

Since ψ_n is an eigenfunction of the operator $\boldsymbol{L}$,

$$\boldsymbol{L} \psi_n = L_n \psi_n. \tag{22.3}$$

With (22.3) and the orthogonality of the functions ψ_m^* and ψ_n, (22.2) becomes

$$\bar{L} = \sum_n \sum_m c_m^* c_n L_n \delta_{mn} = \sum_n c_n^* c_n L_n, \quad \text{i.e.} \quad \bar{L} = \sum_n |c_n|^2 L_n. \tag{22.4}$$

Multiplying (22.1) by (22.1′) and integrating over all space, we obtain

$$1 = \int \psi^* \psi \, \mathrm{d}x = \sum_n \sum_m c_m^* c_n \int \psi_m^* \psi_n \, \mathrm{d}x = \sum_m \sum_n c_m^* c_n \delta_{mn} = \sum_n |c_n|^2.$$

Thus

$$\sum_n |c_n|^2 = 1. \tag{22.5}$$

If $w(L_n)$ denotes the probability that a random quantity L has one of the possible values L_n, then the general definition of the mean gives

$$\bar{L} = \sum_n w(L_n) L_n \tag{22.6}$$

if

$$\sum w(L_n) = 1. \tag{22.7}$$

Comparison of (22.6) and (22.7) with (22.4) and (22.5) shows[10] that

$$w(L_n) = |c_n|^2\,. \tag{22.8}$$

The probability of finding a value of the mechanical quantity L which is one of its possible values L_n is equal to the squared modulus of the amplitude of the eigenstate ψ_n. In other words, this probability is determined by the intensity $|c_n|^2$ with which the eigenstate ψ_n is represented in the state ψ.

To calculate the probability of obtaining a given value of a quantity which has a continuous spectrum we use a procedure entirely analogous to that for a discrete spectrum. The state ψ under consideration is expanded in terms of the eigenfunctions $\psi(x, L)$ of the operator $\boldsymbol{L}$:

$$\psi(x) = \int c(L)\,\psi(x, L)\,\mathrm{d}L\,, \tag{22.9}$$

with $\psi(x, L)$ normalised to a δ function and ψ to unity.

Let us again calculate the mean value of L in the state ψ:

$$\bar{L} = \int \psi^* \cdot \boldsymbol{L}\psi\,\mathrm{d}x = \int\int c^*(L')\,\psi^*(x, L')\,\mathrm{d}L' \cdot \boldsymbol{L} \int c(L)\,\psi(x, L)\,\mathrm{d}L\,\mathrm{d}x\,.$$

Since $\psi(x, L)$ is an eigenfunction,

$$\boldsymbol{L}\psi(x, L) = L\psi(x, L)\,.$$

Substituting this in the above expression for $\bar{L}$ and changing the order of integration, we obtain

$$\bar{L} = \int\int c^*(L')\,c(L)\,\mathrm{d}L'\,\mathrm{d}L \cdot L \int \psi^*(x, L')\,\psi(x, L)\,\mathrm{d}x$$

and, from (21.12),

$$\bar{L} = \int\int c^*(L')\,c(L)\,\mathrm{d}L'\,\mathrm{d}L \cdot L\,\delta(L' - L)\,.$$

Hence, from the properties of the δ function,

$$\bar{L} = \int |c(L)|^2\,L\,\mathrm{d}L\,. \tag{22.10}$$

Similarly,

$$\begin{aligned} 1 &= \int \psi^*\psi\,\mathrm{d}x = \int \mathrm{d}x \int c^*(L')\,\psi^*(x, L')\,\mathrm{d}L' \int c(L)\,\psi(x, L)\,\mathrm{d}L \\ &= \int\int c^*(L')\,c(L)\,\mathrm{d}L'\,\mathrm{d}L \cdot \delta(L' - L) = \int |c(L)|^2\,\mathrm{d}L\,, \end{aligned}$$

i.e.

$$\int |c(L)|^2\,\mathrm{d}L = 1\,. \tag{22.11}$$

If the probability that the value of a continuous random quantity lies between L and $L + \mathrm{d}L$ is $w(L)\,\mathrm{d}L$, the general definition of the mean value gives

$$\bar{L} = \int w(L)\,\mathrm{d}L \tag{22.12}$$

if

$$\int w(L)\,\mathrm{d}L = 1\,. \tag{22.13}$$

[10] For a completely rigorous comparison of (22.6) and (22.4) it is necessary to consider an operator which is a function of L, equal to 1 for $L = L_n$ and 0 for $L \neq L_n$. The mean value of such an operator is $|c_n|^2$ from (22.4) and w_n from (22.6), so that $|c_n|^2 = w_n$.

Comparing (22.12) and (22.13) with (22.10) and (22.11), we obtain

$$w(L)\,\mathrm{d}L = |c(L)|^2\,\mathrm{d}L. \tag{22.14}$$

Thus for the continuous spectrum also we obtain a statistical interpretation of the intensities of the eigenstates $|c(L)|^2$.[11]

The above formulae are valid only for a pure ensemble described by a single wave function $\psi(x)$. For a mixed ensemble they must be somewhat generalised.

Let a mixed ensemble consist of the pure ensembles $\psi_1, \psi_2, \ldots, \psi_\alpha, \ldots$, mixed in the proportions $P_1, P_2, \ldots, P_\alpha, \ldots$. Then, if the probability of finding the value L_n of some quantity in the pure ensemble ψ_α is $w_\alpha(L_n)$, the total probability of finding $L = L_n$ in the mixed ensemble is

$$w(L_n) = \sum_\alpha P_\alpha\, w_\alpha(L_n). \tag{22.15}$$

Similarly, for a quantity with a continuous spectrum,

$$w(L)\,\mathrm{d}L = \sum_\alpha P_\alpha\, w_\alpha(L)\,\mathrm{d}L, \tag{22.16}$$

with

$$w_\alpha(L_n) = |c_{\alpha n}|^2, \qquad w_\alpha(L) = |c_\alpha(L)|^2, \tag{22.17}$$

where $c_{\alpha n}$ and $c_\alpha(L)$ are the amplitudes of the eigenfunctions of the operator $\boldsymbol{L}$($\psi_n(x)$ or $\psi(x, L)$ respectively) in the expansion of $\psi_\alpha(x)$. In accordance with Formulae (22.15) and (22.16) the mean value of the quantity $\bar{L}$ in a mixed ensemble is

$$\bar{L} = \sum_\alpha P_\alpha \bar{L}_\alpha, \tag{22.18}$$

where $\bar{L}_\alpha$ is the mean value of L in the pure ensemble ψ_α:

$$\bar{L}_\alpha = \int \psi_\alpha^* \boldsymbol{L}\, \psi_\alpha\, \mathrm{d}x. \tag{22.19}$$

23. Conditions for a simultaneous measurement of different mechanical quantities to be possible

We have seen that in the quantum region there are no states of particles in which the momentum and the corresponding co-ordinate simultaneously have definite values. Many other quantities have a similar mutually exclusive relationship. In order that states should exist in which two quantities L and M simultaneously have definite values, i.e. $\overline{\Delta L^2} = 0$, $\overline{\Delta M^2} = 0$, it is necessary that the wave function of such a state should be a common eigenfunction of the operators $\boldsymbol{L}$ and $\boldsymbol{M}$. However, the equations for the eigenfunctions of the operators $\boldsymbol{L}$ and $\boldsymbol{M}$:

$$\boldsymbol{L}\psi_L = L\psi_L \qquad \text{and} \qquad \boldsymbol{M}\psi_M = M\psi_M \tag{23.1}$$

[11] It may be noted that Formula (22.14) includes as a particular case Formula (12.4) for the momentum probability distribution: $c(p_x, p_y, p_z)$ is the amplitude of a state ψ_p with a definite momentum, that is, an eigenstate of the momentum operator. Hence $c(p_x, p_y, p_z)$ and $c(L)$ in (22.14) have a similar significance. To obtain (12.4) from (22.14) we need only take L to be the three momentum components p_x, p_y, p_z and $\mathrm{d}L$ correspondingly the product $\mathrm{d}p_x\,\mathrm{d}p_y\,\mathrm{d}p_z$.

have in general different solutions $\psi_L \neq \psi_M$. Hence, in states ψ_L with a definite value of L ($\overline{\Delta L^2} = 0$), the quantity M has no definite value ($\overline{\Delta M^2} > 0$), and in a state ψ_M with a definite value of M ($\overline{\Delta M^2} = 0$) the quantity L has no definite value ($\overline{\Delta L^2} > 0$).

The two quantities L and M simultaneously have definite values only in particular cases; for this it is necessary that $\psi_M = \psi_L$. It can be shown that the condition for two quantities L and M always to be able to have definite values simultaneously is that their operators $\boldsymbol{L}$ and $\boldsymbol{M}$ should commute. That is, the operator equation

$$\boldsymbol{LM} = \boldsymbol{ML} \tag{23.2}$$

must hold.[12] If, however,

$$\boldsymbol{LM} \neq \boldsymbol{ML}, \tag{23.3}$$

then the quantities L and M do not have definite values (except possibly exceptional ones) simultaneously.

Thus two quantities represented by commuting operators can simultaneously have definite values and so can be measured simultaneously, at least in principle. Two quantities represented by non-commuting operators can not simultaneously have definite values and so cannot be measured simultaneously.[13] The measurement of one such quantity L leads to a state ψ_L. When M is measured in this state, we obtain some new state ψ_M which does not coincide with the original state ψ_L. In other words, measurement of one such quantity changes the state of the system in such a way that the value of the other quantity becomes indeterminate.

We see that in general there is an effect of the measuring apparatus on the state of the system similar to that discussed above for the measurement of momentum and co-ordinate (cf. Sections 14, 15). Hence every apparatus used in the quantum region for the measurement of mechanical quantities pertaining to microparticles must be carefully considered as regards an analysis of the significance of the results of measurements with it and the changes which it may cause in the state of the system. Dogmatic judgements not founded on an analysis of the structure of the apparatus may lead to erroneous conclusions.

24. Co-ordinate and momentum operators of a microparticle

Since the wave function is treated as a function of the co-ordinates of a particle, the operator of the co-ordinate x of the particle is the number x itself. The effect of a function $F(x, y, z)$ of the co-ordinates of the particle, as an operator, reduces simply to multiplication of $\psi(x, y, z)$ by $F(x, y, z)$.

With this choice of variables[14] in the wave function, the operators of the momentum components of a particle will be, in accordance with Section 13,

$$\boldsymbol{P}_x = -i\hbar\,\partial/\partial x, \qquad \boldsymbol{P}_y = -i\hbar\,\partial/\partial y, \qquad \boldsymbol{P}_z = -i\hbar\,\partial/\partial z, \tag{24.1}$$

[12] See Appendix IV.

[13] See footnote 16.

[14] The possibility of a different choice of independent variables in the wave function is discussed in Chapter VII.

or, in vector form,

$$\vec{P} = -i\hbar\nabla, \tag{24.1'}$$

where ∇ is the gradient operator (nabla).

The momentum component and co-ordinate operators satisfy certain commutation rules which are useful in calculations involving them.

Let $\psi(x, y, z)$ be the wave function. Then

$$x(\boldsymbol{P}_x\psi) = x(-i\hbar\,\partial\psi/\partial x) = -i\hbar x\,\partial\psi/\partial x\,;$$

$$\boldsymbol{P}_x(x\psi) = -i\hbar\,\partial(x\psi)/\partial x = -i\hbar x\,\partial\psi/\partial x - i\hbar\psi\,.$$

Subtraction gives

$$(x\boldsymbol{P}_x - \boldsymbol{P}_x x)\psi = i\hbar\psi\,, \qquad \text{i.e.} \qquad x\boldsymbol{P}_x - \boldsymbol{P}_x x = i\hbar\,, \tag{24.2}$$

and similarly

$$y\boldsymbol{P}_y - \boldsymbol{P}_y y = i\hbar\,, \tag{24.2'}$$

$$z\boldsymbol{P}_z - \boldsymbol{P}_z z = i\hbar\,. \tag{24.2''}$$

These rules are called *Heisenberg's commutation relations.*

It is easily seen that

$$x\boldsymbol{P}_y - \boldsymbol{P}_y x = 0\,, \tag{24.3}$$

$$y\boldsymbol{P}_z - \boldsymbol{P}_z y = 0\,, \tag{24.3'}$$

$$z\boldsymbol{P}_y - \boldsymbol{P}_y z = 0\,, \tag{24.3''}$$

and so on.

We can similarly establish more general commutation relations for any function $F(x, y, z)$ and the momentum operators:

$$F\boldsymbol{P}_x - \boldsymbol{P}_x F = i\hbar\,\partial F/\partial x\,, \tag{24.4}$$

$$F\boldsymbol{P}_y - \boldsymbol{P}_y F = i\hbar\,\partial F/\partial y\,, \tag{24.4'}$$

$$F\boldsymbol{P}_z - \boldsymbol{P}_z F = i\hbar\,\partial F/\partial z\,. \tag{24.4''}$$

The relations (24.2) and (24.4) show that there are no states in which the momentum and the corresponding co-ordinate simultaneously have definite values. Essentially (24.2) and (24.4) express in operator form the uncertainty relation already discussed above.

Let us now determine the eigenvalues and eigenfunctions of the operator of the momentum component along any axis (for example, OX). According to Section 20, the equation for the eigenfunctions of the momentum operator is

$$\boldsymbol{P}_x\psi = p_x\psi\,, \tag{24.5}$$

where p_x is the eigenvalue. Using the value of $\boldsymbol{P}_x$, we find from this

$$-i\hbar\,\partial\psi/\partial x = p_x\psi\,. \tag{24.5'}$$

This equation is easily integrated to give

$$\psi_{p_x}(x) = Ne^{ip_x x/\hbar}\,, \tag{24.6}$$

where N is a constant. For this solution to be everywhere finite (it is obviously continuous and single-valued), p_x must be some real number. Thus the spectrum of eigenvalues p_x is found to be continuous:

$$-\infty < p_x < \infty . \tag{24.7}$$

The factor N can be chosen so that the function ψ_{p_x} is normalised to the delta function, by putting[15] $N = 1/\sqrt{(2\pi\hbar)}$. Thus the final normalised and orthogonal eigenfunctions of the operator $\boldsymbol{P}_x$ are

$$\psi_{p_x}(x) = \frac{1}{\sqrt{(2\pi\hbar)}} e^{ip_x x/\hbar}, \tag{24.8}$$

$$\int \psi^*_{p'_x}(x)\,\psi_{p_x}(x)\,\mathrm{d}x = \delta(p'_x - p_x), \tag{24.9}$$

i.e. the eigenfunctions of the momentum operator ψ_{p_x} are plane de Broglie waves. This is not unexpected. The fact that a de Broglie wave is a state with a definite value of the momentum of the particle was essentially the starting point of quantum mechanics (Sections 7 and 12).

25. The angular momentum operator of a microparticle

In classical mechanics the angular momentum is the vector product of the radius vector $\mathbf{r}$ from some chosen point (for example, the centre of force) to a particle and its momentum:

$$\mathbf{M} = \mathbf{r} \times \mathbf{p} . \tag{25.1}$$

The significance of this quantity in mechanics arises from the fact that it is an integral of the motion in a central force field. In quantum mechanics the angular momentum is represented by the operator

$$\vec{\boldsymbol{M}} = \mathbf{r} \times \vec{\boldsymbol{P}}, \tag{25.2}$$

where $\vec{\boldsymbol{P}}$ is the vector momentum operator (24.1′) and $\mathbf{r}$ the radius vector. The basis of this choice of the angular momentum operator is not only the outward analogy with the classical expression (25.1) but also the fact that the quantity represented by the operator $\vec{\boldsymbol{M}}$ is also an integral of the motion in a central force field (cf. Section 33) and has properties analogous to those of the angular momentum in classical mechanics.

The operators of the angular momentum components along the co-ordinate axes, corresponding to the value of the vector product (25.2), are

$$\left.\begin{aligned}
\boldsymbol{M}_x &= \boldsymbol{P}_z y - \boldsymbol{P}_y z = i\hbar\left(z\frac{\partial}{\partial y} - y\frac{\partial}{\partial z}\right),\\
\boldsymbol{M}_y &= \boldsymbol{P}_x z - \boldsymbol{P}_z x = i\hbar\left(x\frac{\partial}{\partial z} - z\frac{\partial}{\partial x}\right),\\
\boldsymbol{M}_z &= \boldsymbol{P}_y x - \boldsymbol{P}_x y = i\hbar\left(y\frac{\partial}{\partial x} - x\frac{\partial}{\partial y}\right),
\end{aligned}\right\} \tag{25.3}$$

[15] See Appendix III, Formula (20).

and the operator of the square of the angular momentum is

$$\boldsymbol{M}^2 = \boldsymbol{M}_x^2 + \boldsymbol{M}_y^2 + \boldsymbol{M}_z^2 = -\hbar^2 \left\{ \left(z\frac{\partial}{\partial y} - y\frac{\partial}{\partial z} \right)^2 + \left(x\frac{\partial}{\partial z} - z\frac{\partial}{\partial x} \right)^2 + \left(y\frac{\partial}{\partial x} - x\frac{\partial}{\partial y} \right)^2 \right\}. \tag{25.4}$$

Let us derive the commutation rules for the components of angular momentum. These rules will be needed later, but here they may serve as an illustration of the methods of operator algebra. The commutator $\boldsymbol{G} = \boldsymbol{M}_y\boldsymbol{M}_z - \boldsymbol{M}_z\boldsymbol{M}_y$ is calculated as follows. For $\boldsymbol{M}_y$ and $\boldsymbol{M}_z$ we substitute the expressions (25.3); the term $\boldsymbol{M}_y\boldsymbol{M}_z$ is

$$\begin{aligned} \boldsymbol{M}_y\boldsymbol{M}_z &= (\boldsymbol{P}_z x - \boldsymbol{P}_x z)(\boldsymbol{P}_x y - \boldsymbol{P}_y x) \\ &= \boldsymbol{P}_z x \boldsymbol{P}_x y - \boldsymbol{P}_x z \boldsymbol{P}_x y - \boldsymbol{P}_z x \boldsymbol{P}_y x + \boldsymbol{P}_x z \boldsymbol{P}_y x \\ &= y\boldsymbol{P}_z x \boldsymbol{P}_x - zy\boldsymbol{P}_x^2 - x^2 \boldsymbol{P}_z \boldsymbol{P}_y + z\boldsymbol{P}_y \boldsymbol{P}_x x, \end{aligned}$$

since y and $\boldsymbol{P}_z$, $\boldsymbol{P}_x$ commute, as do z and $\boldsymbol{P}_x$, $\boldsymbol{P}_y$, and x and $\boldsymbol{P}_z$, $\boldsymbol{P}_y$. Similarly

$$\boldsymbol{M}_z\boldsymbol{M}_y = y\boldsymbol{P}_z\boldsymbol{P}_x x - zy\boldsymbol{P}_x^2 - x^2\boldsymbol{P}_z\boldsymbol{P}_y + z\boldsymbol{P}_y x\boldsymbol{P}_x .$$

Subtraction gives

$$\boldsymbol{M}_y\boldsymbol{M}_z - \boldsymbol{M}_z\boldsymbol{M}_y = y\boldsymbol{P}_z(x\boldsymbol{P}_x - \boldsymbol{P}_x x) + z\boldsymbol{P}_y(\boldsymbol{P}_x x - x\boldsymbol{P}_x).$$

From (24.2) we thus obtain

$$\boldsymbol{M}_y\boldsymbol{M}_z - \boldsymbol{M}_z\boldsymbol{M}_y = i\hbar(y\boldsymbol{P}_z - \boldsymbol{P}_y z) = i\hbar\boldsymbol{M}_x .$$

Cyclic interchange of x, y, z gives the three commutators

$$\boldsymbol{M}_y\boldsymbol{M}_z - \boldsymbol{M}_z\boldsymbol{M}_y = i\hbar\boldsymbol{M}_x , \tag{25.5}$$

$$\boldsymbol{M}_z\boldsymbol{M}_x - \boldsymbol{M}_x\boldsymbol{M}_z = i\hbar\boldsymbol{M}_y , \tag{25.5'}$$

$$\boldsymbol{M}_x\boldsymbol{M}_y - \boldsymbol{M}_y\boldsymbol{M}_x = i\hbar\boldsymbol{M}_z . \tag{25.5''}$$

Thus the angular momentum component operators do not commute.

Each component, however, commutes with the square of the total angular momentum:

$$\boldsymbol{M}_x\boldsymbol{M}^2 - \boldsymbol{M}^2\boldsymbol{M}_x = 0, \tag{25.6}$$

$$\boldsymbol{M}_y\boldsymbol{M}^2 - \boldsymbol{M}^2\boldsymbol{M}_y = 0, \tag{25.6'}$$

$$\boldsymbol{M}_z\boldsymbol{M}^2 - \boldsymbol{M}^2\boldsymbol{M}_z = 0. \tag{25.6''}$$

The proof is left to the reader.

It follows from these commutation rules that the angular momentum components M_x, M_y, M_z cannot be measured simultaneously. In a state in which one component has a definite value ($\overline{\Delta M_x^2} = 0$), the other two components do not have definite values ($\overline{\Delta M_y^2} > 0$, $\overline{\Delta M_z^2} > 0$).[16] On the other hand, any component and the square of the total angular momentum can be measured simultaneously.

[16] An exception is the case $\overline{M_x^2} = 0$, when $\overline{M_x^2} = \overline{M_y^2} = \overline{M_z^2} = 0$.

Let us now determine the possible values of the component of the angular momentum in some arbitrary direction, and the possible values of the absolute magnitude of the angular momentum (more precisely, of M^2). To solve this problem, it is convenient to use spherical polar co-ordinates, with a selected direction as the axis OZ. In these co-ordinates

$$x = r \sin\theta \cos\phi\,, \qquad y = r \sin\theta \sin\phi\,, \qquad z = r\cos\theta\,, \tag{25.7}$$

where θ is the angle between the axis OZ and the radius vector, and ϕ the angle measured in the xy plane from the axis OX.

A somewhat lengthy transformation of Formulae (25.3) from Cartesian to spherical polar co-ordinates gives

$$\boldsymbol{M}_x = +\, i\hbar\left(\sin\phi\frac{\partial}{\partial\theta} + \cot\theta\cos\phi\frac{\partial}{\partial\phi}\right), \tag{25.8}$$

$$\boldsymbol{M}_y = -\, i\hbar\left(\cos\phi\frac{\partial}{\partial\theta} - \cot\theta\sin\phi\frac{\partial}{\partial\phi}\right), \tag{25.8'}$$

$$\boldsymbol{M}_z = -\, i\hbar\,\partial/\partial\phi\,, \tag{25.8''}$$

$$\boldsymbol{M}^2 = -\,\hbar^2\,\nabla^2_{\theta,\phi}\,, \tag{25.9}$$

where $\nabla^2_{\theta,\phi}$ is the spherical Laplacian operator:

$$\nabla^2_{\theta,\phi} = \frac{1}{\sin\theta}\frac{\partial}{\partial\theta}\left(\sin\theta\frac{\partial}{\partial\theta}\right) + \frac{1}{\sin^2\theta}\frac{\partial^2}{\partial\phi^2}. \tag{25.10}$$

Since the operators (25.8) and (25.9) act only on the angles θ and ϕ, the wave function need only be considered as a function of these angles, i.e.

$$\psi = \psi(\theta, \phi). \tag{25.11}$$

The equation to determine the eigenvalues of the operator $\boldsymbol{M}^2$ is, according to (20.2) (with $\boldsymbol{L} = \boldsymbol{M}^2$, $L = M^2$),

$$\boldsymbol{M}^2\psi = M^2\psi\,. \tag{25.12}$$

Substituting $\boldsymbol{M}^2$ from (25.9) and putting

$$\lambda = M^2/\hbar^2\,, \tag{25.13}$$

we obtain from (25.12)

$$\frac{1}{\sin\theta}\frac{\partial}{\partial\theta}\left(\sin\theta\frac{\partial\psi}{\partial\theta}\right) + \frac{1}{\sin^2\theta}\frac{\partial^2\psi}{\partial\phi^2} + \lambda\psi = 0\,. \tag{25.14}$$

This equation has to be solved for the whole range of the variables θ and ϕ ($0 \leqslant \theta \leqslant \pi$, $0 \leqslant \phi \leqslant 2\pi$), and the relevant solutions must be finite, continuous and single-valued. Equation (25.14) is well known as the equation of spherical harmonics; details concerning these functions and the solution of Equation (25.14) are given in Appendix V. Here we shall give only a brief summary.

It is found that solutions of this equation satisfying the conditions imposed do not exist for all values of λ, but only when

$$\lambda = l(l+1), \tag{25.15}$$

where l is a positive integer.

For each such value of l there are $2l + 1$ solutions which represent spherical harmonics. We denote them by

$$Y_{lm}(\theta, \phi) = \sqrt{\frac{(l-|m|)!\,(2l+1)}{(l+|m|)!\,4\pi}}\, P_l^{|m|}(\cos\theta)\, e^{im\phi}, \tag{25.16}$$

where m is an integer which takes only the values

$$m = 0, \pm 1, \pm 2, \ldots, \pm l;\; l = 1, 2, 3, \ldots \tag{25.17}$$

($2l + 1$ values in all). The symbol $|m|$ denotes the absolute magnitude of m. The function $P_l^{|m|}(\cos\theta)$ is defined by

$$P_l^{|m|}(\xi) = (1-\xi^2)^{\frac{1}{2}|m|} \frac{\mathrm{d}^{|m|}}{\mathrm{d}\xi^{|m|}} P_l(\xi), \qquad \xi = \cos\theta, \tag{25.18}$$

where $P_l(\xi)$ is the *Legendre polynomial*:

$$P_l(\xi) = \frac{1}{2^l\, l!} \frac{\mathrm{d}^l}{\mathrm{d}\xi^l}\left[(\xi^2 - 1)^l\right]. \tag{25.19}$$

The coefficient of $P_l^{|m|}$ in (25.16) is so chosen that the orthogonal functions Y_{lm} are also normalised to unity over the surface of the sphere, i.e.

$$\int_0^{\pi}\int_0^{2\pi} Y^*_{l'm'}\, Y_{lm} \sin\theta \,\mathrm{d}\theta\, \mathrm{d}\phi = \delta_{l'l}\,\delta_{m'm}. \tag{25.20}$$

(The co-ordinates θ and ϕ specify points on the surface of the sphere, and the surface element is $\sin\theta \;\mathrm{d}\theta\; \mathrm{d}\phi$.)

Let us now apply these results to our problem. As already stated, Equation (25.14) has single-valued finite solutions only for $\lambda = l(l + 1)$. The eigenvalues of the squared angular momentum operator are therefore

$$M_l^2 = \hbar^2\, l(l+1), \qquad l = 0, 1, 2, \ldots, \tag{25.21}$$

and the corresponding eigenfunctions are

$$\psi_{lm}(\theta, \phi) = Y_{lm}(\theta, \phi), \qquad m = 0, \pm 1, \ldots, \pm l. \tag{25.22}$$

The eigenvalue M_l^2 (25.21) corresponds to a total of $2l + 1$ eigenfunctions with different values of m. Thus we have a case of degeneracy (see Section 21). The significance of this degeneracy is easily seen if it is noticed that the eigenfunctions of the squared angular momentum operator $\boldsymbol{M}^2$ are also eigenfunctions of the operator of the angular momentum component $\boldsymbol{M}_z$. The equation for the eigenfunctions of the operator $\boldsymbol{M}_z$ is

$$\boldsymbol{M}_z \psi = M_z \psi\,; \tag{25.23}$$

substituting for M_z from (25.8″), we obtain $-i\hbar\,\partial\psi/\partial\phi = M_z\psi$. Substitution of ψ_{lm}, which is proportional to $e^{im\phi}$, shows that $-i\hbar \cdot im\psi_{lm} = M_z\psi_{lm}$, i.e. Equation (25.23) is satisfied by the function ψ_{lm}, and the eigenvalues of the operator M_z are

$$M_z = \hbar m, \qquad m = 0, \pm 1, \ldots, \pm l. \tag{25.24}$$

Hence it follows that the states ψ_{lm} with a given total angular momentum M_l^2 (i.e. a given l) but with different suffixes m are states with different components of angular momentum along the axis OZ.

This result shows that the possible values of the absolute magnitude of the angular momentum (25.21) and the possible values of the component of the angular momentum along an arbitrary axis OZ (25.24) have quantum values. No values other than those mentioned can occur in Nature. In states in which M^2 and M_z have definite values, the components M_x and M_y do not have definite values (except for $l = 0$, when $M^2 = M_x = M_y = M_z = 0$); for the functions (25.22) are not eigenfunctions of the operators M_x and M_y (25.8), as may be seen by direct calculation. This also follows from the fact that M_x, M_y and M_z do not commute.

The possible values of M_x and M_y are, of course, the same as those of M_z (25.24), since the direction OZ is in no way distinguished, and the validity of this statement may be seen by taking OX or OY as the polar axis. If we measure M_x or M_y, therefore, we shall always obtain one of the values $\hbar m$ ($m = 0, \pm 1, \pm 2, \ldots, \pm l$), but a new state is thereby produced with a definite value of M_x, say. This will be a state with indeterminate M_y and M_z, i.e. simultaneous measurements of the components of angular momentum are mutually exclusive: measurement of one component makes the value of another indeterminate.

The reader's attention should be drawn to certain symmetry properties of the eigenfunctions of the angular momentum operators. Let us replace the co-ordinates x, y, z by $-x, -y, -z$ respectively. This operation of reflection in the origin is called *inversion*. In spherical polar co-ordinates it signifies replacement of the co-ordinates r, θ, ϕ by r, $\pi - \theta$, $\phi + \pi$ respectively. Under this transformation of co-ordinates, $e^{im\phi}$ becomes $e^{im(\phi+\pi)} = (-1)^m e^{im\phi}$, and $P_l^{|m|}(\cos\theta)$ becomes $P_l^{|m|}(-\cos\theta) = (-1)^{l+|m|} \times P_l^{|m|}(\cos\theta)$; see (25.18) and (25.19). Thus $Y_{lm}(\theta, \phi)$ becomes $(-1)^l Y_{lm}(\theta, \phi)$, i.e. is multiplied by $(-1)^l$, whatever the value of m. Thus inversion multiplies the wave function by $+1$ for even l and by -1 for odd l.

The states with $(-1)^l = +1$ (even l) are called *even states* or *states of positive parity*, while those with $(-1)^l = -1$ (odd l) are called *odd states* or *states of negative parity*. The concept of parity of states is more general, however, than the parity of a state with a given angular momentum (see Section 107).

26. The energy operator and the Hamilton's function operator

(a) The kinetic energy operator T. Experiment shows that the kinetic energy of microparticles is related to the momentum in the same way as for macroscopic bodies[17], i.e.

[17] This fact has essentially been utilised already in the fundamental de Broglie relations (see Section 7).

the kinetic energy T of a particle of mass μ and momentum p is

$$T = p^2/2\mu = (p_x^2 + p_y^2 + p_z^2)/2\mu . \tag{26.1}$$

This means that the kinetic energy operator must be written

$$\boldsymbol{T} = \boldsymbol{P}^2/2\mu = (\boldsymbol{P}_x^2 + \boldsymbol{P}_y^2 + \boldsymbol{P}_z^2)/2\mu . \tag{26.2}$$

Substituting the values of the operators $\boldsymbol{P}_x$, $\boldsymbol{P}_y$, $\boldsymbol{P}_z$ from (24.1), we find

$$\boldsymbol{T} = -(\hbar^2/2\mu)\nabla^2 , \tag{26.2'}$$

where ∇^2 is the Laplacian operator ($\nabla^2 = \partial^2/\partial x^2 + \partial^2/\partial y^2 + \partial^2/\partial z^2$). With this choice of the operator $\boldsymbol{T}$ its eigenvalues T are (26.1) if p_x, p_y, p_z are the eigenvalues of the momentum operators $\boldsymbol{P}_x$, $\boldsymbol{P}_y$, $\boldsymbol{P}_z$. For, the equation for the eigenfunctions $\psi(x, y, z)$ of the operator $\boldsymbol{T}$ is

$$\boldsymbol{T}\psi = T\psi , \tag{26.3}$$

and is satisfied by a function which represents a plane de Broglie wave:

$$\psi_T(x, y, z) = \frac{1}{(2\pi\hbar)^{\frac{3}{2}}} e^{i(p_x x + p_y y + p_z z)/\hbar} . \tag{26.4}$$

This function is also an eigenfunction of the momentum operators, so that the kinetic energy T can be measured simultaneously with the momenta p_x, p_y, p_z (the operators $\boldsymbol{T}$, $\boldsymbol{P}_x$, $\boldsymbol{P}_y$, $\boldsymbol{P}_z$ commute, of course). The operator $\boldsymbol{T}$ can easily be written in any system of curvilinear co-ordinates. To do so, it is sufficient to write the Laplacian operator ∇^2 in the appropriate co-ordinate system. In particular, in a spherical polar co-ordinate system the operator ∇^2 has the form

$$\nabla^2 = \frac{1}{r^2}\frac{\partial}{\partial r}\left(r^2 \frac{\partial}{\partial r}\right) + \frac{\nabla^2_{\theta,\phi}}{r^2}, \tag{26.5}$$

where $\nabla^2_{\theta,\phi}$ is given by (25.10).

Substituting ∇^2 from (26.5) in (26.2′) and using (25.9), we obtain

$$\boldsymbol{T} = \boldsymbol{T}_r + \boldsymbol{M}^2/2\mu r^2 , \tag{26.6}$$

where $\boldsymbol{M}^2$ is the operator of the squared angular momentum and $\boldsymbol{T}_r$ is

$$\boldsymbol{T}_r = -\frac{\hbar^2}{2\mu}\frac{1}{r^2}\frac{\partial}{\partial r}\left(r^2\frac{\partial}{\partial r}\right). \tag{26.7}$$

The operator $\boldsymbol{T}_r$ may be regarded as the kinetic energy operator corresponding to motion along the radius vector, and the operator $\boldsymbol{M}^2/2\mu r^2$ as the operator of the kinetic energy in the transverse direction.[18]

(b) The total energy operator $\boldsymbol{H}$. First of all, let us note that the potential energy operator $\boldsymbol{U}$ is simply $U(x, y, z)$, since the potential energy is a function only of the

[18] Equation (26.6) agrees entirely with the representation of the kinetic energy in classical mechanics in the form $T = p_r^2/2\mu + M^2/2\mu r^2$, where p_r is the component of momentum along the radius vector.

co-ordinates x, y, z of the particle. In classical mechanics the total energy is the sum of the potential energy and the kinetic energy.

Similarly, in quantum mechanics, the operator which represents the total energy is the sum of the kinetic and potential energy operators, i.e.

$$\boldsymbol{H} = \boldsymbol{T} + \boldsymbol{U}(x, y, z). \tag{26.8}$$

The form of the potential energy $\boldsymbol{U}(x, y, z)$ is, as in classical mechanics, taken from experiment and represents the field of forces acting on the particle.

It may be noted that in quantum mechanics we cannot say that the total energy is the sum of the kinetic energy and the potential energy. The kinetic energy is a function of momenta, but the potential energy is a function of co-ordinates. As we know, there are no states of quantum ensembles in which particles simultaneously have definite momenta and co-ordinates. Thus it is impossible to measure the total energy of a particle by measuring its kinetic and potential energies separately.[19]

The total energy must be measured directly as a single quantity. The possible values of the total energy depend on the form of $U(x, y, z)$, i.e. on the nature of the particle and on the field of force in which it moves. The finding of these values is one of the most important problems of quantum mechanics and will be discussed below.

The total energy expressed as a function of momenta and co-ordinates is called in classical mechanics *Hamilton's function*. The kinetic energy operator $\boldsymbol{T}$ is expressed here in terms of the momentum operators (using (26.2)), and so we shall call $\boldsymbol{H}$ the Hamilton's function operator or briefly the *Hamiltonian*.

27. The Hamiltonian

The concept of Hamilton's function may be extended to non-conservative systems, and it is therefore a somewhat more general concept of mechanical energy.

In classical mechanics there are simple rules for writing down Hamilton's function. Its form is determined by the nature of the mechanical system, i.e. the nature of the particles and their interaction between themselves and with the external field. Knowing this Hamilton's function, we can easily find the equations of motion in any system of co-ordinates.

Similar rules exist in quantum mechanics to describe the operator of Hamilton's function (the Hamiltonian).

At present we shall consider only the motion of a single particle in an external field, and only later (Section 102) consider the Hamiltonian for a system of particles.

Two important cases must be distinguished, namely those where the forces do not depend on the velocity of the particle, and those where they do. In the former case **F** is a function only of the co-ordinates of the particle and of time, and can be represented as the gradient of some function $U(x, y, z)$, which we call the *force function*[20]:

[19] The operators $\boldsymbol{T}$ and $\boldsymbol{U}$, of course, do not commute, as may easily be seen by using the commutation rule (24.4). Hence it follows that T and U cannot be simultaneously determined for the same state ψ.

[20] The force function in mechanics is often taken as $-U$. It should also be noted that, by representing the force as the gradient of U, we exclude rotational fields (where **curl F** $\neq 0$). Such forces, independent of velocity, are unknown in the mechanics of microparticles.

$$\mathbf{F} = -\nabla U(x, y, z, t). \tag{27.1}$$

If the forces are independent of time, $U(x, y, z)$ is just the potential energy of the particle. In this case Hamilton's function is the same as the total energy of the particle, $T + U(x, y, z)$. The corresponding Hamiltonian is (26.8), and is the same as the total energy operator. In the more general case, Hamilton's function is the sum of the kinetic energy T and the force function U: $H = T + U(x, y, z, t)$. Since here U is not the potential energy, H is not the total energy of the system.

In complete analogy with the classical expression for Hamilton's function, the Hamiltonian is written for this case in quantum mechanics in the form

$$\boldsymbol{H} = \boldsymbol{T} + U(x, y, z, t), \tag{27.2}$$

where U is the force function.

There remains the case where the forces depend on the velocity of the particle. The only known forces of this type in the microuniverse are those which occur in an electromagnetic field (the Lorentz force). It is therefore sufficient to consider the Hamiltonian for the motion of a charged particle (of charge e and mass μ) in an arbitrary electromagnetic field.

It is known from the theory of fields that an arbitrary electromagnetic field $\mathbf{E}$ (electric field), $\mathbf{H}$ (magnetic field) can be described by means of a scalar potential V and a vector potential $\mathbf{A}$, with

$$\mathbf{E} = -\nabla V - \frac{1}{c}\frac{\partial \mathbf{A}}{\partial t}, \tag{27.3}$$

$$\mathbf{H} = \operatorname{curl} \mathbf{A}. \tag{27.4}$$

The classical Hamilton's function H which gives the correct equations of motion in an electromagnetic field is

$$H = \frac{1}{2\mu}\left(\mathbf{p} - \frac{e}{c}\mathbf{A}\right)^2 + eV, \tag{27.5}$$

where $\mathbf{p}$ (p_x, p_y, p_z) is the generalised momentum vector (so that $\mathbf{p} - (e/c)\mathbf{A} = \mu\mathbf{v}$, where $\mathbf{v}$ is the velocity of the particle, but $\mathbf{p} \neq \mu\mathbf{v}$).[21]

It is found that in quantum mechanics we obtain the correct Hamiltonian if $\mathbf{p}$ is replaced by the momentum operator $\vec{\boldsymbol{P}} = -i\hbar\nabla$, i.e. the Hamiltonian operator for this case is

$$\boldsymbol{H} = \frac{1}{2\mu}\left(\vec{\boldsymbol{P}} - \frac{e}{c}\vec{\boldsymbol{A}}\right)^2 + eV. \tag{27.6}$$

If there are, in addition to the electromagnetic forces, other forces described by a force function U, the general expression for the Hamiltonian is

$$\boldsymbol{H} = \frac{1}{2\mu}\left(\vec{\boldsymbol{P}} - \frac{e}{c}\vec{\boldsymbol{A}}\right)^2 + eV + U. \tag{27.7}$$

21 See Appendix VI.

Expanding the operator $(\vec{P} - (e/c)\vec{A})^2$ in explicit form gives

$$\left(\vec{P} - \frac{e}{c}\vec{A}\right)^2 = \left(P_x - \frac{e}{c}A_x\right)^2 + \left(P_y - \frac{e}{c}A_y\right)^2 + \left(P_z - \frac{e}{c}A_z\right)^2. \tag{27.8}$$

By the definition of operator products

$$\left(P_x - \frac{e}{c}A_x\right)^2 = \left(P_x - \frac{e}{c}A_x\right)\cdot\left(P_x - \frac{e}{c}A_x\right)$$

$$= P_x^2 - \frac{e}{c}P_x A_x - \frac{e}{c}A_x P_x + \frac{e^2}{c^2}A_x^2.$$

From (24.4)

$$P_x A_x - A_x P_x = -i\hbar\, \partial A_x/\partial x\,,$$

and hence

$$\left(P_x - \frac{e}{c}A_x\right)^2 = P_x^2 - \frac{2e}{c}A_x P_x + \frac{i\hbar e}{c}\frac{\partial A_x}{\partial x} + \frac{e^2}{c^2}A_x^2.$$

Repeating the calculation for the two remaining terms in (27.8) and adding the results, we find

$$H = \frac{1}{2\mu}\vec{P}^2 - \frac{e}{\mu c}\vec{A}\cdot\vec{P} + \frac{i\hbar e}{2\mu c}\operatorname{div}\vec{A} + \frac{e^2}{2\mu c^2}\vec{A}^2 + eV + U. \tag{27.9}$$

The Hamiltonian operator is seen from the discussion in this and the preceding sections to be determined by two factors: (1) the nature of the particle (or, in the general case, of the system of particles; cf. Section 102), (2) the nature of the fields acting on it.

This operator is fundamental in mechanics, since, when it has been fixed, we have essentially formulated mathematically all the properties of the system concerned. In particular, the number of independent variables in the Hamiltonian is, by definition, equal to the number of degrees of freedom of the system.

Success in solving a problem (in the sense of achieving agreement between theory and experiment) is decided by the correct choice of the Hamiltonian and the proper taking into account of all important interactions.

The independent variables in the Hamiltonian will usually be the Cartesian co-ordinates of the particle, since with this choice of variables the interaction operators (for example, the potential energy) are most simply expressed (being purely multiplicative), and the kinetic energy operator is expressed by a comparatively simple second-order differential operator. Other choices of independent variables are possible, however.[22]

In order to obtain an expression for the Hamiltonian in an arbitrary system of curvilinear co-ordinates q_1, q_2, q_3, it is sufficient to transform to this system the Hamiltonian obtained for a Cartesian co-ordinate system, following the usual rules of the differential calculus. Formula (26.5) gives an example of such a transformation. The form of the Hamiltonian in curvilinear co-ordinates is not so simply related to the

[22] If the particle has 'spin' (cf. Sections 58, 59, 60), the spin variable appears in the Hamiltonian together with the co-ordinates.

classical Hamilton's function as in Cartesian co-ordinates (replacing **p** by the operator $\vec{P}$). This is no accident: the Cartesian system in quantum mechanics is distinguished among all co-ordinate systems by the fact that in the Cartesian system the kinetic energy is expressed by the sum of the squares of the momentum components p_x, p_y, p_z, so that if the momentum is measured the kinetic energy may be calculated.

In curvilinear co-ordinates the kinetic energy is expressed as a quadratic function of the generalised momenta:

$$T = \sum_{i,k=1}^{3} a_{ik}(q_1, q_2, q_3)\, p_i p_k, \tag{27.10}$$

and the coefficients a_{ik} are functions of the co-ordinates. Measurement of p_k ($k = 1, 2, 3$) does not determine the kinetic energy, since the a_{ik} must also be known. The latter are functions of the co-ordinates q_k ($k = 1, 2, 3$) and so cannot be determined simultaneously with the momenta p_k. Thus a measurement of the momenta is at the same time a measurement of the kinetic energy only in Cartesian co-ordinates.[23]

[23] See Appendix VII for the equations of quantum mechanics in curvilinear co-ordinates.

CHAPTER IV

CHANGE OF STATE WITH TIME

28. Schrödinger's equation

Let the wave function $\psi(x, 0)$ which describes the state of an ensemble of particles (x denoting all the co-ordinates of a particle) be given at some instant $t = 0$. By means of this wave function we may calculate the probability of different results of measuring various mechanical quantities at time $t = 0$ for the ensemble of particles in the state $\psi(x, 0)$. In this sense we say that the wave function describes the state of a particle at time $t = 0$.

Let us now assume that we wish to make measurements not at time $t = 0$ but at some later time $t > 0$. During this time the state of the particle (or, generally, of the system of particles) changes and will be represented by some other wave function, which we denote by $\psi(x, t)$. As we know, the wave function changes also as a consequence of measurements (the reduction of a wave packet; see Section 17). We shall here suppose that no measurements are made between $t = 0$ and some instant t, so that only those changes of state are considered which are caused exclusively by the motion of the particle (or system of particles), without any interaction with measuring apparatus. What will then be the relation between the wave functions $\psi(x, 0)$ and $\psi(x, t)$?

Since the wave function fully describes a pure ensemble, it must also determine the subsequent evolution of the ensemble. This requirement expresses the principle of causality as applied to quantum mechanics.[1] Mathematically, this means that from the wave function $\psi(x, 0)$ for $t = 0$ it must be possible to determine the wave function $\psi(x, t)$ at later instants.

Let us consider the function ψ at an instant Δt infinitesimally near to $t = 0$. Then

$$\psi(x, \Delta t) = \psi(x, 0) + \left[\frac{\partial \psi(x, t)}{\partial t}\right]_{t=0} \Delta t + \dots .$$

According to the above discussion $[\partial\psi(x, t)/\partial t]_{t=0}$ must be determined by $\psi(x, 0)$, i.e.

$$[\partial\psi(x, t)/\partial t]_{t=0} = \boldsymbol{L}(x, 0)\, \psi(x, 0),$$

[1] We shall leave open the question whether this generally accepted formulation of the causality principle is the only one possible. A formulation of the problem is also possible in which the solution is not determined by the initial data but selected by conditions relating both to the past and to the future, leading to a problem of finding eigensolutions in space and time.

where $\boldsymbol{L}(x, 0)$ is some operation to be performed on $\psi(x, 0)$ in order to obtain $[\partial\psi/\partial t]_{t=0}$.

Since the time $t = 0$ is entirely arbitrary, it follows that

$$\partial\psi(x,t)/\partial t = \boldsymbol{L}(x,t)\,\psi(x,t). \tag{28.1}$$

The form of the operator $\boldsymbol{L}$, which may be called the *time translation operator*, cannot be determined from the above ideas of quantum mechanics and must be postulated.

According to the principle of superposition of states, this operator must be linear. It cannot contain time derivatives or time integrals; for, if it contained a first derivative with respect to t, $\boldsymbol{L}$ could not be the desired operator, which expresses the first derivative with respect to t in terms of $\psi(x, t)$; and if it contained higher derivatives with respect to t, (28.1) would be an equation for ψ of order higher than the first, and so, in order to determine the state at subsequent instants, it would be necessary to know not only $\psi(x, 0)$ but also the time derivatives of ψ: $(\partial\psi/\partial t)_0$, $(\partial^2\psi/\partial t^2)_0$, ...[2], i.e. the wave function ψ would not determine the state of the system, and this contradicts our fundamental hypothesis. The presence of an integral with respect to t would signify that the value of ψ over an interval of time, i.e. the history of the process, was involved. Thus $\boldsymbol{L}$ can contain t only as a parameter.

Equation (28.1) enables us to find the function $\psi(x, t)$ from the initial wave function $\psi(x, 0)$ and thus to predict the probability of results of various measurements at time t, assuming that during the time interval from 0 to t the system undergoes no additional interactions, and in particular is not subjected to measurement.

The change in the wave function which occurs in measurements (reduction) is not described by any differential equation, but follows directly from the result of the measurement itself (Section 17).

The correct choice of the operator $\boldsymbol{L}$ is suggested by a consideration of free motion with a definite value of the momentum $\mathbf{p}$. The wave function for such a motion is a de Broglie wave:

$$\psi(x, y, z, t) = Ne^{-i(Et - p_x x - p_y y - p_z z)/\hbar},$$

where

$$E = (p_x^2 + p_y^2 + p_z^2)/2\mu.$$

Direct substitution shows that this wave satisfies the equation

$$\frac{\partial\psi}{\partial t} = \frac{i\hbar}{2\mu}\nabla^2\psi.$$

This can also be written

$$\frac{\partial\psi}{\partial t} = \frac{1}{i\hbar}\boldsymbol{H}\psi,$$

[2] For example, the equation for the vibrations of a string is a second-order equation with respect to time. To determine the state of the string at time $t = 0$ it is necessary to know not only the deflection of the string $a(x, t)$ at $t = 0$ but also the velocity of the points of the string $\partial a(x, t)/\partial t$ at $t = 0$.

where the operator $\boldsymbol{H}$ is the Hamiltonian for free motion of a particle:

$$\boldsymbol{H} = \boldsymbol{T} = -(\hbar^2/2\mu)\nabla^2 .$$

Hence it follows that for free motion the time translation operator $\boldsymbol{L} = (1/i\hbar)\,\boldsymbol{H}$.

In quantum mechanics this particular result is generalised by assuming that the operator $\boldsymbol{L}$ is *always* given by

$$\boldsymbol{L} = \frac{1}{i\hbar}\boldsymbol{H}, \tag{28.2}$$

where $\boldsymbol{H}$ is the Hamiltonian operator (the operator of Hamilton's function), whose form in various cases has been discussed in Section 27.

In accordance with this postulate, Equation (28.1) for the wave function can be written

$$i\hbar\,\partial\psi/\partial t = \boldsymbol{H}\psi . \tag{28.3}$$

This is called *Schrödinger's equation.* It is one of the foundations of quantum mechanics[3] and its justification lies in its agreement with experiment rather than in the theoretical and historical reasons which led to its establishment.

The explicit form of Schrödinger's equation (28.3) in the absence of a magnetic field is, in accordance with the value of the operator $\boldsymbol{H}$ in (27.2) and (26.2′),

$$i\hbar\frac{\partial\psi}{\partial t} = -\frac{\hbar^2}{2\mu}\nabla^2\psi + U(x, y, z, t)\,\psi . \tag{28.4}$$

When a magnetic field is present, $\boldsymbol{H}$ must be taken from (27.9).

The most important property of Schrödinger's equation is the presence of the quantity i as a coefficient of the derivative $\partial\psi/\partial t$. In classical physics, equations of the first order in time have no periodic solutions: they describe irreversible processes, such as diffusion and thermal conduction.[4] The presence of i before $\partial\psi/\partial t$ has the result that Schrödinger's equation, though of the first order with respect to time, may have periodic solutions.

For the same reason, the wave function ψ is in general complex. In classical wave theory, waves are also written in the complex form u = constant $\times\ e^{i(\omega t - kx)}$, but ultimately we use only the real or the imaginary part of u. For example, the actual displacement of particles (e.g. a string) is u' = constant $\times \sin(\omega t - kx)$. The use of i is here simply a convenience in calculation.

The situation is not the same in quantum mechanics. If we take, instead of the de Broglie wave ψ, its real or imaginary part, for instance

$$\psi' = N \sin\left[(Et - p_x x - p_y y - p_z z)/\hbar\right],$$

[3] In many textbooks an attempt is made to 'derive' Schrödinger's equation. In reality, it is not a consequence of anything, but a foundation of the new theory, and so we prefer to postulate it after merely advancing the above-mentioned arguments in favour of the postulated form.

[4] The nature of the solution of a differential equation also depends, of course, on the boundary conditions. In the above comparison we are considering cases where both $U(x, y, z)$ and the boundary conditions are independent of time.

ψ' cannot be made to satisfy an equation of the first order in the time which is compatible with de Broglie's relations $\omega = E/\hbar$, $\mathbf{k} = \mathbf{p}/\hbar$.

The statement of the problem 'find $\psi(x, t)$ when $\psi(x, 0)$ is given', which is based on Schrödinger's equation, is meaningful only if $\psi(x, 0)$ can be uniquely related to certain definite physical conditions. This procedure is not trivial, however, since the wave function is by its nature a quantity which cannot be measured (for example, ψ and $\psi' = a\psi$, where a is any constant, represent the same state). Only the values of mechanical quantities L, M, N of a particle (or system of particles) can be measured, together with the probabilities of finding these values for an ensemble of particles (or systems).

We can therefore expect only that from measurements of probabilities for the ensemble it is possible to calculate the wave function to within an unimportant constant factor. This problem of calculating the wave function from measured probabilities is in general by no means simple, since the probabilities determine only $|\psi(x)|^2$ or, in general, the squared moduli $|c_n|^2$ of the amplitudes in an expansion of $\psi(x)$ in terms of eigenfunctions of some operator, while the phase of $\psi(x)$ or c_n remains undetermined.[5] The problem becomes simple or even trivial only in exceptional cases.

For example, it will be shown in Section 29 that the wave function is real in states where there is no flux of particles. In such cases the probability density $w(x) = |\psi(x)|^2 = \psi^2(x)$, and $\psi(x) \cong \sqrt{w(x)}$.

The whole problem of determining $\psi(x, 0)$ is, however, simplified by the fact that, in the great majority of cases of practical interest, we have an ensemble of particles with a definite complete set of mechanical variables L, M, N. If their values are known from measurements at time $t = 0$, we can use the mathematical methods of quantum mechanics to calculate the initial wave function also. For, if at time $t = 0$ the values L, M, N are measured, we can say that the initial wave function is the common eigenfunction of the operators $\mathbf{L}$, $\mathbf{M}$, $\mathbf{N}$ which pertains to the eigenvalues[6] L, M, N.

In this way the problem of determining the wave function is reduced to finding which quantities form a complete set.

It is shown below that these quantities must have the following properties: (1) they can be measured simultaneously; (2) they are equal in number to the degrees of freedom of the system; (3) they are independent.

With a view to subsequent generalisations, we shall suppose that the wave function is a function of f variables (a system with f degrees of freedom).

The desired function is an eigenfunction, and therefore belongs to a complete set of orthogonal functions in a space of f dimensions. Every such function is specified by f parameters $\alpha, \beta, \gamma, \ldots$, which label the functions. If such a function $\psi_{\alpha,\beta,\gamma,\ldots}(x, y, z, \ldots)$ is an eigenfunction of the operators $\mathbf{L}, \mathbf{M}, \mathbf{N}, \ldots$, the eigenvalues $L, M, N, \ldots$ will be functions of these parameters:

$$\left.\begin{aligned} \mathbf{L}\psi_{\alpha,\beta,\gamma,\ldots} &= L(\alpha, \beta, \gamma, \ldots)\, \psi_{\alpha,\beta,\gamma,\ldots}, \\ \mathbf{M}\psi_{\alpha,\beta,\gamma,\ldots} &= M(\alpha, \beta, \gamma, \ldots)\, \psi_{\alpha,\beta,\gamma,\ldots}, \\ \mathbf{N}\psi_{\alpha,\beta,\gamma,\ldots} &= N(\alpha, \beta, \gamma, \ldots)\, \psi_{\alpha,\beta,\gamma,\ldots}, \ldots . \end{aligned}\right\} \tag{28.5}$$

---

[5] Compare the theory of scattering (Chapter XIII).

[6] For example, if the initial state is specified by the momentum $\mathbf{p}$ of the particle (when $L = p_x$, $M = p_y$, $N = p_z$), then $\psi(r, 0) = \psi_p(x)$ is a plane de Broglie wave with momentum $\mathbf{p}$.

These equations are compatible if

$$[L, M] = [L, N] = [M, N] = \dots = 0, \tag{28.6}$$

i.e. if the quantities $L, M, N, \dots$ can be measured simultaneously. In order to determine the parameters $\alpha, \beta, \gamma, \dots$ from the measured $L, M, N, \dots$, it is necessary to solve f such equations:

$$\left.\begin{aligned} L &= L(\alpha, \beta, \gamma, \dots), \\ M &= M(\alpha, \beta, \gamma, \dots), \\ N &= N(\alpha, \beta, \gamma, \dots), \dots, \end{aligned}\right\} \tag{28.7}$$

and none of these must be a consequence of another, i.e. the quantities $L, M, N, \dots$ must be independent.[7]

29. Conservation of number of particles

From Schrödinger's equation we can derive the law of conservation of number of particles, expressed by the equation of continuity:

$$\partial w/\partial t + \operatorname{div} \mathbf{j} = 0, \tag{29.1}$$

where w is the mean number density of particles at the point x, y, z and $\mathbf{j}$ is the mean particle current density.

In order to obtain this equation, we take Schrödinger's equation for the simple case of potential forces (28.4):

$$i\hbar\frac{\partial\psi}{\partial t} = -\frac{\hbar^2}{2\mu}\nabla^2\psi + U\psi. \tag{29.2}$$

The equation for the complex conjugate function is

$$-i\hbar\frac{\partial\psi^*}{\partial t} = -\frac{\hbar^2}{2\mu}\nabla^2\psi^* + U\psi^*. \tag{29.2'}$$

Multiplying Equation (29.2) by ψ^* and (29.2′) by ψ, and subtracting, we have

$$i\hbar\left(\psi^*\frac{\partial\psi}{\partial t} + \psi\frac{\partial\psi^*}{\partial t}\right) = -\frac{\hbar^2}{2\mu}(\psi^*\nabla^2\psi - \psi\nabla^2\psi^*).$$

This can be written

$$\frac{\partial}{\partial t}(\psi\psi^*) = \frac{i\hbar}{2\mu}\operatorname{div}(\psi^*\nabla\psi - \psi\nabla\psi^*); \tag{29.3}$$

$\psi^*\psi$ is the probability density w:

$$w = \psi^*\psi. \tag{29.4}$$

If $\mathbf{j}$ denotes the vector

$$\mathbf{j} = \frac{i\hbar}{2\mu}(\psi\nabla\psi^* - \psi^*\nabla\psi), \tag{29.5}$$

then Equation (29.3) can be written

$$\partial w/\partial t + \operatorname{div} \mathbf{j} = 0. \tag{29.6}$$

[7] These parameters may be continuous or discrete. In the simplest case of separable variables, this function is of the form $\psi_{\alpha,\beta,\gamma,\dots}(x, y, z, \dots) = u_\alpha(x)v_\beta(y)w_\gamma(z)\dots$; see [27].

Hence it follows that $\mathbf{j}$ is the *probability current density vector*. Equation (29.6) has a more evident interpretation if we notice that $w = \psi^*\psi$ can be regarded as the *mean particle density*. Then $\mathbf{j}$ must be regarded as the mean flux of particles per second through an area of 1 cm^2. Accordingly, Equation (29.6) must be interpreted as the law of conservation of the number of particles. In particular, integrating (29.6) over some finite volume V and using Gauss' theorem, we have

$$\frac{\partial}{\partial t}\int_V w\,\mathrm{d}v = -\int_V \operatorname{div}\mathbf{j}\,\mathrm{d}v = -\int_S j_n\,\mathrm{d}s\,, \tag{29.7}$$

where the last integral is taken over a surface S bounding the volume V. On extending the integration to all space ($V \to \infty$) and using the fact that the wave functions ψ, and therefore also the current density $\mathbf{j}$, vanish on an infinitely remote surface[8], we find

$$\frac{\mathrm{d}}{\mathrm{d}t}\int_\infty w\,\mathrm{d}v = \frac{\mathrm{d}}{\mathrm{d}t}\int \psi^*\psi\,\mathrm{d}v = 0\,, \tag{29.8}$$

i.e. the total probability of finding a particle somewhere in space is independent of time. Hence the number of particles remains unchanged. Equation (29.8) also shows that the normalisation of the wave functions does not change in the course of time, a fact already mentioned in Section 10.

If we multiply $\mathbf{j}$ and w by the particle mass μ:

$$\rho_\mu = \mu w = \mu\,|\psi|^2\,, \qquad \mathbf{j}_\mu = \tfrac{1}{2}i\hbar\,(\psi\nabla\psi^* - \psi^*\nabla\psi)\,, \tag{29.9}$$

then ρ_μ is the mean density of matter (mass density) and $\mathbf{j}_\mu$ the mean current density of matter (or mass). From (29.6) it follows that these quantities obey the equation of continuity:

$$\partial\rho_\mu/\partial t + \operatorname{div}\mathbf{j}_\mu = 0\,. \tag{29.10}$$

Thus the change in the mean mass in some infinitesimal region is due to the inflow or outflow of this mass through the surface bounding the region.

In the same way, multiplying w and $\mathbf{j}$ by the particle charge e, we obtain the mean density of electric charge and of electric current:

$$\rho_e = ew = e\,|\psi|^2\,, \qquad \mathbf{j}_e = \frac{ie\hbar}{2\mu}(\psi\nabla\psi^* - \psi^*\nabla\psi)\,, \tag{29.11}$$

for which again we have the equation of continuity

$$\partial\rho_e/\partial t + \operatorname{div}\mathbf{j}_e = 0\,. \tag{29.12}$$

Equations (29.10) and (29.12) express the laws of conservation of mass and electricity in quantum mechanics.

[8] If the functions ψ are not integrable, the integral $\int j_n \mathrm{d}s$ may not be zero even on an infinitely remote surface. Physically, this signifies that there is a flux of particles to or from infinity.

If we write the wave function ψ in the form

$$\psi = ue^{i\Theta}, \tag{29.13}$$

where u is the real amplitude and Θ the real phase, then substitution of (29.13) in (29.5) gives

$$\mathbf{j} = (\hbar/\mu)\, u^2 \nabla\Theta\,. \tag{29.5'}$$

Since u^2 is the density w, the quantity $(\hbar/\mu)\,\nabla\Theta$ can be interpreted as the *mean velocity* at the point x, y, z:

$$\mathbf{v} = (\hbar/\mu)\,\nabla\Theta, \tag{29.14}$$

and $(\hbar/\mu)\,\Theta$ as the *velocity potential.*

From Formula (29.5′) it is particularly clearly seen that the current density $\mathbf{j}$ is non-zero only when the state is described by a complex function ψ.

When a magnetic field $\mathbf{H}$ is present, described by a vector potential $\mathbf{A}$ ($\mathbf{H} = \mathbf{curl}\,\mathbf{A}$), the formula for the current density $\mathbf{j}$ must be modified[9]: in the presence of a magnetic field we have, instead of (29.5), the following expression for the current density:

$$\mathbf{j} = \frac{i\hbar}{2\mu}(\psi\nabla\psi^* - \psi^*\nabla\psi) - \frac{e}{\mu c}\mathbf{A}\psi^*\psi\,. \tag{29.5''}$$

In order to derive this expression, we must substitute in Schrödinger's equation (28.3) the Hamiltonian (27.9) for motion in an arbitrary electromagnetic field. This substitution gives Schrödinger's equation as

$$\begin{aligned} i\hbar\frac{\partial\psi}{\partial t} = -\frac{\hbar^2}{2\mu}\nabla^2\psi + \frac{ie\hbar}{\mu c}\mathbf{A}\cdot\nabla\psi + \frac{ie\hbar}{2\mu c}(\operatorname{div}\mathbf{A})\,\psi\, + \\ + \frac{e^2}{2\mu c^2}A^2\psi + eV\psi + U\psi, \end{aligned} \tag{29.15}$$

and the equation for the conjugate function is

$$\begin{aligned} -i\hbar\frac{\partial\psi^*}{\partial t} = -\frac{\hbar^2}{2\mu}\nabla^2\psi^* - \frac{ie\hbar}{\mu c}\mathbf{A}\cdot\nabla\psi^* - \frac{ie\hbar}{2\mu c}(\operatorname{div}\mathbf{A})\,\psi^*\, + \\ + \frac{e^2}{2\mu c^2}A^2\psi^* + eV\psi^* + U\psi^*. \end{aligned} \tag{29.16}$$

We again multiply the first equation by ψ^*, the second by ψ, and subtract. This gives

$$\begin{aligned} i\hbar\frac{\partial(\psi^*\psi)}{\partial t} = -\frac{\hbar^2}{2\mu}\operatorname{div}(\psi^*\nabla\psi - \psi\nabla\psi^*)\, + \\ + \frac{i\hbar e}{\mu c}\{(\operatorname{div}\mathbf{A})\,\psi^*\psi + \mathbf{A}\cdot(\psi^*\nabla\psi + \psi\nabla\psi^*)\}\,. \end{aligned}$$

[9] The modification arises because, in the presence of the magnetic field, the operators P_x, P_y, P_z are generalised momentum operators, not ordinary momentum operators (mass times velocity). The situation is the same in classical mechanics; see Appendix VI, Formula (10′).

The expression in the braces may be transformed as follows:

$$(\operatorname{div}\mathbf{A})\psi^*\psi + \mathbf{A}\cdot(\psi^*\nabla\psi + \psi\nabla\psi^*) = (\operatorname{div}\mathbf{A})\psi^*\psi + \mathbf{A}\cdot\nabla(\psi^*\psi)$$
$$= \operatorname{div}(\mathbf{A}\psi^*\psi).$$

Substituting this result in the previous equation and dividing by $i\hbar$, we obtain

$$\frac{\partial(\psi^*\psi)}{\partial t} + \operatorname{div}\left\{\frac{i\hbar}{2\mu}[\psi\nabla\psi^* - \psi^*\nabla\psi] - \frac{e}{\mu c}\mathbf{A}\psi^*\psi\right\} = 0. \tag{29.17}$$

This is the equation of continuity in the presence of a magnetic field with vector potential **A**. The expression in the braces must be the current density **j**, and it is the same as (29.5″).

The validity of the equation of continuity is closely related to the self-adjointness of the Hamiltonian $\boldsymbol{H}$. This property of the Hamiltonian has been implicitly used in deriving (29.5) and (29.17). In Appendix VIII this aspect of the problem is considered in more detail, and it is shown how, from the requirement that the operator $\boldsymbol{H}$ is self-adjoint, there follow the conditions on the behaviour of the wave function at infinity and at singular points (Section 20) which ensure that the equation of continuity holds in all space.

30. Stationary states

In the absence of variable external fields, the Hamiltonian $\boldsymbol{H}$ is independent of time and is equal to the total-energy operator $\boldsymbol{H}(x)$. In this case Schrödinger's equation

$$i\hbar\,\partial\psi(x,t)/\partial t = \boldsymbol{H}(x)\,\psi(x,t) \tag{30.1}$$

has important solutions obtained by separating the variables x and t:

$$\psi(x,t) = \psi(x)\,f(t). \tag{30.2}$$

Substituting (30.2) in (30.1) and denoting the separation constant by E, we have

$$i\hbar\,\partial f/\partial t = Ef, \tag{30.3}$$

$$\boldsymbol{H}(x)\,\psi(x) = E\psi(x). \tag{30.4}$$

The first equation can be solved immediately:

$$f(t) = \text{constant} \times e^{-iEt/\hbar}. \tag{30.5}$$

The second equation is seen to be the same as that for the eigenfunctions of the energy operator[10] $\boldsymbol{H}$. If these functions are denoted by $\psi_n(x)$ and the eigenvalues by E_n (for definiteness, we shall take the case of a discrete energy spectrum), the final solution (30.2) becomes

$$\psi_n(x,t) = \psi_n(x)\,e^{-iE_nt/\hbar}. \tag{30.6}$$

Hence it follows that states with definite values of the energy E_n ($\overline{\Delta E^2} = 0$) are harmonically dependent on time, with frequency

$$\omega_n = E_n/\hbar. \tag{30.7}$$

[10] Equation (30.4) is derived from the general Equation (20.2) by putting $\boldsymbol{L} = \boldsymbol{H}$, $L = E$.

This result extends the applicability of de Broglie's relation $E = \hbar\omega$, which originally applied to free motion, to any system.

The states (30.6) with definite values of the energy are called *stationary states*, for reasons which will now be explained. The equation (30.4) is called *Schrödinger's equation for stationary states*. Since Equation (30.1) is linear, its general solution $\psi(x, t)$ can be represented as a superposition of stationary states with arbitrary constant amplitudes:

$$\psi(x,t) = \Sigma c_n \psi_n(x) e^{-iE_n t/\hbar}. \tag{30.8}$$

The amplitudes c_n are determined by the initial function $\psi(x, 0)$. The orthogonality of the functions ψ_n shows that

$$c_n = \int \psi(x,0)\psi_n^*(x)\,dx. \tag{30.9}$$

Let us now calculate the position probability $w_n(x, t)$ of a particle and the probability current density $\mathbf{j}_n(x, t)$ in the nth stationary state. According to (29.4) and (29.5), we have

$$w_n(x,t) = |\psi_n(x,t)|^2 = \psi_n^*(x,t)\psi_n(x,t),$$

$$\mathbf{j}_n(x,t) = \frac{i\hbar}{2\mu}\{\psi_n(x,t)\nabla\psi_n^*(x,t) - \psi_n^*(x,t)\nabla\psi_n(x,t)\}.$$

Substituting $\psi_n(x, t)$ from (30.6), we find that

$$w_n(x,t) = w_n(x,0), \tag{30.10}$$

$$\mathbf{j}_n(x,t) = \mathbf{j}_n(x,0), \tag{30.11}$$

i.e. in stationary states the position probability of a particle and the probability current density are independent of time.

Hence it follows (using (29.11)) that in these states the mean density of electric charge $\overline{\rho_e}$ and the mean electric current density $\overline{\mathbf{j}_e}$ are independent of time.

Thus a system in a state with a definite energy E_n ($\overline{\Delta E^2} = 0$) is, electrically speaking, a system with a static charge distribution and constant currents.

The description of stationary states will be more complete if we draw the reader's attention to the fact that in stationary states the probability $w(L)$ of finding some value L of any mechanical quantity (not explicitly dependent on time) is independent of time. The mean value $\bar{L}$ is also a constant. To prove this, we use Formula (22.14):

$$w(L) = |c(L)|^2,$$

where $c(L)$ is the amplitude in the expansion of $\psi(x, t)$ in eigenfunctions $\psi_L(x)$ of the operator $\boldsymbol{L}$ which represents the quantity L. According to (21.16) we have for the stationary state $\psi_n(x, t)$ (30.6)

$$c(L) = \int \psi_L^*(x)\psi_n(x,t)\,dx = e^{-iE_n t/\hbar}\int \psi_L^*(x)\psi_n(x)\,dx$$

and so

$$w(L) = |c(L)|^2 = |\int \psi_L^*(x)\psi_n(x)\,dx|^2 = \text{constant}. \tag{30.12}$$

CHAPTER V

CHANGE OF MECHANICAL QUANTITIES WITH TIME

31. Time derivatives of operators

Schrödinger's equation allows the establishment of simple rules whereby the change of the mean value of any mechanical quantity in an infinitesimal interval of time may be calculated; that is, the time derivative $d\bar{L}/dt$ of the mean value $\bar{L}$ of some quantity L may be found.

The physical significance of this derivative is as follows. Let us assume that at time t there is a state described by the wave function $\psi(x, t)$, and that we make measurements of the quantity L in this state. We find the results of the individual measurements to be $L', L'', L''', \ldots$. The mean of a large number of measurements will be $\bar{L}(t)$ and is calculated from the formula

$$\bar{L}(t) = \int \psi^*(x, t) \cdot \boldsymbol{L}\psi(x, t)\, dx. \tag{31.1}$$

Let another series of measurements be made at time $t' = t + \Delta t$ close to t, giving another series of results.[1] The mean value of these results will be different, since during the time Δt the state changes and the same results $L', L'', L''', \ldots$ will be obtained with different probabilities. In addition, it may happen that the quantity itself depends explicitly on time, so that the possible values $L', L'', L''', \ldots$ vary in the course of time. Let the mean result of the measurements at time $t + \Delta t$ be $\bar{L}(t + \Delta t)$. Then

$$\frac{d\bar{L}}{dt} = \lim_{\Delta t \to 0} \frac{\bar{L}(t + \Delta t) - \bar{L}(t)}{\Delta t}. \tag{31.2}$$

To calculate this derivative, we differentiate (31.1) with respect to time, obtaining

$$\frac{d\bar{L}}{dt} = \int \psi^* \cdot \frac{\partial \boldsymbol{L}}{\partial t} \psi\, dx + \int \frac{\partial \psi^*}{\partial t} \cdot \boldsymbol{L}\psi\, dx + \int \psi^* \cdot \boldsymbol{L} \frac{\partial \psi}{\partial t}\, dx. \tag{31.3}$$

[1] The measurement of the value of any quantity at exactly a given instant may lead to a change in the state of a microparticle. Hence the making of two series of measurements at times t and $t + \Delta t$ must be imagined to be more precisely as follows. There is an ensemble of a large number N of independent objects in the state $\psi(x, t)$, and we divide N into two large groups N' and N''. At time t we make measurements in the first group of N' particles and obtain $\bar{L}(t)$ (this in general changes the state of these objects, which is then no longer described by the function $\psi(x, t)$). Then, at time $t + \Delta t$, we make measurements in the group of N'' particles unaffected by the first measurement. These measurements give $\bar{L}(t + \Delta t)$.

It is evident that the first term is the mean value of $\partial L/\partial t$, and is zero if L does not depend explicitly on time. The last two terms can be simplified by means of Schrödinger's equation (28.3), which gives

$$\frac{\partial \psi}{\partial t} = \frac{1}{i\hbar} \boldsymbol{H} \psi ,$$
$$\frac{\partial \psi^*}{\partial t} = - \frac{1}{i\hbar} \boldsymbol{H}^* \psi^* .$$

Substituting these expressions in (31.3), we find

$$\frac{\mathrm{d}\bar{L}}{\mathrm{d}t} = \overline{\frac{\partial L}{\partial t}} - \frac{1}{i\hbar} \int (\boldsymbol{H}^* \psi^*) \cdot (\boldsymbol{L} \psi) \,\mathrm{d}x + \frac{1}{i\hbar} \int \psi^* \cdot (\boldsymbol{L} \boldsymbol{H} \psi) \,\mathrm{d}x .$$

The first integral is transformed by using the self-adjointness of the operator $\boldsymbol{H}$. With $\psi^* = u_1^*$, $\boldsymbol{L}\psi = u_2$, the self-adjointness relation (18.7) gives

$$\int (\boldsymbol{H}^* \psi^*) \cdot (\boldsymbol{L} \psi) \,\mathrm{d}x = \int u_2 \cdot \boldsymbol{H}^* u_1^* \,\mathrm{d}x = \int u_1^* \cdot \boldsymbol{H} u_2 \,\mathrm{d}x = \int \psi^* \cdot (\boldsymbol{H} \boldsymbol{L} \psi) \,\mathrm{d}x .$$

Substituting in $\mathrm{d}\bar{L}/\mathrm{d}t$, we find

$$\frac{\mathrm{d}\bar{L}}{\mathrm{d}t} = \overline{\frac{\partial L}{\partial t}} + \frac{1}{i\hbar} \int \psi^* \cdot (\boldsymbol{L} \boldsymbol{H} - \boldsymbol{H} \boldsymbol{L}) \psi \,\mathrm{d}x . \tag{31.4}$$

We shall use the notation

$$[\boldsymbol{H}, \boldsymbol{L}] = \frac{1}{i\hbar} (\boldsymbol{L} \boldsymbol{H} - \boldsymbol{H} \boldsymbol{L}) , \tag{31.5}$$

and refer to this operator as the *quantum Poisson bracket*.[2]

With this notation, (31.4) may be written

$$\frac{\mathrm{d}\bar{L}}{\mathrm{d}t} = \overline{\frac{\partial L}{\partial t}} + \overline{[H, L]} . \tag{31.6}$$

We see that the time derivative of the mean value $\bar{L}$ is the mean value of a quantity represented by the operator $\partial \boldsymbol{L}/\partial t + [\boldsymbol{H}, \boldsymbol{L}]$. Hence this operator must be taken as the operator $\mathrm{d}\boldsymbol{L}/\mathrm{d}t$ of the time derivative $\mathrm{d}L/\mathrm{d}t$ of a quantity L represented by the operator $\boldsymbol{L}$:

$$\frac{\mathrm{d}\boldsymbol{L}}{\mathrm{d}t} = \frac{\partial \boldsymbol{L}}{\partial t} + [\boldsymbol{H}, \boldsymbol{L}] . \tag{31.7}$$

This definition of the operator which represents the time derivative $\mathrm{d}L/\mathrm{d}t$ has the result that

$$\mathrm{d}\bar{L}/\mathrm{d}t = \overline{\mathrm{d}L/\mathrm{d}t} = \int \psi^* \cdot \frac{\mathrm{d}\boldsymbol{L}}{\mathrm{d}t} \psi \,\mathrm{d}x , \tag{31.8}$$

that is, the time derivative of the mean is equal to the mean time derivative.

[2] The term is taken from classical mechanics; see Appendix VI, Formula (4).

If the quantity L does not depend explicitly on time, Formulae (31.6) and (31.7) become simply

$$\overline{\mathrm{d}L/\mathrm{d}t} = \overline{[H, L]}, \tag{31.9}$$

$$\mathrm{d}\boldsymbol{L}/\mathrm{d}t = [\boldsymbol{H}, \boldsymbol{L}]. \tag{31.10}$$

Finally, the reader's attention may also be drawn to the fact that, in calculating the operator of the time derivative of a product or sum of operators, the quantum Poisson bracket can be treated like an ordinary derivative (though the order of factors must be preserved). It is easy to see that, if $\boldsymbol{L} = \boldsymbol{A} + \boldsymbol{B}$, then

$$\frac{\mathrm{d}\boldsymbol{L}}{\mathrm{d}t} = [\boldsymbol{H}, \boldsymbol{A} + \boldsymbol{B}] = [\boldsymbol{H}, \boldsymbol{A}] + [\boldsymbol{H}, \boldsymbol{B}] = \frac{\mathrm{d}\boldsymbol{A}}{\mathrm{d}t} + \frac{\mathrm{d}\boldsymbol{B}}{\mathrm{d}t}, \tag{31.11}$$

and if $\boldsymbol{L} = \boldsymbol{AB}$, then

$$\frac{\mathrm{d}\boldsymbol{L}}{\mathrm{d}t} = [\boldsymbol{H}, \boldsymbol{AB}] = [\boldsymbol{H}, \boldsymbol{A}]\,\boldsymbol{B} + \boldsymbol{A}\,[\boldsymbol{H}, \boldsymbol{B}] = \frac{\mathrm{d}\boldsymbol{A}}{\mathrm{d}t}\boldsymbol{B} + \boldsymbol{A}\frac{\mathrm{d}\boldsymbol{B}}{\mathrm{d}t}. \tag{31.12}$$

32. Equations of motion in quantum mechanics. Ehrenfest's theorems

Let us now find the laws governing the change of momenta and co-ordinates with time.

These are quantities which do not depend explicitly on time. Hence, by (31.10), the operators of the time derivatives of these quantities are simply the quantum Poisson brackets, i.e. are expressed in terms of the operators of the quantities themselves and the Hamiltonian $\boldsymbol{H}$ for the mechanical system concerned.

Let the operators of the Cartesian co-ordinates x, y, z and the corresponding momenta p_x, p_y, p_z be denoted by $\boldsymbol{X}, \boldsymbol{Y}, \boldsymbol{Z}$ and $\boldsymbol{P}_x, \boldsymbol{P}_y, \boldsymbol{P}_z$ respectively.[3]

The Hamiltonian $\boldsymbol{H}$ will be a function of these operators and, in general, of the time t:

$$\boldsymbol{H} = \boldsymbol{H}(\boldsymbol{P}_x, \boldsymbol{P}_y, \boldsymbol{P}_z, \boldsymbol{X}, \boldsymbol{Y}, \boldsymbol{Z}, t). \tag{32.1}$$

Also, let $\mathrm{d}\boldsymbol{X}/\mathrm{d}t$, $\mathrm{d}\boldsymbol{Y}/\mathrm{d}t$, $\mathrm{d}\boldsymbol{Z}/\mathrm{d}t$ denote the operators of the derivatives of the co-ordinates with respect to time, i.e. the operators of the velocity components along the co-ordinate axes, and $\mathrm{d}\boldsymbol{P}_x/\mathrm{d}t$, $\mathrm{d}\boldsymbol{P}_y/\mathrm{d}t$, $\mathrm{d}\boldsymbol{P}_z/\mathrm{d}t$ the operators of the time derivatives of the momentum components.

Substituting for $\boldsymbol{L}$ in (31.10) the operators $\boldsymbol{X}, \boldsymbol{Y}, \boldsymbol{Z}, \boldsymbol{P}_x, \boldsymbol{P}_y, \boldsymbol{P}_z$, we obtain the required operator equations:

$$\mathrm{d}\boldsymbol{X}/\mathrm{d}t = [\boldsymbol{H}, \boldsymbol{X}]; \quad \mathrm{d}\boldsymbol{Y}/\mathrm{d}t = [\boldsymbol{H}, \boldsymbol{Y}]; \quad \mathrm{d}\boldsymbol{Z}/\mathrm{d}t = [\boldsymbol{H}, \boldsymbol{Z}]; \tag{32.2}$$

$$\mathrm{d}\boldsymbol{P}_x/\mathrm{d}t = [\boldsymbol{H}, \boldsymbol{P}_x]; \quad \mathrm{d}\boldsymbol{P}_y/\mathrm{d}t = [\boldsymbol{H}, \boldsymbol{P}_y]; \quad \mathrm{d}\boldsymbol{P}_z/\mathrm{d}t = [\boldsymbol{H}, \boldsymbol{P}_z]. \tag{32.2'}$$

These operator equations are entirely analogous to the classical Hamilton's equations and are therefore called the *quantum Hamilton's equations*.[4]

[3] We shall consider the motion only in Cartesian co-ordinates. See Appendix VII for the equations in curvilinear co-ordinates.

[4] Cf. Appendix VI, Equation (7).

In classical mechanics the first set of equations (derivatives of the co-ordinates) establishes the relation between velocity and momentum, while the second set (derivatives of the momenta) expresses the laws of variation of momentum with time. The quantum Hamilton's equations have the same significance. In order to see this, we must write in explicit form the Poisson brackets in (32.2) and (32.2'). For simplicity we shall consider the case where there are no magnetic forces. In this case the Hamiltonian has the form (see (27.2))

$$H = \frac{1}{2\mu}(P_x^2 + P_y^2 + P_z^2) + U(X, Y, Z, t). \tag{32.3}$$

Regarding the wave function as a function of the co-ordinates x, y, z of the particle and time t, we have the following expressions for the operators:

$$\begin{gathered} X = x, \qquad Y = y, \qquad Z = z, \\ P_x = -i\hbar\frac{\partial}{\partial x}, \qquad P_y = -i\hbar\frac{\partial}{\partial y}, \qquad P_z = -i\hbar\frac{\partial}{\partial z}. \end{gathered} \tag{32.4}$$

Let us now calculate the operator $\mathrm{d}X/\mathrm{d}t$. We have

$$[H, X] = \frac{1}{i\hbar}(XH - HX) = \frac{1}{2\mu i\hbar}(XP_x^2 - P_x^2 X), \tag{32.5}$$

since X commutes with P_y, P_z and $U(x, y, z, t)$. The commutation rule (24.2) for the operators X and P_x gives

$$\begin{aligned} P_x^2 X &= P_x(P_x X) = P_x(XP_x - i\hbar) = (P_x X)P_x - i\hbar P_x \\ &= (XP_x - i\hbar)P_x - i\hbar P_x = XP_x^2 - 2i\hbar P_x. \end{aligned} \tag{32.6}$$

Substituting this expression in (32.5), we find

$$[H, X] = \frac{1}{\mu}P_x. \tag{32.7}$$

For y and z a similar result is obviously obtained, and so

$$\frac{\mathrm{d}X}{\mathrm{d}t} = \frac{P_x}{\mu}, \qquad \frac{\mathrm{d}Y}{\mathrm{d}t} = \frac{P_y}{\mu}, \qquad \frac{\mathrm{d}Z}{\mathrm{d}t} = \frac{P_z}{\mu}, \tag{32.8}$$

i.e. the velocity operator is equal to the momentum operator divided by the mass μ of the particle. In other words, the relation between the velocity and momentum operators is the same as that between the corresponding quantities in classical mechanics.

Let us now find the operator $\mathrm{d}P_x/\mathrm{d}t$. From (32.2') and (24.4) we have

$$[H, P_x] = \frac{1}{i\hbar}(P_x U - UP_x) = -\frac{\partial U}{\partial x}, \tag{32.9}$$

and so

$$\frac{\mathrm{d}P_x}{\mathrm{d}t} = -\frac{\partial U}{\partial x}, \qquad \frac{\mathrm{d}P_y}{\mathrm{d}t} = -\frac{\partial U}{\partial y}, \qquad \frac{\mathrm{d}P_z}{\mathrm{d}t} = -\frac{\partial U}{\partial z}. \tag{32.10}$$

$-\partial U/\partial x$, $-\partial U/\partial y$, $-\partial U/\partial z$ are just the force component operators[5], so that (32.10) may be written

$$\mathrm{d}\boldsymbol{P}_x/\mathrm{d}t = \boldsymbol{F}_x, \qquad \mathrm{d}\boldsymbol{P}_y/\mathrm{d}t = \boldsymbol{F}_y, \qquad \mathrm{d}\boldsymbol{P}_z/\mathrm{d}t = \boldsymbol{F}_z, \tag{32.11}$$

i.e. the operator of the time derivative of the momentum is equal to the force operator. Equations (32.10) may therefore be regarded as Newton's equations in operator form.

If we calculate the mean values of the quantities $\mathrm{d}x/\mathrm{d}t$, $\mathrm{d}p_x/\mathrm{d}t$, etc. in some state ψ, (32.8) and (32.10) give, using (31.8),

$$\overline{\mathrm{d}x/\mathrm{d}t} = \mathrm{d}\bar{x}/\mathrm{d}t = \bar{p}_x/\mu, \tag{32.12}$$

$$\overline{\mathrm{d}p_x/\mathrm{d}t} = \mathrm{d}\bar{p}_x/\mathrm{d}t = -\overline{\partial U/\partial x} = F_x, \tag{32.13}$$

and so on. In other words, the time derivative of the mean co-ordinate $\bar{x}$ is equal to the mean momentum divided by the mass of the particle, and the derivative of the mean momentum $\overline{p_x}$ is equal to the mean force $\overline{F_x}$.

In explicit form, Equations (32.12) and (32.13) are

$$\frac{\mathrm{d}}{\mathrm{d}t}\int \psi^* x\, \psi\, \mathrm{d}x = \frac{1}{\mu}\int \psi^* \cdot \boldsymbol{P}_x \psi\, \mathrm{d}x, \tag{32.12'}$$

$$\frac{\mathrm{d}}{\mathrm{d}t}\int \psi^* \cdot \boldsymbol{P}_x \psi\, \mathrm{d}x = -\int \psi^* \frac{\partial U}{\partial x}\, \psi\, \mathrm{d}x. \tag{32.13'}$$

These are called *Ehrenfest's theorems*. Differentiating (32.12) with respect to time and eliminating $\mathrm{d}\overline{p_x}/\mathrm{d}t$ from (32.12) and (32.13), we get the quantum Newton's equation:

$$\mu \partial^2 \bar{x}/\partial t^2 = -\overline{\partial U/\partial x} = \overline{F_x}. \tag{32.14}$$

33. Integrals of the motion

In quantum mechanics we have the same integrals of the motion as in classical mechanics. The quantity L is an integral of the motion if

$$\frac{\mathrm{d}\boldsymbol{L}}{\mathrm{d}t} = \frac{\partial \boldsymbol{L}}{\partial t} + [\boldsymbol{H}, \boldsymbol{L}] \equiv 0. \tag{33.1}$$

A case of particular importance is that where the quantity L does not depend explicitly on time; then (33.1) becomes

$$\mathrm{d}\boldsymbol{L}/\mathrm{d}t = [\boldsymbol{H}, \boldsymbol{L}] \equiv 0, \tag{33.2}$$

that is, for integrals of the motion (not explicitly time-dependent) the quantum Poisson bracket is zero.

Since $[\boldsymbol{H}, \boldsymbol{L}]$ is the commutator of the operator $\boldsymbol{L}$ and the Hamiltonian operator, any quantity L which does not depend explicitly on time will be an integral of the motion if its operator commutes with the Hamiltonian operator.

From Formulae (33.1) and (33.2) it follows that the mean value of integrals of the

[5] These are simply functions of the co-ordinates.

motion is independent of time:

$$d\bar{L}/dt = 0. \tag{33.3}$$

We shall now show that the probability $w(L_n, t)$ of finding at some instant t a particular value of an integral of the motion, say L_n, is also independent of time.[6]

Since the operators $\boldsymbol{L}$ and $\boldsymbol{H}$ commute, they have common eigenfunctions $\psi_n(x)$:

$$\boldsymbol{L}\psi_n = L_n\psi_n, \tag{33.4}$$

$$\boldsymbol{H}\psi_n = E_n\psi_n. \tag{33.4'}$$

We now expand an arbitrary state $\psi(x, t)$ in terms of the eigenfunctions ψ_n. These functions are stationary-state functions, so that (cf. (30.8))

$$\psi(x, t) = \sum_n c_n\psi_n(x)\, e^{-iE_nt/\hbar}, \tag{33.5}$$

or

$$\psi(x, t) = \sum_n c_n(t)\,\psi_n(x), \tag{33.6}$$

where

$$c_n(t) = c_n e^{-iE_nt/\hbar} = c_n(0)\, e^{-iE_nt/\hbar}. \tag{33.7}$$

The expansion (33.6) is an expansion of $\psi(x, t)$ in terms of eigenfunctions of the operator $\boldsymbol{L}$, and so

$$w(L_n, t) = |c_n(t)|^2 = |c_n(0)|^2 = \text{constant}. \tag{33.8}$$

The form of the integrals of the motion depends on the nature of the force field in which the particle is moving. For *free motion* the force function $U(x, y, z, t) = 0$, and the Hamiltonian is

$$\boldsymbol{H} = \boldsymbol{T} = \frac{1}{2\mu}(\boldsymbol{P}_x^2 + \boldsymbol{P}_y^2 + \boldsymbol{P}_z^2). \tag{33.9}$$

As in classical mechanics, the momentum is in this case an integral of the motion, i.e. a conserved quantity, for

$$[\boldsymbol{H}, \boldsymbol{P}_x] = [\boldsymbol{H}, \boldsymbol{P}_y] = [\boldsymbol{H}, \boldsymbol{P}_z] = 0, \tag{33.10}$$

i.e.

$$d\boldsymbol{P}_x/dt = 0, \qquad d\boldsymbol{P}_y/dt = 0, \qquad d\boldsymbol{P}_z/dt = 0. \tag{33.11}$$

In the field of a *central force* the area law applies; the angular momentum is an integral of the motion. For, in such a field the potential energy U is a function of the distance from the centre: $U = U(r)$. In this case, therefore, the Hamiltonian $\boldsymbol{H}$ may be written (cf. (26.6))

$$\boldsymbol{H} = \boldsymbol{T}_r + \frac{\boldsymbol{M}^2}{2\mu r^2} + U(r). \tag{33.12}$$

The operators of the squared angular momentum $\boldsymbol{M}^2$ and of its components $\boldsymbol{M}_x$, $\boldsymbol{M}_y$,

[6] This refers to the integrals of the motion which do not depend explicitly on time.

M_z depend, according to (25.8), only on the angles θ and ϕ, and so do not act on functions of r. Moreover, the operator M^2 in (33.12) commutes with M_x, M_y and M_z (see (25.6)). Hence all four operators commute with H (33.12), so that

$$[H, M^2] = 0, \qquad dM^2/dt = 0, \tag{33.13}$$

$$\left.\begin{aligned} [H, M_x] = [H, M_y] = [H, M_z] = 0, \\ dM_x/dt = dM_y/dt = dM_z/dt = 0. \end{aligned}\right\} \tag{33.14}$$

Thus the angular momentum is an integral of the motion in a field of central forces.

Let us now apply Equation (33.1) to the Hamiltonian. Putting $L = H$, we obtain

$$dH/dt = \partial H/\partial t + [H, H] = \partial H/\partial t. \tag{33.15}$$

If the Hamiltonian does not depend explicitly on time, then

$$dH/dt = 0. \tag{33.16}$$

In this case, however, the Hamiltonian is the same as the total-energy operator. Hence (33.16) expresses the fact that the total energy is an integral of the motion in a field of forces independent of time. In other words, (33.16) expresses the *law of conservation of energy* in quantum mechanics.

According to the above properties of the integrals of the motion, (33.16) must be taken as signifying that the mean value $\bar{E}$ of the energy and the probabilities of finding various possible values of the energy $E = E_n$ are both independent of time.[7]

[7] See Section 113 regarding the law of conservation of energy in quantum mechanics.

CHAPTER VI

THE RELATION BETWEEN QUANTUM MECHANICS, CLASSICAL MECHANICS AND OPTICS

34. The transition from the quantum equations to Newton's equations

Ehrenfest's theorems proved in Section 32 assert that in any state ψ the mean values of mechanical quantities obey the quantum Newton's equation[1]:

$$\mu \, \mathrm{d}^2 \bar{x}/\mathrm{d}t^2 = -\overline{\partial U/\partial x}. \tag{34.1}$$

Let us suppose that ψ differs appreciably from zero only in a very small region of space Δx. We shall call such a state a *wave packet*. If the mean value of x varied in accordance with the classical Newton's equation and the shape of the wave packet did not change, then the motion of the wave packet $|\psi|^2$ could be regarded as the motion of a point mass obeying Newtonian mechanics. In general no such motion is obtained in quantum mechanics, since firstly the wave packet is broadened out, and secondly, if the motion of the centroid $\bar{x}$ of the wave packet coincided with the motion of a point mass in the field $U(x)$, the equation

$$\overline{\partial U/\partial x} = \partial U(\bar{x})/\partial \bar{x} \tag{34.2}$$

would have to hold. This equation is in general not valid. We shall nevertheless examine the conditions in which the motion of the wave packet approximately coincides with that of a point mass. The mean value $\bar{x}$ of the co-ordinate x, i.e. the co-ordinate of the centroid of the wave packet, is determined by the formula

$$\bar{x} = \int \psi^* x \psi \, \mathrm{d}x. \tag{34.3}$$

The mean value of the force is

$$-\overline{\partial U/\partial x} = -\int \psi^* (\partial U/\partial x) \psi \, \mathrm{d}x. \tag{34.4}$$

If we put $x = \bar{x} + \xi$, then

$$-\overline{\partial U/\partial x} = -\int \psi^*(\bar{x}+\xi) \frac{\partial U(\bar{x}+\xi)}{\partial \bar{x}} \psi(\bar{x}+\xi) \, \mathrm{d}\xi. \tag{34.4'}$$

Let us assume that $U(x)$ is a sufficiently slowly varying function of x in the region where $|\psi|^2$ differs appreciably from zero. Then $\partial U(\bar{x}+\xi)/\partial \bar{x}$ can be expanded as a

[1] We shall consider only one dimension. The generalisation of the discussion to the three-dimensional case offers no difficulty.

power series in ξ. This gives

$$-\frac{\overline{\partial U}}{\partial x} = -\frac{\partial U(\bar{x})}{\partial \bar{x}} \int \psi^* \psi \, d\xi - \frac{1}{1!} \frac{\partial^2 U(\bar{x})}{\partial \bar{x}^2} \int \psi^* \xi \psi \, d\xi - \\ - \frac{1}{2!} \frac{\partial^3 U(\bar{x})}{\partial \bar{x}^3} \int \psi^* \xi^2 \psi \, d\xi - \dots \tag{34.5}$$

But

$$\int \psi^* \psi \, d\xi = \int \psi^* \psi \, dx = 1,$$
$$\int \psi^* \xi \psi \, d\xi = \int \psi^* (x - \bar{x}) \psi \, dx = 0,$$
$$\int \psi^* \xi^2 \psi \, d\xi = \int \psi^* (x - \bar{x})^2 \psi \, dx = \overline{\Delta x^2}.$$

Hence

$$-\frac{\overline{\partial U}}{\partial x} = -\frac{\partial U(\bar{x})}{\partial \bar{x}} - \frac{1}{2} \frac{\partial^3 U(\bar{x})}{\partial \bar{x}^3} \overline{\Delta x^2} - \dots \tag{34.6}$$

From Equation (34.1), we therefore have

$$\mu \frac{d^2 \bar{x}}{dt^2} = -\frac{\partial U(\bar{x})}{\partial \bar{x}} - \frac{1}{2} \frac{\partial^3 U(\bar{x})}{\partial \bar{x}^3} \overline{\Delta x^2} - \dots \tag{34.7}$$

If the force field varies only slowly in space, then by taking the width $\overline{\Delta x^2}$ of the wave packet sufficiently small we can neglect all terms in this equation except the first. This gives Newton's equation for the motion of the centroid $\bar{x}$ of the wave packet:

$$\mu \, d^2 \bar{x}/dt^2 = -\partial U(\bar{x})/\partial \bar{x}, \tag{34.7'}$$

which will be valid for the time interval t during which the terms omitted in (34.7) are small, i.e. at least while

$$\left| \frac{\partial U(\bar{x})}{\partial \bar{x}} \right| \gg \frac{1}{2} \left| \frac{\partial^3 U(\bar{x})}{\partial \bar{x}^3} \right| \overline{\Delta x^2}. \tag{34.8}$$

The quantity $\overline{\Delta x^2}$ which defines the size of the wave packet is a function of time, and in general increases with time (see below), i.e. the wave packet is spread out. Hence, even if the inequality (34.8) holds at the initial instant, it may cease to be valid after some time t. But even if this inequality is satisfied, the state of the particles is not necessarily the same as the classical state.[2]

For, if we take a very narrow wave packet ($\overline{\Delta x^2}$ small), the mean potential energy of the particle according to quantum mechanics is almost equal to the potential energy of a point mass at the centroid of the wave packet:

$$\bar{U} = \int \psi^* U \psi \, dx \approx U(\bar{x}), \tag{34.9}$$

[2] For all functions $U(x)$ of the form $U = a + bx + cx^2$ it follows from (34.7) that the motion of the centroid of the wave packet coincides exactly with the classical motion of a point mass in the field $U(x)$. Such cases include (a) free motion, (b) motion in a uniform field, (c) a harmonic oscillator, and certain others (e.g. the results for a uniform magnetic field are the same as for an oscillator).

but this is not true of the kinetic energy T: we have

$$\bar{T} = \frac{\overline{p^2}}{2\mu} = \frac{1}{2\mu}\overline{(p - \bar{p} + \bar{p})^2} = \frac{\overline{\Delta p^2}}{2\mu} + \frac{\bar{p}^2}{2\mu}, \tag{34.10}$$

and Heisenberg's relation gives

$$\overline{\Delta p^2} \geqslant \hbar^2/4\overline{\Delta x^2},$$

so that in (34.10) the first (quantum) term may be much greater than the classical energy of a particle moving with momentum $\bar{p}$.

The quantum term in (34.10) may be neglected if

$$\frac{\bar{p}^2}{2\mu} \gg \frac{\overline{\Delta p^2}}{2\mu}, \qquad \text{or} \qquad \bar{p}^2 \gg \hbar^2/4\overline{\Delta x^2}. \tag{34.11}$$

Thus the motion of a particle may be regarded as taking place according to the laws of classical mechanics in time t if during this time the inequalities (34.8) and (34.11) can be simultaneously satisfied.

The simultaneous validity of these two inequalities is favoured by the following conditions:

(1) the kinetic energy $\bar{T}$ of the particle being large,

(2) the field $U(x)$ being a slowly varying function of the co-ordinate x.

Thus the transition from the quantum to the Newtonian equations of motion occurs when the kinetic energies of particles are large and fields vary smoothly.

Let us now consider the spreading of a wave packet for a freely moving particle. The mean square deviation $\overline{\Delta x^2}$ is the mean value of $\Delta x^2 = x^2 - \bar{x}^2$, where $\bar{x}$ is the co-ordinate of the centroid of the wave packet. From (34.7) we have

$$\mathrm{d}\bar{x}/\mathrm{d}t = \bar{v}, \qquad \bar{x} = \bar{v}t + \bar{x}_0, \tag{34.12}$$

i.e. the centroid moves uniformly with velocity $\bar{v}$. The time derivatives of Δx^2 are calculated from the general Formula (31.7); putting $L = \Delta x^2$, we find

$$\frac{\mathrm{d}}{\mathrm{d}t}\Delta x^2 = \frac{\partial}{\partial t}\Delta x^2 + [\boldsymbol{H}, \Delta x^2] = -\frac{\mathrm{d}}{\mathrm{d}t}\bar{x}^2 + [\boldsymbol{H}, x^2]$$

and since for free motion the operator $\boldsymbol{H} = (1/2\mu)\,\boldsymbol{P}^2$, we have[3]

$$[\boldsymbol{H}, x^2] = \frac{1}{2\mu}[\boldsymbol{P}^2, x^2] = \frac{1}{2\mu i\hbar}(x^2\boldsymbol{P}^2 - \boldsymbol{P}^2x^2) = (1/\mu)(x\boldsymbol{P} + \boldsymbol{P}x).$$

Thus the operator of $\mathrm{d}(\Delta x^2)/\mathrm{d}t$ is

$$\frac{\mathrm{d}(\Delta x^2)}{\mathrm{d}t} = \frac{x\boldsymbol{P} + \boldsymbol{P}x}{\mu} - \frac{\mathrm{d}}{\mathrm{d}t}\bar{x}^2 = \frac{x\boldsymbol{P} + \boldsymbol{P}x}{\mu} - 2\bar{v}\bar{x}. \tag{34.13}$$

Let us now calculate the second derivative:

$$\frac{\mathrm{d}^2(\Delta x^2)}{\mathrm{d}t^2} = \frac{\partial}{\partial t}\left(\frac{\mathrm{d}(\Delta x^2)}{\mathrm{d}t}\right) + \left[\boldsymbol{H}, \frac{\mathrm{d}(\Delta x^2)}{\mathrm{d}t}\right] = -\frac{\mathrm{d}^2\bar{x}^2}{\mathrm{d}t^2} + \left[\boldsymbol{H}, \frac{x\boldsymbol{P} + \boldsymbol{P}x}{\mu}\right];$$

$$\left[\boldsymbol{H}, \frac{x\boldsymbol{P} + \boldsymbol{P}x}{\mu}\right] = \frac{1}{2i\hbar\mu^2}\{(x\boldsymbol{P} + \boldsymbol{P}x)\boldsymbol{P}^2 - \boldsymbol{P}^2(x\boldsymbol{P} + \boldsymbol{P}x)\} = \frac{2\boldsymbol{P}^2}{\mu^2},$$

[3] In the subsequent calculations we use the formula $\boldsymbol{P}_x x = x\boldsymbol{P}_x - i\hbar$.

i.e.

$$\frac{d^2(\Delta x^2)}{dt^2} = \frac{2P^2}{\mu^2} - \frac{d^2\bar{x}^2}{dt^2} = \frac{2P^2}{\mu^2} - 2\bar{v}^2 . \tag{34.14}$$

Since P^2 commutes with H, all higher derivatives of Δx^2 are zero. Thus the expansion of Δx^2 as a Taylor series in t is

$$\Delta x_t^2 = \Delta x_0^2 + \left(\frac{xP + Px}{\mu} - 2\bar{v}\bar{x}\right) t + \frac{1}{2!}\left(\frac{2P^2}{\mu^2} - 2\bar{v}^2\right) t^2 . \tag{34.15}$$

Replacing the operators by mean values gives

$$\overline{\Delta x_t^2} = \overline{\Delta x_0^2} + \left(\frac{\overline{xp + px}}{\mu} - 2\,\bar{v}\bar{x}\right) t + \left(\frac{\overline{p^2}}{\mu^2} - \bar{v}^2\right) t^2 . \tag{34.16}$$

The quantity $\overline{\Delta x_t^2}$ is necessarily positive, and so (34.16) shows that $\overline{\Delta x_t^2}$ increases without limit (possibly after passing through a minimum) as t increases, i.e. the wave packet is spread out. In many cases (depending on the form of $\psi(x, 0)$) the term in t is absent. Then (34.16) takes the particularly simple form

$$\overline{\Delta x_t^2} = \overline{\Delta x_0^2} + \overline{\Delta v^2}\, t^2 , \tag{34.17}$$

where $\overline{\Delta v^2}$ is the mean square deviation of the velocity:

$$\overline{\Delta v^2} = \frac{\overline{p^2}}{\mu^2} - \bar{v}^2 = \overline{v^2} - \bar{v}^2 .$$

The spreading of such a wave packet corresponds to the dispersal of a cloud of particles in classical mechanics when their initial positions and velocities are distributed about mean values with mean square deviations $\overline{\Delta x_0^2}$ and $\overline{\Delta v^2}$. In classical mechanics, however, a cloud can exist with $\overline{\Delta x_0^2}$ and $\overline{\Delta v^2}$ equal to zero. In quantum mechanics this is impossible, owing to the uncertainty relation. Figure 20 illustrates the above description of the motion and spreading of a wave packet.

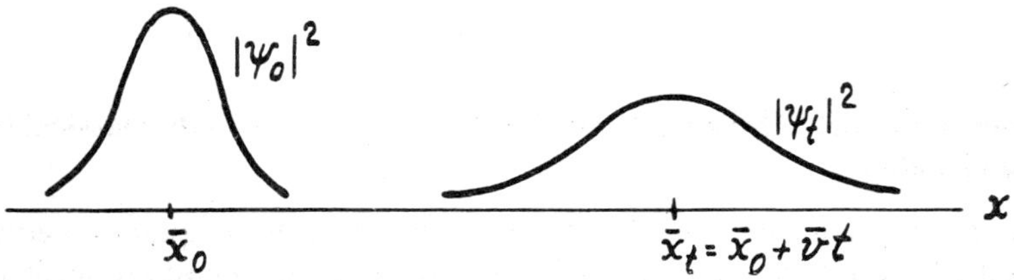

Fig. 20. Motion and spreading of a wave packet in the absence of external forces.

As an example of the application of the theory of motion of a wave packet discussed in this section, let us find the conditions under which the scattering of a particle in the field of an atom may be treated by the methods of classical mechanics. Let the range of the forces of interaction between the atom and a particle passing near it be a. It is clear that, if we can speak of the path of the particle within the atom, the dimensions Δx of the wave packet must be much less than a (Figure 21).

From (34.10) and (34.11) we conclude that the kinetic energy of the particle $\bar{T} = \bar{p}^2/2\mu \gg \hbar^2/8\mu\overline{\Delta x^2} \gg \hbar^2/8\mu a^2$ (since $\Delta x \ll a$). With this condition the wave packet is not appreciably spread during the time of transit of the particle through the atom, which in order of magnitude is $t = a/\bar{v} = a\mu/\bar{p}$; it follows from (34.17) that the spread of the wave packet is $\Delta x' \simeq \Delta v \cdot t = (\Delta p/\mu)(a\mu/\bar{p}) = (\Delta p/\bar{p})\, a$; since, when (34.11) holds, $\Delta p \ll \bar{p}$, we have $\Delta x' \ll a$.

The range of action of the forces is, in order of magnitude, equal to the radius of the atom $a \simeq 10^{-8}$ cm. For an α particle, with typical energy $T = 1$ MeV $= 1.6 \times 10^{-6}$ erg, $\bar{p}_\alpha = \sqrt{(2\mu_\alpha T)} = 4.6 \times 10^{-15}$ g cm/s (the mass of the α particle being $\mu_\alpha = 6.7 \times 10^{-24}$ g). But $\hbar/a = 1 \times 10^{-19}$ g cm/s. Thus for an α particle inequality (34.11) holds. The scattering of an α particle may therefore be treated by the methods of classical mechanics (as was first done by Rutherford in his celebrated theory of the scattering of α particles). However, if the α particle passes near the nucleus, it is necessary to take account of the effect of nuclear forces, for which the range of interaction $a \simeq 10^{-12}$ cm, $\hbar/a = 1 \times 10^{-15}$ and the inequality (34.11) is not well satisfied. Hence the scattering of α particles by nuclear forces ('anomalous' scattering) can not be treated by the methods of classical mechanics.

For electrons ($\mu_e = 9 \times 10^{-28}$ g), for example, with $T = 100$ eV we have $\bar{p}_e = 5.4 \times 10^{-19}$, so that $\bar{p}_e$ is comparable with $\hbar/a$, and classical mechanics cannot be applied to this case.

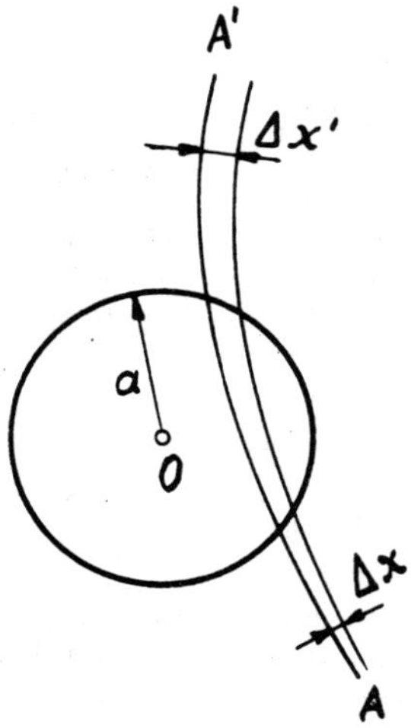

Fig. 21. Scattering of a particle in the field of an atom. O centre of the atom, a range of interaction of forces, AA' path of wave packet spreading from width Δx to $\Delta x'$.

35. The transition from Schrödinger's time-dependent equation to the classical Hamilton-Jacobi equation

In the last section we have established the relation between the quantum equations of motion and Newton's equations, and therefore between quantum mechanics and classical mechanics. This relation may be derived in another manner: it may be shown that the classical Hamilton-Jacobi equation is a limiting case of Schrödinger's time-dependent equation. To show this, let us first recall the Hamilton-Jacobi equation. For simplicity, we shall consider only the motion of a single particle of mass μ in a potential field $U(x, y, z, t)$. The Hamilton-Jacobi equation involves the action function $S_0(x, y, z, t)$, which has the property that

$$p_x = -\partial S_0/\partial x\,, \qquad p_y = -\partial S_0/\partial y\,, \qquad p_z = -\partial S_0/\partial z\,, \tag{35.1}$$

where p_x, p_y, p_z are the components of the particle momentum along the co-ordinate axes. The Hamilton-Jacobi equation itself for the case under consideration is

$$\frac{\partial S_0}{\partial t} = \frac{1}{2\mu}\left\{\left(\frac{\partial S_0}{\partial x}\right)^2 + \left(\frac{\partial S_0}{\partial y}\right)^2 + \left(\frac{\partial S_0}{\partial z}\right)^2\right\} + U(x,y,z,t)\,. \tag{35.2}$$

Since the Hamilton's function $H(p_x, p_y, p_z, x, y, z, t)$ is

$$H(p_x, p_y, p_z, x, y, z, t) = \frac{1}{2\mu}(p_x^2 + p_y^2 + p_z^2) + U(x, y, z, t), \tag{35.3}$$

it follows from (35.1) and (35.2) that the Hamilton-Jacobi equation can be written

$$\frac{\partial S_0}{\partial t} = H\left(-\frac{\partial S_0}{\partial x}, -\frac{\partial S_0}{\partial y}, -\frac{\partial S_0}{\partial z}, x, y, z, t\right). \tag{35.4}$$

If the Hamilton's function does not depend explicitly on time, it is equal to the particle energy E. Then (35.4) gives

$$\partial S_0/\partial t = E, \qquad S_0 = Et - s_0(x, y, z). \tag{35.5}$$

Equations (35.1) show that the paths are orthogonal to the surfaces S_0 = constant. If H does not depend explicitly on time, the form of these surfaces does not vary with time. Figure 22 shows these surfaces and the possible paths of the particle.

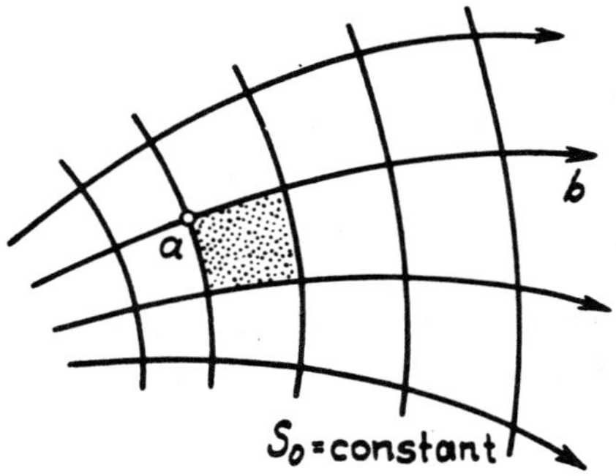

Fig. 22. Paths and surfaces of constant action.

A particle which at time $t = 0$ is at the point a will subsequently move along the path ab. Let us imagine a cloud of particles having various initial co-ordinates x_0, y_0, z_0, and let a volume element ΔV contain $\Delta N = \rho \Delta V$ particles, where ρ is the particle density. At time t all these particles have moved into some other region of space, but their number is, of course, unchanged. Hence, if we follow the motion of the volume element ΔV attached to these particles, the number of particles in it remains constant. Denoting the total derivative by D/Dt, we have

$$\frac{D}{Dt}\Delta N = \Delta V \frac{D\rho}{Dt} + \rho \frac{D}{Dt}\Delta V = 0.$$

It is known that the total derivatives of ρ and ΔV are

$$D\rho/Dt = \partial\rho/\partial t + \mathbf{v}\cdot\nabla\rho,$$
$$D(\Delta V)/Dt = \Delta V \operatorname{div}\mathbf{v},$$

where $\mathbf{v}$ is the velocity of the particles. Combining these results, we have the equation of continuity:

$$\partial\rho/\partial t + \operatorname{div}(\rho\mathbf{v}) = 0. \tag{35.6}$$

From (35.1)

$$\mathbf{v} = \mathbf{p}/\mu = -(1/\mu)\nabla S_0. \tag{35.7}$$

Hence (35.6) can be written

$$\frac{\partial\rho}{\partial t} - \frac{1}{\mu}\operatorname{div}(\rho\nabla S_0) = 0,$$

or

$$\frac{\partial\rho}{\partial t} = \frac{1}{\mu}\{\nabla\rho\cdot\nabla S_0 + \rho\nabla^2 S_0\}. \tag{35.8}$$

Thus the cloud of particles moves like a liquid, and the volume occupied by it is not 'spread out', but merely deformed.

Equation (35.8) also has another interpretation. If the number of particles ΔN in the volume ΔV is divided by the total number of particles N, then $\Delta N/N$ can be regarded as the probability of finding a particle in the volume ΔV, and the density ρ can be regarded as the probability density.

Let us now turn to quantum mechanics. We shall show that Schrödinger's time-dependent equation

$$i\hbar\,\partial\psi/\partial t = H\psi, \qquad H = -(\hbar^2/2\mu)\nabla^2 + U(x, y, z, t) \tag{35.9}$$

leads to approximately the same results as the Hamilton-Jacobi equation discussed above. To prove this, we represent the wave function ψ in the form

$$\psi = e^{-iS/\hbar}, \tag{35.10}$$

where S is some function to be determined. Since

$$\frac{\partial\psi}{\partial x} = -\frac{i}{\hbar}\frac{\partial S}{\partial x}\psi, \qquad \frac{\partial^2\psi}{\partial x^2} = -\frac{1}{\hbar^2}\left(\frac{\partial S}{\partial x}\right)^2\psi - \frac{i}{\hbar}\frac{\partial^2 S}{\partial x^2}\psi,$$

we find on substituting (35.10) in (35.9) the following equation for the function S:

$$\frac{\partial S}{\partial t} = \frac{1}{2\mu}\left\{\left(\frac{\partial S}{\partial x}\right)^2 + \left(\frac{\partial S}{\partial y}\right)^2 + \left(\frac{\partial S}{\partial z}\right)^2\right\} + U(x, y, z, t) + \frac{i\hbar}{2\mu}\nabla^2 S. \tag{35.11}$$

We now expand S in powers of $i\hbar$:

$$S = S_0 + i\hbar S_1 + (i\hbar)^2 S_2 + \dots. \tag{35.12}$$

Substituting (35.12) in (35.11) and equating coefficients of like powers of $\hbar$, we obtain the equations

$$\frac{\partial S_0}{\partial t} = \frac{1}{2\mu}\left\{\left(\frac{\partial S_0}{\partial x}\right)^2 + \left(\frac{\partial S_0}{\partial y}\right)^2 + \left(\frac{\partial S_0}{\partial z}\right)^2\right\} + U(x, y, z, t), \tag{35.13}$$

$$\frac{\partial S_1}{\partial t} = \frac{1}{2\mu}\left\{2\frac{\partial S_0}{\partial x}\frac{\partial S_1}{\partial x} + 2\frac{\partial S_0}{\partial y}\frac{\partial S_1}{\partial y} + 2\frac{\partial S_0}{\partial z}\frac{\partial S_1}{\partial z} + \nabla^2 S_0\right\}$$

$$= \frac{1}{2\mu}\{2\nabla S_0 \cdot \nabla S_1 + \nabla^2 S_0\}\,, \text{etc.} \tag{35.13'}$$

The first of these equations is the same as the Hamilton-Jacobi equation (35.2), and the second is easily seen to be the same as the equation of continuity (35.8). For the probability of finding a particle in the neighbourhood of the point x, y, z is

$$\rho = |\psi|^2 = e^{2S_1+\cdots}. \tag{35.14}$$

Hence

$$\nabla\rho = 2\nabla S_1\, e^{2S_1+\cdots}, \qquad \partial\rho/\partial t = 2(\partial S_1/\partial t)\, e^{2S_1},$$

and on multiplying Equation (35.13′) by $2e^{2S_1}$ we obtain the equation of continuity (35.8).

It remains to examine the range of applicability of the above approximate solution of Schrödinger's equation. In deriving (35.13) from (35.11) we have neglected the term $(i\hbar/2\mu)\nabla^2 S$; this is possible if

$$\frac{1}{2\mu}(\nabla S_0)^2 \gg \left|\frac{i\hbar}{2\mu}\nabla^2 S_0\right|. \tag{35.15}$$

Using (35.1), we can write this inequality as

$$\frac{p^2}{2\mu} \gg \frac{\hbar}{2\mu}|\mathrm{div}\,\mathbf{p}|\,. \tag{35.16}$$

This signifies that the kinetic energy must be large and the changes in momentum $|\mathrm{div}\ \mathbf{p}|$ must be small. For one dimension we have

$$p^2 \gg \hbar\,\mathrm{d}p/\mathrm{d}x\,. \tag{35.16'}$$

In terms of the de Broglie wavelength $\lambda = 2\pi\hbar/p$ this gives

$$\mathrm{d}\lambda/\mathrm{d}x \ll 2\pi\,, \tag{35.17}$$

i.e. the wavelength must be a slowly varying function of the co-ordinates.

36. Quantum mechanics and optics

Historically speaking, quantum mechanics arose partly from the parallels established by Hamilton between geometrical optics and mechanics. These forgotten analogies were brought by de Broglie into modern physics, and with their aid the first steps in quantum (wave) mechanics were taken.[4] It has frequently been said that Schrödinger constructed a mechanics by analogy with wave optics.

Analogies often help in solving a particular physical problem, but they remain no more than analogies. The final equation derived by Schrödinger is not the same as any

[4] See [24].

of the previously known equations of wave propagation. The latter are always of the second order in the time, whereas Schrödinger's equation is of the first order; and there are other differences.

Nevertheless, it is of interest to compare Schrödinger's equation with the equations of wave optics. Let us assume that we have a homogeneous medium in which waves are propagated with velocity v. Then the equation for the displacement f in the propagation of such waves is

$$\nabla^2 f - \frac{1}{v^2}\frac{\partial^2 f}{\partial t^2} = 0. \tag{36.1}$$

For a wave of frequency ω we can put

$$f = ue^{-i\omega t}; \tag{36.2}$$

then (36.1) gives

$$\nabla^2 u + k^2 u = 0, \qquad k^2 = \omega^2/v^2, \tag{36.3}$$

where $k = 2\pi/\lambda$ is the wave number and λ the wavelength. Equation (36.3) is rigorously valid for a homogeneous medium.[5] However, it also describes phenomena of diffraction and interference in the case where the velocity v is a function of co-ordinates. It may therefore be regarded as the wave equation for a heterogeneous medium also. In that case k^2 is a function of co-ordinates, but we shall continue to call k the wave number, and $\lambda = 2\pi/k$ the wavelength.

We also use the refractive index $n(x, y, z)$:

$$n(x, y, z) = k/k_0 = \lambda_0/\lambda, \tag{36.4}$$

where λ_0 is the wavelength *in vacuo*. Then Equation (36.3) can be written

$$\nabla^2 u + k_0^2 n^2 u = 0. \tag{36.5}$$

If the inhomogeneities of the medium are such that the refractive index n varies only slightly over a distance of one wavelength, the wave equation (36.5) gives the fundamental equation of geometrical optics (otherwise the diffraction of waves by these inhomogeneities is involved).

Let us put

$$u = ae^{ik_0\Theta}, \tag{36.6}$$

where a is the amplitude of the wave and $k_0\Theta$ its phase. If the wavelength is small, k_0 is large. We expand a and Θ in inverse powers of k_0:

$$a = a_0 + \frac{1}{k_0}a_1 + \frac{1}{k_0^2}a_2 + \dots, \tag{36.7}$$

$$\Theta = \Theta_0 + \frac{1}{k_0}\Theta_1 + \frac{1}{k_0^2}\Theta_2 + \dots. \tag{36.8}$$

[5] The equation of wave propagation in a heterogeneous medium (for example, propagation of electromagnetic waves in a medium of varying dielectric constant) is in fact more complex than (36.3).

Substituting (36.7) and (36.8) in (36.6), (36.6) in (36.5), and collecting equal powers of k_0, we obtain

$$-k_0^2 a_0 (\nabla \Theta_0)^2 + k_0^2 n^2 a_0 + O(k_0) = 0, \tag{36.9}$$

where $O(k_0)$ denotes terms of order k_0 and below. Neglecting lower powers of k_0, we have from this

$$(\nabla \Theta_0)^2 = n^2. \tag{36.10}$$

This is the fundamental equation of geometrical optics, which determines the surfaces of constant phase

$$\Theta_0(x, y, z) = \text{constant} \tag{36.11}$$

in terms of the refractive index $n(x, y, z)$. The rays are orthogonal to these surfaces.

Let us compare with Equation (36.9) the Hamilton-Jacobi equation (35.2) for the action function S_0. Substituting in (35.2) $S_0 = Et - s_0$, we can write that equation in the form

$$(\nabla s_0)^2 = 2\mu[E - U(x, y, z)]. \tag{36.12}$$

Comparison of this equation with (36.10) shows that the problem of the propagation of short waves (k_0 large) in a heterogeneous medium with refractive index $n(x, y, z)$ may be compared with the problem of the motion of a point mass in a field of force with potential energy $U(x, y, z)$, the refractive index being represented by the quantity $\sqrt{[2\mu(E - U)]}$, and the phase by s_0. The paths of the particles are orthogonal to the surfaces $s_0(x, y, z) = \text{constant}$. Hence the paths coincide with rays of light in a medium whose refractive index n is proportional to $\sqrt{[2\mu(E - U)]}$. Thus the classical mechanics of a point mass is analogous to geometrical optics.

If Equation (36.3) is regarded as an equation of wave optics, we can say that wave (quantum) mechanics is analogous to wave optics: Schrödinger's equation

$$i\hbar \frac{\partial \psi}{\partial t} = -\frac{\hbar^2}{2\mu} \nabla^2 \psi + U(x, y, z)\psi$$

is reduced by the substitution

$$\psi = u e^{-iEt/\hbar} \tag{36.13}$$

to the form

$$\nabla^2 u + (2\mu/\hbar^2)[E - U]u = 0. \tag{36.14}$$

Now let there be no force in some region of space: $U = C = \text{constant}$. Let the wave number in this region be k_0. Then

$$k_0^2 = (2\mu/\hbar^2)(E - C) \tag{36.15}$$

(usually C is taken as zero). In terms of the refractive index of waves for this region of space:

$$n^2 = (E - U)/(E - C) = k^2/k_0^2, \tag{36.16}$$

we can write Equation (36.14) in the form

$$\nabla^2 u + k_0^2 n^2 u = 0, \tag{36.17}$$

which is the same as (36.5).

Some simple problems on the calculation of reflection and refraction of waves are given in Section 96.

In deriving (36.10) from (36.9) we have neglected terms $O(k_0)$. If these are worked out it is easily seen that a term $k_0 \nabla^2 \Theta_0$ has been neglected in comparison with $k_0^2(\nabla\Theta_0)^2$. Taking for simplicity the one-dimensional case, we can write the condition for our approximation to be valid as

$$k_0^2(\partial\Theta_0/\partial x)^2 \gg k_0|\partial^2\Theta_0/\partial x^2| \,. \tag{36.18}$$

Since $k = 2\pi/\lambda = k_0\, \partial\Theta_0/\partial x$, we have

$$|\partial\lambda/\partial x| \ll 2\pi\,, \tag{36.19}$$

which is the same as the condition (35.17) derived previously for the transition from Schrödinger's equation to the Hamilton-Jacobi equation.

It follows from (36.16) that the refractive index n, and therefore also the wavelength $\lambda = 2\pi/k$, vary appreciably only in the region of space where the potential energy U varies appreciably, i.e. within the range of interaction a of the forces. If this range $a \gg \lambda$, then both U and n will vary only slightly over a distance λ (except in some exceptional cases where the potential energy varies extremely rapidly).[6]

Hence for rough calculations the condition (36.19) can be replaced by the simpler condition $\lambda \ll a$. The application of this condition to the examples given in Section 34 is left to the reader.

37. The quasiclassical approximation (the Wentzel–Kramers–Brillouin method)

The relation between quantum mechanics, classical mechanics and optics described in Sections 35 and 36 makes it possible to develop an approximate method of solving Schrödinger's equation, suitable for cases where the condition (36.19) holds, i.e. where the wavelength varies only slowly (in the language of optics, where the refractive index $n(x)$ of the medium varies only slowly in space).

Then, putting

$$\psi = e^{-i(Et-s)/\hbar} \tag{37.1}$$

in accordance with (36.12) and (36.13) and $s = s_0 + i\hbar s_1 + \dots$ (35.12), we obtain

$$\psi = e^{-s_1} e^{-i(Et-s_0)/\hbar}. \tag{37.1'}$$

In what follows we shall consider the case where the potential U depends only on one co-ordinate: $U = U(x)$, when s_0 and s_1 also will depend only on x. Then $\nabla s_0 = (\mathrm{d}s_0/\mathrm{d}x, 0, 0)$ and from (36.12)

$$s_0(x) = \int p(x)\,\mathrm{d}x\,, \tag{37.2}$$

where $p(x)$ is the particle momentum:

$$p(x) = \pm\sqrt{\{2\mu[E - U(x)]\}} = \pm\,|p(x)|. \tag{37.2'}$$

[6] It may seem that, for any microparticles of sufficiently high energy (and therefore short wavelength λ), classical mechanics will always be valid, but this is not so: when the energy of the particle increases, inelastic-collision phenomena appear (ionisation and excitation of atoms, bremsstrahlung, excitation and fission of atomic nuclei, etc.) which cannot be treated without the use of quantum mechanics.

Using (35.13′), we calculate s_1; $\partial s_1/\partial t$ must be equated to zero. The result is

$$2\frac{ds_0}{dx}\frac{ds_1}{dx} - \frac{d^2 s_0}{dx^2} = 0. \tag{37.3}$$

Hence $s_1 = \frac{1}{2}\log p(x) - \log c$, so that

$$\psi = \frac{c}{\sqrt{p(x)}} e^{(i/\hbar)\int p(x)dx}. \tag{37.4}$$

In this approximation the probability of finding the particle in the range x to $x + dx$ is

$$w(x)\,dx = |\psi(x)|^2\,dx = |c|^2\,dx/p(x), \tag{37.5}$$

i.e. it is inversely proportional to the velocity $v(x) = p(x)/\mu$, and therefore directly proportional to the time taken to traverse the interval dx, as it should be according to the classical theory. Taking into account the two possible signs of $p(x)$ in (37.2′), we must write the complete solution as a superposition of two solutions:

$$\psi(x) = \frac{c_1}{\sqrt{p(x)}} \exp\left(\frac{i}{\hbar}\int_a^x |p(x)|\,dx\right) + \frac{c_2}{\sqrt{p(x)}} \exp\left(-\frac{i}{\hbar}\int_a^x |p(x)|\,dx\right). \tag{37.6}$$

The constants c_1, c_2 and a must be determined from the boundary conditions for the wave function $\psi(x)$.[7] It is clear that only two of the three constants are independent.

The case of the *turning points* requires special consideration. These are the points where the total energy E is equal to the potential energy $U(x)$. At such a point the kinetic energy and momentum of the particle become zero: $T = 0, p = 0$.

According to classical mechanics a particle at such a point changes the sign of its velocity and begins to move in the opposite direction; hence the name of *turning* point.

On the wave picture, motion is possible even in a region where $E < U(x)$; this will be discussed in detail in Sections 96, 97. The quantity $p(x)$ (37.2′) will be purely imaginary, and of course no longer signifies the momentum:

$$p(x) = \pm i\sqrt{\{2\mu[U(x) - E]\}} = \pm i|p(x)|. \tag{37.2″}$$

One of the solutions in (37.6) then increases without limit as x increases. Only bounded wave functions are physically significant, and therefore the constant c_2 must be put equal to zero in the region where $E < U(x)$, so that

$$\psi(x) = \frac{c_1}{\sqrt{|p(x)|}} \exp\left(-\frac{1}{\hbar}\int_a^x p(x)\,dx\right). \tag{37.6′}$$

For the subsequent discussion of the turning points it is convenient to take the constant a equal to the value of x at the turning point $E = U(a)$, $p(a) = 0$.

[7] See Appendix VIII.

It is seen from (37.6) and (37.6′) that these approximate solutions become infinite at the turning points. The joining of solutions on the two sides of the turning point therefore requires a consideration of a more accurate solution of Schrödinger's equation in the neighbourhood of the turning point. This is obtained by expressing the potential $U(x)$ in the neighbourhood of $x = a$ in the form $U(x) = U(a) + (\mathrm{d}U/\mathrm{d}x)_a(x - a) + \ldots$ and solving Schrödinger's equation for this linear potential. Here we shall give only the results of the calculation.

We shall suppose that $E < U(x)$ for $x > a$ and $E > U(x)$ for $x < a$; then it is found that the correct choice of constants is such that[8]

$$\psi(x) = \frac{c}{\sqrt{p(x)}} \sin\left\{\frac{1}{\hbar}\int_x^a p(x)\,\mathrm{d}x + \tfrac{1}{4}\pi\right\}, \qquad x < a, \tag{37.7}$$

$$\psi(x) = \frac{c}{2\sqrt{|p(x)|}} \exp\left(-\frac{1}{\hbar}\int_a^x |p(x)|\,\mathrm{d}x\right), \qquad x > a. \tag{37.7′}$$

For the case where $E > U(x)$ for $x > a$,

$$\psi = \frac{c}{\sqrt{p(x)}} \sin\left\{\frac{1}{\hbar}\int_a^x p(x)\,\mathrm{d}x + \tfrac{1}{4}\pi\right\}. \tag{37.7″}$$

Let us now suppose that the range of motion of the particle is bounded and lies between two turning points, $b < x < a$. Then in (37.7″) the limit a must be replaced by b. Evidently the two solutions (37.7) and

$$\psi(x) = \frac{c}{\sqrt{p(x)}} \sin\left\{\frac{1}{\hbar}\int_b^x p(x)\,\mathrm{d}x + \tfrac{1}{4}\pi\right\} \tag{37.8}$$

must coincide in the range $b < x < a$. This can happen only if

$$\frac{1}{\hbar}\int_b^a p(x)\,\mathrm{d}x + \tfrac{1}{2}\pi = (n + 1)\,\pi, \tag{37.9}$$

where n is an integer.

Extending the integral to the whole path of the particle from a to b and back, we have

$$\oint p(x)\,\mathrm{d}x = (n + \tfrac{1}{2})\cdot 2\pi\hbar. \tag{37.10}$$

This is the quantisation condition according to the old semiclassical theory of Bohr. The appearance of the term $\frac{1}{2}$ together with the integer n is unimportant, since the classical approximation is strictly valid only when $n \gg 1$ (the condition of short wavelength).

[8] See, for instance, [58].

CHAPTER VII

BASIC THEORY OF REPRESENTATIONS

38. Different representations of the state of quantum systems

We have seen that a characteristic feature of quantum mechanics is that the simultaneous use of various classical corpuscular quantities (p_x and x, T and U, M_x and M_y, etc.) is entirely meaningless, since ensembles in which these pairs of quantities exist simultaneously do not occur in Nature.

For every quantum system, therefore, all measuring apparatus can be divided into several groups. Apparatus in one such group classifies particles (or, more generally, systems) in an ensemble according to properties which exclude a classification according to properties corresponding to some other group of measuring apparatus. For example, for particles whose centroid has co-ordinates x, y, z, we can easily distinguish two groups of apparatus. The first group may include apparatus which analyses the ensemble of such particles according to the co-ordinates x, y, z or any functions $F(x, y, z)$ of them (for example, the potential energy $U(x, y, z)$), and the second group apparatus which analyses the ensemble according to the momenta p_x, p_y, p_z or any functions $\Phi(p_x, p_y, p_z)$ of them (for example, the kinetic energy $T(p_x, p_y, p_z)$). Other groups of apparatus are also possible.

So far we have represented the state of the particles by the wave function $\psi(x)$, with the co-ordinate x of the particle as variable (for simplicity we shall use below only one co-ordinate x).

The classification of particles according to co-ordinates x is effected by apparatus which excludes classification with respect to p_x (below we shall write simply p instead of p_x). Let us suppose, however, that we are interested in the classification of particles not with respect to their co-ordinates x but with respect to their momenta. Then we must use an apparatus which analyses the ensemble with respect to p and not x. But the wave function ψ which describes the ensemble is taken as a function of x. Can we not describe the state of the ensemble so that the wave function is a function of the momentum p?

In the former case we shall say that the state is related to an apparatus which analyses the ensemble with respect to the co-ordinates x of the particles (the first 'frame of reference'); in the latter case, to an apparatus which analyses the ensemble with respect to the momenta p_x (the second 'frame of reference'). Briefly we say that the state is given in the 'x' representation or in the 'p' representation.[1]

[1] We should say 'co-ordinate representation' and 'momentum representation'.

It is very easy to find the p representation. Let a wave function $\psi(x, t)$ be given in the x representation, and let us expand this function in terms of the eigenfunctions of the momentum operator $\psi_p(x)$ (i.e. as a Fourier integral). Then

$$\psi(x,t) = \int c(p,t)\,\psi_p(x)\,dp\,, \tag{38.1}$$

$$c(p,t) = \int \psi(x,t)\,\psi_p^*(x)\,dx\,. \tag{38.2}$$

If we know the amplitudes $c(p, t)$, we also know $\psi(x, t)$, which is entirely determined if $c(p, t)$ is given. Hence $c(p, t)$ may be regarded as a wave function given as a function of the momentum p and physically representing the same state as the functions $\psi(x, t)$. Formula (38.1) is to be regarded as a transformation of the wave function from the p representation to the x representation, and formula (38.2) as a transformation from the x representation to the p representation.

Let us now consider the representation of a state when the energy E of the particle is taken as the independent variable. For definiteness, let E have a discrete spectrum of values $E_1, E_2, \ldots, E_n, \ldots$. We denote the corresponding eigenfunctions by $\psi_1(x)$, $\psi_2(x), \ldots, \psi_n(x), \ldots$. The wave function $\psi(x, t)$ can be represented in the form of a series:

$$\psi(x,t) = \sum_n c_n(t)\,\psi_n(x)\,, \tag{38.3}$$

$$c_n(t) = \int \psi(x,t)\,\psi_n^*(x)\,dx\,. \tag{38.4}$$

Again $\psi(x, t)$ is entirely determined if all the amplitudes $c_n(t)$ are given, and conversely $c_n(t)$ is determined by $\psi(x, t)$. Hence the set of all the $c_n(t)$ may be regarded as a wave function describing the same state as $\psi(x, t)$, but in a representation in which the independent variable is the energy[2] E.

From this point of view (38.3) is a transformation of the wave function from the E representation to the x representation. Formula (38.4) gives the inverse transformation. It follows from Formulae (38.1)–(38.4) that the probability of finding any value of the independent variable is equal to the squared modulus of the wave function in the corresponding representation. For, if we have some state $\psi(x, t)$, the probability $w(x, t)\,dx$ of finding a value of the co-ordinate between x and $x + dx$ is

$$w(x,t)\,dx = |\psi(x,t)|^2\,dx\,. \tag{38.5}$$

The probability $w(p, t)\,dp$ of finding the momentum p between p and $p + dp$ is

$$w(p,t)\,dp = |c(p,t)|^2\,dp\,. \tag{38.6}$$

The probability $w(E_n, t)$ of finding the energy E_n is

$$w(E_n,t) = |c_n(t)|^2 = |c(E_n,t)|^2\,. \tag{38.7}$$

39. Different representations of operators of mechanical quantities. Matrices

In order to complete the expression of states ψ in terms of different independent variables, it is necessary to find a method of representing operators in the same variables.

[2] By analogy with $c(p,t)$, we can write $c(E,t)(E = E_1, E_2, \ldots, E_n, \ldots)$ instead of $c_n(t)(n = 1, 2, \ldots)$.

So far we have regarded operators $\boldsymbol{L}$ as 'functions' of x, assuming that $\boldsymbol{L}$ has the form $\boldsymbol{L}(-i\hbar\,\partial/\partial x, x)$. In this case the operator $\boldsymbol{L}$ acts on functions of the form $\psi(x)$ and generates a new function $\phi(x)$ according to the formula

$$\phi(x) = \boldsymbol{L}(-i\hbar\,\partial/\partial x, x)\,\psi(x). \tag{39.1}$$

We can therefore say that the operator $\boldsymbol{L}$ has been taken in the x representation.

Let us now find the operator $\boldsymbol{L}$ in the energy representation (E representation), assuming that the energy has a discrete spectrum of values E_n. Let the corresponding eigenfunctions be $\psi_n(x)$. Then the functions ϕ and ψ can be written in the form

$$\psi(x) = \sum_n c_n \psi_n(x), \tag{39.2}$$

$$\phi(x) = \sum_n b_n \psi_n(x). \tag{39.3}$$

The set c_n is ψ in the E representation, and the set b_n is ϕ in the E representation. The operator $\boldsymbol{L}$ generates from ψ a new function ϕ, and from the c_n it generates new amplitudes b_n. If we find an operator which expresses the b_n directly in terms of the c_n, this gives the operator $\boldsymbol{L}$ in the E representation. To do so, we substitute ψ and ϕ from (39.2) and (39.3) in (39.1). The result is

$$\sum_n b_n \psi_n(x) = \sum_n c_n \cdot \boldsymbol{L}\,\psi_n(x). \tag{39.4}$$

Multiplying (39.4) by $\psi_m^*(x)$ and integrating over all x-space, we obtain, using the orthogonality of the functions $\psi_n(x)$,

$$b_m = \sum_n L_{mn}\, c_n, \tag{39.5}$$

where

$$L_{mn} = \int \psi_m^*(x) \cdot \boldsymbol{L}\,\psi_n(x) \cdot \mathrm{d}x. \tag{39.6}$$

If we know all the L_{mn}, Formula (39.5) gives all the amplitudes b_n (the function ϕ in the E representation) from given c_n (i.e. the function ψ in the E representation). Hence the set of all the quantities L_{mn} should be regarded as the operator $\boldsymbol{L}$ in the E representation.

These quantities may be arranged in a square array:

$$\boldsymbol{L} = \left|\begin{matrix} L_{11} & L_{12} & L_{13} & \dots & L_{1n} & \dots \\ L_{21} & L_{22} & L_{23} & \dots & L_{2n} & \dots \\ \dots & \dots & \dots & \dots & \dots & \dots \\ L_{m1} & L_{m2} & L_{m3} & \dots & L_{mn} & \dots \\ \dots & \dots & \dots & \dots & \dots & \dots \end{matrix}\right|, \tag{39.7}$$

which has an infinite number of rows and columns. This array is called a *matrix*, and the quantities L_{mn} themselves are called *matrix elements*. Each matrix element has two suffixes.[3] The first is the number of the row and the second is the number

[3] Other notations for the matrix elements, introduced by Dirac and often used, are

$(m|L|n)$ instead of L_{mn} ,

of the column. The rows and columns in such a matrix may be interchanged, but the same arrangement must, of course, be used throughout any one calculation. We shall arbitrarily number the rows and columns in order of increasing eigenvalues: $E_1 \leqslant E_2 \leqslant E_3 \leqslant \dots \leqslant E_n \leqslant \dots$.

A representation of operators L can also be found in cases where the independent variable has a continuous spectrum of values. As an example, let us consider the p representation. Analogously to (39.2) and (39.3) we have

$$\psi(x) = \int c(p)\,\psi_p(x)\,\mathrm{d}p\,, \tag{39.2$'$}$$

$$\phi(x) = \int b(p)\,\psi_p(x)\,\mathrm{d}p\,, \tag{39.3$'$}$$

where $c(p)$ and $b(p)$ are the functions ψ and ϕ in the p representation. Let us find the relation between $c(p)$ and $b(p)$. Substitution of (39.2′) and (39.3′) in (39.1) gives

$$\int b(p)\,\psi_p(x)\,\mathrm{d}p = \int c(p)\cdot L\psi_p(x)\cdot \mathrm{d}p\,. \tag{39.4$'$}$$

Multiplying this equation by $\psi_{p'}^*(x)$ and integrating with respect to x, and using the orthogonality of the functions $\psi_p(x)$, we have

$$\int b(p)\delta(p'-p)\,\mathrm{d}p = \int c(p)\,\mathrm{d}p \int \psi_{p'}^*(x)\cdot L\psi_p\cdot \mathrm{d}x\,,$$

or

$$b(p') = \int L_{p'p}c(p)\,\mathrm{d}p\,, \tag{39.5$'$}$$

where

$$L_{p'p} = L(p',p) = \int \psi_{p'}^*(x)\cdot L\psi_p(x)\cdot \mathrm{d}x\,. \tag{39.6$'$}$$

The quantity $L_{p'p}$ is the operator L in the p representation. It depends on two variables p' and p, which take the same values. We again call $L_{p'p}$ the *matrix element* of the operator L in the p representation, and the set of all values of $L_{p'p}$ a *matrix*. It is clear that in this case the $L_{p'p}$ can not be put in the form of an array. Nevertheless, we still call p' the *row number* and p the *column number*.

We see that in an arbitrary representation operators are represented by matrices.[4] In the x representation we had operators in the form of differential operators. It may be shown, however (see Section 40), that in this representation also the operators may be written in matrix form.

40. Matrices and operations on them

Among the elements of matrices we distinguish what are called *diagonal elements*. These are the elements whose row and column numbers are equal, i.e. the elements of the form L_{nn}. For a continuous spectrum the diagonal elements are those of the form L_{pp}. If the matrix has only diagonal elements, it is said to be a *diagonal matrix*.

---

or even more explicitly

$$(E_m|L|E_n) \text{ instead of } L_{mn}\,.$$

The latter notation indicates not only the operator (L) to which the matrix element belongs but also the representation (E) in which it is taken, and finally the numbers m and n of the eigenvalues to which the matrix element belongs. This notation is especially convenient when there is degeneracy (Section 21) and the wave functions have several suffixes.

[4] For E or p may be interpreted as any quantity L having respectively a discrete or continuous spectrum of values. More generally, E or p may be taken as a set of independent, simultaneously measurable quantities $L, M, N, \dots$.

A more detailed account of the theory of representations and transformations from one variable to another will be found in [29, 42].

For a discrete spectrum such a matrix has the form

$$L = \begin{vmatrix} L_{11} & 0 & 0 & \dots & 0 & \dots \\ 0 & L_{22} & 0 & \dots & 0 & \dots \\ \dots & \dots & \dots & \dots & \dots & \dots \\ 0 & 0 & 0 & \dots & L_{nn} & \dots \\ \dots & \dots & \dots & \dots & \dots & \dots \end{vmatrix}. \tag{40.1}$$

An important case of a diagonal matrix is the *unit matrix* δ, whose elements δ_{mn} are given by

$$\delta_{mn} = \int \psi_m^* \psi_n \, \mathrm{d}x = \begin{cases} 0 \text{ for } m \neq n, \\ 1 \text{ for } m = n. \end{cases} \tag{40.2}$$

This matrix has the form

$$\delta = \begin{vmatrix} 1 & 0 & 0 & 0 & \dots \\ 0 & 1 & 0 & 0 & \dots \\ 0 & 0 & 1 & 0 & \dots \\ 0 & \dots & \dots & \dots & \dots \\ \dots & \dots & \dots & \dots & \dots \end{vmatrix}. \tag{40.2'}$$

From the definition of the elements (40.2) of the unit matrix it follows that the unit matrix remains the unit matrix in any representation, since Equation (40.2) holds good for any set of orthogonal functions $\psi_n(x)$. The elements of a diagonal matrix L can always be written in the form

$$L_{mn} = L_n \delta_{mn}. \tag{40.3}$$

It is often necessary to consider, as well as a matrix L with elements L_{mn}, various matrices derived from it. Among these are, first of all, the *complex conjugate matrix* L^*. The elements of this matrix are the complex conjugates of the corresponding elements of the original matrix:

$$(L^*)_{mn} = (L_{mn})^*. \tag{40.4}$$

Next, from a given matrix we can form the *transposed matrix* $\tilde{L}$. This is obtained from the original matrix by interchanging rows and columns. The elements of this matrix are given by the formula

$$(\tilde{L})_{mn} = L_{nm}. \tag{40.5}$$

If we take the complex conjugate matrix of the transposed matrix, i.e. $(\tilde{L})^*$, we obtain the *adjoint matrix* of the original one, denoted by L^+. Its elements are given by

$$(L^+)_{mn} = (\tilde{L}^*)_{mn} = L^*_{nm}. \tag{40.6}$$

If the adjoint matrix is the same as the original one:

$$L^+ = L \text{ (i.e. } L_{mn} = L^*_{nm}), \tag{40.7}$$

the matrix is said to be *Hermitian* or *self-adjoint*. This definition is in entire agreement with our previous definition of an Hermitian or self-adjoint operator (18.7): if the operator $\boldsymbol{L}$ is Hermitian, its matrix elements are

$$L_{mn} = \int \psi_m^* \cdot \boldsymbol{L}\, \psi_n \cdot \mathrm{d}x = \int \psi_n \cdot \boldsymbol{L}^* \psi_m^* \,\mathrm{d}x = L_{nm}^* .$$

Let us now consider algebraic operations on matrices, beginning with the *addition* of matrices. Let some operator $\boldsymbol{C}$ be given which is the sum of operators $\boldsymbol{A}$ and $\boldsymbol{B}$. Then the sum of the matrices $\boldsymbol{A}$ and $\boldsymbol{B}$ will be taken to be the matrix of the operator $\boldsymbol{C}$. It is easy to find the elements of this matrix; we have

$$C_{mn} = \int \psi_m^* \cdot \boldsymbol{C} \psi_n \cdot \mathrm{d}x = \int \psi_m^* \cdot \boldsymbol{A}\, \psi_n \cdot \mathrm{d}x + \int \psi_m^* \cdot \boldsymbol{B} \psi_n \cdot \mathrm{d}x , \tag{40.8}$$

so that

$$C_{mn} = A_{mn} + B_{mn} , \tag{40.9}$$

i.e. a matrix element of a sum of operators is equal to the sum of the corresponding elements of each of the operators forming the sum.

The rule of matrix *multiplication* is of great importance in applications. To establish this rule, let us calculate the matrix element of an operator $\boldsymbol{C}$ which is the product of two operators $\boldsymbol{A}$ and $\boldsymbol{B}$. Using the definition of the matrix element, we have

$$C_{mn} = \int \psi_m^* \cdot \boldsymbol{C} \psi_n \cdot \mathrm{d}x = \int \psi_m^* \cdot \boldsymbol{A} (\boldsymbol{B} \psi_n) \cdot \mathrm{d}x . \tag{40.10}$$

The quantity $\boldsymbol{B}\psi_n$ is itself some function and can be expanded in a series of the orthogonal functions $\psi_k(x)$:

$$\boldsymbol{B}\psi_n = \sum_k b_k \psi_k(x) ,$$

where

$$b_k = \int \psi_k^* \cdot \boldsymbol{B} \psi_n \cdot \mathrm{d}x = B_{kn} .$$

Substituting this expansion in (40.10), we obtain

$$C_{mn} = \int \psi_m^* \cdot \boldsymbol{A} \sum_k B_{kn} \psi_k \cdot \mathrm{d}x = \sum_k B_{kn} \int \psi_m^* \cdot \boldsymbol{A} \psi_k \cdot \mathrm{d}x = \sum_k B_{kn} A_{mk} .$$

Hence

$$C_{mn} = \sum_k A_{mk} B_{kn} . \tag{40.11}$$

This gives the rule of matrix multiplication: in order to obtain the element C_{mn} of the matrix which is the product of operators $\boldsymbol{A}$ and $\boldsymbol{B}$, the elements of the mth row of the matrix $\boldsymbol{A}$ must be multiplied by those of the nth column of the matrix $\boldsymbol{B}$, and the products added. The rules of matrix addition (40.9) and matrix multiplication (40.11) enable us to find, from given matrices of operators $\boldsymbol{A}$, $\boldsymbol{B}$, ..., the matrices which represent various functions of $\boldsymbol{A}$, $\boldsymbol{B}$, ...

The multiplication rule also makes possible the derivation of a somewhat different form of (39.5), which gives the result of the action of the operator $\boldsymbol{L}$ on the wave function. This formula may be regarded as a matrix product. For this purpose we put the wave function in the E representation in the form of a matrix of one column:

$$\psi = \begin{vmatrix} c_1 & 0 & 0 & \dots \\ c_2 & 0 & 0 & \dots \\ \dots & \dots & \dots & \dots \\ c_n & \dots & \dots & \dots \\ \dots & \dots & \dots & \dots \end{vmatrix}, \tag{40.12}$$

and similarly for the function ϕ:

$$\phi = \begin{vmatrix} b_1 & 0 & 0 & \dots \\ b_2 & 0 & 0 & \dots \\ \dots & \dots & \dots & \dots \\ b_m & 0 & 0 & \dots \\ \dots & \dots & \dots & \dots \end{vmatrix}. \tag{40.13}$$

It is now easy to see that (39.5) can be written as a matrix product:

$$\phi = \boldsymbol{L}\,\psi\,, \tag{40.14}$$

where ϕ is the matrix (40.13), ψ the matrix (40.12), and $\boldsymbol{L}$ the matrix (39.7). For example, b_m is the element in the mth row and first column of the matrix (40.13). According to (40.11), it must be obtained by multiplying the elements in the mth row of the matrix (39.7) by those in the first column of the matrix ψ (40.12). This gives precisely the Equations (39.5). The adjoint wave function $\overset{*}{c}_1$, $\overset{*}{c}_2$, ..., $\overset{*}{c}_n$ can be written as the matrix adjoint to (40.12), having one row:

$$\psi^+ = \begin{vmatrix} \overset{*}{c}_1 & \overset{*}{c}_2 & \dots & \overset{*}{c}_n & \dots \\ 0 & 0 & \dots & 0 & \dots \\ \dots & \dots & \dots & \dots & \dots \\ \dots & \dots & \dots & \dots & \dots \end{vmatrix}. \tag{40.12'}$$

The writing of wave functions in the matrix form (40.12) occurs in the theory of the magnetic moment of the electron.

We may also note the following result derived from the rule of matrix multiplication. The matrix $\boldsymbol{C}^+$ adjoint to the product $\boldsymbol{C}$ of two matrices $\boldsymbol{A}$ and $\boldsymbol{B}$ must be written as

$$\boldsymbol{C}^+ = (\boldsymbol{AB})^+ = \boldsymbol{B}^+\,\boldsymbol{A}^+\,; \tag{40.15}$$

for, the elements C^+_{mn} are, by definition of the adjoint matrix, equal to $\overset{*}{C}_{nm}$. From (40.11) we have

$$C^+_{mn} = \overset{*}{C}_{nm} = \sum_k \overset{*}{A}_{nk}\,\overset{*}{B}_{km} = \sum_k \tilde{\overset{*}{B}}_{mk}\,\tilde{\overset{*}{A}}_{kn} = \sum_k (B^+)_{mk}(A^+)_{kn}\,.$$

In a precisely similar manner, replacing sums by integrals, and the symbol δ_{mn} by $\delta(p'-p)$, we obtain corresponding formulae for continuous matrices. Instead of (40.2) we have the unit matrix

$$\delta = \delta(p'-p)\,. \tag{40.2''}$$

The elements of a diagonal matrix are now written as

$$L_{p'p} = L(p')\delta(p'-p)\,. \tag{40.3'}$$

The self-adjointness property is expressed by the formula

$$L_{p'p} = L_{pp'}^* . \tag{40.7'}$$

The matrix element of the sum of two matrices A and B is

$$C_{p'p} = A_{p'p} + B_{p'p} , \tag{40.9'}$$

and that of the product of two matrices A and B is

$$C_{p'p} = \int A_{p'p''} B_{p''p} \, dp'' . \tag{40.11'}$$

The following are some examples of continuous matrices. Let us first consider the operator of the co-ordinate x in the p representation. According to the definition of the matrix element we have

$$x_{p'p} = \int \psi_{p'}^* x \psi_p \, dx = \frac{1}{2\pi\hbar} \int e^{-ip'x/\hbar} x e^{ipx/\hbar} \, dx \tag{40.16}$$

$$= \frac{\hbar}{i} \frac{\partial}{\partial p} \frac{1}{2\pi\hbar} \int e^{-i(p'-p)x/\hbar} \, dx = - i\hbar \frac{\partial}{\partial p} \delta(p - p') .$$

From Formula (39.5′), which defines the effect of an operator L, given in matrix form, on the wave function, we find

$$b(p') = \int x_{p'p} c(p) \, dp = - i\hbar \int \frac{\partial}{\partial p} \delta(p' - p) c(p) \, dp .$$

Integration by parts gives

$$b(p') = \Big[- i\hbar\delta(p' - p) c(p) \Big]_{-\infty}^{\infty} + i\hbar \int \delta(p' - p) \frac{\partial c(p)}{\partial p} \, dp ,$$

or

$$b(p) = i\hbar \partial c(p) / \partial p , \tag{40.17}$$

i.e. the operator x in the p representation can be given either as the matrix (40.16) or as the differential operator $i\hbar \, \partial/\partial p$ (40.17). The latter result has already been derived in Section 13.

The operator x in its own representation may be represented by the diagonal matrix

$$x_{x'x} = x' \delta(x - x') , \tag{40.18}$$

and the operator of any function $V(x)$ by the matrix

$$V_{x'x} = V(x') \delta(x - x') ; \tag{40.18'}$$

for, if we replace in (39.5′) b by ϕ, c by ψ and p by x, we have

$$\phi(x') = \int V_{x'x} \psi(x) \, dx = \int V(x') \delta(x - x') \psi(x) \, dx ,$$

or

$$\phi(x) = V(x) \psi(x) , \tag{40.19}$$

i.e. the effect of the function $V(x)$ in the x representation is to multiply $\psi(x)$ by $V(x)$, another known result.

Similarly, the operator P may be put in the matrix form

$$P_{x'x} = i\hbar \frac{\partial}{\partial x} \delta(x - x') ; \tag{40.20}$$

we have

$$\phi(x') = \int P_{x'x} \psi(x) \, dx = i\hbar \int \frac{\partial}{\partial x} \delta(x - x') \psi(x) \, dx ,$$

and integration by parts gives

$$\phi(x) = - i\hbar \frac{\partial}{\partial x} \psi(x) , \tag{40.21}$$

i.e. the matrix representation (40.20) of the operator P is equivalent to the differential representation $P = - i\hbar \, \partial/\partial x$.

On the basis of Formulae (40.18) and (40.20), any operator given in the form $L(-i\hbar\,\partial/\partial x, x) = L(P, x)$ can be written in matrix form so that

$$\phi(x') = \int L_{x'x}\psi(x)\,\mathrm{d}x = L(-i\hbar\,\partial/\partial x', x')\psi(x'). \tag{40.22}$$

To determine the matrix elements $L_{x'x}$ it is sufficient to consider the operators P and x in $L(P, x)$ as matrices (40.18) and (40.20) and to effect the multiplication and addition of these operators according to the rules (40.9′) and (40.11′) for continuous matrices. For example, it is easy to see that in the matrix co-ordinate representation the Hamiltonian

$$H = \frac{P_x^2}{2\mu} + V(x) = -\frac{\hbar^2}{2\mu}\frac{\partial^2}{\partial x^2} + V(x) \tag{40.23}$$

will have matrix elements

$$H_{x'x} = -\frac{\hbar^2}{2\mu}\frac{\partial^2\delta(x-x')}{\partial x^2} + V(x')\delta(x-x'). \tag{40.24}$$

41. Determination of the mean value and spectrum of a quantity represented by an operator in matrix form

Formula (19.1) for the mean value of a quantity represented by an operator $L(-i\hbar\,\partial/\partial x, x)$ in the state $\psi(x, t)$ can easily be put in matrix form. Let $\psi_n(x)$ be the eigenfunction belonging to the nth eigenvalue of the quantity taken as the independent variable (for example, the energy). We express $\psi(x, t)$ and $\psi^*(x, t)$ as series:

$$\psi(x,t) = \sum_n c_n\psi_n(x,t), \tag{41.1}$$

$$\psi^*(x,t) = \sum_m c_m^*\psi_m^*(x,t). \tag{41.1'}$$

Substituting these in the formula

$$\bar{L} = \int \psi^*(x,t)\cdot L\psi(x,t)\cdot\mathrm{d}x,$$

we have

$$\bar{L} = \sum_n\sum_m c_m^* c_n \int \psi_m^*\cdot L\psi_n\cdot\mathrm{d}x,$$

or

$$\bar{L} = \sum_n\sum_m c_m^* L_{mn} c_n. \tag{41.2}$$

This is the expression for the mean value $\bar{L}$ of the quantity L if the operator L which represents this quantity is given in matrix form. Regarding the c_n as a matrix ψ with one column (40.12), and the c_m^* as the adjoint matrix ψ^+ with one row (40.12′), we can use the rule of matrix multiplication to write (41.2) as

$$\bar{L} = \psi^+ L\psi. \tag{41.3}$$

The spectrum of a quantity (that is, its possible values) and the eigenfunctions of the operator L which represents it are given, according to (20.2), by the equation $L\psi_L = L\psi_L$. Substituting in this equation ψ from (41.1), multiplying on the left by ψ_m^* and integrating with respect to x, we obtain

$$\sum_n c_n \int \psi_m^*\cdot L\,\psi_n\cdot\mathrm{d}x = L\sum c_n \int \psi_m^*\psi_n\,\mathrm{d}x$$

or

$$\sum_n L_{mn} c_n = L c_m . \tag{41.4}$$

This is an infinite set of linear homogeneous algebraic equations to determine the amplitudes of the eigenfunction c_n and the eigenvalues L_n of the operator.

It is known from algebra that a set of linear homogeneous equations has a non-zero solution only when the determinant formed by the coefficients in the equations is zero. In our case this determinant has an infinite number of rows and columns[5]:

$$\begin{Vmatrix} L_{11} - L & L_{12} & L_{13} & \cdots & L_{1n} & \cdots \\ L_{21} & L_{22} - L & L_{23} & \cdots & L_{2n} & \cdots \\ \cdots & \cdots & \cdots & \cdots & \cdots & \cdots \\ L_{n1} & L_{n2} & L_{n3} & \cdots & L_{nn} - L & \cdots \\ \cdots & \cdots & \cdots & \cdots & \cdots & \cdots \end{Vmatrix} = 0 . \tag{41.5}$$

This equation places a restriction on the possible values of L. It is an equation of infinite degree in L (a transcendental equation) and has an infinite number of roots $L = L_1, L_2, \ldots, L_\alpha \ldots$. It is known from algebra that the roots of such an equation must be real. The set of values L_α for which the Equations (41.4) have solutions is the set of eigenvalues of the operator $\boldsymbol{L}$. Substituting in (41.4) one root of Equation (41.5), for instance L_α, we find the solution corresponding to this root:

$$L = L_\alpha, c_1 = c_1(L_\alpha), c_2 = c_2(L_\alpha), \ldots, c_n = c_n(L_\alpha), \ldots . \tag{41.6}$$

The set of values $c_1, c_2, \ldots, c_n, \ldots$ found in this way is the eigenfunction of the operator $\boldsymbol{L}$ belonging to the αth eigenvalue $L = L_\alpha$.

The same wave function in the x representation is[6]

$$\psi_\alpha(x) = \sum_n c_n(L_\alpha)\psi_n(x) . \tag{41.6'}$$

Every quantity is represented by a diagonal matrix in its own representation. For, if $\psi_n(x)$ is an eigenfunction of the operator $\boldsymbol{L}$, the matrix of $\boldsymbol{L}$ has the form

$$L_{mn} = \int \psi_m^* \cdot \boldsymbol{L}\, \psi_n \cdot \mathrm{d}x = \int \psi_m^* \cdot L_n \psi_n \cdot \mathrm{d}x = L_n \delta_{mn} , \tag{41.7}$$

where L_n is the nth eigenvalue of the operator $\boldsymbol{L}$. Hence the problem of finding the eigenvalues of the operator $\boldsymbol{L}$ may be regarded as that of reducing the matrix of an operator $\boldsymbol{L}$, given in an arbitrary representation, to the diagonal form (41.7).

Since commuting operators have eigenfunctions in common, their matrices can be simultaneously brought to diagonal form.

[5] Such a determinant must be regarded as the limit as $N \to \infty$ of the determinant for a system of a finite number N of unknowns c_n. Equation (41.5) is meaningful if this limit exists. An example of such an equation is given in [90], Part II, p. 415.

[6] The function $\psi_\alpha(x)$ may be obtained directly by solving the differential equation $\boldsymbol{L}\psi = L\psi$. The solution of Equations (41.4) and (41.5) is usually no simpler than the solution of this differential equation, but in an approximate solution of the equations (see Chapter XI) the equations in matrix form are very convenient.

The corresponding formulae for the case of continuous matrices are obtained from those given above by replacing the sums by integrals. The derivation is so simple that we shall merely give the results. The mean value of the quantity L is

$$\bar{L} = \iint \mathrm{d}p' \, \mathrm{d}p \, c^*(p') L_{p'p} c(p) \tag{41.2'}$$

in the momentum representation and

$$\bar{L} = \iint \mathrm{d}x' \, \mathrm{d}x \, \psi^*(x') L_{x'x} \psi(x) \tag{41.2''}$$

in the co-ordinate representation. Instead of Equation (41.4) we have respectively

$$\int L_{p'p} c(p) \, \mathrm{d}p = L \cdot c(p') \,, \tag{41.4'}$$

$$\int L_{x'x} \psi(x) \, \mathrm{d}x = L \cdot \psi(x') \,. \tag{41.4''}$$

Finally, instead of (41.7),

$$P_{p'p} = p' \cdot \delta(p' - p) \,, \tag{41.7'}$$

$$x_{x'x} = x' \cdot \delta(x' - x) \,. \tag{41.7''}$$

Equations (41.4′) and (41.4″) are either differential or integral equations.

42. Schrödinger's equation and the time dependence of operators in matrix form

Schrödinger's equation (28.3) can be rewritten in matrix form if we expand $\psi(x, t)$ as a series of eigenfunctions $\psi_n(x)$ of any operator. Substituting in (28.3) $\psi(x, t)$ as the series

$$\psi(x, t) = \sum_n c_n(t) \psi_n(x) \,,$$

multiplying on the left by $\psi_m^*(x)$ and integrating with respect to x, we find

$$i\hbar \frac{\mathrm{d}c_m}{\mathrm{d}t} = \sum_n H_{mn} c_n \,, \qquad m = 1, 2, 3, \ldots \,, \tag{42.1}$$

where

$$H_{mn} = \int \psi_m^*(x) \cdot \boldsymbol{H} \psi_n(x) \cdot \mathrm{d}x \tag{42.2}$$

is the matrix element of the Hamiltonian $\boldsymbol{H}$. This equation determines $c_n(t)$ (i.e. $\psi(x, t)$) from $c_n(0)$ (i.e. $\psi(x, 0)$) given at the initial instant.

Let $\boldsymbol{H}$ be the total-energy operator, and let us take as the functions $\psi_n(x)$ the eigenfunctions of the operator $\boldsymbol{H}$. Then the $c_n(t)$ are the amplitudes of the stationary states, and the matrix H_{mn} is diagonal:

$$H_{mn} = \int \psi_m^* \cdot \boldsymbol{H} \psi_n \cdot \mathrm{d}x = E_n \delta_{mn} \,. \tag{42.3}$$

Subsitution of these values of H_{mn} in (42.1) gives Schrödinger's equation for this case:

$$i\hbar \, \mathrm{d}c_m / \mathrm{d}t = E_m c_m \,. \tag{42.4}$$

Hence

$$c_m(t) = c_m(0) \, e^{-iE_m t/\hbar} \,, \tag{42.5}$$

i.e. the amplitudes of stationary states are harmonic functions of time. This is in accordance with the results of Section 30.

Let us now apply Schrödinger's equation in matrix form to the calculation of the time derivative of an operator. Differentiating with respect to time the mean value (41.2), we have

$$\frac{\mathrm{d}\bar{L}}{\mathrm{d}t} = \frac{\overline{\mathrm{d}L}}{\mathrm{d}t} = \sum_m \sum_n c_m^* \frac{\partial L_{mn}}{\partial t} c_n + \sum_m \sum_n \frac{\mathrm{d}c_m^*}{\mathrm{d}t} L_{mn} c_n + \sum_m \sum_n c_m^* L_{mn} \frac{\mathrm{d}c_n}{\mathrm{d}t};$$

(42.1) gives

$$- i\hbar \frac{\mathrm{d}c_m^*}{\mathrm{d}t} = \sum_k H_{mk}^* c_k^*, \qquad i\hbar \frac{\mathrm{d}c_n}{\mathrm{d}t} = \sum_k H_{nk} c_k.$$

Substituting these derivatives in the expression for $\mathrm{d}\bar{L}/\mathrm{d}t$, we obtain

$$\frac{\mathrm{d}\bar{L}}{\mathrm{d}t} = \sum_m \sum_n c_m^* \frac{\partial L_{mn}}{\partial t} c_n + \frac{1}{i\hbar} \sum_m \sum_n \sum_k c_m^* L_{mn} H_{nk} c_k - $$
$$- \frac{1}{i\hbar} \sum_m \sum_n \sum_k H_{mk}^* c_k^* L_{mn} c_n.$$

Since, from the self-adjointness of the operator, $H_{mk}^* = H_{km}$, and since the suffixes m, n and k take the same range of values, we can (interchanging k and n in the second term, and k and m in the third term) rewrite this equation in the form

$$\frac{\mathrm{d}\bar{L}}{\mathrm{d}t} = \sum_m \sum_n c_m^* \frac{\partial L_{mn}}{\partial t} c_n + \frac{1}{i\hbar} \sum_m \sum_n c_m^* \left\{ \sum_k L_{mk} H_{kn} - \sum_k H_{mk} L_{kn} \right\} c_n.$$

Since, from the rule of matrix multiplication,

$$\sum_k L_{mk} H_{kn} = (LH)_{mn}, \qquad \sum_k H_{mk} L_{kn} = (HL)_{mn},$$

we have

$$\frac{\mathrm{d}\bar{L}}{\mathrm{d}t} = \frac{\overline{\mathrm{d}L}}{\mathrm{d}t} = \sum_m \sum_n c_m^* \left\{ \frac{\partial L_{mn}}{\partial t} + \frac{1}{i\hbar} (LH - HL)_{mn} \right\} c_n, \tag{42.6}$$

where

$$\frac{1}{i\hbar} (LH - HL)_{mn} = \frac{1}{i\hbar} \sum_k (L_{mk} H_{kn} - H_{mk} L_{kn}) = [H, L]_{mn} \tag{42.7}$$

is the matrix element of the Poisson bracket. A comparison with Formula (41.2) for the mean value shows that the matrix element of the time derivative of $\boldsymbol{L}$ is

$$\left(\frac{\mathrm{d}L}{\mathrm{d}t} \right)_{mn} = \frac{\partial L_{mn}}{\partial t} + [H, L]_{mn}. \tag{42.8}$$

Formulae (42.6) and (42.8) are the matrix forms of (31.4) and (31.7) respectively.

The following is an important particular case. Let the Hamiltonian $\boldsymbol{H}$ be independent of time, so that $\boldsymbol{H}$ is the total-energy operator, and let us choose the E representation. Then the matrix $\boldsymbol{H}$ is diagonal:

$$H_{kn} = E_k \delta_{kn}, \qquad H_{mk} = E_m \delta_{mk}.$$

Assuming also that the operator $\boldsymbol{L}$ does not depend explicitly on time, we obtain from (42.7) and (42.8)

$$\left(\frac{\mathrm{d}L}{\mathrm{d}t}\right)_{mn} = \frac{1}{i\hbar}(E_n - E_m) L_{mn},$$

or

$$\left(\frac{\mathrm{d}L}{\mathrm{d}t}\right)_{mn} = i\omega_{mn} L_{mn}, \tag{42.9}$$

where

$$\omega_{mn} = (E_m - E_n)/\hbar \tag{42.10}$$

is the Bohr frequency. In particular, the matrix of the velocity operator has elements

$$(\mathrm{d}x/\mathrm{d}t)_{mn} = i\omega_{mn} x_{mn}, \tag{42.11}$$

where x_{mn} are the matrix elements of the co-ordinate x. The relation between the velocity and the co-ordinate is exactly the same as for an oscillator of frequency ω_{mn}.

Formula (42.9) becomes quite obvious if we use what is called *Heisenberg's representation of operators*. This method consists in setting up the matrix of some operator $\boldsymbol{L}$ by means of the wave functions of stationary states taken for time t:

$$\psi_n(x, t) = \psi_n(x)\, e^{-iE_n t/\hbar}.$$

It is clear that this can be done, since the $\psi_n(x, t)$, like the $\psi_n(x)$, form a complete orthogonal set of functions. Thus in Heisenberg's representation the matrix element of the operator $\boldsymbol{L}$ is determined from the formula

$$L_{mn}(t) = \int \psi_m^*(x, t) \cdot \boldsymbol{L} \psi_n(x, t) \cdot \mathrm{d}x = L_{mn} e^{i\omega_{mn} t}. \tag{42.12}$$

Hence we have for an operator not explicitly dependent on time

$$(\mathrm{d}L/\mathrm{d}t)_{mn} = \mathrm{d}L_{mn}/\mathrm{d}t = i\omega_{mn} L_{mn}(t). \tag{42.9'}$$

This formula differs from (42.9) only in that the time dependence is transferred from the wave functions to the operators.

According to (42.12) the matrix elements of operators not explicitly dependent on time are, in Heisenberg's representation, harmonic functions of time with the Bohr frequency ω_{mn}.

For continuous matrices we have instead of (42.1)

$$i\hbar\, \partial c(p')/\partial t = \int H_{p'p}\, c(p)\, \mathrm{d}p, \tag{42.1'}$$

or, in the co-ordinate representation,

$$i\hbar \partial \psi(x')/\partial t = \int H_{x'x} \psi(x)\, \mathrm{d}x, \tag{42.1''}$$

and instead of (42.8)

$$\left(\frac{\mathrm{d}L}{\mathrm{d}t}\right)_{p'p} = \frac{\partial L_{p'p}}{\partial t} + [H, L]_{p'p}, \tag{42.8'}$$

$$\left(\frac{\mathrm{d}L}{\mathrm{d}t}\right)_{x'x} = \frac{\partial L_{x'x}}{\partial t} + [H, L]_{x'x}. \tag{42.8''}$$

The remaining formulae in this section relate specifically to the energy representation.

It is seen from Sections 39–42 that the use of continuous matrices allows us to make the matrix method of writing operators completely uniform, so that all possible representations of operators and wave functions become entirely equivalent. For this reason the matrix notation for operators is especially convenient in discussing general theoretical problems.

In the solution of particular problems, however, the co-ordinate representation is usually employed. This is because the energy of interaction in non-relativistic theory depends only on the co-ordinates, while the kinetic energy is a simple function of momentum ($\boldsymbol{P}^2/2m$). Hence Schrödinger's equation in the co-ordinate representation is a comparatively simple second-order differential equation. In an approximate solution, however, other representations may well be superior to the co-ordinate representation.

43. Unitary transformations

Let us consider the transformation of some operator $\boldsymbol{G}$ from one arbitrary representation to another. Let the operator $\boldsymbol{G}$ be represented in the first representation by the matrix $\boldsymbol{G}'$, whose elements are labelled by the eigenvalues $L = L_1, L_2, \ldots, L_n, \ldots, L_m, \ldots$ of the operator $\boldsymbol{L}$ (the L representation). In the second representation let the same operator $\boldsymbol{G}$ be represented by a matrix $\boldsymbol{G}''$, whose elements are labelled by the eigenvalues $M = M_1, M_2, \ldots, M_\alpha, \ldots, M_\beta, \ldots$ of the operator $\boldsymbol{M}$ (the M representation). For definiteness we shall assume that $\boldsymbol{L}$ and $\boldsymbol{M}$ have discrete spectra. If the operator $\boldsymbol{G}$ is originally given in the x representation ($\boldsymbol{G} = \boldsymbol{G}(-i\hbar\,\partial/\partial x, x)$) and the eigenfunctions of the operators $\boldsymbol{L}$ and $\boldsymbol{M}$ are $\psi_1(x), \psi_2(x), \ldots, \psi_n(x), \ldots, \psi_m(x), \ldots$ and $\phi_1(x), \phi_2(x), \ldots, \phi_\alpha(x), \ldots, \phi_\beta(x), \ldots$ respectively, then the matrix elements of the operator $\boldsymbol{G}$ in the L representation are

$$G_{mn} = \int \psi_m^*(x)\cdot \boldsymbol{G}(-i\hbar\,\partial/\partial x, x)\,\psi_n(x)\cdot \mathrm{d}x\,, \tag{43.1}$$

and in the M representation

$$G_{\alpha\beta} = \int \phi_\alpha^*(x)\cdot \boldsymbol{G}(-i\hbar\,\partial/\partial x, x)\,\phi_\beta(x)\cdot \mathrm{d}x\,. \tag{43.2}$$

We ask what is the relation between the matrix $\boldsymbol{G}'$ with elements G_{mn} and the matrix $\boldsymbol{G}''$ with elements $G_{\alpha\beta}$. Let us expand the eigenfunctions of the operator $\boldsymbol{M}$ in terms of the eigenfunctions of the operator $\boldsymbol{L}$:

$$\phi_\beta(x) = \sum_n \psi_n(x)\,S_{n\beta}\,, \qquad \phi_\alpha^*(x) = \sum_m \psi_m^*(x)\,S_{m\alpha}^*\,, \tag{43.3}$$

where

$$S_{n\beta} = \int \psi_n^*(x)\,\phi_\beta(x)\,\mathrm{d}x\,, \qquad S_{m\alpha}^* = \int \psi_m(x)\,\phi_\alpha^*(x)\,\mathrm{d}x\,. \tag{43.4}$$

Substitution of (43.3) in (43.2), using (43.1), gives

$$G_{\alpha\beta} = \sum_m \sum_n S_{m\alpha}^*\,G_{mn}\,S_{n\beta}\,. \tag{43.5}$$

The set of quantities $S_{n\beta}$ may be regarded as a matrix $\boldsymbol{S}$ whose rows and columns are labelled by the eigenvalues of the quantities L and M respectively. As well as the matrix S we have the adjoint matrix S^+ whose elements are given by

$$(S^+)_{\alpha m} = S^*_{m\alpha}, \tag{43.6}$$

so that $S^+ = \tilde{S}^*$, and therefore the rows and columns of this matrix are labelled by the eigenvalues of M and L respectively. By (43.6), the formula for the transformation from G_{mn} to $G_{\alpha\beta}$ (43.5) can be written

$$G_{\alpha\beta} = \sum_m \sum_n (S^+)_{\alpha m} G_{mn} S_{n\beta}, \tag{43.7}$$

or, by the rule of matrix multiplication, in the matrix form

$$G'' = S^+ G' S. \tag{43.8}$$

Thus the matrix S and its adjoint matrix S^+ may be regarded as matrices by means of which the transformation of an operator is effected from one representation (L) to another (M). The matrix S has an important property. Multiplying the functions $\phi^*_\alpha(x)$ and $\phi_\beta(x)$ and integrating the product with respect to x, we have from the orthogonality of the eigenfunctions

$$\sum_m \sum_n S^*_{m\alpha} S_{n\beta} \delta_{mn} = \delta_{\alpha\beta}, \tag{43.9}$$

or

$$\sum_n (S^+)_{\alpha n} S_{n\beta} = \delta_{\alpha\beta}, \tag{43.10}$$

i.e. in matrix form

$$S^+ S = 1. \tag{43.11}$$

Similarly, by expanding the functions $\psi_n(x)$ in terms of the functions $\phi_\beta(x)$, it may be shown that

$$\sum_\alpha S_{m\alpha} S^+_{\alpha n} = \delta_{mn}, \tag{43.12}$$

i.e.

$$SS^+ = 1. \tag{43.11'}$$

A matrix which satisfies the conditions (43.11) and (43.11′) is said to be *unitary*. Since the product of S^+ and S or of S and S^+ gives the unit matrix, it follows that S^+ is the matrix inverse to S, i.e.

$$S^+ = S^{-1}. \tag{43.13}$$

It should be noted that a unitary matrix is not Hermitian, since for an Hermitian matrix $S^+ = S$ instead of (43.13). From the above discussion we can conclude that the transformation of an operator from one representation to another is achieved by means of a unitary matrix S with elements (43.4). The transformation (43.8) is also said to be *unitary*.

Formula (43.1) may also be regarded as a unitary transformation from the coordinate representation to the L representation. For this purpose we need only write the operator $G(-i\hbar\,\partial/\partial x, x)$ in matrix form. Then we have instead of (43.1)

$$G_{mn} = \int\int \psi_m^*(x')\, G_{x'x}\, \psi_n(x)\, dx\, dx' . \tag{43.1'}$$

Putting $(S^+)_{mx'} = \psi_m^*(x')$ and $S_{xn} = \psi_n(x)$, we bring the transformation (43.1') to the form (43.8). Thus the wave functions $\psi_m^*(x)$, $\psi_n(x)$ are just the matrix elements of the unitary matrices S^+ and S which transform from the co-ordinate representation to the L representation.

It has already been noted (Section 41) that the problem of finding the eigenvalues of any operator may be regarded as one of reducing the matrix representing the operator to diagonal form. In terms of unitary transformations, this problem can be formulated as follows: to find a unitary transformation S which transforms the matrix of the operator G into a diagonal matrix. In order to find this transformation, we multiply equation (43.8) on the left by S. Using (43.11'), we obtain

$$SG'' = G'\,S, \tag{43.14}$$

or in explicit form

$$\sum_\alpha S_{m\alpha}\, G_{\alpha\beta} = \sum_n G_{mn}\, S_{n\beta} . \tag{43.15}$$

If the matrix $G_{\alpha\beta}$ is diagonal, then

$$S_{m\alpha}\, G_{\alpha\alpha} = \sum_n G_{mn}\, S_{n\alpha} . \tag{43.16}$$

Since the eigenvalues $G_{\alpha\alpha}$ are unknown, we must drop the suffix α, obtaining

$$S_m G = \sum_n G_{mn}\, S_n , \tag{43.17}$$

and this is the same as (41.4) if we put $G = L$, $S = c$.

Another important property of a unitary transformation is the following: a unitary transformation does not change the sum of the diagonal elements of a matrix. This sum is called the *trace* (or *spur*) of the matrix and denoted by

$$\mathrm{Tr}\, G = \sum_n G_{nn} . \tag{43.18}$$

From (43.7) we have

$$\begin{aligned} \sum_a G_{\alpha\alpha} &= \sum_\alpha \sum_m \sum_n (S^+)_{\alpha m}\, G_{mn}\, S_{n\alpha} = \sum_m \sum_n G_{mn} \sum_\alpha (S^+)_{\alpha m}\, S_{n\alpha} \\ &= \sum_m \sum_n G_{mn}\, \delta_{mn} = \sum_n G_{nn} , \end{aligned} \tag{43.19}$$

i.e. the trace of the matrix is an invariant of the unitary transformation. This property is often made use of in applications.

44. The unitary transformation from one instant to another

The variation of wave functions with time can also be considered in terms of the unitary transformation.

Let the Hamiltonian H be independent of time. We denote its eigenfunctions by $\psi_n(x)$ and the eigenvalues of the energy by E_n. Then

$$H\psi_n(x) = E_n\psi_n(x)\,. \tag{44.1}$$

Let us apply to the function $\psi_n(x)$ the operator

$$S(t) = \exp\,(-\,iHt/\hbar), \tag{44.2}$$

the significance of which will appear below.

From (44.1) it follows that $H^s\psi_n(x) = E_n{}^s\psi_n(x)$, and therefore

$$S(t)\,\psi_n(x) = \sum_{s=0}^{\infty}\frac{1}{s!}\left(-\frac{it}{\hbar}\right)^s H^s\psi_n(x) = \sum_{s=0}^{\infty}\frac{1}{s!}\left(-\frac{it}{\hbar}\right)^s E_n{}^s\psi_n(x)$$
$$= e^{-iE_nt/\hbar}\psi_n(x)\,. \tag{44.3}$$

Now let an arbitrary initial function $\psi(x, 0)$ be prescribed. We expand it in terms of $\psi_n(x)$:

$$\psi(x, 0) = \Sigma\, c_n\psi_n(x)\,. \tag{44.4}$$

Applying to $\psi(x, 0)$ the operator (44.2), we have from (44.3)

$$S(t)\,\psi(x, 0) = \Sigma\, c_n e^{-iE_nt/\hbar}\psi_n(x)\,. \tag{44.5}$$

The right-hand side of this equation is just the wave function at time t. Hence

$$\psi(x, t) = S(t)\,\psi(x, 0)\,. \tag{44.6}$$

The operator $S(t)$ is unitary. For, if this operator is taken in the energy representation, its matrix elements are seen directly from (44.3) to be simply $S_{mn} = e^{-iE_nt/\hbar}\delta_{mn}$, and the matrix elements $(S^+)_{mn}$ of the adjoint matrix are $e^{iE_nt/\hbar}\delta_{mn}$. Hence[7] $S^+ = S^{-1}$.

It follows from (44.6) that the unitary transformation (44.2) with the operator S is equivalent to Schrödinger's equation.[8] We can say, in other words, that the motion may be regarded as a succession of unitary transformations.[9]

Let us expand the functions $\psi(x, 0)$ and $\psi(x, t)$ in terms of the eigenfunctions $\phi_\alpha(x)$ of some operator L. Then (44.6) becomes

$$c_\beta(t) = \sum_\alpha S_{\beta\alpha}(t)\,c_\alpha(0)\,, \tag{44.6'}$$

where $c_\beta(t)$ and $c_\alpha(0)$ are the amplitudes in the expansions of $\psi(x, t)$ and $\psi(x, 0)$ respectively, and $S_{\beta\alpha}(t)$ is the matrix element of the operator $S(t)$ in the L representation.

Let the quantity L have some definite value L_α at the initial instant. This means that $\psi(x, 0) = \phi_\alpha(x)$, $c_\alpha(0) = 1$, $c_{\alpha'}(0) = 0$ for $\alpha' \neq \alpha$. In this particular (but very important) case we have instead of (44.6′)

$$c_\beta(t) = S_{\beta\alpha}(t)\,. \tag{44.7}$$

According to the general theory (Section 22) the probability of finding $L = L_\beta$ at time t is $|c_\beta(t)|^2$. But it has been assumed that at $t = 0$ the quantity L has the unique value L_α. Thus, with this choice of initial conditions $|c_\beta(t)|^2$ is the probability of finding $L = L_\beta$ at time t if $L = L_\alpha$ at $t = 0$. In other words, $P_{\beta\alpha}(t) = |c_\beta(t)|^2$ is what is called the *quantum transition probability*[10] from the state $L = L_\alpha$ to

---

[7] It can easily be shown in the same way that any unitary operator can be written in the form $S' = e^{i\eta}$, where η is the Hermitian phase operator.

[8] If the Hamiltonian H depends on time, (44.6) is still valid, but in this case $S'(t)$ cannot be written in the form (44.2); instead $S'(t) = e^{i\eta_1(t)}$ and is given by the relation $H = i\hbar(\partial S'/\partial t)S'^+$, i.e. to find S' we need the solution of Schrödinger's equation. In recent years it has been suggested that the phase operator $\eta(t)$ is perhaps more fundamental than the Hamilton's function operator H. In that case Schrödinger's equation would not be needed at all. However, no rules have been discovered for finding η for a given physical system.

[9] Similarly in classical mechanics the motion may be regarded as a succession of contact transformations.

[10] Cf. Sections 84 and 85, where the problem of quantum transitions is discussed in detail and approximate methods for calculating transition probabilities are given.

the state $L = L_\beta$. According to (44.7)

$$P_{\beta\alpha}(t) = |S_{\beta\alpha}(t)|^2 . \tag{44.8}$$

Since the matrix S is unitary, and not Hermitian, $|S_{\beta\alpha}|^2$ is in general not equal to $|S_{\alpha\beta}|^2$, so that the probability of a transition from state α to state β is in general not equal to the probability of the reverse transition from state β to state α.

It should not be concluded from this that quantum mechanics is irreversible. We know from classical mechanics that, if the forces do not depend on the velocities, the reversal of the velocities of all the particles has the result that the same motion is executed in the opposite direction.

It can be shown that under these conditions exactly the same reversibility holds good in quantum mechanics. The probability of going in time t from a state with particle momenta $p_1^0, p_2^0, \ldots$ (state α) to one with momenta $p_1, p_2, \ldots$ (state β) is equal to the probability of going in the same time from a state with the reversed momenta $-p_1, -p_2, \ldots$ (reversed state β) to one with momenta $-p_1^0, -p_2^0, \ldots$ (reversed state α).[11]

It will be seen from this brief account of unitary transformations that the whole mathematical formalism of quantum mechanics can be expressed in terms of operators in matrix form and in terms of unitary transformations.

45. The density matrix

Let an operator L be given in the co-ordinate representation in the form of a matrix $L_{x'x}$. The mean value $\bar{L}_\alpha$ in the state $\psi_\alpha(x)$ is (see (41.2″))

$$\bar{L}_\alpha = \int\int \mathrm{d}x' \, \mathrm{d}x \, \psi_\alpha^*(x') L_{x'x} \psi_\alpha(x) . \tag{45.1}$$

If from the pure ensembles specified by the wave functions ψ_α we form a mixed ensemble such that each pure state is represented with probability P_α, then the mean value $\bar{L}$ in the mixed ensemble is (see (22.18))

$$\bar{L} = \sum_\alpha P_\alpha \bar{L}_\alpha = \sum_\alpha P_\alpha \int\int \mathrm{d}x' \, \mathrm{d}x \, \psi_\alpha^*(x') L_{x'x} \psi_\alpha(x) \tag{45.2}$$

(if $\sum P_\alpha = 1$). Equation (45.2) can be written in the form

$$\bar{L} = \int\int \mathrm{d}x' \, \mathrm{d}x \, \rho_{xx'} L_{x'x} , \tag{45.3}$$

where

$$\rho_{xx'} = \sum_\alpha P_\alpha \psi_\alpha^*(x') \psi_\alpha(x) . \tag{45.4}$$

The operator $\boldsymbol{\rho}$ represented by the matrix whose elements are (45.4) is called the *density matrix*.[12] The expression (45.3) is just the sum of the diagonal elements of the operator $\boldsymbol{\rho} L$. We can therefore write (45.3) as

$$\bar{L} = \mathrm{Tr}(\boldsymbol{\rho} L) . \tag{45.5}$$

In another representation, expanding $\psi_\alpha(x)$ in terms of the eigenfunctions $\phi_n(x)$ (of some operator M having a discrete spectrum of eigenvalues $M_1, M_2, \ldots, M_n, \ldots$), we obtain from (45.2)

$$\bar{L} = \sum_\alpha \sum_m \sum_n P_\alpha c_{\alpha m}^* L_{mn} c_{\alpha n} , \tag{45.6}$$

[11] See [10], where the problem is considered in detail.

[12] Introduced by von Neumann [68].

and

$$\rho_{nm} = \sum_{\alpha} P_{\alpha} c_{\alpha m}^{*} c_{\alpha n}, \tag{45.7}$$

where $c_{\alpha n}$ are the amplitudes in the expansion of $\psi_{\alpha}(x)$ in terms of $\phi_n(x)$. Thus in this representation we have

$$\bar{L} = \sum_{m}\sum_{n} \rho_{nm} L_{mn} = \mathrm{Tr}(\boldsymbol{\rho} \boldsymbol{L}). \tag{45.8}$$

The diagonal element of the matrix $\boldsymbol{\rho}$ has the significance of probability (or probability density). For, putting $x' = x$ in (45.4), we find that

$$\rho_{xx} = \sum_{\alpha} P_{\alpha} |\psi_{\alpha}(x)|^2 = w(x) \tag{45.9}$$

is the probability density distribution for the co-ordinate x in the mixed ensemble. Similarly we have from (45.7)

$$\rho_{nn} = \sum_{\alpha} P_{\alpha} |c_{\alpha n}|^2 = w_n, \tag{45.10}$$

the probability of finding the value $M = M_n$ in the mixed ensemble.

Let us now consider how the operator $\boldsymbol{\rho}$ will vary in the course of time. The matrix (45.4) determines $\boldsymbol{\rho}$ for some instant which we can take as the initial instant ($t = 0$). The mixed ensemble described by this matrix is a set of independent systems, each of which is (with probability P_{α}) in one of the pure states $\psi_{\alpha}(x) = \psi_{\alpha}(x, 0)$. A system which is in the pure state $\psi_{\alpha}(x, 0)$ at time $t = 0$ will also be in the pure state $\psi_{\alpha}(x, t)$ at times $t > 0$, and this state can be found from Schrödinger's equation

$$i\hbar\, \partial\psi_{\alpha}(x, t)/\partial t = \int H_{xx''} \psi_{\alpha}(x'', t)\, \mathrm{d}x'' \tag{45.11}$$

or, for the complex conjugate function $\psi_{\alpha}^{*}(x', t)$, from the complex conjugate equation

$$-\, i\hbar\, \partial\psi_{\alpha}^{*}(x', t)/\partial t = \int H_{x'x''}^{*} \psi_{\alpha}^{*}(x'', t)\, \mathrm{d}x''. \tag{45.11'}$$

Here $H_{x'x''}$ is the matrix element of the Hamiltonian in the x representation.

P_{α} is the probability that at time $t = 0$ the system is in the state $\psi_{\alpha}(x, 0) = \psi_{\alpha}(x)$, and thus, being the probability of initial data, is of course independent of time.[13] Hence at times $t > 0$ the matrix $\boldsymbol{\rho}$ will be

$$\rho_{xx'}(t) = \sum P_{\alpha} \psi_{\alpha}^{*}(x', t) \psi_{\alpha}(x, t). \tag{45.4'}$$

Differentiating this equation with respect to time and expressing the derivatives of the wave functions in terms of the Hamiltonian operator by means of (45.11) and (45.11'), we find

$$\frac{\partial \rho_{xx'}(t)}{\partial t} = \frac{1}{i\hbar}\int H_{xx''} \rho_{x''x'}\, \mathrm{d}x'' - \frac{1}{i\hbar}\int \rho_{xx''} H_{x''x'}\, \mathrm{d}x''; \tag{45.12}$$

here we have used the fact that $H_{x'x''}^{*} = H_{x''x'}$. In operator form

$$\partial\boldsymbol{\rho}/\partial t = -\left[\boldsymbol{H}, \boldsymbol{\rho}\right], \tag{45.13}$$

where $[\boldsymbol{H}, \boldsymbol{\rho}]$ is the quantum Poisson bracket.

13 But may be changed as a result of measurement; see below.

This operator equation enables us to determine the operator $\boldsymbol{\rho}$ for any instant if it is known for $t = 0$.

The advantage of describing the ensemble by means of the operator $\boldsymbol{\rho}$, in comparison with the description by means of the ψ function, is that the operator $\boldsymbol{\rho}$ permits a uniform treatment of both mixed and pure ensembles.

Let us now consider the changes in the operator $\boldsymbol{\rho}$ which occur as a result of measurement. Let a complete measurement be made (measurement of a quantity or set of quantities M). Let the eigenfunctions of the operator $\boldsymbol{M}$ be $\phi_n(x)$. Then the probability of finding a value M_n is given by (45.10). After the measurement, a new mixed ensemble exists, in which new pure states $\phi_n(x)$ are represented with probabilities[14] w_n, i.e. after the measurement we have

$$\rho'_{xx'} = \sum w_n \phi_n^*(x')\, \phi_n(x), \tag{45.14}$$

and we obtain a completely new mixed ensemble.

In quantum statistics, states are never described by a complete set of measurements, and so it is always mixed ensembles which are concerned. In consequence the density operator $\boldsymbol{\rho}$ is of especial importance in quantum statistics.

It is well known that in classical statistics an ensemble of independent systems (usually called a *Gibbs ensemble*) is described by a probability density $D(p, x)$, such that the quantity $D(p, x)\mathrm{d}p\ \mathrm{d}x$ signifies the probability of finding a system with momentum near p and co-ordinate near x.[15] According to Liouville's theorem this density is constant, so that

$$\frac{\mathrm{d}D}{\mathrm{d}t} = \frac{\partial D}{\partial t} + [H, D]_{cl} = 0, \tag{45.15}$$

where $[H, D]_{cl} = (\partial H/\partial p)\,(\partial D/\partial x) - (\partial H/\partial x)\,(\partial D/\partial p)$ is the classical Poisson bracket. From (45.15) it follows that

$$\partial D/\partial t = -\,[H, D]_{cl}\,. \tag{45.15$'$}$$

The analogy between (45.15′) and (45.13) is obvious.

A classical Gibbs ensemble and a quantum mixed ensemble are identical in nature (being assemblies of independent systems). The operator $\boldsymbol{\rho}$ is for this reason called the *density operator* by analogy with the probability density D. The relation between $\boldsymbol{\rho}$ and D can be more fully established if we use instead of $\rho_{xx'}$ the matrix $\boldsymbol{R}(p, x)$ whose rows and columns are labelled by the momentum and the co-ordinate respectively:

$$\boldsymbol{R}(p, x) = \int \rho_{xx'} \frac{e^{ip(x-x')/\hbar}}{2\pi\hbar}\, \mathrm{d}x'\,. \tag{45.16}$$

[14] If, of course, no choice of subsets, say with $M = M_n$, is made. In that case the ensemble obtained after the measurement will be pure, with $\psi = \phi_n(x)$.

[15] We use the notation corresponding to an ensemble of systems with one degree of freedom x. The symbols p and x may be understood to represent the momenta and co-ordinates of all particles in the system.

Then

$$\int \boldsymbol{R}(p,x)\,\mathrm{d}p = \int \rho_{xx'}\,\delta(x-x')\,\mathrm{d}x' = w(x), \tag{45.17}$$

$$\int \boldsymbol{R}(p,x)\,\mathrm{d}x = \int \rho_{xx'}\frac{e^{ip(x-x')/\hbar}}{2\pi\hbar}\,\mathrm{d}x\,\mathrm{d}x' = \rho_{pp} = w(p), \tag{45.17'}$$

where $w(x)$ and $w(p)$ are the probability densities for the co-ordinate x and the momentum[16] p. These formulae are entirely analogous to the classical formulae

$$\int D(p,x)\,\mathrm{d}p = w_{cl}(x), \qquad \int D(p,x)\,\mathrm{d}x = w_{cl}(p). \tag{45.18}$$

Furthermore it may be shown that the matrix $\boldsymbol{R}(p, x)$ satisfies an equation which in certain conditions (for smooth fields and a smooth function $R(p, x)$) becomes the classical equation (45.15').[17] Hence the quantity $R(p, x)$ is entirely analogous to the classical probability (probability density) $D(p, x)$ and may be regarded as a generalisation of the concept of probability to quantities which cannot be measured simultaneously (*quasiprobability*). The quantity $\rho_{xx'}$ corresponds to the Fourier components of the density $D(p, x)$, i.e. to

$$\Delta_{xx'} = \int D(p,x)\,e^{-ip(x-x')/\hbar}\,\mathrm{d}p. \tag{45.19}$$

[16] In order to derive (45.17′) we must use the fact that

$$\int \psi_\alpha^*(x')\,\frac{e^{-ipx'/\hbar}}{\sqrt{(2\pi\hbar)}}\,\mathrm{d}x' = c_\alpha^*(p)\,.$$

[17] This matrix was introduced by the present author; see [8] and also [85].

CHAPTER VIII

THEORY OF THE MOTION OF MICROPARTICLES IN A FIELD OF POTENTIAL FORCES

46. Introductory remarks

In this chapter we shall consider some simple problems of atomic mechanics concerning the motion of particles in the field of potential forces. Generally speaking, if the forces are independent of time, the basic problem of atomic mechanics is that of finding the stationary states of the system. For in this case, according to (30.8), an arbitrary state $\psi(x,t)$ can be represented as a superposition of stationary states with constant amplitudes c_n:

$$\psi(x,t) = \sum_n c_n \psi_n(x,t), \tag{46.1}$$

$$\psi_n(x,t) = \psi_n(x) \cdot e^{-iE_n t/\hbar}, \tag{46.2}$$

where the $\psi_n(x)$ are wave functions of stationary states and the E_n the corresponding energy values. The wave functions $\psi_n(x)$ are the eigenfunctions of the energy operator $\boldsymbol{H}$ and are determined, according to (30.4), from Schrödinger's equation for stationary states:

$$\boldsymbol{H}\psi = E\psi . \tag{46.3}$$

The problem of finding the stationary states is at the same time the problem of finding the spectrum of the energy E.

The particular significance of this problem in atomic mechanics lies in the fact that, in contrast to classical mechanics, quantum mechanics leads in many cases to a quantisation of energy, i.e. to a discrete spectrum of values $E_1, E_2, \ldots, E_n, \ldots$. These values are often called *quantum levels* or *energy levels.*

If a system (for example,an electron in an atom, a molecule, etc.) has such an energy spectrum and is subjected to a weak external interaction, its quantum levels are unchanged (more precisely, are changed only slightly). However, the external interaction may cause a system to go from one level to another, so that the state of the system may be changed considerably. The probabilities of such transitions will be calculated later.

If the possible values of the energy have been found, we can immediately say what are the possible values of the energy of the system under consideration if a weak interaction[1] is set up between it and some other system or an external field. For instance, if the energy levels found are $E_1, E_2, \ldots, E_n, \ldots E_m, \ldots$, energy can be exchanged only in quantities given by

$$\Delta E = E_m - E_n. \tag{46.4}$$

47. A harmonic oscillator

In classical mechanics the Hamilton's function for a one-dimensional harmonic oscillator is

$$H = \tfrac{1}{2}p_x^2/\mu + \tfrac{1}{2}\mu\omega_0^2 x^2. \tag{47.1}$$

Here p_x is the particle momentum, μ the particle mass, x the deviation from the position of equilibrium, and ω_0 the natural (angular) frequency of the oscillator.

It may be noted that, since mechanical oscillations are under consideration, the harmonic oscillator is an idealisation, because the value $U = \frac{1}{2}\mu\omega_0{}^2x^2$ of the potential energy signifies that the force increases without limit as we move away from the equilibrium position. In any actual case there occur considerable deviations from harmonicity above a certain value of the amplitude, and for large x the force of interaction tends to zero (U tends to a constant). For small amplitudes of oscillation x, however, the concept of a harmonic oscillator may certainly be used.

The theory of the one-dimensional harmonic oscillator is of great importance in applications, since by a suitable choice of co-ordinates (*'normal co-ordinates'*) the motion of any system of particles executing small oscillations can be reduced to that of an assembly of independent oscillators.[2]

In quantum mechanics we shall understand by a one-dimensional oscillator a system described by a Hamiltonian operator $\boldsymbol{H}$ such that

$$\boldsymbol{H} = \frac{\boldsymbol{P}_x^2}{2\mu} + \tfrac{1}{2}\mu\omega_0^2\boldsymbol{X}^2 \tag{47.2}$$

(in exact correspondence with (47.1)), where $\boldsymbol{P}_x$ is the momentum operator and $\boldsymbol{X}$ is the co-ordinate operator.[3] Accordingly, Schrödinger's equation in the x representation for stationary states of an oscillator is

$$-\frac{\hbar^2}{2\mu}\frac{d^2\psi}{dx^2} + \tfrac{1}{2}\mu\omega_0^2 x^2\psi = E\psi. \tag{47.3}$$

[1] If the interaction between the systems is strong, we have one large system; if the external field is strong, the levels in the system are considerably changed. The assumption that the interaction is weak is therefore essential.

[2] See Section 109. An important application of the theory of quantum harmonic oscillators occurs in the quantum theory of light [47].

[3] The question may be asked whether it is reasonable to call a system with the Hamiltonian (47.2) a harmonic oscillator. The answer is that a system described by the Hamiltonian (47.2) emits and absorbs only a single frequency ω_0 (see Section 90, A), and as $\hbar \to 0$ it tends to a classical system with the Hamilton's function (47.1) (cf. Sections 34, 35).

To solve this equation we use the dimensionless quantities

$$\left.\begin{aligned} \xi &= x/x_0\,, \\ x_0 &= \sqrt{(\hbar/\mu\omega_0)}\,, \\ \lambda &= 2E/\hbar\omega_0\,. \end{aligned}\right\} \tag{47.4}$$

Denoting differentiation with respect to ξ by a prime and regarding ψ as a function of ξ, we obtain from (47.3) after some elementary transformations

$$\psi'' + (\lambda - \xi^2)\psi = 0\,. \tag{47.5}$$

We have to find solutions of this equation which are finite, continuous and single-valued in the range $-\infty < \xi < \infty$. Such solutions of Equation (47.5) exist not for all values of the parameter λ but only for

$$\lambda = 2n + 1\,, \qquad n = 0, 1, 2, 3, \dots, \tag{47.6}$$

and the corresponding functions ψ_n are

$$\psi_n(\xi) = e^{-\frac{1}{2}\xi^2} H_n(\xi)\,, \tag{47.7}$$

where $H_n(\xi)$ is the Hermite polynomial[4] of order n, given by the formula

$$H_n(\xi) = \frac{(-1)^n}{\sqrt{(2^n n!\sqrt{\pi})}}\, e^{\xi^2} \frac{\mathrm{d}^n e^{-\xi^2}}{\mathrm{d}\xi^n}\,; \tag{47.8}$$

the coefficient in front of e^{ξ^2} is chosen so that the function $\psi_n(\xi)$ is normalised to unity with respect to ξ:

$$\int_{-\infty}^{\infty} \psi_n^2(\xi)\,\mathrm{d}\xi = \int_{-\infty}^{\infty} e^{-\xi^2} H_n^2(\xi)\,\mathrm{d}\xi = 1\,. \tag{47.9}$$

Thus the mere requirements of continuity and finiteness of ψ result in the parameter λ having only the discrete values (47.6). According to (47.4) this parameter determines the energy. A comparison of (47.4) and (47.6) shows that the possible values E_n are

$$E_n = \hbar\omega_0 (n + \tfrac{1}{2})\,, \qquad n = 0, 1, 2, 3, \dots, \tag{47.10}$$

i.e. that the energy E of the oscillator can have only discrete values. The number n which specifies the quantum level is called the *principal quantum number*.

Finally, the eigenfunction belonging to the nth eigenvalue is, in the x representation,

$$\psi_n(x) = \frac{e^{-\frac{1}{2}\xi^2}}{\sqrt{x_0}} H_n(\xi)\,, \tag{47.11}$$

where $\xi = x/x_0$. These functions are normalised so that

$$\int_{-\infty}^{\infty} \psi_n^2(x)\,\mathrm{d}x = 1\,.$$

[4] Details of the solution of Equation (47.5) and in particular the condition (47.6) are given in Appendix IX.

It should be noted that the wave functions of the oscillator are of the same parity as the principal quantum number n; this is easily seen from (47.11) and (47.8).

Using Formula (47.8), we can write out some eigenfunctions of the form (47.11):

$$\psi_0(x) = \frac{1}{\sqrt{(x_0\sqrt{\pi})}} e^{-x^2/2x_0^2}, \qquad n = 0; \tag{47.12}$$

$$\psi_1(x) = \frac{1}{\sqrt{(2x_0\sqrt{\pi})}} e^{-x^2/2x_0^2} \cdot \frac{2x}{x_0}, \qquad n = 1; \tag{47.12'}$$

$$\psi_2(x) = \frac{1}{\sqrt{(2^2 \cdot 2x_0\sqrt{\pi})}} e^{-x^2/2x_0^2} \left(4\frac{x^2}{x_0^2} - 2\right), \qquad n = 2. \tag{47.12''}$$

The first of these functions does not vanish except at $x = \pm\infty$. The second is zero for $x = 0$. A point where the wave function is zero is called a *node*. The third function is zero for $x = \pm x_0/\sqrt{2}$, and thus has two nodes. It may be noted that the number of nodes of each function is equal to n. This property holds good for any[5] n. Thus the quantum number is equal to the number of nodes of the eigenfunction. These wave functions are shown in Figure 23a. The form of the functions $\psi_n(x)$ is similar

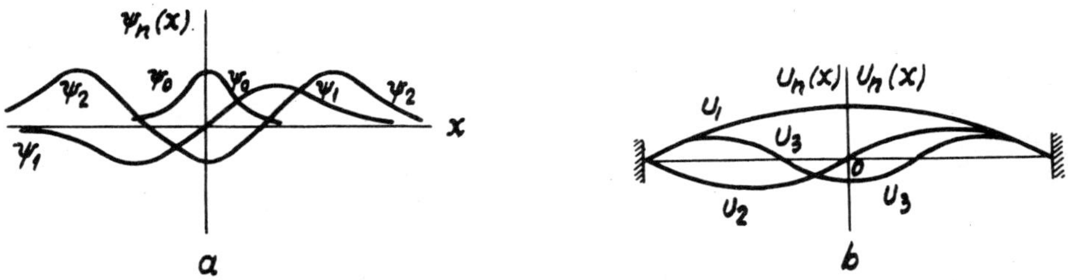

Fig. 23. Wave functions: (a) wave functions of an oscillator for $n = 0$, 1, and 2; (b) vibrations of a string; U_1 fundamental, U_2, U_3 first two harmonics.

to that of the function $U_n(x)$ representing the vibration of a string fixed at both ends. For comparison, Figure 23b shows the function $U_n(x)$ for the fundamental ($n = 0$), the first harmonic ($n = 1$) and the second harmonic ($n = 2$).

This analogy between the vibrations of a string and the wave function of an oscillator is no accident. It is due to two circumstances. Firstly, both systems are one-dimensional; secondly, the vibrations of the string are eigenvibrations. According to a general theorem concerning the nodes of eigenfunctions (see the last footnote), the numbers of nodes of the function $\psi_n(x)$ and of the function $U_n(x)$ must be the same.

In order to afford a still more complete understanding of the quantum states of an oscillator, we show in Figure 24 the potential function of the oscillator:

$$U(x) = \tfrac{1}{2}\mu\omega_0^2 x^2 .$$

[5] The number of the eigenfunction is always equal to the number of nodes. A general proof of this theorem is given in [27], pp. 451–455.

The ordinate is the potential energy and the abscissa is the deviation x. The horizontal lines show the energy levels E_n (47.10) for various values of n. Such diagrams, showing both the energy spectrum and the potential energy, are quite frequently used. They allow a simple comparison with the classical picture of the motion. Let us consider, for example, the level E_1. According to classical mechanics a particle with energy E_1 could be found only in the range AB, since A and B are the points at which the potential energy is equal to the total energy. At these points the kinetic energy T is zero, since

$$E = T + U, \qquad T = E - U. \tag{47.13}$$

The points A and B are called *turning points*. Evidently $OA = OB$ is the amplitude of oscillation of a particle having energy E_1.

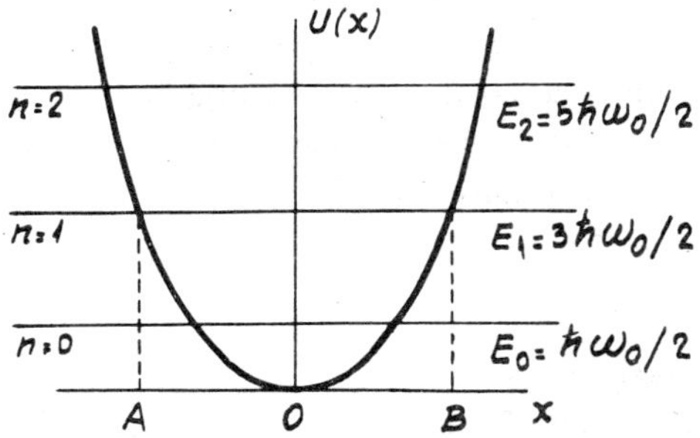

Fig. 24. Diagram of quantum levels (E_n) and potential energy $U(x) = \frac{1}{2}\mu\omega_0{}^2 x^2$ for a harmonic oscillator.

Let us calculate the probability $w(x)\,\mathrm{d}x$ of finding the particle in the range x to $x + \mathrm{d}x$, according to classical mechanics. This probability is proportional to the time $\mathrm{d}t$ during which the particle traverses the distance $\mathrm{d}x$. If the period of oscillation is $T = 2\pi/\omega_0$, we can put

$$w_{cl}(x)\,\mathrm{d}x = \mathrm{d}t/T = (\omega_0/2\pi)\,\mathrm{d}x/v\,, \tag{47.14}$$

where v is the velocity of the particle. Expressing v as a function of x, since

$$x = a \sin \omega_0 t\,, \tag{47.15}$$

where $a = \sqrt{(2E/\mu\omega_0^2)}$ is the amplitude of the oscillations, we have

$$v = a\omega_0 \cos \omega_0 t\,, \tag{47.16}$$

i.e. by (47.15),

$$v = a\omega_0 \sqrt{(1 - x^2/a^2)}\,. \tag{47.17}$$

Hence

$$w_{cl}(x)\,\mathrm{d}x = \frac{1}{2\pi a}\frac{\mathrm{d}x}{\sqrt{(1 - x^2/a^2)}}\,, \qquad -a \leqslant x \leqslant a\,. \tag{47.18}$$

This probability is shown in Figure 25. As we should expect, the greatest probability occurs at the turning points A and B.

The probability of finding the particle in the range x to $x + dx$ according to quantum mechanics is (for $n = 1$)

$$w_{qu}(x)\,dx = \psi_1^2(x)\,dx,$$

with ψ_1 given by (47.12′). Hence

$$w_{qu}(x)\,dx = \frac{2}{\sqrt{\pi}} e^{-x^2/x_0{}^2} \frac{x^2}{x_0^2} \frac{dx}{x_0}. \tag{47.19}$$

This probability also is shown in Figure 25. It is seen that the quantum probability also has maxima near the classical turning points (to be precise, for $E_1 = 3\hbar\omega_0/2$ $OA = OB = \sqrt{(3\hbar/\mu\omega_0)}$ and $OA' = OB' = \sqrt{(\hbar/\mu\omega_0)}$), but unlike the classical case the

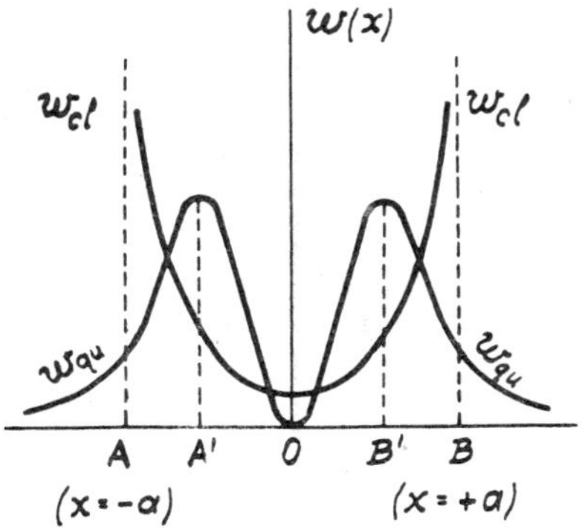

Fig. 25. Comparison of the quantum position probability distribution (w_{qu}) of a particle (for $n = 1$) with the classical distribution (w_{cl}). A, B turning points; A', B' points of maximum w_{qu}.

probability of finding the particle is not zero even beyond the turning points. This result involves no contradiction in quantum mechanics, since Equation (47.13) does not hold good in quantum mechanics: the kinetic energy T and the potential energy U cannot be measured simultaneously.

The difference between the quantum and classical cases is particularly marked if we consider the state of lowest energy. According to the classical theory the least energy of the oscillator is $E = 0$ and corresponds to a particle at rest in the equilibrium position. The probability $w_{cl}(x)$ in this case has the form shown in Figure 26; it is zero everywhere except at the point $x = 0$. According to the quantum theory, the least energy of the oscillator is $E_0 = \frac{1}{2}\hbar\omega_0$; it is called the *zero-point energy*. The probability $w_{qu}(x)$ in this case is

$$w_{qu} = \psi_0^2(x) = \frac{1}{x_0\sqrt{\pi}} e^{-x^2/x_0{}^2}.$$

This also is shown in Figure 26.

Let us examine the properties of the zero-point energy in more detail. It is evident that this energy cannot be removed from the oscillator, since it is essentially the minimum energy which the oscillator can have. It can be removed only by altering the oscillator itself, namely by reducing the frequency ω_0, i.e. by changing the elasticity coefficient. The existence of the zero-point energy is typical of quantum systems, and is a direct consequence of the uncertainty relations

$$\overline{\Delta p^2}\cdot\overline{\Delta x^2} \geqslant \tfrac{1}{4}\hbar^2 . \tag{47.20}$$

The mean values $\bar{p}$ and $\bar{x}$ in a state with a definite value of the energy are zero:

$$\bar{x} = \int \psi_n x \psi_n \,\mathrm{d}x = \int \psi_n^2 x \,\mathrm{d}x = 0 \tag{47.21}$$

(since the integrand is an odd function),

$$\bar{p} = \int \psi_n \cdot P_x \psi_n \cdot \mathrm{d}x = -\,i\hbar \int \psi_n (\partial\psi_n/\partial x)\,\mathrm{d}x = \left[-\tfrac{1}{2} i\hbar\, \psi_n^2(x)\right]_{-\infty}^{\infty} = 0 . \tag{47.22}$$

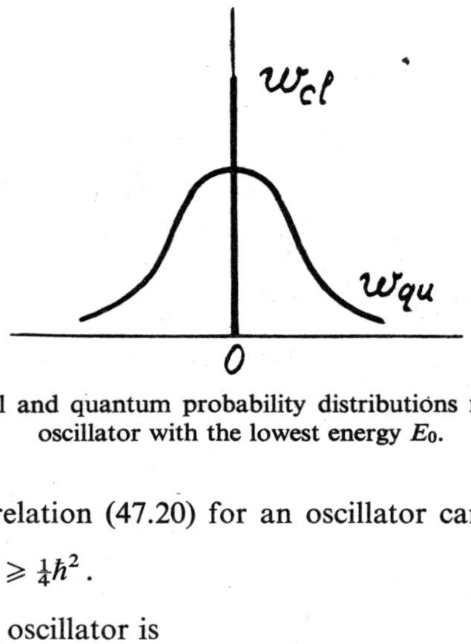

Fig. 26. Classical and quantum probability distributions for the state of an oscillator with the lowest energy E_0.

Hence the uncertainty relation (47.20) for an oscillator can be written

$$\overline{p^2}\cdot\overline{x^2} \geqslant \tfrac{1}{4}\hbar^2 . \tag{47.20'}$$

The mean energy of the oscillator is

$$\bar{E} = \overline{p^2}/2\mu + \tfrac{1}{2}\mu\omega_0^2 \overline{x^2} . \tag{47.23}$$

From a comparison of (47.20′) and (47.23) we see immediately that by decreasing the potential energy we increase the kinetic energy and *vice versa*. In particular, the state with lowest potential energy $\bar{U} = 0$ is the state with infinite kinetic energy $\bar{T} = \infty$. Combination of (47.20′) and (47.23) gives

$$\bar{E} \geqslant \overline{p^2}/2\mu + \mu\omega_0^2 \hbar^2/8\overline{p^2} . \tag{47.24}$$

From this it is easy to find the minimum value of $\bar{E}$: the condition $\partial\bar{E}/\partial\overline{p^2} = 0$ gives

$$\min E \geqslant \tfrac{1}{2}\hbar\omega_0 , \tag{47.25}$$

i.e. the zero-point energy is the lowest energy compatible with the uncertainty relation.

The atoms in a molecule or in a solid afford an example of particles executing small oscillations. The existence of the zero-point energy and zero-point vibrations of the atoms can be shown experimentally by observing the scattering of light by crystals. The scattering of light is due to vibrations of the atoms. As the temperature decreases, the amplitude of oscillations should, according to the classical theory, decrease without limit, and the scattering of light should therefore cease. Experiment [50] shows, however, that the intensity of scattered light tends to a limiting value as the temperature decreases, indicating that the vibrations of the atoms do not cease even at absolute zero. This confirms the existence of zero-point vibrations.

48. An oscillator in the energy representation

Let us take the representation in which the independent variable is the energy E of the oscillator. In this representation the total-energy operator $\boldsymbol{H}$ is a diagonal matrix with elements

$$H_{mn} = E_n \delta_{mn}, \tag{48.1}$$

or, using (47.10),

$$\boldsymbol{H} = \begin{vmatrix} \hbar\omega_0/2 & 0 & 0 & 0 & \cdots \\ 0 & 3\hbar\omega_0/2 & 0 & 0 & \cdots \\ 0 & 0 & 5\hbar\omega_0/2 & 0 & \cdots \\ \cdots & \cdots & \cdots & \cdots & \cdots \end{vmatrix}. \tag{48.2}$$

Any state $\psi(x, t)$ of the oscillator can be represented as a superposition of stationary states (cf. Section 30):

$$\psi(x, t) = \sum c_n(0)\, \psi_n(x)\, e^{-iE_n t/\hbar} = \sum_n c_n(t)\, \psi_n(x), \tag{48.3}$$

where $\psi_n(x)$ is given by (47.11) and E_n by (47.10). The set of all the c_n is the wave function in the E representation. The probability of finding, in the state $\psi(x, t)$, the energy E_n is

$$w(E_n) = |c_n(t)|^2 = |c_n(0)|^2. \tag{48.4}$$

This probability is independent of time, in accordance with the fact that the energy is an integral of the motion.

Let us find the co-ordinate operator $\boldsymbol{X}$ in the E representation. According to the general theory it must be represented by a matrix with elements

$$x_{mn} = \int \psi_m x \psi_n \,\mathrm{d}x. \tag{48.5}$$

Substituting ψ_m and ψ_n from (47.11), we obtain

$$\left.\begin{aligned} x_{mn} &= x_0 \int_{-\infty}^{\infty} e^{-\xi^2} H_m(\xi)\, \xi H_n(\xi)\, \mathrm{d}\xi, \\ x_0 &= \sqrt{(\hbar/\mu\omega_0)}. \end{aligned}\right\} \tag{48.6}$$

This integral can be evaluated:

$$\int_{-\infty}^{\infty} e^{-\xi^2} H_m \xi H_n \,\mathrm{d}\xi = \begin{cases} \sqrt{(\tfrac{1}{2}n)} & \text{for } m = n-1\,, \\ \sqrt{(\tfrac{1}{2}n + \tfrac{1}{2})} & \text{for } m = n+1\,, \\ 0 & \text{otherwise}\,. \end{cases} \tag{48.7}$$

Using this result, we can write (48.6) in terms of the symbol δ_{mn} in the form

$$x_{mn} = x_0 \left[\sqrt{(\tfrac{1}{2}n)}\,\delta_{n-1,m} + \sqrt{(\tfrac{1}{2}n + \tfrac{1}{2})}\,\delta_{n+1,m}\right]. \tag{48.8}$$

The matrix $\boldsymbol{X}$ is seen from (48.8) to have non-zero elements only on the diagonals adjoining the main diagonal:

$$\boldsymbol{X} = x_0 \begin{vmatrix} 0 & \sqrt{(\frac{1}{2})} & 0 & 0 & \cdots \\ \sqrt{(\frac{1}{2})} & 0 & \sqrt{(\frac{2}{2})} & 0 & \cdots \\ 0 & \sqrt{(\frac{2}{2})} & 0 & \sqrt{(\frac{3}{2})} & \cdots \\ \cdots & \cdots & \cdots & \cdots & \cdots \end{vmatrix}. \tag{48.9}$$

In the Heisenberg representation the matrix elements of the operator $\boldsymbol{X}$ are (see (42.12))

$$x_{mn}(t) = x_{mn}\, e^{i\omega_{mn}t}, \tag{48.10}$$

where

$$\omega_{mn} = (E_m - E_n)/\hbar = \omega_0 (m - n). \tag{48.11}$$

Since $x_{mn} \neq 0$ only for $m = n \pm 1$, all the matrix elements of the co-ordinate of the oscillator oscillate with the same frequency, namely the natural frequency ω_0 of the oscillator.

Let us now calculate the mean value $\bar{x}$ of the co-ordinate x of the oscillator for an arbitrary state. The general formula (41.2) gives

$$\overline{x(t)} = \sum_m \sum_n c_m^*(t)\, x_{mn}\, c_n(t) = \sum_m \sum_n c_m^*(0)\, x_{mn}(t)\, c_n(0). \tag{48.12}$$

From the above discussion of the matrix elements $x_{mn}(t)$, the mean value $\bar{x}$ is a harmonic function of time with frequency ω_0, i.e. $\bar{x}$ depends on time in the same way as the co-ordinate of a classical oscillator [6]:

$$\overline{x(t)} = a \cos(\omega_0 t + \phi), \tag{48.13}$$

where a is the amplitude and ϕ the phase.

The matrix of the momentum operator in the E representation can be found either by calculating the integrals

$$p_{mn} = \int \psi_m^* \cdot \boldsymbol{P}_x \psi_n \cdot \mathrm{d}x = -\,i\hbar \int \psi_m^* \cdot \mathrm{d}\psi_n/\mathrm{d}x \cdot \mathrm{d}x\,, \tag{48.14}$$

or more simply from the quantum equations of motion, according to which

$$\boldsymbol{P} = \mu\, \mathrm{d}\boldsymbol{X}/\mathrm{d}t\,, \tag{48.15}$$

[6] The same result can be obtained directly from Ehrenfest's theorem. Equation (34.1) for an oscillator becomes $\mu\, \mathrm{d}^2\bar{x}/\mathrm{d}t^2 = -\,\mu\omega_0{}^2\bar{x}$, whence by integration $\bar{x} = a \cos(\omega_0 t + \phi)$.

i.e.

$$p_{mn} = \mu(\mathrm{d}X/\mathrm{d}t)_{mn}. \tag{48.16}$$

Using Formula (42.11), we find

$$p_{mn} = i\omega_{mn}\,\mu x_{mn}, \tag{48.17}$$

or

$$p_{mn} = i\mu\omega_0(m - n)\,x_{mn}. \tag{48.18}$$

The calculation of the integrals (48.14) naturally gives the same result.

49. Motion in the field of a central force

The field of a central force is distinguished by the fact that the potential energy of a particle in such a field depends only on its distance r from some centre (the centre of force). The laws of motion in the field of a central force are the foundation of atomic mechanics: the solution of the general problem of the motion of electrons in an atom is based to a considerable extent on the results for the motion of a single particle in the field of a central force.

Let $U(r)$ be the potential energy of the particle; then we can write the total-energy operator $\boldsymbol{H}$ (33.12) as

$$\boldsymbol{H} = \boldsymbol{T}_r + \frac{\boldsymbol{M}^2}{2\mu r^2} + U(r), \tag{49.1}$$

where $\boldsymbol{M}^2$ is the operator of the squared angular momentum, and $\boldsymbol{T}_r$ the kinetic-energy operator for radial motion.

It follows from the general theory of the integrals of the motion (Section 33) that the integrals of the motion in the field of a central force are the total energy E and the angular momentum (i.e. M^2, M_x, M_y, M_z). Our problem is to find the stationary states of a particle moving in such a field $U(r)$.

Schrödinger's equation for the stationary states in this case is

$$\boldsymbol{T}_r\psi + \frac{\boldsymbol{M}^2}{2\mu r^2}\psi + U(r)\psi = E\psi. \tag{49.2}$$

It is reasonable to look for the wave function ψ as a function of spherical polar co-ordinates r, θ, ϕ. We have to find single-valued, continuous and finite solutions ψ of Equation (49.2) for the whole range of the variables r, θ, ϕ, i.e. $0 \leqslant r \leqslant \infty$, $0 \leqslant \theta \leqslant \pi$, $0 \leqslant \phi \leqslant 2\pi$. Since the operators $\boldsymbol{H}$ and $\boldsymbol{M}^2$ commute, they must have common eigenfunctions, and so we can write a second equation for ψ:

$$\boldsymbol{M}^2\psi = M^2\psi. \tag{49.3}$$

The eigenvalues of $\boldsymbol{M}^2$ are, according to Section 25, equal to $\hbar^2 l(l+1)$, so that instead of $\boldsymbol{M}^2\psi$ in (49.2) we can put $\hbar^2 l(l+1)\psi$. This gives the equation

$$\boldsymbol{T}_r\psi + \frac{\hbar^2 l(l+1)}{2\mu r^2}\psi + U(r)\psi = E\psi. \tag{49.2'}$$

This equation contain explicitly only one variable r. Putting now

$$\psi(r, \theta, \phi) = R(r) \cdot Y_{lm}(\theta, \phi), \tag{49.4}$$

where $Y_{lm}(\theta, \phi)$ is an eigenfunction of the operator M^2, we can satisfy Equations (49.3) and (49.2′) simultaneously if the function $R(r)$ satisfies the equation

$$T_r R + \frac{\hbar^2 l(l+1)}{2\mu r^2} R + U(r) R = ER, \tag{49.5}$$

obtained by dividing (49.2′) by Y_{lm}. We call this Schrödinger's equation for the radial function $R(r)$.

It may be recalled that the functions Y_{lm} are also eigenfunctions of a component of the angular momentum, namely the component M_z with our choice of co-ordinates. Hence the total energy, the square of the angular momentum, and the component of the angular momentum along some arbitrary direction OZ are quantities which can be measured simultaneously in the field of a central force.

The possible values of the energy E are determined from Equation (49.5), and depend on the form of $U(r)$. They may depend on the value of the angular momentum M^2 (through the number l), but they can not depend on the angular-momentum component M_z (nor therefore on the number m), since M_z does not appear in Equation (49.5). This is because the field in question has central symmetry, so that all directions in space are physically equivalent, and so the energy cannot depend on the orientation of the angular momentum in space.

To draw any further conclusions we must specify the form of $U(r)$ more precisely. In all real physical systems the interaction is infinitely small at infinitely large distances. This means that asymptotically (as $r \to \infty$) the potential energy takes a constant value:

$$U(r)_{r \to \infty} = C, \tag{49.6}$$

where C is an arbitrary constant which determines the level of the potential energy.

We shall see that the nature of the solution of Equation (49.5) depends essentially on whether the total energy E is greater or less than the potential energy at infinity C. Since C is an arbitrary constant, we shall, unless otherwise stated, take it as zero and distinguish two cases: $E > 0$ and $E < 0$.

We must also specify the form of $U(r)$ near the centre of force (as $r \to 0$), and we shall assume that $U(r)$ has at the origin a pole of order less than 2:

$$U(r)_{r \to 0} = A/r^{\alpha}, \qquad \alpha < 2. \tag{49.7}$$

These assumptions about the form of $U(r)$ cover a very wide range of problems in atomic mechanics. For example, in the problem of the motion of a valency electron in an atom we are concerned with the motion of the electron in the field of the nucleus, the latter being surrounded by a shell of electrons closer to it.

At short distances the effect of these electrons is unimportant, and the field is mainly the Coulomb field of the nucleus. The potential energy of the electron in the

Coulomb field is of the form A/r, and so is of the type (49.7). For the interaction of two atoms at short distances the greatest interaction is the repulsion of the nuclei according to Coulomb's law, i.e. the potential energy is again of the form A/r. In both examples U has a first-order pole at $r = 0$.

To discuss the solution of Equation (49.5) we write it in the form

$$R(r) = u(r)/r, \tag{49.8}$$

substitute this in (49.5), and note that by (26.7)

$$T_r R = -\frac{\hbar^2}{2\mu}\frac{1}{r^2}\frac{\partial}{\partial r}\left(r^2\frac{\partial R}{\partial r}\right) = -\frac{\hbar^2}{2\mu}\frac{1}{r}\frac{\mathrm{d}^2 u}{\mathrm{d}r^2}; \tag{49.9}$$

then we obtain for u the equation

$$-\frac{\hbar^2}{2\mu}\frac{\mathrm{d}^2 u}{\mathrm{d}r^2} + \frac{\hbar^2\, l(l+1)}{2\mu r^2}u + Uu = Eu. \tag{49.10}$$

Let us first consider the asymptotic solutions of this equation as $r \to \infty$. Neglecting for large r the term in $1/r^2$ and $U(r)$ (since we take C in (49.6) as zero) we get the simple equation

$$-\frac{\hbar^2}{2\mu}\frac{\mathrm{d}^2 u}{\mathrm{d}r^2} = Eu. \tag{49.11}$$

With the notation

$$k^2 = 2\mu E/\hbar^2 \text{ for } E > 0 \quad \text{and} \quad \lambda^2 = -2\mu E/\hbar^2 \text{ for } E < 0 \tag{49.12}$$

we obtain the general solution of (49.11) in the form

$$u = C_1 e^{ikr} + C_2 e^{-ikr} \qquad (E > 0), \tag{49.13}$$

$$u = C_1 e^{-\lambda r} + C_2 e^{\lambda r} \qquad (E < 0), \tag{49.14}$$

where C_1 and C_2 are arbitrary constants. According to (49.8) the asymptotic solution of Equation (49.5) is

$$R = C_1\frac{e^{ikr}}{r} + C_2\frac{e^{-ikr}}{r} \qquad (E > 0), \tag{49.15}$$

$$R = C_1\frac{e^{-\lambda r}}{r} + C_2\frac{e^{\lambda r}}{r} \qquad (E < 0). \tag{49.16}$$

In the former case ($E > 0$) the solution R is finite and continuous for any values of the constants. It is seen to be a superposition of ingoing and outgoing spherical waves. The probability of finding the particle does not become zero in this case even for large r: the probability of finding the particle between r and $r + \mathrm{d}r$ is proportional to $|R|^2$ and to the volume of the spherical shell $4\pi r^2 \mathrm{d}r$:[7]

$$w(r)\,\mathrm{d}r \approx |R|^2 \cdot 4\pi r^2\,\mathrm{d}r = 4\pi\,|C_1 e^{ikr} + C_2 e^{-ikr}|^2\,\mathrm{d}r.$$

[7] A more detailed examination (see, for example, [37; 58]) shows that to neglect the potential energy $U(r)$ in Equation (49.10), as we have done, is legitimate only if $U(r)$ tends to zero more rapidly

Such states correspond to aperiodic orbits in classical mechanics, when the particle moves from infinity towards the centre and then returns to infinity. Since the state under consideration is a stationary state the flux of particles arriving must be equal to the flux of particles leaving. This means that the amplitudes C_1 and C_2 of the incoming and outgoing waves must be equal in modulus. If we put $C_1 = (1/2i)Ae^{i\alpha}$, $C_2 = -(1/2i)Ae^{-i\alpha}$, where A and α are real, the asymptotic solution (49.15) can be put in the form

$$R = A\frac{\sin(kr + \alpha)}{r}, \tag{49.15'}$$

that is, a stationary spherical wave.

The situation is different if $E < 0$. In this case we must put $C_2 = 0$, as otherwise R becomes infinite for $r \to \infty$. The required solution is therefore

$$R = C_1\frac{e^{-\lambda r}}{r}. \tag{49.16'}$$

For these states

$$w(r)\,dr \approx 4\pi\,|C_1|^2\,e^{-2\lambda r}\,dr,$$

and for large r the quantity $w(r) \to 0$, i.e. the particle can be found only near the centre of force. Such states correspond to periodic orbits in classical mechanics, when the particle moves around the centre of force.

Let us now examine the behaviour of the solutions near the centre ($r \to 0$). We seek $u(r)$ as a power series:

$$u(r) = r^{\gamma}(1 + a_1 r + a_2 r^2 + \ldots), \tag{49.17}$$

and substitute this expression for u in Equation (49.10). Then the lowest power of r is $r^{\gamma-2}$ or $r^{\gamma-\alpha}$. We see that, if $\alpha < 2$, the lowest power is $r^{\gamma-2}$. The term in $r^{\gamma-2}$ is the largest as $r \to 0$, and so, ignoring quantities of higher order, we find that the result of substituting (49.17) in (49.10) is

$$[\gamma(\gamma-1) - l(l+1)]\,r^{\gamma-2} + \text{higher-order terms} = 0. \tag{49.18}$$

If this equation is satisfied identically for all (infinitesimal) values of r, we must have

$$\gamma(\gamma-1) = l(l+1), \tag{49.19}$$

whence

$$\gamma = l + 1 \qquad \text{or} \qquad \gamma = -l. \tag{49.20}$$

Consequently, for $r \to 0$ the solution $R = u/r$ is

$$R = C_1' r^l(1 + a_1 r + a_2 r^2 + \ldots) + C_2' r^{-l-1}(1 + a_1' r + a_2' r^2 + \ldots), \tag{49.21}$$

where C_1' and C_2' are arbitrary constants.

---

than $1/r$ as $r \to \infty$. For a Coulomb field $U(r)_{r\to\infty} = B/r$, and the asymptotic solutions (49.15) and (49.16) are somewhat modified, but not sufficiently to affect the validity of our subsequent discussion.

If this function remains finite, $C_2' = 0$. Thus the eigenfunction R for small r has the form

$$R = C_1' r^l (1 + a_1 r + a_2 r^2 + \dots). \tag{49.22}$$

As $r \to \infty$ this particular solution becomes either (49.15) (if $E > 0$) or (49.16) (if $E < 0$). By putting $C_2' = 0$ we choose a particular solution of Equation (49.10). Hence the coefficients C_1 and C_2 in (49.15) or (49.16) will be in a definite ratio; the absolute values of these coefficients are not significant, since the Equation (49.10) is homogeneous. This ratio now depends only on the parameters of Equation (49.10), in particular E. Consequently, when $C_2' = 0$ we have

$$C_2/C_1 = f(E), \tag{49.23}$$

where f is some function of E which depends on the form of Equation (49.10), i.e. on $U(r)$.

If the energy of the particle $E > 0$, the particular solutions in (49.15) are both finite, and so for any ratio C_2/C_1 the solution (49.15) is admissible, and in particular for the value of C_2/C_1 which is obtained from the requirement that $C_2' = 0$. Thus we need place no further restriction on the ratio C_2/C_1.[8] The parameter E can have any value. Hence, if $E > 0$, the energy is not quantised, but can take all values from zero to infinity.

Thus for $E > 0$ we have a continuous spectrum of energy. The situation is different for $E < 0$. From the requirement of finiteness of the function R at the origin ($C_2' = 0$) it does not follow that $C_2 = 0$, so that in general, when R is finite at the origin, the solution will become infinite at infinity. In order to obtain solutions finite at infinity, we must further require that $C_2 = 0$. This places a restriction on the possible values of the energy E, since (49.23) then gives

$$C_2/C_1 = f(E) = 0. \tag{49.24}$$

This is a transcendental equation for E, and its roots

$$E = E_1, E_2, \dots, E_n, \dots \tag{49.25}$$

will be the eigenvalues of the energy operator, since the solution R is finite both at $r = 0$ and at $r = \infty$ only for these values of E. Consequently, for $E < 0$ we obtain a discrete spectrum of possible energy values. In this case the *quantum levels* (49.25) are obtained.

Let us now consider in more detail some of the most typical forms of the potential

[8] The condition $C_2' = 0$ gives the asymptotic expression (49.15′) for R. By putting $C_2' = 0$, we choose ψ without singularity at the origin. Consequently the conservation equation (29.7) for $\psi^*\psi$ is valid (cf. also Appendix IX). For stationary states (29.7) gives $\int J_N \, dS = 0$ for any closed surface. If we take this surface to be a sphere with centre at the origin, $J_N = J_r$, and (29.5) and (49.4) give

$$J_r = \tfrac{1}{2} i\hbar \{\psi \partial\psi^*/\partial r - \psi^* \partial\psi/\partial r\} = \tfrac{1}{2} i\hbar Y_{lm} Y_{lm}^* \{R \partial R^*/\partial r - R^* \partial R/\partial r\}.$$

Substitution in the previous formula, using the results that $dS = r^2 \, d\Omega$, $\int Y_{lm} Y_{lm}^* \, d\Omega = 1$, gives $R \partial R^*/\partial r = R^* \partial R/\partial r$. It is easily seen that this equation cannot hold if $|C_1| \neq |C_2|$.

energy $U(r)$. In every case we shall suppose that the potential energy has either no pole at $r = 0$, or a pole of order less than $1/r^2$. The potential energy at infinity will be taken to be zero. Figure 27 shows the potential energy U as a function of the distance r from the centre for the case of repulsion of a particle. Here the total energy of the particle is negative.[9] For $E > 0$ the energy spectrum is continuous. In the case of repulsive forces, therefore, all values of the energy from 0 to $+\infty$ are possible. This is shown in the diagram by hatching. Figure 28 shows the potential energy for the case of attraction. Here we have to distinguish two possibilities: $E > 0$ and $E < 0$. In the first case the spectrum is continuous (the hatched part of the diagram);

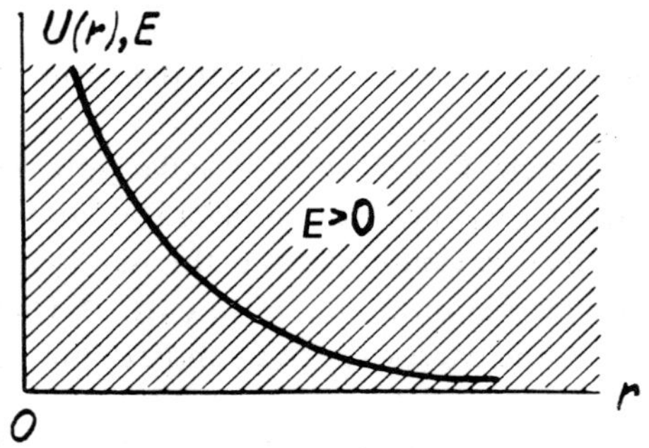

Fig. 27. Potential energy for the case of repulsion from a centre. The energy spectrum ($E > 0$) is continuous.

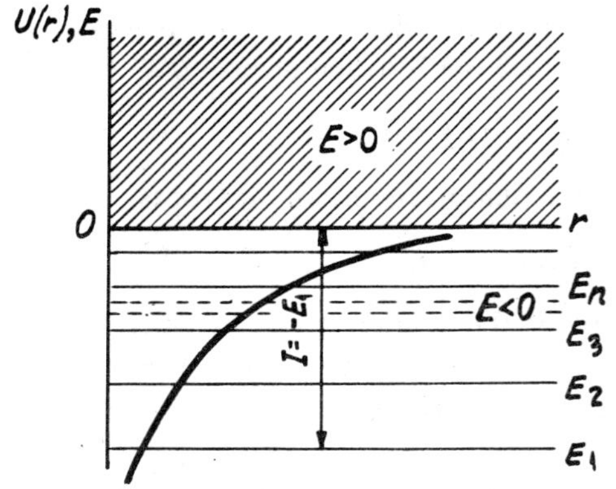

Fig. 28. Potential energy for the case of attraction to a centre. The energy spectrum for $E > 0$ is continuous; for $E < 0$ it consists of separate energy levels $E_1, E_2, \ldots, E_n, \ldots$ I is the ionisation energy.

[9] In classical mechanics this follows from the fact that the kinetic energy $T > 0$, and if $U > 0$ then $E > 0$. In quantum mechanics the situation is exactly the same:

$$E = (1/2\mu) \int \psi^* \cdot \mathbf{P}^2 \psi \cdot dv + \int \psi^* U \psi \, dv .$$

The first term is the kinetic energy and is necessarily positive, since the eigenvalues of the operator $\mathbf{P}^2$ are positive. If $U > 0$, then $E > 0$.

in the second case we have a discrete spectrum of values $E_1, E_2, ..., E_n, ...$. These quantum levels are shown in Figure 28 by horizontal lines. This spectrum consisting of a discontinuous and a continuous part is in fact the energy spectrum of an electron interacting with a nucleus or positive ion (attraction according to Coulomb's law).

The discrete levels correspond, as stated above, to the motion of an electron in an atom (the probability of finding the electron far from the atom is vanishingly small). The continuous spectrum, on the other hand, corresponds to an ionised atom, since the electron in this case may be at any distance from the atom. The energy necessary for ionisation, called the ionisation energy I, can easily be derived from the diagram. The energy of the electron in the normal non-excited state of the atom is E_1. In order that the atom should be ionised it is necessary that the energy of its electron should be greater than zero, and so the minimum work which must be done to ionise the atom in its ground state is

$$I = 0 - E_1 = - E_1 . \tag{49.26}$$

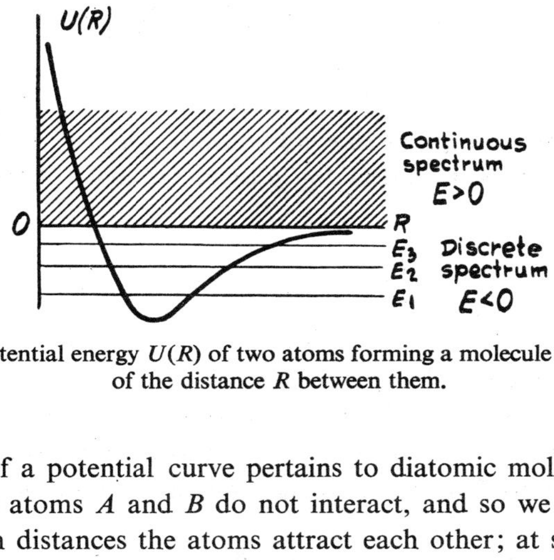

Fig. 29. Potential energy $U(R)$ of two atoms forming a molecule as a function of the distance R between them.

Another example of a potential curve pertains to diatomic molecules AB. At large distances apart the atoms A and B do not interact, and so we can put $U = 0$ for $r = \infty$. At medium distances the atoms attract each other; at small distances they repel because of the repulsion of the nuclei and the electron shells when one atom penetrates into the other. The potential energy therefore has the form shown in Figure 29. For $E > 0$ we again have a continuous spectrum. The probability $w(r)$ remains finite even as $r \to \infty$; the atoms A and B can be at any distance apart (the dissociated molecule). For $E < 0$ we obtain a series of discrete levels $E_1, E_2, ..., E_n, ...$. In this case $w(r) \to 0$ as $r \to \infty$. The atoms are close together and form the molecule AB.

For dissociation of a molecule in the normal (lowest) state, the work of dissociation D is necessary:

$$D = - E_1 . \tag{49.27}$$

It may be noted that according to classical theory this work would be $D' = - U_{min}$, where $U_{\min}$ is the lowest potential energy. D is less than D' by the zero-point energy $\frac{1}{2}\hbar\omega_0$.

These examples show that, if we know the potential energy $U(r)$, we can deduce the nature of the energy spectrum without solving Schrödinger's equation.

50. Motion in a Coulomb field

The simplest problem of atomic mechanics is that of the motion of an electron in the Coulomb field of a nucleus. This occurs in the hydrogen atom H, in the helium ion He^+, in doubly ionised lithium Li^{++} and in similar ions, which are said to be *hydrogen-like*. Denoting the charge on the nucleus by $+Ze$, where e is the unit of charge and Z the atomic number of the nucleus, we find that the potential energy of the electron in the field of such a nucleus is, according to Coulomb's law,

$$U(r) = -Ze^2/r. \tag{50.1}$$

In order to find the quantum levels for this motion of the electron, we have to solve Schrödinger's equation for the radial function R. Putting

$$R = u/r, \tag{50.2}$$

we get Equation (49.10) for u, as shown in Section 49.

Substituting U from (50.1) and taking μ to be the mass of the electron, we have to solve the equation

$$-\frac{\hbar^2}{2\mu}\frac{d^2u}{dr^2} + \frac{\hbar^2}{2\mu}\frac{l(l+1)}{r^2}u - \frac{Ze^2}{r}u = Eu. \tag{50.3}$$

This is a case of attraction (Figure 28), and so according to the general theory of motion in a field of central forces we have a continuous energy spectrum for $E > 0$ and a discrete one for $E < 0$. We shall derive this discrete spectrum and the corresponding eigenfunctions R. For convenience, we replace r and E by the dimensionless quantities

$$\rho = r/a \quad \text{and} \quad \varepsilon = E/E_1, \tag{50.4}$$

where

$$\left.\begin{aligned} a &= \hbar^2/\mu e^2 = 0.529 \times 10^{-8}\,\text{cm}, \\ E_1 &= \mu e^4/2\hbar^2 = e^2/2a = 13.55\,\text{eV}. \end{aligned}\right\} \tag{50.5}$$

Substitution of (50.4) in (50.3) has the result that the atomic constants μ, e, $\hbar$ no longer appear in the equation, which becomes

$$\frac{d^2u}{d\rho^2} + \left[\varepsilon + \frac{2Z}{\rho} - \frac{l(l+1)}{\rho^2}\right]u = 0. \tag{50.6}$$

In accordance with the discussion in Section 49 of the asymptotic behaviour of u, we seek u in the form

$$u(\rho) = e^{-\alpha\rho} f(\rho), \qquad \alpha = \sqrt{-\varepsilon}, \tag{50.7}$$

where $f(\rho)$ is a new function to be found.

Substituting $u(\rho)$ from (50.7) in (50.6), we obtain an equation for $f(\rho)$, after some easy calculations:

$$\frac{d^2 f}{d\rho^2} - 2\alpha \frac{df}{d\rho} + \left[\frac{2Z}{\rho} - \frac{l(l+1)}{\rho^2}\right] f = 0. \tag{50.8}$$

We shall seek the solution of this equation in the form of a power series in ρ. We know from the general theory that the solution of Equation (50.3) which is finite at $r = 0$ is such that the power series in r begins with the term in r^{l+1}. It then follows from (50.7) that the solution of (50.8) finite at the origin must begin with ρ^{l+1}. We therefore seek $f(\rho)$ in the form

$$f(\rho) = \rho^{l+1} \sum_{\nu=0}^{\infty} a_\nu \rho^\nu, \tag{50.9}$$

where a_ν are some coefficients as yet unknown.

The series (50.9) must be such that the function $R(r)$, which we can now, from (50.2) and (50.7), write in the form

$$R(\rho) = \frac{e^{-\alpha\rho} f(\rho)}{\rho}, \tag{50.2'}$$

does not become infinite when $\rho \to \infty$. To find the coefficients a_ν, we substitute (50.9) in (50.8) and collect powers of ρ. This gives

$$\begin{aligned} \sum_\nu \{a_{\nu+1} [(\nu + l + 2)(\nu + l + 1) - l(l+1)] + \\ + a_\nu [2Z - 2\alpha(\nu + l + 1)]\} \rho^{\nu+l} = 0. \end{aligned} \tag{50.10}$$

If the series (50.9) is a solution of Equation (50.8), it is necessary that (50.10) should be satisfied identically for all ρ from 0 to ∞. This can happen only if the coefficient of each power of ρ is zero, i.e. if

$$\begin{aligned} a_{\nu+1} [(\nu + l + 2)(\nu + l + 1) - l(l+1)] + \\ + a_\nu [2Z - 2\alpha(\nu + l + 1)] = 0 \end{aligned} \tag{50.11}$$

for all ν. This gives a recurrence relation between a_ν and $a_{\nu+1}$:

$$a_{\nu+1} = \frac{2\alpha(\nu + l + 1) - 2Z}{[(\nu + l + 2)(\nu + l + 1) - l(l+1)]} a_\nu, \qquad \nu = 0, 1, 2, 3, \ldots . \tag{50.12}$$

The first coefficient a_0 is, of course, arbitrary, since the equation is homogeneous. If a_0 is given some value, (50.12) gives a_1, then a_2, and so on. By calculating all the a_ν we obtain the required solution in the form of a power series in ρ.

It is easy to see that the series obtained will converge for all values of ρ, but for large ρ it increases so rapidly that $R = e^{-\alpha\rho}f/\rho$ tends to infinity with ρ.[10] Thus, as follows also from the general theory of Section 49, the solution finite for $\rho = 0$ will not in general be finite for $\rho = \infty$. However, the solution will obviously be finite for $\rho = \infty$ if the series terminates at some term. Then $f(\rho)$ is a polynomial, and R will tend to zero as $\rho \to \infty$. Such a solution is an eigenfunction of the equation, since it is finite and single-valued throughout the range from $\rho = 0$ to $\rho = \infty$.

It is easy to see that the termination of the series at, for example, the term with $\nu = n_r$ can be effected only for some particular value of the parameter α in the equation. For, suppose that coefficient a_{n_r} is not zero. If the following coefficient a_{n_r+1} vanishes, we must have

$$2\alpha(n_r + l + 1) - 2Z = 0,$$

i.e.

$$\alpha = Z/(n_r + l + 1). \tag{50.13}$$

It is clear that, if this condition holds, not only a_{n_r+1} but all subsequent coefficients are zero, since all are proportional to a_{n_r+1}. Thus (50.13) is a necessary and sufficient condition for the solution $f(\rho)$ to reduce to a polynomial, and the function $R(\rho)$ therefore to remain everywhere finite. Putting

$$n = n_r + l + 1 \tag{50.14}$$

and substituting in (50.13) the value of α from (50.7), we obtain

$$\varepsilon = -Z^2/n^2. \tag{50.13'}$$

Thus, using the expression for E in terms of ε (50.4), we find that finite and single-valued solutions R exist only for the following values of the electron energy:

$$E_n = -\frac{Z^2 e^4 \mu}{2\hbar^2}\cdot\frac{1}{n^2}, \tag{50.15}$$

where the number n, according to (50.14), takes the values

$$n = 1, 2, 3, \ldots, \qquad n_r = 0, 1, 2, 3, \ldots. \tag{50.16}$$

[10] Putting $\lambda = Z/\alpha$, $s = 2l + 1$, we can write (50.12) as

$$a_{\nu+1} = \frac{2\alpha}{\nu + 1}\,\frac{\nu + \frac{1}{2}s + \frac{1}{2} - \lambda}{\nu + s + 1}\,a_\nu .$$

This shows that the ratio $a_{\nu+1}/a_\nu \to 2\alpha/(\nu + 1)$ as $\nu \to \infty$. We can also take some value ν' of ν such that

$$\frac{\nu' + \frac{1}{2}s + \frac{1}{2} - \lambda}{\nu' + s + 1} > \tfrac{1}{2}(1 + \varepsilon),$$

where $\varepsilon > 0$, $\frac{1}{2}(1 + \varepsilon) < 1$. From this value of ν onwards the coefficients a_ν increase more rapidly than those in the series defined by the recurrence formula

$$b_{\nu+1} = \frac{\alpha(1 + \varepsilon)}{\nu + 1}\,b_\nu .$$

This series gives $f_1(\rho) = e^{\alpha(1+\varepsilon)\rho}$, and so, since $f(\rho)$ increases more rapidly than $f_1(\rho)$, the function (50.2′) must tend to infinity with ρ.

The number n is seen to determine the energy of the electron, and is called the *principal quantum number*.

This formula for the quantum levels E_n of an electron moving in a Coulomb field was first derived by Bohr, using semi-classical quantum theory. In that theory, where quantisation was of the nature of an artificial device, it was necessary to specify that the value $n = 0$ was impossible. In quantum mechanics the exclusion of this value follows automatically, since l takes the values 0, 1, 2, ... and n_r is the number of a term in the series (50.9), its lowest value being zero.

Before going on to discuss in detail the quantum levels E_n, let us consider the form of the eigensolutions $R(\rho)$. For the eigensolutions $\alpha = Z/n$, and so Formula (50.12) is simplified to

$$a_{\nu+1} = -\frac{2Z}{n}\frac{n-(l+\nu+1)}{(\nu+1)(2l+\nu+2)}a_\nu. \tag{50.12'}$$

Calculating the coefficients successively and substituting in (50.9), we obtain

$$\begin{aligned} f(\rho) = a_0\rho^{l+1}\Bigg\{1 &- \frac{n-l-1}{1!(2l+2)}\left(\frac{2Z\rho}{n}\right) + \\ &+ \frac{(n-l-1)(n-l-2)}{2!(2l+2)(2l+3)}\left(\frac{2Z\rho}{n}\right)^2 + \dots + \\ &+(-1)^{n_r}\times\frac{(n-l-1)(n-l-2)\dots 1}{(n_r)!(2l+2)(2l+3)\dots(2l+n_r+1)}\left(\frac{2Z\rho}{n}\right)^{n_r}\Bigg\}. \end{aligned} \tag{50.17}$$

Hence we see that it is convenient to use a new variable

$$\xi = 2Z\rho/n = (2Z/na)r. \tag{50.18}$$

Combining all the constant coefficients into one factor N_{nl}, we find from (50.9′) that the function $R_{nl}(\rho)$ belonging to the quantum numbers n and l is

$$R_{nl}(\xi) = N_{nl}e^{-\frac{1}{2}\xi}\xi^l L_{n+l}^{2l+1}(\xi), \tag{50.19}$$

where L_{n+l}^{2l+1} denotes the polynomial in the braces in Formula (50.17). This notation is in accordance with that usual in mathematics, since the polynomial in (50.17) can be expressed in terms of derivatives of the Laguerre polynomials defined by the formula

$$L_k(\xi) = e^\xi\frac{d^k}{d\xi^k}(e^{-\xi}\xi^k). \tag{50.20}$$

Then the polynomial $L_k^s(\xi)$ is defined as

$$L_k^s(\xi) = \frac{d^s}{d\xi^s}L_k(\xi). \tag{50.21}$$

Putting $k = n + l$ and $s = 2l + 1$, we easily see that the polynomial in the braces in Formula (50.17) is obtained.

Formulae (50.20) and (50.21) make possible an easy calculation of the function R_{nl}. The factor N_{nl} in (50.19) will be so chosen that the function R_{nl} is normalised to unity:

$$\int_0^\infty R_{nl}^2 r^2 \, \mathrm{d}r = 1 . \tag{50.22}$$

The complete eigenfunction, according to (49.4), is equal to the product of R_{nl} and the eigenfunction of the angular momentum operator, i.e.

$$\psi_{nlm}(r, \theta, \phi) = R_{nl}(r)\, Y_{lm}(\theta, \phi) . \tag{50.23}$$

The energy E_n is seen from (50.15) to depend only on the principal quantum number n. If this number is given, it follows from (50.14) that the orbital number l can have only the values

$$l = 0, 1, 2, \ldots, n-1 \qquad (n_r = n-1, n-2, \ldots, 0) . \tag{50.24}$$

Also, as we know, for given l the magnetic number m takes the values

$$m = 0, \pm 1, \pm 2, \ldots, \pm l . \tag{50.25}$$

Let us now calculate how many different wave functions belong to the quantum level E_n. For each l we have $2l + 1$ functions with different values of m; and l takes values from 0 to $n - 1$, so that the total number of functions is

$$\sum_{l=0}^{n-1} (2l + 1) = n^2 . \tag{50.26}$$

Thus n^2 different states belong to each quantum level E_n. This is a case of n^2-fold degeneracy.

51. The spectrum and wave functions of the hydrogen atom

Substituting in Formula (50.15) the values of the universal constants e, μ and $\hbar$, we can calculate the quantum levels of an electron moving in the Coulomb field of a nucleus of atomic number Z. Figure 30 shows these levels for the hydrogen atom ($Z = 1$).

The left-hand ordinate scale gives the energy of the levels in electron-volts (the energy being measured not from 0 but from the lowest level E_1). It is seen that, as the principal quantum number n increases, the levels lie closer together, and for $n = \infty$, $E_\infty = 0$; then follows the continuous spectrum $E > 0$, corresponding to the ionised atom. The ionisation energy of the hydrogen atom is

$$I = -E_1 = \mu e^4/2\hbar^2 = 13.55\,\mathrm{eV} . \tag{51.1}$$

In order to see the significance of the right-hand ordinate scale, it should be recalled that the frequency of light ω emitted in a transition from the level E_{nlm} to the level $E_{n'l'm'}$ is, according to the quantum theory of light, given by Bohr's equation[11]

$$\hbar\omega = E_{nlm} - E_{n'l'm'} . \tag{51.2}$$

[11] This will be proved below. For the present we refer to the discussion in Section 2.

Substituting the energy E_{nlm} from (50.15), we obtain

$$\omega = \frac{Z^2 e^4 \mu}{2\hbar^3}\left(\frac{1}{n'^2} - \frac{1}{n^2}\right), \qquad n' < n. \tag{51.3}$$

This formula (for $Z = 1$) gives the frequency of light emitted or absorbed by a hydrogen atom. The quantity $E_{nlm}/\hbar$ is called a *spectral term*. The differences between the terms give the frequencies. For the hydrogen atom the term is

$$E_n/\hbar = (e^4\mu/2\hbar^3)\cdot 1/n^2, \qquad n = 1, 2, 3, \ldots. \tag{51.4}$$

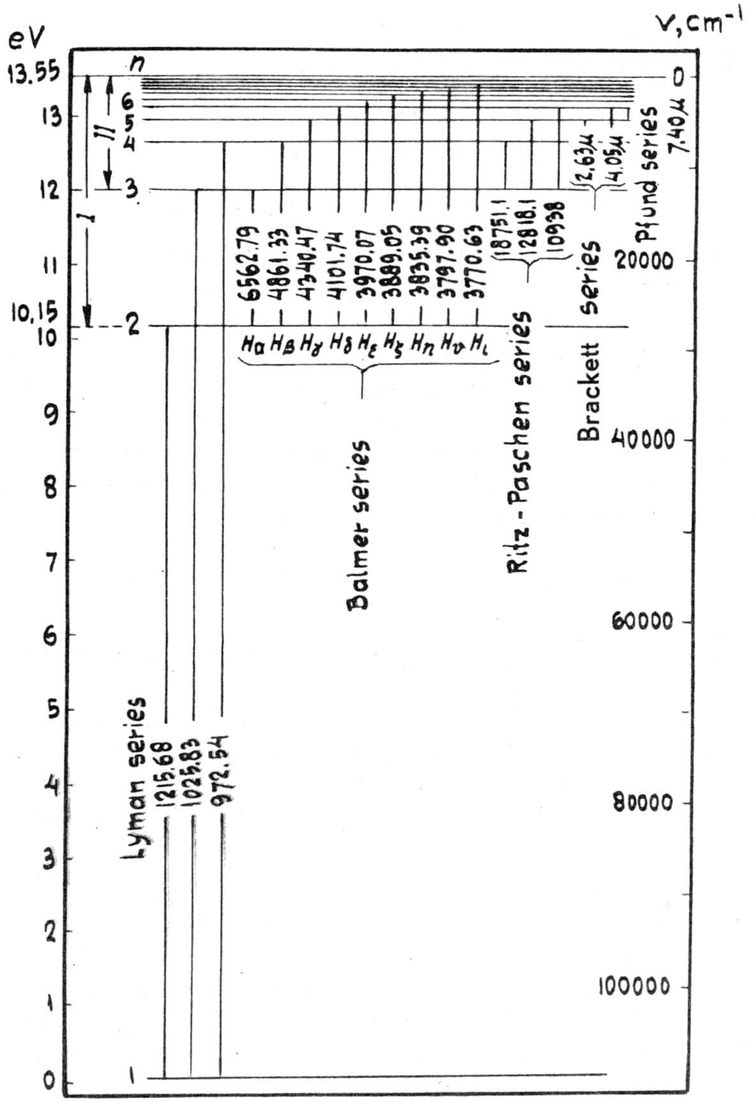

Fig. 30. Diagram of quantum levels of the hydrogen atom.

The quantity

$$R = e^4\mu/4\pi\hbar^3 = 3.27 \times 10^{15}\,\text{sec}^{-1} \tag{51.4'}$$

is called the *Rydberg-Ritz constant* and was first calculated theoretically by Bohr. In spectroscopy the values of the terms are often given not as frequencies $E/\hbar$ but as wave numbers which show how many times the wave length λ divides into 1 cm. If the angular frequency of the light is ω, the ordinary frequency is $\nu = \omega/2\pi$. This frequency also is usually measured as $1/\lambda$, so that the spectroscopic frequency (wave number) is equal to the ordinary frequency ν divided by the velocity of light c:

$$\nu_{\text{spectr}} = 1/\lambda = \nu/c = (\omega/2\pi c)\,\text{cm}^{-1}.$$

The Rydberg-Ritz constant in terms of wave numbers is

$$R = e^4\mu/4\pi\hbar^3 c = 109737.40\,\text{cm}^{-1}. \tag{51.4''}$$

The hydrogen terms in the same units are

$$R/n^2 = 1.09 \times 10^5/n^2, \qquad n = 1, 2, 3, \ldots. \tag{51.4'''}$$

The right-hand ordinate scale on the hydrogen level diagram (Figure 30) gives the values of the spectral terms in reciprocal centimetres. The lines joining the levels have lengths proportional to the quantum energy of the light emitted or absorbed in the transition of an electron between the levels. The numbers beside these lines give the wavelengths λ of the light in Ångströms.

The frequencies belonging to transitions which have the same lower level form what is called a *spectral series*. The most important series of hydrogen are as follows. Transitions to the level $n = 1$ (the lowest level) form the *Lyman series*. The frequencies of this series can be calculated from the formula

$$\nu = R\left(\frac{1}{1^2} - \frac{1}{n^2}\right), \qquad n = 2, 3, \ldots. \tag{51.5}$$

Of these, the longest wavelength, $\lambda = 1215.68$ Å, occurs for the line with $n = 2$. This is in the ultraviolet region of the spectrum.

Transitions to the level $n = 2$ correspond to emission of visible light. These spectral lines form the *Balmer series*. The frequencies of this series are

$$\nu = R\left(\frac{1}{2^2} - \frac{1}{n^2}\right), \qquad n = 3, 4, \ldots. \tag{51.6}$$

Formula (51.6) was found by Balmer in 1885 from an analysis of empirical data on the hydrogen spectrum. Later, this formula played a very important part in the elucidation of spectra, and served as a crucial test of the quantum theory of the atom. The spectral lines of the Balmer series are denoted by the letters $H_\alpha (n = 3)$, $H_\beta (n = 4)$, $H_\gamma (n = 5)$, and so on. In addition to the Balmer and Lyman series, Figure 30 also shows other series corresponding to transitions to levels with $n = 3$, 4, and 5 (the Ritz-Paschen, Brackett and Pfund series respectively). The lines of these series are in the infrared region of the spectrum.

The spectra of the hydrogen-like ions He^+, Li^{++} etc. have the same form as the hydrogen spectrum discussed above, but all the lines are shifted to shorter wavelengths, since the Rydberg constant is multiplied by Z^2. According to (51.3) and (51.4″), the frequencies for these ions are obtained from the formula

$$\nu = Z^2 R\left(\frac{1}{n'^2} - \frac{1}{n^2}\right), \qquad n' < n. \tag{51.7}$$

Let us now examine in more detail the quantum states and the corresponding eigenfunctions $\psi_{nlm}(r, \theta, \varphi)$ (50.23). Any particular state specified by the three quantum numbers n, l, m is an eigenstate of three quantities that can be measured simultaneously: the energy, the square of the angular momentum, and the component of the angular momentum in some direction OZ. All three of these quantities have definite values in the state ψ_{nlm}, namely

$$E_n = -\frac{Z^2 e^4 \mu}{2\hbar^2} \cdot \frac{1}{n^2}, \tag{51.8}$$

$$M_l^2 = \hbar^2\, l(l+1), \qquad l = 0, 1, 2, \ldots, n-1, \tag{51.9}$$

$$M_z = \hbar m, \qquad m = 0, \pm 1, \pm 2, \ldots, \pm l. \tag{51.10}$$

Thus the dynamical significance of the quantum numbers n, l, m is that the principal number n gives the energy E_n, the orbital number l gives the angular momentum M_l^2 and the magnetic number m gives the component M_z of the angular momentum in some arbitrary direction OZ.[12]

The three quantities E, M^2, M_z completely determine the wave function ψ_{nlm}, and therefore form a complete set of quantities. The number of quantities is three, equal to the number of degrees of freedom, as it should be (cf. Section 14).

The squared absolute magnitude of $\psi_{nlm}(r, \theta, \phi)$ (the co-ordinate representation) gives the probability that when the position of an electron in the quantum state n, l, m is determined it will be found in the neighbourhood of the point r, θ, ϕ. More precisely, this probability is given by

$$w_{nlm}(r, \theta, \phi)\, r^2\, \mathrm{d}r \sin\theta\, \mathrm{d}\theta\, \mathrm{d}\phi = |\psi_{nlm}(r, \theta, \phi)|^2\, r^2\, \mathrm{d}r \sin\theta\, \mathrm{d}\theta\, \mathrm{d}\phi. \tag{51.11}$$

In order to exhibit more clearly the nature of this probability, Figure 31 shows a spherical polar co-ordinate system. The polar axis OZ is distinguished by being the direction in which the component $M_z = \hbar m$ of the angular momentum is taken. Denoting by $\mathrm{d}\Omega$ the element of solid angle $\sin\theta\, \mathrm{d}\theta\, \mathrm{d}\phi$ around θ and ϕ and using Formula (50.23) for ψ_{nlm}, we can write the probability (51.11) in the form

$$w_{nlm}(r, \theta, \phi)\, r^2\, \mathrm{d}r\, \mathrm{d}\Omega = R_{nl}^2(r)\, r^2\, \mathrm{d}r\, |Y_{lm}(\theta, \phi)|^2\, \mathrm{d}\Omega. \tag{51.12}$$

12 The number l is called the *orbital quantum number* because in the old Bohr theory it determined the shape of the orbit for a given energy; m is called the *magnetic quantum number* because it plays an important part in magnetic phenomena (Sections 74, 75, 129, 130).

If we integrate (51.12) over all angles Ω, we obtain the probability of finding the electron between two spheres of radii r and $r + \mathrm{d}r$. We denote this probability by

$$w_{nl}(r)\,\mathrm{d}r = R_{nl}^2(r)\,r^2\,\mathrm{d}r\,; \tag{51.13}$$

it is shown in Figure 32 for various states. The numbers on the curves are the values

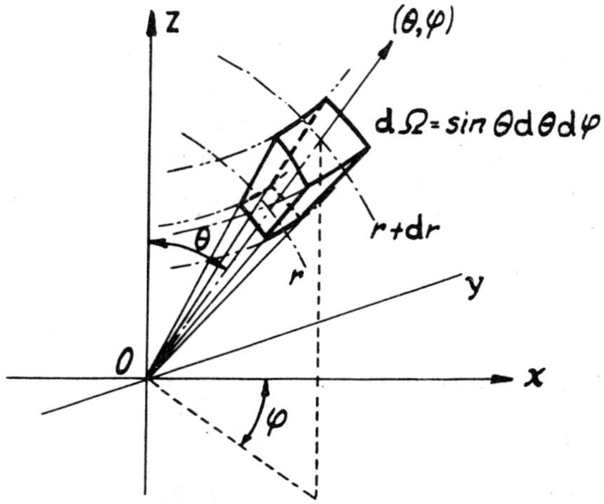

Fig. 31. Spherical polar co-ordinates.

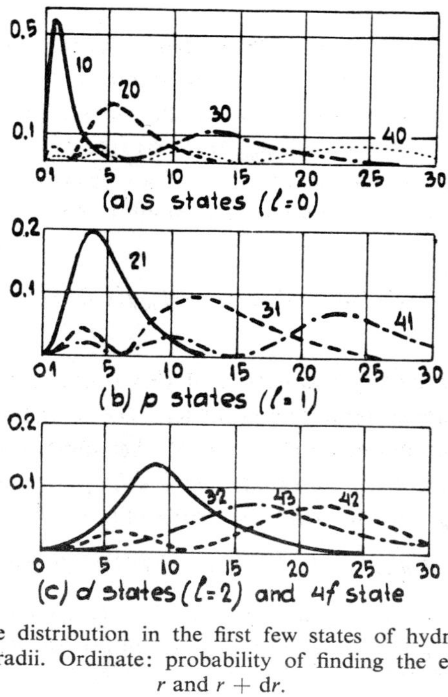

Fig. 32. Charge distribution in the first few states of hydrogen. Abscissa: r in hydrogen radii. Ordinate: probability of finding the electron between r and $r + \mathrm{d}r$.

of n and l ($n_r = n - l - 1$). For example, 31 signifies $n = 3$, $l = 1$ ($n_r = 1$). The abscissa is the distance from the centre, $\rho = r/a$ (50.4). It is easily seen from the graphs that the number n_r (called the *radial quantum number*) is equal to the number of nodes of the wave function R_{nl}. These are not point nodes, but nodal surfaces, since R_{nl} is zero for some $r = r'$, and this represents the surface of a sphere of radius r'. Thus in a state described by the numbers n, l, there are $n_r = n - l - 1$ nodal surfaces having the form of spheres.

Let us now find the significance of the length a defined earlier. Equation (50.19) shows that for large r ($\rho \to \infty$) the radial function has the form

$$R_{nl}(\rho) = N_{nl}\, e^{-Zr/na} (2Zr/na)^{n-1} + \ldots. \tag{51.14}$$

For large r, therefore, the probability $w_{nl}(r)$ is

$$w_{nl}(r) = N_{nl}^2\, e^{-2Zr/na} (2Zr/na)^{2n}. \tag{51.15}$$

Hence it follows that $na/2Z$ is a length representing the dimensions of the atom, since for $r \gg na/2Z$ the probability $w_{nl}(r)$ is practically zero.

A more detailed calculation may be given for the lowest quantum state ($n = 1$). In this case (50.19) gives

$$R_{10}(\rho) = N_{10}\, e^{-Z\rho}. \tag{51.16}$$

Hence

$$w_{10}(r) = N_{10}^2\, a^2\, e^{-2Zr/a} (r/a)^2. \tag{51.17}$$

The maximum value of this probability is obtained when $Z\rho = Zr/a = 1$. Hence, in the state $n = 1$ ($l = m = 0$) the most probable result is to find the electron at

$$r_0 = a/Z = h^2/\mu e^2 Z = 0.529 \times 10^{-8}/Z\,\text{cm}. \tag{51.18}$$

This is just the *radius of the first Bohr orbit*, the value of which was first derived by Bohr in 1913 using the old quantum theory.

Since the lowest orbit is, according to Bohr's theory, a circle, the probability of

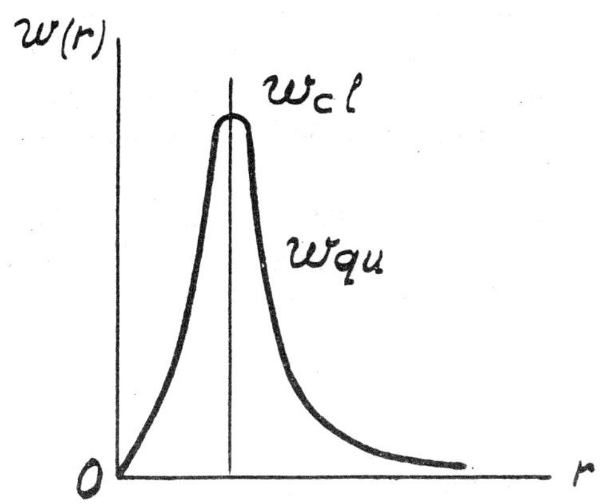

Fig. 33. Comparison of $w_{cl}(r)$ and $w_{qu}(r)$ for state with $n = 1$ ($l = m = 0$).

finding the electron in the state $n = 1$ should be zero except on a sphere of radius $r = r_0$. According to the later quantum mechanics, however, the probability is non-zero in all space. Figure 33 compares the probabilities according to the older theory (w_{cl}) and the later theory (w_{qu}) for the state $n = 1$ of the hydrogen atom. The correlation shown here between w_{cl} and w_{qu} is found for other states also: it is far from complete, as we see from the mere fact that in quantum mechanics the angular momentum in the lowest state is $M_l^2 = 0$ $(l = 0)$, while according to the old theory $M_l^2 = \hbar^2$ in this state. Despite the lack of correspondence, the probability distribution pattern is easier to understand and indicates a relation between quantum and classical mechanics which indeed exists (cf. Chapter VI).

Let us now consider the angular distribution. If (51.11) is integrated over r from 0 to ∞, we obtain the probability $w_{lm}(\theta, \phi)\,\mathrm{d}\Omega$ that the electron is found to lie somewhere in the solid angle $\mathrm{d}\Omega$ (see Figure 31) around the direction (θ, ϕ). Using the normalisation of the functions R_{nl}, we have

$$w_{lm}(\theta, \phi)\,\mathrm{d}\Omega = |Y_{lm}(\theta, \phi)|^2\,\mathrm{d}\Omega . \tag{51.19}$$

The form of the function $Y_{lm}(\theta, \phi)$ shows that the probability is independent of the angle ϕ, and is[13]

$$w_{lm}(\theta)\,\mathrm{d}\Omega = N_{lm}^2\left[P_l^{|m|}(\cos\theta)\right]^2\mathrm{d}\Omega . \tag{51.20}$$

Hence the angular distribution has rotational symmetry about the axis along which the component of angular momentum is determined (the axis OZ in our case).

Figure 34 shows graphs of the probability distribution w_{lm} for various states l, m. A polar system of co-ordinates θ, w_{lm} has been used, with w_{lm} along the radius vector. For comparison the Bohr orbits are shown in the appropriate position. For $l = 0$, $m = 0$ the probability is

$$w_{00}(\theta) = \left[P_0^0\right]^2 = 1/4\pi \tag{51.21}$$

and does not depend on the angle θ, so that we have spherical symmetry. A state in which the angular momentum is zero $(l = 0)$ is called an *s state*, and the corresponding term is called an *s term*; thus an *s* state has spherical symmetry. There is no corresponding Bohr orbit. This was one of the difficulties of Bohr's theory, since states with $l = 1 (m = 0, \pm 1)$ had to be correlated with the optical *s* term, whereas experiment clearly showed that the electron in the *s* term has no orbital angular momentum (or magnetic moment).

A state with $l = 1 (m = 0, \pm 1)$ is called a *p state*, and the corresponding term a *p term*. The probability in this case is given by the functions $P_1^1(\cos\theta)$ and $P_1^0(\cos\theta)$. Substituting the values of these functions from (25.18), (25.19), we have

$$w_{1,\pm 1}(\theta) = (3/8\pi)\sin^2\theta , \tag{51.22}$$

$$w_{10}(\theta) = (3/4\pi)\cos^2\theta . \tag{51.22'}$$

Figure 34 shows the probabilities $w_{1,\pm 1}$, w_{10} and also the corresponding orbits ac-

[13] N_{lm} is the normalisation factor (Appendix V).

cording to Bohr's theory. It is seen that, whereas according to Bohr's theory, in the case $m = \pm 1$, for example, the probability of finding the electron is non-zero only in the orbital plane ($\theta = \frac{1}{2}\pi$), according to the quantum theory it is non-zero for other values of θ also (on cones θ = constant). The point of agreement is that the maximum probability is at $\theta = \frac{1}{2}\pi$. A similar agreement occurs for $m = 0$ (maximum at $\theta = 0$).

A state with $l = 2$ ($m = 0, \pm 1, \pm 2$) is called a *d state*, and the term is a *d term*.

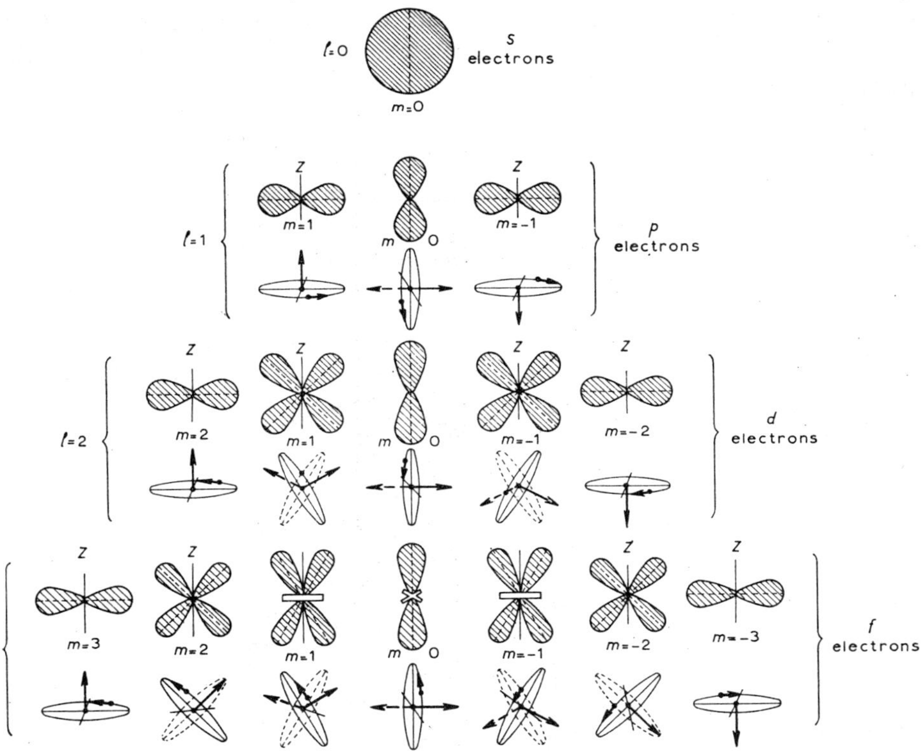

Fig. 34. Angular distribution of electrons $w_{lm}(\theta)$ for s, p, d and f states.

Figure 34 also shows the probability w_{21} for $l = 2$, $m = 1$. From the formulae for the spherical harmonics (25.16) we easily obtain

$$w_{21}(\theta) = N_{21}^2 \left[P_2^1(\cos\theta)\right]^2 = (15/8\pi)\sin^2\theta\cos^2\theta\,. \tag{51.23}$$

For $l = 2$ and $m = 1$ Bohr's theory gives a set of orbits, the normals to which form a cone with axis OZ and semi-vertical angle 60°. This cone also has the maximum probability, according to Bohr's theory. According to quantum mechanics the maximum is at angle of 45°.

The form of the probabilities $w_{lm}(\theta)$ (Figure 34) enables us to obtain an idea of the shape of the atom in various states. This shape is determined by the value of the orbital number l, while the magnetic number m is seen to determine the orientation of the atom in space.

From the above expressions for the probabilities $w_{lm}(\theta)$ we see that the function P_l^m with $l = 0$ has no node, that with $l = 1$ and $m = 0$ has one nodal surface (the plane $\theta = \frac{1}{2}\pi$); that with $l = 2$ and $m = 1$ also has one nodal surface (the plane $\theta = \frac{1}{2}\pi$). In general the equation $P_l^m(\cos\theta) = 0$ has $l - |m|$ real roots $\theta_1, \theta_2, \ldots, \theta_{l-|m|}$. These angles are the semi-vertical angles of the cones θ = constant which form the nodal surfaces. The part of the wave function ψ_{nlm} which depends on the angle ϕ, namely $e^{im\phi}$, has no node, but its real part $\cos m\phi$ and its imaginary part $i \sin m\phi$ each have m nodes $\phi_1, \phi_2, \ldots, \phi_m$, which give nodal planes passing through the polar axis.

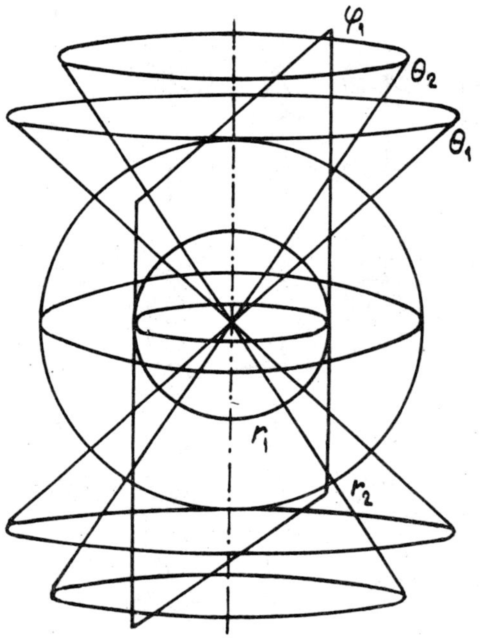

Fig. 35. Nodal surfaces of the real part of the function $\psi_{nlm}(r, \theta, \phi)$: $n_r = n - l - 1$ spheres; $l - |m|$ cones; $|m|$ planes.

Figure 35 shows a family of nodal surfaces of the function ψ_{nlm} consisting of spheres (nodes of the function R_{nl}), cones (nodes of the function P_l^m) and planes (nodes of the function $\cos m\phi$ or $\sin m\,\phi$). The number of spheres is n_r, of cones $l - |m|$ and of planes $|m|$. Altogether there are $n_r + l - |m| + |m| = n_r + l = n - 1$ nodal surfaces. This again illustrates the general theorem mentioned above.

The nodal surfaces shown in Figure 35 have the same geometry as those of a vibrating sphere. The functions $\psi_{nlm}(r, \theta, \phi)$ are therefore similar to the functions which represent the vibrati n of a sphere, just as the eigenfunctions $\psi_n(x)$ of an oscillator are similar to the functions which represent the vibration of a string.

52. Motion of an electron in univalent atoms

There are a number of atoms which have one valency electron: these are the atoms of the alkali metals Li, Na, K, ... We shall call these *hydrogen-like*. In such atoms there is a group of inner electrons, and the outer (valency) electron moves in the field of the nucleus and these inner electrons.

Strictly speaking, this is a multi-electron problem, but in the atoms mentioned there is one feature whereby the problem may be reduced to that of the motion of a single electron in the field of central forces: if the valency electron is removed from such an atom, the remaining electrons form an electron shell of the inert-gas type. For example the ion Li^+ has an electron shell similar to that of the He atom. Both experiment and theory show that the electron shell of an inert gas forms a very stable system which is spherically symmetrical and is not much deformed by external interactions. Thus we can proceed approximately as follows: we assume the outer valency electron does not affect the inner electrons, and so the motion of the outer electron in the field of the nucleus and the inner electrons may be considered.

Owing to the spherically symmetrical distribution of the latter, the field due to them is a central field.[14] Let us find the potential energy $U(r)$ of the outer valency electron in the field of the atomic nucleus and the inner electrons. If $V(r)$ is the potential of this field, then

$$U(r) = -eV(r). \tag{52.1}$$

Next, let $\rho(r)$ be the mean density of electric charge due to the inner electrons.[15] Then the total electron charge $-eN(r)$ within a sphere of radius r is

$$-eN(r) = 4\pi \int_0^r \rho(r) r^2 \, dr. \tag{52.2}$$

Including also the charge on the nucleus $+eZ$, we can write the total charge in this sphere as

$$eZ^*(r) = e[Z - N(r)], \tag{52.3}$$

where eZ^* denotes the effective charge of the nucleus at a distance r. Hence Gauss's theorem shows that the field $\mathscr{E}_r$ is

$$\mathscr{E}_r(r) = eZ^*(r)/r^2, \tag{52.4}$$

and the potential $V(r)$ is

$$V(r) = -e \int_\infty^r \frac{Z^*(r)}{r^2} dr. \tag{52.5}$$

[14] We may again emphasise that this is really only an approximation, since the outer electron will in fact polarise the inner electron shell.

[15] The probability $\rho(r)$ may be calculated by the methods of quantum mechanics. For example, for Li^+ we have two electrons moving in the field of the nucleus. Here the problem is the same as for the He atom, which is discussed in Section 121. $\rho(r)$ can also be measured experimentally (see Section 79).

It follows from (52.3) that the effect of the electron shell reduces to a screening of the field eZ/r^2 of the nucleus, the screening being different at different distances from the nucleus. Near the nucleus the field is not screened. For, as $r \to 0$,

$$\lim_{r \to 0} \frac{eN(r)}{r^2} = -4\pi\rho(0) \lim_{r \to 0} \frac{1}{r^2} \int_0^r r^2 \, dr .$$

In this range, therefore, $\mathscr{E}_r = eZ/r^2$, and the potential is

$$V(r) = eZ/r + \text{constant}. \tag{52.6}$$

In the range $r \gg a$, on the other hand, where a is the radius of the electron shell, $N(r)_{r \gg a} = N$, the total number of electrons in the shell, and $\mathscr{E}_r = e(Z - N)/r^2$; the potential is

$$V(r) = e(Z - N)/r, \tag{52.7}$$

corresponding to the potential of the nucleus with its charge decreased by the charge of the shell electrons.

Often a still rougher approximation is used, in which the dependence of the effective charge $eZ^*(r)$ on r is neglected, and some suitable constant value is taken for

$$Z^* = Z - N(r_0). \tag{52.8}$$

This approximation is very inaccurate, however, and does not give good results.[16] The potential energy $U(r) = -eV(r)$ for the valency electron of a hydrogen-like atom is of the kind discussed in Section 50 (with a pole of order $1/r$). Since $N < Z$, we have a case of attraction. Hence it follows that the energy spectrum of a hydrogen-like atom will consist of a continuous spectrum ($E > 0$) corresponding to the ionised atom and a discrete spectrum ($E < 0$) formed by the quantum levels of the atom.

We shall not discuss the solution of the radial Equation (49.5) for this form of the potential energy. It can be solved only by numerical integration, and we shall merely describe the results.

The most important fact is that in this case the energy E depends not only on the principal quantum number n but also on the radial number n_r. This is easily understood. Equation (49.5) for the functions R, which determines the quantum levels E_n, involves the orbital quantum number l. E will therefore depend, in general, on the number l. The value of E also depends on the number of the eigenfunction of Equation (49.5), i.e. on the radial number n_r. Thus in general the eigenvalues of E depend on two quantum numbers n_r and l, and since $n = n_r + l + 1$ we can say that they depend on n and l. The complete numbering of the levels and eigenfunctions will thus be

$$\left.\begin{aligned} &\psi_{nlm}(r, \theta, \phi) = R_{nl}(r)\, Y_{lm}, \\ &E = E_{nl}, \qquad l = 0, 1, 2, \ldots, n-1, \\ &m = 0, \pm 1, \ldots, \pm l, \qquad n = 1, 2, 3, \ldots, \end{aligned}\right\} \tag{52.9}$$

[16] Of course, the applicability or otherwise of any particular approximation depends also on the degree of accuracy desired.

and not E_n as in the case of a Coulomb field. The fact that in a Coulomb field the energy depends only on n is a special property of this field, the reason for which is explained in [39]. In a Coulomb field the numbers n_r and l appear in the expression for the energy only in the sum $n = n_r + l + 1$.

Thus, in a Coulomb field, as already mentioned, there is *degeneracy* (l degeneracy) consisting in the fact that the energy for a given principal number n is independent of the angular momentum (l). In the general case of a central field $U(r)$ this l degeneracy is removed, and terms with the same principal quantum number n but with different orbital numbers l have different values. Figure 36 shows the levels for the

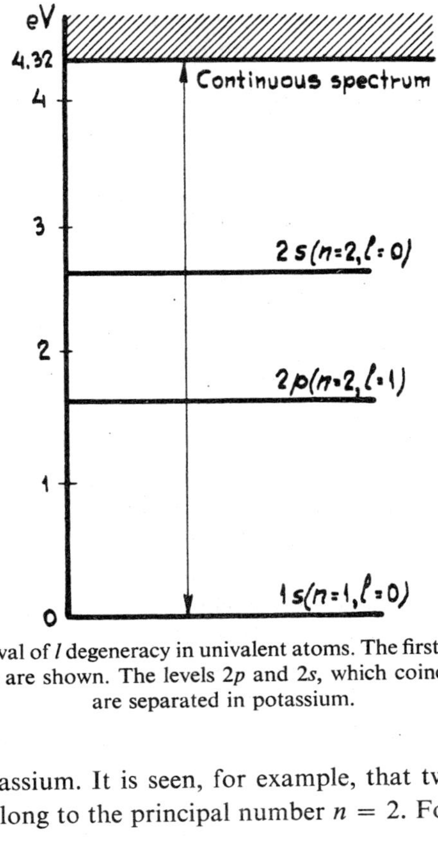

Fig. 36. Removal of l degeneracy in univalent atoms. The first three levels of the potassium atom are shown. The levels $2p$ and $2s$, which coincide in hydrogen, are separated in potassium.

univalent atom of potassium. It is seen, for example, that two levels $l = 0$ (s term) and $l = 1$ (p term) belong to the principal number $n = 2$. For hydrogen these levels coincide.

The magnetic quantum number m, as already explained, determines the orientation of the atom in space, and so the energy of the atom (in the absence of external fields) cannot depend on this number.

53. Currents in atoms. The magneton

Let us calculate the electric current density flowing in an atom if an electron is in a stationary state with a definite value of the angular momentum component $M_z = \hbar m$.

The wave function of such a state is

$$\psi_{nlm}(r, \theta, \phi) = R_{nl}(r)\, P_l^{|m|}(\cos\theta)\, e^{im\phi}\,. \tag{53.1}$$

According to (29.11) the electric current density in the state ψ_{nlm} is given by

$$\mathbf{J} = -\frac{ie\hbar}{2\mu}\{\psi_{nlm}\nabla\psi^*_{nlm} - \psi^*_{nlm}\nabla\psi_{nlm}\}\,; \tag{53.2}$$

here we take the charge on the electron to be $-e$, with $e = 4.778 \times 10^{-10}$ absolute e.s.u. It is convenient to derive the vector $\mathbf{J}$ in spherical polar co-ordinates r, θ, ϕ. To do so, we note that in these co-ordinates the components of the gradient operator ∇ are $\partial/\partial r$, $(1/r)\,\partial/\partial\theta$, $(1/r\sin\theta)\,\partial/\partial\phi$. The components of the vector $\mathbf{J}$ along the radius, the meridian and the circle of latitude are respectively

$$J_r = -\frac{i\hbar e}{2\mu}\left\{\psi_{nlm}\frac{\partial\psi^*_{nlm}}{\partial r} - \psi^*_{nlm}\frac{\partial\psi_{nlm}}{\partial r}\right\} = 0\,, \tag{53.3}$$

$$J_\theta = -\frac{i\hbar e}{2\mu r}\left\{\psi_{nlm}\frac{\partial\psi^*_{nlm}}{\partial\theta} - \psi^*_{nlm}\frac{\partial\psi_{nlm}}{\partial\theta}\right\} = 0\,, \tag{53.4}$$

$$\begin{aligned} J_\phi &= -\frac{i\hbar e}{2\mu r\sin\theta}\left\{\psi_{nlm}\frac{\partial\psi^*_{nlm}}{\partial\phi} - \psi^*_{nlm}\frac{\partial\psi_{nlm}}{\partial\phi}\right\} \\ &= -\frac{e\hbar m}{\mu r\sin\theta}|\psi_{nlm}|^2\,. \end{aligned} \tag{53.5}$$

The first two of these equations are obtained immediately by noting that $P_l^{|m|}$ and R_{nl} are real functions of the variables θ and r, and the last because ψ_{nlm} is proportional to $e^{im\phi}$. Thus in stationary states the radial and meridional current components are zero (as is evident also from geometrical considerations: for instance, if $J_r \neq 0$, the charges will either disperse or accumulate), and the current flows only along circles of latitude (Figure 37). This flow corresponds exactly to the mean current according to classical mechanics for a family of orbits with the same total angular momentum M^2 and the same component M_z of the angular momentum along the axis OZ.

Now, using Formula (53.5) for the current density, it is easy to find the magnetic moment $\mathfrak{M}_z$ of the atom. The current $\mathrm{d}I$ flowing through an area $\mathrm{d}\sigma$ in a meridional plane (Figure 37) is

$$\mathrm{d}I = J_\phi\,\mathrm{d}\sigma\,. \tag{53.6}$$

The magnetic moment due to this current is

$$\mathrm{d}\mathfrak{M}_z = \mathrm{d}I\cdot S/c = J_\phi S\,\mathrm{d}\sigma/c\,, \tag{53.7}$$

where S is the area round which the current $\mathrm{d}I$ flows; $S = \pi r^2\sin^2\theta$ (see Figure 37). Hence

$$\mathrm{d}\mathfrak{M}_z = \frac{\pi r^2\sin^2\theta}{c}J_\phi\,\mathrm{d}\sigma = -\frac{\pi r^2\sin^2\theta}{c}\frac{e\hbar m}{\mu r\sin\theta}|\psi_{nlm}|^2\,\mathrm{d}\sigma\,. \tag{53.8}$$

In order to obtain the total moment $\mathfrak{M}_z$, we must sum over all the current tubes, obtaining

$$\mathfrak{M}_z = -\frac{e\hbar m}{2\mu c}\int 2\pi r \sin\theta \, \mathrm{d}\sigma \, |\psi_{nlm}|^2 . \qquad (53.9)$$

But $2\pi r \sin\theta \, \mathrm{d}\sigma$ is the volume of the tube. Since the quantity $|\psi_{nlm}|^2$ is constant within the tube, the integral in (53.9) is just the integral of $|\psi_{nlm}|^2$ over the whole volume, which by the normalisation condition is equal to unity. Thus the component of the magnetic moment along the axis is

$$\mathfrak{M}_z = -e\hbar m/2\mu c = -\mathfrak{M}_B m , \qquad (53.10)$$

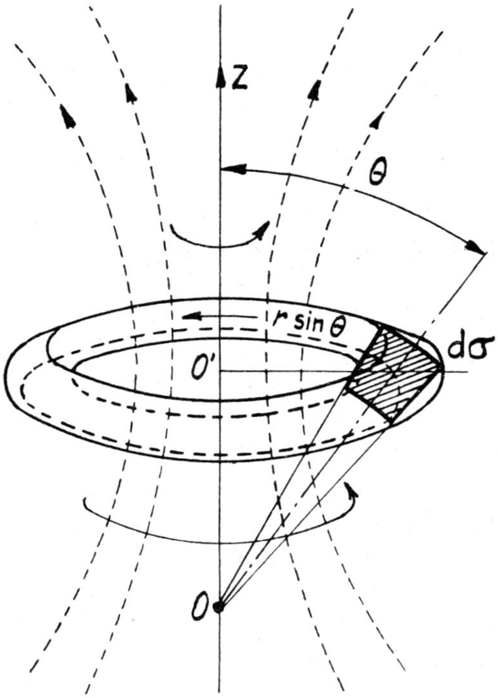

Fig. 37. Currents in an atom for a given angular momentum M^2 and angular momentum component M_z.

where

$$\mathfrak{M}_B = e\hbar/2\mu c = 9 \times 10^{-21} \, \mathrm{erg/G} , \qquad (53.11)$$

i.e. this component has a quantised value equal to an integral number of Bohr magnetons $\mathfrak{M}_B$ (see Section 3). The minus sign is due to the negative charge on the electron.

The above calculation shows that in states with $M_z \neq 0$ an electric current flows in the atom. This current causes a magnetic moment (53.10), so that the whole atom is a magnetic dipole. The ratio of the component $\mathfrak{M}_z$ of the magnetic moment to the component M_z of the angular momentum is

$$\mathfrak{M}_z/M_z = -e/2\mu c \tag{53.12}$$

and is exactly the same as the ratio of these quantities in classical theory for a charge $-e$ of mass μ moving in a closed orbit [4,82]. It may be noted that, since the axis OZ is in no way distinguished, the same relation is obtained for the components of $\mathfrak{M}$ and $\mathbf{M}$ in any direction. Hence (53.12) must be interpreted as meaning that the ratio of the magnetic moment vector $\mathfrak{M}$ to the angular momentum vector $\mathbf{M}$ is $-e/2\mu c$.

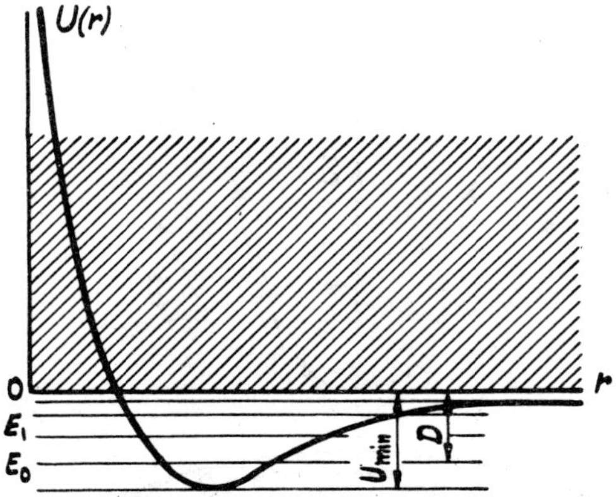

Fig. 38. Potential energy for the atoms of a diatomic molecule, and energy spectrum: continuous for $E > 0$, a series of levels E_0, E_1, ... for $E < 0$.

54. Quantum levels of the diatomic molecule

Let us now consider a molecule consisting of two atoms A and B, of masses m_A and m_B. Let the potential energy as a function of the distance r between the atoms be $U(r)$; it has the form shown in Figure 38. We shall consider only the relative motion of the atoms A and B. It is known from classical mechanics that the relative motion of two particles with energy of interaction $U(r)$ is the same as the motion of a material particle with the reduced mass μ, where

$$1/\mu = 1/m_A + 1/m_B, \tag{54.1}$$

in the field $U(r)$ of a central force, while the common translatory motion is the same as the free motion of a material particle of mass $m_A + m_B$. This situation also occurs in quantum mechanics, as will be shown in Section 104. On this basis we can write the total-energy operator for the relative motion of the atoms A and B as

$$H = T_r + \frac{M^2}{2\mu r^2} + U(r), \tag{54.2}$$

where r is the distance between the atoms, and the angles θ and ϕ (which appear in M^2) define the direction of the line AB.

Schrödinger's equation for stationary states is the same as (49.2). The wave function may again be sought in the form

$$\psi(r, \theta, \phi) = R(r)\, Y_{lm}(\theta, \phi), \qquad R = u/r, \tag{54.3}$$

and for u we obtain the equation

$$-\frac{\hbar^2}{2\mu}\frac{d^2u}{dr^2} + \left[\frac{\hbar^2\, l(l+1)}{2\mu r^2} + U(r)\right]u = Eu. \tag{54.4}$$

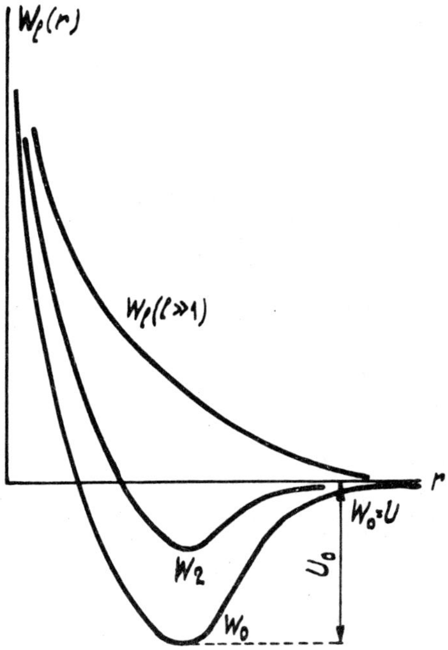

Fig. 39. Relation between vibration and rotation in a diatomic molecule.

The term $\hbar^2 l(l+1)/2\mu r^2$ may be regarded as an additional potential energy, so that the total potential energy for radial motion may be taken as

$$W_l(r) = U(r) + \hbar^2\, l(l+1)/2\mu r^2 \tag{54.5}$$

and Equation (54.4) may be written

$$-\frac{\hbar^2}{2\mu}\frac{d^2u}{dr^2} + W_l(r)\, u = Eu. \tag{54.4'}$$

Figure 39 shows a graph of the function $W_l(r)$ for various l.

In the absence of rotation ($l = 0$), $W_0(r) = U(r)$, and we have the case discussed in Section 49 (Figure 29). If the rotation is not rapid (l small), $W_l(r)$ still does not differ greatly from $U(r)$: the curve of $U(r)$ is only slightly modified. Finally, if l is very large, the curve of $W_l(r)$ takes the form shown in Figure 39 ($l \gg 1$). We know that for $l = 0$ the molecule has a discrete spectrum for $E < 0$ and a continuous spectrum for $E > 0$. In rapid rotation $W_l(r)$ is everywhere positive. It then follows from the theorem proved in Section 49 that $E > 0$, and the spectrum is therefore continuous. The molecule dissociates into the atoms A and B. This dissociation is due to the action of the centrifugal force resulting from the rotation of the molecule.

Let us consider the case where rotation is slow, so that W_l differs only slightly from $U(r)$, at least in the region of the minimum of $U(r)$ ($r = r_l$). We expand $W_l(r)$ in powers of the deviation $r - r_l$ from the equilibrium position. The equilibrium position r_l depends on l and is given by the minimum of $W_l(r)$:

$$\frac{\mathrm{d}W_l}{\mathrm{d}r} = \frac{\mathrm{d}U}{\mathrm{d}r} - \frac{\hbar^2 l(l+1)}{\mu r^3} = 0. \tag{54.6}$$

This gives $r = r_l$. Next,

$$W_l(r) = W_l(r_l) + \tfrac{1}{2}\left(\frac{\mathrm{d}^2 W_l(r)}{\mathrm{d}r^2}\right)_{r_l}(r - r_l)^2, \tag{54.7}$$

with

$$W_l(r_l) = U(r_l) + \hbar^2 l(l+1)/2\mu r_l^2. \tag{54.8}$$

With the notation

$$(\mathrm{d}^2 W_l/\mathrm{d}r^2)_{r=r_l} = \mu\omega_l^2, \qquad \mu r_l^2 = I_l, \qquad x = r - r_l, \tag{54.9}$$

substitution of $W_l(r)$ from (54.7) in (54.4′) gives

$$-\frac{\hbar^2}{2\mu}\frac{\mathrm{d}^2 u}{\mathrm{d}x^2} + \left[U(r_l) + \frac{\hbar^2 l(l+1)}{2I_l} + \tfrac{1}{2}\mu\omega_l^2 x^2\right]u = Eu. \tag{54.4″}$$

Putting

$$E' = E - U(r_l) - \hbar^2 l(l+1)/2I_l, \tag{54.10}$$

this becomes

$$-\frac{\hbar^2}{2\mu}\frac{\mathrm{d}^2 u}{\mathrm{d}x^2} + \tfrac{1}{2}\mu\omega_l^2 x^2 u = E'u. \tag{54.4‴}$$

This is the equation (47.3) for the stationary states of an oscillator having the eigenfrequency ω_l. According to (47.10) the eigenvalues of E' are

$$E'_n = \hbar\omega_l(n + \tfrac{1}{2}), \qquad n = 0, 1, 2, \ldots, \tag{54.11}$$

and according to (47.11) the eigenfunctions are

$$u_n(x) = e^{-\frac{1}{2}\xi^2} H_n(\xi), \qquad \xi = x/x_0. \tag{54.12}$$

The total internal energy of the molecule is (from (54.10))

$$E_{nl} = U(r_l) + \hbar\omega_l(n + \tfrac{1}{2}) + \hbar^2 l(l+1)/2I_l, \tag{54.13}$$

$$n = 0, 1, 2, \ldots, \qquad l = 0, 1, 2, \ldots. \tag{54.13′}$$

The eigenfunctions of the molecule are

$$\psi_{nlm}(r,\theta,\phi) = \frac{1}{r}\, u_n(r)\, Y_{lm}(\theta,\phi)\,. \qquad (54.14)$$

These wave functions describe the rotation and vibrations of the molecule. The energy E_{nl} of the molecule is equal to the sum of the energy of vibrations with frequency ω_l and the energy of rotation of the molecule

$$E_l = \hbar^2\, l(l+1)/2I_l\,. \qquad (54.15)$$

Since $\hbar^2 l(l+1)$ is the squared angular momentum M_l^2, we see that the expression for the energy of rotation of the molecule in quantum mechanics is the same as in classical mechanics, since according to (54.9) I_l is the moment of inertia of the molecule.[17] Formula (54.15) shows that the energy of rotation is quantised, and the distance between adjoining levels l and $l+1$ is

$$\Delta E_l = \hbar^2 (l+1)/I \qquad (54.16)$$

if we neglect the slight dependence of the moment of inertia on l, i.e. the extension of the molecule owing to the centrifugal force.

The solutions derived above are, of course, only approximate. We have neglected anharmonic vibrations of the molecule by omitting higher terms in the expansion of $W_l(r)$ in powers of $r - r_l$. This is permissible if the deviations $r - r_l$ are small in comparison with the distance r_l (or r_0) between the atoms. The theory of the oscillator shows that the mean value $\overline{x^2} = (\hbar/\mu\omega_0)(n+\frac{1}{2})$; to see this, it is sufficient to calculate the matrix element $x^2{}_{mn}$ using the matrix x_{mn} (48.8). Hence

$$\overline{|x|} = \sqrt{\overline{x^2}} = \sqrt{(\hbar/\mu\omega_0)}\sqrt{(n+\tfrac{1}{2})}\,,$$

and the condition for our approximation to be valid may be written in the form

$$\overline{|x|}/r_0 \ll 1 \qquad \text{or} \qquad \sqrt{(\hbar/\mu\omega_0)}\sqrt{(n+\tfrac{1}{2})} \ll r_0\,, \qquad (54.17)$$

i.e. the approximation is best when the masses of the atoms in the molecule are large, the vibration frequency ω_0 is large, and the distance r_0 between the atoms is large. In addition, the level of the vibrations must not be too high (n must be small). For large n and l the vibrations and rotation of the molecule are strongly coupled, and our whole approximation ceases to be valid. For small n and l, on the other hand, we can neglect the dependence of r_l on l and replace I_l and ω_l by the values I_0 and ω_0 for $l = 0$.

The values of I_0 and ω_0 are usually such that the 'quantum' of vibrational energy $\hbar\omega_0$ is much greater than the 'quantum' of rotational energy $\hbar^2/2I$. Thus, for example, for the hydrogen molecule $\hbar\omega_0 = 8.75 \times 10^{-13}$ erg, $\hbar^2/2I = 1.15 \times 10^{-14}$ erg. Accordingly, the energy spectrum of the molecule consists of a series of vibrational levels (various values of the number n) and rotational levels (various l), the latter being very close together. Figure 40 is a diagram of the energy spectrum of the

[17] In classical mechanics the energy of rotation is $M^2/2I$.

molecule. The broken line at the boundary of the continuous spectrum is $E = 0$ and corresponds to the energy of the dissociating molecule. This energy value can be reached for any n when l is sufficiently large.

The dissociation energy D of a molecule in the ground state ($n = l = 0$) is, as shown in Section 49,

$$D = U_0 - \tfrac{1}{2}\hbar\omega_0 . \tag{54.18}$$

The most important field of phenomena in which the quantisation of the motion of the molecule is observed is that of molecular spectra. Let the possible levels of the energy of an electron in the molecule be E_N. Then the total energy of the molecule and its optical electron is

$$E = E_N + \hbar\omega_0(n + \tfrac{1}{2}) + \hbar^2\, l(l+1)/2I + \text{constant}. \tag{54.19}$$

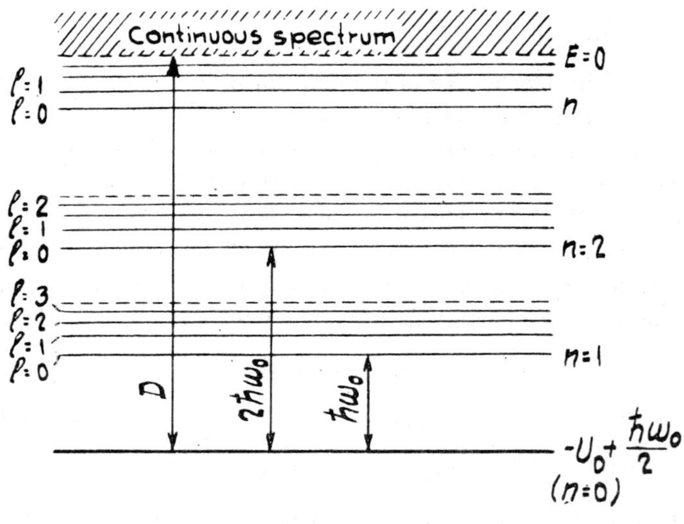

Fig. 40. Diagram of vibrational (n) and rotational (l) levels of a diatomic molecule.

By writing the energy in this form we assume that the coupling between the motion of the electrons and that of the atoms is only slight, so that we can approximately represent the energy as the sum of the energies of the electron and the atoms. This coupling nevertheless exists, and even with weak coupling a change in the state of the electron (from a level E_N to another level $E_{N'}$) will be accompanied by a change in the state of the atoms. Hence, if the molecule absorbs a quantum of light $\hbar\omega$, part of this energy is used for excitation of the electron and the other part for excitation of motion of the atoms in the molecule. Conversely, a quantum of frequency $\hbar\omega$ can be emitted not only from the energy of the electron but also from the energy of motion of the atoms in the molecule. In order to obtain the frequencies ω of the light emitted and absorbed by the molecule, therefore, we must take E in Bohr's frequency rule $\hbar\omega = E' - E$ to be the energy of the entire molecule. Substituting E from (54.19), we find

$$\hbar\omega = E_{N'} - E_N + \hbar\omega_0(n' - n) + \hbar^2[l'(l'+1) - l(l+1)]/2I. \tag{54.20}$$

Denoting the frequency $(E_{N'} - E_N)/\hbar$ due to electron transitions by $\nu^0_{N'N}$, we can write (54.20) as

$$\omega = \nu^0_{N'N} + \omega_0(n' - n) + \frac{\hbar}{2I}[(l' + \tfrac{1}{2})^2 - (l + \tfrac{1}{2})^2]. \tag{54.21}$$

$\nu^0_{N'N}$ is usually much greater than ω_0, and still more so than $\hbar/2I$. Hence, together with the spectral line corresponding to the purely electron transition (frequency $\nu^0_{N'N}$), observation in the spectroscope shows a number of very close lines which almost coalesce.[18] This is called a *band spectrum*. It is typical of diatomic molecules; atoms have a spectrum consisting of lines fairly far apart, although these sometimes split into a small number of neighbouring lines. Lines in bands are due to changes in the rotational motion of molecules. These bands are therefore often called *rotational bands*. In addition to the lines due to a change in rotation (the number l), there will be lines due to a change in the vibrational motion (the number n). These are often called *vibrational lines*.

Thus the complexity of molecular spectra is due to the fact that, in the exchange of energy between the molecule and light, generally the molecule as a whole participates: not only the states of the optical electron but also those of vibration and rotation of the molecule undergo change. The theory of molecular spectra now forms a widely developed but by no means complete branch of atomic mechanics.[19]

The quantum nature of the motion of the molecule is seen not only in molecular spectra but also in the specific heat of diatomic gases. According to classical theory the specific heat per degree of freedom is $\frac{1}{2}k$, where k is Boltzmann's constant, 1.38×10^{-16} erg/deg. The diatomic molecule has altogether six degrees of freedom, and so, according to classical theory, the specific heat should have the constant value $7k/2$.[20] Experiment shows, however, that at medium temperatures the specific heat is indeed constant but equal to $5k/2$, falling to $3k/2$ at low temperatures. These results are fully explained by quantum theory.

If at temperature T the mean energy of translational motion of the molecule $3kT/2 < \hbar\omega_0$, vibrations of the molecule are not excited (more precisely, are rarely excited). The molecule can in this case be regarded as rigid and its number of degrees of freedom as 5 instead of 6. The vibration is said to be 'frozen'. The 'freezing point' T_v is evidently given by the inequality

$$3kT_v/2 \leqslant \hbar\omega_0. \tag{54.22}$$

18 Whether they do or not depends, of course, on the resolving power of the spectroscope.

19 Details are given in [55, 69]; see also Section 125.

20 One degree of freedom is vibrational and, since the kinetic and potential energies are equal, corresponds to $2 \cdot \frac{1}{2}k$, not $\frac{1}{2}k$.

For hydrogen the value of T_v is 4300°. This high value of T_v explains the fact that at ordinary temperatures the specific heat of diatomic gases is $5k/2$.

As the temperature decreases, a point is reached at which the translational energy is less than the 'rotational quantum' $\hbar^2/2I$; the rotation is then no longer excited and plays no part in the heat balance. The rotation is 'frozen' at a temperature T_r given by the inequality

$$3kT_r/2 \leqslant \hbar^2/2I\,. \tag{54.23}$$

For $T \ll T_r$ the specific heat of rotation is zero, and only that of the translational motion, $3k/2$, remains.

Figure 41 shows the specific heat of rotation c_r as a function of temperature. The agreement between quantum theory and experiment is seen to be complete. The broken line shows the specific heat according to classical theory, which is in contradiction with experiment at low temperatures.[21]

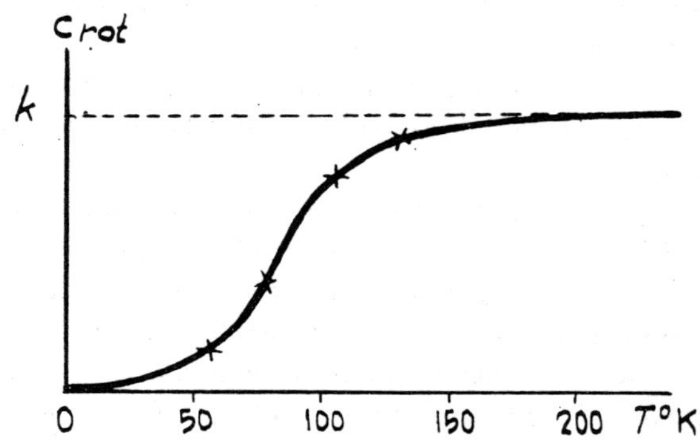

Fig. 41. Specific heat of the hydrogen molecule due to the rotational degrees of freedom.

55. Motion of an electron in a periodic field

One important case is that of the motion of an electron in a periodic potential field $U(x, y, z)$. If the field has periods a, b, c in the directions OX, OY, OZ respectively, the property of periodicity can be expressed by the equations

$$U(x + a, y, z) = U(x, y, z), \tag{55.1}$$

$$U(x, y + b, z) = U(x, y, z), \tag{55.1'}$$

$$U(x, y, z + c) = U(x, y, z). \tag{55.1''}$$

Such a periodic field occurs within ideal crystals, where the ions, and so also the mean electric charge, have a periodic distribution. The electric field potential is, of course, also a periodic function of the co-ordinates x, y, z. If an electron is brought into such a system, it has a periodic potential energy of the form (55.1).

[21] Details concerning the specific heat of diatomic gases are given in [40].

Strictly speaking, this is a multi-electron problem. The replacement of such a problem by the simpler problem of the motion of a single electron in an external field is an approximation, although it is valid when the velocity of the electron considered is large and if inelastic collisions of the electron are unimportant. No justification has so far been given for the use of this approximation to treat the motion of the electrons of the crystal itself, although the theoretical results allow an interpretation of numerous phenomena.

Figure 42 shows the potential energy of an electron in a crystal as a function of x when the axis OX passes through the centres of the atoms forming the crystal. These centres are at the points ..., $-2a$, $-a$, 0, a, $2a$, ..., and at these points U has a pole of the first order $(-e^2Z/r)$.

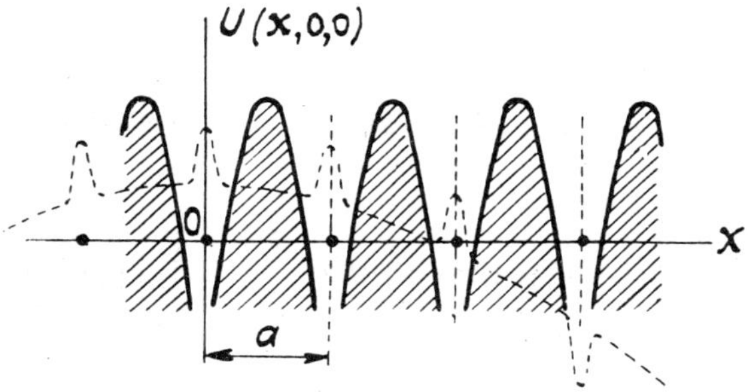

Fig. 42. Curve of potential energy of an electron in a crystal (as a function of one co-ordinate x). The broken line is the wave function (modulated wave).

To ascertain the possible energy levels of an electron in a periodic field and the eigenfunctions of the energy we have to solve Schrödinger's equation, which we take first in the x representation:

$$-(\hbar^2/2\mu)\nabla^2\psi + U\psi = E\psi, \tag{55.2}$$

where μ is the mass of the electron and U the potential energy, which satisfies the periodicity condition (55.1). Our purpose being merely to ascertain the fundamental properties of motion in a periodic field, we shall consider only one dimension. Then (55.1) and (55.2) become

$$U(x+a) = U(x), \tag{55.1'''}$$

$$-\frac{\hbar^2}{2\mu}\frac{d^2\psi}{dx^2} + U(x)\psi = E\psi. \tag{55.2'}$$

To investigate this equation we change to the p representation, putting

$$\psi(x) = \int_{-\infty}^{\infty} c(k)\frac{e^{ikx}}{\sqrt{(2\pi)}}dk, \qquad k = k_x = p_x/\hbar, \tag{55.3}$$

where p_x is the momentum along the axis OX. The potential energy U is correspondingly expanded as a Fourier series:

$$U(x) = \sum_{-\infty}^{\infty} U_n e^{-2\pi i n x/a}, \qquad U_n = U^*_{-n}. \tag{55.4}$$

The coefficients U_n in this series are just $U(x)$ in the p representation. Substitution of (55.3) and (55.4) in (55.2′) gives

$$\frac{\hbar^2}{2\mu}\int_{-\infty}^{\infty} k^2 c(k)\frac{e^{ikx}}{\sqrt{(2\pi)}}\,\mathrm{d}k + \sum_{-\infty}^{\infty} U_n \int_{-\infty}^{\infty} c(k)\frac{e^{i(k-2\pi n/a)x}}{\sqrt{(2\pi)}}\,\mathrm{d}k = E\int_{-\infty}^{\infty} c(k)\frac{e^{ikx}}{\sqrt{(2\pi)}}\,\mathrm{d}k. \tag{55.5}$$

Multiplication by $e^{-ik'x}/\sqrt{(2\pi)}$ and integration with respect to x from $-\infty$ to ∞ gives δ functions:

$$\frac{\hbar^2}{2\mu}\int_{-\infty}^{\infty} k^2 c(k)\,\delta(k-k')\,\mathrm{d}k + \sum_{-\infty}^{\infty} U_n \int_{-\infty}^{\infty} c(k)\,\delta\!\left(k - \frac{2\pi n}{a} - k'\right)\mathrm{d}k = E\int_{-\infty}^{\infty} c(k)\,\delta(k-k')\,\mathrm{d}k. \tag{55.5′}$$

Finally, effecting the integration over k and interchanging k' and k, we have

$$\frac{\hbar^2}{2\mu}k^2 c(k) + \sum_{-\infty}^{\infty} U_n c\left(k + \frac{2\pi n}{a}\right) = E c(k). \tag{55.6}$$

This equation is just Equation (55.2′) in the p representation. Its particular property is that it involves only $c(k)$ whose arguments differ by $2\pi n/a$ ($n = 0, \pm 1, \pm 2, \ldots$).

The quantities $c(k)$, $c(k + 2\pi n/a)$ are unknown quantities which have to be calculated. They are all interrelated by equations of the form (55.6), which are easily obtained by replacing k in (55.6) by $k + 2\pi m/a$, where m is an integer. Taking the term in E to the left-hand side in (55.6), we easily find the following equations for all the functions $c(k + 2\pi m/a)$:

$$\begin{aligned} &\cdots\cdots\cdots\cdots\cdots\cdots\cdots\cdots\cdots\cdots\cdots\cdots \\ m = 1, \quad &\left[\frac{\hbar^2}{2\mu}\left(k + \frac{2\pi}{a}\right)^2 - E\right] c\left(k + \frac{2\pi}{a}\right) + \\ &+ \sum_{-\infty}^{\infty} U_n c\left(k + \frac{2\pi}{a} + \frac{2\pi n}{a}\right) = 0, \end{aligned} \tag{55.7}$$

$$\left.\begin{aligned}
m = 0, \quad & \left[\frac{\hbar^2}{2\mu}k^2 - E\right]c(k) + \sum_{-\infty}^{\infty} U_n c\left(k + \frac{2\pi n}{a}\right) = 0\,, \\
m = -1, \quad & \left[\frac{\hbar^2}{2\mu}\left(k - \frac{2\pi}{a}\right)^2 - E\right]c\left(k - \frac{2\pi}{a}\right) + \\
& + \sum_{-\infty}^{\infty} U_n c\left(k - \frac{2\pi}{a} + \frac{2\pi n}{a}\right) = 0\,, \\
& \ldots\ldots\ldots\ldots\ldots\ldots\ldots\ldots\ldots\ldots\ldots\ldots
\end{aligned}\right\} \tag{55.7}$$

This is a set of linear homogeneous algebraic equations for an infinite number of unknowns $c(k + 2\pi m/a)$, $m = 0, \pm 1, \pm 2, \ldots$. If these equations have non-zero solutions, their determinant Δ must be zero. This determinant depends on E and k (and on all the coefficients U_n) and is a transcendental function of E. The equation

$$\Delta(E, k) = 0 \tag{55.8}$$

therefore has an infinite number of roots $E = E_1, E_2, \ldots, E_j, \ldots$, each of which is a function of the wave number k. Hence it follows that the energy spectrum of a particle moving in a periodic field consists of separate regions

$$E = E_j(k)\,, \qquad j = 1, 2, 3, \ldots\,, \tag{55.9}$$

in each of which the energy is a function of the wave number k. These regions are called *allowed energy zones* or just *zones*.

We shall show that within each zone the energy is a periodic function of the wave number k, with period $2\pi/a$. To prove this, we replace k in the equations (55.7) by $k \pm 2\pi/a$. It is then seen at once from (55.7) that such a replacement merely rearranges the equations, i.e. the set of equations as a whole is unchanged. The roots E_j are therefore also unchanged, so that

$$E_j(k \pm 2\pi/a) = E_j(k)\,. \tag{55.10}$$

Thus the energy is in fact a periodic function of k, and can therefore be expressed by a Fourier series:

$$E_j(k) = \sum_{m=0}^{\infty} E_{jm} \cos mak\,, \tag{55.11}$$

where the coefficients E_{jm} depend only on the form of the potential energy $U(x)$, i.e. on the U_n.[22]

Figure 43 shows typical curves of $E_j(k)$ for the first two zones, E_1 and E_2. In the first zone the energy varies from a minimum value E_1' to a maximum E_1'', and in the second zone from E_2' to E_2''. The range of E from E_1'' to E_2' does not occur and forms a *forbidden* zone. Thus the spectrum consists of sections of the continuous

[22] We have written a cosine series; the general Fourier series contains both cosines and sines, but it is easily seen from (55.7) that replacing k by $-k$ cannot alter the coefficients in the equations, which are again changed into one another. E must therefore be an even function of k.

spectrum (bands) from E_1' to E_1'', from E_2' to E_2'', and so on. The forbidden zones usually become narrower as the number of the zone increases, until finally a continuous spectrum is reached in the limit as $j \to \infty$.

The general form of the eigenfunctions is also easily derived. To each eigenvalue $E = E_j(k)$ there belongs a definite solution of Equations (55.7) and $c_j(k)$ with values of k differing from a fixed value only by an integral multiple of $2\pi/a$. To write the $c_j(k)$ as a single function we can use δ functions:

$$c_{jk}(k') = c_j(k') \sum_{n=-\infty}^{\infty} \delta\left(k + \frac{2\pi n}{a} - k'\right). \tag{55.12}$$

This is the solution belonging to the eigenvalue $E_j(k)$ and taken in the p representation (since $k' = p'/\hbar$). Hence we find ψ in the x representation:

$$\begin{aligned}\psi_{jk}(x) &= \int_{-\infty}^{\infty} c_{jk}(k') \frac{e^{ik'x}}{\sqrt{(2\pi)}} dk' \\ &= \int_{-\infty}^{\infty} \sum_{n=-\infty}^{\infty} c_j(k')\,\delta\left(k + \frac{2\pi n}{a} - k'\right) \frac{e^{ik'x}}{\sqrt{(2\pi)}} dk'.\end{aligned}$$

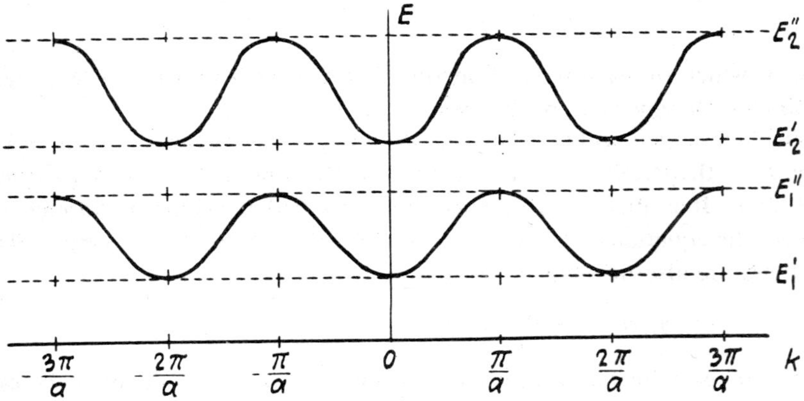

Fig. 43. Energy spectrum and energy as functions of the wave number k for an electron moving in a periodic field.

Effecting the integration with respect to k', we obtain

$$\psi_{jk}(x) = \sum_{n=-\infty}^{\infty} c_j\left(k + \frac{2\pi n}{a}\right) \frac{e^{i(k + 2\pi n/a)x}}{\sqrt{(2\pi)}}. \tag{55.13}$$

Taking e^{ikx} outside the sum, we have

$$\psi_{jk}(x) = e^{ikx} u_{jk}(x), \tag{55.14}$$

where $u_{jk}(x)$ is some periodic function of x with period a:

$$u_{jk}(x + a) = u_{jk}(x). \tag{55.15}$$

$\psi_{jk}(x)$ in Equation (55.14) is the eigenfunction of the energy operator in the x representation which belongs to the eigenvalue $E_j(k)$, i.e. to the jth zone and the wave number k. It is a plane wave (e^{ikx}) modulated with the periodicity of the potential energy. The broken curve in Figure 42 shows the real part of such a function. The points on the axis OX mark the positions of the atomic nuclei (poles of the function $U(x)$). Near these points the function $\psi_{jk}(x)$ is close to those for the isolated atoms.

It follows immediately from the solution (55.13) that states with definite values of the energy ($\overline{\Delta E^2} = 0$) are (as always when a field is present) states with indefinite values of the momentum p. For, in a state with energy $E_j(k)$, the possible values of the momentum p are

$$p = \hbar(k + 2\pi n/a), \qquad n = 0, \pm 1, \pm 2, \ldots, \tag{55.16}$$

with probability

$$w(p_n) = \frac{1}{2\pi\hbar}\left|c\left(k + \frac{2\pi n}{a}\right)\right|^2 \tag{55.17}$$

for $p_n = \hbar(k + 2\pi n/a)$. The mean value $\bar{p}$ of the momentum in the state ψ_{jk} is in general not zero.

We shall now prove a theorem concerning the motion of a wave packet in a periodic field, similar to that for the motion of a wave packet in the absence of a field (Section 7). The time dependence of the functions $\psi_{jk}(x)$, which represent stationary states, is harmonic with frequency $\omega = E_j(k)/\hbar$:

$$\psi_{jk}(x, t) = \psi_{jk}(x)\, e^{-iE_j(k)t/\hbar}. \tag{55.18}$$

From these states we form a wave packet, taking only those functions which belong to a particular zone j, and accordingly drop the suffix j.

The definition of a wave packet gives

$$\psi(x, t) = \int_{k_0 - \Delta k}^{k_0 + \Delta k} c(k)\, e^{i(kx - \omega t)}\, u_k(x)\, dk, \tag{55.19}$$

where Δk is a small range. Putting

$$k = k_0 + \delta, \qquad \omega(k) = \omega(k_0) + (d\omega/dk)_0\, \delta + \ldots$$

and assuming $c(k)$ and $u_k(x)$ to be slowly varying functions of k (in the range $k_0 \pm \Delta k$), we obtain instead of (55.19)

$$\psi(x, t) = c(k_0)\, u_{k_0}(x)\, e^{i(k_0 x - \omega_0 t)} \int_{-\Delta k}^{\Delta k} e^{i[x - (d\omega/dk)_0 t]\delta}\, d\delta. \tag{55.19'}$$

The factors taken outside the integral are rapidly varying functions of x and t. The integral over δ, however, varies only slowly if Δk is small. This integral may therefore be treated, as in Section 7, as the amplitude of the wave packet $\psi(x, t)$.

Repeating precisely the arguments of Section 7, we find that the maximum amplitude (the centroid of the wave packet) moves with the group velocity

$$v = (d\omega/dk)_0 = (1/\hbar)(dE(k)/dk)_0. \tag{55.20}$$

Hence it follows that the mean momentum of such a wave packet is

$$\bar{p} = \mu v = (\mu/\hbar)(\mathrm{d}E/\mathrm{d}k)_0 . \tag{55.21}$$

Using the expression (55.11) for E, we can derive an expression for the mean momentum of a group of states in the jth zone near $k = k_0$:

$$\bar{p} = -\frac{\mu a}{\hbar} \sum_{m=1}^{\infty} E_{jm}\, m \sin mak . \tag{55.22}$$

This shows that at the boundaries of the zone ($k = \pm\, n\pi/a$) the mean momentum $\bar{p} = 0$. It is easy to see from the form of the functions $\psi_{jk}(x)$ (55.13) that in these cases we have modulated stationary waves. For $k \neq n\pi/a$ the mean momentum is in general not zero. Consequently, states with a given energy in a periodic field are states with a mean momentum which is in general not zero.

If we take in the series (55.11) only the first two terms ($m = 0$ and $m = 1$), we obtain

$$E_j(k) = E_{j0} + E_{j1} \cos ka . \tag{55.11'}$$

In the centre of the zone (near $k = 0$, Figure 43) we can expand (55.11′) in powers of k, so that

$$E_j(k) = E_{j0} + E_{j1}(1 - \tfrac{1}{2}k^2a^2 + \ldots). \tag{55.11''}$$

For free motion the energy is

$$E_k = \text{constant} + \hbar^2 k^2/2\mu \tag{55.11'''}$$

(see Section 7). Hence (55.11″) can be written

$$E_j(k) = \text{constant} + \hbar^2 k^2/2\mu^* , \tag{55.23}$$

where μ^* is the *effective mass*, given by

$$\frac{1}{\mu^*} = -\frac{E_{j1}\, a^2}{\hbar^2} = \frac{1}{\hbar^2}\left(\frac{\mathrm{d}^2 E_j(k)}{\mathrm{d}k^2}\right)_{k=0} . \tag{55.24}$$

Accordingly, the momentum is

$$\bar{p} = \mu \hbar\, k/\mu^* , \tag{55.25}$$

which differs from that of a free particle by the coefficient μ/μ^*. The energy at the boundaries of the zone ($k = \pm\, \pi/a$) can be represented similarly. Let us consider, for example, the neighbourhood of the point $k = \pi/a$.

Putting $k = \pi/a - \xi$, we have

$$\cos ak = \cos(\pi - \xi a) = -\cos \xi a ,$$

and in this range

$$E_j(k) = E_{j0} - E_{j1}(1 - \tfrac{1}{2}\xi^2 a^2 + \ldots),$$

$$E_j(k) = \text{constant} + \hbar^2 \xi^2/2\mu^{**} , \qquad \xi = \pi/a - k , \tag{55.23'}$$

where μ^{**} is the effective mass at the boundary of the zone. From (55.24), $\mu^{**} = -\mu^*$, so that the effective masses at the centre and at the boundary of the zone have opposite signs.

The theorems proved in this section are of extreme importance in the modern theory of metals.[23] No details can be given here of this theory, which is now very extensive, and we shall merely make some very brief comments.[24] The theorem on the motion of a wave packet in a periodic field shows that in such a field an electron moves with a constant mean momentum which is in general not zero, a result due to F. Bloch (1927). The ohmic resistance of a metal can therefore be due only to the fact that an actual metal is not a medium with a perfectly periodic field. Deviations from strict periodicity of the field cause scattering of the electron waves $\psi_{jk}(x)$ and lead to a change in the mean momentum $\bar{p}_{jk}$ of the electron, and this gives rise to the ohmic resistance. These deviations from periodicity have two causes: (1) the thermal vibrations of the metal atoms, (2) the presence of foreign inclusions in the crystal, and chance microdeformations. As the temperature of the metal decreases, so does the amplitude of the vibrations of the atoms, and so also the scattering of electron waves, so that the resistance falls. In a well-prepared crystal the second cause may be of little importance, and hence the resistance of the metal will tend to zero (or to a very small value) as the temperature decreases.[25] According to the classical theory the resistance should increase ('freezing of the electron gas').

The quantitative theory of the ohmic resistance of metals based on this qualitative picture gives good agreement with experiment [79].

Another interesting fact is the following. Although Tolman's experiments have definitely shown that the conductivity of metals is due to the motion of electrons, it has been found that in certain metals the sign of the Hall effect is as if the conductivity were due to positively charged particles. This anomaly is entirely explained by quantum mechanics: it can be shown that, if the conductivity of a metal is due to electrons at the boundary of the zone, the result will be as if they were positively charged particles and not electrons.

Let us suppose that an electric field $\mathscr{E}$ acts on an electron at the boundary of the zone. The force on the electron is $e\mathscr{E}$. This force causes a change in the mean momentum which, according to Ehrenfest's theorem, is $\mathrm{d}\mathbf{p}/\mathrm{d}t = e\mathscr{E}$. From (55.21),

$$\frac{\mathrm{d}\bar{p}}{\mathrm{d}t} = \frac{\mathrm{d}}{\mathrm{d}t}\left(\frac{\mu}{h}\frac{\mathrm{d}E}{\mathrm{d}k}\right) = \frac{\mu}{h}\frac{\mathrm{d}^2 E}{\mathrm{d}k^2}\frac{\mathrm{d}k}{\mathrm{d}t}.$$

The work done by the field in 1 sec is

$$\frac{\mathrm{d}E}{\mathrm{d}t} = \frac{\mathrm{d}E}{\mathrm{d}k}\frac{\mathrm{d}k}{\mathrm{d}t} = e\mathscr{E}\cdot v = e\mathscr{E}\cdot\frac{1}{h}\frac{\mathrm{d}E}{\mathrm{d}k}.$$

[23] We should have to generalise these theorems to three dimensions, but this generalisation amounts to a trivial increase in the number of variables (x, y, z instead of x; k_x, k_y, k_z instead of k), all the theorems remaining valid.

[24] A detailed account of the modern theory of metals is given in [79].

[25] This decrease in the resistance of metals should not be confused with the phenomenon of superconductivity, which consists in a sharp discontinuous disappearance of the resistance of certain metals with decreasing temperature.

Hence $dk/dt = e\mathscr{E}/\hbar$. Since, by (55.23′),

$$d^2E/dk^2 = d^2E/d\xi^2 = \hbar^2/\mu^{**},$$

we obtain

$$d\bar{p}/dt = e\mathscr{E}\mu/\mu^{**}. \tag{55.26}$$

In general μ^* is positive, as is seen from the fact that, as the periodic field U decreases to zero, i.e. for the case of free motion, $\mu^* \rightarrow \mu$. From (55.25) it follows that $\mu^{**} = -\mu^* < 0$. Consequently, according to (55.26), an electron at the boundary of the zone moves as if it had a charge $e' = e\mu/\mu^{**}$ which is opposite in sign to e, since $\mu/\mu^{**} < 0$.

CHAPTER IX

MOTION OF A CHARGED MICROPARTICLE IN AN ELECTROMAGNETIC FIELD

56. An arbitrary electromagnetic field

Let us now consider the motion of particles with charge e and mass μ in an arbitrary electromagnetic field. Let the electric field strength be $\mathscr{E}$, and the magnetic field strength be $\mathscr{H}$. These can be expressed in terms of the scalar potential V and the vector potential $\mathbf{A}$:

$$\mathscr{E} = -\frac{1}{c}\frac{\partial \mathbf{A}}{\partial t} - \nabla V, \tag{56.1}$$

$$\mathscr{H} = \mathbf{curl}\,\mathbf{A}. \tag{56.2}$$

The Hamiltonian in this case is (27.9):

$$H = \frac{1}{2\mu}\vec{P}^2 - \frac{e}{\mu c}\mathbf{A}\cdot\vec{P} + \frac{i\hbar e}{2\mu c}\operatorname{div}\mathbf{A} + \frac{e^2}{2\mu c^2}\mathbf{A}^2 + eV + U, \tag{56.3}$$

where U is the force function and appears if other forces act in addition to electromagnetic forces.

We shall not here seek stationary states, since in an arbitrary electromagnetic field they do not always exist; we shall merely derive the equations of motion and from them some general conclusions.

To obtain the equations of motion we can use the general theory of Section 32. According to (32.2) and (32.2′) we have only to calculate the quantum Poisson brackets for the co-ordinates x, y, z and the momenta P_x, P_y, P_z, the operator H being given by (56.3).[1]

Let us first find the velocity operator $\mathrm{d}X/\mathrm{d}t$ (from which $\mathrm{d}Y/\mathrm{d}t$ and $\mathrm{d}Z/\mathrm{d}t$ are written down by analogy). We have

$$\frac{\mathrm{d}X}{\mathrm{d}t} = [H, x] = \frac{1}{2\mu}[\vec{P}^2, x] - \frac{e}{\mu c}[\mathbf{A}\cdot\vec{P}, x]. \tag{56.4}$$

The first bracket has already been calculated in (32.5), and the corresponding term

[1] The following calculation is similar to the classical one given in Appendix VI.

is $\boldsymbol{P}_x/\mu$. In the second term

$$[\mathbf{A}\cdot\vec{\boldsymbol{P}}, x] = [A_x\boldsymbol{P}_x, x] = \frac{1}{i\hbar}(xA_x\boldsymbol{P}_x - A_x\boldsymbol{P}_x x) = \frac{1}{i\hbar}[xA_x\boldsymbol{P}_x - A_x(x\boldsymbol{P}_x - i\hbar)] = A_x. \tag{56.5}$$

Hence

$$\left.\begin{aligned} \mathrm{d}\boldsymbol{X}/\mathrm{d}t &= [\boldsymbol{H}, x] = \frac{1}{\mu}\left(\boldsymbol{P}_x - \frac{e}{c}A_x\right);\\ \mathrm{d}\boldsymbol{Y}/\mathrm{d}t &= [\boldsymbol{H}, y] = \frac{1}{\mu}\left(\boldsymbol{P}_y - \frac{e}{c}A_y\right);\\ \mathrm{d}\boldsymbol{Z}/\mathrm{d}t &= [\boldsymbol{H}, z] = \frac{1}{\mu}\left(\boldsymbol{P}_z - \frac{e}{c}A_z\right). \end{aligned}\right\} \tag{56.6}$$

These operator equations are exactly the same as the second group of the classical Hamilton's equations (see Appendix VI, Formula (10′)), if $\vec{\boldsymbol{P}}$ is regarded as a quantity and not an operator.

The first group of equations is obtained somewhat less directly. We have

$$\mathrm{d}\boldsymbol{P}_x/\mathrm{d}t = [\boldsymbol{H}, \boldsymbol{P}_x] = -\frac{e}{\mu c}[\mathbf{A}\cdot\vec{\boldsymbol{P}}, \boldsymbol{P}_x] + \frac{i\hbar e}{2\mu c}[\operatorname{div}\mathbf{A}, \boldsymbol{P}_x] + \frac{e^2}{2\mu c^2}[\mathbf{A}^2, \boldsymbol{P}_x] + [eV + U, \boldsymbol{P}_x]. \tag{56.7}$$

The terms are, starting with the last,

$$\left.\begin{aligned} [eV + U, \boldsymbol{P}_x] &= -e\,\partial V/\partial x - \partial U/\partial x,\\ \frac{e^2}{2\mu c^2}[\mathbf{A}^2, \boldsymbol{P}_x] &= -\frac{e^2}{2\mu c^2}\frac{\partial \mathbf{A}^2}{\partial x}\\ &= -\frac{e^2}{\mu c^2}\left(A_x\frac{\partial A_x}{\partial x} + A_y\frac{\partial A_y}{\partial x} + A_z\frac{\partial A_z}{\partial x}\right),\\ \frac{i\hbar e}{2\mu c}[\operatorname{div}\mathbf{A}, \boldsymbol{P}_x] &= -\frac{i\hbar e}{2\mu c}\frac{\partial \operatorname{div}\mathbf{A}}{\partial x}\\ &= -\frac{i\hbar e}{2\mu c}\left(\frac{\partial^2 A_x}{\partial x^2} + \frac{\partial^2 A_y}{\partial y\,\partial x} + \frac{\partial^2 A_z}{\partial z\,\partial x}\right),\\ -\frac{e}{\mu c}[\mathbf{A}\cdot\vec{\boldsymbol{P}}, \boldsymbol{P}_x] &= \frac{e}{\mu c}\left(\frac{\partial A_x}{\partial x}\boldsymbol{P}_x + \frac{\partial A_y}{\partial x}\boldsymbol{P}_y + \frac{\partial A_z}{\partial x}\boldsymbol{P}_z\right). \end{aligned}\right\} \tag{56.8}$$

Hence

$$\frac{\mathrm{d}\boldsymbol{P}_x}{\mathrm{d}t} = -\frac{\partial U}{\partial x} - e\frac{\partial V}{\partial x} + \frac{e}{\mu c}\left\{\frac{\partial A_x}{\partial x}\left(\boldsymbol{P}_x - \frac{e}{c}A_x\right) + \frac{\partial A_y}{\partial x}\left(\boldsymbol{P}_y - \frac{e}{c}A_y\right) + \frac{\partial A_z}{\partial x}\left(\boldsymbol{P}_z - \frac{e}{c}A_z\right)\right\} - \frac{i\hbar e}{2\mu c}\frac{\partial \operatorname{div}\mathbf{A}}{\partial x}. \tag{56.9}$$

In order to obtain the derivative not of the generalised momentum but of the ordinary momentum, which according to (56.6) is

$$\mu \frac{\mathrm{d}\boldsymbol{X}}{\mathrm{d}t} = \boldsymbol{P}_x - \frac{e}{c} A_x, \tag{56.10}$$

we must subtract $(e/c)\,\mathrm{d}A_x/\mathrm{d}t$ from (56.9). Now

$$\frac{e}{c}\frac{\mathrm{d}A_x}{\mathrm{d}t} = \frac{e}{c}\frac{\partial A_x}{\partial t} + \frac{e}{c}[\boldsymbol{H}, A_x]. \tag{56.11}$$

Substitution of $\boldsymbol{H}$ from (56.3) gives

$$\frac{e}{c}[\boldsymbol{H}, A_x] = \frac{e}{2\mu c}[\vec{\boldsymbol{P}}^2, A_x] - \frac{e^2}{\mu c^2}[\mathbf{A}\cdot\vec{\boldsymbol{P}}, A_x]. \tag{56.12}$$

The brackets are

$$[\vec{\boldsymbol{P}}^2, A_x] = 2\left(\frac{\partial A_x}{\partial x}\boldsymbol{P}_x + \frac{\partial A_x}{\partial y}\boldsymbol{P}_y + \frac{\partial A_x}{\partial z}\boldsymbol{P}_z\right) - i\hbar\nabla^2 A_x, \tag{56.13}$$

$$[\mathbf{A}\cdot\vec{\boldsymbol{P}}, A_x] = A_x\frac{\partial A_x}{\partial x} + A_y\frac{\partial A_x}{\partial y} + A_z\frac{\partial A_x}{\partial z}. \tag{56.14}$$

Hence we have

$$\frac{e}{c}\frac{\mathrm{d}A_x}{\mathrm{d}t} = \frac{e}{c}\frac{\partial A_x}{\partial t} + \frac{e}{\mu c}\left\{\frac{\partial A_x}{\partial x}\left(\boldsymbol{P}_x - \frac{e}{c}A_x\right) + \right.$$
$$\left. + \frac{\partial A_x}{\partial y}\left(\boldsymbol{P}_y - \frac{e}{c}A_y\right) + \frac{\partial A_x}{\partial z}\left(\boldsymbol{P}_z - \frac{e}{c}A_z\right)\right\} - \frac{i\hbar e}{2\mu c}\nabla^2 A_x. \tag{56.15}$$

Now subtracting $(e/c)\,\mathrm{d}A_x/\mathrm{d}t$ (56.15) from $\mathrm{d}\boldsymbol{P}_x/\mathrm{d}t$ (56.9), we find

$$\frac{\mathrm{d}}{\mathrm{d}t}\left(\boldsymbol{P}_x - \frac{e}{c}A_x\right)$$
$$= -\frac{\partial U}{\partial x} - e\left(\frac{1}{c}\frac{\partial A_x}{\partial t} + \frac{\partial V}{\partial x}\right) + \frac{e}{\mu c}\left(\frac{\partial A_y}{\partial x} - \frac{\partial A_x}{\partial y}\right)\left(\boldsymbol{P}_y - \frac{e}{c}A_y\right) -$$
$$- \frac{e}{\mu c}\left(\frac{\partial A_x}{\partial z} - \frac{\partial A_z}{\partial x}\right)\left(\boldsymbol{P}_z - \frac{e}{c}A_z\right) + \frac{i\hbar e}{2\mu c}\left(\nabla^2 A_x - \frac{\partial\,\mathrm{div}\,\mathbf{A}}{\partial x}\right). \tag{56.16}$$

But

$$-\frac{1}{c}\frac{\partial A_x}{\partial t} - \frac{\partial V}{\partial x} = \mathscr{E}_x, \qquad \frac{\partial A_y}{\partial x} - \frac{\partial A_x}{\partial y} = \mathscr{H}_z, \qquad \frac{\partial A_x}{\partial z} - \frac{\partial A_z}{\partial x} = \mathscr{H}_y,$$

$$\nabla^2 A_x - \frac{\partial\,\mathrm{div}\,\mathbf{A}}{\partial x} = \frac{\partial}{\partial y}\left(\frac{\partial A_x}{\partial y} - \frac{\partial A_y}{\partial x}\right) - \frac{\partial}{\partial z}\left(\frac{\partial A_z}{\partial x} - \frac{\partial A_x}{\partial z}\right) = -\,\mathrm{curl}_x\mathscr{H}.$$

Using also (56.10), we obtain from (56.16)

$$\mu\frac{\mathrm{d}^2\boldsymbol{X}}{\mathrm{d}t^2} = -\frac{\partial U}{\partial x} + e\mathscr{E}_x + \frac{e}{c}\left\{\mathscr{H}_z\frac{\mathrm{d}\boldsymbol{Y}}{\mathrm{d}t} - \mathscr{H}_y\frac{\mathrm{d}\boldsymbol{Z}}{\mathrm{d}t}\right\} - \frac{i\hbar e}{2\mu c}\,\mathrm{curl}_x\mathscr{H}. \tag{56.17}$$

The velocity operators $\mathrm{d}\boldsymbol{Y}/\mathrm{d}t$ and $\mathrm{d}\boldsymbol{Z}/\mathrm{d}t$ do not commute with the field $\mathscr{H}$ (if it is not uniform). It is therefore better to symmetrise (56.17). We have

$$\mathscr{H}_z\frac{\mathrm{d}\boldsymbol{Y}}{\mathrm{d}t}=\frac{1}{\mu}\mathscr{H}_z\left(\boldsymbol{P}_y-\frac{e}{c}A_y\right)=\frac{1}{\mu}\left(\boldsymbol{P}_y-\frac{e}{c}A_y\right)\mathscr{H}_z+\frac{i\hbar}{\mu}\frac{\partial\mathscr{H}_z}{\partial y},$$

$$\mathscr{H}_y\frac{\mathrm{d}\boldsymbol{Z}}{\mathrm{d}t}=\frac{1}{\mu}\mathscr{H}_y\left(\boldsymbol{P}_z-\frac{e}{c}A_z\right)=\frac{1}{\mu}\left(\boldsymbol{P}_z-\frac{e}{c}A_z\right)\mathscr{H}_y+\frac{i\hbar}{\mu}\frac{\partial\mathscr{H}_y}{\partial z}.$$

Thus

$$\mathscr{H}_z\frac{\mathrm{d}\boldsymbol{Y}}{\mathrm{d}t}-\mathscr{H}_y\frac{\mathrm{d}\boldsymbol{Z}}{\mathrm{d}t}=\frac{1}{2}\left\{\mathscr{H}_z\frac{\mathrm{d}\boldsymbol{Y}}{\mathrm{d}t}+\frac{\mathrm{d}\boldsymbol{Y}}{\mathrm{d}t}\mathscr{H}_z-\mathscr{H}_y\frac{\mathrm{d}\boldsymbol{Z}}{\mathrm{d}t}-\frac{\mathrm{d}\boldsymbol{Z}}{\mathrm{d}t}\mathscr{H}_y\right\}+ \\ +\frac{i\hbar}{2\mu}\operatorname{curl}_x\mathscr{H}. \tag{56.18}$$

Substitution of (56.18) in (56.17) gives

$$\mu\frac{\mathrm{d}^2\boldsymbol{X}}{\mathrm{d}t^2}=-\frac{\partial U}{\partial x}+e\mathscr{E}_x+\frac{e}{2c}\left\{\mathscr{H}_z\frac{\mathrm{d}\boldsymbol{Y}}{\mathrm{d}t}+\frac{\mathrm{d}\boldsymbol{Y}}{\mathrm{d}t}\mathscr{H}_z- \\ -\mathscr{H}_y\frac{\mathrm{d}\boldsymbol{Z}}{\mathrm{d}t}-\frac{\mathrm{d}\boldsymbol{Z}}{\mathrm{d}t}\mathscr{H}_y\right\}. \tag{56.19}$$

The expression

$$F_x=e\mathscr{E}_x+\frac{e}{2c}\left\{\left(\mathscr{H}_z\frac{\mathrm{d}\boldsymbol{Y}}{\mathrm{d}t}+\frac{\mathrm{d}\boldsymbol{Y}}{\mathrm{d}t}\mathscr{H}_z\right)-\left(\mathscr{H}_y\frac{\mathrm{d}\boldsymbol{Z}}{\mathrm{d}t}+\frac{\mathrm{d}\boldsymbol{Z}}{\mathrm{d}t}\mathscr{H}_y\right)\right\} \tag{56.20}$$

must be regarded as the operator of the Lorentz force on a particle of charge e in a field $\mathscr{E},\mathscr{H}$, since the classical expression for the Lorentz force is

$$F_x=e\mathscr{E}_x+\frac{e}{c}\left(\mathscr{H}_z\frac{\mathrm{d}y}{\mathrm{d}t}-\mathscr{H}_y\frac{\mathrm{d}z}{\mathrm{d}t}\right).$$

The two remaining equations for the y and z components are evidently obtained by cyclic interchange of x, y and z.

On going from the operator equation (56.19) to the equation for mean values, for which purpose we multiply (56.19) on the left by $\psi^*(x, y, z, t)$ and on the right by $\psi(x, y, z, t)$ and integrate over all space, we derive Ehrenfest's theorem for motion in an electromagnetic field:

$$\mu\overline{\frac{\mathrm{d}^2X}{\mathrm{d}t^2}}=-\overline{\frac{\partial U}{\partial x}}+e\overline{\mathscr{E}}_x+\frac{e}{2c}\left\{\overline{\left(\mathscr{H}_z\frac{\mathrm{d}Y}{\mathrm{d}t}+\frac{\mathrm{d}Y}{\mathrm{d}t}\mathscr{H}_z\right)}- \\ -\overline{\left(\mathscr{H}_y\frac{\mathrm{d}Z}{\mathrm{d}t}+\frac{\mathrm{d}Z}{\mathrm{d}t}\mathscr{H}_y\right)}\right\}. \tag{56.21}$$

This equation is entirely analogous to the classical Newton's equation

$$\mu\frac{\mathrm{d}^2x}{\mathrm{d}t^2}=-\frac{\partial U}{\partial x}+e\mathscr{E}_x+\frac{e}{c}\left\{\mathscr{H}_z\frac{\mathrm{d}y}{\mathrm{d}t}-\mathscr{H}_y\frac{\mathrm{d}z}{\mathrm{d}t}\right\}. \tag{56.21'}$$

Let us now consider the special case of motion in uniform electric and magnetic fields. Here $\mathscr{E}$ and $\mathscr{H}$ are independent of the co-ordinates and therefore commute with the operators $\mathrm{d}\boldsymbol{X}/\mathrm{d}t$, $\mathrm{d}\boldsymbol{Y}/\mathrm{d}t$, $\mathrm{d}\boldsymbol{Z}/\mathrm{d}t$. For uniform fields we therefore have instead of (56.21)

$$\mu\frac{\mathrm{d}^2\bar{x}}{\mathrm{d}t^2} = e\mathscr{E}_x + \frac{e}{c}\left\{\mathscr{H}_z\frac{\mathrm{d}\bar{y}}{\mathrm{d}t} - \mathscr{H}_y\frac{\mathrm{d}\bar{z}}{\mathrm{d}t}\right\}; \tag{56.22}$$

$\bar{x}$, $\bar{y}$, $\bar{z}$ are the co-ordinates of the centroid of the wave packet. Comparison with (56.21′) shows that in a uniform electromagnetic field the centroid of the wave packet moves in accordance with the laws of classical mechanics like a particle of charge e and mass μ.

If there is no magnetic field, (56.22) becomes

$$\mu\,\mathrm{d}^2\bar{x}/\mathrm{d}t^2 = e\mathscr{E}_x, \qquad \bar{x} = (e\mathscr{E}_x/2\mu)\,t^2 + \bar{v}_0 t + \bar{x}_0, \tag{56.23}$$

i.e. we have a uniformly accelerated motion of the centroid of the wave packet. It may be noted that in a uniform electric field there are no stationary solutions; the corresponding wave functions become infinite at $x = \pm\infty$ (depending on the direction of the field $\mathscr{E}_x$). For, according to (56.23) the centroid of the wave packet must be at infinity for $t = \infty$: the field 'blows' the particles in the direction in which the potential energy decreases.

In a magnetic field, stationary solutions exist (see Section 57). They also exist in the simultaneous presence of electric and magnetic fields if these are at right angles.

It follows from (56.1) and (56.2) that, if we replace the potentials $\mathbf{A}$ and V by other potentials $\mathbf{A}'$ and V' such that

$$\mathbf{A}' = \mathbf{A} + \nabla f, \tag{56.24}$$

$$V' = V - \frac{1}{c}\frac{\partial f}{\partial t}, \tag{56.25}$$

where f is any function of the co-ordinates and time, then the potentials $\mathbf{A}'$ and V' represent the same field as $\mathbf{A}$ and V, since

$$\mathscr{E}' = -\frac{1}{c}\frac{\partial \mathbf{A}'}{\partial t} - \nabla V' = \mathscr{E} - \frac{1}{c}\frac{\partial}{\partial t}\nabla f + \frac{1}{c}\nabla\frac{\partial f}{\partial t} = \mathscr{E},$$

$$\mathscr{H}' = \mathbf{curl}\,\mathbf{A}' = \mathscr{H} + \mathbf{curl}\,\nabla f = \mathscr{H}.$$

Thus the quantities $\mathbf{A}$ and V are arbitrary within the limits indicated by (56.24) and (56.25). But these potentials appear in the Hamiltonian $\boldsymbol{H}$, and it might therefore seem that the physical results would be affected by this arbitrariness in the choice of $\mathbf{A}$ and V. This is not so, however. The physical results depend only on the field $\mathscr{E}$, $\mathscr{H}$, and not on the potentials $\mathbf{A}$, V. In particular, the equation of motion (56.21) involves only the fields, and not the potentials. This is an example illustrating the correctness of the above statement.

The reader may verify by direct substitution that if a solution of Schrödinger's equation

$$i\hbar\,\partial\psi/\partial t = \boldsymbol{H}\psi \tag{56.26}$$

has been found, where $\boldsymbol{H}$ is the Hamiltonian (56.3), then a solution ψ' of Schrödinger's equation

$$i\hbar\partial\psi'/\partial t = \boldsymbol{H}'\psi', \tag{56.26′}$$

where $\boldsymbol{H}'$ differs from $\boldsymbol{H}$ in that $\mathbf{A}$ and V are replaced by $\mathbf{A}'$ and V' in accordance with (56.24) and (56.25), is obtained from ψ by means of the formula

$$\psi' = \psi e^{ief/\hbar c}; \tag{56.27}$$

since f is a real function, we have

$$|\psi'|^2 = |\psi|^2, \tag{56.28}$$

$$\mathbf{J}' = \frac{i\hbar}{2\mu}\{\psi'\nabla\psi'^* - \psi'^*\nabla\psi'\} - \frac{e}{\mu c}\mathbf{A}'|\psi'|^2$$

$$= \frac{i\hbar}{2\mu}\{\psi\nabla\psi^* - \psi^*\nabla\psi\} - \frac{e}{\mu c}\mathbf{A}|\psi|^2 = \mathbf{J}, \tag{56.29}$$

since $\nabla\psi' = \nabla\psi \cdot e^{ief/\hbar c} + (ie/\hbar c)\nabla f \cdot \psi'$.

Thus the position probability distribution of the particle and the current density are unchanged when the potentials are altered according to (56.24) and (56.25), which does not affect the electromagnetic field. Similarly, all other physical quantities are likewise unchanged.

This property of Schrödinger's equation is called *electromagnetic* or *gauge invariance*.[2]

57. Motion of a free charged particle in a uniform magnetic field

Let the axis OZ be in the direction of the magnetic field. Then the field components are $\mathscr{H}_x = \mathscr{H}_y = 0$, $\mathscr{H}_z = \mathscr{H}$.

We take the vector potential $\mathbf{A}$ as

$$A_x = -\mathscr{H}y, \qquad A_y = A_z = 0. \tag{57.1}$$

From Equations (57.1) we in fact obtain the desired field:

$$\mathscr{H}_x = 0, \qquad \mathscr{H}_y = 0, \qquad \mathscr{H}_z = -\partial A_x/\partial y = \mathscr{H}, \tag{57.2}$$

and this justifies the above choice of $\mathbf{A}$. We assume other fields absent, and so from (56.3) Schrödinger's equation for stationary states is

$$-\frac{\hbar^2}{2\mu}\nabla^2\psi - \frac{i\hbar e}{\mu c}\mathscr{H}y\frac{\partial\psi}{\partial x} + \frac{e^2}{2\mu c^2}\mathscr{H}^2y^2\psi = E\psi. \tag{57.3}$$

In this equation we can immediately separate the variables by putting

$$\psi(x, y, z) = e^{i(\alpha x + \beta z)}\phi(y), \tag{57.4}$$

where α and β are some constants. Substitution of (57.4) in (57.3) gives the following equation for $\phi(y)$:

$$-\frac{\hbar^2}{2\mu}\frac{d^2\phi}{dy^2} + \frac{e\hbar\alpha}{\mu c}\mathscr{H}y\phi + \frac{e^2\mathscr{H}^2}{2\mu c^2}y^2\phi = \left[E - \frac{\hbar^2\alpha^2}{2\mu} - \frac{\hbar^2\beta^2}{2\mu}\right]\phi. \tag{57.5}$$

This equation is easily reduced to that of an oscillator by putting

$$y = y' - \hbar\alpha c/e\mathscr{H}, \tag{57.6}$$

$$\omega_0 = e\mathscr{H}/\mu c, \tag{57.6'}$$

$$\varepsilon = E - \hbar^2\beta^2/2\mu. \tag{57.6''}$$

[2] The same property holds for the classical Hamilton's equations (see Appendix VI).

Then, after some easy transformations, we obtain

$$-\frac{\hbar^2}{2\mu}\frac{d^2\phi}{dy'^2} + \tfrac{1}{2}\mu\omega_0^2 y'^2\phi = \varepsilon\phi . \tag{57.7}$$

This is the equation of an oscillator of mass μ and frequency ω_0 (see (47.3)).

Hence, using the known solutions for the oscillator, we can write down immediately the solutions required in the present case:

$$\phi_n(y') = e^{-\frac{1}{2}\xi^2} H_n(\xi), \tag{57.8}$$

$$\xi = \sqrt{(\mu\omega_0/\hbar)}\, y' = \sqrt{(\mu\omega_0/\hbar)}(y + \hbar\alpha c/e\mathscr{H}), \tag{57.9}$$

$$\varepsilon = \hbar\omega_0(n + \tfrac{1}{2}), \qquad n = 0, 1, 2, \ldots . \tag{57.10}$$

The eigenfunctions of a particle in the field are thus

$$\psi_{n\alpha\beta}(x, y, z) = e^{i(\alpha x + \beta z)} e^{-\frac{1}{2}\xi^2} H_n(\xi) , \tag{57.11}$$

and the quantum levels are given by the formula

$$E_n(\beta) = \frac{e\hbar\mathscr{H}}{\mu c}(n + \tfrac{1}{2}) + \frac{\hbar^2\beta^2}{2\mu} , \tag{57.12}$$

where $n = 0, 1, 2, \ldots$. The last term is just the kinetic energy of the motion in the direction of OZ (along the field) while the first part,

$$E_n(0) = (e\hbar\mathscr{H}/\mu c)(n + \tfrac{1}{2}), \tag{57.12'}$$

is the energy of the motion in the xy plane perpendicular to the magnetic field. This energy can be written as the potential energy of a current having magnetic moment $\mathfrak{M}$ in a magnetic field $\mathscr{H}$ (0, 0, $\mathscr{H}$):

$$E_n(0) = -\mathfrak{M}\cdot\mathscr{H} = -\mathfrak{M}_z\mathscr{H} = \mathfrak{M}_B(2n + 1)\mathscr{H} . \tag{57.13}$$

From this formula it is seen that the component $\mathfrak{M}_z$ of the magnetic moment in the direction of the magnetic field is an integral multiple of the Bohr magneton $\mathfrak{M}_B$.

The resulting quantisation of the energy of a free particle moving in a magnetic field is an important deduction of quantum mechanics, since it leads to the diamagnetism of an electron gas, whereas according to classical theory an electron gas should have no diamagnetic properties.[3]

The eigenfunctions (57.11) are entirely in accordance with the classical law of motion in a magnetic field, according to which there is a circular motion in the xy plane with frequency ω_0 (it is this part of the energy which is quantised) and free motion along the axis OZ.[4] For the wave function (57.11) signifies that the generalised momentum along the axis OX is $p_x^0 = \hbar\alpha$, and that along the axis OZ is $p_z^0 = \hbar\beta$. Along the axis OY we have a harmonic motion with frequency ω_0 about the equilibrium position $y_0 = cp_x^0/e\mathscr{H}$. According to classical mechanics the mo-

[3] See, e.g. [79]; [7], where the theory due to L. D. Landau is described.

[4] See Appendix X, where the corresponding calculation according to classical mechanics is given.

mentum along the axis OX is also constant, and this does not contradict the fact that a harmonic motion also occurs along this axis about some equilibrium position x_0, since $p_x = \mu v_x + eA_x/c$, not μv_x.

The generalised momentum p_x^0 determines the equilibrium position y_0, and the energy of the motion $E_n(\beta)$ is therefore independent of p_x^0.

The fact that the quantum-mechanical solution seems to give a harmonic motion only along the axis OY, while the classical circular motion corresponds to harmonic motions along both OY and OX (with phase difference $\frac{1}{2}\pi$), is due to the fact that the wave function $\psi_{n\alpha\beta}(x, y, z)$ (57.11) describes a state in which the equilibrium position x_0 of oscillations along the axis OX is indeterminate.

Since the energy $E_n(\beta)$ is independent of α, we have an infinite degeneracy corresponding to the various possible positions of the equilibrium point x_0. The energy $E_n(\beta)$ therefore belongs not only to the solution $\psi_{n\alpha\beta}$ found above but also to all wave functions of the form

$$\psi_{n\beta}(x, y, z) = \int_{-\infty}^{\infty} c(\alpha)\, e^{i(\alpha x + \beta z)}\, e^{-\frac{1}{2}\xi^2}\, H_n(\xi)\, d\alpha,$$

with $c(\alpha)$ an arbitrary function of α. In particular, $c(\alpha)$ may be chosen so that the solution $\psi_{n\beta}$ corresponds to a definite position of the equilibrium point x_0 on the axis OX.

CHAPTER X

INTRINSIC ANGULAR MOMENTUM AND MAGNETIC MOMENT OF THE ELECTRON. SPIN

58. Experimental proofs of the existence of electron spin

The theory of the motion of a charged particle in a magnetic field as given in the preceding chapter is by no means complete. The reason is that, in addition to the angular momentum and magnetic moment due to the motion of the centre of mass of the electron, it is also necessary to ascribe to the electron intrinsic angular momentum and magnetic moment, as if it were not a point mass but a rotating charged body.[1] These are called the *spin* angular momentum and magnetic moment, as opposed to those due to the motion of the centre of mass of the electron, which we shall now call *orbital*. The phenomenon itself is called the *spin* of the electron.

We shall briefly describe the experimental facts which prove the existence of electron spin.[2] One of the simplest and most direct indications of the existence of electron spin is obtained from the experiments of Stern and Gerlach on spatial quantisation (Section 3). These authors observed that a beam of hydrogen atoms known to be in the s state split into two. In this state the angular momentum is zero, and therefore so is the orbital magnetic moment. The fact that the beam of atoms undergoes deviation in a magnetic field shows, however, that these atoms have a magnetic moment in the s state. The fact that it is split into only two beams shows that the component of this magnetic moment can have only two values. The results of measurement show that the magnitude of the magnetic moment is equal to the Bohr magneton $\mathfrak{M}_B$. Thus, in the s state of an atom having only one electron, there is a magnetic moment $\boldsymbol{\mathfrak{M}}$ whose component in the direction of the magnetic field takes only two values, $\pm\,\mathfrak{M}_B$.

The existence of this magnetic moment in a state where there is certainly no orbital moment can be explained by supposing that it is intrinsic to the electron itself. This supposition is also suggested by the following important results. The spectral lines even of atoms which have a single optical electron are more complex than would follow from our earlier discussion of the theory of the motion of an electron in a field of central forces. For example, in the sodium atom, instead of one spectral line a (Figure 44) corresponding to the transition $2p \rightarrow 1s$, two very close lines are

[1] The classical theory of the rotating electron was first developed by Ya. I. Frenkel' [41].

[2] A detailed account is given in [87]; [99], Ch. IV.

observed (b, c) starting from two neighbouring levels. This is called the *sodium doublet* (the lines 5895.93 Å and 5889.96 Å).

Thus the p term of sodium must be regarded as consisting of two neighbouring levels. A similar structure of spectral lines is also observed in other atoms and is called the *multiplet structure* of spectra.

The theory of the motion of an electron in a field of central forces shows that the $2p$ term ($n = 2$, $l = 1$) consists of three coincident levels ($m = 0, \pm 1$), not of two neighbouring levels. The three levels can be split only in an external field, whereas the doublet b, c is observed in the absence of a field.

The hypothesis that the electron has an intrinsic magnetic moment $\mathfrak{M}_B$ enables us to explain immediately the twofold splitting of the terms of univalent atoms. Electric currents exist in an atom in every state (p, d, ...) except the s state, where the orbital angular momentum is zero (cf. Section 53). These currents create an internal magnetic field. Depending on the orientation of the spin magnetic moment

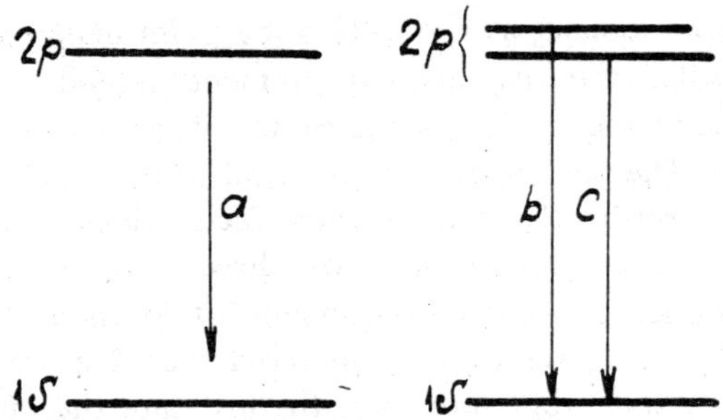

Fig. 44. Multiplet structure of the $2p$ level. Transitions b and c form two close lines (a doublet).

of the electron (parallel or antiparallel to this field), two states are obtained with slightly different energies, so that each of the levels p, d, ... is split into two neighbouring levels (see Section 62).

As we shall see, the splitting of the spectral lines of atoms in a magnetic field (the Zeeman effect, Section 74) also demands the hypothesis that there exists a spin of the electron and can be explained only in this way.

Let us now consider the intrinsic angular momentum of the electron, denoting it by $\mathbf{s}$. If the component s_z of this angular momentum in any direction OZ were given by an integral number of times Planck's constant, $m_s\hbar$ (as is the case for the orbital angular momentum), we should expect at least three orientations of the spin ($m_s = 0, \pm 1$). In fact, the result of Stern and Gerlach's experiment mentioned above, as well as the twofold splitting of the p, d, ... levels, indicates that only two orientations of the electron spin are possible. These facts led the Dutch physicists Uhlenbeck and Goudsmit to the hypothesis that the component s_z of the angular momentum of the electron in any direction is measured by one-half of Planck's constant and can therefore take only two values:

$$s_z = \pm \tfrac{1}{2}\hbar . \tag{58.1}$$

This hypothesis was supplemented by Uhlenbeck and Goudsmit, in accordance with experimental results, by the assumption that the electron has an intrinsic magnetic moment $\mathfrak{M}$ whose component $\mathfrak{M}_z$ in any direction can take only two values:

$$\mathfrak{M}_z = \pm \mathfrak{M}_B = \pm e\hbar/2\mu c . \tag{58.2}$$

It follows from (58.1) and (58.2) that the ratio of the spin magnetic moment to the spin angular momentum is $-e/\mu c$:

$$\mathfrak{M} = -(e/\mu c)\mathbf{s} , \tag{58.3}$$

whereas the ratio of the orbital quantities is $-e/2\mu c$ (see Section 53).

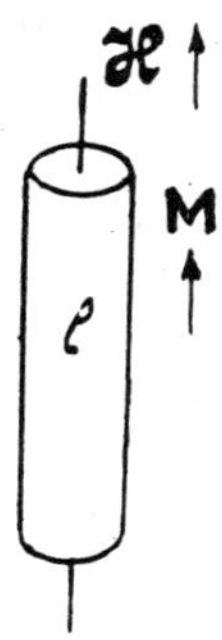

Fig. 45. Diagram of Einstein and de Haas' experiment.

The existence of the relation (58.3) between the magnetic moment and the angular momentum was discovered as early as 1915 in an experiment by Einstein and de Haas. The principle of this experiment is briefly as follows. A ferromagnetic rod l (Figure 45) is suspended on threads so as to be able to rotate about its axis. If the direction of the longitudinal magnetic field $\mathscr{H}$ is changed, then so is the direction of magnetisation of the rod, i.e. its magnetic moment $\mathfrak{M}$. Since the magnetic moment is proportional to the angular momentum,

$$\mathfrak{M} = -(e/2\mu c)\mathbf{M} , \tag{58.4}$$

the angular momentum $\mathbf{M}$ of all the electrons in the rod is also changed.[3] In consequence the rod begins to rotate and the threads are twisted. From this torsion we can determine $\mathbf{M}$, and so verify the ratio $\mathfrak{M}/\mathbf{M}$. For electrons this ratio must be negative, since the charge on the electron is $-e$, and this was in fact found by experiment, showing that the magnetisation of a ferromagnetic body is due to the

[3] It may be noted that we here give Formula (58.4) in terms of the total moment of all the electrons. Since it is valid for every electron in the rod, it must also be valid for the electrons as a whole; see [82, 87].

motion of the electrons. The ratio was, however, found to be $-e/\mu c$, not $-e/2\mu c$. For the orbital motion, on the most general assumptions, both classical and quantum theory give $-e/2\mu c$. The result of the experiment was therefore puzzling. If we suppose that the magnetisation is due to the spin of the electron, not its orbital motion, the ratio $\mathfrak{M}/\mathbf{M}$ should be $-e/\mu c$, as found in the experiment. This hypothesis not only explained the results of Einstein and de Haas' experiment but also laid the foundations of the modern theory of ferromagnetism (see Section 130).

It may be mentioned that the existence of the spin of the electron may now be regarded as a consequence of Dirac's relativistic theory of the electron. A discussion of that theory is, however, outside the scope of this book.[4]

59. The electron spin operator

Let us now consider the mathematical formulation of Uhlenbeck and Goudsmit's hypothesis. In accordance with the general principles of quantum mechanics, the corresponding angular momentum of the electron (which for brevity we shall call simply the electron spin) must be represented by a linear self-adjoint operator. Let the operators of the components of the spin along the co-ordinate axes be $\boldsymbol{s}_x, \boldsymbol{s}_y, \boldsymbol{s}_z$. In order to determine the form of these operators, we impose the condition that they should obey the same commutation rules as the orbital angular momentum components $\boldsymbol{M}_x, \boldsymbol{M}_y, \boldsymbol{M}_z$. Then, replacing $\vec{\boldsymbol{M}}$ in (25.5) by $\vec{\boldsymbol{s}}$, we obtain[5]

$$\left.\begin{aligned} \boldsymbol{s}_x\boldsymbol{s}_y - \boldsymbol{s}_y\boldsymbol{s}_x &= i\hbar\, \boldsymbol{s}_z\,, \\ \boldsymbol{s}_y\boldsymbol{s}_z - \boldsymbol{s}_z\boldsymbol{s}_y &= i\hbar\, \boldsymbol{s}_x\,, \\ \boldsymbol{s}_z\boldsymbol{s}_x - \boldsymbol{s}_x\boldsymbol{s}_z &= i\hbar\, \boldsymbol{s}_y\,. \end{aligned}\right\} \tag{59.1}$$

The component of the spin in any direction can (by the original hypothesis) take the two values $\pm\frac{1}{2}\hbar$. Hence the operators $\boldsymbol{s}_x, \boldsymbol{s}_y, \boldsymbol{s}_z$ must each be represented by a two-rowed matrix, since such a matrix, when put in a diagonal form, has only two diagonal elements and therefore only two eigenvalues. Putting

$$\boldsymbol{s}_x = \tfrac{1}{2}\hbar\boldsymbol{\sigma}_x\,, \qquad \boldsymbol{s}_y = \tfrac{1}{2}\hbar\boldsymbol{\sigma}_y\,, \qquad \boldsymbol{s}_z = \tfrac{1}{2}\hbar\boldsymbol{\sigma}_z\,, \tag{59.2}$$

we can say that the operators $\boldsymbol{\sigma}_x, \boldsymbol{\sigma}_y, \boldsymbol{\sigma}_z$ (the *spin matrices*) must be two-rowed matrices of the form

$$\boldsymbol{\sigma}_x = \begin{vmatrix} a_{11} & a_{12} \\ a_{21} & a_{22} \end{vmatrix}, \qquad \boldsymbol{\sigma}_y = \begin{vmatrix} b_{11} & b_{12} \\ b_{21} & b_{22} \end{vmatrix}, \qquad \boldsymbol{\sigma}_z = \begin{vmatrix} c_{11} & c_{12} \\ c_{21} & c_{22} \end{vmatrix}, \tag{59.3}$$

whose eigenvalues are ± 1. Substituting (59.2) in (59.1) and cancelling $\frac{1}{4}\hbar^2$, we find

$$\boldsymbol{\sigma}_x\boldsymbol{\sigma}_y - \boldsymbol{\sigma}_y\boldsymbol{\sigma}_x = 2i\boldsymbol{\sigma}_z\,, \tag{59.4}$$

[4] See Dirac's book ([29], Section 72). Dirac showed that, from the relativistic equation for the motion of an electron, it necessarily follows that the electron must have the magnetic moment (58.2) and the angular momentum (58.1). This provided a theoretical justification of Uhlenbeck and Goudsmit's hypothesis.

[5] It can be shown by means of group theory that the rules (59.1) are the only possible ones; see, for example, [71].

$$\boldsymbol{\sigma}_y \boldsymbol{\sigma}_z - \boldsymbol{\sigma}_z \boldsymbol{\sigma}_y = 2\, i \boldsymbol{\sigma}_x , \tag{59.4'}$$

$$\boldsymbol{\sigma}_z \boldsymbol{\sigma}_x - \boldsymbol{\sigma}_x \boldsymbol{\sigma}_z = 2\, i \boldsymbol{\sigma}_y . \tag{59.4''}$$

Since the eigenvalues of $\boldsymbol{\sigma}_x$, $\boldsymbol{\sigma}_y$ and $\boldsymbol{\sigma}_z$ are ± 1, the eigenvalues of the operators $\boldsymbol{\sigma}_x^2$, $\boldsymbol{\sigma}_y^2$, $\boldsymbol{\sigma}_z^2$ are $+ 1$. In their own representation, therefore, the latter matrices must be

$$\boldsymbol{\sigma}_x^2 = \begin{vmatrix} 1 & 0 \\ 0 & 1 \end{vmatrix}, \quad \boldsymbol{\sigma}_y^2 = \begin{vmatrix} 1 & 0 \\ 0 & 1 \end{vmatrix}, \quad \boldsymbol{\sigma}_z^2 = \begin{vmatrix} 1 & 0 \\ 0 & 1 \end{vmatrix}, \tag{59.5}$$

i.e. they are unit matrices $\boldsymbol{\delta}$:

$$\boldsymbol{\delta} = \begin{vmatrix} 1 & 0 \\ 0 & 1 \end{vmatrix}. \tag{59.6}$$

The unit matrix remains a unit matrix in every representation (see Section 40). The matrices $\boldsymbol{\sigma}_x^2$, $\boldsymbol{\sigma}_y^2$, $\boldsymbol{\sigma}_z^2$ therefore have the form (59.5) in all possible representations. Let us now consider the combination

$$2\, i (\boldsymbol{\sigma}_x \boldsymbol{\sigma}_y + \boldsymbol{\sigma}_y \boldsymbol{\sigma}_x) = 2\, i \boldsymbol{\sigma}_x \boldsymbol{\sigma}_y + \boldsymbol{\sigma}_y \cdot 2\, i \boldsymbol{\sigma}_x .$$

From (59.4), this can be written as

$$(\boldsymbol{\sigma}_y \boldsymbol{\sigma}_z - \boldsymbol{\sigma}_z \boldsymbol{\sigma}_y)\, \boldsymbol{\sigma}_y + \boldsymbol{\sigma}_y (\boldsymbol{\sigma}_y \boldsymbol{\sigma}_z - \boldsymbol{\sigma}_z \boldsymbol{\sigma}_y)$$
$$= \boldsymbol{\sigma}_y \boldsymbol{\sigma}_z \boldsymbol{\sigma}_y - \boldsymbol{\sigma}_z \boldsymbol{\sigma}_y^2 + \boldsymbol{\sigma}_y^2 \boldsymbol{\sigma}_z - \boldsymbol{\sigma}_y \boldsymbol{\sigma}_z \boldsymbol{\sigma}_y = \boldsymbol{\sigma}_y^2 \boldsymbol{\sigma}_z - \boldsymbol{\sigma}_z \boldsymbol{\sigma}_y^2 \;;$$

but $\boldsymbol{\sigma}_y^2 = \boldsymbol{\delta}$ is the unit matrix, so that $\boldsymbol{\sigma}_y^2 \boldsymbol{\sigma}_z = \boldsymbol{\sigma}_z \boldsymbol{\sigma}_y^2$. Hence

$$\boldsymbol{\sigma}_x \boldsymbol{\sigma}_y = - \boldsymbol{\sigma}_y \boldsymbol{\sigma}_x \;; \tag{59.7}$$

the matrices $\boldsymbol{\sigma}_x$ and $\boldsymbol{\sigma}_y$ are said to *anticommute*.

Combining (59.7) with (59.4) and using a cyclic permutation of $\boldsymbol{\sigma}_x$, $\boldsymbol{\sigma}_y$, $\boldsymbol{\sigma}_z$, we find

$$\left.\begin{aligned} \boldsymbol{\sigma}_x \boldsymbol{\sigma}_y &= - \boldsymbol{\sigma}_y \boldsymbol{\sigma}_x = i \boldsymbol{\sigma}_z , \\ \boldsymbol{\sigma}_y \boldsymbol{\sigma}_z &= - \boldsymbol{\sigma}_z \boldsymbol{\sigma}_y = i \boldsymbol{\sigma}_x , \\ \boldsymbol{\sigma}_z \boldsymbol{\sigma}_x &= - \boldsymbol{\sigma}_x \boldsymbol{\sigma}_z = i \boldsymbol{\sigma}_y . \end{aligned}\right\} \tag{59.8}$$

Let us now find the explicit form of the matrices $\boldsymbol{\sigma}_x$, $\boldsymbol{\sigma}_y$, $\boldsymbol{\sigma}_z$. Suppose that the matrix $\boldsymbol{\sigma}_z$, say, has been brought to diagonal form. Since its eigenvalues are ± 1, this will be

$$\boldsymbol{\sigma}_z = \begin{vmatrix} 1 & 0 \\ 0 & -1 \end{vmatrix}. \tag{59.9}$$

It can be shown that in the same representation the other two matrices $\boldsymbol{\sigma}_x$ and $\boldsymbol{\sigma}_y$ will be

$$\boldsymbol{\sigma}_x = \begin{vmatrix} 0 & 1 \\ 1 & 0 \end{vmatrix}, \quad \boldsymbol{\sigma}_y = \begin{vmatrix} 0 & -i \\ i & 0 \end{vmatrix}. \tag{59.9'}$$

To prove this, we form the products $\boldsymbol{\sigma}_z \boldsymbol{\sigma}_x$ and $\boldsymbol{\sigma}_x \boldsymbol{\sigma}_z$. The rule of matrix multiplication

(Section 40) gives

$$\boldsymbol{\sigma}_z\boldsymbol{\sigma}_x = \begin{vmatrix} 1 & 0 \\ 0 & -1 \end{vmatrix} \begin{vmatrix} a_{11} & a_{12} \\ a_{21} & a_{22} \end{vmatrix} = \begin{vmatrix} a_{11} & a_{12} \\ -a_{21} & -a_{22} \end{vmatrix},$$

$$\boldsymbol{\sigma}_x\boldsymbol{\sigma}_z = \begin{vmatrix} a_{11} & a_{12} \\ a_{21} & a_{22} \end{vmatrix} \begin{vmatrix} 1 & 0 \\ 0 & -1 \end{vmatrix} = \begin{vmatrix} a_{11} & -a_{12} \\ a_{21} & -a_{22} \end{vmatrix}.$$

From (59.8) we have

$$\begin{vmatrix} a_{11} & a_{12} \\ -a_{21} & -a_{22} \end{vmatrix} = - \begin{vmatrix} a_{11} & -a_{12} \\ a_{21} & -a_{22} \end{vmatrix} = \begin{vmatrix} -a_{11} & a_{12} \\ -a_{21} & a_{22} \end{vmatrix},$$

or $a_{11} = -a_{11}$, $a_{12} = a_{12}$, $-a_{21} = -a_{21}$, $-a_{22} = a_{22}$, i.e. $a_{11} = 0$, $a_{22} = 0$. The matrix $\boldsymbol{\sigma}_x$ is therefore of the form

$$\boldsymbol{\sigma}_x = \begin{vmatrix} 0 & a_{12} \\ a_{21} & 0 \end{vmatrix}. \tag{59.10}$$

Now we form $\boldsymbol{\sigma}_x^2$:

$$\boldsymbol{\sigma}_x^2 = \begin{vmatrix} 0 & a_{12} \\ a_{21} & 0 \end{vmatrix} \begin{vmatrix} 0 & a_{12} \\ a_{21} & 0 \end{vmatrix} = \begin{vmatrix} a_{12}a_{21} & 0 \\ 0 & a_{12}a_{21} \end{vmatrix}.$$

Comparison with (59.5) shows that $a_{12}a_{21} = 1$. The matrix must be self-adjoint, i.e. $a_{12} = a_{21}^*$. Thus $|a_{12}|^2 = 1$. This gives

$$\boldsymbol{\sigma}_x = \begin{vmatrix} 0 & e^{i\alpha} \\ e^{-i\alpha} & 0 \end{vmatrix}, \tag{59.11}$$

where α is a real number. Similarly

$$\boldsymbol{\sigma}_y = \begin{vmatrix} 0 & e^{i\beta} \\ e^{-i\beta} & 0 \end{vmatrix}. \tag{59.11'}$$

Now multiplying $\boldsymbol{\sigma}_x$ by $\boldsymbol{\sigma}_y$ and $\boldsymbol{\sigma}_y$ by $\boldsymbol{\sigma}_x$ and using (59.8), we find

$$\begin{vmatrix} e^{i(\alpha-\beta)} & 0 \\ 0 & e^{-i(\alpha-\beta)} \end{vmatrix} = - \begin{vmatrix} e^{-i(\alpha-\beta)} & 0 \\ 0 & e^{i(\alpha-\beta)} \end{vmatrix},$$

whence

$$e^{i(\alpha-\beta)} = -e^{-i(\alpha-\beta)},$$

i.e. $\alpha - \beta = \frac{1}{2}\pi$. Then all the relations are satisfied for any value of α. We can therefore take without restriction $\alpha = 0$, $\beta = -\frac{1}{2}\pi$, and substitution in (59.11) and (59.11′) gives (59.9′).

According to (59.2) we can obtain from (59.9) and (59.9′) the matrices of the operators s_x, s_y, s_z in the representation in which s_z is diagonal (the s_z representation):

$$s_x = \begin{vmatrix} 0 & \frac{1}{2}h \\ \frac{1}{2}h & 0 \end{vmatrix}, \quad s_y = \begin{vmatrix} 0 & -\frac{1}{2}ih \\ \frac{1}{2}ih & 0 \end{vmatrix}, \quad s_z = \begin{vmatrix} \frac{1}{2}h & 0 \\ 0 & -\frac{1}{2}h \end{vmatrix}. \tag{59.12}$$

It may be noted that the suffixes 1 and 2 which number the elements of the matrices $\boldsymbol{\sigma}$ and $\boldsymbol{s}$ now acquire a significance (the representation having been chosen): the suffix 1 relates to the first eigenvalue of $s_z(+\frac{1}{2}\hbar)$ and 2 to the second eigenvalue $(-\frac{1}{2}\hbar)$.

Let us now derive the operator of the square of the electron spin. From (59.12)

$$\begin{aligned}\vec{s}^2 &= s_x^2 + s_y^2 + s_z^2 \\ &= \tfrac{3}{4}\hbar^2 \begin{vmatrix} 1 & 0 \\ 0 & 1 \end{vmatrix} = \tfrac{3}{4}\hbar^2 \boldsymbol{\delta}.\end{aligned} \tag{59.13}$$

Using the quantum numbers m_s and l_s which define respectively the value of the spin component in any direction and the square of the spin, we can write the spin quantisation formulae in a way entirely analogous to Formulae (51.9), (51.10) for the orbital angular momentum:

$$s^2 = \hbar^2 l_s(l_s+1), \qquad l_s = \tfrac{1}{2}, \tag{59.14}$$

$$s_z = \hbar m_s, \qquad m_s = \pm\tfrac{1}{2}. \tag{59.15}$$

60. Spin functions

We see that in quantum mechanics the spin state must be described by two quantities: the absolute value $|\mathbf{s}|$ (or s^2) and the component s_z of the spin in some direction. The first quantity (s^2) is assumed to be the same for all electrons, and we are therefore concerned with only one variable, s_z. Thus, in addition to the three variables which define the motion of the centre of mass of the electron (x, y, z or p_x, p_y, p_z, etc.) there is a further variable s_z which defines the spin of the electron. We can therefore say that the electron has four degrees of freedom.[6]

Accordingly, the wave function ψ which defines the state of the electron must be regarded as a function of the four variables of which three relate to the centre of mass of the electron and one to the spin (s_z). For example, in the co-ordinate representation for the electron we must put

$$\psi = \psi(x, y, z, s_z, t). \tag{60.1}$$

Since the spin variable has only two values ($\pm\frac{1}{2}\hbar$), we can say that instead of one function we have two:

$$\psi_1 = \psi(x, y, z, +\tfrac{1}{2}\hbar, t), \tag{60.2}$$

$$\psi_2 = \psi(x, y, z, -\tfrac{1}{2}\hbar, t). \tag{60.2'}$$

[6] The spin variables s_x, s_y, s_z differ from all the quantum quantities previously discussed, in that they take only two values. This is due to the assumption that the squared spin angular momentum s^2 of the electron has a strictly defined value.

If s^2 were regarded as variable, we should have not one but two degrees of freedom, as for a top. It might therefore be thought that the electron spin considered in present-day physics is only one possible state of rotation of the electron (other values of s^2 might, for example, require a large energy of rotation). This problem must remain undecided at present. It is of particular interest with respect to mesons, for which a whole 'spectrum' of masses is observed.

We shall sometimes write these functions as single-column matrices:

$$\Psi = \begin{vmatrix} \psi_1 & 0 \\ \psi_2 & 0 \end{vmatrix} \equiv \begin{vmatrix} \psi_1 \\ \psi_2 \end{vmatrix}, \tag{60.3}$$

and the adjoint function as a single-row matrix:

$$\Psi^+ = \begin{vmatrix} \psi_1^* & \psi_2^* \\ 0 & 0 \end{vmatrix} \equiv \begin{vmatrix} \psi_1^* & \psi_2^* \end{vmatrix}. \tag{60.3'}$$

This type of notation enables us to use the rule (40.11).

It is clear that the wave functions ψ_1 and ψ_2 will be different only if there is an actual relation between the spin and the motion of the centre of mass. Such a relation exists, and represents the interaction of the spin magnetic moment and the magnetic field of the currents due to the motion of the centre of mass of the charged electron. This interaction causes the multiplet structure of spectra (see Section 58). Hence, if we ignore the multiplet structure of spectra, we can neglect the interaction between the spin and the orbital motion. In this approximation

$$\psi_1(x, y, z, t) = \psi_2(x, y, z, t) = \psi(x, y, z, t). \tag{60.4}$$

However, in order still to mark the fact that a particle possessing spin is involved, we write the function (60.1) in a form corresponding to a separation of the variables:

$$\psi(x, y, z, s_z, t) = \psi(x, y, z, t)\, S_\alpha(s_z), \tag{60.5}$$

where $S_\alpha(s_z)$ denotes the *spin function*. This is essentially just a symbol indicating the spin state of the particle.

The value of this symbol, or spin function, is as follows. The suffix α takes two values, which are usually taken as $+\frac{1}{2}$ and $-\frac{1}{2}$ (instead of 1 and 2). The first value, $+\frac{1}{2}$ (or 1) denotes that the component of the spin along some chosen direction OZ is $+\frac{1}{2}\hbar$. The second value of the suffix α denotes the spin state with the other possible value of the spin component in that direction, namely $-\frac{1}{2}\hbar$. The 'argument' s_z of the 'function' S_α is regarded as an independent variable, which can take the two values $\pm\frac{1}{2}\hbar$. Then

$$S_{+\frac{1}{2}}(\tfrac{1}{2}\hbar) = 1, \qquad S_{+\frac{1}{2}}(-\tfrac{1}{2}\hbar) = 0, \tag{60.6}$$

since, from the meaning of the symbol, in the state $\alpha = +\frac{1}{2}$, $s_z = +\frac{1}{2}\hbar$, and in that state we cannot have $s_z = -\frac{1}{2}\hbar$, so that the corresponding function is zero. Similarly

$$S_{-\frac{1}{2}}(\tfrac{1}{2}\hbar) = 0, \qquad S_{-\frac{1}{2}}(-\tfrac{1}{2}\hbar) = 1. \tag{60.6'}$$

The notation (60.1) and, in the particular case where spin and orbital motion do not interact, (60.5) enables us to regard the spin s_z as a dynamical variable like any other mechanical quantity.

The spin 'wave' functions $S_\alpha(s_z)$ defined above are orthogonal and normalised. To see this, we take the product $S_\alpha^*(s_z)\, S_\beta(s_z)$, where S^* denotes, as usual, the complex conjugate function to S, and $\alpha, \beta = \pm \frac{1}{2}$. We sum this product over all possible values of the spin variable s_z (of which there are only two, $\pm \frac{1}{2}\hbar$). Then it follows directly from (60.6) and (60.6′) (since $S^* = S$) that

$$\sum_{s_z} S_\alpha^*(s_z)\, S_\beta(s_z) = \delta_{\alpha\beta}\,. \tag{60.7}$$

The function $S_\alpha(s_z)$ can also be written in the matrix form (60.3):

$$\boldsymbol{S}_{+\frac{1}{2}} = \begin{vmatrix} 1 & 0 \\ 0 & 0 \end{vmatrix}, \qquad \boldsymbol{S}_{-\frac{1}{2}} = \begin{vmatrix} 0 & 0 \\ 1 & 0 \end{vmatrix}, \tag{60.8}$$

$$\boldsymbol{S}^+_{+\frac{1}{2}} = \begin{vmatrix} 1 & 0 \\ 0 & 0 \end{vmatrix}, \qquad \boldsymbol{S}^+_{-\frac{1}{2}} = \begin{vmatrix} 0 & 1 \\ 0 & 0 \end{vmatrix}. \tag{60.8′}$$

Let us now calculate the effect of any spin operator of the type

$$\boldsymbol{L} = \begin{vmatrix} L_{11} & L_{12} \\ L_{21} & L_{22} \end{vmatrix} \tag{60.9}$$

on the wave function. If the operator $\boldsymbol{L}$ is taken in the s_z representation, the suffixes 1 and 2 denote the numbers of the eigenvalues $s_z(\pm \frac{1}{2}\hbar)$. According to Formula (39.5), which defines the effect of an operator given in matrix form on the wave function, the operator $\boldsymbol{L}$ changes the function $\Psi(\psi_1, \psi_2)$ into another function $\Phi(\phi_1, \phi_2)$ such that

$$\phi_1 = L_{11}\psi_1 + L_{12}\psi_2, \tag{60.10}$$

$$\phi_2 = L_{21}\psi_1 + L_{22}\psi_2\,. \tag{60.10′}$$

The difference between (60.10) and (39.5) is merely that in (60.10) we have two-rowed matrices and correspondingly a function of two components, while in (39.5) we had a matrix L_{mn} with an infinite number of elements and a function ψ with an infinite number of components $c_n(c_1, c_2, \ldots)$.

Putting Ψ in the form of the (column) matrix (60.3), we can write the two equations (60.10) and (60.10′) as a single matrix equation

$$\boldsymbol{\Phi} = \boldsymbol{L}\boldsymbol{\Psi} \tag{60.11}$$

(see (40.14)), since in explicit form this gives

$$\boldsymbol{\Phi} = \begin{vmatrix} \phi_1 & 0 \\ \phi_2 & 0 \end{vmatrix} = \begin{vmatrix} L_{11} & L_{12} \\ L_{21} & L_{22} \end{vmatrix} \begin{vmatrix} \psi_1 & 0 \\ \psi_2 & 0 \end{vmatrix} = \begin{vmatrix} L_{11}\psi_1 + L_{12}\psi_2 & 0 \\ L_{21}\psi_1 + L_{22}\psi_2 & 0 \end{vmatrix}, \tag{60.11′}$$

which is the same as (60.10) and (60.10′). In what follows, if the operator is spin-dependent, we shall take the symbol $\boldsymbol{L}\boldsymbol{\Psi}$ to denote a product of this type, which essentially represents two equations (60.10), (60.10′) in the form of one matrix equation.

The mean value of any spin quantity L in a state ψ_1, ψ_2 is, according to the general formula (41.2),

$$\bar{L}(x, y, z, t) = \psi_1^* L_{11}\psi_1 + \psi_1^* L_{12}\psi_2 + \psi_2^* L_{21}\psi_1 + \psi_2^* L_{22}\psi_2\,. \tag{60.12}$$

Since the functions ψ_1 and ψ_2 depend also on the co-ordinates of the centre of mass of the electron, we have written $\bar{L}(x, y, z, t)$, bearing in mind that the mean value obtained from (60.12) is the mean value of L for a given position of the centre of mass of the electron. The mean value in the state ψ_1, ψ_2 for any position of the electron is given by the formula

$$\bar{L}(t) = \int \bar{L}(x, y, z, t)\,\mathrm{d}x\,\mathrm{d}y\,\mathrm{d}z\,. \tag{60.13}$$

Formulae (60.12) and (60.13) can be written

$$\bar{L}(x, y, z, t) = \Psi^+ \boldsymbol{L} \Psi\,, \tag{60.12'}$$

$$\bar{L}(t) = \int \Psi^+ \boldsymbol{L} \Psi \mathrm{d}x\,\mathrm{d}y\,\mathrm{d}z \tag{60.13'}$$

if we represent Ψ as a single-column matrix. In particular,

$$\begin{aligned} \bar{\sigma}_x(x, y, z, t) = \Psi^+ \boldsymbol{\sigma}_x \Psi &= \begin{vmatrix} \psi_1^* & \psi_2^* \\ 0 & 0 \end{vmatrix} \begin{vmatrix} 0 & 1 \\ 1 & 0 \end{vmatrix} \begin{vmatrix} \psi_1 & 0 \\ \psi_2 & 0 \end{vmatrix} \\ &= \begin{vmatrix} \psi_1^*\psi_2 + \psi_2^*\psi_1 & 0 \\ 0 & 0 \end{vmatrix} = \psi_1^*\psi_2 + \psi_2^*\psi_1\,. \end{aligned} \tag{60.14}$$

Similarly

$$\bar{\sigma}_y(x, y, z, t) = \Psi^+ \boldsymbol{\sigma}_y \Psi = -i\psi_1^*\psi_2 + i\psi_2^*\psi_1\,, \tag{60.14'}$$

$$\bar{\sigma}_z(x, y, z, t) = \Psi^+ \boldsymbol{\sigma}_z \Psi = \psi_1^*\psi_1 - \psi_2^*\psi_2\,. \tag{60.14''}$$

61. Pauli's equation

Let us consider the motion of an electron in an electromagnetic field, taking spin into account. According to the fundamental hypothesis (Section 58) the electron has a magnetic moment

$$\mathfrak{M}_B = -(e/\mu c)\,\mathbf{s}\,. \tag{61.1}$$

Owing to the existence of this magnetic moment, an electron in a magnetic field $\mathscr{H}(\mathscr{H}_x, \mathscr{H}_y, \mathscr{H}_z)$ acquires an additional potential energy equal to the energy of a magnetic dipole in a field $\mathscr{H}$:

$$\Delta U = -\mathfrak{M}_B \cdot \mathscr{H}\,. \tag{61.2}$$

The operator of this energy, according to (61.1), is

$$\begin{aligned} \Delta U = (e/\mu c)\vec{\mathbf{s}}\cdot\mathscr{H} &= (e\hbar/2\mu c)\vec{\boldsymbol{\sigma}}\cdot\mathscr{H} \\ &= (e\hbar/2\mu c)(\boldsymbol{\sigma}_x\mathscr{H}_x + \boldsymbol{\sigma}_y\mathscr{H}_y + \boldsymbol{\sigma}_z\mathscr{H}_z)\,, \end{aligned} \tag{61.3}$$

where $\vec{\boldsymbol{\sigma}}$ is the vector operator with components $\boldsymbol{\sigma}_x$, $\boldsymbol{\sigma}_y$, $\boldsymbol{\sigma}_z$ (59.9) and (59.9'). The Hamiltonian (27.7) for the motion of a charged particle in an electromagnetic field must therefore be supplemented, when spin is taken into account, by the term (61.3), becoming

$$\boldsymbol{H} = \frac{1}{2\mu}\left(\vec{\boldsymbol{P}} + \frac{e}{c}\mathbf{A}\right)^2 - eV + U + \frac{e\hbar}{2\mu c}\vec{\boldsymbol{\sigma}}\cdot\mathscr{H} \tag{61.4}$$

(the electron charge is taken as $-e$).

Schrödinger's equation for the wave function $\Psi(\psi_1, \psi_2)$ now becomes

$$i\hbar \frac{\partial \Psi}{\partial t} = \frac{1}{2\mu}\left(\vec{P} + \frac{e}{c}\mathbf{A}\right)^2 \Psi - eV\Psi + U\Psi + \frac{e\hbar}{2\mu c}\vec{\boldsymbol{\sigma}}\cdot\mathscr{H}\Psi. \tag{61.5}$$

This is called *Pauli's equation*. It may be noted that Ψ signifies the column matrix (60.3), and (61.5) therefore represents essentially two equations for the two functions ψ_1 and ψ_2 in the form of one matrix equation.

Let us now determine the current density. To do so, we write (61.5) in the form

$$i\hbar \frac{\partial \Psi}{\partial t} = \boldsymbol{H}_0 \Psi + \frac{e\hbar}{2\mu c}(\vec{\boldsymbol{\sigma}}\cdot\mathscr{H})\Psi, \tag{61.6}$$

where $\boldsymbol{H}_0$ denotes the terms which do not contain the operators $\boldsymbol{\sigma}$. The equation for the adjoint function Ψ^+, which we regard as the row matrix (60.3′), is

$$-i\hbar \frac{\partial \Psi^+}{\partial t} = \boldsymbol{H}_0^* \Psi^+ + \frac{e\hbar}{2\mu c}[(\vec{\boldsymbol{\sigma}}\cdot\mathscr{H})\Psi]^+ . \tag{61.6′}$$

The symbol [ ]$^+$ signifies that the rows and columns in the relevant matrix are interchanged and the complex conjugate elements are taken.

Now multiplying (61.6) on the left by Ψ^+ and (61.6′) on the right by Ψ and subtracting, we obtain

$$\begin{aligned} i\hbar \frac{\partial}{\partial t}(\Psi^+\Psi) &= \Psi^+(\boldsymbol{H}_0\Psi) - (\boldsymbol{H}_0^*\Psi^+)\Psi + \\ &+ \frac{e\hbar}{2\mu c}\{\Psi^+(\vec{\boldsymbol{\sigma}}\cdot\mathscr{H})\Psi - [(\vec{\boldsymbol{\sigma}}\cdot\mathscr{H})\Psi]^+\Psi\}. \end{aligned} \tag{61.7}$$

From (40.15)

$$[(\vec{\boldsymbol{\sigma}}\cdot\mathscr{H})\Psi]^+ = \Psi^+(\vec{\boldsymbol{\sigma}}^+\cdot\mathscr{H}) \tag{61.8}$$

since the operator $\vec{\boldsymbol{\sigma}}^+ = \vec{\boldsymbol{\sigma}}$ is self-adjoint. The expression in the braces in (61.7) is therefore zero. The remaining terms, which do not contain the operators $\boldsymbol{\sigma}$, give, after calculations entirely analogous to those for the current density in Section 29,[7]

$$\begin{aligned} &i\hbar \frac{\partial}{\partial t}(\psi_1^*\psi_1 + \psi_2^*\psi_2) \\ &= -\frac{\hbar^2}{2\mu}\operatorname{div}\{\psi_1^*\nabla\psi_1 - \psi_1\nabla\psi_1^* + \psi_2^*\nabla\psi_2 - \psi_2\nabla\psi_2^*\} - \\ &\quad - \frac{i\hbar e}{\mu c}\operatorname{div}\{\mathbf{A}(\psi_1^*\psi_1 + \psi_2^*\psi_2)\}. \end{aligned} \tag{61.9}$$

[7] When using the matrix notation we are always operating with four functions ψ_1^*, ψ_2^*, ψ_1, ψ_2 at once. The reader is advised, after acquainting himself with matrix methods, to write equations (61.6) and (61.6′) in explicit form (four equations), and to derive the same result by multiplying the first pair by ψ_1^* and ψ_2^* and the second pair by ψ_1 and ψ_2.

Rewriting this equation as an equation of continuity for the probability density w and the particle current density $\mathbf{J}$, we find

$$w(x, y, z, t) = \psi_1^* \psi_1 + \psi_2^* \psi_2 , \tag{61.10}$$

$$\mathbf{J} = \frac{i\hbar}{2\mu}[(\psi_1 \nabla \psi_1^* - \psi_1^* \nabla \psi_1) + (\psi_2 \nabla \psi_2^* - \psi_2^* \nabla \psi_2)] - \\ - \frac{e}{\mu c} \mathbf{A} (\psi_1^* \psi_1 + \psi_2^* \psi_2), \tag{61.11}$$

or

$$w(x, y, z, t) = \Psi^+ \Psi, \qquad \mathbf{J} = \frac{i\hbar}{2\mu}[\Psi \nabla \Psi^+ - \Psi^+ \nabla \Psi] - \frac{e}{\mu c} \mathbf{A} \Psi^+ \Psi. \tag{61.12}$$

These formulae show that the position probability distribution and the current density are obtained additively from two parts, each of which relates to electrons with a single definite spin orientation. The formula for normalising the probability is

$$\int (\psi_1^* \psi_1 + \psi_2^* \psi_2) \mathrm{d}x \, \mathrm{d}y \, \mathrm{d}z = 1 \quad \text{or} \quad \int \Psi^+ \Psi \mathrm{d}x \, \mathrm{d}y \, \mathrm{d}z = 1 . \tag{61.13}$$

The quantities

$$w_1 (x, y, z, t) = \psi_1^* \psi_1 , \qquad w_2 (x, y, z, t) = \psi_2^* \psi_2 \tag{61.14}$$

are the probability densities for finding an electron at the point x, y, z at time t with $s_z = + \frac{1}{2}\hbar$ and $- \frac{1}{2}\hbar$ respectively. The quantities

$$\left. \begin{aligned} w_1 &= \int \psi_1^* \psi_1 \, \mathrm{d}x \, \mathrm{d}y \, \mathrm{d}z , \\ w_2 &= \int \psi_2^* \psi_2 \, \mathrm{d}x \, \mathrm{d}y \, \mathrm{d}z \end{aligned} \right\} \tag{61.15}$$

are the probabilities of finding an electron with spin $s_z = + \frac{1}{2}\hbar$ and $- \frac{1}{2}\hbar$ respectively. The mean density of electric charge ρ_e and the mean electric current density $\mathbf{J}_e$ are, by (61.12),

$$\rho_e = - e \Psi^+ \Psi, \qquad \mathbf{J}_e = \frac{i\hbar e}{2\mu} [\Psi^+ \nabla \Psi - \Psi \nabla \Psi^+] + \frac{e^2}{\mu c} \mathbf{A} \Psi^+ \Psi ; \tag{61.16}$$

ρ_e and $\mathbf{J}_e$ do not completely describe all electromagnetic field sources in the case of the electron. The magnetic moment of the electron (61.1), which creates a magnetic field, must also be taken into account. From (61.1) and the general formula (60.12), we obtain an expression for the mean magnetic moment density (magnetisation $\mathbf{I}$):

$$\mathbf{I}(x, y, z, t) = - \frac{e\hbar}{2\mu c} \Psi^+ \vec{\boldsymbol{\sigma}} \Psi . \tag{61.17}$$

According to Maxwell's equations for the magnetic field we have

$$\mathbf{curl}\, \mathscr{H} = 4\pi \mathbf{J}_e / c , \qquad \operatorname{div} \mathbf{B} = 0 , \qquad \mathbf{B} = \mathscr{H} + 4\pi \mathbf{I} . \tag{61.18}$$

From these equations we can find the magnetic field due to an electron in the state Ψ, if $\mathbf{J}_e$ and $\mathbf{I}$ are taken as (61.16) and (61.17). Expressing $\mathscr{H}$ in the first equation (61.18) in terms of the induction $\mathbf{B}$, we obtain

$$\mathbf{curl\,B} = (4\pi/c)\{\mathbf{J}_e + c\,\mathbf{curl\,I}\}. \tag{61.18'}$$

Thus, instead of the magnetisation $\mathbf{I}$ we can consider an equivalent current given by

$$\mathbf{J}_s = c\,\mathbf{curl\,I} = -(e\hbar/2\mu)\mathbf{curl}(\Psi^+ \vec{\boldsymbol{\sigma}}\Psi), \qquad \operatorname{div}\mathbf{J}_s = 0. \tag{61.19}$$

The total electric current corresponding to both orbital and spin motion is

$$\mathbf{J}'_e = \frac{i\hbar e}{2\mu}[\Psi^+\nabla\Psi - \Psi(\nabla\Psi^+)] + \frac{e^2}{\mu c}\mathbf{A}\Psi^+\Psi - \frac{e\hbar}{2\mu}\mathbf{curl}(\Psi^+ \vec{\boldsymbol{\sigma}}\Psi). \tag{61.20}$$

To calculate the components of the spin current $\mathbf{J}_s$ we must use Formulae (60.14), (60.14′) and (60.14″).

62. Splitting of spectral lines in a magnetic field

Let us consider an atom with one valency electron in a uniform external magnetic field. This electron will be simultaneously subject to the interaction of the magnetic field and that of the electric field of the nucleus and the inner electrons. The electric field will be assumed central, and the potential energy of the electron in that field will be denoted by $U(r)$.

Let the magnetic field be directed along the axis OZ, and let us take the vector potential $\mathbf{A}$ in the form

$$A_x = -\tfrac{1}{2}\mathscr{H}_y, \qquad A_y = \tfrac{1}{2}\mathscr{H}_x, \qquad A_z = 0. \tag{62.1}$$

The magnetic field is then correctly given by the formula $\mathscr{H} = \mathbf{curl\,A}$:

$$\mathscr{H}_x = \mathscr{H}_y = 0, \qquad \mathscr{H}_z = \mathscr{H}. \tag{62.2}$$

Substituting this value of $\mathbf{A}$ in the Hamiltonian (61.4), we obtain Pauli's equation:

$$\begin{aligned} i\hbar\frac{\partial\Psi}{\partial t} = & -\frac{\hbar}{2\mu}\nabla^2\Psi + U(r)\Psi - \frac{i\hbar e}{2\mu c}\mathscr{H}\left(x\frac{\partial\Psi}{\partial y} - y\frac{\partial\Psi}{\partial x}\right) + \\ & + \frac{e^2}{8\mu c^2}\mathscr{H}^2(x^2+y^2)\Psi + \frac{e\hbar}{2\mu c}(\boldsymbol{\sigma}_z\mathscr{H})\Psi. \end{aligned} \tag{62.3}$$

The term in $\mathscr{H}^2$ can be neglected for weak fields.[8] The operator

$$-i\hbar\left(x\frac{\partial}{\partial y} - y\frac{\partial}{\partial x}\right) = -i\hbar\frac{\partial}{\partial\phi} = M_z \tag{62.4}$$

is the operator of the orbital angular momentum component. Also, denoting by

$$H^0 = -(\hbar^2/2\mu)\nabla^2 + U(r) \tag{62.5}$$

[8] As will be shown in Section 129, the term neglected gives rise to weak diamagnetic phenomena.

the Hamiltonian of the electron in the absence of the magnetic field, we have

$$i\hbar\frac{\partial\Psi}{\partial t} = \boldsymbol{H}^0\Psi + \frac{e\mathscr{H}}{2\mu c}(\boldsymbol{M}_z + \hbar\boldsymbol{\sigma}_z)\Psi. \tag{62.6}$$

From this equation it follows that, since $\mathscr{H}^2$ is neglected, the term representing the effect of the magnetic field may be regarded as the potential energy ΔU of a magnetic dipole with moment $-(e/2\mu c)(\vec{\boldsymbol{M}} + \hbar\vec{\boldsymbol{\sigma}})$ in a magnetic field $\mathscr{H}$:

$$\Delta U = -\mathscr{H}\cdot\mathfrak{M} = (e\mathscr{H}/2\mu c)(\boldsymbol{M}_z + \hbar\boldsymbol{\sigma}_z). \tag{62.7}$$

We shall look for stationary states, representing the wave function in the form

$$\Psi(x, y, z, t) = \Psi(x, y, z)\, e^{-iEt/\hbar}, \tag{62.8}$$

where E is the energy of the stationary state. Substituting this in (62.6), we find

$$\boldsymbol{H}^0\Psi + \frac{e\mathscr{H}}{2\mu c}(\boldsymbol{M}_z + \hbar\boldsymbol{\sigma}_z)\Psi = E\Psi. \tag{62.6'}$$

We take a representation in which the matrix $\boldsymbol{\sigma}_z$ is diagonal (the s_z representation); then

$$\boldsymbol{\sigma}_z\Psi = \begin{vmatrix} 1 & 0 \\ 0 & -1 \end{vmatrix}\begin{vmatrix} \psi_1 \\ \psi_2 \end{vmatrix} = \begin{vmatrix} \psi_1 \\ -\psi_2 \end{vmatrix}, \tag{62.9}$$

and so Equation (62.6′) is resolved into two equations for ψ_1 and ψ_2 separately:

$$\boldsymbol{H}^0\psi_1 + \frac{e\mathscr{H}}{2\mu c}(\boldsymbol{M}_z + \hbar)\psi_1 = E\psi_1, \tag{62.10}$$

$$\boldsymbol{H}^0\psi_2 + \frac{e\mathscr{H}}{2\mu c}(\boldsymbol{M}_z - \hbar)\psi_2 = E\psi_2. \tag{62.10'}$$

The solution of these equations is obtained immediately if we note that in the absence of a magnetic field there are two solutions:

$$\Psi'_{nlm} = \begin{pmatrix} \psi_{nlm} \\ 0 \end{pmatrix}, \qquad E = E^0_{nl} \qquad \text{for spin } s_z = +\tfrac{1}{2}\hbar, \tag{62.11}$$

$$\Psi''_{nlm} = \begin{pmatrix} 0 \\ \psi_{nlm} \end{pmatrix}, \qquad E = E^0_{nl} \qquad \text{for spin } s_z = -\tfrac{1}{2}\hbar, \tag{62.11'}$$

with

$$\psi_{nlm} = R_{nl}(r)\cdot Y_{lm}(\theta, \phi). \tag{62.12}$$

Since $\boldsymbol{M}_z\psi_{nlm} = \hbar m\psi_{nlm}$, these solutions are also solutions of Equations (62.10) and (62.10′), but belong to different eigenvalues. Substituting (62.11) and (62.11′) in (62.10) and (62.10′), we find two solutions:

$$\Psi'_{nlm}, \quad E = E'_{nlm} = E^0_{nl} + (e\hbar\mathscr{H}/2\mu c)(m+1), \quad s_z = +\tfrac{1}{2}\hbar, \tag{62.13}$$

$$\Psi''_{nlm}, \quad E = E''_{nlm} = E^0_{nl} + (e\hbar\mathscr{H}/2\mu c)(m-1), \quad s_z = -\tfrac{1}{2}\hbar, \tag{62.13'}$$

i.e. the wave functions are unchanged (since the term in $\mathscr{H}^2$ has been neglected): the atom is not deformed by the magnetic field. The energy, however, now depends on the orientation of the moment relative to the field, i.e. on the magnetic number m: levels which coincide in the absence of the magnetic field are now separated (the m degeneracy is removed).

Figure 46 shows the splitting of the s and p terms. The splitting of the p term is obtained from (62.13) and (62.13′) by considering the various possible values of m for $l = 1$ (i.e. $m = \pm 1, 0$). The splitting of the s term ($l = 0$, $m = 0$) is due only to the spin of the electron. This is an important result of spin theory: it was just this splitting which was observed by Stern and Gerlach in their experiments.

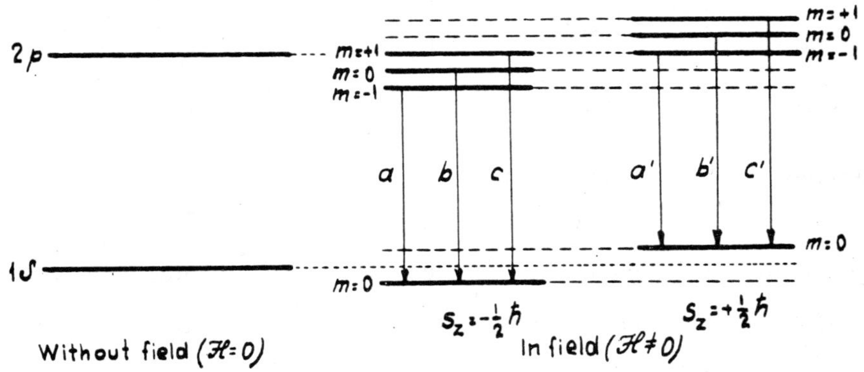

Fig. 46. Splitting of s and p terms in a strong magnetic field (taking spin into account).

Owing to the splitting of the levels the number of possible transitions is increased, and therefore so is the number of observable spectral lines. This is called the *normal Zeeman effect* (as opposed to the *anomalous Zeeman effect* discussed in Section 74). It will be shown in Section 90B that in optical transitions the number m can change only by ± 1 or 0. In addition, the spin magnetic moment has only a very weak interaction with the field of light waves. The calculation therefore involves only those transitions in which the spin is unchanged. These transitions are shown in Fig. 46 by the lines a, b, c and a', b', c'. The frequencies of these transitions are calculated from the formula

$$\omega_{n'l'm',\,n''l''m''} = \frac{E_{n'l'm'} - E_{n''l''m''}}{h} = \frac{E^0_{n'l'} - E^0_{n''l''}}{h} + \frac{e\mathscr{H}}{2\mu c}(m' - m''). \tag{62.14}$$

Denoting the frequencies in the absence of the field by ω^0, and those in the presence of the field by ω, we obtain

$$\omega = \omega^0 + (e\mathscr{H}/2\mu c)(m' - m''). \tag{62.15}$$

Since $m' - m'' = \pm 1$ or 0, we have three frequencies, one unchanged and two shifted by $\pm e\mathscr{H}/2\mu c$. This splitting into three lines (the *normal Zeeman triplet*) is just what is given by the classical theory of the Zeeman effect. In the classical theory ([59, 77, 62]; [95], p. 7–37; [94], p. 443), the Zeeman effect is explained by the precession of an orbit in the magnetic field with frequency equal to the Larmor frequency $O_L = e\mathscr{H}/2\mu c$. The quantum formula (62.15) does not involve Planck's constant $\hbar$, and so there must be agreement with the classical result (since no change results from putting $\hbar = 0$). This agreement is in fact found.

We shall show that in quantum mechanics also the Zeeman effect is due to a precessional motion of the angular momentum round the direction of the magnetic field. To do so, we calculate the time derivatives of the orbital and spin angular momenta. The general formula (31.10) gives

$$\frac{\mathrm{d}M_x}{\mathrm{d}t} = [H, M_x], \qquad \frac{\mathrm{d}M_y}{\mathrm{d}t} = [H, M_y], \qquad \frac{\mathrm{d}M_z}{\mathrm{d}t} = [H, M_z], \tag{62.16}$$

$$\frac{\mathrm{d}s_x}{\mathrm{d}t} = [H, s_x], \qquad \frac{\mathrm{d}s_y}{\mathrm{d}t} = [H, s_y], \qquad \frac{\mathrm{d}s_z}{\mathrm{d}t} = [H, s_z]. \tag{62.17}$$

Substituting the Hamiltonian from (62.6)

$$H = H^0 + (e\mathscr{H}/2\mu c)(M_z + \hbar\sigma_z) = H^0 + O_L M_z + 2\,O_L s_z, \tag{62.18}$$

and noting that H^0 commutes with $\vec{M}$ and $\vec{s}$, which also commute with each other (since $\vec{M}$ acts on functions of θ and ϕ, and $\vec{s}$ on functions of s_x, s_y, s_z), we find

$$\frac{\mathrm{d}M_x}{\mathrm{d}t} = \frac{O_L}{i\hbar}(M_x M_z - M_z M_x),$$

$$\frac{\mathrm{d}M_y}{\mathrm{d}t} = \frac{O_L}{i\hbar}(M_y M_z - M_z M_y), \qquad \frac{\mathrm{d}M_z}{\mathrm{d}t} = 0,$$

$$\frac{\mathrm{d}s_x}{\mathrm{d}t} = \frac{2\,O_L}{i\hbar}(s_x s_z - s_z s_x), \qquad \frac{\mathrm{d}s_y}{\mathrm{d}t} = \frac{2\,O_L}{i\hbar}(s_y s_z - s_z s_y), \qquad \frac{\mathrm{d}s_z}{\mathrm{d}t} = 0.$$

Using (25.5) and (59.1) we obtain

$$\frac{\mathrm{d}M_x}{\mathrm{d}t} = -\,O_L M_y, \qquad \frac{\mathrm{d}M_y}{\mathrm{d}t} = O_L M_x, \qquad \frac{\mathrm{d}M_z}{\mathrm{d}t} = 0, \tag{62.19}$$

$$\frac{\mathrm{d}s_x}{\mathrm{d}t} = -\,2\,O_L s_y, \qquad \frac{\mathrm{d}s_y}{\mathrm{d}t} = 2\,O_L s_x, \qquad \frac{\mathrm{d}s_z}{\mathrm{d}t} = 0. \tag{62.20}$$

On going from these operator formulae to the mean values, and bearing in mind that O_L is just a number, we find

$$\frac{\mathrm{d}\bar{M}_x}{\mathrm{d}t} = -\,O_L \bar{M}_y, \qquad \frac{\mathrm{d}\bar{M}_y}{\mathrm{d}t} = O_L \bar{M}_x, \qquad \frac{\mathrm{d}\bar{M}_z}{\mathrm{d}t} = 0, \tag{62.21}$$

$$\frac{\mathrm{d}\bar{s}_x}{\mathrm{d}t} = -\,2\,O_L \bar{s}_y, \qquad \frac{\mathrm{d}\bar{s}_y}{\mathrm{d}t} = 2\,O_L \bar{s}_x, \qquad \frac{\mathrm{d}\bar{s}_z}{\mathrm{d}t} = 0. \tag{62.22}$$

From these equations it follows that the components of the orbital and spin angular momenta in the direction of the magnetic field are both integrals of the motion. The component of the orbital angular momentum perpendicular to the magnetic field rotates with the Larmor frequency O_L, and the corresponding component of the spin angular momentum rotates with twice the frequency, $2\,O_L$ (owing to the anomalous ratio of the magnetic moment and the angular momentum; cf. (61.1)). For (62.21) gives

$$d^2\bar{M}_x/dt^2 = -O_L\,d\bar{M}_y/dt = -O_L^2\bar{M}_x, \qquad \bar{M}_y = -(1/O_L)\,d\bar{M}_x/dt. \tag{62.23}$$

Hence

$$\bar{M}_x = A\sin(O_Lt+\alpha), \quad \bar{M}_y = -A\cos(O_Lt+\alpha), \quad \bar{M}_z = \text{constant}. \tag{62.23'}$$

Similarly, from (62.22)

$$\bar{s}_x = B\sin(2\,O_Lt+\beta), \quad \bar{s}_y = -B\cos(2\,O_Lt+\beta), \quad \bar{s}_z = \text{constant}. \tag{62.24}$$

63. Motion of the spin in a variable magnetic field

In a variable magnetic field the intrinsic angular momentum of a particle will not be an integral of the motion, and therefore transitions are possible from one quantum state to another. In this section we shall consider a case of the motion of the spin of a particle in a variable field whose theory finds an important application in the measurement of the magnetic moments of atomic nuclei by Rabi's method. Figure 47 shows a diagram of Rabi's experiment.

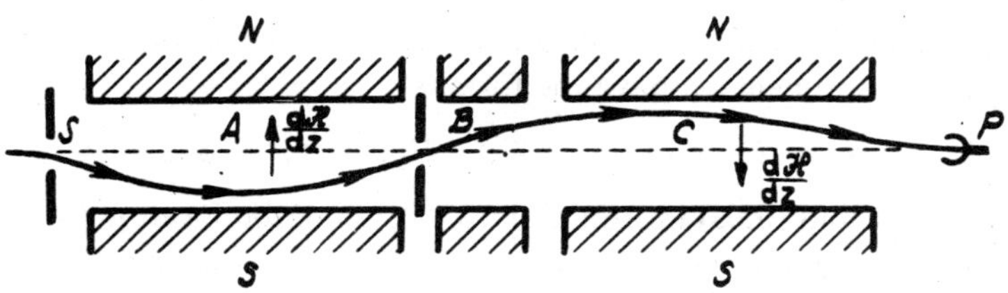

Fig. 47. Diagram of experiment to measure the magnetic moments of atomic nuclei (Rabi). S source of particle beam (slit), A first region of constant non-uniform magnetic field, C second region, B region of variable field, P particle collector.

The magnets A and C produce a non-uniform field constant in time, as in Stern and Gerlach's experiment, but the directions of the gradients of the fields in the magnets A and C are opposite. On passing through the non-uniform field in A a particle is deflected so that it cannot reach the collector P. This deflection is corrected by the field in C, which causes the particle to deviate in the opposite direction. As a result, the particle reaches the collector P as if it had simply moved in a straight line (as in the absence of the fields).

An additional variable field $\mathscr{H}_1$, able to reverse the magnetic moment of the particle, is applied in a small region B lying between A and C. If the magnetic moment of the particle is reversed as it passes through this field the deflection by the field in A will not be compensated by that due to the field in C, and the 'reversed' particles will not enter the collector P.

The frequency ω and strength $\mathscr{H}_1$ of the added variable field are so chosen that the probability of reversal of the magnetic moment is a maximum, and therefore the flux of particles to the collector P is a minimum. It will be shown below that, if we know ω and $\mathscr{H}_1$ corresponding to the maximum probability of reversal, we can determine the magnetic moment of the particle. This method of measurement of the magnetic moment is very accurate. Since we are concerned only with the motion of the spin (that of the centre of mass can be described by the methods of classical mechanics[9]), we need only write down Schrödinger's equation for the spin function $\mathbf{S}$ (60.5). This equation is[10]

$$i\hbar\, \mathrm{d}\mathbf{S}/\mathrm{d}t = -(\mathscr{H}\cdot\mathfrak{M})\,\mathbf{S}\,. \tag{63.1}$$

For simplicity we shall suppose that the particle has spin $\frac{1}{2}\hbar$. Then the magnetic moment $\mathfrak{M}$ is represented by a two-rowed matrix:

$$\mathfrak{M}_x = \mu\boldsymbol{\sigma}_x\,, \qquad \mathfrak{M}_y = \mu\boldsymbol{\sigma}_y\,, \qquad \mathfrak{M}_z = \mu\boldsymbol{\sigma}_z\,, \tag{63.2}$$

where $\boldsymbol{\sigma}_x$, $\boldsymbol{\sigma}_y$, $\boldsymbol{\sigma}_z$ are the Pauli matrices (59.9), (59.9′), and μ is the magnitude of the component of the magnetic moment in some direction. There is no such simple relation between the angular momentum $\mathbf{S}$ and the magnetic moment $\mathfrak{M}$ for nuclear particles, even the nucleons (the proton and the neutron), as there is for the electron (58.3). We shall therefore regard μ as merely some constant characteristic of the particle. The magnetic field in region B is assumed, in accordance with the arrangement of Rabi's experiment, to have the form[11]

$$\mathscr{H}_x = H_1 \cos\omega t\,, \qquad \mathscr{H}_y = H_1 \sin\omega t\,, \qquad \mathscr{H}_z = H_0\,. \tag{63.3}$$

Substituting (63.2) and (63.3) in Equation (63.1), using the Pauli matrices (59.9) and (59.9′) and the rule giving the effect of these matrices on the spin functions, we find an equation for the components S_1 and S_2 of the spin function (the first belonging to $\mathfrak{M}_z = +\mu$ and the second to $\mathfrak{M}_z = -\mu$):

$$i\hbar \frac{\mathrm{d}S_1}{\mathrm{d}t} = -\mu H_0 S_1 - \mu H_1 e^{-i\omega t} S_2\,, \tag{63.4}$$

$$i\hbar \frac{\mathrm{d}S_2}{\mathrm{d}t} = \mu H_0 S_2 - \mu H_1 e^{i\omega t} S_1\,. \tag{63.4′}$$

[9] This can be done for heavy particles (nuclei and atoms) but not for electrons. Bohr has shown that the Stern-Gerlach method can not be used to measure the magnetic moment of a free electron (see, for example, [67], p. 61).

[10] This equation does not contain the kinetic-energy operator, which in this case would have to be the kinetic energy of the intrinsic rotation of the particle. However, since S^2 remains constant, this energy should be regarded as constant, and so it need not be included in the equation.

[11] In Rabi's actual experiments the variable component of the magnetic field was linearly polarised, but in the calculations it is more convenient to take a field rotating in the xy plane; the results are not essentially different.

We shall suppose that at the time when the particle enters the variable field ($t = 0$) its magnetic moment is in the direction OZ, so that for $t = 0$ we have $S_1 = 1$, $S_2 = 0$. Putting

$$\nu = 2\mu H_0/\hbar, \qquad H_1/2H_0 = \Delta, \tag{63.5}$$

we can rewrite Equations (63.4) and (63.4′) in the form

$$dS_1/dt = \tfrac{1}{2} i\nu S_1 + i\nu \Delta e^{-i\omega t} S_2, \tag{63.6}$$

$$dS_2/dt = -\tfrac{1}{2} i\nu S_2 + i\nu \Delta e^{i\omega t} S_1. \tag{63.6′}$$

Differentiating (63.6′) with respect to time and using (63.6), we can eliminate the function S_1. The variable coefficient $e^{-i\omega t}$ also disappears. A simple calculation gives the following equation for S_2:

$$\frac{d^2 S_2}{dt^2} = -\left(\tfrac{1}{2}\omega\nu + \nu^2 \Delta^2 + \tfrac{1}{4}\nu^2\right) S_2 + i\omega \frac{dS_2}{dt}. \tag{63.7}$$

This is solved by substituting $S_2 = ae^{i\Omega t}$. The characteristic equation for the frequency Ω is then

$$\Omega^2 - \Omega\omega - \left(\tfrac{1}{2}\omega\nu + \nu^2 \Delta^2 + \tfrac{1}{4}\nu^2\right) = 0. \tag{63.8}$$

If we put

$$q^2 = \frac{\nu^2 + 4\nu^2 \Delta^2}{\omega^2} = \frac{1}{\omega^2}\left(\frac{\mu^2 H_0^2}{I^2} + \frac{\mu^2 H_1^2}{I^2}\right), \tag{63.9}$$

where $I = \frac{1}{2}\hbar$ is the spin component, and $\tan\theta = H_1/H_0$, it is easily seen that Equation (63.8) gives

$$\Omega = \tfrac{1}{2}\omega \pm \tfrac{1}{2}\omega\left(1 + q^2 + 2q\cos\theta\right)^{\frac{1}{2}} = \tfrac{1}{2}\omega \pm \delta. \tag{63.10}$$

The general solution for S_2 is

$$S_2(t) = a_1 e^{\frac{1}{2}i\omega t + i\delta t} + a_2 e^{\frac{1}{2}i\omega t - i\delta t}. \tag{63.11}$$

In accordance with the initial conditions we must put $a_1 = -a_2 = A/2i$, so that

$$S_2(t) = A e^{\frac{1}{2}i\omega t} \sin\delta t. \tag{63.11′}$$

The amplitude A is given by the condition $S_1(0) = 1$. Substituting (63.11′) in (63.6′) with $t = 0$ gives $A = i\nu\Delta/\delta$. Hence

$$S_2(t) = (i\nu\Delta/\delta)\, e^{\frac{1}{2}i\omega t} \sin\delta t. \tag{63.12}$$

The probability of finding at time t a magnetic moment $\mathfrak{M}_z = -\mu$ is

$$\begin{aligned} P(t) &= |S_2(t)|^2 = (\nu^2 \Delta^2/\delta^2)\sin^2\delta t \\ &= \frac{q^2 \sin^2\theta}{1 + q^2 + 2q\cos\theta} \sin^2\left\{\tfrac{1}{2}\omega\left(1 + q^2 + 2q\cos\theta\right)^{\frac{1}{2}} t\right\}. \end{aligned} \tag{63.13}$$

The time t in Rabi's experiment is equal to the time during which the particle passes through the region B. If the particle velocity is v and the length of the region B is l, then $t = l/v$.

In the experiment we take $q = 1$ and $\delta t = \frac{1}{2}\pi$ (in order to obtain the maximum reversal probability $P(t)$). Hence it is easy to estimate that for $v \approx 10^5$ cm/sec and $l = 1$ cm the frequency ω of the variable field is 10^6 c/s.

In order to judge the accuracy of this remarkable method, it may be mentioned that Rabi's procedure has been used to measure the magnetic moments μ of the proton (p) and the neutron (n), the results being $\mu_p = 2.7896 \pm 0.0002$, $\mu_n = 1.935 \pm 0.02$ (the unit being the Bohr nuclear magneton $e\hbar/2Mc$, with M the mass of the proton. This magneton is less than the electron magnetic moment by a factor 1842).

64. Properties of the total angular momentum

We have seen that both the orbital angular momentum **M** and the spin angular momentum **s** are quantities which take only discrete quantum values. Let us now consider the total angular momentum, which is the sum of the orbital and spin angular momenta.

The operator of the total angular momentum is given by the sum of the operators $\vec{M}$ and $\vec{s}$:

$$\vec{J} = \vec{M} + \vec{s}, \tag{64.1}$$

$$J_x = M_x + s_x, \qquad J_y = M_y + s_y, \qquad J_z = M_z + s_z. \tag{64.1'}$$

We shall show that the operators of the total angular momentum components obey the same commutation rules (25.5) as those of the orbital angular momentum M_x, M_y, M_z. To do so, we note that $\vec{M}$ and $\vec{s}$ commute, since the operator $\vec{M}$ acts on the coordinates but the operator $\vec{s}$ does not. Hence

$$\begin{aligned} J_xJ_y - J_yJ_x &= (M_x + s_x)(M_y + s_y) - (M_y + s_y)(M_x + s_x) \\ &= M_xM_y - M_yM_x + s_xs_y - s_ys_x = i\hbar M_z + i\hbar s_z, \end{aligned} \tag{64.2}$$

the last expression following from (25.5) and (59.1). Thus

$$J_xJ_y - J_yJ_x = i\hbar J_z, \tag{64.3}$$

$$J_yJ_z - J_zJ_y = i\hbar J_x, \tag{64.3'}$$

$$J_zJ_x - J_xJ_z = i\hbar J_y; \tag{64.3''}$$

the last two equations are obtained from the first by cyclic interchange.

Let us now find the operator J^2 of the squared total angular momentum. We have

$$\begin{aligned} J^2 &= (\vec{M} + \vec{s})^2 = \vec{M}^2 + \vec{s}^2 + 2\vec{M}\cdot\vec{s} \\ &= M^2 + s^2 + 2(M_xs_x + M_ys_y + M_zs_z). \end{aligned} \tag{64.4}$$

The operator J^2 commutes with any component of $\vec{J}$. Let us consider, for example, the component along OZ, $J_z = M_z + s_z$. Since M_z commutes with M^2, s^2 and s_z

with M^2, s^2, we have

$$J^2 J_z - J_z J^2 = 2(M_x s_x + M_y s_y + M_z s_z)(M_z + s_z) - \\ - 2(M_z + s_z)(M_x s_x + M_y s_y + M_z s_z).$$

Expansion of the parentheses gives

$$J^2 J_z - J_z J^2 = 2\{(M_x M_z - M_z M_x) s_x + (M_y M_z - M_z M_y) s_y + \\ + M_x (s_x s_z - s_z s_x) + M_y (s_y s_z - s_z s_y)\},$$

and, substituting here for the expressions in parentheses from (25.5) and (59.1),

$$J^2 J_z - J_z J^2 = 2\{- i\hbar M_y s_x + i\hbar M_x s_y + M_x(- i\hbar s_y) + \\ + M_y(+ i\hbar s_x)\} = 0.$$

The statement is similarly proved for the other two components. Thus

$$J^2 J_x - J_x J^2 = 0, \tag{64.5}$$

$$J^2 J_y - J_y J^2 = 0, \tag{64.5'}$$

$$J^2 J_z - J_z J^2 = 0; \tag{64.5''}$$

these equations have the same form as (25.6). Hence it follows that the operator J^2 and the operator of any one component, for instance J_z, can be brought to diagonal form simultaneously, and so the quantities J^2 and J_z can be simultaneously measured.

It is easy to see also that the operator J^2 commutes with the operators M^2 and s^2. For, using Formula (64.4), we immediately see this from the fact that M^2 commutes with M^2, M_x, M_y, M_z, s_x, s_y, s_z, s^2. Likewise s^2, which is a unit matrix (multiplied by $\frac{3}{4}\hbar^2$; see (59.13)), commutes with s_x, s_y and s_z. Hence

$$J^2 M^2 - M^2 J^2 = 0, \tag{64.6}$$

$$J^2 s^2 - s^2 J^2 = 0. \tag{64.6'}$$

Consequently J^2, M^2 and s^2 are also simultaneously measurable quantities. From (64.4)

$$\vec{M}\cdot\vec{s} = \tfrac{1}{2}(J^2 - M^2 - s^2). \tag{64.7}$$

Since **M·s** is derived from simultaneously measurable quantities, this scalar product can be measured simultaneously with J^2, M^2 and s^2.

Since

$$\vec{M}\cdot\vec{s} + s^2 = \vec{J}\cdot\vec{s}, \tag{64.8}$$

we obtain also from (64.7)

$$\vec{J}\cdot\vec{s} = \tfrac{1}{2}(J^2 - M^2 + s^2). \tag{64.9}$$

We shall show below that the squared total angular momentum J^2 and its component J_z in any direction are quantised like the orbital angular momentum, but the quantum numbers take half-integral values:

$$J^2 = \hbar^2 j(j+1), \qquad j = \tfrac{1}{2}, \tfrac{3}{2}, \tfrac{5}{2}, \dots ; \tag{64.10}$$

$$J_z = \hbar m_j, \qquad m_j = \pm\tfrac{1}{2}, \pm\tfrac{3}{2}, \dots, \pm j ; \tag{64.11}$$

the quantum number j, which gives the eigenvalues of the total angular momentum, can be expressed in terms of the orbital number l and the spin number l_s (59.14) by

$$j = l + l_s \qquad \text{or} \qquad j = |l - l_s| . \tag{64.12}$$

From the formulae for the eigenvalues of J^2 (64.10), M^2 (25.21) and s^2 (59.14) we obtain the following relations, which are important in spectroscopy, for the eigenvalues of $\vec{M}\cdot\vec{s}$ and $\vec{J}\cdot\vec{s}$:

$$\mathbf{M}\cdot\mathbf{s} = \tfrac{1}{2}\hbar^2 [j(j+1) - l(l+1) - l_s(l_s+1)] , \tag{64.13}$$

$$\mathbf{J}\cdot\mathbf{s} = \tfrac{1}{2}\hbar^2 [j(j+1) - l(l+1) + l_s(l_s+1)] . \tag{64.14}$$

We shall later apply these formulae in the theory of the anomalous Zeeman effect.

Let us now consider the proof of Formulae (64.10) and (64.11). The equation for the eigenvalues of J^2 is

$$J^2 \Psi = J^2 \Psi , \tag{64.15}$$

where Ψ signifies the column matrix

$$\Psi = \begin{vmatrix} \psi_1 \\ \psi_2 \end{vmatrix} . \tag{64.16}$$

Using (64.4), (59.13) and (59.12), (64.15) becomes in explicit form

$$M^2 \begin{vmatrix} \psi_1 \\ \psi_2 \end{vmatrix} + \tfrac{3}{4}\hbar^2 \begin{vmatrix} 1 & 0 \\ 0 & 1 \end{vmatrix} \begin{vmatrix} \psi_1 \\ \psi_2 \end{vmatrix} + 2\left\{ M_x \cdot \tfrac{1}{2}\hbar \begin{vmatrix} 0 & 1 \\ 1 & 0 \end{vmatrix} + M_y \cdot \tfrac{1}{2}\hbar \begin{vmatrix} 0 & -i \\ i & 0 \end{vmatrix} + \right.$$
$$\left. + M_z \cdot \tfrac{1}{2}\hbar \begin{vmatrix} 1 & 0 \\ 0 & -1 \end{vmatrix} \right\} \begin{vmatrix} \psi_1 \\ \psi_2 \end{vmatrix} = J^2 \begin{vmatrix} \psi_1 \\ \psi_2 \end{vmatrix} . \tag{64.17}$$

Effecting the multiplication and addition of matrices, we find

$$\begin{vmatrix} M^2\psi_1 + \tfrac{3}{4}\hbar^2\psi_1 + \hbar M_z\psi_1 + \hbar(M_x - iM_y)\psi_2 & 0 \\ M^2\psi_2 + \tfrac{3}{4}\hbar^2\psi_2 - \hbar M_z\psi_2 + \hbar(M_x + iM_y)\psi_1 & 0 \end{vmatrix} = \begin{vmatrix} J^2\psi_1 & 0 \\ J^2\psi_2 & 0 \end{vmatrix} , \tag{64.18}$$

and thus finally the two equations

$$M^2\psi_1 + \tfrac{3}{4}\hbar^2\psi_1 + \hbar M_z\psi_1 + \hbar(M_x - iM_y)\psi_2 = J^2\psi_1 , \tag{64.19}$$

$$M^2\psi_2 + \tfrac{3}{4}\hbar^2\psi_2 - \hbar M_z\psi_2 + \hbar(M_x + iM_y)\psi_1 = J^2\psi_2 . \tag{64.19'}$$

These are easily solved by putting

$$\psi_1 = aY_{lm}(\theta, \phi) , \qquad \psi_2 = bY_{l,\,m+1}(\theta, \phi) , \tag{64.20}$$

where $Y_{lm}(\theta, \phi)$ is a spherical harmonic and a and b are undetermined coefficients. Then

$$M^2\psi_1 = \hbar^2 l(l+1)\psi_1 , \qquad M_z\psi_1 = \hbar m\psi_1 , \tag{64.21}$$

$$M^2\psi_2 = \hbar^2 l(l+1)\psi_2 , \qquad M_z\psi_2 = \hbar(m+1)\psi_2 , \tag{64.21'}$$

and

$$(M_x - iM_y)Y_{lm} = -\hbar\sqrt{[(l+m)(l-m+1)]}\,Y_{l,m-1}\,, \tag{64.22}$$

$$(M_x + iM_y)Y_{lm} = -\hbar\sqrt{[(l-m)(l+m+1)]}\,Y_{l,m+1}\,. \tag{64.22'}$$

The two latter equations are derived from the properties of spherical harmonics.[12]

Substituting ψ_1 and ψ_2 from (64.20) in Equations (64.19) and using (64.21) and (64.22) we obtain, after cancelling $\hbar^2 Y_{lm}$ in the first equation and $\hbar^2 Y_{l,m+1}$ in the second,

$$[l(l+1) + \tfrac{3}{4} + m]a - \sqrt{[(l+m+1)(l-m)]}b = \lambda a\,, \tag{64.23}$$

$$[l(l+1) + \tfrac{3}{4} - m - 1]b - \sqrt{[(l+m+1)(l-m)]}a = \lambda b\,, \tag{64.23'}$$

where

$$\lambda = J^2/\hbar^2\,. \tag{64.24}$$

In order that these equations should have non-zero solutions it is necessary that their determinant should be zero. This gives an equation for λ:

$$\begin{vmatrix} l(l+1) + \tfrac{3}{4} + m - \lambda & -\sqrt{[(l+m+1)(l-m)]} \\ -\sqrt{[(l+m+1)(l-m)]} & l(l+1) + \tfrac{3}{4} - m - 1 - \lambda \end{vmatrix} = 0\,. \tag{64.25}$$

This gives the two roots

$$\lambda = (l + \tfrac{1}{2})^2 \pm (l + \tfrac{1}{2})\,. \tag{64.26}$$

Then from (64.24) we find the required eigenvalues

$$J^2 = \hbar^2(l + \tfrac{1}{2})(l + \tfrac{3}{2})\,, \tag{64.27}$$

$$J^2 = \hbar^2(l - \tfrac{1}{2})(l + \tfrac{1}{2})\,. \tag{64.27'}$$

The first value corresponds to addition of the orbital and spin angular momenta, and the second value to their subtraction. Substituting the value of λ in Equations (64.23) and solving, we find a and b, and thus also the eigenfunctions (64.20). They may also be normalised so that $a^2 + b^2 = 1$.

After some simple calculations we obtain the following functions: for the eigenvalue (64.27),

$$\psi_1 = \sqrt{\frac{l+m+1}{2l+1}}\,Y_{lm}\,, \qquad \psi_2 = -\sqrt{\frac{l-m}{2l+1}}\,Y_{l,m+1}\,; \tag{64.28}$$

for the eigenvalue (64.27')

$$\psi_1 = \sqrt{\frac{l-m}{2l+1}}\,Y_{lm}\,, \qquad \psi_2 = \sqrt{\frac{l+m+1}{2l+1}}\,Y_{l,m+1}\,. \tag{64.28'}$$

The solutions are seen to be degenerate: for a given l we can take various $m = 0, \pm 1, \pm 2, \ldots, \pm l$, and the eigenvalue J^2 does not depend on m. The reason for this degeneracy is that for a given magnitude J^2 of the angular momentum it can have various orientations in space. In order to see this, we may show that the solutions (64.28) and (64.28') are also eigenfunctions of the operator $\boldsymbol{J}_z$ of the component along OZ of the total angular momentum $\mathbf{J}$. The equation for the eigenfunctions of the operator $\boldsymbol{J}_z$ is

$$\boldsymbol{J}_z\Psi = J_z\Psi\,, \tag{64.29}$$

or explicitly

$$(\boldsymbol{M}_z + \boldsymbol{s}_z)\Psi = \left\{\boldsymbol{M}_z + \tfrac{1}{2}\hbar\begin{vmatrix} 1 & 0 \\ 0 & -1 \end{vmatrix}\right\}\begin{vmatrix} \psi_1 \\ \psi_2 \end{vmatrix} = J_z\begin{vmatrix} \psi_1 \\ \psi_2 \end{vmatrix}\,.$$

Hence, using (64.21), we find

$$\begin{vmatrix} (\hbar m + \tfrac{1}{2}\hbar)\psi_1 & 0 \\ (\hbar m + \hbar - \tfrac{1}{2}\hbar)\psi_2 & 0 \end{vmatrix} = \hbar(m + \tfrac{1}{2})\begin{vmatrix} \psi_1 \\ \psi_2 \end{vmatrix} = J_z\begin{vmatrix} \psi_1 \\ \psi_2 \end{vmatrix}\,, \tag{64.30}$$

i.e. the solutions found belong to the eigenvalue

$$J_z = \hbar(m + \tfrac{1}{2})\,. \tag{64.31}$$

---

[12] See Appendix V, Formulae (33) and (34).

Considering the solutions (64.28) and (64.28′) we see that in the first solution m can take the values $m = -(l+1)$ (when $\psi_1 = 0$), $-l, -l+1, \ldots, 0, 1, 2, \ldots, l$, and in the second solution the values $m = -l, -l+1, \ldots, 0, 1, 2, \ldots, l-1$ (for $m = l$, $\psi_1 = \psi_2 = 0$). Using the quantum number $j = l + \frac{1}{2} = l + l_s$ or $j = |l - l_s| = |l - \frac{1}{2}|$, we can write (64.27) and (64.27′) in the form (64.10). Finally, putting $m_j = m + \frac{1}{2}$ by reason of the above-mentioned possible values of m for a given l, we obtain (64.11).

65. Labelling of atomic terms having regard to the electron spin. Multiplet structure of spectra

The state of an electron in a field of central forces has been described by means of three quantum numbers n, l, m. The quantum levels E_{nl} of such an electron are defined by two quantum numbers n and l. We have entirely ignored the spin of the electron. If the spin is taken into account, each state $\psi_{nlm}(r, \theta, \phi)$ is in reality double, since two orientations of the spin are possible:

$$s_z = \hbar m_s, \qquad m_s = \pm \tfrac{1}{2}. \tag{65.1}$$

Thus, in addition to the three quantum numbers which define the state of the centre of mass of the electron, we have a fourth, m_s, which defines the spin of the electron. Let the wave function of the electron, taking spin into account, be denoted by $\psi_{nlmm_s}(r, \theta, \phi, s_z)$. Since the interaction of the spin with the orbital motion is for the moment neglected, according to (60.5) this function can be written in the form

$$\psi_{nlmm_s}(r, \theta, \phi, s_z) = \psi_{nlm}(r, \theta, \phi)\, S_{m_s}(s_z) \tag{65.2}$$

(with the suffix α to S being replaced by m_s). The corresponding quantum level is

$$E = E_{nl}. \tag{65.3}$$

The four quantum numbers can take the following values:

$$n = 1, 2, 3, \ldots, \qquad 0 \leqslant l \leqslant n-1, \qquad -l \leqslant m \leqslant l, \qquad m_s = \pm \tfrac{1}{2}. \tag{65.4}$$

For each term E_{nl} we have $2l + 1$ states differing in the orientation of the orbital angular momentum, each of which in turn consists of two states of different spin, making in all $2(2l + 1)$ states. Thus there is $2(2l + 1)$ – fold degeneracy.

If we now take into account the weak interaction of the spin with the magnetic field of the orbital currents, the energy of the state will also depend on the orientation of the spin **s** relative to the orbital angular momentum **M**. We shall not describe the calculation of this interaction, since the correction for the interaction of the spin with the orbital motion is of the same order as the correction resulting from the velocity dependence of the mass of the electron. A correct calculation of the splitting of the levels therefore involves in this case the relativistic equation for the motion of the electron, and a consideration of this lies outside the scope of the present book.[13] We shall merely give a qualitative analysis of this splitting and an estimate

[13] See [29]. A calculation of the splitting is given in [5].

of its magnitude. The magnetic moment $\mathfrak{M}_B$ of the electron is in the field $\mathscr{H}_l$ of the orbital current. Its energy in this field is

$$\Delta E = -\mathfrak{M}_B \cdot \mathscr{H}_l. \tag{65.5}$$

We can estimate the magnetic field $\mathscr{H}_l$ as that of a dipole equivalent to the orbital currents, i.e. a dipole with moment $\mathfrak{M}_l$. This field is[14]

$$\mathscr{H}_l = \frac{3(\mathfrak{M}_l \cdot \mathbf{r})\,\mathbf{r}}{r^5} - \frac{\mathfrak{M}_l}{r^3}, \tag{65.6}$$

where $\mathbf{r}$ is the radius vector between the dipoles $\mathfrak{M}_l$ and $\mathfrak{M}_B$. Since we want only the order of magnitude of ΔE, we can take $\mathscr{H}_l \approx \mathfrak{M}_l/a^3$, where a is a length of the order of interatomic distances (10^{-8} cm). Then

$$\Delta E \approx \frac{\mathfrak{M}_B \mathfrak{M}_l}{a^3} \cos(\mathfrak{M}_B, \mathscr{H}_l). \tag{65.7}$$

The quantities $\mathfrak{M}_l$, $\mathfrak{M}_B$ are of the same order as the Bohr magneton (9×10^{-21} erg/G), and by the properties of spin $\cos(\mathfrak{M}, \mathscr{H})$ can take only the two values ± 1 (depending on whether the spin is parallel or antiparallel to the field $\mathscr{H}_l$). Substituting in (65.7) the numerical values, we find $\Delta E \approx \pm 8 \times 10^{-15}$ erg. This is small in comparison with the energy difference between levels with different n and l, and so the resulting spectral lines are close together. In particular, for the sodium doublet mentioned in Section 58 (wavelengths 5896 Å and 5890 Å), $\Delta E = 2.8 \times 10^{-15}$ erg.

Thus the difference in orientation of the spin magnetic moment relative to the internal magnetic field of the atom can explain the occurrence of the multiplet structure of spectral lines.

It follows from the above discussion that for atoms with a single optical electron only *doublets* (pairs of lines) are possible, corresponding to the two orientations of the electron spin. This theoretical deduction is fully confirmed by examination of spectral lines.[15]

Let us now consider the labelling of the levels of an atom when the multiplet structure is taken into account. When allowance is made for the spin-orbit interaction neither the orbital angular momentum $\mathbf{M}$ nor the spin angular momentum $\mathbf{s}$ has a definite value in a state with a definite energy (they do not commute with the Hamiltonian operator). According to classical mechanics we should have a precession of the vectors $\mathbf{M}$ and $\mathbf{s}$ round the total angular momentum vector $\mathbf{J}$:

$$\mathbf{J} = \mathbf{M} + \mathbf{s}, \tag{65.8}$$

as shown in Figure 48.[16] The total angular momentum $\mathbf{J}$ remains constant. A corresponding situation occurs in quantum mechanics also. When the spin interaction is taken into account, only the total angular momentum $\mathbf{J}$ has a definite value in a

[14] See, for example, [82], p. 269; [101], p. 184.
[15] Details are given in [44]; [99], Ch. IV.
[16] Details regarding the semiclassical theory are given in [23].

state with a given energy (it commutes with the Hamiltonian operator $\boldsymbol{H}$). Hence, when the spin-orbit interaction is taken into account, the classification must be by values of the total angular momentum **J**.

As shown in Section 64, the total angular momentum is quantised by the same rules as the orbital angular momentum: in terms of a quantum number j which determines the total angular momentum **J** we have

$$J^2 = \hbar^2 j(j+1), \tag{65.9}$$

and the component of **J** in an arbitrary direction OZ has the values

$$J_z = \hbar m_j ; \tag{65.10}$$

here

$$j = l + l_s, \qquad l_s = \tfrac{1}{2} \tag{65.11}$$

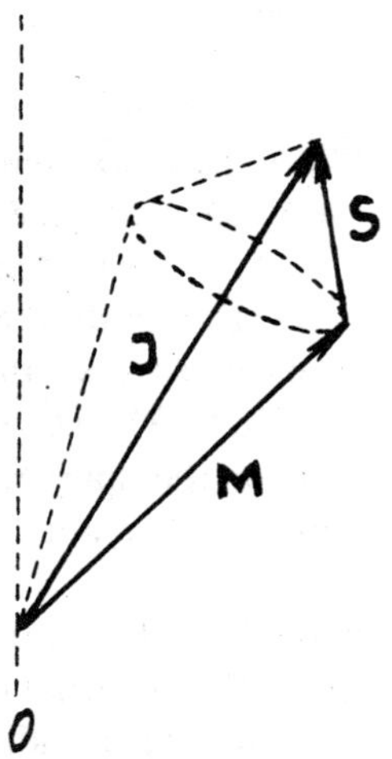

Fig. 48. Addition of spin and orbital angular momenta and their precession round the direction of the total angular momentum **J**.

if the spin and orbital angular momenta are parallel, and

$$j = |l - l_s| \tag{65.12}$$

if they are antiparallel. Similarly the quantum number m_j which determines the component J_z is

$$m_j = m + m_s, \qquad m_s = \pm\tfrac{1}{2}. \tag{65.13}$$

Since l and m are integers, while l_s and m_s are half-integers, we have

$$j = \tfrac{1}{2}, \tfrac{3}{2}, \tfrac{5}{2}, \ldots, \qquad m_j = \pm\tfrac{1}{2}, \pm\tfrac{3}{2}, \ldots, \pm j. \tag{65.14}$$

Depending on the orientation of the spin, the term energy will be different for $j = l + \frac{1}{2}$ and $j = |l - \frac{1}{2}|$. In this case, therefore, the energy levels must be described by the values of the principal number n, the value of the orbital number l, and the number j which gives the total angular momentum, i.e. in this case

$$E = E_{nlj}. \tag{65.15}$$

The wave functions will depend on the spin variable s_z and are different for different j:

$$\psi_{nljm_j} = \psi_{nljm_j}(r, \theta, \phi, s_z). \tag{65.16}$$

(In this case the variables r, θ, ϕ, s_z do not separate.) The quantum levels for a given l and different j are close together, the separation in energy being just the difference in the spin-orbit interaction energy for the two different orientations of the spin. The four numbers n, l, j, m_j can take the following values:

$$n = 1, 2, 3, \dots, \tag{65.17}$$

$$0 \leqslant l \leqslant n-1, \tag{65.17'}$$

$$j = l + l_s \quad \text{or} \quad |l - l_s|, \quad l_s = \tfrac{1}{2}, \tag{65.17''}$$

$$-j \leqslant m_j \leqslant j. \tag{65.17'''}$$

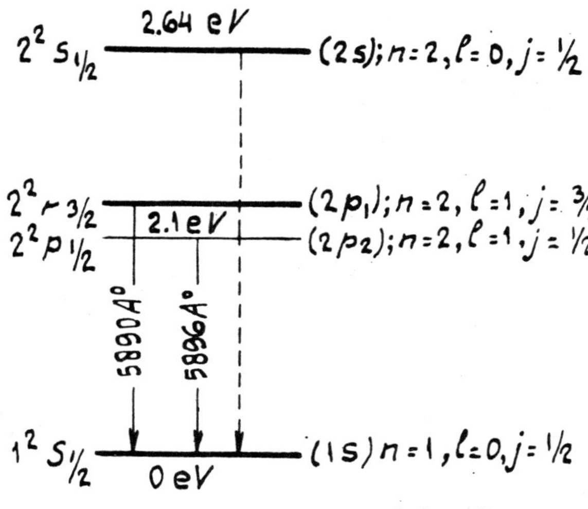

Fig. 49. Multiplet structure of the $2p$ term of the sodium atom. The lines 5889.963 Å and 5895.930 Å form the well-known sodium doublet, the yellow lines D_2 and D_1. The $2s$ term is far removed from the $2p$ terms, as is to be expected in hydrogen-like atoms (the l degeneracy is removed).

The orbital angular momentum l is denoted in spectroscopy by letters (as already mentioned): $s(l=0)$, $p(l=1)$, $d(l=2)$, $f(l=3)$, … . The principal quantum number n is placed before the letter. The number j is shown at the bottom right. Thus, for example, the level (or term) with $n = 3$, $l = 1$, $j = \frac{3}{2}$ is denoted by $3p_{\frac{3}{2}}$. Sometimes a further figure is added, as $3\,^2p_{\frac{3}{2}}$; the 2 at the top left indicates that the term $3\,^2p_{\frac{3}{2}}$ is a doublet term. Where there is only one optical electron this indication is superfluous, since all the levels are doublet levels ($j = l + l_s$ and $j = |l - l_s|$, except of course the s levels, where $l = 0$).

For helium we find a more complex multiplet structure. Owing to the presence of two electrons there are singlet and triplet terms (see Section 122). In order to distinguish these cases the figure indicating the multiplicity of the level is retained. For instance, the level denoted usually by $E_{3,1,\frac{1}{2}}$ (65.15) is denoted in spectroscopy by $3\,^2p_{\frac{1}{2}}$.

Figure 49 shows a level diagram for a hydrogen-like atom (i.e. an atom with a single optical electron), taking into account the multiplet structure. The diagram also shows the quantum numbers and the spectroscopic nomenclature.

Each of the levels E_{nlj} corresponds to $2j + 1$ states with different m_j, i.e. a different orientation of the total angular momentum **J** in space. These coincident levels are separated only when an external field is applied (see the theory of the anomalous Zeeman effect, Section 74). In the absence of such a field there is $(2j + 1)$-fold degeneracy. For instance, the $2s_{\frac{1}{2}}$ term has degeneracy 2: two states, differing in the orientation of the spin. The $2p_{\frac{3}{2}}$ term has fourfold degeneracy, corresponding to the orientations of **J**: $m_j = \pm\frac{1}{2}, \pm\frac{3}{2}$.

CHAPTER XI

PERTURBATION THEORY

66. Statement of the problem

The problem of finding the quantum levels of a system (i.e. of finding the eigenvalues and eigenfunctions of the energy operator $\boldsymbol{H}$) can be solved by means of known mathematical functions in only a very few cases. In most problems of atomic mechanics such simple solutions do not exist. Considerable importance therefore attaches to a wide range of cases where the problem concerned can be approximately reduced to that of a simpler system for which the eigenvalues E_n^0 and eigenfunctions ψ_n^0 are known. This possibility arises when the energy operator $\boldsymbol{H}$ of the system under consideration differs only slightly from the operator $\boldsymbol{H}^0$ of the simpler system.

The exact meaning of the words 'differs only slightly' will be explained below. Here we shall mention the types of problem which can be approximately solved. Let us assume that we know the wave functions and the quantum levels of the electrons moving in an atom, and wish to find how the quantum levels and wave functions are changed if the atom is placed in an external electric or magnetic field.

The fields attained in experiments are usually small in comparison with the Coulomb field within the atom.[1] The interaction of the external field can be regarded as a small correction or, as we shall call it, a *perturbation* (a term taken from celestial mechanics, and originally used to describe the effect of one planet on the orbit of another). A similar procedure can be used to take account of the weak interactions of electrons within atoms, for example magnetic interactions and sometimes even Coulomb interactions. The general methods of solution of such problems form the subject of perturbation theory.

We shall first of all consider only cases where the energy operator $\boldsymbol{H}$ has a discrete spectrum. Let the given Hamiltonian be

$$\boldsymbol{H} = \boldsymbol{H}^0 + \boldsymbol{W}. \tag{66.1}$$

We regard the difference $\boldsymbol{W}$ as being small, and call it the *perturbation energy*, or sometimes simply the *perturbation*; and we further assume that the eigenvalues E_n^0 and eigenfunctions ψ_n^0 of the operator $\boldsymbol{H}^0$ are known, so that

$$\boldsymbol{H}^0 \psi_n^0 = E_n^0 \psi_n^0. \tag{66.2}$$

[1] An electric field may reach values comparable with the fields within the atom; cf. Section 101.

The problem is to find the eigenvalues E_n of the operator $\boldsymbol{H}$ and its eigenfunctions. This problem, as we know, reduces to that of solving Schrödinger's equation

$$\boldsymbol{H}\psi = E\psi . \tag{66.3}$$

This differs from (66.2) only by the term $\boldsymbol{W}\psi$, which is assumed small.

For an approximate solution of the problem by the method of perturbation theory, we first of all write Equation (66.3) in a representation in which the fundamental variable is the eigenvalues E_n^0 of the operator $\boldsymbol{H}^0$, i.e. Equation (66.3) is taken in the E^0 representation. If the operator $\boldsymbol{H}$ (66.1) and therefore Equation (66.3) are initially given in the co-ordinate representation, as is usually the case, we must change from this to the E^0 representation. This is done as follows. We shall everywhere write only one co-ordinate x explicitly; if necessary, x may be taken as representing any number of variables, just as the symbol n in the wave function ψ_n may be taken as representing several quantum numbers. Let the eigenfunctions of the operator $\boldsymbol{H}^0$ in the co-ordinate representation (the x representation) be $\psi_n^0(x)$. We expand the required function $\psi(x)$ in terms of the functions $\psi_n^0(x)$:

$$\psi(x) = \sum_n c_n \psi_n^0(x) . \tag{66.4}$$

Then the c_n form the function ψ in the E^0 representation.

Substituting (66.4) in Equation (66.3), multiplying by $\psi_m^{0*}(x)$ and integrating with respect to x, we obtain

$$\sum_n H_{mn} c_n = E c_m , \tag{66.5}$$

where H_{mn} is a matrix element of the operator $\boldsymbol{H}$ in the E^0 representation:

$$H_{mn} = \int \psi_m^{0*}(x) \cdot \boldsymbol{H}\psi_n^0 \cdot \mathrm{d}x . \tag{66.6}$$

The matrix formed by the elements H_{mn} is the operator $\boldsymbol{H}$ in the E^0 representation. Using (66.1) and (66.2), we find

$$\begin{aligned} H_{mn} &= \int \psi_m^{0*} \cdot (\boldsymbol{H}^0 + \boldsymbol{W}) \psi_n^0 \cdot \mathrm{d}x \\ &= \int \psi_m^{0*} \cdot \boldsymbol{H}^0 \psi_n^0 \cdot \mathrm{d}x + \int \psi_m^{0*} \cdot \boldsymbol{W}\psi_n^0 \cdot \mathrm{d}x = E_n^0 \delta_{mn} + W_{mn} , \end{aligned} \tag{66.6$'$}$$

where W_{mn} is the matrix element of the perturbation energy in the E^0 representation:

$$W_{mn} = \int \psi_m^{0*} \cdot \boldsymbol{W}\psi_n^0 \cdot \mathrm{d}x . \tag{66.7}$$

The matrix formed by the elements W_{mn} is the operator $\boldsymbol{W}$ in the same representation. Substituting (66.6′) in (66.5), we have

$$\sum_n [E_n^0 \delta_{mn} + W_{mn}] c_n = E c_m . \tag{66.8}$$

Taking all the terms to the left-hand side gives

$$[E_m^0 + W_{mm} - E] c_m + \sum_{n \neq m} W_{mn} c_n = 0 , \tag{66.9}$$

where n and m take all values corresponding to the labelling of the functions ψ_n^0 of the unperturbed system.

So far we have made no use of the assumption that W is small, and Equation (66.9) is rigorously true. The problem of perturbation theory is to utilise the assumption that the quantities W_{mn} are small. In order to make the degree of smallness of W explicit we put

$$W = \lambda w, \tag{66.10}$$

where λ is a small parameter; for $\lambda = 0$ the operator H becomes H^0. Then Equation (66.9) becomes

$$[E_m^0 + \lambda w_{mm} - E]\, c_m + \lambda \sum_{n \neq m} w_{mn}\, c_n = 0. \tag{66.11}$$

This can be solved in powers of λ, regarded as a small quantity. For $\lambda = 0$, Equation (66.11), becomes simply (66.2) in the E^0 representation:

$$(E_m^0 - E)\, c_m = 0, \tag{66.12}$$

whose solutions are

$$E^{(0)} = E_m^0, \qquad c_m^{(0)} = 1. \tag{66.13}$$

For small values of λ we should naturally expect that the solutions of Equations (66.11) would be close to those of (66.12), i.e. to (66.13). This can be explicitly stated by representing the eigenfunctions c_m of Equation (66.11) and its eigenvalues E_m as series of powers of the small parameter λ:

$$c_m = c_m^{(0)} + \lambda c_m^{(1)} + \lambda^2 c_m^{(2)} + \dots \tag{66.14}$$

and

$$E_m = E_m^{(0)} + \lambda E^{(1)} + \lambda^2 E^{(2)} + \dots. \tag{66.15}$$

For $\lambda = 0$, (66.14) and (66.15) become (66.13), and $E_m^{(0)}$ must be E_m^0. It is found that the solution of Equations (66.11) depends considerably on whether the states of the system H^0 are degenerate or not. If they are degenerate, several eigenfunctions ψ_m^0 belong to each eigenvalue E_m^0, but if they are not there is only one such function. We shall discuss these two cases separately.

67. Perturbation in the absence of degeneracy

Suppose that only one eigenfunction ψ_n^0, and correspondingly one amplitude c_n^0, belongs to each eigenvalue E_n^0 of the unperturbed equation (66.2). We substitute the series (66.14) and (66.15) in Equation (66.11) and collect like powers. This gives

$$\begin{aligned}
&[E_m^0 - E^{(0)}]\, c_m^{(0)} + \lambda \{[w_{mm} - E^{(1)}]\, c_m^{(0)} + \\
&\quad + [E_m^0 - E^{(0)}]\, c_m^{(1)} + \sum_{n \neq m} w_{mn}\, c_n^{(0)}\} + \\
&\quad + \lambda^2 \{[w_{mm} - E^{(1)}]\, c_m^{(1)} - E^{(2)}\, c_m^{(0)} + \\
&\quad + [E_m^0 - E^{(0)}]\, c_m^{(2)} + \sum_{n \neq m} w_{mn}\, c_n^{(1)}\} + \dots = 0.
\end{aligned} \tag{67.1}$$

This representation of Equation (66.11) makes it easy to obtain a solution by successive approximations. The zeroth-order approximation is obtained by putting $\lambda = 0$, which gives

$$[E_m^0 - E^{(0)}]\, c_m^{(0)} = 0\,, \qquad m = 1, 2, 3, \ldots, k, \ldots. \tag{67.2}$$

This is the equation for the unperturbed system $\boldsymbol{H}^0$. Let us consider the change in the level E_k^0 and the eigenfunction ψ_k^0 due to the perturbation $\boldsymbol{W}$. Then we take the kth solution of (67.2):

$$E^{(0)} = E_k^0\,, \qquad c_m^{(0)} = \delta_{mk}\,, \tag{67.3}$$

i.e. all the $c_m^{(0)}$ are zero except $c_k^{(0)}$, which is unity.

We call the solution (67.3) the *zeroth-order approximation*, and substitute it in Equation (67.1) in order to find the next (first-order) approximation. The substitution gives

$$\lambda\left\{[w_{mm} - E^{(1)}]\,\delta_{mk} + (E_m^0 - E_k^0)\, c_m^{(1)} + \sum_{n \neq m} w_{mn}\,\delta_{nk}\right\} + O(\lambda^2) = 0\,, \tag{67.4}$$

where $O(\lambda^2)$ denotes terms of order λ^2 and higher. In the first approximation we must neglect these terms as being small. This gives

$$[w_{mm} - E^{(1)}]\,\delta_{mk} + (E_m^0 - E_k^0)\, c_m^{(1)} + \sum_{n \neq m} w_{mn}\,\delta_{nk} = 0\,. \tag{67.4$'$}$$

The equation with $m = k$ shows that

$$w_{kk} - E^{(1)} = 0\,, \tag{67.4$''$}$$

whence the correction in the first approximation to E_k^0 is

$$E^{(1)} = w_{kk}\,. \tag{67.5}$$

The equations with $m \neq k$ give the corrections $c_m^{(1)}$ to the amplitudes: from (67.4′)

$$(E_m^0 - E_k^0)\, c_m^{(1)} + w_{mk} = 0\,. \tag{67.4$'''$}$$

Hence

$$c_m^{(1)} = w_{mk}/(E_k^0 - E_m^0)\,, \qquad m \neq k\,. \tag{67.6}$$

Let us now find the second approximation, in which terms in λ^2 must be taken into account. Substitution of the first approximation (67.5) and (67.6) in (67.1) gives

$$\begin{aligned} \lambda^2 \Bigg\{ &(w_{mm} - w_{kk}) \frac{w_{mk}}{E_k^0 - E_m^0} - E^{(2)}\,\delta_{mk} + \\ &+ (E_m^0 - E_k^0)\, c_m^{(2)} + \sum_{n \neq k} w_{mn} \frac{w_{nk}}{E_k^0 - E_n^0} \Bigg\} + O(\lambda^3) = 0\,, \end{aligned} \tag{67.7}$$

where $O(\lambda^3)$ denotes terms of order λ^3 and higher. Neglecting these terms, we obtain equations to determine $E^{(2)}$ and $c_m^{(2)}$ (second approximation). The equation with

$m = k$ is

$$- E^{(2)} + \sum_{k \neq n} \frac{w_{kn} w_{nk}}{E_k^0 - E_n^0} = 0. \tag{67.7$'$}$$

Thus the correction to the energy in the second approximation is

$$E^{(2)} = \sum_{k \neq n} \frac{w_{kn} w_{nk}}{E_k^0 - E_n^0}. \tag{67.8}$$

The equations with $m \neq k$ give $c_m^{(2)}$:

$$c_m^{(2)} = - \frac{w_{kk} w_{mk}}{(E_m^0 - E_k^0)^2} + \sum_n \frac{w_{mn} w_{nk}}{(E_k^0 - E_n^0)(E_k^0 - E_m^0)}, \quad m \neq k, \quad n \neq k. \tag{67.9}$$

The same procedure can be continued to higher and higher approximations, but we shall not go beyond the second approximation. The results are, using (66.14), (66.15), (67.3), (67.5), (67.6), (67.8) and (67.9),

$$E_k = E_k^0 + \lambda w_{kk} + \lambda^2 \sum_{n \neq k} \frac{w_{kn} w_{nk}}{E_k^0 - E_n^{(0)}} + O(\lambda^3), \tag{67.10}$$

$$\left. \begin{aligned} c_k = 1, \qquad c_m = \lambda \frac{w_{mk}}{E_k^0 - E_m^0} + \\ + \lambda^2 \left\{ \sum_n \frac{w_{mn} w_{nk}}{(E_k^0 - E_n^0)(E_k^0 - E_m^0)} - \frac{w_{kk} w_{mk}}{(E_m^0 - E_k^0)^2} \right\} + O(\lambda^3). \end{aligned} \right\} \tag{67.11}$$

These formulae show that the assumption that the operator $\boldsymbol{W}$ is small in comparison with $\boldsymbol{H}^0$ signifies that the ratio

$$|\lambda w_{nm}/(E_n^0 - E_m^0)| \ll 1, \qquad n \neq m; \tag{67.12}$$

when this condition is satisfied the correction terms in (67.10) and (67.11) are small and the eigenvalues E_k of the operator $\boldsymbol{H}$ and its eigenfunctions $c_m(k)$ are close to the eigenvalues and eigenfunctions of the operator $\boldsymbol{H}^0$. The condition (67.12) is the condition for perturbation theory to be applicable. From (66.10), this condition can also be written

$$|W_{mn}|/|E_m - E_n| \ll 1, \qquad n \neq m, \tag{67.13}$$

where W_{mn} are the matrix elements of the perturbation operator.

Using (66.4), (67.6) and (67.5) we can write the above solution in the x representation as

$$\psi_k(x) = \psi_k^0(x) + \sum_{m \neq k} \frac{W_{mk}}{E_k^0 - E_m^0} \psi_m^0(x) + \ldots, \tag{67.14}$$

$$E_k = E_k^0 + W_{kk} + \ldots, \qquad W_{kk} = \int \psi_k^{0*} \cdot \boldsymbol{W} \psi_k^0 \cdot \mathrm{d}x. \tag{67.15}$$

The latter formula shows that the correction to the levels in the first approximation is equal to the mean value of the perturbation energy in the unperturbed state (ψ_k^0).

From the condition (67.13) of applicability of the method of perturbation theory it is seen immediately that the success of the approximate calculation depends on which quantum level is being considered. For example, in a Coulomb field the energy differences of neighbouring levels are given by the formula

$$E_n{}^0 - E_{n\pm 1}{}^0 = -E_1{}^0\left(\frac{1}{n^2} - \frac{1}{(n\pm 1)^2}\right) = \frac{\mp 2n - 1}{n^2(n\pm 1)^2}E_1{}^0 .$$

For small n this quantity may be much greater than $W_{n,\,n\pm 1}$, but for large n it tends to zero as $1/n^3$, and the condition (67.13) may not be satisfied. The method of perturbation theory may therefore be valid for calculating the corrections for lower quantum levels but not for the higher quantum levels. This fact must be borne in mind in using perturbation theory for specific problems.

Another point which should be noted is that there are some special cases when the condition (67.13) is satisfied but nevertheless the quantum states of the systems $\boldsymbol{H}$ and $\boldsymbol{H}^0$ are totally different. The reason is that the perturbation energy W may be such that it causes an appreciable change in the asymptotic behaviour of the potential energy $U(x)$. Let us assume that a perturbation $W = \lambda x^3$ is applied to a harmonic oscillator. Then Schrödinger's equation is

$$-\frac{\hbar^2}{2\mu}\frac{\mathrm{d}^2\psi}{\mathrm{d}x^2} + \tfrac{1}{2}\mu\omega_0{}^2x^2\psi + \lambda x^3\psi = E\psi . \tag{67.16}$$

For $\lambda = 0$ we have the equation of the harmonic oscillator, with a discrete energy spectrum $E_n = \hbar\omega_0(n + \frac{1}{2})$. The perturbation matrix elements $w_{mn} = \lambda(x^3)_{mn}$ for small λ may be arbitrarily small in comparison with $E_m - E_n = \hbar\omega_0(m - n)$. Nevertheless, for any λ other than $\lambda = 0$ Equation (67.16) has a *continuous* spectrum of eigenvalues. For the potential energy $U(x) = \frac{1}{2}\mu\omega_0{}^2x^2 + \lambda x^3$ has the form shown in Figure 50. For any E, $U(x) < E$ for large negative x, i.e. the asymptotic value of the potential energy is less than E, and the energy spectrum must therefore be continuous.

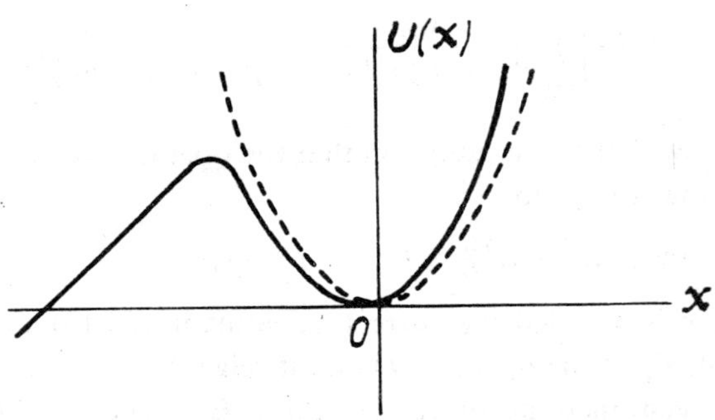

Fig. 50. Curve of the potential energy $U(x) = \frac{1}{2}\mu\omega_0{}^2x^2 + \lambda x^3$. The broken curve shows $U_0(x) = \frac{1}{2}\mu\omega_0{}^2x^2$.

The question arises as to the significance in this case of the approximate functions $\psi_n(x)$ and levels E_n which we can calculate from $\psi_n{}^0$ and $E_n{}^0$ by the method of perturbation theory, using the fact that the parameter λ is small. It is found that for small λ the functions $\psi_n(x)$ found by perturbation theory are large near the potential well of $U(x)$ and small outside it. Figure 51 shows again the curve of the potential energy $U(x)$ from Figure 50 and also the squared modulus $|\psi(x)|^2$ of the wave function. Figure 51a corresponds to the case where the energy $E = E_n \approx E_n{}^0$. If the energy E is not equal to E_n, the wave function $\psi_E(x)$ increases far from the potential well of $U(x)$ (Figure 51b). In the former case we can say that the particles are near the equilibrium position $x = 0$ (so to speak 'in the atom') while in the latter case they are mainly outside the atom at an infinite distance. Stationary states can occur only if waves exist both going to and coming from infinity so that the flux of particles through a surface surrounding the atom is zero. This case is of little interest; usually we have only outgoing waves (see Section 99). There will then be no steady states. If we require that only outgoing waves should exist, the functions $\psi_n(x)$ found by perturbation theory describe the behaviour of particles only during a certain time t, but this time increases as the parameter λ decreases and may be quite long. Such states $\psi_n(x)$ and the corresponding levels $E_n{}^0$ will be called *quasistationary*.

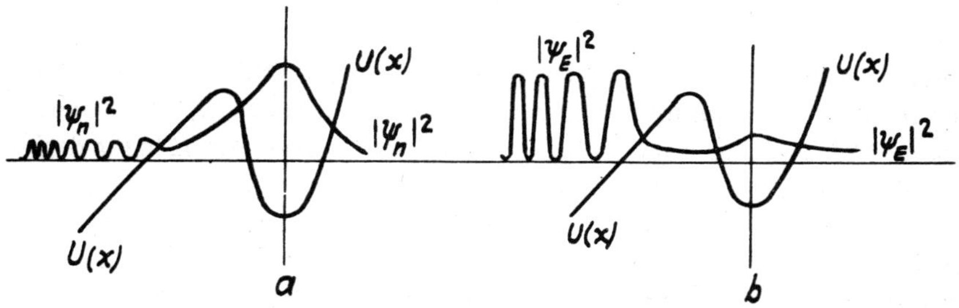

Fig. 51. Potential energy $U(x) = \frac{1}{2}\mu\omega_0{}^2x^2 + \lambda x^3$ and probability density $|\psi|^2$: (a) for $E = E_n$, (b) for $E \neq E_n$.

68. Perturbation in the presence of degeneracy

In most problems important in applications, degeneracy occurs, that is, in the unperturbed system ($\boldsymbol{H}^0$) not just one state ψ_n^0 but several states $\psi_{n1}^0, \psi_{n2}^0, \ldots, \psi_{n\alpha}^0, \ldots, \psi_{nf}^0$ belong to the eigenvalue $E = E_n^0$. If some perturbation $\boldsymbol{W}$ now acts, we cannot say without a special investigation which of the functions $\psi_{n\alpha}^0$ will be the zeroth-order approximation to the eigenfunctions of the operator $\boldsymbol{H} = \boldsymbol{H}^0 + \boldsymbol{W}$. For, instead of the functions $\psi_{n1}^0, \ldots, \psi_{n\alpha}^0, \ldots, \psi_{nf}^0$ which belong to the eigenvalue E_n^0, we can take new functions $\phi_{n1}^0, \phi_{n2}^0, \ldots, \phi_{n\alpha}^0, \ldots, \phi_{nf}^0$ obtained from the former by the linear orthogonal transformation

$$\phi_{n\alpha}^0 = \sum_{\beta=1}^{f} a_{\alpha\beta}\psi_{n\beta}^0, \tag{68.1}$$

$$\sum_{\beta=1}^{f} a_{\alpha\beta}a_{\alpha'\beta}^* = \delta_{\alpha\alpha'}. \tag{68.2}$$

The functions $\phi_{n\alpha}^0$, being linear combinations of the functions $\psi_{n\beta}^0$, will also be solutions of Schrödinger's equation

$$\boldsymbol{H}^0\phi_n = E_n^0\phi_n, \tag{68.3}$$

belonging to the eigenvalue E_n^0, and with the additional condition (68.2) they will be orthogonal if the functions $\psi_{n\alpha}^0$ are orthogonal. The functions $\phi_{n\alpha}^0$ are therefore also possible zeroth-order approximation functions, but we do not know what coefficients $a_{\alpha\beta}$ must be taken in order to obtain the correct zeroth-order approximation.

To resolve this problem, we return to Equation (66.9), but we must now make the notation a little more precise. When there is degeneracy, the eigenfunctions of the operator have at least two suffixes (n, α). In this case, therefore, (66.4) must be written in a more detailed form by replacing the suffix n by n and α. This gives

$$\psi(x) = \sum_{n,\alpha} c_{n\alpha}\psi_{n\alpha}^0(x). \tag{68.4}$$

Accordingly (66.9) becomes, replacing n by n, α and m by m, β,

$$[E_m^0 + W_{m\beta,m\beta} - E]\, c_{m\beta} + \sum_{n,\,\alpha \neq m,\,\beta} W_{m\beta,n\alpha}\, c_{n\alpha} = 0\,, \tag{68.5}$$

where

$$W_{m\beta,\,n\alpha} = \int \psi_{m\beta}^{0*} \cdot W \psi_{n\alpha}^{0} \cdot \mathrm{d}x \tag{68.6}$$

is the matrix element of the perturbation energy and is obtained from (66.7) by increasing the number of quantum numbers describing the states. E_m^0 is the energy of the mth quantum level for the unperturbed problem. This energy, owing to the degeneracy, does not depend on the quantum number α.

Let us assume that we now wish to find the quantum level E_k of the perturbed system which is close to E_k^0 and the corresponding eigenfunctions $\psi_{k\alpha}(x)$. We shall solve this problem only in the first approximation for the levels and the zeroth-order approximation for the functions.

In the absence of degeneracy we have assumed that the zeroth-order approximation functions are just the same as the unperturbed functions, and accordingly put in that approximation $c_{n\alpha}^0 = 1$ and all other c zero. This cannot be done when degeneracy is present, since, on omitting the perturbation W in the zeroth-order approximation, we have from (68.5)

$$[E_k^0 - E]\, c_{k\beta} = 0;$$

this shows that, for $E = E_k^0$, $c_{k\beta}$ is not necessarily zero, and this true not only of one $c_{k\beta}$ but of all those with $\beta = 1, 2, \ldots, f$ belonging to the eigenvalue E_k^0. Thus in the zeroth-order approximation not one but several amplitudes are non-zero. The correct zeroth-order approximation for the functions of the kth level is therefore

$$\left.\begin{aligned} c_{k\alpha}^{(0)} &= c_{k\alpha}^{0}(\neq 0)\,, \qquad \alpha = 1, 2, \ldots, f, \\ c_{n\alpha}^{(0)} &= 0(n \neq k)\,. \end{aligned}\right\} \tag{68.7}$$

In this approximation we take from Equations (68.5) those which contain non-zero $c_{k\alpha}$, namely

$$[E_k^0 + W_{k\beta,\,k\beta} - E]\, c_{k\beta}^{(0)} + \sum_{\alpha \neq \beta} W_{k\beta,\,k\alpha}\, c_{k\alpha}^{(0)} = 0\,. \tag{68.8}$$

Since we are considering only the zeroth-order approximation to the kth level, we can drop the suffix k, retaining it mentally, and put

$$W_{\beta\alpha} = W_{k\beta,\,k\alpha} = \int \psi_{k\beta}^{0*} \cdot W \psi_{k\alpha}^{0} \cdot \mathrm{d}x\,, \tag{68.9}$$

$$c_{\alpha}^{(0)} = c_{k\alpha}^{(0)}\,, \qquad \alpha = 1, 2, \ldots, f_k\,. \tag{68.9'}$$

Then Equations (68.8) become

$$[E_k^0 + W_{\beta\beta} - E]\, c_{\beta}^{(0)} + \sum_{\alpha \neq \beta}^{f_k} W_{\beta\alpha}\, c_{\alpha}^{(0)} = 0\,, \qquad \beta = 1, 2, \ldots, f_k\,. \tag{68.10}$$

We retain the suffix in E_k^0 in order to emphasise still that we are considering the group of f_k states belonging to the level E_k^0.

If Equations (68.10) have non-zero solutions, their determinant must be zero:

$$\Delta(E) = \begin{vmatrix} E_k^0 + W_{11} - E & W_{12} & \cdots & W_{1f_k} \\ W_{21} & E_k^0 + W_{22} - E & \cdots & W_{2f_k} \\ \cdots & \cdots & \cdots & \cdots \\ W_{f_k 1} & \cdots & \cdots & E_k^0 + W_{f_k f_k} - E \end{vmatrix} = 0. \tag{68.11}$$

This is an algebraic equation of degree f_k to determine E, often called the *secular equation.*[2] From it we obtain f_k roots:

$$E = E_{k1}, E_{k2}, \ldots, E_{k\alpha}, \ldots, E_{kf_k}. \tag{68.12}$$

Since the matrix elements $W_{\beta\alpha}$ are assumed small, these roots will be almost equal. We thus have the important result that, when a perturbation is applied, a degenerate level (E_k^0) is split into a number of close levels (68.12). The degeneracy is removed. If some of the roots (68.12) are equal, the degeneracy is only partly removed.

For each of the roots $E_{k\alpha}$ (68.12) we obtain a different solution for the amplitudes $c_\beta^{(0)}$ from Equation (68.10). In order to indicate that the solution $c_1^{(0)}, c_2^{(0)}, \ldots, c_\beta^{(0)}, \ldots, c_{f_k}^{(0)}$ belongs to the level $E_{k\alpha}$, we add a further suffix α, so that the solution of Equations (68.10) for $E_{k\alpha}$ is

$$E = E_{k\alpha}, \quad c = c_{\alpha 1}^{(0)}, c_{\alpha 2}^{(0)}, \ldots, c_{\alpha\beta}^{(0)}, \ldots, c_{\alpha f_k}^{(0)}, \quad \alpha = 1, 2, \ldots, f_k. \tag{68.13}$$

If the suffix k were still retained, the full notation for $c^{(0)}$ would be $c_{k\alpha\beta}^{(0)}$. Equation (68.13) is the zeroth-order approximate wave function of the operator $\boldsymbol{H}$ in the E_k^0 representation. In the x representation the solution (68.13) is

$$\phi_{k\alpha} = \sum_{\beta=1}^{f_k} c_{\alpha\beta}^{(0)} \psi_{k\beta}^0(x). \tag{68.13'}$$

Thus each level $E = E_{k\alpha}$ now has a different function $\phi_{k\alpha}$, which is the zeroth-order approximation function for the perturbed system $(\boldsymbol{H})$.

The difference between the functions (68.13′) and (68.1) is that in (68.1) the coefficients $a_{\alpha\beta}$ are arbitrary (apart from the orthogonality condition (68.2)) while in (68.13′) the coefficients $c_{\alpha\beta}^{(0)}$ are determinate. Accordingly, the zeroth-order approximation functions $\phi_{k\alpha}$ are a particular case of the functions $\phi_{k\alpha}^0$ of the unperturbed problem. If further approximations are calculated it is easily seen that the condition for the method of perturbation theory to be valid is again (67.13), which now has the form

$$|W_{m\beta,\, n\alpha}| \ll |E_m^0 - E_n^0| \tag{68.14}$$

for the case where there is degeneracy.

It has been shown in Section 41 that the problem of finding the eigenvalues and eigenfunctions of any operator $\boldsymbol{L}$ given in matrix form amounts to the solution of Equations (41.4) and (41.5). Taking the operator $\boldsymbol{L}$ in (41.4) to be the total-energy operator $\boldsymbol{H}$, we must replace the suffix n in that equation by

[2] This name is taken from astronomy.

two suffixes n, α in the case of degeneracy, and m likewise by m, β. This gives the equations

$$\sum_{n,\,\alpha} H_{m\beta,\,n\alpha} c_{n\alpha} = E c_{m\beta}\,, \tag{68.15}$$

which are identical with (68.5), since

$$H_{m\beta,\,n\alpha} = E_m{}^0 \delta_{mn} + W_{m\beta,\,n\alpha}\,. \tag{68.16}$$

The equation (41.5) corresponding to (41.4) has a somewhat more complex form in the present case, since the rows and columns of the matrix of the operator H are numbered by the two quantum numbers n and α: for each n there are f_n different values of α (for f_n-fold degeneracy). The number f_n increases with n; for the first level $f_1 = 1$ and the term 'degeneracy' is not used.

It is not difficult to arrange the elements $H_{m\beta,\,n\alpha}$ in a matrix. For example, we can number a column $(n, 1)$ and the following columns $(n, 2), (n, 3), \ldots, (n, f_n)$, after which come columns numbered $(n+1, 1)$, $(n+1, 2)$, $\ldots$ up to $(n+1, f_{n+1})$ and so on. Similarly we have rows numbered $(m, 1)$, $(m, 2), \ldots, (m, f_m)$ etc. With this numbering of the elements of the matrix $H_{m\beta,\,n\alpha}$ the equation for the eigenvalues of E can be written in the following form (corresponding to Equation (41.5)):

$$\begin{vmatrix}
\boxed{H_{11,11}-E} & H_{11,21} & \ldots & H_{11,2f_2} & \ldots & H_{11,k1} & \ldots & H_{11,kf_k} & \ldots \\
H_{21,11} & H_{21,21}-E & \ldots & H_{21,2f_2} & \ldots & H_{21,k1} & \ldots & H_{21,kf_k} & \ldots \\
\ldots & \ldots & \ldots & \ldots & \ldots & \ldots & \ldots & \ldots & \ldots \\
H_{2f_2,11} & H_{2f_2,21} & \ldots & H_{2f_2,2f_2}-E & \ldots & H_{2f_2,k1} & \ldots & H_{2f_2,kf_k} & \ldots \\
\ldots & \ldots & \ldots & \ldots & \ldots & \ldots & \ldots & \ldots & \ldots \\
H_{k1,11} & H_{k1,21} & \ldots & \ldots & \ldots & H_{k1,k1}-E & \ldots & H_{k1,kf_k} & \ldots \\
\ldots & \ldots & \ldots & \ldots & \ldots & \ldots & \ldots & \ldots & \ldots \\
H_{k\beta,11} & H_{k\beta,21} & \ldots & \ldots & \ldots & H_{k\beta,k1} & \ldots & H_{k\beta,kf_k} & \ldots \\
\ldots & \ldots & \ldots & \ldots & \ldots & \ldots & \ldots & \ldots & \ldots \\
H_{kf_k,11} & H_{kf_k,21} & \ldots & \ldots & \ldots & H_{kf_k,k1} & \ldots & H_{kf_k,kf_k}-E & \ldots \\
\ldots & \ldots & \ldots & \ldots & \ldots & \ldots & \ldots & \ldots & \ldots
\end{vmatrix} = 0. \tag{68.17}$$

The matrix elements enclosed in rectangles belong to the same quantum number. For example, the single element in the first rectangle belongs to the level $k = 1$, the elements in the second rectangle to the level $k = 2$, and those in the third rectangle to the kth level. If we neglect matrix elements belonging to different levels, i.e. elements of the type $H_{m\beta,\,n\alpha} (m \neq n)$ (which according to (68.16) are equal to $W_{m\beta,\,n\alpha}$), Equation (68.17) becomes simply

$$\begin{vmatrix}
\boxed{H_{11,11}-E} & 0 & \ldots & \ldots & \ldots & \ldots & \ldots & \ldots & \ldots \\
0 & H_{21,21}-E & \ldots & H_{21,2f_2} & 0 & \ldots & \ldots & \ldots & \ldots \\
0 & \ldots & \ldots & \ldots & \ldots & \ldots & \ldots & \ldots & \ldots \\
0 & H_{2f_2,21} & \ldots & H_{2f_2,2f_2}-E & 0 & \ldots & \ldots & \ldots & \ldots \\
\ldots & \ldots & \ldots & \ldots & \ldots & \ldots & \ldots & \ldots & \ldots \\
0 & \ldots & \ldots & \ldots & \ldots & H_{k1,k1}-E & \ldots & H_{k1,kf_k} & \ldots \\
0 & \ldots & \ldots & \ldots & \ldots & \ldots & \ldots & \ldots & \ldots \\
0 & \ldots & \ldots & \ldots & \ldots & H_{kf_k,k1} & \ldots & H_{kf_k,kf_k}-E & \ldots \\
\ldots & \ldots & \ldots & \ldots & \ldots & \ldots & \ldots & \ldots & \ldots
\end{vmatrix} = 0. \tag{68.18}$$

Such a matrix is called a *block matrix*. Its determinant $\Delta^0(E)$ is a product of determinants of lower order[3]:

$$\Delta^0(E) = |H_{11,11} - E| \cdot \begin{vmatrix} H_{21,21}-E & \ldots & \ldots \\ \ldots & \ldots & \ldots \\ \ldots & \ldots & H_{2f_2,2f_2}-E \end{vmatrix} \ldots \begin{vmatrix} H_{k1,k1}-E & \ldots & \ldots \\ \ldots & \ldots & \ldots \\ \ldots & \ldots & H_{kf_k,kf_k}-E \end{vmatrix} \ldots = 0. \tag{68.19}$$

[3] This result is obtained immediately if the determinant (68.18) is expanded according to the usual rule as the products of elements and minors.

Denoting the determinants which appear here by $\Delta_{f_k}(E)$, we have

$$\Delta^0(E) = \Delta_{f_1}(E)\Delta_{f_2}(E) \ldots \Delta_{f_k}(E) \ldots = 0. \tag{68.20}$$

This equation is satisfied if any $\Delta_{f_k}(E) = 0$. The roots of these equations give in the first approximation the energies of the various levels. The equation

$$\Delta_{f_k}(E) = 0 \tag{68.21}$$

is identical with Equation (68.11) obtained in a different manner.

In Section 41 we have shown that the problem of finding the eigenvalues of an operator can be regarded as that of bringing its matrix to diagonal form. From the above discussion it is seen that the first approximation of perturbation theory consists in neglecting the matrix elements which belong to different levels, and thus reducing the problem of bringing an infinite matrix to diagonal form to the corresponding problem for finite matrices, namely the sub-matrices in the block matrix (68.18).

69. Splitting of levels in the case of twofold degeneracy

Let us consider the particular case of removal of degeneracy by a perturbation when the relevant level of the unperturbed system is doubly degenerate. Let the two functions ψ_{k1}^0 and ψ_{k2}^0 belong to the eigenvalue E_k^0 of the operator $\boldsymbol{H}^0$ ($f_k = 2$). Any two functions ϕ_{k1}^0 and ϕ_{k2}^0 obtained from ψ_{k1}^0 and ψ_{k2}^0 by an orthogonal transformation will also be eigenfunctions of the operator $\boldsymbol{H}^0$ belonging to the level E_k^0. We can write this transformation in the form (cf. (68.1))

$$\phi_{k1}^0 = a_{11}\psi_{k1}^0 + a_{12}\psi_{k2}^0, \tag{69.1}$$

$$\phi_{k2}^0 = a_{21}\psi_{k1}^0 + a_{22}\psi_{k2}^0. \tag{69.1'}$$

In order to satisfy the orthogonality condition (68.2) we put

$$\left.\begin{aligned} a_{11} &= \cos\theta\cdot e^{i\beta}, & a_{12} &= \sin\theta\cdot e^{-i\beta}, \\ a_{21} &= -\sin\theta\cdot e^{i\beta}, & a_{22} &= \cos\theta\cdot e^{-i\beta}, \end{aligned}\right\} \tag{69.2}$$

where θ and β are two arbitrary angles. Thus

$$\left.\begin{aligned} \phi_{k1}^0 &= \cos\theta\cdot e^{i\beta}\psi_{k1}^0 + \sin\theta\cdot e^{-i\beta}\psi_{k2}^0, \\ \phi_{k2}^0 &= -\sin\theta\cdot e^{i\beta}\psi_{k1}^0 + \cos\theta\cdot e^{-i\beta}\psi_{k2}^0 \end{aligned}\right\} \tag{69.3}$$

are the most general expressions for the wave functions belonging to the doubly degenerate level E_k^0.

The orthogonality and normalisation of these functions are easily confirmed directly, and it is also seen that the coefficients $a_{\alpha\beta}$ (69.2) satisfy the orthogonality condition (68.2). For $\beta = \theta = 0$, (69.3) give the original functions ψ_{k1}^0 and ψ_{k2}^0. Now let some perturbation $\boldsymbol{W}$ be applied. The zeroth-order approximation is given by the functions of the unperturbed system, i.e. (69.1) and (69.1′), but with definite values of the coefficients: that is, the angles θ and β depend on the form of the perturbation $\boldsymbol{W}$. To determine these angles we shall seek directly the coefficients c_1 and c_2 in the superposition

$$\phi = c_1\psi_{k1}^0 + c_2\psi_{k2}^0. \tag{69.4}$$

According to the foregoing theory these coefficients are given by Equation (68.10), which in the particular case considered has the form

$$\left.\begin{aligned}[E_k^0 + W_{11} - E]\,c_1 + W_{12}\,c_2 = 0\,,\\ [E_k^0 + W_{22} - E]\,c_2 + W_{21}\,c_1 = 0\,,\end{aligned}\right\} \qquad (69.5)$$

where W_{11}, W_{12}, W_{21}, W_{22} are the matrix elements of the perturbation:

$$W_{11} = \int \psi_{k1}^{0*} \cdot W \psi_{k1}^0 \cdot \mathrm{d}x\,, \qquad (69.6)$$

$$W_{22} = \int \psi_{k2}^{0*} \cdot W \psi_{k2}^0 \cdot \mathrm{d}x\,, \qquad (69.6')$$

$$W_{12} = W_{21}^* = \int \psi_{k1}^{0*} \cdot W \psi_{k2}^0 \cdot \mathrm{d}x\,. \qquad (69.6'')$$

Then the secular Equation (68.11) becomes

$$\Delta_2(E) = \begin{vmatrix} W_{11} - \varepsilon & W_{12} \\ W_{21} & W_{22} - \varepsilon \end{vmatrix} = 0\,, \qquad (69.7)$$

where ε is the correction to the energy of the kth level:

$$\varepsilon = E - E_k^0\,. \qquad (69.8)$$

Expanding the determinant (69.7) and solving the resulting quadratic equation, we find two roots:

$$\varepsilon_{1,2} = \tfrac{1}{2}(W_{11} + W_{22}) \pm \sqrt{[\tfrac{1}{4}(W_{11} - W_{22})^2 + |W_{12}|^2]}\,. \qquad (69.9)$$

Equations (69.5) give

$$c_1/c_2 = W_{12}/(\varepsilon - W_{11})\,. \qquad (69.10)$$

Putting

$$W_{12} = |W_{12}|\, e^{2i\beta} \qquad (69.11)$$

and substituting in (69.10) the first root ε_1 (with the plus sign), we have

$$\frac{c_1}{c_2} = \frac{|W_{12}|\, e^{2i\beta}}{\tfrac{1}{2}(W_{22} - W_{11}) + \sqrt{[\tfrac{1}{4}(W_{11} - W_{22})^2 + |W_{12}|^2]}} = \cot\theta \cdot e^{2i\beta}\,, \qquad (69.12)$$

and for the second root ε_2 (with the minus sign)

$$\frac{c_1}{c_2} = \frac{|W_{12}|\, e^{2i\beta}}{\tfrac{1}{2}(W_{22} - W_{11}) - \sqrt{[\tfrac{1}{4}(W_{11} - W_{22})^2 + |W_{12}|^2]}} = -\tan\theta \cdot e^{2i\beta}\,. \qquad (69.12')$$

Thus we have the following solutions (in the x representation):

$$\left.\begin{aligned} E_{k1} &= E_k^0 + \tfrac{1}{2}(W_{11} + W_{22}) + \sqrt{[\tfrac{1}{4}(W_{11} - W_{22})^2 + |W_{12}|^2]}\,,\\ \phi_{k1} &= \cos\theta \cdot e^{i\beta}\, \psi_{k1}^0 + \sin\theta \cdot e^{-i\beta}\, \psi_{k2}^0 \end{aligned}\right\} \qquad (69.13)$$

and

$$\left.\begin{aligned} E_{k2} &= E_k^0 + \tfrac{1}{2}(W_{11} + W_{22}) - \sqrt{[\tfrac{1}{4}(W_{11} - W_{22})^2 + |W_{12}|^2]}\,,\\ \phi_{k2} &= -\sin\theta \cdot e^{i\beta}\, \psi_{k1}^0 + \cos\theta \cdot e^{-i\beta}\, \psi_{k2}^0\,, \end{aligned}\right\} \qquad (69.13')$$

with

$$-\tan\theta = \frac{|W_{12}|}{\frac{1}{2}(W_{22} - W_{11}) - \sqrt{[\frac{1}{4}(W_{11} - W_{22})^2 + |W_{12}|^2]}}, \tag{69.14}$$

$$e^{2i\beta} = W_{12}/|W_{12}|. \tag{69.15}$$

A very important particular case is that where

$$W_{11} = W_{22}, \qquad W_{12} = W_{21}. \tag{69.16}$$

In this case

$$\left.\begin{aligned} E_{k1} &= E_k^0 + W_{11} + W_{12},\\ \phi_{k1} &= \frac{1}{\sqrt{2}}(\psi_{k1}^0 + \psi_{k2}^0), \end{aligned}\right\} \tag{69.17}$$

$$\left.\begin{aligned} E_{k2} &= E_k^0 + W_{11} - W_{12},\\ \phi_{k2} &= -\frac{1}{\sqrt{2}}(\psi_{k1}^0 - \psi_{k2}^0). \end{aligned}\right\} \tag{69.17'}$$

The transformation (69.3) is a rotation. We can obtain a direct geometrical analogy by taking $\beta = 0$ (this implies that $W_{12} = W_{21}$). Then the coefficients a are real. The coefficients c, which are particular values of the coefficients a, are also real. Instead of (69.4) we can put, with $c_1 = \xi$, $c_2 = \eta$,

$$\phi = \xi\psi_1^0 + \eta\psi_2^0 \tag{69.18}$$

retaining the suffix k mentally. If we impose the condition

$$\xi^2 + \eta^2 = 1, \tag{69.19}$$

then the mean value of the perturbation energy W in the state (69.18) is

$$\bar{W} = \int (\xi\psi_1^{0*} + \eta\psi_2^{0*}) \cdot W(\xi\psi_1^0 + \eta\psi_2^0)\, dx. \tag{69.20}$$

According to (69.6)

$$\bar{W} = W_{11}\xi^2 + 2W_{12}\xi\eta + W_{22}\eta^2. \tag{69.21}$$

This equation may be regarded as that of a second-degree curve in the $\xi\eta$ plane. Thus the mean value of W is a quadratic form in the amplitudes ξ, η which represent the state ϕ.

Let us now replace the co-ordinates ξ, η by new co-ordinates ξ', η' which differ from the former by a rotation through an angle θ:

$$\xi = \cos\theta \cdot \xi' - \sin\theta \cdot \eta', \; \eta = \sin\theta \cdot \xi' + \cos\theta \cdot \eta'. \tag{69.22}$$

Substitution in (69.18) gives

$$\left.\begin{aligned} \phi &= \xi'\phi_1^0 + \eta'\phi_2^0,\\ \phi_1^0 &= \cos\theta \cdot \psi_1^0 + \sin\theta \cdot \psi_2^0,\\ \phi_2^0 &= -\sin\theta \cdot \psi_1^0 + \cos\theta \cdot \psi_2^0. \end{aligned}\right\} \tag{69.23}$$

The matrix W must be diagonal with respect to the functions ϕ_1^0 and ϕ_2^0, since

$$\left.\begin{aligned} W_{11}' &= \int \phi_1^{0*} \cdot W\phi_1^0 \cdot dx = \varepsilon_1,\\ W_{22}' &= \int \phi_2^{0*} \cdot W\phi_2^0 \cdot dx = \varepsilon_2,\\ W_{12}' &= \int \phi_1^{0*} \cdot W\phi_2^0 \cdot dx = W_{21}' = 0. \end{aligned}\right\} \tag{69.24}$$

The mean value of W in the state ϕ therefore takes the form

$$\bar{W} = \int \phi^* \cdot W\phi \cdot dx = \varepsilon_1\xi'^2 + \varepsilon_2\eta'^2, \tag{69.25}$$

i.e. in the new variables ξ', η' the mean energy is a second-degree curve referred to its principal axes (Figure 52).

Thus the problem of reducing the matrix W to diagonal form is the same as the geometrical problem of reducing a second-degree curve to canonical form (i.e. referring it to its principal axes). In the more general case ξ and η are complex, so that the problems do not correspond entirely, but the analogy is preserved if ξ and η are still regarded as co-ordinates of a point.

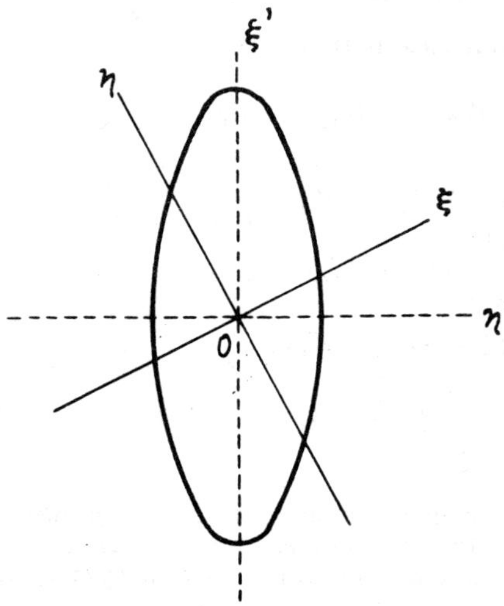

Fig. 52. Geometrical illustration of the reduction of a second-order matrix to diagonal form.

70. Comments on the removal of degeneracy

We have shown that when a perturbation is applied the degeneracy of the unperturbed system is removed: the coincident levels are separated. What is the reason for this? To answer this question, let us first consider the causes of degeneracy.

We have seen that, for example, the levels of an electron in a field of central forces are $(2l + 1)$-fold degenerate (ignoring spin degeneracy). This degeneracy arises because the energy of the electron in a field of central forces is independent of the orientation of the angular momentum relative to the field. Mathematically this is expressed by the fact that the Hamiltonian in this case has spherical symmetry: the Hamiltonian

$$\boldsymbol{H}^0 = -\frac{\hbar^2}{2\mu}\nabla^2 + U(r) \qquad (r = \sqrt{(x^2 + y^2 + z^2)}) \tag{70.1}$$

remains unchanged when the co-ordinate system is rotated from x, y, z to x', y', z'. For we have

$$x^2 + y^2 + z^2 = x'^2 + y'^2 + z'^2 \tag{70.2}$$

and

$$\nabla^2 = \frac{\partial^2}{\partial x} + \frac{\partial^2}{\partial y} + \frac{\partial^2}{\partial z^2} = \frac{\partial^2}{\partial x'^2} + \frac{\partial^2}{\partial y'^2} + \frac{\partial^2}{\partial z'^2}; \tag{70.2'}$$

the latter equation follows most simply from the fact that $\nabla^2 = (\nabla)^2$, since ∇ is a vector operator and the square of a vector is unchanged by a rotation. Thus

$$\boldsymbol{H}^0(x, y, z) = \boldsymbol{H}^0(x', y', z'). \tag{70.3}$$

If the perturbation applied does not have spherical symmetry, the energy of the electron will depend on the orientation of the angular momentum and the levels will be split. Moreover, Equation (70.3) will not be valid for the operator $\boldsymbol{H}$. This example shows that the presence of degeneracy is due to some symmetry of the field, and the removal of degeneracy is due to the violation of this symmetry.

Another example is the following. Suppose that we have an oscillator in the *xy* plane with equal frequencies ω_0 for oscillations in the directions OX and OY. Schrödinger's equation for such an oscillator is

$$-\frac{\hbar^2}{2\mu}\left(\frac{\partial^2 \Psi}{\partial x^2} + \frac{\partial^2 \Psi}{\partial y^2}\right) + \tfrac{1}{2}\mu\omega_0^2(x^2 + y^2)\Psi = E\Psi. \tag{70.4}$$

The Hamiltonian in this equation remains unchanged by a rotation of the coordinates about the axis OZ. Thus it has rotational symmetry. From the above discussion we should expect degeneracy, and this is in fact found. For Equation (70.4) is solved immediately by separating the variables:

$$\Psi(x, y) = \psi_1(x)\psi_2(y), \qquad E = E_1 + E_2. \tag{70.5}$$

Substitution of (70.5) in (70.4) gives by the usual method the two equations:

$$-\frac{\hbar^2}{2\mu}\frac{\partial^2\psi_1}{\partial x^2} + \tfrac{1}{2}\mu\omega_0^2 x^2\psi_1 = E_1\psi_1, \tag{70.6}$$

$$-\frac{\hbar^2}{2\mu}\frac{\partial^2\psi_2}{\partial y^2} + \tfrac{1}{2}\mu\omega_0^2 y^2\psi_2 = E_2\psi_2. \tag{70.6'}$$

These equations for oscillators have known eigenfunctions and eigenvalues:

$$\psi_1(x) = \psi_{n_1}(x), \qquad E_1 = \hbar\omega_0(n_1 + \tfrac{1}{2}), \qquad n_1 = 0, 1, 2, \ldots, \tag{70.7}$$

$$\psi_2(y) = \psi_{n_2}(y), \qquad E_2 = \hbar\omega_0(n_2 + \tfrac{1}{2}), \qquad n_2 = 0, 1, 2, \ldots. \tag{70.7'}$$

Hence

$$\Psi_{n_1 n_2}(x, y) = \psi_{n_1}(x)\psi_{n_2}(y), \qquad E_{n_1 n_2} = \hbar\omega_0(n_1 + n_2 + 1). \tag{70.8}$$

In terms of the 'principal quantum number'

$$n = n_1 + n_2 + 1, \qquad n_2 = n - n_1 - 1, \tag{70.9}$$

we have

$$\Psi_{n n_1}(x, y) = \psi_{n_1}(x)\psi_{n-n_1-1}(y), \quad E_n = \hbar\omega_0 n, \quad n = 1, 2, \ldots. \tag{70.10}$$

To each level E_n there will correspond n functions with $n_1 = 0, 1, \ldots, n - 1$. Thus there is in fact degeneracy.

Let us now assume that the perturbation $\boldsymbol{W}$ consists in a change in the coefficient of elasticity for oscillations along OY. Then the frequency of the oscillations along this axis is altered, becoming ω_1, say. The Hamiltonian of the perturbed system is

$$\boldsymbol{H} = -\frac{\hbar^2}{2\mu}\left(\frac{\partial^2}{\partial x^2} + \frac{\partial^2}{\partial y^2}\right) + \tfrac{1}{2}\mu\omega_0^2 x^2 + \tfrac{1}{2}\mu\omega_1^2 y^2 ,$$
$$W(y) = \tfrac{1}{2}\mu(\omega_1^2 - \omega_0^2)\, y^2 .$$

Here W is the perturbation. In the example considered an exact solution can evidently be obtained for the perturbed system, by replacing ω_0 in (70.7′) by ω_1. The solution is then

$$\left.\begin{aligned} \Psi_{n_1 n_2}(x, y) &= \psi_{n_1}(x)\,\psi_{n_2}(y), \\ E_{n_1 n_2} &= \hbar\omega_0 n_1 + \hbar\omega_1 n_2 + \tfrac{1}{2}\hbar\omega_0 + \tfrac{1}{2}\hbar\omega_1 \end{aligned}\right\} \qquad (70.8')$$

or

$$\left.\begin{aligned} \Psi_{nn_1}(x, y) &= \psi_{n_1}(x)\,\psi_{n-n_1-1}(y), \\ E_{nn_1} &= \hbar\omega_0 n_1 + \hbar\omega_1 (n - n_1 - 1) + \tfrac{1}{2}\hbar\omega_0 + \tfrac{1}{2}\hbar\omega_1 . \end{aligned}\right\} \qquad (70.10')$$

It is seen that the levels with different values of n_1 and the same n will have different energies. A single level E_n of the unperturbed system is split into a total of n levels $E_{n0}, E_{n1}, \ldots, E_{n,\,n-1}$. The degeneracy is removed.

The conclusion to be drawn from these examples is the following. If the Hamiltonian $\boldsymbol{H}^0(x, y, z)$ remains invariant with respect to some transformation of co-ordinates $(x, y, z) \rightarrow (x', y', z')$, then the eigenvalues E^0 are degenerate. If this invariance is destroyed by the perturbation, then the degeneracy is, at least in part, removed.

CHAPTER XII

SIMPLE APPLICATIONS OF PERTURBATION THEORY

71. The anharmonic oscillator

A harmonic oscillator is an idealisation of actual mechanical systems. The real potential energy of particles is never represented by the function $\frac{1}{2}\mu\omega_0^2x^2$ but by a much more complicated function $U(x)$. The former expression is valid only when x is small. In order to make more precise the expression for the potential energy $U(x)$ we can take into account, besides the term $\frac{1}{2}\mu\omega_0^2x^2$, also higher terms in the expansion of $U(x)$ in powers of the displacement x:

$$U(x) = \tfrac{1}{2}\mu\omega_0^2 x^2 + \lambda x^3 + \dots. \tag{71.1}$$

If the added terms remain small we have a harmonic oscillator slightly perturbed by the presence of deviations from the curve for an ideal harmonic oscillator. We shall call this an *anharmonic oscillator*.

Let us find the quantum levels of the anharmonic oscillator, regarding the additional terms in (71.1) as small (λ small). We solve this problem by the method of perturbation theory, using the known solutions for the harmonic oscillator. The perturbation W will be given by the additional terms in the expression for the potential energy[1]:

$$W(x) = \lambda x^3 + \dots. \tag{71.2}$$

The quantum levels of the unperturbed system ($\lambda = 0$) are those of the harmonic oscillator; its eigenvalues and eigenfunctions will be denoted by

$$E_n^0 = \hbar\omega_0(n + \tfrac{1}{2}), \qquad \psi_n^0(x). \tag{71.3}$$

In this case there is no degeneracy: only one state ψ_n^0 belongs to each level. The matrix element of the perturbation energy W is

$$W_{mn} = \int \psi_m^0 \cdot W\psi_n^0\,\mathrm{d}x = \lambda \int \psi_m^0 x^3 \psi_n^0\,\mathrm{d}x = \lambda(x^3)_{mn}, \tag{71.4}$$

where $(x^3)_{mn}$ denotes the matrix elements of x^3.

According to Formula (67.10) the energy of the kth level of the perturbed system is

1 We can suppose that the spectrum of the perturbed system remains discrete, since λx^3 is a correction term and does not apply for large x. Thus the form of the correction (71.2) does not justify the conclusion that the asymptotic behaviour of $U(x)$ is fundamentally changed. In Section 67 this conclusion was deduced by formally regarding the additional term λx^3 as valid for large x also.

in the second approximation

$$E_k = E_k^0 + \lambda (x^3)_{kk} + \lambda^2 \sum_{n \neq k} \frac{(x^3)_{nk} (x^3)_{kn}}{E_k^0 - E_n^0}. \tag{71.5}$$

Thus we have only to calculate the matrix $(x^3)_{mn}$. This matrix could be calculated directly from (71.4), using the functions ψ_n^0 (see (47.11)), but we shall use a simpler procedure. The matrix x_{mn} is known (see (48.8)). The rule of matrix multiplication enables us to calculate from this the matrix $(x^3)_{mn}$:

$$(x^3)_{kn} = \sum_l x_{kl} (x^2)_{ln} = \sum_l x_{kl} \sum_m x_{lm} x_{mn} = \sum_l \sum_m x_{kl} x_{lm} x_{mn}. \tag{71.6}$$

Substituting the values of the matrix elements x_{kl}, x_{lm}, x_{mn} from (48.8), we obtain

$$\begin{aligned}(x^3)_{kn} = (\hbar/\mu\omega_0)^{\frac{3}{2}} \sum_l \sum_m [\sqrt{(\tfrac{1}{2}k)}\, \delta_{k-1,l} + \sqrt{(\tfrac{1}{2}k + \tfrac{1}{2})}\, \delta_{k+1,l}] \times \\ \times [\sqrt{(\tfrac{1}{2}l)}\, \delta_{l-1,m} + \sqrt{(\tfrac{1}{2}l + \tfrac{1}{2})}\, \delta_{l+1,m}] \times \\ \times [\sqrt{(\tfrac{1}{2}m)}\, \delta_{m-1,n} + \sqrt{(\tfrac{1}{2}m + \tfrac{1}{2})}\, \delta_{m+1,n}].\end{aligned} \tag{71.7}$$

The double series in l and m is easily summed by means of the δ symbols, and the result is

$$\begin{aligned}(x^3)_{kn} = \left(\frac{\hbar}{\mu\omega_0}\right)^{\frac{3}{2}} \Bigg\{ \sqrt{\frac{k(k-1)(k-2)}{8}}\, \delta_{k-3,n} + \sqrt{\frac{9k^3}{8}}\, \delta_{k-1,n} + \\ + \sqrt{\frac{9(k+1)^3}{8}}\, \delta_{k+1,n} + \sqrt{\frac{(k+1)(k+2)(k+3)}{8}}\, \delta_{k+3,n} \Bigg\}.\end{aligned} \tag{71.8}$$

Hence it follows that $(x^3)_{kk} = 0$, and the correction to E_k^0 in the first approximation is therefore zero. The second-approximation correction, which involves a sum over n, is also easily calculated, since (71.8) shows that only the four terms with $n = k \pm 3, k \pm 1$ remain. Moreover, $(x^3)_{kn} = (x^3)_{nk}$. Substituting (71.8) in (71.5) and using (71.3), we then find

$$E_k = \hbar\omega_0 (k + \tfrac{1}{2}) - \frac{\lambda^2}{\hbar\omega_0} \left(\frac{\hbar}{\mu\omega_0}\right)^3 \cdot \tfrac{15}{4} (k^2 + k + \tfrac{11}{30}), \qquad k = 0, 1, 2, \ldots. \tag{71.9}$$

This is the required approximate expression for the energy of the quantum levels of the oscillator with allowance for the anharmonic correction term λx^3.

The condition for this approximation to be valid is easily found. The matrix element $\lambda (x^3)_{kn}$ of the perturbation for large quantum numbers k is in order of magnitude, from (71.8),

$$W_{kn} \approx \lambda (\hbar/\mu\omega_0)^{\frac{3}{2}} k^{\frac{3}{2}}.$$

The difference of the levels $E_k^0 - E_n^0 \approx \hbar\omega_0$. Thus the condition (67.13) for perturbation theory to be applicable reduces to

$$\lambda (\hbar/\mu\omega_0)^{\frac{3}{2}} k^{\frac{3}{2}} / \hbar\omega_0 \ll 1. \tag{71.10}$$

Our approximation is therefore valid for not too high levels, such that

$$k \ll (\hbar\omega_0/\lambda)^{\frac{2}{3}}\,\mu\omega_0/\hbar. \tag{71.10'}$$

In the language of classical mechanics, this condition signifies that the amplitude of the oscillation must not be too great.

Formula (71.9) is used in calculating the vibrational levels of a molecule. When considering the diatomic molecule in Section 54 we took only the second term in the expansion of the potential energy $U(x)$ in powers of the displacement x from the equilibrium position, and accordingly we obtained harmonic vibrations of the molecule. If the following term in the expansion were taken into account, as in general it must be, the vibrational levels of the molecule would be determined by Formula (71.9) and not (71.3).

72. Splitting of spectral lines in an electric field

It has been found by experiment that in an electric field the spectral lines of atoms undergo splitting; this is called the *Stark effect*. Figure 53 shows the splitting pattern for the hydrogen lines H_β, H_γ, H_δ, H_ε and H_ζ of the Balmer series. Experiment shows that the effect of an electric field on the hydrogen atom is very different from its effect on other atoms. In hydrogen, the splitting of the spectral lines is proportional to the first power of the electric field $\mathscr{E}$, but in all other atoms it is proportional to the square of the field ($\mathscr{E}^2$). In strong fields (of the order of 10^5 V/cm) a further splitting appears which is proportional to higher powers of $\mathscr{E}$. In addition, as the field increases, the spectral lines are observed to be broadened and finally to disappear. The latter phenomenon will be considered in detail in Section 101; for the present we shall consider fields below 10^5 V/cm.

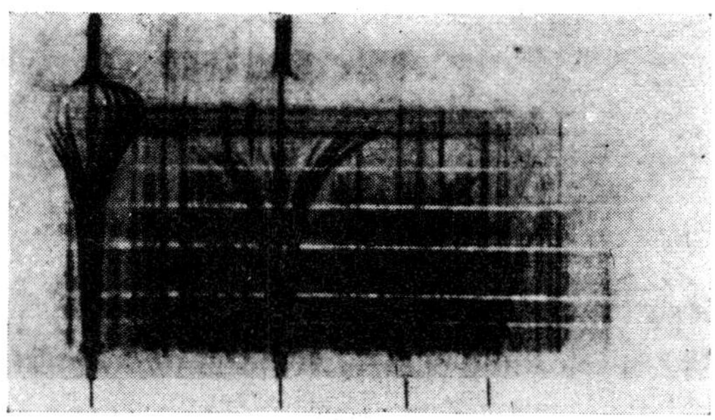

H_β H_γ H_δ H_ε H_ζ

Fig. 53. Splitting of spectral lines of the Balmer series in strong electric fields.*

* The field increases upwards, its maximum value being 1.14 MV/cm, and the white lines are lines of constant field. The unperturbed (zero field) hydrogen lines are also shown by the central line in each splitting pattern, which is almost straight. On comparing the Stark lines adjoining undisplaced lines it is clear that the red (left-hand) line is always much further than the violet line from the undisplaced line (the quadratic Stark effect). This is particularly evident for the H_β line. It is also seen that all the lines cease to exist at a certain critical field which is less for H_ε than for H_δ, less for H_δ than for H_γ, and so on, and less for the red components of each line than for the violet components. The phenomenon of the disappearance of lines is explained in Section 101.

From a comparison of the electric field within the atom $\mathscr{E}_0 = e/a^2 = 5.13 \times 10^9$ V/cm (where a is the radius of the first Bohr orbit) with the external field ($\mathscr{E} < 10^5$ V/cm) it follows that the action of the external field may generally be regarded as a perturbation. We shall make use of this fact to find the quantum levels and wave functions of the atomic electron in the presence of the external field $\mathscr{E}$. Let the potential energy of the optical electron in the atom be denoted by $U(r)$. If there is also a uniform external field $\mathscr{E}$, the electron will have some additional potential energy W, which is easily calculated. Let us take the axis OZ in the direction of the electric field $\mathscr{E}$. Then the potential energy of the electron in the field is

$$W = e\mathscr{E}z = -D_z\mathscr{E}, \tag{72.1}$$

where $D_z = -ez$ is the component of the electric moment in the direction of OZ.[2] The total potential energy of the electron is

$$U'(r) = U(r) + e\mathscr{E}z. \tag{72.2}$$

Schrödinger's equation for the stationary states is

$$-(\hbar^2/2\mu)\nabla^2\psi + [U(r) + e\mathscr{E}z]\psi = E\psi. \tag{72.3}$$

The perturbation W is of the type discussed in Section 67: even when the field $\mathscr{E}$ is arbitrarily small it changes the asymptotic behaviour of the potential energy. If $\mathscr{E} = 0$, $U' \to 0$ as $z \to \pm\infty$, but if $\mathscr{E} \neq 0$, $U' \to \pm\infty$ as $z \to \pm\infty$. We can therefore apply perturbation theory (for small λ) only in the sense explained in Section 67. Thus by using perturbation theory we find the quantum energy values E_n for which the electron is near the atom for a fairly long time ('quasistationary' states). Regarding W as a perturbation in this sense, we shall assume the states of the electron in the atom in the absence of the external field to be known.

Let us first consider a hydrogen-like atom. The energy of the quantum levels of the atom in the absence of the field may be denoted by

$$E = E^0_{nl}, \qquad 0 \leqslant l \leqslant n-1, \qquad n = 1, 2, 3, \ldots, \tag{72.4}$$

and the corresponding wave functions by

$$\psi^0_{nlm} = R_{nl}(r)\,P^m_l(\cos\theta)\,e^{im\phi}, \qquad -l \leqslant m \leqslant l. \tag{72.5}$$

Each level E^0_{nl} is $(2l+1)$-fold degenerate owing to the different possibilities for the orientation of the orbital angular momentum M_z. Since we are considering a definite level n, l we can omit these suffixes and retain only m. Then the functions belonging to the level E^0_{nl} can be denoted by

$$\psi^0_{-l}, \psi^0_{-l+1}, \ldots, \psi^0_m, \ldots, \psi^0_l. \tag{72.6}$$

The most general function representing a state with energy E^0_{nl} is

$$\phi = \sum_{m=-l}^{l} c_m \psi^0_m. \tag{72.7}$$

[2] We denote the charge on the electron by $-e$ and take the origin at the centre of the atom.

Let us calculate the mean value of the component D_z of the electric moment in such a state. We have

$$\begin{aligned} \overline{D_z} &= \int \phi^* \cdot D_z \phi \cdot \mathrm{d}v = \sum_m \sum_{m'} c_m^* c_{m'} \int \psi_m^{0*} \cdot D_z \psi_{m'}^0 \cdot \mathrm{d}v \\ &= \sum_m \sum_{m'} c_m^* c_{m'} (D_z)_{mm'}, \end{aligned} \tag{72.8}$$

where

$$(D_z)_{mm'} = \int \psi_m^{0*} \cdot D_z \psi_{m'}^0 \cdot \mathrm{d}v \tag{72.9}$$

is the matrix element of the electric moment D_z. It follows from (72.1) that the matrix elements of the perturbation energy are

$$W_{mm'} = -(D_z)_{mm'} \mathscr{E}. \tag{72.10}$$

Let us now calculate $(D_z)_{mm'}$. Substituting in (72.9) the wave functions ψ_{nlm}^0 from (72.5) and bearing in mind that $z = r \cos\theta$, we obtain

$$(D_z)_{mm'} = -e \int_0^\infty R_{nl}^2 r^3 \,\mathrm{d}r \int_0^\pi P_l^m P_l^{m'} \cos\theta \sin\theta \,\mathrm{d}\theta \int_0^{2\pi} e^{i(m-m')\phi} \,\mathrm{d}\phi. \tag{72.11}$$

If $m \neq m'$ this integral is zero, since $e^{i(m-m')\phi}$ is a periodic function of ϕ; if $m = m'$, the second integral in (72.11) is an even function of $\cos\theta$ and is therefore zero. Thus $(D_z)_{mm'} = 0$, and therefore the mean value $\overline{D_z}$ (72.8) of the electric moment in any state belonging to the level E_{nl}^0 is zero. According to (72.10) the perturbation energy is also zero. Hence it follows that in hydrogen-like atoms in an electric field there can be no splitting of the levels which is proportional to the field, since the mean electric moment is zero. The splitting proportional to higher powers of the field will of course occur. For the electron wave functions in the field will be different from ψ_{nlm}^0 (the zeroth-order approximation). In the first approximation we can put

$$\psi_{nlm} = \psi_{nlm}^0 + u_{nlm} + \dots, \tag{72.12}$$

where u_{nlm} is some additional term proportional to the first power of the field $\mathscr{E}$.

Calculation shows that in this approximation, when the deformation of the atom is taken into account, the mean electric moment $\overline{D_z}$ is not zero, but is proportional to the field $\mathscr{E}$:

$$\overline{D_z} = \alpha \mathscr{E}. \tag{72.13}$$

This is the result of the polarisation of the atom in the field. The potential energy of this moment in the field $\mathscr{E}$ is

$$W = -\tfrac{1}{2}\alpha \mathscr{E}^2, \tag{72.14}$$

corresponding to the work of polarisation

$$W = -\int_0^{\mathscr{E}} \mathscr{E} \, d(D_z)$$

when the field is increased from zero to $\mathscr{E}$.

The shift of the quantum levels will therefore be proportional to the square of the field, $\mathscr{E}^2$. We shall not discuss the calculation of the quantity α, which is called the *polarisability*.[3]

The situation is different in the hydrogen atom, where in addition to the degeneracy due to various orientations of the orbital angular momentum there is also l degeneracy. To each quantum level E_n^0 there belong n^2 functions of the type (72.5), which differ both in the number $l(l = 0, 1, ..., n - 1)$ and in the number m. Retaining the level number n mentally, we can write the functions belonging to the level E_n^0 in the form

$$\psi_{lm}^0, \qquad l = 0, 1, 2, ..., n - 1; \qquad m = 0, \pm 1, ..., \pm l, \tag{72.15}$$

making in all n^2 such functions. The most general state belonging to the level E_n^0 will now be

$$\phi = \sum_{l=0}^{n-1} \sum_{m=-l}^{l} c_{lm} \psi_{lm}^0 . \tag{72.16}$$

The mean electric moment $\overline{D_z}$ in the state ϕ is not zero, because the superposition (72.16) involves functions with different values of l (see the calculation in the next section). The mean additional energy in the field $\mathscr{E}$ in the state ϕ,

$$\bar{W} = - \mathscr{E} \overline{D_z}, \tag{72.17}$$

is therefore in general non-zero and proportional to the field. As a result the shift of the levels should be proportional to the field, and this is in fact observed. Thus the essence of the difference in the behaviour of the hydrogen atom and the hydrogen-like atoms in an electric field is that in the former there is an electric moment in the group of states belonging to the level E_n^0, while in the latter there is no electric moment in the group of states of the atom belonging to the level E_{nl}^0, and such a moment appears only as a result of the polarisation (deformation) of the atom.

73. Splitting of spectral lines of the hydrogen atom in an electric field

The derivation of the general formula for the splitting of the levels of hydrogen in an electric field will be found in many textbooks, for instance [58]. We shall here discuss only one example, which demonstrates the essential features, namely the splitting of the second quantum level of the hydrogen atom ($n = 2$) (the first level is not degenerate and is therefore not split). Thus we shall take the simplest case.

The quantum level in question has four states, with the following wave functions:

$$\begin{aligned} \psi_{200} &= R_{20}(r) \cdot Y_{00} \quad (s \text{ term}); \\ \psi_{210} &= R_{21}(r) \cdot Y_{10} \\ \psi_{211} &= R_{21}(r) \cdot Y_{11} \quad (p \text{ term}). \\ \psi_{21,-1} &= R_{21}(r) \cdot Y_{1,-1} \end{aligned} \tag{73.1}$$

[3] See, for example, [5], pp. 339–349, where a calculation of α for helium is given. It may be noted that the formula for the polarisability α can be derived from dispersion theory (Section 92) by substituting for the frequency of the external field $\omega = 0$.

According to (25.16)

$$Y_{00} = \frac{1}{\sqrt{(4\pi)}}, \qquad Y_{10} = \sqrt{\frac{3}{4\pi}} \cos\theta, \qquad Y_{1,\pm 1} = \sqrt{\frac{3}{8\pi}} \sin\theta \cdot e^{\pm i\phi}. \tag{73.2}$$

From (50.19) we have the radial functions R_{nl}:

$$\left.\begin{aligned} R_{20} &= \frac{1}{\sqrt{(2a^3)}} e^{-r/2a}\left(1 - \frac{r}{2a}\right), \\ R_{21} &= \frac{1}{\sqrt{(6a^3)}} e^{-r/2a} \frac{r}{2a}, \end{aligned}\right\} \tag{73.3}$$

where a is the radius of the Bohr orbit and $1/\sqrt{(2a^3)}$ and $1/\sqrt{(6a^3)}$ are normalising factors. Using the formulae $x = r \sin\theta \cos\phi$, $y = r \sin\theta \sin\phi$, $z = r\cos\theta$, we can write the functions (73.1) as

$$\left.\begin{aligned} \psi_{200} &= \psi_1^0 = \frac{1}{\sqrt{(4\pi)}} R_{20} = f(r), \\ \psi_{210} &= \psi_2^0 = \sqrt{\frac{3}{4\pi}} R_{21} \frac{z}{r} = F(r) \cdot z, \\ \psi_{211} &= \psi_3^0 = \sqrt{\frac{3}{8\pi}} R_{21} \frac{x + iy}{r} = F(r) \cdot \frac{x + iy}{\sqrt{2}}, \\ \psi_{21,-1} &= \psi_4^0 = \sqrt{\frac{3}{8\pi}} R_{21} \frac{x - iy}{r} = F(r) \cdot \frac{x - iy}{\sqrt{2}}. \end{aligned}\right\} \tag{73.4}$$

The most general state belonging to the level E_2^0 will be

$$\phi = \sum_{\alpha=1}^{4} c_\alpha \psi_\alpha^0 . \tag{73.5}$$

In order to determine approximately the quantum levels and the wave functions in the presence of an external electric field $\mathscr{E}$ by means of perturbation theory, we have to solve Equations (68.10), which in our case are

$$[E_2^0 - E + W_{\beta\beta}] c_\beta + \sum_{\alpha \neq \beta} W_{\beta\alpha} c_\alpha = 0, \qquad \alpha, \beta = 1, 2, 3, 4, \tag{73.6}$$

$$W_{\beta\alpha} = \int \psi_\beta^{0*} e\mathscr{E} z \psi_\alpha^0 \, \mathrm{d}v . \tag{73.7}$$

From the form (73.4) of the functions it is easily seen that all the integrals (73.7) except two are zero because the integrand is an odd function of z. The two exceptions are

$$W_{12} = W_{21} = e\mathscr{E} \int f(r) F(r) z^2 \, \mathrm{d}v , \tag{73.8}$$

and these are easily calculated in polar co-ordinates. From (73.3) and (73.4) we have

$$W_{12} = \frac{e\mathscr{E}\sqrt{3}}{4\pi} \cdot \frac{1}{\sqrt{12}\, a^3} \int_0^\infty \int_0^\pi \int_0^{2\pi} e^{-r/2a} \left(1 - \frac{r}{2a}\right) \frac{e^{-r/2a}}{r} \cdot \frac{r}{2a} \times$$
$$\times z^2 r^2 \sin\theta \, \mathrm{d}r \, \mathrm{d}\theta \, \mathrm{d}\phi .$$

Now

$$\int_0^{\pi}\int_0^{2\pi} z^2 \sin\theta \, d\theta \, d\phi = r^2 \int_0^{\pi}\int_0^{2\pi} \cos^2\theta \sin\theta \, d\theta \, d\phi = 4\pi r^2/3 .$$

Using the variable $\xi = r/a$, we obtain finally

$$W_{12} = W_{21} = \frac{e\mathscr{E}a}{12}\int_0^{\infty} e^{-\xi}(1 - \tfrac{1}{2}\xi)\,\xi^4 \, d\xi = -\,3e\mathscr{E}a . \tag{73.8'}$$

Let us now write down the equations (73.6) in explicit form. From the above discussion of the matrix elements $W_{\alpha\beta}$ we have

$$\left.\begin{aligned} (E_2^0 - E)\,c_1 + W_{12}\,c_2 &= 0 ,\\ (E_2^0 - E)\,c_2 + W_{21}\,c_1 &= 0 ,\\ (E_2^0 - E)\,c_3 &= 0 ,\\ (E_2^0 - E)\,c_4 &= 0 . \end{aligned}\right\} \tag{73.6'}$$

The determinant $\Delta_2(E)$ of these equations must be zero (see Section 68):

$$\Delta_2(E) = \begin{vmatrix} E_2^0 - E & W_{12} & 0 & 0 \\ W_{21} & E_2^0 - E & 0 & 0 \\ 0 & 0 & E_2^0 - E & 0 \\ 0 & 0 & 0 & E_2^0 - E \end{vmatrix} = (E_2^0 - E)^2\,[(E_2^0 - E)^2 - W_{12}^2] = 0 . \tag{73.9}$$

Hence we find the roots E_1, E_2, E_3, E_4, which are equal to the energies of the perturbed levels:

$$E_1 = E_2^0 + W_{12} , \qquad E_2 = E_2^0 - W_{12} , \qquad E_3 = E_4 = E_2^0 . \tag{73.10}$$

Thus the degeneracy is only partly removed: the fourfold level is split into only three different levels.[4] Figure 54 shows the pattern of this splitting.

Consequently, instead of a single spectral line corresponding to the transition $E_2^0 \to E_1^0$ (shown by the arrow a in the diagram), we have three lines corresponding to

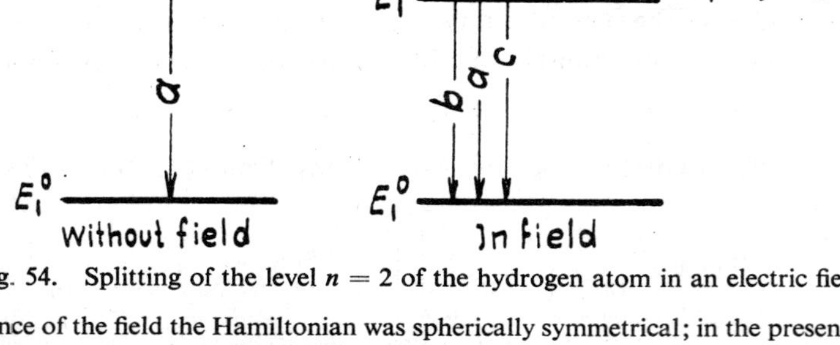

Fig. 54. Splitting of the level $n = 2$ of the hydrogen atom in an electric field.

[4] In the absence of the field the Hamiltonian was spherically symmetrical; in the presence of the field there remains a symmetry about the direction of the field.

the transitions

$$(a)\quad E_3, E_4 \to E_1^0$$
$$(b)\quad E_1 \to E_1^0$$
$$(c)\quad E_2 \to E_1^0.$$

This is the phenomenon of the splitting of spectral lines in an electric field. It may be noted that for simplicity we have calculated the splitting of the first line of the ultraviolet Lyman series, whereas Stark himself examined that of the Balmer series lines in the visible spectrum.

It follows from (73.10) and (73.8′) that the difference ΔE between the energy levels E_1 and E_2 is $6e\mathscr{E}a$, i.e. $\Delta E \approx 3 \times 10^{-8}\ \mathscr{E}$ eV, if $\mathscr{E}$ is in V/cm. The splitting is slight: even if $\mathscr{E} = 10^4$ V/cm, $\Delta E = 3 \times 10^{-4}$ eV, whereas the difference $E_2^0 - E_1^0 \approx 10$ eV.

Let us now calculate in the zeroth-order approximation the wave functions ϕ which belong to the levels E_1, E_2, E_3 and E_4. To do so, we must find the amplitudes c_a from Equations (73.6′). Substituting in these equations $E = E_3 = E_4 = E_2^0$, we find that c_3 and $c_4 \neq 0$ but $c_1 = c_2 = 0$. Hence the most general state for the undisplaced levels is described by the function

$$\phi = c_3\psi_3^0 + c_4\psi_4^0, \qquad E = E_2^0; \tag{73.11}$$

c_3 and c_4 are arbitrary (the degeneracy is not removed). Substituting now in (73.6′) $E = E_1 = E_2^0 + W_{12}$, we have $c_3 = c_4 = 0$, $c_1 = c_2$. Hence the level E_1 has a corresponding wave function

$$\phi_1 = \frac{1}{\sqrt{2}}(\psi_1^0 + \psi_2^0), \qquad E_1 = E_2^0 + W_{12}. \tag{73.12}$$

Similarly for $E = E_2$: $c_3 = c_4 = 0$ and $c_1 = -c_2$, and the wave function is

$$\phi_2 = \frac{1}{\sqrt{2}}(\psi_1^0 - \psi_2^0), \qquad E_2 = E_2^0 - W_{12}. \tag{73.12′}$$

(The factor $1/\sqrt{2}$ serves to normalise ϕ_1 and ϕ_2 to unity.) Thus in the presence of the field $\mathscr{E}$ the wave functions of the stationary states[5] will be ϕ_1, ϕ_2, $\phi_3 = \psi_3^0$ and $\phi_4 = \psi_4^0$. It is left to the reader to verify that the perturbation matrix W in the new representation,

$$W'_{\alpha\beta} = e\mathscr{E}\int \phi_\alpha^* z \phi_\beta \, dv, \tag{73.13}$$

is a diagonal matrix

$$W' = \begin{vmatrix} 3ea\mathscr{E} & 0 & 0 & 0 \\ 0 & -3ea\mathscr{E} & 0 & 0 \\ 0 & 0 & 0 & 0 \\ 0 & 0 & 0 & 0 \end{vmatrix}, \tag{73.14}$$

as it should be according to the general theory. From this it follows that the level splitting pattern can also be interpreted as follows. The levels E_3 and E_4 are not

[5] More precisely 'almost stationary'; see Sections 99, 101.

shifted, because in the states ϕ_3 and ϕ_4 the electric moment is zero. The shifts of the levels E_1 and E_2 are determined by the fact that in the states ϕ_1 and ϕ_2 this moment is respectively $3ae\mathscr{E}$ and $-3ae\mathscr{E}$, i.e. in the former case it is orientated opposite to the field and in the latter case it is along the field.

74. Splitting of spectral lines in a weak magnetic field

The theory of the splitting of spectral lines in a magnetic field considered in Section 62 is by no means complete, since it neglects the multiplet structure of spectral lines. We shall now take this structure into account.

The Hamiltonian $\boldsymbol{H}$ of an atomic electron in a magnetic field is, according to (62.6),

$$\boldsymbol{H} = \boldsymbol{H}^0 + (e\mathscr{H}/2\mu c)(\boldsymbol{M}_z + \hbar\boldsymbol{\sigma}_z) = \boldsymbol{H}^0 + (e\mathscr{H}/2\mu c)(\boldsymbol{M}_z + 2\boldsymbol{s}_z), \tag{74.1}$$

terms in $\mathscr{H}^2$ being regarded as negligibly small. $\boldsymbol{H}^0$ is the Hamiltonian in the absence of the external magnetic field:

$$\boldsymbol{H}^0 = -(\hbar^2/2\mu)\nabla^2 + U(r). \tag{74.2}$$

To take into account the multiplet structure of the spectrum we must add to this Hamiltonian the terms giving the spin-orbit interaction energy (which, as explained in Section 65, bring about the structure of the spectra). We must also recall the comment in Section 65 that the correction for the velocity dependence of the electron mass (the relativistic effect) is of the same order as the spin-orbit interaction. We denote all these additional terms in the electron energy which lead to the multiplet structure by

$$\boldsymbol{W}^0 = \boldsymbol{W}^0(x, y, z, \boldsymbol{s}, -i\hbar\,\partial/\partial x, -i\hbar\,\partial/\partial y, -i\hbar\,\partial/\partial z). \tag{74.3}$$

We shall not write this operator explicitly, but merely indicate the arguments on which it depends. The appearance in $\boldsymbol{W}^0$ of the electron momentum operators is evident from the fact that the internal magnetic field $\mathscr{H}_l$ created by the orbital motion of the electron depends on the velocity of the electron and therefore on its momentum.[6] Thus the total Hamiltonian has the form

$$\boldsymbol{H} = \boldsymbol{H}^0 + \boldsymbol{W}^0 + \boldsymbol{W}, \qquad \boldsymbol{W} = (e\mathscr{H}/2\mu c)(\boldsymbol{M}_z + 2\boldsymbol{s}_z). \tag{74.4}$$

We shall distinguish two cases: one where the magnetic field is so large that the energy W of the electron in the external field is much greater than the energy W^0 which brings about the multiplet splitting, and the other where the energy W in the external field is much less than the energy W^0 (small magnetic fields).

Let us make more precise the terms 'strong field' and 'weak field'. The energy W^0, which is neglected, is equal in order of magnitude to the energy difference of the levels of the doublet (see Figure 46). Let this quantity be denoted by

$$\Delta E_{jj'} = E^0_{nlj} - E^0_{nlj'}. \tag{74.5}$$

[6] According to Biot and Savart's law this field is $\mathscr{H}_l = (e/c)(\mathbf{v}\times\mathbf{r})/r^3$, where $\mathbf{v}$ is the velocity of the electron and $\mathbf{r}$ the radius vector from the electron to the point where the field $\mathscr{H}_l$ is observed.

The splitting due to the magnetic field is, according to (62.13), in order of magnitude $e\hbar\mathscr{H}/2\mu c$. Hence the approximation considered in Section 62 corresponds to the condition

$$e\hbar\mathscr{H}/2\mu c \gg |\Delta E_{jj'}| . \tag{74.6}$$

If, for example, $\Delta E_{jj'} = 5.3 \times 10^{-15}$ erg (the D_1 and D_2 lines of sodium; see Figure 49), then (74.6) gives $\mathscr{H} > 5 \times 10^4$ Oe. A weak field, on the other hand, satisfies the condition $W^0 \gg W$, i.e.

$$e\hbar\mathscr{H}/2\mu c \ll |\Delta E_{jj'}| , \qquad \mathscr{H} \ll |2\mu c\Delta E_{jj'}/e\hbar| . \tag{74.7}$$

In the former case (strong fields) we can neglect the quantity W^0 in comparison with W. We then have the case already considered in Section 62 (the normal Zeeman effect). For weak fields the separation of the levels in the multiplet, $\Delta E_{jj'}$, is much greater than $e\hbar\mathscr{H}/2\mu c$, and so in the zeroth-order approximation we can neglect the energy W of the electron in the external field in comparison with W^0 and take as the Hamiltonian of the unperturbed system

$$\boldsymbol{H} = \boldsymbol{H}^0 + \boldsymbol{W}^0 , \tag{74.8}$$

with $\boldsymbol{W}$ a perturbation. The resulting pattern of level splitting and corresponding line splitting is much more complex than that considered in Section 62. The phenomenon itself is called the *anomalous Zeeman effect.*

In order to discuss this splitting, we note that the quantum levels E^0_{nlj} of the unperturbed system (with Hamiltonian (74.8)) are, as shown in Section 65, $(2j + 1)$-fold degenerate in accordance with the possible orientations of the total angular momentum **J**. In the presence of an external field such a level must be split, since different orientations of **J** will correspond to different energies of the magnetic moment in the external field $\mathscr{H}$. In order to find the splitting we must determine the eigenvalues of the perturbation energy $\boldsymbol{W}$. To do this we use the fact (cf. Section 65) that the states of the unperturbed system, taking into account the multiplet structure, are described by four quantum numbers n, l, j, m_j. The matrix elements of the perturbation energy $\boldsymbol{W}$ will therefore have the form $W_{nljm_j,\, n'l'j'\, m_{j'}}$. If we take only the zeroth-order approximation, then, as explained in Section 68, we must neglect the matrix elements of the perturbation energy which belong to different levels of the unperturbed system. Since such levels are labelled by n, l, j in our problem, in the zeroth-order approximation we need consider only the elements

$$W_{m_j m_{j'}} = W_{nljm_j, nljm_{j'}} . \tag{74.9}$$

The validity of this approximation is ensured by the smallness of the magnetic field. Since the matrix elements $W_{m_j m_{j'}}$ are of the order of $e\hbar\mathscr{H}/2\mu c$, the condition (74.7) can be written as

$$\left|\frac{W_{nljm_j,\, nlj'm_{j'}}}{E_{nlj} - E_{nlj'}}\right| \ll 1 , \tag{74.10}$$

which is just the condition for perturbation theory to be valid. Here we have taken the energy difference within the multiplet (different j and j' but the same n and l). It is clear that for different n and l (74.10) is satisfied if it is for the same n and l.

From the above it is seen that the problem amounts to that of reducing the matrix $W_{m_j m_{j'}}$ to diagonal form. To do this, we express the perturbation energy W in terms of the component J_z of the total angular momentum $\mathbf{J}$ along the axis OZ. We have

$$W = (e\mathscr{H}/2\mu c)(M_z + 2s_z) = O_L(J_z + s_z), \tag{74.11}$$

where O_L is the Larmor frequency. Let us now consider the product $s_z J^2$. This can be put in the form

$$\begin{aligned} s_z J^2 &= s_z(J_x^2 + J_y^2 + J_z^2) \\ &= J_z(s_x J_x + s_y J_y + s_z J_z) + (s_z J_x - J_z s_x) J_x + (s_z J_y - J_z s_y) J_y \\ &= J_z(\vec{s}\cdot\vec{J}) + Q, \end{aligned} \tag{74.12}$$

$$Q = (s_z J_x - J_z s_x) J_x + (s_z J_y - J_z s_y) J_y. \tag{74.13}$$

The theorem of composition of angular momenta together with (64.9) shows that (74.12) may be written

$$s_z J^2 = J_z \cdot \tfrac{1}{2}(J^2 - M^2 + s^2) + Q. \tag{74.12'}$$

If we now take a representation in which J^2 is a diagonal matrix, (74.12′) may be divided by J^2 (since a diagonal matrix behaves as an ordinary quantity, not as an operator). In this representation, therefore, (74.12′) becomes

$$s_z = \frac{J_z}{2J^2}\{J^2 - M^2 + s^2\} + \frac{Q}{J^2}, \tag{74.13'}$$

and so the perturbation energy W may be written

$$W = O_L J_z \left\{1 + \frac{J^2 - M^2 + s^2}{2J^2}\right\} + O_L \frac{Q}{J^2}. \tag{74.14}$$

The matrix elements of the operator Q are zero except when $j \neq j'$.

For the operator Q can be put in the form

$$Q = \gamma_x J_y - \gamma_y J_x, \tag{74.15}$$

where

$$\gamma_x = J_y s_z - J_z s_y, \qquad \gamma_y = J_z s_x - J_x s_z, \qquad \gamma_z = J_x s_y - J_y s_x \tag{74.16}$$

(the suffixes being permuted cyclically). Using the rules of commutation of the angular-momentum components (Section 64), it is easily shown that

$$J_x\gamma_x + J_y\gamma_y + J_z\gamma_z = 0, \tag{74.17}$$

$$J_z\gamma_x - \gamma_x J_z = i\hbar\gamma_y, \qquad J_z\gamma_y - \gamma_y J_z = -i\hbar\gamma_x, \qquad J_z\gamma_z - \gamma_z J_z = 0; \tag{74.18}$$

the last three equations lead to others by cyclic permutation of x, y, z. If we now take the three orbital angular momentum components M_x, M_y, M_z and the three co-ordinates x, y, z, it is easy to see that they satisfy entirely similar relations:

$$M_x x + M_y y + M_z z = 0, \tag{74.17'}$$

$$M_z x - x M_z = i\hbar y, \qquad M_z y - y M_z = -i\hbar x, \qquad M_z z - z M_z = 0. \tag{74.18'}$$

A comparison of (74.17′) and (74.18′) with (74.17) and (74.18) shows that the structure of the matrices $\boldsymbol{J}_x, \boldsymbol{J}_y, \boldsymbol{J}_z$ with respect to $\gamma_x, \gamma_y, \gamma_z$ is the same as that of $\boldsymbol{M}_x, \boldsymbol{M}_y, \boldsymbol{M}_z$ with respect to x, y, z. It is shown in Section 90B that the only non-zero matrix elements of x, y, z are $x_{l,l\pm1}$, $y_{l,l\pm1}$, $z_{l,l\pm1}$ (where l is the orbital quantum number). The diagonal elements x_{ll}, y_{ll}, z_{ll} are zero. But l is just the number of the eigenvalue $M^2{}_l$. Thus the diagonal matrix elements of x, y, z are zero in a representation in which $\boldsymbol{M}^2$ is diagonal. The diagonal elements of $\gamma_x, \gamma_y, \gamma_z$ are therefore zero in a representation in which $\boldsymbol{J}^2$ is diagonal, i.e.

$$(\gamma_x)_{jm_j, jm_{j'}} = 0\,, \qquad (\gamma_y)_{jm_j, jm_{j'}} = 0\,, \qquad (\gamma_z)_{jm_j, jm_{j'}} = 0\,. \tag{74.19}$$

Since moreover $\boldsymbol{J}_x, \boldsymbol{J}_y, \boldsymbol{J}_z$ commute with $\boldsymbol{J}^2$, their non-zero matrix elements are

$$(J_x)_{jm_j, jm_{j'}}\,, \qquad (J_y)_{jm_j, jm_{j'}}\,, \qquad (J_z)_{jm_j, jm_{j'}}\,. \tag{74.20}$$

It follows from (74.19) and (74.20) that the matrix elements of $\boldsymbol{Q}$ of the form $Q_{jm_j, jm_{j'}}$ are zero (as is easily seen by deriving $\boldsymbol{Q}$ from γ and $\boldsymbol{J}$, using the rule of matrix multiplication).

Thus the operator $\boldsymbol{Q}$ does not affect the perturbation matrix under consideration, whose elements belong to the same value of the total angular momentum J. In other words, all the elements of the matrix $W_{m_j m_{j'}}$ arise from the part of $\boldsymbol{W}$ which does not contain $\boldsymbol{Q}$, i.e. from the operator

$$\boldsymbol{W}' = O_L \boldsymbol{J}_z \left\{1 + \frac{\boldsymbol{J}^2 - \boldsymbol{M}^2 + \boldsymbol{s}^2}{2\boldsymbol{J}^2}\right\}. \tag{74.21}$$

Since $\boldsymbol{J}_z$, $\boldsymbol{M}^2$, $\boldsymbol{s}^2$ and $\boldsymbol{J}^2$ commute with one another, their matrices can be simultaneously brought to diagonal form. The matrix $\boldsymbol{W}'$ (with elements $W'_{m_j m_{j'}}$) is then also brought to diagonal form. In order to obtain its diagonal elements, we need only replace the operators $\boldsymbol{J}_z$, $\boldsymbol{M}^2$, $\boldsymbol{s}^2$ and $\boldsymbol{J}^2$ by their eigenvalues. Since

$$\left.\begin{aligned} J_z &= \hbar m_j\,, \qquad J^2 = \hbar^2 j(j+1)\,,\\ M^2 &= \hbar^2 l(l+1)\,, \qquad s^2 = \hbar^2 l_s(l_s+1)\,, \end{aligned}\right\} \tag{74.22}$$

we find

$$W' = \hbar\, O_L\, m_j \left\{1 + \frac{j(j+1) - l(l+1) + l_s(l_s+1)}{2j(j+1)}\right\}. \tag{74.23}$$

This formula gives the splitting in a weak magnetic field of the quantum level specified by the numbers j, l; since one electron is involved, $l_s = \frac{1}{2}$.[7]

Denoting now the correction W' to the level energy E_{nlj} by ΔE_{jlm_j}, we can write (74.23) as

$$\Delta E_{jlm_j} = \hbar\, O_L\, m_j\, g\,, \tag{74.24}$$

where g denotes the *Landé factor*:

$$g = 1 + \frac{j(j+1) - l(l+1) + l_s(l_s+1)}{2j(j+1)}\,. \tag{74.25}$$

Since m_j takes all values from $-j$ to $+j$, it is seen from (74.24) that each level E_{nlj} is split into $2j + 1$ levels in a weak magnetic field.

[7] Cf. Section 105, where W' is given for a system of electrons.

Figure 55 shows the splitting pattern of the levels ${}^2S_{\frac{1}{2}}$ $(j = \frac{1}{2}, l = 0)$, ${}^2P_{\frac{1}{2}}$ $(j = \frac{1}{2}, l = 1)$ and ${}^2P_{\frac{3}{2}}$ $(j = \frac{3}{2}, l = 1)$.

When the field $\mathscr{H}$ is large the splitting pattern is simplified and becomes that derived previously (Figure 46). This simplification of the splitting of spectral lines in a magnetic field when the field becomes strong has been observed experimentally [87].

75. A diagrammatic interpretation of the splitting of levels in a weak magnetic field (the vector model)

The formula (74.23) derived above for the splitting of quantum levels in weak magnetic fields has a diagrammatic interpretation in terms of a vector model. In a magnetic field the squared total angular momentum J^2 and its component J_z along the magnetic field

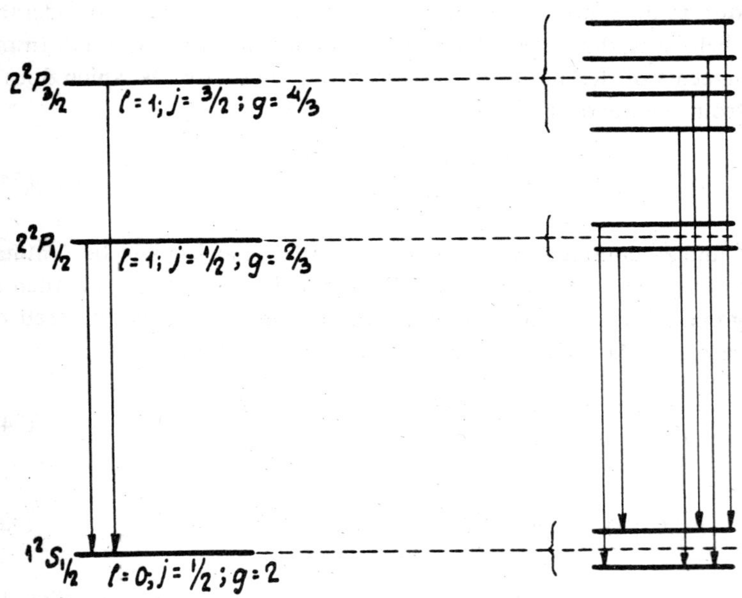

Fig. 55. Splitting of the levels ${}^2S_{\frac{1}{2}}$, ${}^2P_{\frac{1}{2}}$, ${}^2P_{\frac{3}{2}}$ in a weak magnetic field.

are integrals of the motion. The total angular momentum vector $\mathbf{J}$, however, is not an integral of the motion, but precesses about the direction of the magnetic field as indicated in Figure 56.

If the coupling between the orbital motion and the spin is great, the relative orientation of the spin vector $\mathbf{s}$ and the orbital angular momentum vector $\mathbf{M}_l$ is preserved, but both precess about the total angular momentum $\mathbf{J}$. The additional energy W in the magnetic field is equal to the energy of magnetic dipoles with moments $-(e/2\mu c)\,\mathbf{M}_l$ and $-(e/\mu c)\,\mathbf{s}$ in the field $\mathscr{H}$:

$$W = \frac{e}{2\mu c}\,\mathbf{M}_l \cdot \mathscr{H} + \frac{e}{\mu c}\,\mathbf{s} \cdot \mathscr{H} = O_L(J_z + s_z)\,. \tag{75.1}$$

We wish to find the mean value of W. The value of J_z is constant, but s_z is variable, and so, in order to calculate the mean value $\overline{W}$, we must find the mean value $\overline{s_z}$, bearing in mind that the vector **s** takes part in two precessional motions: about the vector **J**, and together with **J** about the direction of the magnetic field (OZ).

Since

$$s_z = s\cos(\mathscr{H}, \mathbf{s}), \tag{75.2}$$

we have to calculate the mean value of cos ($\mathscr{H}$, **s**). It is seen from Figure 56 that

$$\overline{\cos(\mathscr{H}, \mathbf{s})} = \cos(\mathbf{s}, \mathbf{J})\cos(\mathbf{J}, \mathscr{H}), \tag{75.3}$$

i.e.

$$\overline{s_z} = s\cos(\mathbf{s}, \mathbf{J})\cos(\mathbf{J}, \mathscr{H}). \tag{75.4}$$

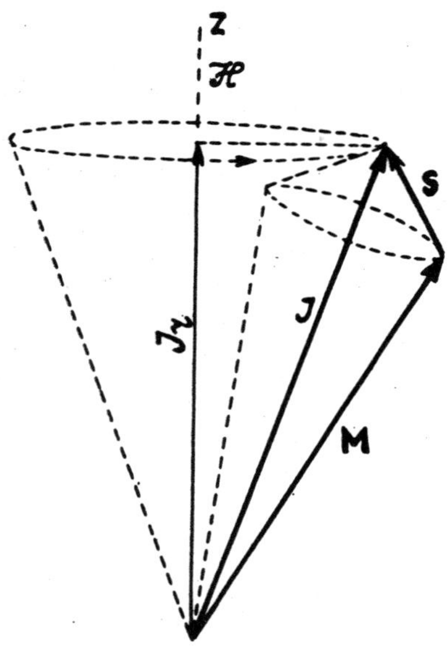

Fig. 56. Precession of the total angular momentum **J** about the direction of the magnetic field.

But

$$\cos(\mathbf{J}, \mathscr{H}) = J_z/J, \tag{75.5}$$

and from the triangle with sides **J**, **M**, **s** we have

$$sJ\cos(\mathbf{s}, \mathbf{J}) = \mathbf{s}\cdot\mathbf{J} = \tfrac{1}{2}(J^2 - M^2 + s^2). \tag{75.6}$$

From these formulae

$$\overline{s_z} = (J_z/2J^2)(J^2 - M^2 + s^2). \tag{75.7}$$

Substitution of $\overline{s_z}$ in (75.1) gives for the energy $\overline{W}$

$$\overline{W} = O_L(J_z + \overline{s_z}) = O_L J_z\left(1 + \frac{J^2 - M^2 + s^2}{2J^2}\right). \tag{75.8}$$

If J_z, J^2, M^2 and s^2 in this formula are replaced by their quantum values (74.22), then (75.8) becomes the quantum formula (74.23).

76. Perturbation theory for the continuous spectrum

Let us now consider the case where the unperturbed system has a continuous energy spectrum. We denote the Hamiltonian of this system by $\boldsymbol{H}^0$, and the eigenfunctions belonging to the energy level E by ψ_E^0. In this notation Schrödinger's equation is

$$\boldsymbol{H}^0 \psi_E^0 = E\psi_E^0. \tag{76.1}$$

We now assume that a perturbation W acts on the system. Schrödinger's equation for the perturbed system is then

$$\boldsymbol{H}\psi \equiv (\boldsymbol{H}^0 + \boldsymbol{W})\psi = E\psi. \tag{76.2}$$

If the perturbation is such that it does not affect the continuity of the spectrum of the operator $\boldsymbol{H}^0$, i.e. if the operator $\boldsymbol{H}$ also has a continuous spectrum, then the effect of the perturbation is merely to change the form of the eigenfunctions belonging to the level E. In this case the problem of perturbation theory is simply to find the functions ψ_E, which can differ only slightly from the function ψ_E^0, if the perturbation $\boldsymbol{W}$ is small. Another case is possible, however, where the perturbation $\boldsymbol{W}$ leads to the occurrence of gaps in the continuous spectrum. Then the problem of perturbation theory involves the determination not only of the wave functions but also of the position and extent of the gaps in the originally continuous energy spectrum.

We shall consider both cases for the simple example of a particle moving freely along the axis OX. Schrödinger's equation has the form

$$-\frac{\hbar^2}{2\mu}\frac{d^2\psi^0}{dx^2} = E\psi^0, \tag{76.3}$$

and has the eigenfunctions and eigenvalues

$$\psi_E^0 = \frac{e^{ipx/\hbar}}{\sqrt{(2\pi\hbar)}}, \qquad E(p) = p^2/2\mu. \tag{76.4}$$

The perturbed equation is

$$-\frac{\hbar^2}{2\mu}\frac{d^2\psi}{dx^2} + W(x)\psi = E'\psi, \tag{76.5}$$

so that $W(x)$ is an additional potential energy. The prime to E is added if the spectrum of the perturbed system is different. We can write quite generally

$$E' = E + \varepsilon, \tag{76.6}$$

$$\psi = \psi_p^0(x) + u(x). \tag{76.7}$$

However, if we assume that perturbation theory can be applied to solve Equation (76.5), this means that $|\varepsilon| \ll E$, $|u(x)| \ll |\psi_p^0(x)|$, and we can neglect the products εW,

εu and uW as being of the second order of smallness. Then substitution of (76.6) and (76.7) in (76.5), using (76.3), gives

$$-\frac{\hbar^2}{2\mu}\frac{d^2u}{dx^2} - Eu = [\varepsilon - W(x)]\,\psi_p^0(x). \tag{76.8}$$

We write $u(x)$ as a superposition of unperturbed states:

$$u(x) = \int_{-\infty}^{\infty} u(p'')\,\psi_{p''}^0(x)\,dp''. \tag{76.9}$$

We now substitute (76.9) in (76.8), multiply by $\psi_{p'}^{0*}(x)$ and integrate over x. Since

$$\int \psi_{p'}^{0*}\psi_p^0\,dx = \delta(p'-p),$$

the result is

$$\frac{p'^2}{2\mu}u(p') - Eu(p') = \varepsilon\delta(p'-p) - W_{p'p}. \tag{76.10}$$

This is Equation (76.8) in the p representation, with

$$W_{p'p} = \int \psi_{p'}^{0*}(x)\cdot W(x)\,\psi_p^0(x)\cdot dx = \frac{1}{2\pi\hbar}\int W(x)\,e^{i(p-p')x/\hbar}\,dx \tag{76.11}$$

the matrix element in the p representation. From (76.10)

$$u(p') = \frac{\varepsilon\delta(p'-p) - W_{p'p}}{E(p') - E(p)}. \tag{76.12}$$

At the point $p' = p$ the denominator in (76.12) is zero. If we take $\varepsilon \neq 0$ we have $u(p') \approx \infty \cdot \delta(p'-p)$, and the solution can never be an approximation to ψ_p^0. We must therefore put $\varepsilon = 0$, i.e.

$$u(p') = -\frac{W_{p'p}}{E(p') - E(p)} = -\frac{2\mu W_{p'p}}{(p'+p)(p'-p)}. \tag{76.13}$$

Substituting this value of $u(p')$ in (76.9) and (76.7), we find

$$\psi_p(x) = \psi_p^0(x) - P\int \frac{W_{p'p}\,\psi_{p'}^0(x)\,dp'}{E(p') - E(p)}. \tag{76.14}$$

The letter P before the integral sign signifies that the point $p' = p$ must be excluded from the integration, since at that point Formula (76.13) becomes meaningless. The function $\psi_p^0(x)$ $(p' = p)$ has in any case been taken outside the integral.[8]

A necessary condition for this method of solution to be valid is that the additional term in (76.14) should be small, i.e.

$$|\psi_p(x) - \psi_p^0(x)| \ll |\psi_p^0(x)|. \tag{76.15}$$

[8] The precise meaning of the symbol P can be defined as follows:

$$P\int F(p,p')\,dp' = \lim_{\Delta\to 0}\left\{\int_{-\infty}^{p-\Delta} F(p,p')\,dp' + \int_{p+\Delta}^{\infty} F(p,p')\,dp'\right\}.$$

The limit thus defined is called the *principal value* of the integral.

It is seen from (76.13) that $u(p')$ near the resonance point $p = p'$ is small when p is large, i.e. when the particle energy E is large. Accordingly our approximation is valid for large particle energies.

In the above calculation we have assumed that the matrix element $W_{p'p}$ is a finite quantity. This will be so if $W(x)$ decreases sufficiently rapidly as $|x| \to \infty$, i.e. the perturbation must be concentrated in a finite region of space (Figure 57). In this case the calculation shows that the energy spectrum remains continuous[9] if the quantity u is small.

If the perturbation $W(x)$ extends over all space and $W_{p'p}$ is infinite, gaps may result in an originally continuous spectrum.

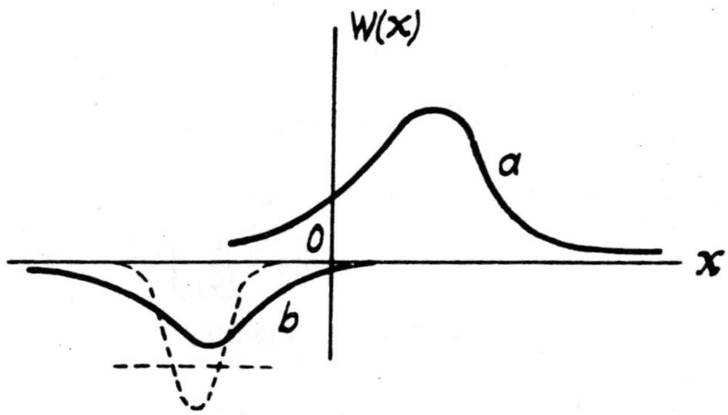

Fig. 57. Curves for the perturbation energy $W(x)$.

As an example, let us take a perturbation of the form

$$W(x) = \lambda \cos(2qx/\hbar) = \tfrac{1}{2}\lambda \{e^{i\cdot 2qx/\hbar} + e^{-i\cdot 2qx/\hbar}\}, \tag{76.16}$$

where λ and q are some parameters. On calculating the matrix elements $W_{p'p}$ from Formula (76.11), we find

$$W_{p'p} = \tfrac{1}{2}\lambda \{\delta(2q + p - p') + \delta(-2q + p - p')\}. \tag{76.17}$$

Substituting this value of $W_{p'p}$ in (76.14), because of the presence of the delta functions we immediately effect the integration to obtain

$$\psi_p(x) = \psi_p^0(x) - \tfrac{1}{2}\lambda \left\{\frac{\psi_{p+2q}^0(x)}{E(p+2q) - E(p)} + \frac{\psi_{p-2q}^0(x)}{E(p-2q) - E(p)}\right\}. \tag{76.18}$$

For small λ this is a valid approximation, but it will not serve at the points

$$E(p \pm 2q) = E(p), \qquad p = \mp q, \tag{76.19}$$

since at these points the term added to ψ_p^0 becomes infinite for all λ.

[9] If the perturbation is represented by the curve b (Figure 57), discrete levels may occur when the minimum is sufficiently deep (as indicated by the broken lines in Figure 57). Our approximate method does not give these levels, since it is valid only for large energies E.

In order to obtain an approximate solution for $p = \pm q$, we use the fact that two functions ψ_p^0 and ψ_{-p}^0 always belong to the level $E(p)$. The most general solution belonging to the level $E(p)$ is

$$\phi^0 = \alpha\psi_p^0 + \beta\psi_{-p}^0, \tag{76.19'}$$

where α and β are undetermined coefficients. If we substitute in (76.7) ϕ^0 instead of ψ_p^0 and repeat the calculations, (76.8) becomes

$$-\frac{\hbar^2}{2\mu}\frac{d^2u}{dx^2} - Eu = (\varepsilon - W)\phi^0. \tag{76.8'}$$

Substituting u from (76.9), multiplying by $\psi_{p'}^{0*}$ and integrating over x, we find instead of (76.10)

$$\begin{aligned} u(p')[E(p') - E(p)] = \varepsilon[\alpha\delta(p - p') + \beta\delta(p + p')] - \\ - (\alpha W_{p'p} + \beta W_{p',-p}) \end{aligned} \tag{76.10'}$$

and finally, with the values of $W_{p'p}$ and $W_{p',-p}$ from (76.17),

$$\begin{aligned} & u(p')[E(p') - E(p)] = \varepsilon[\alpha\delta(p - p') + \beta\delta(p + p')] - \\ & - \tfrac{1}{2}\lambda[\alpha\{\delta(2q + p - p') + \delta(-2q + p - p')\} + \\ & + \beta\{\delta(2q - p - p') + \delta(-2q - p - p')\}]. \end{aligned} \tag{76.10''}$$

If $p \neq \pm q$ we can put $\varepsilon = 0$ and take either $\alpha = 1, \beta = 0$ or $\alpha = 0, \beta = 1$. In the former case we obtain the previous solution, (76.18), and in the latter case the solution ψ_{-p} which is close to ψ_{-p}^0.

For $p = +q$ we have from (76.10″)

$$\begin{aligned} & u(p')[E(p') - E(q)] = \varepsilon[\alpha\delta(q - p') + \beta\delta(q + p')] - \\ & - \tfrac{1}{2}\lambda[\alpha\{\delta(3q - p') + \delta(-q - p')\} + \\ & + \beta\{\delta(q - p') + \delta(-3q - p')\}]. \end{aligned} \tag{76.10'''}$$

For $p' = q$ the left-hand side is zero and the right-hand side must also be zero. Since for $\xi \neq 0$ $\delta(\xi) = 0$, we have

$$\delta(0)[\varepsilon\alpha - \tfrac{1}{2}\lambda\beta] = 0 \tag{76.20}$$

and for $p' = -q$

$$\delta(0)[\varepsilon\beta - \tfrac{1}{2}\lambda\alpha] = 0. \tag{76.20'}$$

Cancelling $\delta(0)$ gives the equations

$$\varepsilon\alpha - \tfrac{1}{2}\lambda\beta = 0, \qquad \varepsilon\beta - \tfrac{1}{2}\lambda\alpha = 0 \tag{76.21}$$

to determine α and β. It is easily seen that for $p = -q$ (76.10″) again gives Equations (76.21). These equations are homogeneous, and equating their determinant to zero gives

$$\varepsilon = \pm\tfrac{1}{2}\lambda, \tag{76.22}$$

the corresponding solutions for α and β being

$$\alpha = \beta \qquad \text{for} \qquad \varepsilon = +\tfrac{1}{2}\lambda \tag{76.23}$$

and

$$\alpha = -\beta \qquad \text{for} \qquad \varepsilon = -\tfrac{1}{2}\lambda. \tag{76.23'}$$

Thus for momentum $p = \pm q$ the solutions are

$$E = E(\pm q) + \tfrac{1}{2}\lambda, \qquad \psi_{\pm q}(x) = \alpha(\psi^0_{\pm q} + \psi^0_{\mp q}), \tag{76.24}$$

$$E = E(\pm q) - \tfrac{1}{2}\lambda, \qquad \psi_{\pm q}(x) = \alpha(\psi^0_{\pm q} - \psi^0_{\mp q}). \tag{76.24'}$$

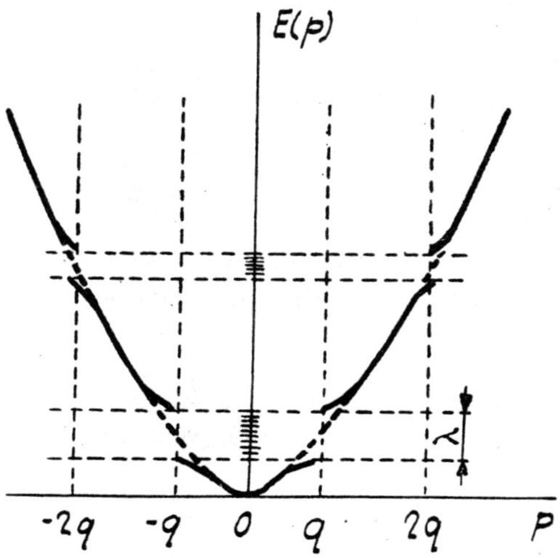

Fig. 58. Formation of gaps (forbidden bands) in a continuous spectrum when a periodic perturbation is applied.

That is to say, the energy has a gap at $p = \pm q$. For momenta far from $p = \pm q$ it has been shown above that $\varepsilon = 0$ and so $E = E(p)$. Figure 58 gives the curve of the energy E as a function of p: the broken line is for the unperturbed motion and the continuous line for the perturbed motion. At the points $p = \pm q$ a discontinuity λ occurs. The other discontinuities at $p = \pm 2q$ shown in the diagram are obtained from a calculation in the second approximation. In general, discontinuities occur at the points $p = \pm nq$, $n = 1, 2, 3, \ldots$.

Thus a spectrum is obtained of the type considered in Section 55, consisting of allowed energy zones from $E = 0$ to $E = E(q) - \tfrac{1}{2}\lambda$, from $E = E(q) + \tfrac{1}{2}\lambda$ to the next discontinuity, and so on, and forbidden energy zones from $E = E(q) - \tfrac{1}{2}\lambda$ to $E = E(q) + \tfrac{1}{2}\lambda$, etc. These forbidden ranges of energy are shown by hatching on the ordinate axis. When the perturbing field is small ($\lambda \to 0$) the gaps become very narrow and so the spectrum of a particle moving in a periodic field of low amplitude is, as it were, the reverse of the discrete spectrum typical of an atom. In the discrete spectrum only certain energy values E_1, E_2, ... are allowed, and all other values are forbidden.

In the case considered here, wide ranges of energy are allowed and certain narrow strips excluded.

Figure 58 shows, besides the gaps calculated above in the continuous spectrum of E, the smooth variation of E as a function of p near these points of discontinuity. This variation could also be obtained from the foregoing calculation by using the fact that the solution (76.18) is invalid not only at the points $p = \pm q$, where it simply becomes infinite, but also at all points where

$$|E(p \pm 2q) - E(p)| \approx \lambda, \tag{76.25}$$

since in this range of momenta the term added to E is large though not infinite. Thus it would be necessary to examine the behaviour of the solutions in the neighbourhood of the points $p = \pm q$, but we shall not do this here.[10]

It is important to note that the existence of these gaps affects the form of the function $E(p)$ near them and so changes the number of states ψ_p (which we can regard as proportional to $\mathrm{d}p$) belonging to an energy range $\mathrm{d}E$. For the unperturbed problem $\mathrm{d}p/\mathrm{d}E = \mu/p$, while for the perturbed problem $\mathrm{d}p/\mathrm{d}E = \infty$ at points where the energy is discontinuous. This result can be obtained immediately: in Section 55 we have shown generally that for a particle moving in a periodic field the group velocity $v = (1/\hbar)\,\mathrm{d}E/\mathrm{d}k = \mathrm{d}E/\mathrm{d}p$ is zero at the zone boundaries. That this is so in our example follows simply from the fact that at the zone boundaries we have stationary waves (76.24) and (76.24′) and not travelling waves ($e^{ipx/\hbar}$).

[10] The calculation is given, for example, in [22], Ch. VIII.

CHAPTER XIII

COLLISION THEORY

77. Statement of the problem in collision theory of microparticles

The theory of the collision of microparticles has become a very extensive branch of atomic mechanics.[1] It is not possible to give a detailed account of this theory in the present book, and we shall merely describe the statement of the problem in collision theory and the simplest methods of solution.

Let us consider some particle A, for definiteness an atom, and a flux of particles B,

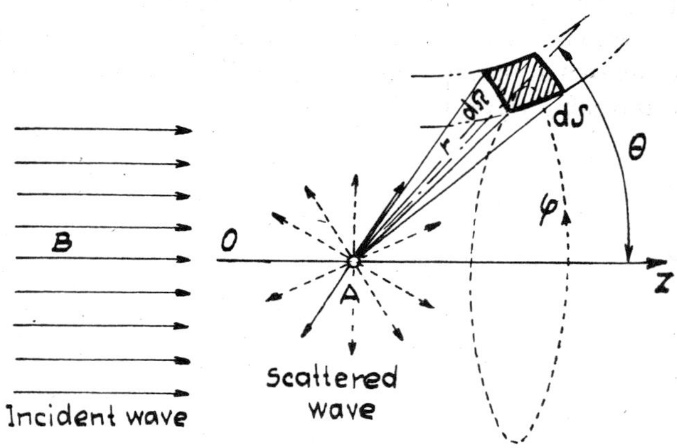

Fig. 59. Collision of particles in quantum mechanics. A scattering atom, B incident beam of particles.

say electrons, incident on it in the direction OZ (Figure 59). When the electrons B collide with the atom they may undergo a change of state in two respects. Firstly they may change their direction of motion, and secondly they may transfer some part ε of their energy E to the atom A.[2] In the latter case we speak of an *inelastic collision* or *inelastic scattering*. If $\varepsilon = 0$, the collision is said to be *elastic* (*elastic scattering*). Experimentally we are concerned with the number of electrons (or particles B) passing

[1] See, for example, [67].

[2] If the atom A is originally in an excited state, a case may also occur in which it transfers its energy to the electron B, and the original energy of the electron is increased by some amount ε. Such a collision is said to be *of the second kind.*

per second through an area dS (Figure 59) placed perpendicular to the line from the centre of the scatterer A. Let the flux of particles passing through this area and having energy $E - \varepsilon$ be dN_ε. This number dN_ε is proportional to the area dS (since it is small) and inversely proportional to the square of the distance r from the scatterer, and is also evidently proportional to the flux N of particles in the primary beam. Thus

$$dN_\varepsilon = Nq(\varepsilon, \theta, \phi)\, dS/r^2, \tag{77.1}$$

where N is the number of particles passing through 1 cm^2 per second in the primary beam, and $q(\varepsilon, \theta, \phi)$ is some proportionality factor between dN_ε and N. The quantity dS/r^2 is the solid angle $d\Omega$ subtended by the area dS at the centre of the scatterer A. The ratio dN_ε/N gives the probability of scattering into the angle $d\Omega$ with loss of energy ε. This ratio is

$$dN_\varepsilon/N = q(\varepsilon, \theta, \phi)\, d\Omega. \tag{77.2}$$

It follows from (77.1) that q has the dimensions of area, since $[dN_\varepsilon] = 1/T$ and $[N] = 1/TL^2$, so that $[q] = L^2$; q is called the *differential cross-section* (of the atom A) for inelastic scattering into the angle $d\Omega$ with loss of energy ε.

The quantity

$$Q_\varepsilon = N_\varepsilon/N = \int q(\varepsilon, \theta, \phi)\, d\Omega, \tag{77.3}$$

where the integral is taken over the whole solid angle 4π, gives what is called the *total cross-section* for inelastic collision with energy loss ε. $N_\varepsilon = Q_\varepsilon N$ is the number of particles per second which lose energy ε by collision when the primary flux is N particles per cm^2 per second.

If the energy loss ε can take continuous values, then for an energy loss between ε and $\varepsilon + d\varepsilon$ (77.2) must be replaced by

$$dN_\varepsilon/N = q(\varepsilon, \theta, \phi)\, d\varepsilon\, d\Omega. \tag{77.2'}$$

In this case $q(\varepsilon, \theta, \phi)\, d\varepsilon$ will signify the differential cross-section for inelastic scattering into the angle $d\Omega$ with energy loss between ε and $\varepsilon + d\varepsilon$. The quantity $q(\varepsilon, \theta, \phi)$ will again be called the differential cross-section for inelastic scattering with respect to an element of solid angle $d\Omega$ and a range of energy $d\varepsilon$. The usual nomenclature is 'cross-section per unit solid angle and per unit energy'.

It may be noted that the cross-section may depend not only on ε, θ and ϕ but also on other parameters pertaining to the collision, such as the spin of the particles. In every case the differential cross-section gives a complete statistical description of the collision process. Thus the problem of collision theory amounts to a calculation of the cross-section $q(\varepsilon, \theta, \phi)$. As we shall see, this quantity in turn is completely determined by the amplitude of the scattered waves.

Leaving aside for the present methods of calculating this quantity in quantum mechanics, let us consider in which cases quantum mechanics and classical mechanics should respectively be used for collision calculations.

To do so let us examine how a collision occurs according to the laws of classical mechanics. Figure 60 shows an atom A with centre O. A sphere of radius a is drawn round it, outside which the forces between the atom A and the incident particle B are small. We call this the *sphere of interaction*.[3] The particle B, moving originally along the axis BZ and incident on this sphere, will be deflected as shown in Figure 60 (for the case of repulsion between A and B). Let ρ be the length of the perpendicular from the centre of the atom to the original direction of motion BZ of the particle. This is called the *impact parameter*. A particle with a given impact parameter is deflected through a definite angle θ, so that $\rho = \rho(\theta)$ and $\theta = \theta(\rho)$. Particles with impact parameters between ρ and $\rho + \mathrm{d}\rho$ are deflected through angles between θ and $\theta + \mathrm{d}\theta$. (We do not consider here the angle ϕ, assuming that the field

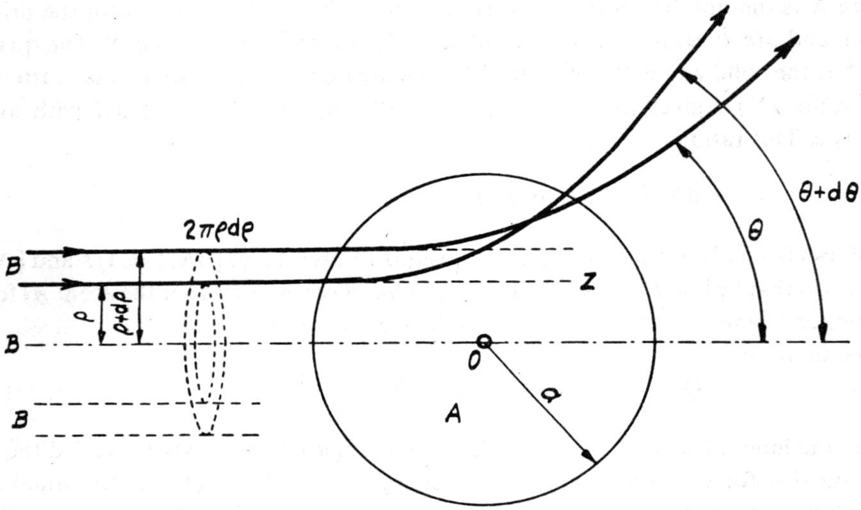

Fig. 60. Collision of particle B with atom A according to classical mechanics (case of repulsion).

of the atom A is spherically symmetrical.) Of the flux of primary particles through an area of 1 cm^2, those which pass through a ring formed by circles of radii ρ and $\rho + \mathrm{d}\rho$ are deflected through angles between θ and $\theta + \mathrm{d}\theta$. The area of this ring is $2\pi\rho\,\mathrm{d}\rho$ (Figure 60).

Thus all those particles in the primary beam which pass through the area $2\pi\rho\,\mathrm{d}\rho$ (Figure 60) are deflected through angles between θ and $\theta + \mathrm{d}\theta$. It follows that $2\pi\rho\,\mathrm{d}\rho$ is the cross-section for deflection through angles between θ and $\theta + \mathrm{d}\theta$. Expressing $\mathrm{d}\rho$ in terms of $\mathrm{d}\theta$, we find the differential cross-section

$$q(\theta) = \rho\,\mathrm{d}\rho/\mathrm{d}\theta\,. \tag{77.4}$$

This classical expression for $q(\theta)$ will not always be applicable to micro-collisions, for the error $\Delta\rho$ in the determination of the impact parameter must be less than ρ itself, and the determination of ρ with error $\Delta\rho$ involves an uncertainty Δp_1 in the momentum perpendicular to the original motion such that $\Delta p_1 \sim \hbar/\Delta\rho$, and consequently an uncertainty $\Delta\theta$ in the angle of deflection such that $\Delta\theta = \Delta p_1/p = \hbar/p\Delta\rho$ (where p is the original momentum of the particle). Since $\theta > \Delta\theta$, $\rho > \Delta\rho$, this gives

$$\theta > \lambda/\rho\,. \tag{77.5}$$

Hence we see that the application of the methods of classical mechanics to consider small deflections is not meaningful. To consider deflections which satisfy the condition (77.5) the general condition for classical mechanics to be valid must be satisfied, namely the change in the potential $U(r)$ over the

[3] This sphere cannot always be defined. For instance, Coulomb's law gives $U = \text{constant}/r$ and no such sphere can be said to exist. The sphere of interaction can be defined only when the forces decrease sufficiently rapidly.

wavelength λ must be small:

$$\frac{1}{U(r)}\frac{\mathrm{d}U(r)}{\mathrm{d}r}\lambda \ll 1 . \tag{77.6}$$

Suppose that the potential of the scattering particle varies appreciably over a distance a, so that a is in order of magnitude the range of interaction of the potential, or the radius of the sphere of interaction. Then the condition (77.6) can be replaced by the less stringent condition

$$\lambda \ll a; \tag{77.7}$$

see Section 36. The wavelength λ of electrons whose energy is of the order of 1 eV is, in order of magnitude, 10^{-8} cm, the same as the dimensions of the region a within which the potential in the atom varies appreciably. In collisions of electrons with atoms the condition (77.7) is therefore not satisfied, and quantum mechanics must be employed. In collisions of α particles ($\lambda \sim 10^{-13}$ cm) with an atom the condition (77.7) is satisfied and a classical treatment of the problem is adequate, but in collisions of α particles (or, of nucleons, or of heavy particles in general) with a nucleus, for which the radius of the sphere of interaction $a \sim 10^{-13}$ cm, we again have $\lambda \sim a$, i.e. quantum mechanics must be used to discuss the problem.

The consideration of a collision with only one atom, instead of with an assembly of atoms forming a gas or a liquid or a solid, is itself an abstraction by no means always valid. By considering only one atom we assume that the particle moves freely before colliding with the atom. This is the essence of the problem stated in terms of two-particle collisions. In order to decide when such a formulation of the problem is possible, we may consider the mean free path traversed by the particle B without collision in the assembly of atoms forming the body.

For definiteness we shall consider only elastic collisions, and define the criterion according to which the particle B is not interacting with the atom A (i.e. is moving freely) in terms of some angle of deflection θ_0. If the angle of deflection $\theta < \theta_0$, we shall suppose that the particle is moving freely and is not deflected; if $\theta > \theta_0$, we shall suppose that an interaction occurs. The cross-section Q_0 for deviation through angles exceeding θ_0 is

$$Q_0 = \int_{\Omega_0} q(\varepsilon, \theta, \phi)\, \mathrm{d}\Omega ; \tag{77.8}$$

here Ω_0 signifies that we exclude small deflections ($\theta < \theta_0$) from the integration. Let us now imagine a flux of N particles B passing through an area of 1 cm². In traversing a distance dx this flux passes through a volume 1 cm² $\times$ dx. If n denotes the number of atoms in 1 cm³ of the body (gas, liquid or solid), the particle B will encounter in this volume $n \times 1\ \mathrm{cm}^2 \times \mathrm{d}x$ atoms A. The probability of collision of one particle B with one atom A when the distance dx is traversed is

$$\frac{Q_0}{1\ \mathrm{cm}^2} n \cdot 1\ \mathrm{cm}^2 \cdot \mathrm{d}x = Q_0 n\, \mathrm{d}x . \tag{77.9}$$

Let $N(x)$ be the flux of undeflected particles at a depth x in the substance. According to (77.9) the decrease in this flux in traversing the distance from x to $x + \mathrm{d}x$ is

$$\mathrm{d}N(x)/\mathrm{d}x = -N(x) Q_0 n . \tag{77.10}$$

Hence

$$N(x) = N_0 e^{-Q_0 n x} , \tag{77.11}$$

and therefore the quantity

$$w(x) = e^{-Q_0 n x} \tag{77.12}$$

is the probability of traversing the distance x without collision. The mean free path l is therefore

$$l = Q_0 n \int_0^\infty e^{-Q_0 n x} x\, \mathrm{d}x = 1/Q_0 n . \tag{77.13}$$

In order that a particle traversing a distance l may in fact be regarded as freely moving relative to any atom in the body, it is necessary that the free path should be larger than the radius a of the sphere of interaction. Otherwise the particle will always be within the sphere of interaction of the atom with which it is about to collide. Thus the condition for the theory of two-particle collisions to be applicable is, both in classical and in quantum mechanics,

$$l \gg a . \tag{77.14}$$

If the sphere of interaction a cannot be defined, the use of two-particle collision theory becomes at least dubious (certainly for collisions where l is small).

In quantum mechanics the condition (77.14) must be supplemented by a further, specifically quantum condition. We are concerned with the changes in momentum (and energy) of a particle in collisions. A state with a definite momentum p is a de Broglie wave with wavelength $\lambda = 2\pi\hbar/p$.

The condition (77.14) shows that we have to consider the motion of a free particle over the free path l, i.e. a wave packet of dimensions not exceeding l. In such a wave packet $\overline{\Delta p^2} \neq 0$ in general and the state has no definite momentum. In order to be able to neglect this uncertainty (and use a monochromatic wave) it is necessary that

$$l \gg \lambda . \tag{77.15}$$

When the conditions (77.14) and (77.15) are not satisfied it is necessary to consider the collision with the whole assembly of atoms A or to seek special procedures whereby the difficulties of this direct statement of the problem may be surmounted.

78. Calculation of elastic scattering by the Born approximation

If only elastic scattering is considered we may disregard the internal structure of the atom A.[4] The interaction of the atom A with the incident particles B may then be considered as equivalent to that of a centre of force. If the atom is spherically symmetrical, the field due to the atom will be a central force field. Taking this case, let us denote by $U(r)$ the potential energy of the particle B in the field of the atom A, where r is the distance from the centre A to B, and let E be the energy of the particle B. If we assume that $U(r) = 0$ for $r = \infty$, we must take $E > 0$, since we are interested in a case where the particle B with energy E moves from infinity towards the atom A. According to the general theory of motion in a field of central forces, such states of the particle B are possible only when $E > 0$.

Denoting the wave function of the particle B by $\psi(x, y, z)$, we can write the corresponding Schrödinger's equation as

$$-(\hbar^2/2\mu)\nabla^2\psi + U(r)\psi = E\psi , \tag{78.1}$$

where μ is the mass of the particle B. The potential energy $U(r)$ will be assumed to decrease sufficiently rapidly with increasing distance r from the atom A. In terms of the wave number k given by

$$k^2 = 2\mu E/\hbar^2 = p^2/\hbar^2 , \tag{78.2}$$

where p is the momentum of the particle, and the quantity

$$V(r) = 2\mu U(r)/\hbar^2 , \tag{78.3}$$

Equation (78.1) may be written

$$\nabla^2\psi + k^2\psi = V(r)\psi . \tag{78.1'}$$

The solutions of this equation for a given energy E are highly degenerate and have a great variety of forms.

We must take solutions which would correspond to the physical problem in question.

[4] In calculations of inelastic collisions, on the other hand, it is necessary to consider the structure of the atom A, since the quantum state of this atom is changed in an inelastic collision.

Thus for large distances from the atom A the solution ψ must consist of the plane wave representing the flux of incident particles B and the outgoing wave representing the scattered particles (the general solution of Equation (78.1′) might, for example, include ingoing waves also).

We accordingly represent ψ as a superposition:

$$\psi = \psi^0 + u\,, \tag{78.4}$$

where ψ^0 represents the flux of incident particles and u the flux of scattered particles. Assuming that the incident particles move along the axis OZ, we take ψ^0 as

$$\psi^0 = e^{ikz}/\sqrt{L^3}\,, \qquad L^3 = 1\,\mathrm{cm}^3\,. \tag{78.5}$$

The normalisation of the function ψ^0 chosen here signifies a density of incident particles $|\psi^0|^2 = 1\,\mathrm{cm}^{-3}$ (one particle per unit volume). According to Formula (29.5), the flux is

$$N = J_z = (k\hbar/\mu)\,|\psi^0|^2 = v\,|\psi^0|^2 = v\,\mathrm{sec}^{-1}\,\mathrm{cm}^{-2}\,, \tag{78.6}$$

where $v = k\hbar/\mu = p/\mu$ is the velocity of the particles. The function u, which represents the state of the scattered particles, must have the form of an outgoing wave for large distances r from the centre of the atom:

$$\underset{r\to\infty}{u(r,\theta)} = A(\theta)\,e^{ikr}/r\,, \tag{78.7}$$

where $A(\theta)$ is the amplitude of the scattered wave and θ the angle between r and OZ, i.e. the angle of scattering.

Let us now calculate the flux of scattered particles at a large distance from the atom. From Formula (29.5) for the particle flux and from (78.7) it follows that the flux of scattered particles is[5]:

$$J_r = \frac{i\hbar}{2\mu}\left\{u\,\frac{\partial u^*}{\partial r} - u^*\frac{\partial u}{\partial r}\right\} = \frac{k\hbar}{\mu}\,|A(\theta)|^2\cdot\frac{1}{r^2} = v\,|A(\theta)|^2/r^2\,. \tag{78.8}$$

Hence the flux through an area $\mathrm{d}S$ is

$$\mathrm{d}N = J_r\,\mathrm{d}S = v\,|A(\theta)|^2\,d\Omega\,. \tag{78.9}$$

Thus from (78.9) and (78.6) we have

$$q(\theta)\,\mathrm{d}\Omega = \mathrm{d}N/N = |A(\theta)|^2\,\mathrm{d}\Omega\,. \tag{78.10}$$

Thus to calculate the cross-section $q(\theta)$ it is sufficient to know the amplitude $A(\theta)$ of the scattered wave. In order to find the scattered wave u, we shall regard $V(r)$ in (78.1′) as a perturbation and use the methods of perturbation theory to solve Equation (78.1′).[6]

[5] See (53.3). The remaining components J_θ and J_ϕ are zero in a field of central forces ($A(\theta)$ being real). It may also be noted that if e^{-ikr} were taken in (78.7) instead of e^{ikr} an ingoing flux would be obtained.

[6] We shall also assume that $V(r)$ decreases with increasing distance more rapidly than $1/r$ (see the first footnote to Section 49). The matrix element of $V(r)$ will be supposed finite, so that, from the discussion in Section 76, the spectrum of E remains continuous.

Substituting (78.4) in (78.1′) and neglecting the term Vu as being of the second order of smallness, we obtain

$$\nabla^2 u + k^2 u = V\psi^0 . \tag{78.11}$$

We have now to find a solution of this equation which has the asymptotic form (78.7). Instead of expanding u in terms of the unperturbed functions, we apply a more direct method to solve (78.11). Let us consider the function

$$\Phi(\mathbf{r}, t) = \Phi_0(\mathbf{r})\, e^{-i\omega t}, \tag{78.12}$$

where $\mathbf{r}$ is the radius vector of the point x, y, z, and t is regarded as the time and ω correspondingly as some frequency. We shall regard Φ as a scalar potential due to electric charges distributed in space with density

$$\rho(\mathbf{r}, t) = \rho_0(\mathbf{r})\, e^{-i\omega t}. \tag{78.13}$$

It is known from electrodynamics that the potential satisfies d'Alembert's equation

$$\nabla^2\Phi - \frac{1}{c^2}\frac{\partial^2\Phi}{\partial t^2} = -4\pi\rho, \tag{78.14}$$

where c is the velocity of propagation of electromagnetic waves. The solution of Equation (78.14) is known: if we take waves emitted by a charge $\rho(\mathbf{r}', t)\, \mathrm{d}v'$ (where $\mathrm{d}v' = \mathrm{d}x'\, \mathrm{d}y'\, \mathrm{d}z'$) at the point $\mathbf{r}'$, then the electric potential at the point $\mathbf{r}$ at time t is

$$\Phi(\mathbf{r}, t) = \int \frac{\rho(\mathbf{r}', t - |\mathbf{r}' - \mathbf{r}|/c)}{|\mathbf{r}' - \mathbf{r}|}\, \mathrm{d}v', \tag{78.15}$$

where $|\mathbf{r}' - \mathbf{r}|$ is the distance from the position $\mathbf{r}'$ of the charge $\rho\, \mathrm{d}v'$ to the point of observation $\mathbf{r}$. Substituting in (78.15) Φ from (78.12) and ρ from (78.13) and cancelling $e^{-i\omega t}$, we obtain

$$\Phi_0(\mathbf{r}) = \int \frac{\rho_0(\mathbf{r}')\, e^{i\omega|\mathbf{r}' - \mathbf{r}|/c}}{|\mathbf{r}' - \mathbf{r}|}\, \mathrm{d}v'. \tag{78.16}$$

If we substitute in d'Alembert's equation Φ from (78.12) and ρ from (78.13) and cancel $e^{-i\omega t}$, we have

$$\nabla^2\Phi_0 + (\omega^2/c^2)\, \Phi_0 = -4\pi\rho_0 . \tag{78.17}$$

Comparing this equation with (78.11), we see that (78.11) and (78.17) are the same if we put

$$\Phi_0 = u, \qquad \omega/c = k, \qquad \rho_0 = -V\psi^0/4\pi . \tag{78.18}$$

Hence, from (78.16),

$$u(\mathbf{r}) = -\frac{1}{4\pi}\int \frac{V(\mathbf{r}')\, \psi^0(\mathbf{r}')\, e^{ik|\mathbf{r}' - \mathbf{r}|}}{|\mathbf{r}' - \mathbf{r}|}\, \mathrm{d}v' \tag{78.19}$$

is the solution of Equation (78.11). We have already taken account of the fact that u contains only outgoing waves, since the solution (78.15) is that for waves emitted, not absorbed, by the charges.

Let us now derive the form of $u(\mathbf{r})$ far from the atom A. To do so, we denote the unit vector in the direction of the incident beam (the axis OZ) by $\mathbf{n}_0$, and that in the direction of the vector $\mathbf{r}$ by $\mathbf{n}$. Let us first transform $|\mathbf{r}' - \mathbf{r}|$. From the triangle shown in Figure 61 we have

$$|\mathbf{r}' - \mathbf{r}|^2 = r^2 + r'^2 - 2(\mathbf{n}\cdot\mathbf{r}')r,$$

where $r = |\mathbf{r}|$, $r' = |\mathbf{r}'|$. Hence, for $r \gg r'$,

$$|\mathbf{r}' - \mathbf{r}| = r - \mathbf{n}\cdot\mathbf{r}' + O(r'/r), \tag{78.20}$$

where $O(r'/r)$ denotes terms of order r'/r and higher.

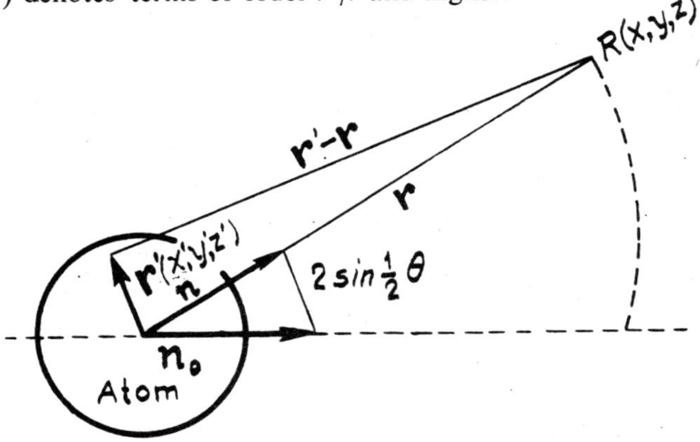

Fig. 61. Illustrating the choice of vectors. $\mathbf{r}'$ radius vector from the centre of the atom to the electron, $\mathbf{r}$ radius vector from the centre of the atom to the point of observation $R(x, y, z)$, θ angle of scattering, $\mathbf{n}_0$ unit vector in the direction of the primary beam, $\mathbf{n}$ unit vector in the direction of the scattered beam.

Substituting $|\mathbf{r}' - \mathbf{r}|$ from (78.20) in (78.19) and neglecting $\mathbf{n} \cdot \mathbf{r}'$ in the denominator in comparison with r, we obtain an expression for u valid for large[7] distances r from the atom:

$$u(\mathbf{r}) = -\frac{1}{4\pi}\frac{e^{ikr}}{r}\int e^{-ik\mathbf{n}\cdot\mathbf{r}'}V(\mathbf{r}')\psi^0(\mathbf{r}')\,\mathrm{d}v'. \tag{78.19'}$$

Substituting $\psi^0(\mathbf{r}')$ from (78.5) and using the fact that $z' = \mathbf{r}' \cdot \mathbf{n}_0$ gives

$$u(\mathbf{r}) = -\frac{1}{4\pi}\frac{e^{ikr}}{r}\int e^{ik(\mathbf{n}_0 - \mathbf{n})\cdot\mathbf{r}'}V(\mathbf{r}')\,\mathrm{d}v'. \tag{78.21}$$

Comparison of (78.21) with (78.7) shows that the amplitude of the scattered wave is

$$A = -\frac{1}{4\pi}\int e^{ik(\mathbf{n}_0 - \mathbf{n})\cdot\mathbf{r}'}V(\mathbf{r}')\,\mathrm{d}v'. \tag{78.22}$$

In terms of the vector

$$\mathbf{K} = k(\mathbf{n}_0 - \mathbf{n}), \qquad K = k|\mathbf{n}_0 - \mathbf{n}| = 2k\sin\tfrac{1}{2}\theta = (4\pi/\lambda)\sin\tfrac{1}{2}\theta \tag{78.23}$$

[7] Namely $r \gg a$, where a is the radius of the sphere of interaction.

we have, using (78.3),

$$A_\theta = -\frac{1}{4\pi}\frac{2\mu}{\hbar^2}\int e^{i\mathbf{K}\cdot\mathbf{r}'}\, U(\mathbf{r}')\,\mathrm{d}v', \tag{78.24}$$

i.e. the amplitude of the scattered wave is proportional to the Fourier component in the expansion of the potential in terms of plane waves $e^{i\mathbf{K}\cdot\mathbf{r}}$. Substituting this value of $A(\theta)$ in (78.10), we find the cross-section:

$$q(\theta) = \frac{1}{16\pi^2}\left(\frac{2\mu}{\hbar^2}\right)^2 |\int e^{i\mathbf{K}\cdot\mathbf{r}'}\, U(\mathbf{r}')\,\mathrm{d}v'|^2. \tag{78.25}$$

This formula is seen from its derivation to be approximate. In collision theory this approximation (the first approximation of perturbation theory) is usually called the *Born approximation*. We cannot discuss in detail the accuracy of this approximation and its suitability in various cases[8], but shall merely mention that the intensity of the scattered wave $|u(\mathbf{r})|^2$ near the scattering centre must be small compared with that of the incident wave $|\psi^0(\mathbf{r})|^2$. From Formula (78.19) we can easily estimate the ratio of $|u|^2$ to $|\psi^0|^2$ at the centre of the atom ($r = 0$). Assuming the forces to be central, so that $V(\mathbf{r}') = V(r')$, and putting in (78.19) $\mathbf{r} = 0$, $\mathrm{d}v' = r'^2\,\mathrm{d}r' \sin\theta'\,\mathrm{d}\theta'\,\mathrm{d}\phi'$, $\mathbf{k}\cdot\mathbf{r}' = kr'\cos\theta'$, after a simple integration over the angles θ' and ϕ' we have

$$\left|\frac{u}{\psi^0}\right| = \frac{2\mu}{\hbar^2}\left|\int_0^\infty U(r')\frac{\sin kr'}{kr'}e^{ikr'}\,r'\,\mathrm{d}r'\right|. \tag{78.26}$$

When $k \to \infty$ the integral on the right tends to zero. When the particle energy is sufficiently large (k large), therefore, Born's method will always be valid.

79. Elastic scattering of fast charged microparticles by atoms

The formula derived above for the differential cross-section $q(\theta)$ can be used to calculate the elastic scattering of sufficiently fast particles. Our derivation has, moreover, implicitly assumed that the atom is at rest both before and after the collision. If the velocity of the incident particles is large and that of the atom before the collision is the thermal velocity, the latter may be neglected. The velocity after the collision, however, may be neglected only if the mass μ of the incident particle is much less than the mass M of the atom. Assuming that all these conditions are met, let us calculate the scattering of particles with mass μ and charge e_1. Let $-e\rho(\mathbf{r}'') = -e\rho(r'')$ denote the density of electric charge due to the atomic electrons at the point $\mathbf{r}''$ (this density being assumed spherically symmetrical), and Z the atomic number. Then the electric potential at the point $\mathbf{r}$ will be

$$\phi(r) = \frac{Ze}{r} - e\int\frac{\rho(r'')\,\mathrm{d}v''}{|\mathbf{r}''-\mathbf{r}|}, \tag{79.1}$$

and the potential energy of a particle in this field will be

$$U(r) = e_1\phi(r) = \frac{Zee_1}{r} - ee_1\int\frac{\rho(r'')\,dv''}{|\mathbf{r}''-\mathbf{r}|}. \tag{79.2}$$

[8] See [67].

Substituting this value of $U(r)$ in (78.24), we obtain

$$A(\theta) = -\frac{2\mu Zee_1}{\hbar^2 4\pi}\int \frac{e^{i\mathbf{K}\cdot\mathbf{r}'}}{r'}dv' + \frac{2\mu ee_1}{\hbar^2 4\pi}\int e^{i\mathbf{K}\cdot\mathbf{r}'}dv'\int\frac{\rho(r'')\,dv''}{|\mathbf{r}''-\mathbf{r}'|}. \tag{79.3}$$

We shall consider the integrals separately. The integral

$$\phi(\mathbf{r}'') = \int \frac{e^{i\mathbf{K}\cdot\mathbf{r}'}}{|\mathbf{r}''-\mathbf{r}'|}dv' \tag{79.4}$$

may be regarded as the potential at the point $\mathbf{r}''$ due to electric charges distributed in space with density $\rho(\mathbf{r}') = e^{i\mathbf{K}\cdot\mathbf{r}'}$.

The potential $\phi(\mathbf{r}')$ satisfies Poisson's equation:

$$\nabla^2\phi(\mathbf{r}') = -4\pi\rho(\mathbf{r}') = -4\pi e^{i\mathbf{K}\cdot\mathbf{r}'} \tag{79.5}$$

This immediately gives

$$\phi(\mathbf{r}') = 4\pi e^{i\mathbf{K}\cdot\mathbf{r}'}/|\mathbf{K}|^2, \qquad |\mathbf{K}|^2 = K_x^2 + K_y^2 + K_z^2. \tag{79.6}$$

A comparison of (79.4) with the first integral in (79.3) shows that

$$I_1 = \int \frac{e^{i\mathbf{K}\cdot\mathbf{r}'}}{r'}dv' = \int \frac{e^{i\mathbf{K}\cdot\mathbf{r}'}}{|\mathbf{r}'|}dv' = \frac{4\pi}{|\mathbf{K}|^2}. \tag{79.7}$$

The second (double) integral is

$$\begin{aligned} I_2 &= \int e^{i\mathbf{K}\cdot\mathbf{r}'}\,\mathrm{d}v'\int\frac{\rho(\mathbf{r}'')\,dv''}{|\mathbf{r}''-\mathbf{r}'|} = \int \mathrm{d}v''\,\rho(\mathbf{r}'')\int\frac{e^{i\mathbf{K}\cdot\mathbf{r}'}}{|\mathbf{r}''-\mathbf{r}'|}dv' \\ &= \int \mathrm{d}v''\,\rho(\mathbf{r}'')\frac{4\pi e^{i\mathbf{K}\cdot\mathbf{r}''}}{|\mathbf{K}|^2} = \frac{4\pi}{|\mathbf{K}|^2}\int \mathrm{d}v\,\rho(r)\,e^{i\mathbf{K}\cdot\mathbf{r}}. \end{aligned} \tag{79.8}$$

To effect the integration in (79.8) we use spherical polar co-ordinates with the axis parallel to $\mathbf{K}$. Then $\mathrm{d}v = r^2\,\mathrm{d}r\sin\theta\,\mathrm{d}\theta\,\mathrm{d}\phi$, $\mathbf{K}\cdot\mathbf{r} = Kr\cos\theta$. This gives

$$\int \mathrm{d}v\,\rho(r)\,e^{i\mathbf{K}\cdot\mathbf{r}} = \int_0^\infty \rho(r)\,r^2\,\mathrm{d}r\int_0^{2\pi}\mathrm{d}\phi\int_0^\pi e^{iKr\cos\theta}\sin\theta\,\mathrm{d}\theta.$$

With the new variable $\cos\theta = \xi$ the integration over ξ and ϕ is easily carried out, and the result is

$$\int \mathrm{d}v\,\rho(r)\,e^{i\mathbf{K}\cdot\mathbf{r}} = 4\pi\int_0^\infty\frac{\sin Kr}{Kr}\rho(r)\,r^2\,\mathrm{d}r. \tag{79.9}$$

Substituting (79.9) in (79.8) and (79.7) in (79.3), we obtain the final expression for $A(\theta)$:

$$A(\theta) = -\frac{2\mu ee_1}{4\pi\hbar^2}\cdot\frac{4\pi}{K^2}\left\{Z - 4\pi\int_0^\infty\frac{\sin Kr}{Kr}\rho(r)\,r^2\,\mathrm{d}r\right\}. \tag{79.10}$$

Since $K^2 = 4k^2 \sin^2 \frac{1}{2}\theta = (4\mu^2 v^2/\hbar^2) \sin^2 \frac{1}{2}\theta$, where v is the velocity of the particle, we have, putting

$$F(\theta) = 4\pi \int_0^\infty \frac{\sin Kr}{Kr} \rho(r) r^2 \, dr, \tag{79.11}$$

the formula

$$A(\theta) = -\frac{ee_1}{2\mu v^2} \{Z - F(\theta)\} \operatorname{cosec}^2 \tfrac{1}{2}\theta. \tag{79.12}$$

The quantity $F(\theta)$ is called the *atomic factor*. This quantity is seen to determine the scattering of electrons at various angles. The scattering of X-rays is determined by the same quantity.

From (79.12) we find the differential cross-section for elastic scattering of electrons with energy E into the neighbourhood of the angle θ:

$$q(\theta) = \frac{e^2 e_1^2}{4\mu^2 v^4} \{Z - F(\theta)\}^2 \operatorname{cosec}^4 \tfrac{1}{2}\theta. \tag{79.13}$$

To make this formula more specific we shall use a simple assumption regarding the charge density $e\rho$ of the electron cloud, namely (in accordance with the results of quantum mechanics) that ρ decreases exponentially with increasing distance from the centre of the atom:

$$\rho = \rho_0 e^{-r/a}, \tag{79.14}$$

where a is the 'radius' of the atom. The atom as a whole is neutral, and so

$$\int \rho \, dv = Z, \tag{79.15}$$

whence $\rho_0 = Z/8\pi a^3$. Thus

$$\rho = (Z/8\pi a^3) e^{-r/a}. \tag{79.16}$$

The atomic factor is

$$F(\theta) = 4\pi \int_0^\infty \rho(r) \frac{\sin Kr}{Kr} r^2 \, dr = \frac{Z}{2a^3 K^3} \int_0^\infty e^{-\xi/Ka} \sin \xi \cdot \xi \, d\xi,$$

where $\xi = Kr$. The last integral is easily found:

$$\int_0^\infty e^{-\xi/Ka} \sin \xi \cdot \xi \, d\xi = \frac{1}{2i} \int_0^\infty e^{-\xi/Ka} (e^{i\xi} - e^{-i\xi}) \xi \, d\xi = \frac{2a^3 K^3}{(1 + K^2 a^2)^2}.$$

Hence

$$F(\theta) = \frac{Z}{(1 + K^2 a^2)^2} = \frac{Z}{(1 + 4k^2 a^2 \sin^2 \frac{1}{2}\theta)^2}, \tag{79.17}$$

and therefore

$$q(\theta) = \frac{e_1^2 e^2 Z^2}{4\mu^2 v^4}\left[1 - \frac{1}{(1 + 4k^2 a^2 \sin^2 \frac{1}{2}\theta)^2}\right]^2 \operatorname{cosec}^4 \tfrac{1}{2}\theta. \tag{79.18}$$

For fast particles $ka \gg 1$, and so in (79.18) for scattering angles which are not too small we can neglect the second term in the square brackets in comparison with unity. This gives

$$q(\theta) = \frac{e_1^2 e^2 Z^2}{4\mu^2 v^4} \operatorname{cosec}^4 \tfrac{1}{2}\theta. \tag{79.19}$$

This is the same as the formula for elastic scattering of particles of charge e and mass μ in the Coulomb field of a nucleus of charge Ze. It was first derived by Rutherford on the basis of classical mechanics.

The result is quite different for small scattering angles. Whereas (79.19) gives $q(\theta)$ infinite for $\theta = 0$, from (79.18) we find that $q(0)$ is finite.

The fact that for large scattering angles the result is the same as for the Coulomb field of a bare nucleus can be visualised as follows. Large deflections occur for particles which pass close to the nucleus, in which case the field of the electrons has no effect. Small deflections, on the other hand, occur for particles with large impact parameters. In this case the charge on the nucleus is almost entirely screened by the negative charge of the electron cloud. The field then differs greatly from the Coulomb field.

A. SCATTERING OF α PARTICLES

For α particles the charge $e_1 = +2e$, the mass $\mu = 4\mu_H = 6.64 \times 10^{-24}$ g, where μ_H is the mass of the hydrogen atom. If the atomic weight of the atom A is much greater than 4, we can apply the above formulae immediately to calculate the scattering of α particles by atoms. The α particles emitted by radioactive elements have a velocity $v \approx 10^9$ cm/sec. From (78.2) we therefore obtain a wave number $k \approx 10^{12}$ to 10^{13} cm^{-1}. The dimension of the atom $a \approx 10^{-8}$ cm. Thus $ka \approx 10^4$, so that For-

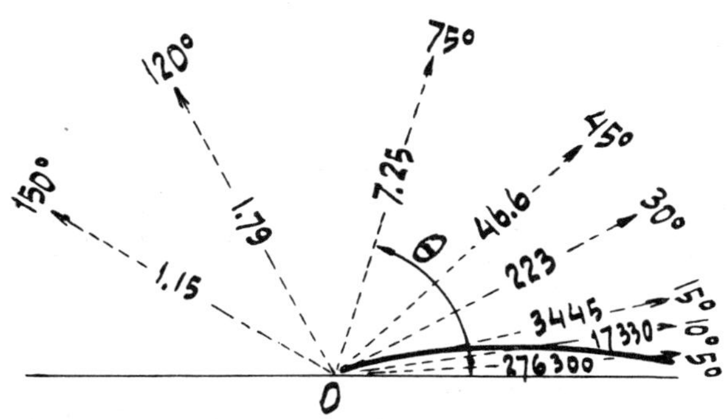

Fig. 62. Scattering of α particles on passing through a gold foil of thickness 0.001 mm. The continuous curve gives a polar plot of $q(\theta)$. The numbers on the radial lines show the results of observations of the numbers of scattered particles.

mula (79.19) can be used instead of (79.18) down to very small angles ($\sin \frac{1}{2}\theta \approx 10^{-4}$ to 10^{-5}). Hence we have for α particles

$$q_{\alpha}(\theta) = (e^4 Z^2/\mu^2 v^4) \operatorname{cosec}^4 \tfrac{1}{2}\theta \tag{79.20}$$

for $\sin \frac{1}{2}\theta \gg 1/ka$. Figure 62 is for scattering by gold, and shows the number of α particles scattered at various angles θ.

As already mentioned, Formula (79.20) was first derived by Rutherford on the basis of classical mechanics by considering hyperbolic orbits of α particles in the Coulomb field of the atomic nucleus. This formula served as the key to the discovery of the existence of the nucleus[9] and is called *Rutherford's formula*. Since the screening of the charge on the nucleus by the electron cloud is unimportant down to very small angles θ ($F(\theta) \approx 0$), Formula (79.20) is the quantum formula for scattering of α particles in a purely Coulomb field of a point charge Ze. Thus scattering in a Coulomb field is the same in quantum mechanics as in classical mechanics.[10]

B. SCATTERING OF ELECTRONS

For electrons $\mu \cong 10^{-27}$ g, so that the Born approximation is valid only for electrons whose energy is a few hundred eV or higher. For 500 eV the electron velocity

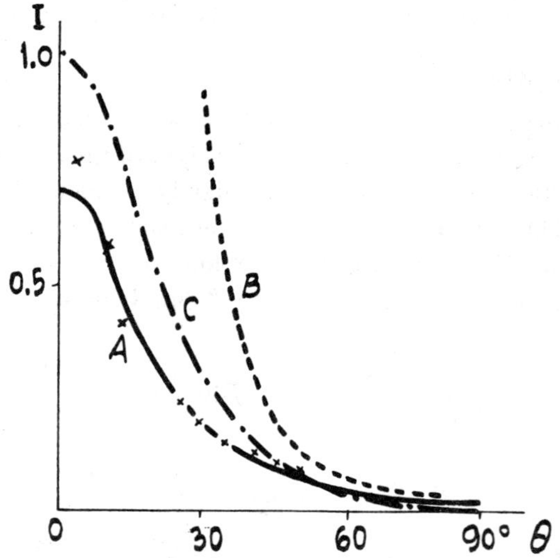

Fig. 63. Elastic scattering in helium. *A* theoretical curve with allowance for screening, *B* Rutherford scattering, *C* scattering of X-rays of the same wavelength, crosses: results of Dymond's measurements.

[9] See the classic work described in [96], p. 205.

[10] Formula (79.20) has been derived here in the Born approximation and for $\sin \frac{1}{2}\theta \gg 1/ka$. A rigorous formula for scattering in a Coulomb field can be derived (see [67], Chapter III). It is found to agree with Rutherford's formula for all angles. For like particles, e.g. the scattering of α particles in helium, quantum mechanics gives a different result, which is confirmed by experiment (see [67], Chapter III, Sections 4 and 4.1).

$v = 1.3 \times 10^9$ cm/sec and $k = 1.3 \times 10^9$ cm^{-1}, i.e. $ka \approx 1$. The atomic factor in (79.18) cannot, therefore, be neglected. In this case the cross-section is

$$q(\theta) = \frac{e^4}{4\mu^2 v^4} \{Z - F(\theta)\}^2 \operatorname{cosec}^4 \tfrac{1}{2}\theta . \tag{79.21}$$

Figure 63 shows the curve for scattering of electrons in helium, calculated theoretically, and the results of measurements by Dymond.

It is noteworthy that the distribution of electric charge in the atom can be determined from observations of electron scattering. For, by observing the scattering of electrons for various velocities v and angles θ, we find the differential cross-section $q(\theta)$ and then from (79.21) the atomic factor $F(\theta)$, which is a function of the number $K = (2\mu v/\hbar) \sin \tfrac{1}{2}\theta$ (see (79.11)). Accordingly we shall regard F as a function of K. From (79.11)

$$\frac{KF(K)}{4\pi} = \int_0^\infty \sin Kr\, \rho(r)\, r\, \mathrm{d}r . \tag{79.22}$$

Fourier's theorem gives

$$4\pi r^2 \rho(r) = \frac{2r}{\pi} \int_0^\infty KF(K) \sin Kr\, \mathrm{d}K \tag{79.23}$$

(using the fact that $KF(K)$ is an odd function of K).

Having determined the atomic factor $F(K)$ from experiment, we can find $\rho(r)$ from (79.23). This quantity is the mean density of electric charge in the atom due to the electron cloud, which can therefore be found experimentally. It may also be calculated theoretically, since the probability of a given position of an electron in the atom is defined by the wave function as $|\psi|^2$. As already noted, the atomic factor $F(K)$ can also be found from experiments on X-ray scattering, and this again allows a determination of ρ.

It is of great interest to compare the predictions of quantum mechanics with the

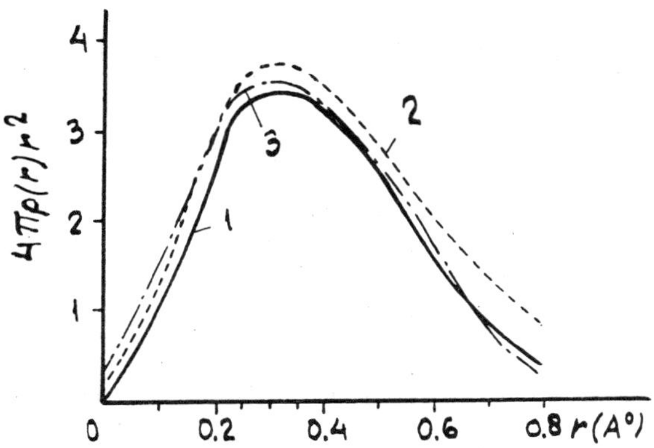

Fig. 64. Density of electric charge in helium ($4\pi\rho r^2$) as a function of the distance r: 1 from electron scattering, 2 from X-ray scattering, 3 theoretical curve.

experimental results for a quantity as sensitive as the mean charge distribution in the atom. Experiment gives an excellent confirmation of theory.[11] As an illustration, Figure 64 shows the quantity $4\pi\rho^2$ according to measurements of the scattering of X-rays and electrons in helium, together with the theoretical curve for this quantity, derived from the wave function ψ for helium (see Section 122). The coincidence of the maxima and of the exponential decrease of ρ as $r \to \infty$ is remarkably good.

Knowing the density of electrons within the atom, we can use (79.2) to determine the interaction energy $U(\mathbf{r})$ between the atom and the scattered electron. Thus from experiments on the elastic scattering of particles we can determine the nature of the forces acting on these particles.

This conclusion follows still more directly from Formula (78.24). The amplitude $A(\theta)$ of the scattered waves depends on θ only through the vector $\mathbf{K}$ (78.23), and can therefore be regarded as a function of $\mathbf{K}$, i.e. $A = A(\mathbf{K})$. Inversion of the Fourier integral (78.24) gives

$$U(\mathbf{r}) = -\frac{2\pi\hbar^2}{\mu}\cdot\frac{1}{(2\pi)^3}\int\!\!\int\!\!\int_{-\infty}^{\infty} e^{-i\mathbf{K}\cdot\mathbf{r}}\, A(\mathbf{K})\, \mathrm{d}K_x\, \mathrm{d}K_y\, \mathrm{d}K_z\,. \tag{79.24}$$

Thus if $A(\mathbf{K})$ is known from experiment we can find $U(\mathbf{r})$, i.e. the interaction energy.

The following point should also be borne in mind here. In the experiment we do not determine $A(\mathbf{K})$ directly, but the cross-section $q(\theta) = |A(\mathbf{K})|^2$. Hence, knowing $q(\theta)$, we can find $A(\mathbf{K})$ only if the amplitude $A(\mathbf{K})$ is real. Otherwise the phase of $A(\mathbf{K})$ remains unknown. It is seen from (78.24) that $A(\mathbf{K})$ is real if $U(\mathbf{r}) = U(-\mathbf{r})$ and in particular for central forces. Moreover, the inversion of the integral (78.24) requires integration over K_x, K_y, K_z from $-\infty$ to $+\infty$, and so, in order to find $U(\mathbf{r})$, we have to know the scattering for infinitely large momenta of the scattered particles (since $0 \leqslant K \leqslant 2p/\hbar = 4\pi/\lambda$). Considering only momentum p (energy $E = p^2/2\mu$) we can calculate only a part of the integral (79.24):

$$\tilde{U}(\mathbf{r}) = -\frac{2\pi\hbar^2}{\mu}\cdot\frac{1}{(2\pi)^3}\int\!\!\int\!\!\int_{-2p/\hbar}^{2p/\hbar} e^{-i\mathbf{K}\cdot\mathbf{r}}\, A(\mathbf{K})\, \mathrm{d}K_x\, \mathrm{d}K_y\, \mathrm{d}K_z\,. \tag{79.24'}$$

If the remaining part of the integral is small, the true potential energy $U(\mathbf{r})$ is replaced by a modified function $\tilde{U}(\mathbf{r})$, so that from an experiment on the scattering of particles with momentum p (wavelength $\lambda = 2\pi\hbar/p$) we cannot deduce anything about variations of $U(\mathbf{r})$ on a scale of the order of λ, since the integral (79.24′) does not include harmonics $e^{-i\mathbf{K}\cdot\mathbf{r}}$ with $K > 4\pi/\lambda = 2p/\hbar$. This is one way of expressing the well-known fact that one cannot obtain an image of details of an object that are smaller than the wavelength of the illumination used.[12]

80. The exact theory of scattering. The phase shift of the scattered waves and the cross-section

Let us now consider the exact solution of Equation (78.1′):

$$\nabla^2\psi + k^2\psi = V(r)\psi\,. \tag{80.1}$$

This equation differs from Equation (49.2), which has been examined in detail in discussing the general theory of motion in the field of a central force, only by the factor $-2\mu/\hbar^2$ and by the arrangement of the terms. The eigensolution of Equation (80.1) belonging to the energy $E = \hbar^2k^2/2\mu$, squared angular momentum

[11] See the review [65].

[12] All these remarks also apply, of course, to the determination of $\rho(r)$ by means of the integral (79.23).

$M = \hbar^2 l(l+1)$ and angular momentum component $M_z = \hbar m$ will therefore be, by (49.4),

$$\psi_{lm}(r, \theta, \phi) = R_l(r) Y_{lm}(\theta, \phi), \tag{80.2}$$

and, if we put $R_l = u_l/r$, (80.1) gives the following equation for u_l:

$$\frac{d^2 u_l}{dr^2} + \left[k^2 - \frac{l(l+1)}{r^2}\right] u_l = V(r) u_l, \tag{80.3}$$

which is essentially the same as (49.10). The general solution of Equation (80.1) belonging to the energy $E = \hbar^2 k^2/2\mu$ can be written as an expansion in terms of the orthogonal functions $\psi_{lm}(r, \theta, \phi)$:

$$\psi(r, \theta, \phi) = \sum_{l=0}^{\infty} \sum_{m=-l}^{l} C_{lm} R_l(r) Y_{lm}(\theta, \phi). \tag{80.4}$$

By representing the solution in the form (80.4) we seek it as a superposition of states with various values of the angular momentum (the number l) and of its component along the axis OZ (the number m).

For the scattering problem, as shown in Section 78, we have to find a particular solution having the asymptotic form

$$\psi_{r\to\infty} = e^{ikz} + A(\theta) e^{ikr}/r, \tag{80.5}$$

i.e. representing a combination of the primary plane wave and the scattered wave. This solution has rotational symmetry about the axis OZ and is therefore independent of the angle ϕ. A particular solution independent of ϕ is obtained from (80.4) by omitting all terms with $m \neq 0$. Since $Y_{l0}(\theta, \phi)$ differs only by a factor from the Legendre polynomial $P_l(\cos\theta)$[13], we can write the desired solution as

$$\psi(r, \theta) = \sum_{l=0}^{\infty} C_l R_l(r) P_l(\cos\theta). \tag{80.6}$$

The problem now is to determine the amplitudes C_l. Let us consider the asymptotic form of the function (80.6). According to (49.15′), as $r \to \infty$, $R_l(r)$ has the asymptotic form $A \sin(kr + \alpha_l)/r$. For convenience in the subsequent calculations we put $\alpha_l = -\frac{1}{2}\pi l + \eta_l$ and normalise the function $R_l(r)$ in such a way that $A = 1/k$. Then

$$R_l(r)_{r\to\infty} = \frac{\sin(kr - \frac{1}{2}\pi l + \eta_l)}{kr}. \tag{80.7}$$

With this choice of normalisation, the asymptotic expression for the function $\psi(r, \theta)$ becomes

$$\psi(r, \theta)_{r\to\infty} = \sum_{l=0}^{\infty} C_l P_l(\cos\theta) \left\{ \frac{e^{ikr - \frac{1}{2}i\pi l + i\eta_l}}{2ikr} - \frac{e^{-ikr + \frac{1}{2}i\pi l - i\eta_l}}{2ikr} \right\}. \tag{80.8}$$

We must now choose the C_l so that (80.8) is the same as (80.5). To do this, we expand

[13] See (25.16).

the plane wave $e^{ikz} = e^{ikr\cos\theta}$ in Legendre polynomials. The expansion is[14]

$$e^{ikz} = \sum_{l=0}^{\infty} (2l+1)\, e^{\frac{1}{2}i\pi l} \sqrt{\frac{\pi}{2kr}}\, J_{l+\frac{1}{2}}(kr)\, P_l(\cos\theta), \tag{80.9}$$

where $J_{l+\frac{1}{2}}(kr)$ is the Bessel function of order $l + \frac{1}{2}$. Physically, this expansion represents the plane wave as a superposition of stationary spherical waves, i.e. it is an expansion in terms of states with various angular momenta about the origin (the point $r = 0$). Each term of the sum (80.9) is itself a solution of Equation (80.1) with $V(r) = 0$, i.e. for free motion, belonging to a given angular momentum (the number l). For large r we have

$$J_{l+\frac{1}{2}}(kr)_{r\to\infty} = \sqrt{\frac{2}{\pi kr}} \sin(kr - \tfrac{1}{2}\pi l). \tag{80.10}$$

Putting also

$$A(\theta) = \sum_{l=0}^{\infty} \frac{A_l}{2ik} P_l(\cos\theta), \tag{80.11}$$

we can express the asymptotic form (80.5) of $\psi(r, \theta)$ as

$$\psi(r,\theta)_{r\to\infty} = \sum_{l=0}^{\infty} P_l(\cos\theta) \left\{ (2l+1)\, e^{\frac{1}{2}i\pi l} \left[\frac{e^{ikr - \frac{1}{2}i\pi l}}{2ikr} - \frac{e^{-ikr + \frac{1}{2}i\pi l}}{2ikr} \right] + \frac{A_l e^{ikr}}{2ikr} \right\}. \tag{80.12}$$

A term-by-term comparison of (80.12) and (80.8) gives

$$C_l e^{-i\eta_l} = (2l+1)\, e^{\frac{1}{2}i\pi l}, \tag{80.13}$$

$$C_l e^{-\frac{1}{2}i\pi l + i\eta_l} = (2l+1) + A_l, \tag{80.13'}$$

whence

$$A_l = (2l+1)(e^{2i\eta_l} - 1). \tag{80.14}$$

Thus the amplitude of the scattered wave is

$$A(\theta) = \frac{1}{2ik} \sum_{l=0}^{\infty} (2l+1)(e^{2i\eta_l} - 1) P_l(\cos\theta). \tag{80.15}$$

The required cross-section is, according to (78.10), simply $|A(\theta)|^2$:

$$q(\theta) = \frac{1}{4k^2} \left|\sum_{l=0}^{\infty} (2l+1)(e^{2i\eta_l} - 1) P_l(\cos\theta)\right|^2. \tag{80.16}$$

---

[14] See, for example, [78], Vol. III; [106], p. 128.

The total cross-section for elastic scattering is[15]

$$Q = \int q(\theta)\,d\Omega = \frac{4\pi}{k^2}\sum_{l=0}^{\infty}(2l+1)\sin^2\eta_l. \tag{80.17}$$

Hence we see that both the differential and the total cross-section are entirely determined by the phase shifts η_l of the scattered waves. The part of the total cross-section given by

$$Q_l = (4\pi/k^2)(2l+1)\sin^2\eta_l \tag{80.18}$$

is the cross-section for particles having a squared angular momentum $M^2 = \hbar^2 l(l+1)$ relative to the centre of force. Q_l is often called a *partial cross-section*. The terminology used for discrete states may be extended to scattering. Thus we speak of *s*-wave scattering ($l = 0$), *p*-wave scattering ($l = 1$), and so on. The *s*-wave scattering is spherically symmetrical, and the *p*-wave scattering has dipole symmetry. By analogy with classical mechanics we can say that scattering of order l corresponds to particles passing at a distance ρ_l (the impact parameter) from the centre of force, where

$$\rho_l = \hbar\sqrt{[l(l+1)]}/p = \lambda\sqrt{[l(l+1)]}/2\pi, \tag{80.19}$$

p being the momentum of the particle and λ its wavelength.[16]

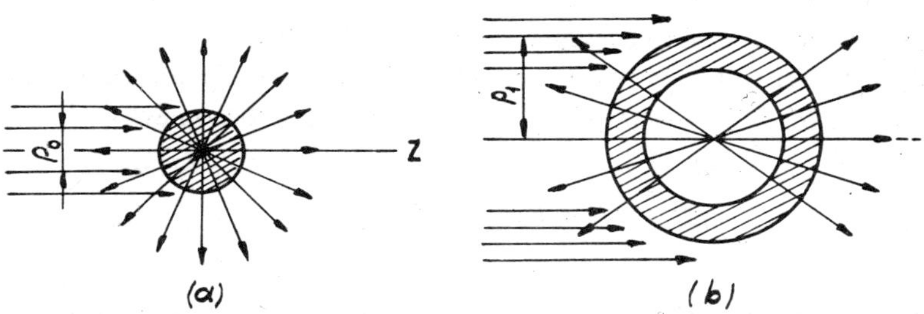

Fig. 65. (a) *s*-wave scattering, $\rho_0 = 0$; (b) *p*-wave scattering, $\rho_1 = (\lambda/2\pi)\sqrt{2}$. The hatched regions are those where $R_l^2(r)$ is appreciably different from zero.

In quantum mechanics a state with a definite angular momentum does not correspond to any definite impact parameter ρ, but the radial wave functions $R_l(r)$ have a maximum near $r = \rho_l$. In Figure 65 the hatched regions are those where $R_l^2(r)$ is appreciably different from zero.

It follows from (80.16) and (80.17) that, to determine the scattering, it is sufficient to know the phase shifts η of the scattered waves, and to find these we require a solution of Equation (80.3) with the asymptotic behaviour (80.7). This is a difficult problem, and in general numerical integration is necessary.[17]

15 Since $\int_{4\pi} P_l^2(\cos\theta)\,d\Omega = 4\pi/(2l+1)$, $\int_{4\pi} P_l(\cos\theta)\,P_{l'}(\cos\theta)\,d\Omega = 0$ for $l \neq l'$.

16 According to classical mechanics $M = p\rho$, $\rho = M/p$.

17 The series (80.15) can be summed in closed form only for a Coulomb field, and the result is Rutherford's formula.

If the number of important phase shifts is small, it is reasonable to represent the experimentally determined cross-section $q(\theta)$ in terms of these phase shifts. This method of treating the experimental data is called *phase-shift analysis*.

It is seen from Formula (80.18) that the maximum value of a partial cross-section is $(4\pi/k^2)(2l + 1) = (\lambda^2/\pi)(2l + 1)$. If the phase shift η_l is small, then $Q_l = (4\pi/k^2)(2l + 1)\,\eta_l^2$. When all phase shifts $\eta_l \ll \frac{1}{2}\pi$, Born's method is appropriate and $A(\theta)$ may be calculated directly (or determined from experiment).

Let us now consider the partial waves belonging to orbital angular momentum l at large distances from the scattering centre. From (80.8) it is seen that such a partial wave may be represented as a superposition of the primary ingoing wave $(1/r)e^{-i(kr - \frac{1}{2}\pi l)}$ and the scattered outgoing wave $(1/r)e^{i(kr - \frac{1}{2}\pi l)}$:

$$\psi_l(r, \theta)_{r\to\infty} = -\frac{(2l + 1)\,P_l(\cos\theta)}{2ik}\, e^{\frac{1}{2}i\pi l}\left\{\frac{1}{r}e^{-i(kr - \frac{1}{2}\pi l)} - \frac{S_l}{r}e^{i(kr - \frac{1}{2}\pi l)}\right\}, \tag{80.20}$$

where

$$S_l = e^{2i\eta_l}. \tag{80.21}$$

The quantity S_l evidently determines the ratio of the amplitude of the outgoing waves to that of the ingoing primary waves, and is called the *scattering matrix*.

In the present case it is in diagonal form:

$$\underset{lm,\,l'm'}{S(E)} = e^{2i\eta_l(E)}\,\delta_{ll'}\,\delta_{mm'}, \tag{80.22}$$

where l, m are the orbital and magnetic quantum numbers.

If $S(E)$ is taken for negative values of E, i.e. for purely imaginary values of k $(k = -i\sqrt{(2\mu|E|/\hbar^2)} = -i\kappa,\ \kappa > 0)$, then $\psi_l(r, \theta)_{r\to\infty}$ becomes

$$\psi_l(r, \theta)_{r\to\infty} = -\frac{(2l + 1)\,P_l(\cos\theta)}{2\kappa}\, e^{\frac{1}{2}i\pi l}\left\{\frac{1}{r}e^{-\kappa r + \frac{1}{2}i\pi l} - \frac{1}{r}e^{\kappa r - \frac{1}{2}i\pi l + 2i\eta_l}\right\}. \tag{80.23}$$

We know that for bound states there is only an exponentially decreasing function $e^{-\kappa r}/r$. For bound states, therefore,

$$e^{2i\eta_l(E)} = 0 \quad \text{or} \quad \eta_l(E) = i\infty. \tag{80.24}$$

Thus bound states with discrete energy values $E = E_1, E_2, \ldots, E_n, \ldots$ lead to the condition

$$S(E) = 0 \tag{80.25}$$

for $E < 0$.[18]

The concept of the scattering matrix is a very wide one, and in general it may be defined as a matrix which transforms waves coming from infinity into waves going to infinity. The significance of this matrix has been especially emphasised by Heisenberg,

[18] In particular, this condition can be used to derive the Balmer term for hydrogen on the basis of the corresponding scattering matrix [9]. Later, Ma showed for one example that the condition $S(E) = 0$ may give not only the correct levels but also some extra roots $E_1', E_2', \ldots$. The problem of the additional conditions which should eliminate these extra roots is not yet resolved.

who has proposed to use it as the basis of quantum mechanics in place of the wave function.[19]

81. The general case of scattering

Using the concept of the scattering matrix, we can generalise the results of Section 80 to the case of inelastic scattering of particles. We shall regard inelastic scattering phenomenologically as the absorption of a beam of primary particles at a scattering centre, since every inelastic interaction of a particle with a target removes the particle from those which are elastically scattered. Thus in this case the amplitude S_l of the elastically scattered wave is less than the amplitude of the incident wave, i.e. $|S_l| < 1$, and so the phase shifts of S_l are complex:

$$\delta_l = \eta_l + i\gamma_l, \tag{81.1}$$

where $\gamma_l(E)$ describes the 'absorption' of particles.

It is easy to see that the partial elastic scattering cross-section is now of the form

$$Q_{l,e} = (\pi/k^2)(2l+1)|1 - S_l|^2, \tag{81.2}$$

and is the same as (80.18) when $\gamma_l = 0$. To calculate the partial cross-section $Q_{l,r}$ for inelastic processes we note that the total number of particles absorbed (or undergoing reaction) per unit time is evidently equal to the incoming flux. This flux is

$$I_r = \frac{i\hbar}{2\mu}\int\left(\psi_l \frac{\partial \psi_l^*}{\partial r} - \frac{\partial \psi_l}{\partial r}\psi_l^*\right) \mathrm{d}s, \tag{81.3}$$

where the integral is taken over a surface surrounding the scattering centre ($\mathrm{d}s = r^2\, \mathrm{d}\Omega$), and ψ_l is the difference between the outgoing and ingoing partial waves (80.20). Substituting in (81.3) and integrating, we find

$$I_r = \frac{\pi h}{\mu k}(2l+1)(1 - |S_l|^2). \tag{81.4}$$

The incident flux $I_0 = \hbar k/\mu$, and the partial inelastic cross-section $Q_{l,r}$ is I_r/I_0:

$$Q_{l,r} = (\pi/k^2)(2l+1)(1 - |S_l|^2). \tag{81.5}$$

The corresponding total cross-sections are obtained by summation over l:

$$Q_e = (\pi/k^2)\sum(2l+1)|1 - S_l|^2, \tag{81.6}$$

$$Q_r = (\pi/k^2)\sum(2l+1)(1 - |S_l|^2) \tag{81.7}$$

[19] In principle $\boldsymbol{S} = \boldsymbol{S}(\boldsymbol{H})$ or $\boldsymbol{\eta} = \boldsymbol{\eta}(\boldsymbol{H})$ may be found, but these operator functions are extremely complex. Nevertheless their existence indicates that the Hamiltonian $\boldsymbol{H}$ could be replaced by the operator $\boldsymbol{S}$ or $\boldsymbol{\eta}$. Some years ago Heisenberg put forward the interesting hypothesis that in relativistic quantum mechanics the wave function may have no physical meaning when the distance between particles is small. Only the wave functions at infinity remain physically meaningful [46]. Since the operator $\boldsymbol{S}$(or $\boldsymbol{\eta}$) determines just the behaviour of the wave functions at infinity, Heisenberg considered that the phase-shift operator is more fundamental than the Hamiltonian. Heisenberg's idea has not been developed in detail. It seems that, unless the theory of relativity itself is modified for space-time on a small scale, there is no need to replace the theory based on the Hamiltonian approach by any ther [11].

and the total cross-section for all processes (elastic and inelastic) is

$$Q_t = Q_e + Q_r = (2\pi/k^2)\sum(2l+1)(1 - Re\, S_l), \tag{81.8}$$

where $Re\, S_l$ denotes the real part of S_l. Thus the inelastic scattering can be described by the use of complex phase shifts.

Formally, this can be regarded as equivalent to defining a complex potential $W(r) = U(r) + iV(r)$, so that the refractive index of the medium $n(r) = \sqrt{(1 + W/E)}$ is also complex.

This treatment of a complex atomic system such as the nucleus is called the *optical model.*

Finally, we shall derive an important *optical theorem* which gives the relation between the imaginary part of the scattering amplitude for a scattering angle $\theta = 0$ and the total cross-section. From (80.15) and (80.21) it is seen that

$$Im\, A(0) = \frac{1}{2k}\sum_0^{\infty}(2l+1)(1 - Re\, S_l), \tag{81.9}$$

where *Im A* denotes, as usual, the imaginary part. Comparison with (81.8) gives

$$Im\, A(0) = kQ_t/4\pi\,. \tag{81.10}$$

This is the optical theorem, which can be used to determine the imaginary part of the scattering amplitude for $\theta = 0$ from the total cross-section for all processes.

We shall now consider two important cases of the formulae just derived.

A. DIFFRACTION SCATTERING

Let us assume that for impact parameters $\rho < R$ total absorption occurs. Thus R will represent the radius of a completely absorbing black sphere. We also suppose that $R \gg \lambda/2\pi$, where λ is the wavelength of the particles being scattered. Simple formulae can be obtained only when this assumption is made.

In accordance with Formula (80.19) this means that total absorption occurs for all $l < L$, where $L(L+1) = (2\pi R/\lambda)^2$, $L \sim 2\pi R/\lambda = kR$. But total absorption of a partial wave corresponds to $|S_l| = 0$ (there being no flux away from the centre of the system). We can therefore write Formulae (81.6) and (81.7) for the total elastic and inelastic cross-sections in the form

$$\begin{aligned} Q_e = Q_r &= \frac{\pi}{k^2}\sum_0^L (2l+1) \\ &= \frac{\pi}{k^2}\sum_0^L (2l+1)\,\Delta l \\ &= \frac{\pi}{k^2}\int_0^{kR} 2l\,dl = \pi R^2, \end{aligned} \tag{81.11}$$

i.e. the cross-sections are constant and equal to the cross-sectional area of the black sphere.

The angular distribution is calculated from Formula (80.15), which now reads

$$A(\theta) = \frac{1}{2ik} \sum_{0}^{L} (2l + 1) \Delta l \, P_l(\cos\theta). \tag{81.12}$$

For small angles θ and for large l (which is the important range here)[20], $P_l(\cos\theta) \cong J_0(\theta l)$, where $J_0(z)$ is the Bessel function. Hence

$$A(\theta) = \frac{1}{2ik} \int_{0}^{kR} J_0(\theta l) \, 2l \, \mathrm{d}l = (R/i\theta) \, J_1(kR\theta), \tag{81.13}$$

so that

$$q(\theta) = (R^2/\theta^2) \, J_1^2(kR\theta). \tag{81.14}$$

This gives a curve with a sharp maximum at $\theta = 0$ and slight minima and maxima away from $\theta = 0$.

Such angular distributions are observed in the scattering of neutrons by nuclei (under the condition $\lambda \ll 2\pi R$) or of pions by nucleons (cf. Figure 13 (Section 8)). In both cases there is a strong inelastic interaction.

For $\lambda \ll 2\pi r_0 A^{\frac{1}{3}}$ (where $r_0 = 1.2 \times 10^{-13}$ cm, A is the mass number, and $R = r_0 A^{\frac{1}{3}}$ is the radius of the nucleus) and for an impact parameter $\rho < R$, the neutron is 'trapped' in the nucleus, which thus acts as a black body. In the scattering of pions by nucleons, for large pion energies ($E \gg mc^2$, where m is the pion mass), there is strong inelastic interaction for impact parameters $\rho < \hbar/\mu c = 1.4 \times 10^{-13}$ cm. Almost every pion then undergoes inelastic scattering (producing new pions and losing energy).

The elastic scattering in this case is similar in pattern to diffraction by a black sphere.

B. RESONANCE SCATTERING

In the interaction of complex systems with particles, resonance phenomena may occur, that is, a very great increase in the cross-section may be observed at a certain particle energy E_{res}. This is, for example, quite typical of neutron-nucleus interactions (cf. Figure 4 (Section 3)).

Let us consider, as an important example, a resonance in the s state. Here the wave function can be written in the form

$$\psi_0(r) = e^{-ikr}/r - S_0 e^{ikr/r}, \tag{81.15}$$

where S_0 is the scattering matrix element for $l = 0$. It is clear that in the case of resonance S_0 varies greatly with k (i.e. with the energy of the particle). It is found that S_0 can be expressed in terms of quantities which vary only slightly near the resonance. To do this we express S_0 in terms of the logarithmic derivative of the wave function on the surface of the system (for instance, the nucleus), i.e. for $r = R$. We assume that for $r > R$ there is practically no interaction. The derivative may therefore be calculated by

[20] See [90], p. 367.

means of the asymptotic function $\psi_0(r)$; and it is also determined by the internal properties of the system. Hence

$$\left[\frac{\mathrm{d}\,[r\psi_0(r)]/\mathrm{d}r}{r\psi_0(r)}\right]_{r=R}\cdot R = -\,ix\,\frac{1+S_0e^{2ix}}{1-S_0e^{2ix}} = f(E), \tag{81.16}$$

where the left-hand member is the logarithmic derivative of the function $r\psi_0(r)$, $x = kR$, and $f(E)$ denotes the value of this derivative as a function of energy expressed in terms of the internal parameters of the system (e.g. a nucleus). Hence

$$S_0 = -\,e^{-2ix}\frac{(x-h)-if_0}{(x+h)+if_0}, \tag{81.17}$$

where we have put $f(E) = f_0(E) - ih(E)$. If, for some value $E = E_{\text{res}}$, $f_0(E_{\text{res}}) = 0$, there is a *resonance*. For, in this range, we can put

$$f_0(E) = (\mathrm{d}f_0/\mathrm{d}E)_{E=E_{\text{res}}}(E-E_{\text{res}}), \qquad h(E) = h(E_{\text{res}}). \tag{81.18}$$

With the notation

$$\left.\begin{aligned} \Gamma_e &= -\,\frac{2kR}{(\mathrm{d}f_0/\mathrm{d}E)_{E=E_{\text{res}}}}\,. \\ \Gamma_r &= -\,\frac{2h(E_{\text{res}})}{(\mathrm{d}f_0/\mathrm{d}E)_{E=E_{\text{res}}}} \end{aligned}\right\} \tag{81.19}$$

we find

$$S_0 = -\,e^{-2ikR}\frac{\tfrac{1}{2}(-\Gamma_e+\Gamma_r) - i(E-E_{\text{res}})}{-\tfrac{1}{2}(\Gamma_e+\Gamma_r) + i(E-E_{\text{res}})}. \tag{81.20}$$

Substituting in the formulae for the elastic cross-section

$$Q_{0,e} = \sigma_e = (\pi/k^2)\,|1-S_0|^2$$

and the inelastic cross-section

$$Q_{0,r} = \sigma_r = (\pi/k^2)\{1-|S_0|^2\},$$

we obtain

$$Q_{0,r} = \sigma_r = \frac{\pi}{k^2}\frac{\Gamma_r\Gamma_e}{(E-E_{\text{res}})^2+\tfrac{1}{4}\Gamma^2}, \tag{81.21}$$

$$Q_{0,e} = \sigma_e = \frac{\pi}{k^2}\left|\frac{\Gamma_e}{E-E_{\text{res}}+\tfrac{1}{2}i\Gamma} + 2e^{ikR}\sin kR\right|^2. \tag{81.22}$$

In these formula $\Gamma = \Gamma_e + \Gamma_r$ is the total *resonance half-width* (the cross-section being less by a factor of two for $|E - E_{\text{res}}| = \tfrac{1}{2}\Gamma$). The quantity Γ_e is called the *elastic scattering partial half-width*, and Γ_r the *reaction* (or *inelastic scattering*) *partial half-width*. The elastic scattering amplitude consists of two terms, the *resonance scattering* (the term inversely proportional to $E - E_{\text{res}} + \tfrac{1}{2}i\Gamma$) and the *potential scattering* (the term proportional to $\sin kR$). This part of the scattering is independent of the internal parameters of the nucleus, and depends only on its dimension R and on the energy of the particle.

Formulae (81.21) and (81.22) were first derived by Breit and Wigner and describe the scattering near resonance. They are similar to the well-known formulae of optics for scattering near a resonance line of the spectrum.

Figure 4 (Section 3) shows resonances in the cross-section for interaction of neutrons with the oxygen nucleus. Each of the maxima in the figure can be adequately described by Breit and Wigner's formulae if there is no other maximum in the vicinity.

It may be noted that resonance is typically a quantum phenomenon. The formulae show that for $E = E_{\text{res}}$ the total cross-section is

$$Q_{0,t} = \sigma_t = \sigma_e + \sigma_r = (4\pi/k^2)\Gamma_e/\Gamma = (\lambda^2/\pi)\Gamma_e/\Gamma$$

and can take very large values $\sim \lambda^2$ (since $\Gamma_e \sim \Gamma$), many times greater than the area of interaction of nuclear forces ($\sim \pi R^2$).

For example, the resonance in the absorption of thermal neutrons by ${}^{135}_{54}Xe$ has a cross-section 100 000 times the geometrical cross-sectional area of the nucleus. This resonance is of great practical importance in the operation of nuclear reactors.[21]

82. Scattering of a charged particle in a Coulomb field

The motion of a charged particle in a Coulomb field has been deduced in Section 50, but we were there concerned with bound states ($E < 0$) and did not consider the case $E > 0$ which occurs in the scattering of particles.

By the following the method of Section 50 we could also find the radial functions $R_l(\rho)$ ($\rho = r/a$, and l is the orbital number) for the case $E > 0$. In this case, however, it would be necessary to seek a complicated linear combination of these functions in order to obtain an asymptotic solution of the type (80.5). In the scattering problem it is therefore more convenient to select a method which is more direct and better suited to the problem.

To do this, we start from Equation (49.2) with the variables not separated, and put there $U(r) = e^2Z_1Z_2/r$, where eZ_1 and eZ_2 are the charges of the particles and r the distance between them. We use the notation $k^2 = 2\mu E/\hbar^2$, $\beta = 2\mu e^2Z_1Z_2/\hbar^2$, so that Equation (49.2) becomes

$$\nabla^2\psi + \left(k^2 - \frac{\beta}{r}\right)\psi = 0. \tag{82.1}$$

Let us now seek ψ in the form

$$\psi = e^{ikz}F(r - z). \tag{82.2}$$

Then it is easily seen that F satisfies the equation

$$\zeta\frac{d^2F}{d\zeta^2} + \frac{dF}{d\zeta}(1 - ik\zeta) - \tfrac{1}{2}\beta F = 0, \tag{82.3}$$

[21] See [13].

where $\zeta = r - z$. Writing $F(\zeta)$ as a series

$$F(\zeta) = \zeta^{\gamma}(1 + a_1\zeta + a_2\zeta^2 + \ldots), \tag{82.4}$$

we find by the usual method that $\gamma^2 = 0$, so that $F(\zeta)$ is regular at the origin.

We can also use recurrence formulae to calculate the coefficients in the series (82.4). It is found that $F(\zeta) = {}_1F_1(-i\alpha, 1, ik\zeta)$, where $\alpha = \beta/2k$ and is related to Whittaker's confluent hypergeometric function.[22]

The asymptotic expansion of this function is known[23] to be

$$\begin{aligned} {}_1F_1(-i\alpha, 1, ik\zeta) = {} & \frac{e^{\frac{1}{2}\pi\alpha}}{\Gamma(1+i\alpha)}\left(1 - \frac{\alpha^2}{ik\zeta}\right)e^{i\alpha\log k\zeta} - \\ & - \frac{ie^{\frac{1}{2}\pi\alpha}}{\Gamma(1-i\alpha)}\frac{e^{ik\zeta}}{k\zeta}e^{-i\alpha\log k\zeta} + \ldots, \end{aligned} \tag{82.5}$$

where $\Gamma(z)$ is the gamma function. Taking $\psi(r, \theta)$ in the form

$$\psi(r, \theta) = e^{-\frac{1}{2}\pi\alpha}\Gamma(1 + i\alpha)e^{ikz}\,{}_1F_1(-i\alpha, 1, ik\zeta), \tag{82.6}$$

where

$$\alpha = e^2 Z_1 Z_2/\hbar v, \qquad \zeta = r - z = r(1 - \cos\theta) \tag{82.7}$$

and $v = p/\mu$ is the particle velocity, we find from (82.6) and (82.5) for $r, \zeta \to \infty$

$$\psi(r, \theta)_{r,\zeta\to\infty} = I + A(\theta)S, \tag{82.8}$$

where

$$I = \left[1 - \frac{\alpha^2}{ik(r-z)} + \ldots\right]e^{ikz + i\alpha\log k(r-z)}, \tag{82.9}$$

$$S = \frac{1}{r}e^{ikr - i\alpha\log kr} \tag{82.9'}$$

and

$$A(\theta) = \frac{e^2 Z_1 Z_2}{2\mu v^2}\operatorname{cosec}^2\tfrac{1}{2}\theta\cdot e^{-i[\alpha\log(1-\cos\theta) - \pi - 2\eta_0]}, \tag{82.9''}$$

with $e^{2i\eta_0} = \Gamma(1 + i\alpha)/\Gamma(1 - i\alpha)$. A comparison of these formulae with the usual formulae of scattering theory shows that both the incident wave e^{ikz} and the scattered wave e^{ikr}/r are affected by logarithmic factors $e^{i\alpha\log k(r-z)}$ and $e^{-i\alpha\log kr}$. This is a peculiarity of the Coulomb field, which decreases slowly with increasing distance, and so affects the waves at arbitrarily large distances. There are therefore no solutions in a Coulomb field in the form of plane waves or ordinary spherical waves. The differential cross-section $q(\theta)$ for scattering into the angle θ is equal to $|A(\theta)|^2$:

$$q(\theta) = (e^4 Z_1^2 Z_2^2/4\mu^2 v^4)\operatorname{cosec}^4\tfrac{1}{2}\theta, \tag{82.10}$$

and is the same as that previously calculated by Born's method (79.19). The amplitudes

[22] See [90], Chapter XVI.
[23] See [67], p. 48.

$A(\theta)$ in (79.12) and (82.9″) are, however, different in phase. The difference is slight if $\alpha = e^2 Z_1 Z_2/\hbar v \ll 1$. This is the condition for Born's method to be applicable in the present case.

Thus we have shown that the classical Rutherford's formula for the scattering of particles in a Coulomb field follows rigorously from quantum mechanics without correction terms.

CHAPTER XIV

THEORY OF QUANTUM TRANSITIONS

83. Statement of the problem

One of the most important problems of quantum mechanics is to calculate the probabilities of transitions from one quantum state to another. This problem may be formulated as follows. Suppose that at time $t = 0$ we have a pure ensemble of systems which is such that some mechanical quantity L has a definite value $L = L_n$. Such an ensemble will be described by a wave function $\psi_n(x)$ which is an eigenfunction of the operator $\boldsymbol{L}$ belonging to the eigenvalue[1] $L = L_n$. Systems belonging to such an ensemble are said to be in the quantum state n.

In the course of time the state of the systems may change owing to the interaction of external fields or because of internal causes. Thus at time t the ensemble will be described by some new wave function, which we denote by $\psi_n(x, t)$. This newly derived ensemble will in general have no definite value of the quantity L.[2]

If the systems of this ensemble are now classified according to the value of L, i.e. if a spectral resolution is made with respect to L, a new (mixed) ensemble is obtained (cf. Section 17). Some of the systems will have $L = L_m$ and form a pure ensemble described by the wave function $\psi_m(x)$ $[\boldsymbol{L}\psi_m(x) = L_m\psi_m(x)]$, others will have $L = L_{m'}$ and will form a pure ensemble $\psi_{m'}(x)$, and so on. Systems belonging to an ensemble with $L = L_m (m \neq n)$ are said to have made the quantum transition from state n to state m.

This can be illustrated diagrammatically:

$$\begin{array}{lll} t=0 & t & \ldots\rightarrow \psi_m(x)\ L = L_m \\ \psi = \psi_n(x) \longrightarrow & \psi' = \psi_n(x,t) = \sum\limits_m c_{mn}(t)\,\psi_m(x)\ldots & \ldots\rightarrow \psi_{m'}(x)\ L = L_{m'} \\ & & \ldots\rightarrow \psi_{m''}(x)\ L = L_{m''} \\ L = L_n & L\ \text{indefinite} & \ldots\rightarrow \ldots\ldots\ldots\ldots\ldots \end{array}$$

Here the continuous arrow shows the change of the ensemble which occurs spontaneously, without the effect of measurement, i.e. without the spectral resolution with respect to L. This change of the ensemble can be found from Schrödinger's equation. The diagram shows that this new state of the ensemble is a superposition of states with

[1] In general the state will be described not by one but by several mechanical quantities $L, M, N, \ldots$. Accordingly the wave function will have several suffixes: $\psi_{n,r,s,\ldots}$.

[2] The case where L is an integral of the motion forms an exception. Then $\psi_n(x, t) = \psi_n(x)e^{-iE_nt/\hbar}$, and in the state $\psi_n(x, t)$ L still has the unique value L_n.

various values of L (the sum over m). The broken arrows show the changes of the ensemble which occur when the spectral resolution of the ensemble is carried out at time t. As we know (cf. Section 17), this resolution occurs, in particular, when a measurement is made. In other words, the broken arrow represents a 'reduction of the wave packet' (cf. Section 17), in which the superposition $\psi_n(x, t)$ becomes one of the partial states $\psi_m(x)$. Only after this reduction can we speak of a quantum transition from the state $L = L_n$ to, say, the state $L = L_m$.

Thus the concept of a quantum transition necessarily presupposes not only a definite initial state (n) but also a definite final state (m). We emphasise the latter point because defining the final state changes the state of the systems of the ensemble. This process can occur in any interaction which is selective with respect to the characteristic L, i.e. which effects a spectral resolution of the ensemble $\psi_n(x, t)$ in terms of $\psi_m(x)$, and in particular in a measurement of the quantity L.

Let us now consider the concept of the probability of transition from state n to state m. According to the general theory (Section 22) the quantity $P_{mn}(t) = |c_{mn}(t)|^2$ is the probability of finding $L = L_m$ in the state $\psi_n(x, t)$ (see the diagram).[3] Since for $t = 0$ $P_{mn}(0) = 0$ for $m \neq n$ and $P_{mn}(0) = 1$ for $m = n$, the probability $P_{mn}(t) (m \neq n)$ is called the probability of transition from the state $\psi_n(x)$ with $L = L_n$ to the state $\psi_m(x)$ with $L = L_m$ in time t. For, when $m \neq n$ $P_{mn}(t)$ gives the probability of finding at time t the value $L = L_m$, which at $t = 0$ did not exist in the ensemble, since $P_{mn}(0) = 0$. The most important problems in the theory of quantum transitions are the calculation of the probability of a transition from a state with energy E_n to a state with another energy E_m, or the probability of transition from one quantum level to another. Here it may be noted that if a particle (or in general a system) is interacting with an external field depending on time, the concept of potential energy, and therefore of total energy, has no meaning (this is not true of the kinetic energy). In general, therefore, the problem of the transition of a particle from one quantum level to another is meaningful only when the agency causing the transition acts over a finite interval of time, say from $t = 0$ to $t = T$. Outside this interval the total energy is an integral of the motion and can be determined by suitable measurements (see Sections 111 and 112). The solution of Schrödinger's equation, which determines $\psi(x, t)$ from $\psi(x, 0)$, is very difficult. Results of general value can be obtained only when transitions from one level to another are caused by weak interactions which may be regarded as a perturbation. In this case Schrödinger's equation may be written

$$i\hbar\, \partial\psi/\partial t = \boldsymbol{H}^0(x)\psi + \boldsymbol{W}(x, t)\psi, \tag{83.1}$$

where $\boldsymbol{H}^0(x)$ is the total-energy operator of the system in the absence of the perturbation, and $\boldsymbol{W}(x, t)$ is the perturbation. When the perturbation is small, the operator $\boldsymbol{H}^0(x)$ may be regarded as the total-energy operator, and so in this particular case the inclusion or omission of $\boldsymbol{W}(x, t)$ is of secondary importance.

To find the transition probability $P_{mn}(t)$ from the level E_n to the level E_m it is convenient to write Equation (83.1) in the E representation. We expand $\psi(x, t)$ in terms of

[3] The additional suffix n in c_{mn} indicates the initial state. This was not shown in Section 22.

the eigenfunctions $\psi_k(x)$ of the total-energy operator $\boldsymbol{H}^0$:

$$\psi(x,t) = \sum_k c_k(t)\,\psi_k(x)\,e^{-iE_kt/\hbar}\,; \tag{83.2}$$

substituting ψ in this form in (83.1), multiplying by $\psi_m^*(x)e^{iE_mt/\hbar}$ and integrating with respect to x, we obtain in the usual way Equation (83.1) in the E representation:

$$i\hbar\,\mathrm{d}c_m/\mathrm{d}t = \sum_k W_{mk}(t)\,e^{i\omega_{mk}t}\,c_k(t) \tag{83.3}$$

(using the fact that $\boldsymbol{H}^0\psi_k = E_k\psi_k$). Hence $W_{mk}(t)$ is the matrix element of the perturbation energy:

$$W_{mk}(t) = \int \psi_m^*(x)\cdot \boldsymbol{W}(x,t)\,\psi_k(x)\cdot \mathrm{d}x\,, \tag{83.4}$$

and ω_{mk} is the Bohr frequency $(E_m - E_k)/\hbar$ for the transition $E_m \to E_k$. At the initial instant the system is assumed to be in the state $E = E_n$. At $t = 0$, therefore,

$$c_k(0) = 1 \text{ if } k = n\,, \qquad c_k(0) = 0 \text{ if } k \neq n\,. \tag{83.5}$$

The probability of finding the system in the state $E = E_m$ at time t is $|c_m(t)|^2$.[4] Hence the transition probability from E_n to E_m at time t is

$$P_{mn} = |c_m(t)|^2\,. \tag{83.6}$$

Thus we have to determine the quantities $c_k(t)$ from Equations (83.3) with the initial conditions (83.5).

We shall regard $\boldsymbol{W}(x, t)$ as a small perturbation. To solve Equation (83.3) we note that if $\boldsymbol{W}$ is ignored completely the quantities $c_k(t)$ are constants. Hence as a zeroth-order approximation $c_k^{(0)}(t)$ we can take the initial value (83.5):

$$c_k^{(0)}(t) = \delta_{nk}\,. \tag{83.7}$$

Substituting these values on the right-hand side of (83.3), we find the equation for the first approximation $c_m^{(1)}(t)$:

$$i\hbar\,\mathrm{d}c_m^{(1)}(t)/\mathrm{d}t = \sum_k W_{mk}(t)\,e^{i\omega_{mk}t}\,c_k^{(0)} = W_{mn}(t)\,e^{i\omega_{mn}t}\,. \tag{83.8}$$

Hence

$$c_m^{(1)}(t) = \frac{1}{i\hbar}\int_0^t W_{mn}(\tau)\,e^{i\omega_{mn}\tau}\,\mathrm{d}\tau + \delta_{mn}\,. \tag{83.9}$$

Substituting this first approximation for $c_m^{(1)}(t)$ on the right-hand side of (83.3), we find the equation for the second approximation:

$$i\hbar\,\mathrm{d}c_m^{(2)}(t)/\mathrm{d}t = \sum_k W_{mk}(t)\,e^{i\omega_{mk}t}\,c_k^{(1)}(t)\,. \tag{83.10}$$

Since the $c_k^{(1)}(t)$ are known functions of time (83.9), integration of (83.10) with respect to time gives $c_m^{(2)}(t)$, i.e. the second approximation. This procedure can be continued

[4] See Section 22.

further, and leads to the exact solution for $c_m(t)$. In general, however, a large number of approximations are necessary, or else only short intervals of time can be considered. If $W(x, t)$ is small, however, the first or second approximation is sufficient.

Below we shall discuss some special perturbations and systems.

84. Transition probabilities under a time-dependent perturbation

Let us now determine the transition probability from a quantum level E_n to E_m for a system subject to a time-dependent perturbation $W(x, t)$. We shall assume that the perturbation is zero for $t < 0$ and for $t > T$. Assuming that the $W_{mn}(t)$ are so small that the first approximation is adequate even for $t = T$, we obtain from (83.9) the amplitude $c_m^{(1)}(t)$ for $t \geqslant T$ in the form

$$c_m^{(1)} = \frac{1}{i\hbar}\int_0^T W_{mn}(\tau)\, e^{i\omega_{mn}\tau}\,\mathrm{d}\tau = \frac{1}{i\hbar}\int_{-\infty}^{\infty} W_{mn}(\tau)\, e^{i\omega_{mn}\tau}\,\mathrm{d}\tau\,, \quad m \neq n\,. \tag{84.1}$$

For $t > T$, $c_m^{(1)}$ is independent of time, since the energy is an integral of the motion.

The expression for $c_m^{(1)}(t)$ has a simple significance. The perturbation $W(x, t)$ can be expanded as a Fourier integral:

$$W(x, t) = \int_{-\infty}^{\infty} W(x, \omega)\, e^{-i\omega t}\,\mathrm{d}\omega\,. \tag{84.2}$$

Hence, by Fourier's theorem,

$$W(x, \omega) = \frac{1}{2\pi}\int_{-\infty}^{\infty} W(x, t)\, e^{i\omega t}\,\mathrm{d}t\,. \tag{84.3}$$

The perturbation matrix element (83.4) can be written, from (84.2), as

$$\begin{aligned} W_{mn}(t) &= \int \psi_m^*(x)\cdot W(x, t)\,\psi_n(x)\cdot\mathrm{d}x \\ &= \int_{-\infty}^{\infty} e^{-i\omega t}\,\mathrm{d}\omega \int_{-\infty}^{\infty} \psi_m^*(x)\cdot W(x, \omega)\,\psi_n(x)\cdot\mathrm{d}x \\ &= \int_{-\infty}^{\infty} e^{-i\omega t}\, W_{mn}(\omega)\,\mathrm{d}\omega\,, \end{aligned} \tag{84.4}$$

where $W_{mn}(\omega)$ is the matrix element of the Fourier component of frequency ω. Applying Fourier's theorem to (84.4), we find

$$W_{mn}(\omega) = \frac{1}{2\pi}\int_{-\infty}^{\infty} W_{mn}(t)\, e^{i\omega t}\,\mathrm{d}t\,. \tag{84.5}$$

Comparison with the integral in (84.1) shows that

$$c_m^{(1)} = (2\pi/i\hbar)\, W_{mn}(\omega_{mn})\,. \tag{84.6}$$

The approximation is valid if $c_m^{(1)}$ is small; this is a necessary condition, since $c_m^{(0)}(0) = 0$.

According to (83.6) and (84.6) the transition probability from the state E_n to the state E_m is

$$P_{mn} = (4\pi^2/\hbar^2)|W_{mn}(\omega_{mn})|^2 . \tag{84.7}$$

This formula leads to an important result. It shows that $P_{mn} \neq 0$ only if $W_{mn}(\omega_{mn}) \neq 0$, i.e. the transition from E_n to E_m is possible only if the perturbation spectrum contains the frequency $\omega_{mn} = (E_m - E_n)/\hbar$. Thus the transition is of resonance type. The position is as if a quantum system were an assembly of oscillators with eigenfrequencies equal to the Bohr frequencies ω_{mn}. When a variable external interaction occurs, only those oscillators are excited whose frequencies are equal to frequencies present in the external interaction. We shall later describe important applications of Formula (84.7) to optical problems.

Formula (84.7) has been derived for transitions in a discrete spectrum. For transitions in a continuous spectrum it must be somewhat modified. We shall consider the necessary changes for transitions from a discrete to a continuous spectrum, assuming that the system has both (for instance, the spectrum of an atom is of this type).

The states of the continuous spectrum are labelled by continuous parameters, which we denote by α, β, γ. These may be, for example, the three components p_x, p_y, p_z of the particle momentum. We shall for the present write only one of these parameters (α) explicitly. The energy will be a function $E = E(\alpha)$. The corresponding wave function will be $\psi_\alpha(x)$. Then in (83.2) an integral over the states of the continuous spectrum (over α) is added to the sum over states of the discrete spectrum:

$$\psi(x, t) = \sum_k c_k(t)\,\psi_k(x)\,e^{-iE_k t/\hbar} + \int c_\alpha(t)\,\psi_\alpha(x)\,e^{-iE(\alpha)t/\hbar}\,d\alpha . \tag{84.8}$$

Assuming that the functions $\psi_\alpha(x)$ are normalised to $\delta(\alpha - \alpha')$ and repeating the calculations leading from (83.1) to (83.8), we find that

$$c_\alpha^{(1)}(t) = \frac{1}{i\hbar}\int_0^t W_{\alpha n}(\tau)\,e^{i[E(\alpha) - E_n]\tau/\hbar}\,d\tau , \tag{84.9}$$

if the system was originally in the state E_n, with

$$W_{\alpha n}(t) = \int \psi_\alpha(x)\cdot W(x, t)\,\psi_n(x)\,dx . \tag{84.10}$$

The subsequent calculations depend on assumptions concerning the way in which $W(x, t)$ depends on time. We shall suppose that $W(x, t)$ is monochromatic; in transitions in the discrete spectrum it is necessary to take account of the non-monochromaticity of actual perturbations, but for transitions to the continuous spectrum this is not necessary and an actual perturbation may be considered monochromatic. Thus we shall suppose that

$$W(x, t) = W(x)\,e^{i\omega t} + W^*(x)\,e^{-i\omega t} . \tag{84.11}$$

Then

$$W_{\alpha n}(t) = W_{\alpha n}\,e^{i\omega t} + W_{\alpha n}^*\,e^{-i\omega t} , \tag{84.12}$$

where $W_{\alpha n}$ and $W_{\alpha n}^*$ are the matrix elements of the Fourier components of $W(x, t)$. Substituting (84.12) in (84.9) and effecting the integration with respect to time, we find

$$c_\alpha^{(1)} = \frac{1}{i\hbar} W_{\alpha n} \frac{e^{i[E(\alpha) - E_n + \hbar\omega]t/\hbar} - 1}{(i/\hbar)[E(\alpha) - E_n + \hbar\omega]} + \frac{1}{i\hbar} W_{\alpha n}^* \frac{e^{i[E(\alpha) - E_n - \hbar\omega]t/\hbar} - 1}{(i/\hbar)[E(\alpha) - E_n - \hbar\omega]}.$$

Since $\omega > 0$, $E(\alpha) > 0$ and $E_n < 0$, the first term is small; the second term is large when $E(\alpha) = E_n + \hbar\omega$. We shall therefore omit the first term, and thus find for the transition probability from E_n to the range α to $\alpha + d\alpha$ in time t

$$|c_\alpha|^2 \, d\alpha = |W_{\alpha n}|^2 \left| \frac{e^{i[E(\alpha) - E_n - \hbar\omega]t/\hbar} - 1}{E(\alpha) - E_n - \hbar\omega} \right|^2 d\alpha. \tag{84.13}$$

The transition probability from E_n to $(\alpha, \alpha + d\alpha)$ per second is

$$P_{\alpha n} \, d\alpha = \frac{d\,|c_\alpha|^2}{dt} \, d\alpha = 2 \frac{|W_{\alpha n}|^2}{h} \frac{\sin \{[E(\alpha) - E_n - \hbar\omega] \, t/\hbar\}}{E(\alpha) - E_n - \hbar\omega} d\alpha. \tag{84.14}$$

The last factor in (84.14) differs from a δ function for large t only by a factor $1/\pi$. The probability $P_{\alpha n} \, d\alpha$ may therefore be written in the form

$$P_n(\alpha) \, d\alpha = (2\pi/h) \, |W_{\alpha n}|^2 \, \delta[E(\alpha) - E_n - h\omega] \, d\alpha. \tag{84.15}$$

If the state of the continuous spectrum is labelled by several parameters α, β, γ, we obtain similarly for the transition probability from the state E_n to the range $\alpha, \alpha + d\alpha$; $\beta, \beta + d\beta$; $\gamma, \gamma + d\gamma$ per second

$$\begin{aligned} &P_n(\alpha, \beta, \gamma) \, d\alpha \, d\beta \, d\gamma \\ &= (2\pi/\hbar) \, |W_{\alpha\beta\gamma,n}|^2 \, \delta[E(\alpha, \beta, \gamma) - E_n - \hbar\omega] \, d\alpha \, d\beta \, d\gamma. \end{aligned} \tag{84.16}$$

It is easy to find also the transition probability in the continuous spectrum. Taking the initial state $\psi_{\alpha_0\beta_0\gamma_0}$ (i.e. $c_{\alpha\beta\gamma}(0) = \delta(\alpha - \alpha_0) \, \delta(\beta - \beta_0) \, \delta(\gamma - \gamma_0)$) we similarly find for the transition probability per second from α_0, β_0, γ_0 to the range $\alpha, \alpha + d\alpha$; $\beta, \beta + d\beta$; $\gamma, \gamma + d\gamma$

$$\begin{aligned} &P_{\alpha_0\beta_0\gamma_0}(\alpha, \beta, \gamma) \, d\alpha \, d\beta \, d\gamma \\ &= (2\pi/\hbar) \, | \, W_{\alpha\beta\gamma, \, \alpha_0\beta_0\gamma_0} |^2 \, \delta \, [E(\alpha, \beta, \gamma) - E(\alpha_0, \beta_0, \gamma_0) - \hbar\omega] \, d\alpha \, d\beta \, d\gamma. \end{aligned} \tag{84.17}$$

These formulae again show the resonance nature of the transition, since the probabilities found are non-zero only for transitions for which

$$\hbar\omega = E(\alpha, \beta, \gamma) - E_n = \hbar\omega_{\alpha\beta\gamma, \, n} \tag{84.18}$$

or

$$\hbar\omega = E(\alpha, \beta, \gamma) - E(\alpha_0, \beta_0, \gamma_0) = \hbar\omega_{\alpha\beta\gamma, \, \alpha_0\beta_0\gamma_0}, \tag{84.18'}$$

i.e. the frequency of the external interaction is equal to the Bohr frequency for a possible transition.

At the point of resonance the probabilities calculated above become infinite, but in the neighbourhood of that point they are zero.[5] Hence the transition probability to an arbitrarily narrow energy range containing the resonance point is finite. In order to see this we need only replace the parameters α, β, γ which label the states of the continuous spectrum by some new parameters which include the energy. Let these be E, a, b; they are functions of α, β, γ. We have

$$d\alpha\, d\beta\, d\gamma = \rho(E, a, b)\, dE\, da\, db\,. \tag{84.19}$$

$\rho\ (E,\ a,\ b)$ is called the *density of states* per unit interval of energy, a and b. Substituting this value of $d\alpha\, d\beta\, d\gamma$ in the expression (84.16) or (84.17) for the probability and integrating with respect to E, we obtain zero if the range of integration does not contain the resonance point and a finite number if it does. From (84.16) and (84.17) we have

$$P_n(E, a, b)\, da\, db = (2\pi/\hbar)\, |W_{Eabn}|^2\, \rho(E, a, b)\, da\, db\,, \tag{84.20}$$

$$P_{\alpha_0\beta_0\gamma_0}(E, a, b)\, da\, db = (2\pi/\hbar)\, |W_{Eab,\alpha_0\beta_0\gamma_0}|^2\, \rho(E, a, b)\, da\, db\,, \tag{84.21}$$

where of course the value of E is that given by the resonance conditions (84.18) and (84.18′) respectively.

In the particular case where the parameters α, β, γ are taken to be the three components p_x, p_y, p_z of the particle momentum, it is convenient to consider the momentum of the final state in spherical polar co-ordinates p, θ, ϕ. Then

$$dp_x\, dp_y\, dp_z = p^2\, dp\, d\Omega\,, \qquad d\Omega = \sin\theta\, d\theta\, d\phi\,. \tag{84.22}$$

The energy of the particle is $E = p^2/2\mu$, so that $p^2\ dp = p^2\, (dp/dE)\ dE = \mu p\ dE$. Substituting in (84.22) and comparing with (84.19), we find

$$\rho(E, \theta, \phi) = \rho(E) \sin\theta\,, \qquad \rho(E) = \mu p = \tfrac{1}{2}(2\mu)^{\frac{3}{2}}\, E^{\frac{1}{2}}\,. \tag{84.23}$$

Substitution in (84.20) and (84.21) gives

$$P_n(E, \theta, \phi)\, d\Omega = (2\pi/\hbar)\, |W_{E\theta\phi n}|^2\, \rho(E)\, d\Omega\,, \tag{84.24}$$

$$P_{\alpha_0\beta_0\gamma_0}(E, \theta, \phi)\, d\Omega = (2\pi/\hbar)\, |W_{E\theta\phi,\, \alpha_0\beta_0\gamma_0}|^2\, \rho(E)\, d\Omega\,. \tag{84.25}$$

These formulae give the transition probability per second from the state n or α_0, β_0, γ_0 to the state with energy E, the particle momentum being then in the solid angle $d\Omega$.

85. Transitions due to a time-independent perturbation

If the perturbation is independent of time, we can seek a stationary solution $\psi(x) e^{-iE_n t/\hbar}$ of Schrödinger's equation, and so reduce the problem to that of solving the equation

$$H^0 \psi(x) + W(x)\, \psi(x) = E\psi(x)\,,$$

for which methods of approximate solution have already been discussed. The problem

[5] This is not quite accurate, since according to (84.14) we have not a δ function but only an approximation to it; see Section 112.

can also be stated, however, in terms of the theory of quantum transitions. The two statements lead to the same results.[6]

In order to obtain the transition probability due to the interaction of a perturbation independent of time, we need only put $\omega = 0$ in Formulae (84.16) and (84.17). Then the conditions (84.18) and (84.18′) become

$$E(\alpha, \beta, \gamma) = E_n \qquad \text{or} \qquad E(\alpha, \beta, \gamma) = E(\alpha_0, \beta_0, \gamma_0), \tag{85.1}$$

i.e. only transitions without change of energy are possible. This follows from the general theory, since the energy is in this case an integral of the motion. Hence transitions under a perturbation independent of time can only be such that energy is redistributed between parts of the system or some other mechanical quantities are changed (for example, the direction of the particle momentum).

In the continuous spectrum the formula for the transition probability per second from the state $E(\alpha_0, \beta_0, \gamma_0)$ to the state E_0; $a, a + \mathrm{d}a$; $b, b + \mathrm{d}b$ is thus obtained directly from (84.21):

$$P_{\alpha_0\beta_0\gamma_0}(E_0, a, b)\,\mathrm{d}a\,\mathrm{d}b = (2\pi/\hbar)\,|W_{E_0ab,\alpha_0\beta_0\gamma_0}|^2\,\rho(E_0, a, b)\,\mathrm{d}a\,\mathrm{d}b, \tag{85.2}$$

and, if α, β, γ are the momenta,

$$\begin{aligned} &P_{p_{x0}, p_{y0}, p_{z0}}(E_0, \theta, \phi)\,\mathrm{d}\Omega \\ &= (2\pi/\hbar)\,|W_{E_0\theta\phi, p_{x0}p_{y0}p_{z0}}|^2\,\rho(E_0)\,\mathrm{d}\Omega. \end{aligned} \tag{85.3}$$

These formulae are of the same form as (84.21) and (84.25) and differ from them only through the resonance condition (85.1), which expresses the law of conservation of energy.

It may be noted that for a time-independent perturbation there is little value in considering transitions only between discrete states, since the condition that the energies of the initial and final states are equal can then be satisfied only in exceptional cases.

[6] Cf. Section 112, where collisions are treated by the transition method, and Section 78, where the same problem is solved by the method of stationary states.

CHAPTER XV

EMISSION, ABSORPTION AND SCATTERING OF LIGHT BY ATOMIC SYSTEMS

86. Introductory remarks

Problems concerning the interaction of light and microparticles are in some respects outside the scope of quantum mechanics. They cannot be discussed without the use of supplementary ideas relating to laws of creation and annihilation of electromagnetic fields. We can, however, make considerable progress on the basis of Einstein's semi-phenomenological theory of radiation, which is essentially founded on the laws of conservation of energy and momentum in interactions between quantum systems and the electromagnetic radiation field. For the behaviour of a quantum system in a given electromagnetic field is entirely a mechanical problem. We can therefore use the theory of quantum transitions to calculate the probability that an atom enters an excited state, or returns from an excited state to a lower one, under the action of incident light. In the former case the energy of the atom increases by an amount $E_m - E_n$, where E_n is the energy of the initial state and E_m that of the excited state, and in the latter case it decreases by this amount. Let us first consider the former process.

If we suppose that the additional energy $E_m - E_n$ of the atom is taken from the electromagnetic field, then the transition probability of the atom from E_n to E_m is identified with the probability of absorption of an amount of light energy $E_m - E_n$, which is just the light-quantum absorption probability which appears in Einstein's theory. For this interpretation to be possible (not in contradiction with quantum mechanics) it is necessary that the transition of the atom from E_n to E_m should be able to occur only when the energy difference $E_m - E_n$ is equal to the energy of the light quantum, i.e. when Bohr's frequency condition holds:

$$\hbar\omega = E_m - E_n\,. \tag{86.1}$$

We know from the theory of quantum transitions that this is in fact so, since the transition $E_n \to E_m$ is possible only when the spectrum of the external interaction contains the frequency $\omega = (E_m - E_n)/\hbar = \omega_{mn}$. In the present case this means that the spectrum of the incident light must contain this frequency; in other words, it must contain light quanta of energy

$$\varepsilon = \hbar\omega = E_m - E_n\,. \tag{86.2}$$

We also know that the transition $E_n \to E_m$ is entirely brought about by the part of the perturbation which is a harmonic function of time with frequency ω_{mn}. Thus if we imagine the incident light to be resolved into an assembly of monochromatic waves, the transition $E_n \to E_m$ is entirely due to the wave whose frequency is ω_{mn} and whose quanta are accordingly $\varepsilon = \hbar\omega_{mn}$.

The transition of an atom from an excited state E_m to a lower state E_n under the effect of light must be regarded as the emission of a light quantum $\varepsilon = E_m - E_n$, if we again apply the law of conservation of energy. The probability of this transition also can be calculated, and it is the same as the probability of stimulated emission in Einstein's theory (the probability of emission due to the interaction of radiation).

We cannot, however, discuss in terms of mechanics the third process, namely spontaneous emission by an atom, which occurs in the absence of an external interaction – in the absence, that is, of incident light. If the atom is in an excited state in the absence of an external interaction, quantum mechanics asserts that it will remain for an arbitrarily long time in that state. States of definite energy are stationary states (Section 30), and the energy is an integral of the motion. Yet experiment shows that the atom will spontaneously emit light and make the transition to the normal state.

This contradiction should not cause surprise. We have from the start considered a purely mechanical problem, namely the motion of an electron in a given external field (for example, in the electrostatic field of the nucleus), and have neglected the fact that a moving electron creates an electromagnetic field which in turn acts on it. In short, we have neglected the reaction of the field of the electron on the electron itself.

A similar situation is encountered in classical mechanics. If we consider the motion of a charged particle, for example, under the action of a quasielastic force, the result is that a particle which initially has energy E can retain this value of the energy. If, however, we take into account the fact that a moving charged particle produces an electromagnetic field which acts on it, we find that the particle will in fact lose its energy and emit light. Classical theory[1] gives the following formula for the energy $\mathrm{d}E/\mathrm{d}t$ emitted per second by a particle oscillating harmonically with frequency ω_0 and having electric moment $\mathbf{D}_{cl}$:

$$\frac{\mathrm{d}E}{\mathrm{d}t} = \frac{2\omega_0^4}{3c^3}\overline{(\mathbf{D}_{cl})^2}, \tag{86.3}$$

where $\overline{(\mathbf{D}_{cl})^2}$ denotes the time average of $(\mathbf{D}_{cl})^2$. The reaction of this radiation retards the particle, which thus gradually comes to rest.

This problem of emission with allowance for reaction is essentially outside the scope of quantum mechanics, and belongs to quantum electrodynamics. In the present book we shall not discuss the problems of quantum electrodynamics, which are far from being completely solved.[2] We shall avoid the difficulty by postulating, in accordance with Einstein's theory, that such spontaneous emission exists.

Since quantum mechanics can be used to calculate the probability of absorption of light, we can calculate the probability of spontaneous emission by using the general

[1] See, for example, [82], Section 99; [98], p. 631.

[2] The quantum theory of radiation enables Einstein's theory to be justified; see [29,47].

relation (5.11) between this probability and that of absorption, which is established in Einstein's theory, and this we shall now do.

87. Absorption and emission of light

To solve the problem of absorption or emission of light according to the discussion in Section 86, we must calculate the probability that an atom will go from one quantum level to another under the influence of incident light. To do so, we must first determine the interaction of an optical electron in the atom with a light wave.

Let us suppose that the light is polarised and its electric vector is $\mathscr{E}(x, t)$. In addition to the electric field there is also a magnetic field $\mathscr{H}(x, t)$, but the effect of the latter on the electron can be neglected in comparison with that of the electric field.[3] The effect of the electric field differs considerably according as the field $\mathscr{E}(x, t)$ varies appreciably over the extent of the atom or not. It is easy to state a criterion to distinguish these two cases. Let the incident wave be monochromatic (or almost so) and of wavelength λ. Then

$$\mathscr{E}(x, t) = \mathscr{E}_0 \cos(\omega_0 t - 2\pi x/\lambda), \tag{87.1}$$

where $\omega_0 = 2\pi c/\lambda$. We are concerned not with the field in all space, of course, but only with that within the atom. Let the dimension of the atom be a,[4] and the origin be at the centre of the atom. Then the wave phase $2\pi x/\lambda$ varies within the atom by an amount of the order $\pm 2\pi a/\lambda$, and if the size of the atom is much less than the wavelength of the incident light the variation of the phase within the atom can be neglected, so that the field within the atom at any instant can be described by the expression

$$\mathscr{E}(x, t) = \mathscr{E}_0 \cos \omega_0 t \tag{87.1'}$$

and is therefore the same everywhere within the atom.

The condition that the atom should be small in comparison with the wavelength is almost always satisfied; we need only have $\lambda \gg 10^{-8}$ cm (since $a \approx 10^{-8}$ cm). Ultraviolet and visible light have wavelengths thousands of times greater than 10^{-8} cm, so that the condition $\lambda \gg a$ is certainly satisfied for such light. The situation is different for X-rays, since in this range the wavelength is by no means always greater than the size of the atom.[5] The problem of the effect of such radiation is then more complex. Let us first consider the former case, where the wavelength is much greater than the size of the atom. We shall drop the special assumption that the light is monochromatic, though still supposing that the wavelengths present in the spectrum are long compared with the size of the atom. Then the electric field of the light which acts within the atom will be the same throughout the atom, but will depend on time:

$$\mathscr{E} = \mathscr{E}(t). \tag{87.2}$$

[3] The force exerted on the electron by the magnetic field is the Lorentz force $\mathbf{F} = e\mathbf{v} \times \mathscr{H}/c$, where $\mathbf{v}$ is the electron velocity and c the velocity of light. The force exerted by the electric field is $e\mathscr{E}$. In a light wave $\mathscr{E}$ and $\mathscr{H}$ are the same, and so the effect of the magnetic field is less by a factor v/c. The velocity of an electron in an atom is 100 times less than c, and so the magnetic interaction is 100 times weaker.

[4] a is the radius of the region where the wave functions are appreciably different from zero.

[5] The effect of X-rays on inner electrons (the K shell) is often of interest. The size of the K shell for

On the above assumptions it is easy to determine the form of interaction of the electron with the electric field (87.2) of the light, without using the general Hamiltonian for an electron in an external electromagnetic field. The field (87.2) is derived from a scalar potential $\phi(\mathbf{r}, t) = -\boldsymbol{\mathscr{E}} \cdot \mathbf{r} = -(\mathscr{E}_x x + \mathscr{E}_y y + \mathscr{E}_z z)$, so that the force function for an electron at a point $\mathbf{r}$ is in this field

$$W(\mathbf{r}, t) = -e\phi = e(\boldsymbol{\mathscr{E}} \cdot \mathbf{r}) = -\boldsymbol{\mathscr{E}} \cdot \mathbf{D}, \tag{87.3}$$

where $\mathbf{D} = -e\mathbf{r}$ is the electric moment of the electron, $\mathbf{r}$ being the radius vector from the nucleus to the electron.[6] Using also the unit vector $\mathbf{1}$ parallel to the field:

$$\boldsymbol{\mathscr{E}}(t) = \mathbf{1} \cdot \mathscr{E}(t),$$

we can write (87.3) as

$$W(\mathbf{r}, t) = -\mathscr{E}(t) \mathbf{1} \cdot \mathbf{D}. \tag{87.4}$$

If $\boldsymbol{H}^0$ denotes the total-energy operator of the electron, Schrödinger's equation for the wave function $\psi(\mathbf{r}, t)$ is

$$i\hbar\, \partial\psi/\partial t = \boldsymbol{H}^0\psi + W(\mathbf{r}, t)\psi. \tag{87.5}$$

The quantity $W(\mathbf{r}, t)$ will be regarded as a perturbation; this is possible for any intensity of radiation which can be obtained in practice.[7]

Let us now calculate the transition probability of an atom from a quantum level $E_n(\psi = \psi_n)$ to the level $E_m(\psi = \psi_m)$ under the effect of a light field. In order to be able to apply to this problem the theory of quantum transitions described in Section 84, we make the assumption that the light flux begins to act at time $t = 0$ and ceases at time $t = T$. If T is much longer than the period of oscillation of the light waves, this switching on and off of the interaction will not affect the spectral composition of the incident light.

According to (84.7) the transition probability P_{mn} from the state E_n to the state E_m in a time t equal to or greater than T is

$$P_{mn} = (4\pi^2/\hbar^2) \,|W_{mn}(\omega_{mn})|^2, \tag{87.6}$$

where $W_{mn}(\omega_{mn})$ is the Fourier coefficient, for frequency ω_{mn}, of the matrix element of the perturbation energy $W(\mathbf{r}, t)$. According to (87.4),

$$\begin{aligned} W_{mn}(t) &= \int \psi_m^* \cdot W(\mathbf{r}, t)\, \psi_n \cdot \mathrm{d}v \\ &= -\mathscr{E}(t) \int \psi_m^* \mathbf{1} \cdot \mathbf{D}\, \psi_n \cdot \mathrm{d}v = -\mathscr{E}(t) \mathbf{1} \cdot \mathbf{D}_{mn}, \end{aligned} \tag{87.7}$$

where $\mathbf{D}_{mn}$ is the matrix element of the electric moment vector, whose components are

$$\left.\begin{aligned} D_{mn,x} &= -e \int \psi_m^* x \psi_n \,\mathrm{d}v, \\ D_{mn,y} &= -e \int \psi_m^* y \psi_n \,\mathrm{d}v, \\ D_{mn,z} &= -e \int \psi_m^* z \psi_n \,\mathrm{d}v. \end{aligned}\right\} \tag{87.8}$$

---

elements of high atomic number is much less than that of the shell formed by the outer electrons. This extends the range of wavelengths for which the variation of the field phase can be neglected.

[6] The direction of the electric moment is reckoned from the negative to the positive charge, and the vector $\mathbf{r}$ in the opposite direction, from the positive nucleus to the negative electron.

[7] For example, the field of sunlight is about 0.1 e.s.u./cm^2, whereas the atomic field $\mathscr{E}$ is $e/a^2 \approx 10^7$ e.s.u./cm^2.

It follows from (87.7) that the Fourier component of $W_{mn}(t)$ is equal to that of $\mathscr{E}(t)$ multiplied by $-\mathbf{1}\cdot\mathbf{D}_{mn}$ (since $\mathbf{D}_{mn}$ is independent of time). Thus we find that

$$W_{mn}(\omega_{mn}) = -\mathscr{E}(\omega_{mn})\cdot\mathbf{1}\cdot\mathbf{D}_{mn}, \tag{87.9}$$

where $\mathscr{E}(\omega_{mn})$ denotes the Fourier component of $\mathscr{E}(t)$ for frequency ω_{mn}, i.e.

$$\mathscr{E}(\omega_{mn}) = \frac{1}{2\pi}\int_{-\infty}^{\infty}\mathscr{E}(t)\,e^{-i\omega_{mn}t}\,\mathrm{d}t = \frac{1}{2\pi}\int_{0}^{T}\mathscr{E}(t)\,e^{-i\omega_{mn}t}\,\mathrm{d}t. \tag{87.10}$$

Thus the transition probability from E_n to E_m is, according to (87.6),

$$P_{mn} = (4\pi^2/\hbar^2)\,|\mathscr{E}(\omega_{mn})|^2\,|\mathbf{1}\cdot\mathbf{D}_{mn}|^2. \tag{87.11}$$

The squared Fourier component of the electric field, $|\mathscr{E}(\omega_{mn})|^2$, can be expressed in terms of the amount of energy passing through in time T. The density of electromagnetic energy is $\mathscr{E}^2(t)/4\pi$ (the denominator is 4π, not 8π, since there are equal amounts of electric and magnetic energy). The energy flux is $c\mathscr{E}^2(t)/4\pi$ (where c is the velocity of light). Hence the total energy passing through 1 cm^2 is

$$E = \frac{c}{4\pi}\int_{-\infty}^{\infty}\mathscr{E}^2(t)\,\mathrm{d}t = \frac{c}{4\pi}\int_{-\infty}^{\infty}\mathrm{d}t\int_{-\infty}^{\infty}\mathscr{E}(\omega)\,e^{i\omega t}\,\mathrm{d}\omega\int_{-\infty}^{\infty}\mathscr{E}^*(\omega')\,e^{-i\omega' t}\,\mathrm{d}\omega'. \tag{87.12}$$

Integrating first with respect to t and noting that

$$\int_{-\infty}^{\infty} e^{i(\omega-\omega')t}\,\mathrm{d}t = 2\pi\delta(\omega-\omega'),$$

we find that

$$E = \frac{c}{4\pi}\cdot 2\pi\iint_{-\infty}^{\infty}\mathscr{E}(\omega)\,\mathscr{E}^*(\omega')\,\delta(\omega-\omega')\,\mathrm{d}\omega\,\mathrm{d}\omega'$$

$$= \tfrac{1}{2}c\int_{-\infty}^{\infty}|\mathscr{E}(\omega)|^2\,\mathrm{d}\omega = c\int_{0}^{\infty}|\mathscr{E}(\omega)|^2\,\mathrm{d}\omega,$$

since $\mathscr{E}(\omega) = \mathscr{E}^*(-\omega)$ because $\mathscr{E}(t)$ is real. If $E(\omega)$ denotes the energy passing through in the frequency range $\mathrm{d}\omega$, then

$$E = \int_{0}^{\infty} E(\omega)\,\mathrm{d}\omega;$$

comparison with the preceding formula shows that

$$E(\omega) = c\,|\mathscr{E}(\omega)|^2. \tag{87.13}$$

Thus

$$P_{mn} = (4\pi^2/\hbar^2)\,|\mathbf{1}\cdot\mathbf{D}_{mn}|^2\,E(\omega_{mn})/c. \tag{87.14}$$

The amount of energy $E(\omega)$ is evidently equal to the density $\rho(\omega)$ of radiant energy per unit frequency range, multiplied by the velocity of light and the time T during which energy passes:

$$E(\omega) = \rho(\omega) cT. \tag{87.15}$$

From (87.14) and (87.15) we can determine the transition probability p_{mn} from the state E_n to E_m per unit time. To do so, we must divide P_{mn} by the time during which the light acts, i.e. T:

$$p_{mn} = P_{mn}/T.$$

From (87.15) the transition probability per unit time is

$$p_{mn} = (4\pi^2/\hbar^2)|\mathbf{1}\cdot\mathbf{D}_{mn}|^2 \rho(\omega_{mn}). \tag{87.16}$$

If the angle between the electric moment vector $\mathbf{D}_{mn}$ and the direction $\mathbf{1}$ of polarisation of the light field is denoted by Θ_{mn}, we can finally write p_{mn} as

$$p_{mn} = (4\pi^2/\hbar^2)|\mathbf{D}_{mn}|^2 \cos^2 \Theta_{mn} \rho(\omega_{mn}). \tag{87.16'}$$

From this formula we see that to calculate the transition probability it is sufficient to know the electric moment matrix $\mathbf{D}_{mn}$, which is entirely determined by the properties of the atomic system considered. We shall return to this important result later, but next establish the relation between the probability p_{mn} just calculated and the Einstein coefficients discussed in Section 5.

88. Emission and absorption coefficients

According to Einstein's theory the probability per second of absorption of a light quantum $\hbar\omega = E_m - E_n$ with polarisation α and propagated in the solid angle $d\Omega$ is (see (5.2))

$$dW_a = b^m_{n\alpha} \rho_\alpha(\omega, \Omega) d\Omega. \tag{88.1}$$

The probability p_{mn} has been derived on the assumption that the wave is plane and propagated in some definite direction. Accordingly our formula for the probability involves only the spectral distribution and not the angular distribution. The general relation between $\rho_\alpha(\omega)$ and $\rho_\alpha(\omega, \Omega)$ is

$$\rho_\alpha(\omega) = \int \rho_\alpha(\omega, \Omega) d\Omega. \tag{88.2}$$

Since $\rho_\alpha(\omega)$ is finite, and $\rho_\alpha(\omega, \Omega)$ in our case is zero except for a particular direction, the density $\rho_\alpha(\omega, \Omega)$ must be a δ function of the angle Ω:

$$\rho_\alpha(\omega, \Omega) = \rho_\alpha(\omega) \delta(\Omega). \tag{88.3}$$

By integrating (88.3) with respect to Ω and using (88.1) we find the absorption probability per second with respect to a wave propagated in a definite direction (with no divergence):

$$W_a = b^m_{n\alpha} \rho_\alpha(\omega). \tag{88.4}$$

From the law of conservation of energy, the probability of absorption of a light quantum $\hbar\omega_{mn}$ must be equal to the probability for the transition of an atom from the state E_n to E_m, i.e. $W_a = p_{mn}$. Comparing (87.16′) and (88.4), we find that Einstein's coefficient $b^m_{n\alpha}$ for absorption of light is

$$b^m_{n\alpha} = (4\pi^2/h^2)\,|\mathbf{D}_{mn}|^2 \cos^2 \Theta_{mn}\,. \tag{88.5}$$

We must now consider separately the possible polarisations of the light. The formula for the transition probability p_{mn} (87.16′) has been derived on the assumption that the light is polarised in a direction $\mathbf{l}$ which makes an angle Θ_{mn} with the direction of the electric moment $\mathbf{D}_{mn}$, but in Einstein's coefficients $b^m_{n\alpha}$ the suffix $\alpha\,(\alpha = 1, 2)$ indicates that the polarisation is one of two selected as independent ($\mathbf{l}_1$ or $\mathbf{l}_2$). We can, without loss of generality, take as the first direction $\mathbf{l}_1\,(\alpha = 1)$ the direction perpendicular to the ray and lying in the plane containing the ray and the vector $\mathbf{D}_{mn}$, and as the second direction $\mathbf{l}_2\,(\alpha = 2)$ the direction perpendicular to this plane (see Figure 66).

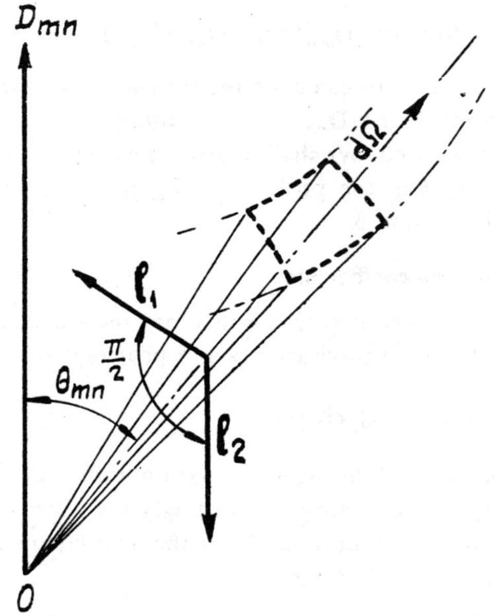

Fig. 66. Choice of independent polarisations l_1, l_2.

Putting $\mathbf{l} = \mathbf{l}_1$, we have $\Theta_{mn} = \frac{1}{2}\pi - \theta_{mn}$, where θ_{mn} is the angle between the polarisation vector $\mathbf{D}_{mn}$ and the direction of propagation of the absorbed radiation. From (88.5) we then have

$$b^m_{n1} = (4\pi^2/h^2)\,|\mathbf{D}_{mn}|^2 \sin^2 \theta_{mn}\,. \tag{88.5′}$$

With $\mathbf{l} = \mathbf{l}_2$ we obtain $\Theta_{mn} = \frac{1}{2}\pi$, i.e.

$$b^m_{n2} = 0\,. \tag{88.5″}$$

Using Formula (5.11), which determines the ratio of the spontaneous emission coefficient $a^n_{m\alpha}$ to the stimulated emission coefficient $b^n_{m\alpha} = b^m_{n\alpha}$ (see (5.7)), we can derive the probability dW'_r of spontaneous emission of a light quantum $\hbar\omega = E_m - E_n$ of polarisation α into the solid angle $d\Omega$:

$$dW'_r = a^n_{m\alpha}\,d\Omega = (\hbar\omega^3/8\pi^3c^3)\,b^n_{m\alpha}\,d\Omega = (\hbar\omega^3/8\pi^3c^3)\,b^m_{n\alpha}\,d\Omega\,, \quad (88.6)$$

where $\omega = (E_m - E_n)/\hbar = \omega_{mn}$. From (88.6) and (88.5'),

$$dW'_{r1} = (\omega^3_{mn}/2\pi c^3\hbar)\,|\mathbf{D}_{mn}|^2 \sin^2\theta_{mn}\,d\Omega \quad (88.7)$$

for light polarised parallel to $\mathbf{l}_1$, and

$$dW'_{r2} = 0 \quad (88.7')$$

for light polarised parallel to $\mathbf{l}_2$.

In order to obtain the total probability of spontaneous emission in the transition from the state E_m to the state E_n, we must integrate dW'_{r1} over all directions of propagation. This gives

$$W'_{r1} = (4\omega^3_{mn}/3\hbar c^3)\,|\mathbf{D}_{mn}|^2\,. \quad (88.8)$$

If the levels E_m and E_n are degenerate, the same frequency ω_{mn} may be emitted in various transitions from E_m to E_n. Summing (88.8) over all such transitions, we obtain the total probability per second of emission of the frequency ω_{mn}, which we denote by

$$A^n_m = (4\omega^3_{mn}/3\hbar c^3)\sum|\mathbf{D}_{mn}|^2\,. \quad (88.9)$$

The quantity $A_m{}^n$ is also called Einstein's coefficient for spontaneous emission of the frequency ω_{mn}. Together with $A_m{}^n$ we define a corresponding coefficient for absorption of isotropic unpolarised radiation of frequency ω_{mn}:

$$B_n{}^m = \frac{1}{8\pi f_n}\sum\int_{4\pi} b_{n\alpha}{}^m\,d\Omega\,, \quad (88.10)$$

where the sum is taken over the two polarisations ($\alpha = 1, 2$) and over all transitions from the level E_n to the level E_m. The quantity f_n signifies the degree of degeneracy of the level E_n. The integral is taken over all directions of propagation of the light. Similarly we can define the coefficient $B_m{}^n$ for stimulated emission:

$$B_m{}^n = \frac{1}{8\pi f_m}\sum\int_{4\pi} b_{m\alpha}{}^n\,d\Omega\,, \quad (88.10')$$

where f_m is the degree of degeneracy of the level E_m. By using the properties of $b_{m\alpha}{}^n$, $b_{n\alpha}{}^m$ and $a_{m\alpha}{}^n$, it is easily shown that

$$f_m B_m{}^n = f_n B_n{}^m,\ A_m{}^n = (\hbar\omega_{mn}{}^3/\pi^2c^3)\,B_m{}^n\,. \quad (88.11)$$

The quantity A^n_m determines the lifetime of the atom in the excited state. If at time t we have N_m atoms in the excited state E_m, the mean number of atoms which make a spontaneous transition to the lower state E_n in time dt is

$$dN_m = -\,A^n_m N_m\,dt\,,$$

whence

$$N_m = N^0_m\,e^{-A_m{}^n t} = N^0_m\,e^{-t/\tau_{mn}}\,, \quad (88.12)$$

where

$$\tau_{mn} = 1/A^n_m. \tag{88.13}$$

These formulae show that τ_{mn} is the mean lifetime of an atom at the excited level E_m.

From (88.9) we obtain

$$\tau_{mn} = \frac{3c^3\hbar}{4\omega^3_{mn}\sum|\mathbf{D}_{mn}|^2}. \tag{88.14}$$

An estimate of this quantity for visible light, with $\omega_{mn} \approx 4 \times 10^{15}$ and $\mathbf{D}_{mn}$ given in order of magnitude by $-ea$ (a being the size of the atom) so that $|\mathbf{D}_{mn}| \approx 2 \times 10^{-18}$, leads to the result $\tau_{mn} \approx 10^{-8}$ sec, i.e. $\tau_{mn} \gg T_{mn} = 2\pi/\omega_{mn} \approx 10^{-15}$ sec.[8]

Let us now calculate the mean energy emitted per second into the solid-angle element $\mathrm{d}\Omega$ by the transition $m \to n$. Since energy $\hbar\omega_{mn} = E_m - E_n$ is emitted in each transition, the mean energy emitted into the angle $\mathrm{d}\Omega$ per second is

$$\mathrm{d}(\mathrm{d}E/\mathrm{d}t) = \mathrm{d}W'_r \cdot \hbar\omega_{mn} = (\omega^4_{mn}/2\pi c^3)|\mathbf{D}_{mn}|^2 \sin^2\theta_{mn}\,\mathrm{d}\Omega, \tag{88.15}$$

and the total energy emitted per second is given by integration over all angles Ω:

$$\mathrm{d}E/\mathrm{d}t = (4\omega^4_{mn}/3c^3)|\mathbf{D}_{mn}|^2. \tag{88.16}$$

Both the angular distribution of the emitted energy (88.15) and the total energy emitted per second are the same as the corresponding formulae for a classical oscillator with eigenfrequency $\omega_0 = \omega_{mn}$ and mean electric moment

$$\overline{(\mathbf{D}_{cl})^2} = 2|\mathbf{D}_{mn}|^2. \tag{88.17}$$

The polarisation of the light is also the same as for a classical oscillator, namely only light with polarisation $\mathbf{l}_1$ is emitted; see Figure 66.

89. The correspondence principle

Let us consider the emission by a charged particle (of charge $-e$) moving in accordance with the laws of classical mechanics, and take for simplicity only the one-dimensional case. Let the period of the motion be $\tau_0 = 2\pi/\omega_0$. Denoting the co-ordinate of the particle by $x(t)$, let us expand it as a Fourier series:

$$x(t) = \sum_{k=-\infty}^{\infty} x_k e^{i\omega_k}, \tag{89.1}$$

$$\omega_k = \omega_0 k, \qquad k = \pm 1, \pm 2, \ldots, \qquad x_k = x^*_{-k};$$

ω_0 is the fundamental frequency and ω_k are the harmonic frequencies. Putting

$$x_k = |x_k|\,e^{i\phi_k}, \tag{89.2}$$

we can write (89.1) in the form

$$x(t) = \sum_{k=1}^{\infty} 2|x_k|\cos(\omega_k t + \phi_k). \tag{89.1'}$$

[8] This fact enables us to regard the excited states of the atom as stationary states (at least approximately); cf. Section 113.

The electric moment D of the particle is $ex(t)$, i.e.

$$D(t) = \sum_{k=-\infty}^{\infty} D_k e^{i\omega_k t} = \sum_{k=1}^{\infty} 2|D_k| \cos(\omega_k t + \phi_k), \tag{89.3}$$

where $D_k = ex_k$. The intensity of radiation of frequency ω_k, its distribution in space and its polarisation are given by the term

$$D_{cl} = 2|D_k| \cos(\omega_k t + \phi_k). \tag{89.4}$$

The mean energy emitted by such a dipole into the solid angle $d\Omega$ is

$$d(dE/dt) = (\omega_k^4/4\pi c^3)\overline{(D_{cl})^2} \sin^2\theta \, d\Omega, \tag{89.5}$$

and the total emission is

$$dE/dt = (2\omega_k^4/3c^3)\overline{(D_{cl})^2}, \tag{89.6}$$

where

$$\overline{(D_{cl})^2} = 4|D_k|^2 \overline{[\cos(\omega_k t + \phi_k)]^2} = 2|D_k|^2. \tag{89.7}$$

Thus instead of (89.5) and (89.6) we have

$$d(dE/dt) = (\omega_k^4/2\pi c^3)|D_k|^2 \sin^2\theta \, d\Omega, \tag{89.5'}$$

$$dE/dt = (4\omega_k^4/3c^3)|D_k|^2. \tag{89.6'}$$

A comparison of these formulae with (88.15) and (88.16) shows that the electric moment matrix element $\mathbf{D}_{mn}$ is fully analogous to the classical Fourier components. This analogy can be extended by considering the variation with time of the electric moments $\mathbf{D}_{mn}$ in the Heisenberg representation. We have supposed that $\mathbf{D}_{mn}$ is independent of time, including the time-dependence in the wave functions. We can, however, also regard the wave functions as independent of time and transfer the time dependence to the operators (matrices); this has been discussed in a general form for any mechanical quantity in Section 42. Then we have

$$\mathbf{D}_{mn}(t) = \mathbf{D}_{mn}(0)\, e^{i\omega_{mn}t} = \mathbf{D}_{mn} e^{i\omega_{mn}t}. \tag{89.8}$$

The corresponding representation in the classical theory signifies that the time factors $e^{i\omega_k t}$ in (89.3) are included in D_k:

$$D_k(t) = D_k(0)\, e^{i\omega_k t} = D_k e^{i\omega_k t}. \tag{89.8'}$$

Thus a particle moving classically can be described, as regards the field which it emits, by a single sequence of harmonically oscillating dipoles (89.8'):

$$D_1 e^{i\omega_1 t}, D_2 e^{i\omega_2 t}, \ldots, D_n e^{i\omega_n t}, \ldots \tag{89.9}$$

with frequencies

$$\omega_1 = \omega_0, \qquad \omega_2 = 2\omega_0, \ldots, \qquad \omega_n = n\omega_0, \ldots, \tag{89.9'}$$

representing the fundamental and the harmonics of the system.

A quantum system, on the other hand, is described as regards emission also by a set of harmonically oscillating dipoles but one forming a much larger manifold. The assembly of these dipoles can be represented by the electric moment matrix

$$\mathbf{D}(t) = \begin{vmatrix} \mathbf{D}_{11} & \mathbf{D}_{12}e^{i\omega_{12}t} & \cdots & \mathbf{D}_{1n}e^{i\omega_{1n}t} & \cdots \\ \mathbf{D}_{21}e^{i\omega_{21}t} & \mathbf{D}_{22} & \cdots & \mathbf{D}_{2n}e^{i\omega_{2n}t} & \cdots \\ \cdots & \cdots & \cdots & \cdots & \cdots \\ \mathbf{D}_{m1}e^{i\omega_{m1}t} & \mathbf{D}_{m2}e^{i\omega_{m2}t} & \cdots & \mathbf{D}_{mn}e^{i\omega_{mn}t} & \cdots \\ \cdots & \cdots & \cdots & \cdots & \cdots \end{vmatrix} \tag{89.10}$$

with frequencies

$$\omega_{mn} = (E_m - E_n)/\hbar, \tag{89.10'}$$

which also form a matrix:

$$\omega = \begin{vmatrix} 0 & \omega_{12} & \cdots & \omega_{1n} & \cdots \\ \omega_{21} & 0 & \cdots & \omega_{2n} & \cdots \\ \cdots & \cdots & \cdots & \cdots & \cdots \\ \omega_{m1} & \omega_{m2} & \cdots & \omega_{mn} & \cdots \\ \cdots & \cdots & \cdots & \cdots & \cdots \end{vmatrix}. \tag{89.10''}$$

The diagonal elements $\mathbf{D}_{nn}(t)$ of the matrix $\mathbf{D}(t)$ are independent of time, since $\omega_{nn} = 0$, and give the mean electric moment of the atom in the nth quantum state. The non-diagonal elements give the emission of the atom and oscillate with the Bohr frequencies.

Thus we arrive at *Ritz's combination principle*, given by (89.10′), which states that the frequencies of atoms are the differences of the terms $E_m/\hbar$, in contrast to the result of classical theory that all the frequencies ω_k are multiples of some fundamental frequency ω_0.

Long before the advent of quantum mechanics, Bohr put forward the hypothesis that the amplitudes $\mathbf{D}_n$ of the classical oscillators can be used to determine the intensities and polarisation of the radiation from quantum systems. This was called the *correspondence principle*. However, before quantum mechanics was devised the application of the principle was far from definite and was at best ambiguous. In Bohr's theory, quantum motions were regarded as motions in quantised orbits. The classical amplitudes $\mathbf{D}_n$, however, pertain to motion in any one definite orbit. They are derived by expanding as a Fourier series the radius vector $\mathbf{r}(t)$ of a particle moving in the nth orbit. Emission occurs, however, in a transition from one quantum state to another or, in the language of the old Bohr theory, in a transition from one orbit (n) to another orbit (m). Which of the two motions should be expanded as a Fourier series in order to obtain the Fourier coefficients $\mathbf{D}_k$ determining the emission was a question which could not be answered.

However, the application of the correspondence principle to transitions between levels with large quantum numbers ($n \gg 1$) accompanied by small changes in the quantum number ($|n - m| = |k| \ll n$) was entirely reasonable. For large quantum numbers n, the quantum orbits are very close together, forming an almost continuous sequence of classical unquantised orbits. For transitions between such orbits, since the change in the number n is small, it was possible to make unequivocal use of the correspondence principle by assuming that the intensity of radiation is determined by the classical Fourier components $\mathbf{D}_k$, because, owing to the smallness of the difference between the nth and mth orbits, it does not matter which of the two motions is resolved into harmonic components to determine the amplitudes of the fundamentals and harmonics, i.e. the quantities $\mathbf{D}_k$.

An important difficulty in Bohr's theory was the impossibility of calculating the intensity of radiation for small quantum numbers and for large changes of quantum number. In this typically quantum range of transitions the correspondence principle was of no use, and attempts to extend it to small values of n led to ambiguous results which at best permitted only qualitative, not quantitative, predictions concerning the nature of the emission.

In the above discussion we have started from Einstein's theory and reached the conclusion that a quantum system absorbs and emits like an assembly of classical harmonic oscillators with Fourier components of the electric moment equal to $\mathbf{D}_{mn}\, e^{i\omega_{mn}t}$. Consequently, to calculate the absorption or emission of light by a quantum system it is necessary to find the absorption or emission by classical oscillators with moments $\mathbf{D}_{mn}\, e^{i\omega_{mn}t}$. By calculating the energy absorbed or emitted per second and dividing by the absorbed or emitted light quantum $\hbar\omega = E_m - E_n$, we obtain the probability per second of the corresponding quantum transition.

This statement may be regarded as the modern form of the principle of correspondence between the quantum and classical theories of radiation.[9]

90. Selection rules for dipole radiation

It may happen that some of the electric moments $\mathbf{D}_{mn}$ are zero. In that case the transition $m \to n$ does not occur under the effect of light, and the corresponding frequency ω_{mn} is neither absorbed nor emitted, despite the fact that the levels E_m and E_n exist. This is described in terms of a *selection rule*, i.e. a rule which as it were selects among all conceivable transitions $E_m \rightleftarrows E_n$ only those which actually occur. It should be borne in mind that a transition is forbidden by the selection rule only with respect to perturbations W whose matrix elements are proportional to $\mathbf{D}_{mn}$. For example, a transition $m \rightleftarrows n$ impossible under the action of light may well occur as a result of collision with an electron.

Let us now consider the properties of the matrices $\mathbf{D}_{mn}$ for important cases and derive the selection rules for absorption and emission of light.

A. SELECTION RULES FOR AN OSCILLATOR

Consider an oscillator of mass μ, eigenfrequency ω_0 and charge e. The quantum levels E_n of such an oscillator are given by the formula

$$E_n = \hbar\omega_0 (n + \tfrac{1}{2}), \qquad n = 0, 1, 2, 3, \ldots. \tag{90.1}$$

The electric moment matrix elements must be

$$D_{mn} = ex_{mn} e^{i\omega_{mn}t} = ex_{mn} e^{i\omega_0 (m-n)t}, \tag{90.2}$$

where x_{mn} are the matrix elements of the co-ordinate. In Section 48 we have calculated the matrix of the co-ordinate and found that its elements are zero except for $m = n \pm 1$.

[9] A more detailed and general formulation of this principle is given by Pauli [71]. It can also be shown ([71], Section 12) that for large quantum numbers m and n and for $|k| = |m - n| \ll m, n$ the matrix elements $\mathbf{D}_{mn}{}^{(x)}(t) = \mathbf{D}_{m,\,m-k}{}^{(x)}(t)$ are approximately equal to the classical Fourier components: $-ex_k(t) = -\, ex_k(0) e^{ik\omega t}$, so that the old form of the correspondence principle is fully incorporated in the new one. A philosophical analysis of the correspondence principle is given in [57].

We thus obtain the selection rule

$$D_{mn} \neq 0 \qquad \text{only for} \qquad m = n \pm 1, \tag{90.3}$$

and the corresponding frequencies are $\omega_{mn} = \omega_0(m - n) = \pm\,\omega_0$, i.e. are equal to the eigenfrequency of the oscillator.

Using (48.9) and putting $D_0 = ex_0 = e\sqrt{(\hbar/\mu\omega_0)}$, we can write the matrix $D(t)$ in the Heisenberg representation as

$$D(t) = \begin{vmatrix} 0 & D_0 e^{i\omega_0 t}\sqrt{\frac{1}{2}} & 0 & 0 & \cdots \\ D_0 e^{-i\omega_0 t}\sqrt{\frac{1}{2}} & 0 & D_0 e^{i\omega_0 t}\sqrt{\frac{2}{2}} & 0 & \cdots \\ 0 & D_0 e^{-i\omega_0 t}\sqrt{\frac{2}{2}} & 0 & D_0 e^{i\omega_0 t}\sqrt{\frac{3}{2}} & \cdots \\ \cdots & \cdots & \cdots & \cdots & \cdots \\ \cdots & \cdots & \cdots & \cdots & \cdots \end{vmatrix}. \tag{90.4}$$

Thus an oscillator can absorb and emit only its eigenfrequency ω_0 (as in classical mechanics).

This selection rule is not always valid. It must be remembered that our calculations of the interaction with light have been based on the assumption that the wavelength λ of the light is much greater than the size a of the system. The interaction is given in terms of the electric moment matrix only when this condition is satisfied. The size of the oscillator is determined by its amplitude, and in order of magnitude is $\sqrt{(\hbar/\mu\omega_0)}\sqrt{(n+\frac{1}{2})}$. Hence the selection rule (90.3) is valid only if

$$\lambda \gg \sqrt{(\hbar/\mu\omega_0)}\sqrt{(n+\tfrac{1}{2})}, \tag{90.5}$$

i.e. when the amplitude of the oscillation is not too large.

It should be noted that actual oscillators become anharmonic at high amplitudes (large n), and this in itself may lead to violation of the simple selection rule.

B. SELECTION RULES FOR THE OPTICAL ELECTRON OF AN ATOM

Let us consider the electric moment matrix for an electron moving in a field of central forces. In this case the wave functions of the stationary states have the form

$$\psi_{nlm}(r,\theta,\phi) = R_{nl}(r)\,P_l^m(\cos\theta)\,e^{im\phi}. \tag{90.6}$$

We have to calculate the electric moment matrix with respect to this set of functions. Since the matrices of the electric moment components differ from those of the electron co-ordinates only by a factor $-e$, we shall calculate the latter. It is also convenient to calculate the matrices not of x, y, z but of the combinations

$$\xi = x + iy = r\sin\theta\cdot e^{i\phi}, \qquad \eta = x - iy = r\sin\theta\cdot e^{-i\phi}, \qquad \zeta = z. \tag{90.7}$$

Using the functions (90.6), we obtain

$$\xi_{nlm,n'l'm'} = \int_0^\infty R_{nl}R_{n'l'}\,r^3\,\mathrm{d}r \int_0^\pi P_l^m P_{l'}^{m'}\sin^2\theta\,\mathrm{d}\theta \int_0^{2\pi} e^{i(m-m')\phi+i\phi}\mathrm{d}\phi,$$

$$\eta_{nlm,n'l'm'} = \int_0^\infty R_{nl} R_{n'l'} r^3 \,\mathrm{d}r \int_0^\pi P_l^m P_{l'}^{m'} \sin^2\theta \,\mathrm{d}\theta \int_0^{2\pi} e^{i(m-m')\phi - i\phi} \,\mathrm{d}\phi,$$

$$z_{nlm,n'l'm'} = \int_0^\infty R_{nl} R_{n'l'} r^3 \,\mathrm{d}r \int_0^\pi P_l^m P_{l'}^{m'} \sin\theta\cos\theta \,\mathrm{d}\theta \int_0^{2\pi} e^{i(m-m')\phi} \,\mathrm{d}\phi. \tag{90.8}$$

The integrals with respect to ϕ are obvious:

$$\int_0^{2\pi} e^{i(m-m')\phi \pm i\phi} \,\mathrm{d}\phi = 2\pi\delta_{m'\mp 1,m}; \qquad \int_0^{2\pi} e^{i(m-m')\phi} \,\mathrm{d}\phi = 2\pi\delta_{m'm}. \tag{90.9}$$

With the notation

$$\int_0^\infty R_{nl} R_{n'l'} r^3 \,\mathrm{d}r = I_{nl,n'l'}, \tag{90.10}$$

$$\int_0^\pi P_l^m P_{l'}^{m'} \sin^2\theta \,\mathrm{d}\theta = S_{ll'}^{mm'}, \tag{90.11}$$

$$\int_0^\pi P_l^m P_{l'}^{m'} \sin\theta\cos\theta \,\mathrm{d}\theta = C_{ll'}^{mm'}, \tag{90.12}$$

we can write the matrix elements (90.8) as

$$\xi_{nlm,n'l'm'} = 2\pi I_{nl,n'l'} \cdot S_{ll'}^{mm'} \cdot \delta_{m,m'-1}, \tag{90.13}$$

$$\eta_{nlm,n'l'm'} = 2\pi I_{nl,n'l'} \cdot S_{ll'}^{mm'} \cdot \delta_{m,m'+1}, \tag{90.14}$$

$$z_{nlm,n'l'm'} = 2\pi I_{nl,n'l'} \cdot C_{ll'}^{mm'} \cdot \delta_{mm'}. \tag{90.15}$$

These formulae immediately give the selection rules for the change in the magnetic number m. The matrix elements of ξ are zero except for $m' = m + 1$, those of η except for $m' = m - 1$ and those of z except for $m' = m$. Thus only those transitions are possible for which the magnetic number m changes according to the rule

$$m' - m = \pm 1 \qquad \text{or} \qquad 0.$$

By examining the integrals $S_{ll'}^{mm'}$ and $C_{ll'}^{mm'}$ we can also set up the selection rule for the orbital quantum number l. To do this, we must establish the conditions under which these integrals are not zero. Let us first consider the integral $C_{ll'}^{mm'}$. We are interested only in the case where $m = m'$:

$$C_{ll'}^{mm} = \int_0^\pi P_{l'}^m P_{l'}^m \cos\theta \sin\theta \,\mathrm{d}\theta. \tag{90.16}$$

With the variable $x = \cos\theta$, we obtain

$$C_{ll'}^{mm} = \int_{-1}^{1} P_l^m(x) \cdot P_{l'}^m(x)\, x \,\mathrm{d}x. \tag{90.16$'$}$$

The properties of the spherical harmonic functions show that

$$x P_l^m(x) = a_{lm} P_{l+1}^m(x) + b_{lm} P_{l-1}^m(x), \tag{90.17}$$

where a_{lm} and b_{lm} are certain coefficients.[10]

[10] See Appendix V, Formula (30).

Bearing in mind that the functions P_l^m are orthogonal and substituting (90.17) in (90.16′), we find that

$$C_{ll'}^{mm} = a_{lm}\,\delta_{l',\,l+1} + b_{lm}\,\delta_{l',l-1}\,, \tag{90.18}$$

and so $C_{ll'}^{mm}$ is non-zero for $l' = l \pm 1$.

Similarly for the integrals $S_{ll'}^{mm'}$ (90.11) we have (with $m' = m \pm 1$)

$$S_{ll'}^{m,m\pm1} = \int_{-1}^{1} P_{l'}^{m\pm1}(x)\sqrt{(1-x^2)}\,P_l^m(x)\,dx\,. \tag{90.16″}$$

Using the formula for the spherical harmonic functions[11]

$$(1-x^2)^{\frac{1}{2}}\,P_l^m(x) = \alpha_{lm}\,P_{l-1}^{m-1}(x) + \beta_{lm}P_{l+1}^{m-1}(x) \tag{90.17′}$$

we find that

$$S_{ll'}^{m,\,m+1} = \alpha_{lm}\,\delta_{l-1,\,l'} + \beta_{lm}\,\delta_{l+1,\,l'}\,. \tag{90.19}$$

Applying the above formula for $(1-x^2)^{\frac{1}{2}}P_{l'}^{m-1}(x)$ we find in a similar way

$$S_{ll'}^{m,\,m-1} = \alpha_{l,\,m-1}\,\delta_{l,\,l'-1} + \beta_{l,\,m-1}\,\delta_{l,\,l'+1}\,. \tag{90.19′}$$

These formulae show that $S_{ll'}^{mm'} \neq 0$ only for $l' = l \pm 1$.

Thus we obtain the selection rule for the orbital quantum number:

$$l' - l = \pm 1\,. \tag{90.20}$$

There are no selection rules for the radial number $n_r = n - l - 1$. The selection rule just found indicates that optical transitions (with $\lambda \gg a$, i.e. for dipole radiation) are possible only between states which are adjoining states as regards the change in the angular momentum $M^2 = \hbar^2 l(l+1)$.

It has been explained that in spectroscopy the state with $l = 0$ is called an s term, with $l = 1$ a p term, with $l = 2$ a d term, and so on. It has long been known to spectroscopists that optical transitions occur only between s and p, p and d, d and f terms. It is seen that quantum mechanics affords an explanation of this fact: the electric moments (dipoles) $\mathbf{D}_{mn}$ are non-zero only for such transitions.

Let us consider more closely the selection rule for the magnetic number m as applied to the normal Zeeman effect. It has been shown in Section 62 that the quantum levels of atoms in a magnetic field are split, and if the field $\mathscr{H}$ is along the axis OZ, the emission frequencies possible *a priori* are determined by Formula (62.15):

$$\omega_{nlm,\,n'l'm'} = \omega_0 + O_L(m' - m)\,, \tag{90.21}$$

where ω_0 is the frequency in the absence of the field $\mathscr{H}$. The functions corresponding to the states E_{nlm} are given by ψ_{nlm} (90.6) (an atom in a magnetic field is not deformed, to a first approximation). The matrix elements $\mathbf{D}_{nlm,\,n'l'm'}$ are therefore also the same as in the absence of the external field $\mathscr{H}$. We can therefore apply to optical transitions, in the presence of the magnetic field, the selection rules which we have derived on the assumption that there is no external field. These rules show that emission and absorption are possible not for all frequencies given by Formula (90.21) but only for three frequencies:

$$\omega = \omega_0 \pm O_L \text{ if } m' - m = \pm 1, \text{ and } \omega = \omega_0 \text{ if } m' = m\,. \tag{90.22}$$

[11] See Appendix V, Formula (31).

This is just the normal Zeeman triplet already discussed in Section 62. We shall now determine the polarisation of the corresponding spectral lines.

For the undisplaced line ($m' = m$) only the electric moment along the axis OZ is different from zero, and so the emission of the undisplaced frequency is due to a dipole along the magnetic field $\mathscr{H}$. The electric vector of the dipole emission is coplanar with the dipole itself. The emission in this frequency will therefore be polarised so that the plane of polarisation passes through the direction of the magnetic field. For $m' = m + 1$ the matrix elements of z and η are zero (see (90.13), (90.14) and (90.15)). Then (90.7) gives

$$y_{nlm,\, n'l',\, m+1} = x_{nlm,\, n'l',\, m+1} \cdot e^{-\frac{1}{2}i\pi} . \qquad (90.23)$$

Similarly for $m' = m - 1$ we obtain

$$y_{nlm,\, n'l',\, m-1} = x_{nlm,\, n'l',\, m-1} \cdot e^{\frac{1}{2}i\pi} . \qquad (90.23')$$

These formulae show that the phase of the dipole along the axis OY is shifted by $\pm \frac{1}{2}\pi$ in comparison with that of the dipole along OX. The transition $m \to m + 1$ corresponds to the excitation of oscillations which are right-hand circularly polarised, while those for $m \to m - 1$ are left-hand circularly polarised. Accordingly the emission with frequency $\omega = \omega_0 + O_L$ is right-hand circularly polarised and that with $\omega = \omega_0 - O_L$ is left-hand circularly polarised.

Thus both the frequencies and the polarisations for the normal Zeeman effect are the same as in the classical Lorentz theory. The advantage of the quantum theory in this problem is that it provides, in addition to these data, the relative value (and, if the excitation conditions are specified, the absolute value) of the intensities for all the components of the Zeeman triplet: $\omega = \omega_0$ and $\omega = \omega_0 \pm O_L$.

91. Intensities in the emission spectrum

If an atom is in the excited state (m), a spontaneous transition to a lower level (n) is possible, with emission of a light quantum $\hbar\omega_{mn}$. In Section 88 we have derived an expression (88.16) for the energy $\mathrm{d}E/\mathrm{d}t$ emitted by an excited atom in unit time. In order to obtain the total observed intensity of emission, this quantity must be multiplied by the number of atoms N_m in the excited state (m). This number depends on the conditions of excitation. If, for example, the excitation is thermal and the emitting substance is in thermal equilibrium at temperature T, then

$$N_m = C(T)\, e^{-E_m/kT} , \qquad (91.1)$$

where C is some function of temperature which depends on the nature of the emitter. If the excitation is by electron impact and equilibrium is established, the number N_m is found from the conditions of equilibrium: the number of transitions per second to excited states under electron impact must be equal to the number of transitions per second to lower states which occur owing to spontaneous emission and also owing to collisions with electrons.

In general, without specifying the particular form of N_m, we can write the intensity I_{mn} of the radiation of frequency ω_{mn} caused by the transition from state m to state n as

$$I_{mn} = N_m \cdot (4\omega_{nm}^4/3c^3)\, |\mathbf{D}_{mn}|^2 . \qquad (91.2)$$

92. Dispersion

The problem of dispersion theory is to calculate the scattering of light. On interaction with a medium, light is not only absorbed but also scattered, changing its direction of propagation and, in general, also its frequency.

One of the simplest problems of dispersion theory is to calculate the refractive index for a gas. According to the classical theory of fields, Maxwell's relation gives the refractive index n of a medium as equal to $\sqrt{\varepsilon}$, where ε is the dielectric constant (permittivity). The latter in turn is related to the polarisability α of the medium: $\varepsilon = 1 + 4\pi\alpha$, so that

$$n^2 - 1 = 4\pi\alpha . \tag{92.1}$$

If N is the number of atoms per cm^3 and β the polarisability of a single atom, then $\alpha = N\beta$, and so

$$n^2 - 1 = 4\pi N\beta . \tag{92.2}$$

The atomic polarisability coefficient β is given by

$$\mathbf{p} = \beta \mathscr{E} , \tag{92.3}$$

where $\mathbf{p}$ is the electric moment of the atom and $\mathscr{E}$ the variable electric field of the light wave. The problem is to calculate β.

In the classical theory an optical electron was regarded as a particle moving under a quasielastic force. Accordingly the polarisability coefficient β was found to be

$$\beta = \frac{e^2}{\mu} \frac{1}{\omega_0^2 - \omega^2}, \tag{92.4}$$

where e is the charge on the electron, μ its mass, ω_0 the eigenfrequency of the optical electron and ω the frequency of the external field.[12] If the atom contains electrons with various eigenfrequencies $\omega_0, \omega_1, \omega_2, \ldots, \omega_k, \ldots$ and the number of electrons with frequency ω_k is f_k, then (92.4) must be replaced by the more general formula

$$\beta = \frac{e^2}{\mu} \sum_k \frac{f_k}{\omega_k^2 - \omega^2} . \tag{92.5}$$

The number f_k can also be regarded as the number of oscillators in the atom having eigenfrequency ω_k. The formula correctly describes the dispersion as regards the dependence of β (and therefore of the refractive index) on the frequency ω of the incident light. However, it was surprising to find that the experimental results gave values of the numbers f_k which were less than unity.

Let us now consider the quantum theory of dispersion. This leads, for coherent scattering, to the same Formula (92.5) as the classical theory, but the quantities f_k are no longer numbers of electrons of the kth kind, but have a quite different meaning. We shall therefore give the f_k a different name, and follow the established terminology in calling them *oscillator strengths.*

The quantum theory enables us to calculate the oscillator strengths f_k in entire agreement with the experimental data.

The problem of the dispersion of light in the quantum theory can be stated in close analogy with the quantum theory of emission and absorption of light. In the latter we seek the probability of absorption of a light quantum; in the dispersion problem we

[12] See [4], Section 25.

seek the probability that an original light quantum (in the incident beam) changes the direction of its momentum, and in general also its energy, on interacting with an atom.

We shall, however, use the correspondence principle to follow a simpler method which is closer to the classical theory: we shall find the electric moment $\mathbf{p}(t)$ produced in an atom which is in the variable field of a light wave. The light is assumed monochromatic of frequency ω. Again taking only the case where the wavelength λ is much greater than the dimension a of the quantum system, we can write the electric field $\mathscr{E}(t)$ of the light wave within the system (atom or molecule) as

$$\mathscr{E} = \mathscr{E}_0 \cos \omega t. \tag{92.6}$$

Suppose that, before the light field is applied, the atom is in one of its quantum levels E_n, and let the eigenfunction corresponding to this state be $\psi_n^0(\mathbf{r}, t)$.

When the light field is present, the state of the atom will be different, since forced vibrations will occur. Let this state be described by the wave function $\psi_n(\mathbf{r}, t)$. This function must satisfy Schrödinger's equation

$$i\hbar \partial \psi_n / \partial t = \boldsymbol{H}^0 \psi_n + \boldsymbol{W} \psi_n, \tag{92.7}$$

where $\boldsymbol{H}^0$ is the total-energy operator of the system (in the absence of the light field) and $\boldsymbol{W}$ the perturbation caused by the light wave. According to (92.6) $\boldsymbol{W}$ is

$$W = e\mathscr{E}_0 \cdot \mathbf{r} \cos \omega t. \tag{92.8}$$

To solve Equation (92.7) we write ψ_n in the form

$$\psi_n(\mathbf{r}, t) = \psi_n^0(\mathbf{r})\, e^{-i\omega_n t} + u_n(\mathbf{r})\, e^{-i(\omega_n - \omega)t} + v_n(\mathbf{r})\, e^{-i(\omega_n + \omega)t}, \tag{92.9}$$

where $\omega_n = E_n/\hbar$, and u_n and v_n are the required corrections to ψ_n^0. The function ψ_n^0 is the wave function of the stationary state of the unperturbed system:

$$\boldsymbol{H}^0 \psi_n^0 = E_n \psi_n^0. \tag{92.10}$$

Substituting (92.9) in Equation (92.7) and neglecting in a first approximation the products $\boldsymbol{W}u_n$, $\boldsymbol{W}v_n$ (since they are proportional to $\mathscr{E}^2$ and thus belong to the second approximation), we obtain

$$\begin{aligned} &\hbar(\omega_n - \omega)\, u_n e^{i\omega t} + \hbar(\omega_n + \omega)\, v_n e^{-i\omega t} \\ &= \boldsymbol{H}^0 u_n e^{i\omega t} + \boldsymbol{H}^0 v_n e^{-i\omega t} + e\mathscr{E}_0 \cdot \mathbf{r} \cdot \tfrac{1}{2}(e^{i\omega t} + e^{-i\omega t})\, \psi_n^0. \end{aligned} \tag{92.11}$$

Equating here the coefficients of the Fourier components, we obtain equations for u_n and v_n:

$$\hbar(\omega_n - \omega)\, u_n = \boldsymbol{H}^0 u_n + \tfrac{1}{2} e\mathscr{E}_0 \cdot \mathbf{r}\, \psi_n^0, \tag{92.12}$$

$$\hbar(\omega_n + \omega)\, v_n = \boldsymbol{H}^0 v_n + \tfrac{1}{2} e\mathscr{E}_0 \cdot \mathbf{r}\, \psi_n^0. \tag{92.12'}$$

To solve these equations we expand u and v as series in the orthogonal functions ψ_n^0:

$$u_n = \sum_l A_{nl}\, \psi_l^0, \tag{92.13}$$

$$v_n = \sum_l B_{nl}\, \psi_l^0. \tag{92.13'}$$

Substituting these expressions for u_n and v_n in (92.12) and (92.12′) and using the fact that the functions ψ_l^0 satisfy the equation $H^0\psi_l^0 = E_l\psi_l^0$, we find

$$\hbar\sum_l A_{nl}(\omega_n - \omega_l - \omega)\psi_l^0 = \tfrac{1}{2}e\mathscr{E}_0\cdot\mathbf{r}\psi_n^0, \tag{92.14}$$

$$\hbar\sum_l B_{nl}(\omega_n - \omega_l + \omega)\psi_l^0 = \tfrac{1}{2}e\mathscr{E}_0\cdot\mathbf{r}\psi_n^0. \tag{92.14′}$$

We multiply these equations by ψ_k^{0*} and integrate over all space. Then the orthogonality of the functions ψ_l^0 and ψ_k^{0*} gives the results

$$\hbar(\omega_n - \omega_k - \omega)A_{nk} = \tfrac{1}{2}e\int\psi_k^{0*}(\mathscr{E}_0\cdot\mathbf{r})\psi_n^0\,dv, \tag{92.15}$$

$$\hbar(\omega_n - \omega_k + \omega)B_{nk} = \tfrac{1}{2}e\int\psi_k^{0*}(\mathscr{E}_0\cdot\mathbf{r})\psi_n^0\,dv. \tag{92.15′}$$

Hence we find A_{nk} and B_{nk}:

$$A_{nk} = -\frac{\mathscr{E}_0\cdot\mathbf{D}_{kn}}{2\hbar(\omega_{nk} - \omega)}, \tag{92.16}$$

$$B_{nk} = -\frac{\mathscr{E}_0\cdot\mathbf{D}_{kn}}{2\hbar(\omega_{nk} + \omega)}, \tag{92.16′}$$

where $\omega_{nk} = \omega_n - \omega_k = (E_n - E_k)/\hbar$ are the eigenfrequencies of the atom and $\mathbf{D}_{kn}$ is the matrix element of the electric moment vector.

It follows from (92.16) and (92.16′) that the method used for the solution of Equations (92.14) and (92.14′) is valid only when the frequency ω of the incident light does not coincide with any of the eigenfrequencies ω_{nk} of the atom, i.e. far from resonance. The necessary distance from ω_{nk} is given by the condition $|\mathscr{E}_0\cdot\mathbf{D}_{kn}| \ll 2\hbar|\omega_{nk} \pm \omega|$. Only with this condition are A_{nk} and $B_{nk} \ll 1$. In order to deal with the resonance region the damping of the oscillators $\mathbf{D}_{kn}\,e^{i\omega_{kn}t}$ must also be taken into account.

Substituting the values found for A_{nk} and B_{nk} in (92.13) and (92.13′), and u_n and v_n in (92.9), we have the approximate expression for $\psi_n(\mathbf{r}, t)$:

$$\begin{aligned}\psi_n(\mathbf{r}, t) = \psi_n^0(\mathbf{r})e^{-i\omega_n t} &- \frac{e^{-i(\omega_n-\omega)t}}{2\hbar}\sum_k\frac{\mathscr{E}_0\cdot\mathbf{D}_{kn}}{\omega_{nk} - \omega}\psi_k^0(\mathbf{r}) - \\ &- \frac{e^{-i(\omega_n+\omega)t}}{2\hbar}\sum_k\frac{\mathscr{E}_0\cdot\mathbf{D}_{kn}}{\omega_{nk} + \omega}\psi_k^0(\mathbf{r}).\end{aligned} \tag{92.17}$$

Let us now calculate in the first approximation the electric moment $\mathbf{p}_{nn}(t)$ induced by the field $\mathscr{E}(t)$ in the state ψ_n^0. This state becomes $\psi_n(\mathbf{r}, t)$ when the field is applied. The mean electric moment in this state is

$$\mathbf{p}_{nn} = -e\int\psi_n^*(\mathbf{r}, t)\mathbf{r}\cdot\psi_n(\mathbf{r}, t)\,dv = -e\int|\psi_n(\mathbf{r}, t)|^2\cdot\mathbf{r}\cdot dv. \tag{92.18}$$

According to (92.17), $|\psi_n(\mathbf{r}, t)|^2$ is, to within terms of the second order in $\mathscr{E}_0$,

$$|\psi_n(\mathbf{r}, t)|^2 = |\psi_n^0|^2 - \frac{e^{i\omega t}}{2\hbar}\sum_k \frac{\mathscr{E}_0 \cdot \mathbf{D}_{kn}}{\omega_{nk} - \omega}\psi_n^{0*}\psi_k^0 - \frac{e^{-i\omega t}}{2\hbar}\sum_k \frac{\mathscr{E}_0 \cdot \mathbf{D}_{kn}}{\omega_{nk} + \omega}\psi_n^{0*}\psi_k^0 - \frac{e^{-i\omega t}}{2\hbar}\sum_k \frac{\mathscr{E}_0 \cdot \mathbf{D}_{kn}^*}{\omega_{nk} - \omega}\psi_k^{0*}\psi_n^0 - \frac{e^{i\omega t}}{2\hbar}\sum_k \frac{\mathscr{E}_0 \cdot \mathbf{D}_{kn}^*}{\omega_{nk} + \omega}\psi_k^{0*}\psi_n^0 .$$

Substituting in (92.18) and noting that

$$-e\int \psi_k^{0*}\,\mathbf{r}\,\psi_n^0\,dv = \mathbf{D}_{kn},$$

we obtain

$$\mathbf{p}_{nn}(t) = \mathbf{D}_{nn} - \frac{e^{i\omega t}}{2\hbar}\sum_k\left\{\frac{(\mathscr{E}_0 \cdot \mathbf{D}_{kn})\mathbf{D}_{kn}^*}{\omega_{kn} - \omega} + \frac{(\mathscr{E}_0 \cdot \mathbf{D}_{kn}^*)\mathbf{D}_{kn}}{\omega_{kn} + \omega}\right\} - \frac{e^{-i\omega t}}{2\hbar}\sum_k\left\{\frac{(\mathscr{E}_0 \cdot \mathbf{D}_{kn})\mathbf{D}_{kn}^*}{\omega_{kn} + \omega} + \frac{(\mathscr{E}_0 \cdot \mathbf{D}_{kn}^*)\mathbf{D}_{kn}}{\omega_{kn} - \omega}\right\}. \tag{92.19}$$

We see that the electric moment $\mathbf{p}_{nn}(t)$ consists of two parts, the moment $\mathbf{D}_{nn}$ which is independent of time and the additional induced moment which depends linearly on the field. $\mathbf{D}_{nn}$ is just the mean electric moment of the atom (or molecule) in the state n. Since it is independent of time, it plays no part in the dispersion of light. The induced moment varies periodically with time, the frequency being the frequency ω of the incident light. Moreover, the phase of oscillation of this moment is related to that of the electric vector of the incident light. This additional moment is responsible for the coherent scattering (dispersion). Denoting it by $\mathbf{p}'_{nn}(t)$, we have

$$\mathbf{p}'_{nn} = \mathbf{p}_{nn} - \mathbf{D}_{nn}.$$

According to (92.19) this induced moment can be written in components as

$$\left.\begin{aligned}
(p'_{nn})_x &= Re\{\beta_{xx}\mathscr{E}_{0x}e^{i\omega t} + \beta_{xy}\mathscr{E}_{0y}e^{i\omega t} + \beta_{xz}\mathscr{E}_{0z}e^{i\omega t}\},\\
(p'_{nn})_y &= Re\{\beta_{yx}\mathscr{E}_{0x}e^{i\omega t} + \beta_{yy}\mathscr{E}_{0y}e^{i\omega t} + \beta_{yz}\mathscr{E}_{0z}e^{i\omega t}\},\\
(p'_{nn})_z &= Re\{\beta_{zx}\mathscr{E}_{0x}e^{i\omega t} + \beta_{zy}\mathscr{E}_{0y}e^{i\omega t} + \beta_{zz}\mathscr{E}_{0z}e^{i\omega t}\},
\end{aligned}\right\} \tag{92.20}$$

where Re denotes the real part of the expression following it.

The quantities such as β_{xy} form the atomic polarisability tensor

$$\beta = \begin{vmatrix} \beta_{xx} & \beta_{xy} & \beta_{xz} \\ \beta_{yx} & \beta_{yy} & \beta_{yz} \\ \beta_{zx} & \beta_{zy} & \beta_{zz} \end{vmatrix}, \tag{92.21}$$

whose typical component is

$$\beta_{xy} = -\frac{1}{\hbar}\sum_k \left\{\frac{(D_{kn})_y (D_{nk})_x}{\omega_{nk} - \omega} + \frac{(D_{nk})_y (D_{kn})_x}{\omega_{nk} + \omega}\right\}, \tag{92.22}$$

where $(D_{nk})_x$, $(D_{kn})_y$ etc. are the components of the vectors $\mathbf{D}_{nk}$ and $\mathbf{D}_{kn}$ along the axes OX and OY. The remaining components of the tensor β are obtained from (92.22) by replacing x, y by all possible pairs from x, y, z. Since $D_{kn} = D^*_{nk}$, the tensor (92.22) is Hermitian:

$$\beta_{xy} = \beta^*_{yx}, \tag{92.23}$$

and the diagonal elements β_{xx}, β_{yy}, β_{zz} are therefore real.

In the general case where β_{xy}, β_{xz}, β_{yz} are complex, the phase of the induced moment $\mathbf{p}'_{nn}$ and its direction do not coincide with those of the electric field $\mathscr{E}(t)$ of the light wave. If all the components of the tensor β are real, the direction of $\mathbf{p}'_{nn}$ does not coincide with that of the field, but their phases are the same.

For comparison with the classical theory let us consider the particular but very important case where the tensor β reduces to a single scalar, i.e. where $\beta_{xy} = \beta_{xz} = \beta_{yz} = 0$, $\beta_{xx} = \beta_{yy} = \beta_{zz} = \beta$. Under these conditions both the phase and the direction of the induced moment are the same as those of the field of the light wave.

In this special case the fundamental difference from the classical theory of dispersion is very clear. From (92.22), with the above assumption, we have (since $\omega_{kn} = -\omega_{nk}$)

$$\beta = \beta_{xx} = \beta_{yy} = \beta_{zz} = \frac{2}{\hbar}\sum_k \frac{\omega_{kn}\,|(D_{nk})_x|^2}{\omega^2_{nk} - \omega^2}, \tag{92.24}$$

where

$$(D_{nk})_x = -e\int \psi^{0*}_n x \psi^0_k \cdot dv$$

and it is assumed that the system is isotropic, i.e. that

$$|(D_{nk})_x|^2 = |(D_{nk})_y|^2 = |(D_{nk})_z|^2 .$$

The formula (92.24) for the polarisability β can be written in a form entirely analogous to the classical formula (92.5):

$$\beta = \frac{e^2}{\mu}\sum_k \frac{f_{nk}}{\omega^2_{nk} - \omega^2}, \tag{92.5'}$$

where

$$f_{nk} = 2\mu\,|x_{nk}|^2\,\omega_{nk}/\hbar = 2\mu\,|\mathbf{D}_{nk}|^2\,\omega_{kn}/e^2\hbar . \tag{92.25}$$

In the quantum theory the quantity f_{nk} is usually called the *oscillator strength*. It is related in a simple manner to the spontaneous transition probability A^k_n: from (88.9) we have

$$f_{nk} = 3\mu c^3 A^k_n / 2e^2\omega^2_{kn} .$$

Thus the oscillator strength f_{nk} determines the intensity of spontaneous emission,

The quantities f_{nk} may be calculated if the wave functions of the system are known.[13]

We see that in the quantum theory the quantities f_{nk} have a quite different significance from that in the classical theory, where the corresponding quantity f_k represents the number of electrons of the kth kind, and is therefore an integer. The oscillator strengths f_{nk} are not integers, in agreement with experiment. It can be shown[14] that their sum is unity. According to the quantum theory, as follows from (92.5′), the sum of the dispersion terms of the form $1/(\omega_{nk}^2 - \omega^2)$ occurs for even a single electron in the state ψ_n^0. This is directly related to the fact that a quantum system behaves, as regards interaction with light, as an assembly of oscillators with moments $\mathbf{D}_{mn}e^{i\omega_{mn}t}$, even if only one particle is involved.

If the atom can be not only in the state ψ_n^0 but also in other states (a mixed ensemble), then in order to obtain the total polarisability β it is necessary to multiply the polarisability due to atoms in the state ψ_n^0 by the probability of finding the atom in that state, and add the resulting expressions. Denoting by w_n the probability that the atom is in the state ψ_n^0, with

$$\sum_n w_n = 1,$$

we obtain for the polarisability α per cm^3 of gas the expression

$$\alpha = \frac{e^2 N}{\mu} \sum_n \sum_k w_n \frac{f_{nk}}{\omega_{nk}^2 - \omega^2} = N\beta, \tag{92.26}$$

where N is the number of atoms per cm^3. The refractive index as a function of the frequency of the incident light is, according to (92.2) and (92.26),

$$n^2(\omega) = 1 + \frac{4\pi e^2 N}{\mu} \sum_n \sum_k w_n \frac{f_{nk}}{\omega_{nk}^2 - \omega^2}. \tag{92.27}$$

Often one or more of the terms in the sum in (92.27) are predominant. This occurs when the frequency ω is not too far from the resonance frequency ω_{nk}.

The oscillator strength f_{nk} may even be negative. If the atom is in the excited state n, the states k will include some for which $\omega_{kn} < 0$ (i.e. $E_k < E_n$). In this case the dispersion curve is unusual, and *negative dispersion* occurs. On the left of Figure 67 is the dispersion curve in the region of anomalous dispersion for the classical case ($f_{nk} > 0$). This dispersion was considered in a number of papers, the most thorough investigations being those of Rozhdestvenskiĭ.[15] Figure 67 also shows, on the right, the curve for negative dispersion ($f_{nk} < 0$), a case not predicted by the classical theory. The phenomenon of negative dispersion was discovered by Ladenburg [54].

The experimental determination of the values of oscillator strengths is a difficult problem [25].

To illustrate the agreement between theory and experiment we may cite the results

[13, 14] See [5].

[15] Rozhdestvenskiĭ used the special method of 'hooks'; see [76].

of Ladenburg and Carst [25] for the oscillator strengths of the hydrogen lines of the Balmer series H_α and H_β. They found that $5.9:1 > f_\alpha : f_\beta > 4.66:1$. The theoretical result is $f_\alpha : f_\beta = 5.37:1$.

93. Raman scattering

We have calculated in Section 92 the electric moment $\mathbf{p}'_{nn}$ induced by light in the nth state of the atom. Let us now consider what additional electric moment $\mathbf{p}_{mn}$ is induced by light in a quantum system when it goes from one state m to another state n. This problem is easily solved by means of the results of Section 92. Formula (92.17) gives the state $\psi_n(\mathbf{r}, t)$ which results from $\psi_n^0(\mathbf{r})e^{-i\omega_n t}$ under the effect of light. An exactly

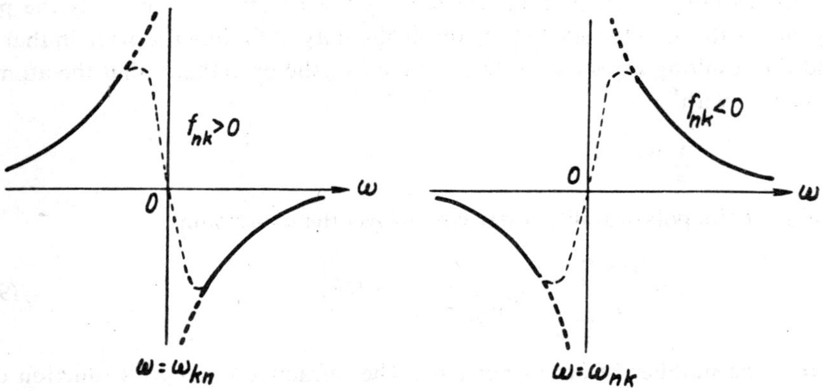

Fig. 67. Dispersion curves for positive and negative dispersion.

similar formula can be written for the state $\psi_m(\mathbf{r}, t)$ which results from the state $\psi_m^0(\mathbf{r})e^{-i\omega_m t}$ under the effect of the same light. Instead of (92.18) we now have the following formula for the moment $\mathbf{p}_{mn}(t)$ corresponding to the transition from m to n:

$$\mathbf{p}_{mn}(t) = -e\int \psi_m^*(\mathbf{r}, t)\,\mathbf{r}\,\psi_n(\mathbf{r}, t)\,\mathrm{d}v\,. \tag{93.1}$$

Substituting here the value of $\psi_n(\mathbf{r}, t)$ from (92.17) and that of $\psi_m^*(\mathbf{r}, t)$ obtained from (92.17) by replacing the suffix n by m, we obtain

$$\mathbf{p}_{mn}(t) = \mathbf{D}_{mn}e^{i\omega_{mn}t} + e^{i(\omega_{mn}+\omega)t}\mathbf{D}_{mn}^{(+)} + e^{i(\omega_{mn}-\omega)t}\mathbf{D}_{mn}^{(-)}, \tag{93.2}$$

where

$$\mathbf{D}_{mn}^{(+)} = -\frac{1}{2\hbar}\sum_k \left\{\frac{(\mathscr{E}_0\cdot\mathbf{D}_{kn})\mathbf{D}_{mk}}{\omega_{nk}-\omega} + \frac{(\mathscr{E}_0\cdot\mathbf{D}_{mk})\mathbf{D}_{kn}}{\omega_{mk}+\omega}\right\}, \tag{93.3}$$

$$\mathbf{D}_{mn}^{(-)} = -\frac{1}{2\hbar}\sum_k \left\{\frac{(\mathscr{E}_0\cdot\mathbf{D}_{kn})\mathbf{D}_{mk}}{\omega_{nk}+\omega} + \frac{(\mathscr{E}_0\cdot\mathbf{D}_{mk})\mathbf{D}_{kn}}{\omega_{mk}-\omega}\right\}. \tag{93.3'}$$

Thus we see that, in addition to the electric moment $\mathbf{D}_{mn}$ already considered, depending periodically on time with frequency ω_{mn}, there appear two other electric moments (93.3) and (93.3′) induced by the light, whose frequencies are the *combination frequencies* $\omega = \omega_{mn} \pm \omega$. The electric moment $\mathbf{D}_{mn}$, as we know, determines the emission and absorption for transitions $E_m \rightleftarrows E_n$. The additional moments $\mathbf{D}_{mn}^{(+)}$ and $\mathbf{D}_{mn}^{(-)}$ cause scattering of the incident light but with a change in frequency, the new frequency being the sum or difference of the frequency ω of the incident light and one of the eigenfrequencies of the system $\omega_{mn} = (E_m - E_n)/\hbar$.

In order to determine the intensity of this scattered light, we can apply the correspondence principle, according to which the atom emits and absorbs light like an assembly of oscillators. According to (93.2) three such oscillators are involved here. The first has already been considered in Section 88, and the other two,

$$\mathbf{D}_{mn}^{(+)} e^{i(\omega_{mn}+\omega)t} \quad \text{and} \quad \mathbf{D}_{mn}^{(-)} e^{i(\omega_{mn}-\omega)t}, \tag{93.4}$$

according to Formula (88.16) for the mean energy emitted by an oscillator per second, give the following intensities for the emission of frequencies $\omega' = \omega_{mn} + \omega$ and $\omega'' = \omega_{mn} - \omega$ respectively:

$$\frac{dE'}{dt} = \frac{4(\omega_{mn} + \omega)^4}{3c^3} |\mathbf{D}_{mn}^{(+)}|^2, \tag{93.5}$$

$$\frac{dE''}{dt} = \frac{4(\omega_{mn} - \omega)^4}{3c^3} |\mathbf{D}_{mn}^{(-)}|^2, \tag{93.5'}$$

where $\mathbf{D}_{mn}^{(+)}$ and $\mathbf{D}_{mn}^{(-)}$ are given by (93.3) and (93.3′) and depend on the intensity of the incident light. Using the law of conservation of energy, we can interpret the resultant scattering with change of frequency on the basis of the concept of light quanta. Let the atom be in the state n with energy E_n, and suppose that a light quantum of frequency ω (energy $\varepsilon = \hbar\omega$) 'collides' with the atom. As a result of the collision, part of the energy

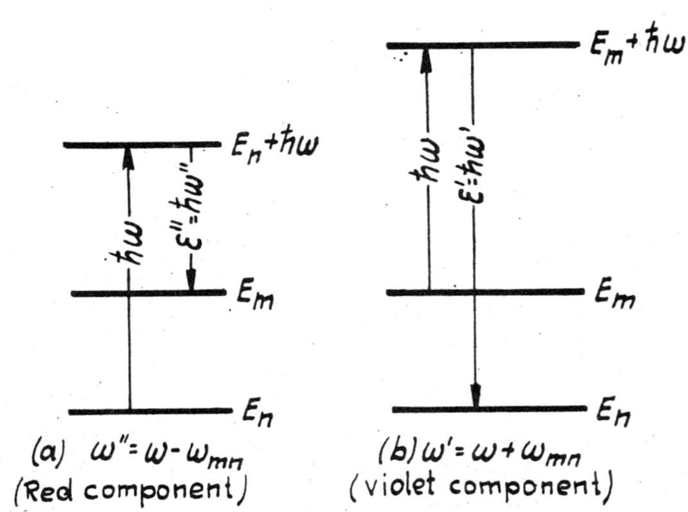

Fig. 68. Diagram of transitions in Raman scattering.

of the quantum may go to excite the atom (by a transition to the state $E_m > E_n$). Then the scattered quantum will have energy $\varepsilon'' = \hbar\omega'' = \hbar\omega - (E_m - E_n)$ (Figure 68a), and frequency $\omega'' = \omega - \omega_{mn}$, $\omega > \omega_{mn} > 0$. If the atom is in the state $E_m > E_n$, the scattered quantum can acquire energy from the atom, which enters the lower state E_n. In this case the energy of the quantum of scattered light ε' will be (Figure 68b) $\varepsilon' = \hbar\omega' = \hbar\omega + (E_m - E_n)$, and its frequency $\omega' = \omega + \omega_{mn}$, where $\omega_{mn} > 0$. The intensities at frequency ω' and ω'' are given by Formulae (93.5) and (93.5′). We see that the application of the law of conservation of energy between the quantum system and the radiation does not permit scattering of frequencies $\omega < \omega_{mn}$. This conclusion does not necessarily follow from (93.5′) and is a special requirement, since we have used only the correspondence principle.[16]

In order to determine the absolute intensities of scattering of frequencies ω' and ω'', (93.5) must be multiplied by the number N_m of atoms in the state m, and (93.5′) by the number N_n of atoms in the state n. The frequencies $\omega' > \omega$, and therefore are often called the 'violet' components of the scattered radiation, while $\omega'' < \omega$ are called the 'red' components. Thus we have finally for the intensities of the violet components

$$I' = N_m \frac{4(\omega + \omega_{mn})^4}{3c^3} |\mathbf{D}_{mn}^{(+)}|^2, \tag{93.6}$$

and for those of the red components

$$I'' = N_n \frac{4(\omega - \omega_{mn})^4}{3c^3} |\mathbf{D}_{mn}^{(-)}|^2. \tag{93.6'}$$

The ratio of these intensities is

$$\frac{I'}{I''} = \frac{N_m (\omega + \omega_{mn})^4 |\mathbf{D}_{mn}^{(+)}|^2}{N_n (\omega - \omega_{mn})^4 |\mathbf{D}_{mn}^{(-)}|^2}. \tag{93.7}$$

Raman scattering was experimentally demonstrated by Landsberg and Mandel'shtam for solids and by Raman for liquids. In both cases, the frequencies ω_{mn} were vibrational frequencies, in Raman's experiments those of the molecules of the liquid. In Mandel'shtam and Landsberg's experiments the frequencies ω_{mn} were those of molecular vibrations of the crystal. For these experiments a particularly important deduction from the formula for the ratio I'/I'' is that the intensity of the violet components must increase with temperature, since the number of excited vibrational states N_m of the crystal increases with temperature according to $N_m \sim 1/(e^{\hbar\omega_{mn}/kT} - 1)$; accordingly the intensity of the violet components in the Raman spectrum must also increase. This theoretical deduction is entirely confirmed by experiment.

The vibrational frequencies of a molecule are determined by its structure. The study of molecular vibrations is therefore a valuable means of investigating molecular structure. These frequencies are in the infrared range, and many molecular vibrations involve no change in the electric moment; these are optically inactive vibrations. Both

[16] In the quantum theory of radiation this result can be deduced; see, for example, [75].

these factors greatly complicate the direct study of the frequencies of molecular vibrations. Raman scattering helps considerably in overcoming this difficulty. By examining the Raman scattering we can make use of visible light and determine from the change in its frequency the frequencies of the molecular vibrations, whether or not they are optically active. The study of Raman scattering by molecules now forms an extensive field of physics. Details of the phenomenon are given in [75].

94. Allowance for change of phase of the electromagnetic field of the wave within the atom. Quadrupole radiation

All our calculations above have assumed that the wavelength λ of the light is large compared with the dimension a of the system.

It is easy to modify the whole theory of interaction of an atom with light in such a way as to eliminate the assumption that $\lambda \gg a$. To do so, we must start from the Hamiltonian (27.9), which describes the behaviour of the electron in an arbitrary electromagnetic field (and we can neglect the small interaction of the electron spin with the field of the light wave).

The vector potential of the light wave can always be so chosen that div $\mathbf{A} = 0$ and the scalar potential $V = 0$. Thus the field of the light wave is given by the formulae

$$\mathscr{E} = -\frac{1}{c}\frac{\partial \mathbf{A}}{\partial t}, \mathscr{H} = \mathbf{curl}\,\mathbf{A}. \tag{94.1}$$

Neglecting also the quantity $\mathbf{A}^2$ in (27.9) as being a second-order small quantity, we can write the Hamiltonian (27.9) as

$$H = \frac{P^2}{2\mu} + U + \frac{e}{\mu c}\mathbf{A}\cdot\vec{P} = H^0 + \frac{e}{\mu c}\mathbf{A}\cdot\vec{P}. \tag{94.2}$$

The perturbation is (in the first approximation)

$$W(\mathbf{r}, t) = (e/\mu c)\,\mathbf{A}\cdot\vec{P} = -(i\hbar e/\mu c)\,\mathbf{A}\cdot\nabla. \tag{94.3}$$

The vector potential can be represented as a Fourier integral:

$$\mathbf{A}(\mathbf{r}, t) = \int \mathbf{A}_0(\omega)\, e^{-i(\omega t - \mathbf{k}\cdot\mathbf{r})}\, d\omega, \tag{94.3'}$$

where $\mathbf{k}$ is the wave vector.[17] Then the Fourier component of the matrix element of the perturbation belonging to the frequency ω_{mn} is

$$W_{mn}(\omega_{mn}) = -\frac{i\hbar e}{\mu c}\mathbf{A}_0(\omega_{mn})\cdot\int \psi_m^* e^{i\mathbf{k}\cdot\mathbf{r}}\cdot\nabla\psi_n\cdot dv. \tag{94.4}$$

From (94.1)

$$\mathbf{A}_0(\omega_{mn}) = \frac{ic}{\omega_{mn}}\mathscr{E}_0(\omega_{mn})\,\mathbf{1},$$

17 We shall suppose that the directions and polarisations of the various partial waves in (94.3′) are the same.

where $\mathscr{E}_0(\omega_{mn})\mathbf{l}$ is the Fourier component of the electric field. Hence

$$|W_{mn}(\omega_{mn})|^2 = |\mathscr{E}_0(\omega_{mn})|^2 \frac{\hbar^2 e^2}{\mu^2 \omega_{mn}^2} |\mathbf{l} \cdot \int \psi_m^* e^{i\mathbf{k}\cdot\mathbf{r}} \nabla \psi_n \cdot dv|^2 . \tag{94.5}$$

Substituting this expression in the formula for the transition probability (87.6) and changing from $|\mathscr{E}_0(\omega_{mn})|^2$ to the radiation density as in Section 89, we find that the transition probability per second is

$$p_{mn} = (4\pi^2/\hbar^2)\,|\mathbf{l} \cdot \mathbf{D}_{mn}(\mathbf{k})|^2 \rho(\omega_{mn}), \tag{94.6}$$

where

$$\mathbf{D}_{mn}(\mathbf{k}) = \frac{\hbar e}{\mu \omega_{mn}} \int \psi_m^* e^{i\mathbf{k}\cdot\mathbf{r}} \cdot \nabla \psi_n \cdot dv . \tag{94.7}$$

Formula (94.6) is entirely similar to (87.16), and can be used to derive Einstein's coefficients b_{na}^m, b_{ma}^n and a_{ma}^n for the short-wave case.

The difference between (87.16) and (94.6) is that in the former $\mathbf{D}_{mn}$ signifies the electric moment, independent of the nature of the radiation and determined by the properties of the atomic system, whereas the vector $\mathbf{D}_{mn}(\mathbf{k})$ depends on the wave vector $\mathbf{k}$ of the radiation. Hence Einstein's coefficients are found to be different from their values for dipole radiation; their general properties established in Section 5 remain unchanged, of course. The angular distribution of the radiation, its polarisation and the frequency dependence are also changed.

The conclusion derived in Section 89 that a quantum system interacts with radiation like an assembly of oscillators remains entirely valid for radiation of any wavelength. The only difference between long waves ($\lambda \gg a$) and short waves ($\lambda \lesssim a$) is that in the former case the quantum system can be regarded as an assembly of dipoles with moments $\mathbf{D}_{mn}e^{i\omega_{mn}t}$, whereas for short waves the variation of the wave phase within the system can not be ignored, and as regards interaction with radiation a quantum system resembles an assembly of oscillators with frequencies ω_{mn}, whose size is not less than the wavelength. In this case it is more appropriate to speak of an assembly of currents and charges distributed in space and periodically dependent on time with frequency ω_{mn}. For long waves we can neglect the change in phase within the atom and expand $e^{i\mathbf{k}\cdot\mathbf{r}}$ in Equation (94.7) in powers of $\mathbf{k}\cdot\mathbf{r}$: $e^{i\mathbf{k}\cdot\mathbf{r}} = 1 + i\mathbf{k}\cdot\mathbf{r} + \ldots$. Since the functions ψ_m^0 and ψ_n^0 differ appreciably from zero only within the atom, this is an expansion in powers of $ka = 2\pi a/\lambda$, i.e. of the ratio of the dimension a to the wavelength λ. From (94.7) we then have

$$\mathbf{D}_{mn}(\mathbf{k}) = \frac{\hbar e}{\mu \omega_{mn}} \int \psi_m^{0*} \cdot \nabla \psi_n^0 \cdot dv + \frac{i\hbar e}{\mu \omega_{mn}} \int \psi_m^{0*} \cdot (\mathbf{k}\cdot\mathbf{r}) \nabla \psi_n^0 \cdot dv + \ldots . \tag{94.8}$$

The first term $\mathbf{D}_{mn}^{(1)}$ is

$$\mathbf{D}_{mn}^{(1)} = \frac{\hbar e}{\mu \omega_{mn}} \int \psi_m^{0*} \cdot \nabla \psi_n^0 \cdot dv = -\frac{e}{i\mu\omega_{mn}} \vec{P}_{mn}, \tag{94.9}$$

where $\vec{P}_{mn}$ is the matrix element of the momentum operator. The quantum equations of motion give

$$\vec{P}_{mn} = i\mu\omega_{mn}\mathbf{r}_{mn}, \qquad (94.10)$$

where $\mathbf{r}_{mn}$ is the matrix element of the radius vector. Hence

$$\mathbf{D}_{mn}^{(1)} = \mathbf{D}_{mn}, \qquad (94.11)$$

i.e. for long waves we obtain as a first approximation from (94.6) the formula (87.16) for dipole radiation. If $\mathbf{D}_{mn} \neq 0$, the following term $\mathbf{D}_{mn}^{(2)}$ can be neglected. Where $\mathbf{D}_{mn} = 0$ owing to the selection rules, the second term in (94.8) need not be zero. When $\mathbf{D}_{mn} = 0$ the emission will be determined by the second term $\mathbf{D}_{mn}^{(2)}$. We shall now show that the emission due to this additional term consists of electric quadrupole and magnetic dipole emission.

According to (94.8), $\mathbf{D}_{mn}^{(2)}$ can be written as

$$\mathbf{D}_{mn}^{(2)} = -\frac{e}{\omega_{mn}}\left\{(\mathbf{k}\cdot\mathbf{r})\frac{\vec{P}}{\mu}\right\}_{mn}, \qquad (94.12)$$

i.e. it is given in terms of the matrix element of the operator[18]

$$(\mathbf{k}\cdot\mathbf{r})\frac{\vec{P}}{\mu} = (\mathbf{k}\cdot\mathbf{r})\frac{\mathrm{d}\mathbf{r}}{\mathrm{d}t}.$$

This operator can be written identically as

$$(\mathbf{k}\cdot\mathbf{r})\frac{\mathrm{d}\mathbf{r}}{\mathrm{d}t} = \tfrac{1}{2}\frac{\mathrm{d}}{\mathrm{d}t}[(\mathbf{k}\cdot\mathbf{r})\mathbf{r}] - \tfrac{1}{2}\left[\mathbf{k}\times\left(\mathbf{r}\times\frac{\mathrm{d}\mathbf{r}}{\mathrm{d}t}\right)\right]. \qquad (94.13)$$

Taking matrix elements and using the fact that

$$\mathbf{r}\times\mathrm{d}\mathbf{r}/\mathrm{d}t = (1/\mu)\mathbf{r}\times\vec{P} = (1/\mu)\vec{M},$$

where $\vec{M}$ is the angular momentum operator, we obtain

$$\frac{1}{\mu}\left\{(\mathbf{k}\cdot\mathbf{r})\vec{P}\right\}_{mn} = \tfrac{1}{2}i\omega_{mn}\left\{(\mathbf{k}\cdot\mathbf{r})\mathbf{r}\right\}_{mn} - \frac{1}{2\mu}\left\{\mathbf{k}\times\vec{M}\right\}_{mn}. \qquad (94.14)$$

Substituting this result in (94.12) and noting that $\mathbf{k}/\omega_{mn} = \mathbf{n}/c$, where $\mathbf{n}$ is a unit vector in the direction of propagation of the radiation (since $k/\omega = 1/c$ and $\omega = \omega_{mn}$), and that $-(e/2\mu c)\mathbf{M} = \mathfrak{M}$, the magnetic moment of the atom, we find

$$\mathbf{D}_{mn}^{(2)} = -\tfrac{1}{2}ie\{(\mathbf{k}\cdot\mathbf{r})\mathbf{r}\}_{mn} - \{\mathbf{n}\times\mathfrak{M}\}_{mn}. \qquad (94.15)$$

Here the first term can be written as the product of the vector $-i\mathbf{k}$ and the matrix element of the second-order tensor

$$\vec{Q} = \begin{vmatrix} \tfrac{1}{2}ex^2 & \tfrac{1}{2}exy & \tfrac{1}{2}exz \\ \tfrac{1}{2}eyx & \tfrac{1}{2}ey^2 & \tfrac{1}{2}eyz \\ \tfrac{1}{2}ezx & \tfrac{1}{2}ezy & \tfrac{1}{2}ez^2 \end{vmatrix}. \qquad (94.16)$$

[18] $\{L\}_{mn}$ will denote the matrix element of the operator L.

In terms of this tensor, called the *quadrupole moment* of the atom, (94.15) can be written

$$\mathbf{D}^{(2)}_{mn} = -i\{\mathbf{k}\cdot\vec{Q}\}_{mn} - \{\mathbf{n}\times\mathfrak{M}\}_{mn}. \tag{94.17}$$

The first term causes the electric quadrupole radiation and the second term the magnetic dipole radiation.

Using the selection rule $l' = l \pm 1$ for dipole radiation (cf. Section 90) and the rule of matrix multiplication, we easily derive the selection rules for quadrupole radiation: since

$$\{x^2\}_{l'l} = \sum_{l''} \{x\}_{l'l''}\{x\}_{l''l}$$

and $l'' = l \pm 1$, $l' = l'' \pm 1$, it follows that $l' = l$ or $l \pm 2$.

The same result is obtained for the other components of the tensor. Thus the selection rule for quadrupole radiation is $l' = l$ or $l \pm 2$. For the magnetic radiation, the matrix of the operator $\mathfrak{M}$ is diagonal with respect to l, and magnetic radiation occurs in transitions with change of the magnetic number m, i.e. the selection rule will be $l' = l$, $m' = m \pm 1$.

The intensity of the quadrupole radiation is much less than that of the dipole radiation if the latter exists, since $\mathbf{D}^{(2)}_{mn}$ is less than the non-vanishing dipole moments by a factor of about $2\pi a/\lambda$. The quadrupole radiation transition probability is therefore in order of magnitude $(2\pi a/\lambda)^2$ times less than the dipole radiation transition probability. Accordingly the lifetime of an atom in the excited state, if the dipole emission is impossible, is $(\lambda/2\pi a)^2$ times the lifetime for an allowed dipole transition, which in Section 88 we estimated as about 10^{-8} sec. Hence, for visible light with $\lambda \sim 5 \times 10^3$ Å, and $a \sim 1$ Å, the lifetime τ in an excited state from which a transition to a lower state is possible only by quadrupole emission is about 10^{-2} sec. Such states of an atom are said to be *metastable*.

Since the magnetic moment of an atom is considerably less than the electric moment, the magnetic radiation also gives a very low transition probability, i.e. metastable levels.

Thus in atoms the quadrupole radiation and the magnetic radiation are of importance only when the dipole radiation is forbidden by the selection rules.

In atomic nuclei which emit γ rays the dipole emission usually is forbidden, and so the emission of γ rays is often due to the quadrupole or magnetic moment of the nucleus.[19]

95. The photoelectric effect

In this section we shall consider the theory of the photoelectric effect in atoms. The problem is to calculate the probability of ionisation of an atom by the action of a light wave and to determine the angular distribution of the electrons liberated. Thus we are concerned with a transition of the electron from the normal state (the lowest level of the discrete spectrum) to levels of the continuous spectrum.

[19] Details are given in [2]; [93], Chapter XII.

The energy of the normal state will be denoted by $E_0(<0)$ and the corresponding wave function by $\psi_0(\mathbf{r})$. The wave functions of the continuous spectrum belonging to the energy E can be taken in many different ways owing to the high degree of degeneracy, provided that they form a complete set of orthogonal functions. We shall take the functions found in the theory of elastic collisions, i.e. a superposition of a plane wave, with a definite momentum $\mathbf{p}(p_x, p_y, p_z)$ of the electron, and the wave scattered by the atom. At large distances from the atom such wave functions will have the form (cf. Section 78)

$$\psi_{p_xp_yp_z}(\mathbf{r}) = \text{constant} \times \{e^{i(p_xx+p_yy+p_zz)/\hbar} + f_{p_xp_yp_z}(\theta,\phi)\, e^{-ikr}/r\}, \tag{95.1}$$

where k is the wave number. Such functions are one possible form for the wave functions of the stationary states of the continuous spectrum. The energy E of the state (95.1) is

$$E = p^2/2\mu = (p_x^2 + p_y^2 + p_z^2)/2\mu. \tag{95.2}$$

The functions (95.1) will be assumed normalised to $\delta(p_x - p_x')$, $\delta(p_y - p_y')$, $\delta(p_z - p_z')$. The perturbation which causes the transitions will be taken, in accordance with (94.3), in the form

$$W(\mathbf{r}, t) = -(i\hbar e/\mu c)\,\mathbf{A}\cdot\nabla, \tag{95.3}$$

where $\mathbf{A}$ is the vector potential of the light wave. The wave is assumed monochromatic, and $\mathbf{A}$ will be taken to have the form

$$\mathbf{A} = \tfrac{1}{2}\mathbf{A}_0 e^{i(\omega t - \mathbf{k}\cdot\mathbf{r})} + \tfrac{1}{2}\mathbf{A}_0 e^{-i(\omega t - \mathbf{k}\cdot\mathbf{r})}, \tag{95.4}$$

where $\mathbf{k}$ is the wave vector. Since the wave is transverse, div $\mathbf{A} = 0$, i.e.

$$\mathbf{A}_0\cdot\mathbf{k} = 0. \tag{95.5}$$

To calculate the required transition probability we can apply Formula (84.24) directly, since it was derived for transitions from the discrete to the continuous spectrum due to a perturbation varying harmonically with time.

Taking E in (84.24) to be the energy E_0 of the normal state of the atom, and the momentum p_x, p_y, p_z (p, θ, ϕ) to be that of the photoelectron, we can use (95.3), (95.4) and (84.12) to obtain the matrix element of the perturbation as

$$W_{E\theta\phi,\,0} = W_{p_xp_yp_z,0} = -\frac{i\hbar e}{2\mu c}\mathbf{A}_0\cdot\int \psi^*_{p_xp_yp_z} e^{i\mathbf{k}\cdot\mathbf{r}}\cdot\nabla\psi_0\cdot \mathrm{d}x\,\mathrm{d}y\,\mathrm{d}z. \tag{95.6}$$

Then the transition probability per second from the state E_0 of the electron to the state $E = E_0 + \hbar\omega$ with momentum lying in the solid angle $\mathrm{d}\Omega$ is

$$P_0(E,\theta,\phi)\,\mathrm{d}\Omega = \frac{2\pi}{\hbar}\frac{(2\mu)^{\frac{3}{2}}}{2}(E_0 + \hbar\omega)^{\frac{1}{2}}\,|W_{p_xp_yp_z,0}|^2\cdot\mathrm{d}\Omega, \tag{95.7}$$

where only those values of the momentum p_x, p_y, p_z are included which satisfy the resonance condition

$$E = p^2/2\mu = (p_x^2 + p_y^2 + p_z^2)/2\mu = E_0 + \hbar\omega. \tag{95.8}$$

Transitions to other levels E are impossible. Noting that $E_0 = -I$, where I is the ionisation potential, we can write (95.8) as

$$p^2/2\mu = \hbar\omega - I. \tag{95.9}$$

This is Einstein's equation for the atomic photoelectric effect.

In order to derive the final expression for $P_0(E, \theta, \phi)$ we must calculate the matrix element (95.6). To do this it is necessary to know the wave function ψ_0 of the initial state and the functions $\psi_{p_x p_y p_z}$ of the continuous spectrum. Let us assume that the photoelectric effect from then K shell is considered; then $-E_0 = I$ is the ionisation potential of the K shell. This shell is close to the nucleus of the atom, and so (ignoring the interaction of the two K electrons) we can take for ψ_0 the wave function of the lowest level E_0 for motion in a Coulomb field, with $n = 1, l = m = 0$. This is

$$\psi_0 = \psi_{100} = (Z^3/\pi a^3)^{\frac{1}{2}} e^{-Zr/a}, \tag{95.10}$$

where Z is the atomic number and a the radius of the first Bohr orbit.

Such a wave function will be a very close approximation to the true one. We shall use only a very rough approximation for the wave functions of the continuous spectrum, simply ignoring the change in the plane wave near the atom due to the action of the field of the atom, and accordingly replacing the exact wave function by the plane wave unperturbed by the field of the atom:

$$\psi_{p_x p_y p_z} = \frac{e^{i(p_x x + p_y y + p_z z)/\hbar}}{(2\pi\hbar)^{\frac{3}{2}}} \tag{95.11}$$

(which is normalised to a δ function with respect to $\mathbf{p}$). Such an approximation is not suitable for an exact calculation, but retains the essential features of the phenomenon. Its accuracy will increase with the energy of the photoelectron, i.e. it is valid for $E \gg -E_0 = I$. With this assumption concerning the wave functions of the continuous spectrum, the matrix element (95.6) can quite easily be calculated. Substituting (95.10) and (95.11) in (95.6), we obtain

$$W_{p_x p_y p_z, 0} = -\frac{i\hbar e}{2\mu c} \frac{1}{(2\pi\hbar)^{\frac{3}{2}}} \left(\frac{Z^3}{\pi a^3}\right)^{\frac{1}{2}} \times \\ \times \int e^{i(\mathbf{k}\cdot\mathbf{r} - \mathbf{p}\cdot\mathbf{r}/\hbar)} \mathbf{A}_0 \cdot (\nabla e^{-Zr/a})\, dx\, dy\, dz. \tag{95.12}$$

Let the wave be propagated in the direction of the axis OX, and let the electric vector (the polarisation) be along the axis OZ. Then OX is the direction of the vector $\mathbf{k}$, and OZ that of the vector $\mathbf{A}_0$. Thus $\mathbf{A}_0 = (0, 0, A_0)$, and so

$$W_{p_x p_y p_z, 0} = \frac{i\hbar e}{2\mu c} \frac{1}{(2\pi\hbar)^{\frac{3}{2}}} \left(\frac{Z^3}{\pi a^3}\right)^{\frac{1}{2}} A_0 \frac{Z}{a} \int e^{i(\mathbf{k} - \mathbf{p}/\hbar)\cdot\mathbf{r}} \frac{z}{r} e^{-Zr/a}\, dx\, dy\, dz. \tag{95.12'}$$

Figure 69 shows the position of the vectors $\mathbf{k}$, $\mathbf{p}$ and $\mathbf{A}_0$. To carry out the integration in (95.12′) we take the vector $\hbar\mathbf{k} - \mathbf{p}$ as the polar axis of a system of spherical polar co-ordinates Θ, Φ.

If the axis OZ in this system has angles Θ', Φ', then $z = r_z = r\cos(OZ, \mathbf{r})$. The cosine of the angle between OZ and $\mathbf{r}$, where $\mathbf{r}$ has polar co-ordinates Θ, Φ, is given by

$$\cos(OZ, \mathbf{r}) = \cos\Theta\cos\Theta' + \sin\Theta\sin\Theta'\cos(\Phi' - \Phi).$$

The angle between $\hbar\mathbf{k} - \mathbf{p}$ and $\mathbf{r}$ is Θ. Hence (95.12′) can be written

$$W_{p_x p_y p_z, 0} = \frac{i\hbar e}{2\mu c}\frac{1}{(2\pi\hbar)^{\frac{3}{2}}}\left(\frac{Z^3}{\pi a^3}\right)^{\frac{1}{2}} A_0 \frac{Z}{a} J, \tag{95.12″}$$

where

$$J = \int_0^\infty r^2\,\mathrm{d}r \int_0^\pi \int_0^{2\pi} \sin\Theta\,\mathrm{d}\Theta\,\mathrm{d}\Phi\, e^{i|\mathbf{k}-\mathbf{p}/\hbar| r\cos\Theta - Zr/a} \times \\ \times \{\cos\Theta\cos\Theta' + \sin\Theta\sin\Theta'\cos(\Phi' - \Phi)\}. \tag{95.13}$$

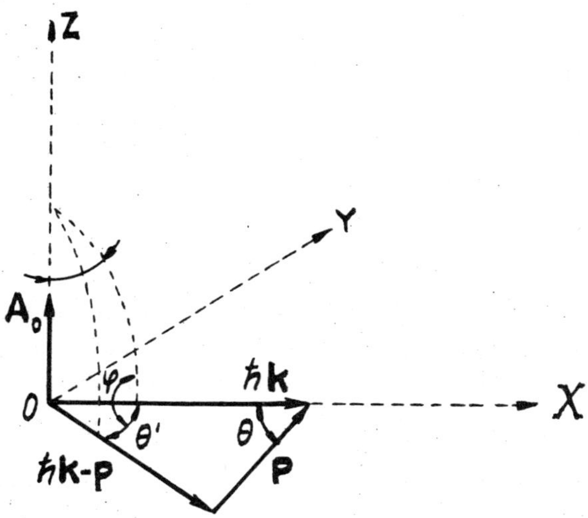

Fig. 69. Position of the vectors $\mathbf{A}_0$, $\mathbf{k}$ and $\mathbf{p}$ in the photoelectric effect.

The integral of $\cos(\Phi' - \Phi)$ with respect to Φ is evidently zero, and so

$$J = 2\pi\cos\Theta' \int_0^\infty r^2\,\mathrm{d}r \int_0^\pi \sin\Theta\,\mathrm{d}\Theta\, e^{i|\mathbf{k}-\mathbf{p}/\hbar| r\cos\Theta - Zr/a} \cdot \cos\Theta. \tag{95.13′}$$

With the variable $\xi = \cos\Theta$, and denoting $|\mathbf{k} - \mathbf{p}/\hbar| r$ by q, we have

$$J = 2\pi\cos\Theta' \int_0^\infty r^2\,\mathrm{d}r \cdot \int_{-1}^{1} \xi\,\mathrm{d}\xi\, e^{iq\xi - Zr/a},$$

and after straightforward integration

$$J = \cos\Theta' \frac{8\pi i |\mathbf{k} - \mathbf{p}/\hbar|}{[(Z^2/a^2) + |\mathbf{k} - \mathbf{p}/\hbar|^2]^2}. \tag{95.13″}$$

It remains to express $\cos\Theta'$ in terms of the angles in the co-ordinate system in which the direction of propagation of the light (i.e. of the vector **k**, the axis OX) is taken as the polar axis. Let ϕ be the angle between the plane of the vectors **p** and $\mathbf{k} - \mathbf{p}/\hbar$ and the plane ZOX (Figure 69), and let the angle between $\hbar\mathbf{k}$ and $\hbar\mathbf{k} - \mathbf{p}$ be θ'. Denoting also the angle between OX and **p** by θ, we obtain from the spherical triangle with sides Θ', θ' and $\frac{1}{2}\pi$

$$\cos\Theta' = \sin\theta' \cos\phi$$

and from the triangle with sides $\hbar\mathbf{k}$, $\mathbf{p}$, $\hbar\mathbf{k} - \mathbf{p}$,

$$\sin\theta' = \sin\theta \frac{p}{|\hbar\mathbf{k} - \mathbf{p}|}.$$

Hence

$$J = \sin\theta\cos\phi \frac{8\pi i p}{\hbar[(Z^2/a^2) + |\mathbf{k} - \mathbf{p}/\hbar|^2]^2}. \tag{95.14}$$

From (95.12″)

$$W_{p_x p_y p_z, 0} = \frac{-8\pi e}{2\mu c}\frac{1}{(2\pi\hbar)^{\frac{3}{2}}}\left(\frac{Z^3}{\pi a^3}\right)^{\frac{1}{2}} A_0 \frac{Z}{a}\frac{p}{[(Z^2/a^2) + |\mathbf{k} - \mathbf{p}/\hbar|^2]^2}\sin\theta\cos\phi. \tag{95.15}$$

Also

$$|\mathbf{k} - \mathbf{p}/\hbar|^2 = k^2 + (p^2/\hbar^2) - (2kp/\hbar)\cos\theta.$$

The law of conservation of energy (95.9), with the assumption that $p^2/2\mu \gg I$ (the condition for our approximation to be valid), gives $p^2/2\mu c = \hbar\omega/c = \hbar k$. Denoting the electron velocity p/μ by v, we find $\hbar k = vp/2c$, and hence

$$|\mathbf{k} - \mathbf{p}/\hbar|^2 = (p^2/\hbar^2)[1 - (v/c)\cos\theta + v^2/4c^2].$$

We are using the non-relativistic theory, and so the formulae are valid in a range which is restricted not only towards low velocities by the condition $\frac{1}{2}\mu v^2 \gg I$, but also towards high velocities: the velocity of the photoelectron must be much less than the velocity of light c. The terms of order v^2/c^2 should therefore be neglected (to include them would be to go beyond the range of applicability of the non-relativistic theory). Hence

$$|\mathbf{k} - \mathbf{p}/\hbar|^2 = (p^2/\hbar^2)[1 - (v/c)\cos\theta]. \tag{95.16}$$

The term Z^2/a^2 also can be neglected in comparison with $|\mathbf{k} - \mathbf{p}/\hbar|^2$, since the latter term is $\approx p^2/\hbar^2$, and $a = \hbar^2/\mu e^2$, so that $Z^2/a^2 = Z^2\mu^2 e^4/\hbar^4 = (2\mu/\hbar^2)Z^2\mu e^4/2\hbar^2$. Balmer's formula shows that $Z^2\mu e^4/2\hbar^2 = -E_0 = I$, so that the condition $Z^2/a^2 \ll p^2/\hbar^2$ is equivalent to $I \ll p^2/2\mu$. Thus when fast photoelectrons are considered we must omit the term Z^2/a^2 in the denominator of (95.15).

Substituting (95.16) in (95.15), we obtain the final expression for the required matrix element:

$$W_{p_x p_y p_z, 0} = -\frac{4\pi e}{\mu c}\frac{1}{(2\pi\hbar)^{\frac{3}{2}}}\left(\frac{Z^3}{\pi a^3}\right)^{\frac{1}{2}} A_0 \frac{Z}{a}\frac{\hbar^4\sin\theta\cos\phi}{p^3[1 - (v/c)\cos\theta]^2}. \tag{95.17}$$

Substituting this value of the matrix element in the expression (95.7) for the probability, we find[20]

$$P_0(E, \theta, \phi)\,\mathrm{d}\Omega = \mathrm{d}\Omega \frac{2e^2 (2\mu)^{\frac{3}{2}} \hbar^4}{\pi \mu^2 c^2} A_0^2 \left(\frac{Z}{a}\right)^5 \frac{(\hbar\omega)^{\frac{1}{2}} \sin^2\theta \cos^2\phi}{p^6 [1 - (v/c)\cos\theta]^4}. \tag{95.18}$$

The quantity A_0^2 may be replaced by the light energy flux. From (95.4) the electric field is

$$\mathscr{E} = -\frac{1}{c}\frac{\partial \mathbf{A}}{\partial t} = \frac{\omega}{c} \mathbf{A}_0 \sin(\omega t - \mathbf{k} \cdot \mathbf{r}).$$

The magnetic field $\mathscr{H}$ is the same, and since it is perpendicular to $\mathscr{E}$ the magnitude of the Poynting vector $\mathbf{S}$ is

$$S = \frac{c}{4\pi} \mathscr{E}\mathscr{H} = \frac{c}{4\pi}\frac{\omega^2}{c^2} A_0^2 \sin^2(\omega t - \mathbf{k} \cdot \mathbf{r}).$$

Its mean value is

$$\bar{S} = \omega^2 A_0^2 / 8\pi c, \; A_0^2 = 8\pi c \bar{S}/\omega^2. \tag{95.19}$$

Substitution in (95.18) gives

$$P_0(E, \theta, \phi)\,\mathrm{d}\Omega = \mathrm{d}\Omega \frac{16e^2 (2\mu)^{\frac{3}{2}} \hbar^4}{\mu^2 c} \frac{Z^5}{a^5} \frac{(\hbar\omega)^{\frac{1}{2}}}{\omega^2 p^6} \frac{\sin^2\theta \cos^2\phi}{[1 - (v/c)\cos\theta]^4} \bar{S}. \tag{95.20}$$

Combining the constants and using the fact that $p^6 = (2\mu E)^3 = (2\mu\hbar\omega)^3$, we obtain

$$P_0(E, \theta, \phi)\,\mathrm{d}\Omega = b\omega^{-\frac{9}{2}} \frac{\sin^2\theta \cos^2\phi}{[1 - (v/c)\cos\theta]^4} \bar{S}\,\mathrm{d}\Omega, \tag{95.21}$$

where

$$b = 4\sqrt{2}\frac{e^2}{\hbar c} \hbar^{\frac{3}{2}} \mu^{-\frac{7}{2}} \left(\frac{Z}{a}\right)^5. \tag{95.22}$$

The formula derived above gives the main properties of the photoelectric emission. Firstly, the number of photoelectrons is proportional to the intensity S of the incident light, while according to (95.9) their velocity depends only on the frequency ω of the incident light, i.e. we obtain precisely the features of the photoelectric effect which present fundamental difficulties when considered from the viewpoint of classical ideas. Next, Formula (95.21) gives the angular distribution of the photoelectrons. Since the angle θ is measured from the direction of propagation of the light, and ϕ from the electric vector, and the maximum photoelectric emission occurs at $\theta = \pm \frac{1}{2}\pi$, $\phi = 0$, this means that the greatest number of photoelectrons move in the direction OZ, i.e. in the direction of the electric vector of the light wave.

When the frequency of the incident light increases, the velocity of the photoelectrons also increases and the factor $[1 - (v/c)\cos\theta]^{-4}$ in (95.21) begins to be of importance, as a result of which the maximum photoelectric emission is displaced

[20] In (95.7) we neglect the initial energy E_0 of the electron in comparison with $\hbar\omega$.

towards smaller values of θ, i.e. in the direction of propagation of the light. This is in accordance with experiment, which gives the results shown in Figure 70. The ordinate is the cosine of the angle θ_{max} between the direction of propagation of the light and the direction of maximum emission; the abscissa is the photoelectron velocity, with the velocity of light taken as unity. $\cos\theta_{max} = 0$ corresponds to the direction of the electric vector of the wave, and $\cos\theta_{max} = 1$ to the direction of the light ray. The theoretical results are seen to agree well with the experimental data (shown by the circles). Using Formula (95.21) we can also derive the absolute magnitude of the photoelectric effect. In such cases the absorption coefficient τ for the incident light is usually calculated; it may be found as follows.

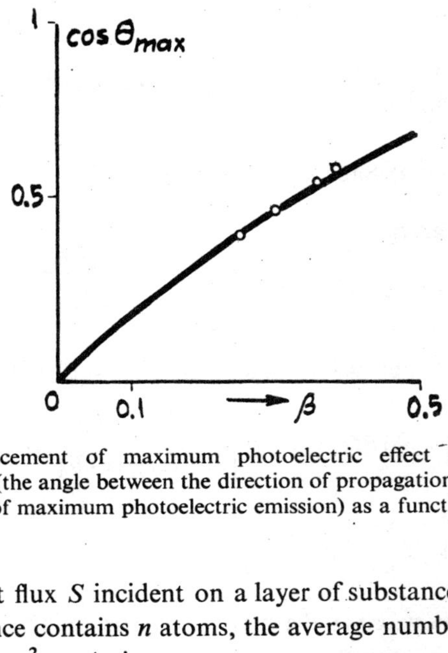

Fig. 70. Displacement of maximum photoelectric effect in the forward direction: θ_{max} (the angle between the direction of propagation of the light and the direction of maximum photoelectric emission) as a function of $\beta = v/c$.

Let us consider a light flux S incident on a layer of substance of thickness Δx. Then if 1 cm^3 of the substance contains n atoms, the average number of atoms ionised per second in a volume 1 cm^2 × Δx is

$$1\,\mathrm{cm}^2 \cdot \Delta x \cdot n \int P_0(E, \theta, \phi)\, d\Omega .$$

The energy absorbed will be this quantity multiplied by $\hbar\omega$ (since a light quantum $\hbar\omega$ is absorbed at each ionisation). The amount of energy entering the layer per second is $S \times 1$ cm^2. Thus the decrease of the energy flux S in passing through a thin layer Δx is

$$\Delta S = -\hbar\omega \cdot n\Delta x \int P_0(E, \theta, \phi)\, d\Omega .$$

Substituting $P_0(E, \theta, \phi)$ from (95.21), we find

$$\Delta S = -\hbar\omega \cdot n\Delta x \cdot b\omega^{-\frac{7}{2}} S \int \frac{\sin^2\theta \cos^2\phi}{[1 - (v/c)\cos\theta]^4}\, d\Omega .$$

Putting

$$\tau = bn\hbar\omega^{-\frac{7}{2}} \int \frac{\sin^2\theta \cos^2\phi}{[1-(v/c)\cos\theta]^4} \mathrm{d}\Omega, \tag{95.23}$$

we have

$$\Delta S = -\tau \bar{S} \Delta x,$$

so that τ is the absorption coefficient. The number of atoms in unit volume is proportional to the density ρ of the substance: $n = 6.06 \times 10^{23}\, \rho/A$, where A is the atomic weight of the substance. Substitution in (95.23), with

$$b' = \frac{6.06 \times 10^{23}}{A} b\hbar \int \frac{\sin^2\theta \cos^2\phi}{[1-(v/c)\cos\theta]^4} \mathrm{d}\Omega,$$

gives the *mass absorption coefficient*

$$\tau/\rho = b'/\omega^{\frac{7}{2}}. \tag{95.24}$$

This frequency dependence also is confirmed by experiments on X-ray absorption. However, it should be borne in mind that (95.24) has been derived for *K*-shell absorption, whereas absorption actually occurs in several shells simultaneously. We shall not consider the resulting complications, but merely refer to the relevant literature [80, 81].

CHAPTER XVI

THE PASSAGE OF MICROPARTICLES THROUGH POTENTIAL BARRIERS

96. Statement of the problem and simplest cases

If there are two regions of space in which the potential energy of a particle is less than on the surface separating the regions, then such regions are separated by a *potential barrier.*

A simple example of such a potential barrier is the barrier in one dimension shown in Figure 71. The ordinate is the potential energy $U(x)$ as a function of the co-

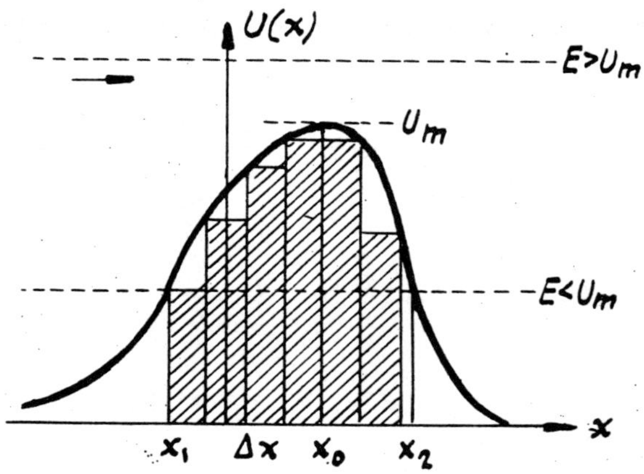

Fig. 71. Potential barrier in one dimension.

ordinate x of the particle. At the point x_0 the potential energy has a maximum U_m. The whole of space $-\infty < x < \infty$ is divided by this point into two regions $x < x_0$ and $x > x_0$ in which $U < U_m$. The significance of the term 'potential barrier' is now clear if we consider the motion of a particle in the field $U(x)$ in terms of classical mechanics. The total energy E of the particle is

$$E = p^2/2\mu + U(x), \tag{96.1}$$

where p is the momentum of the particle and μ its mass. Solving (96.1) for the momentum gives

$$p(x) = \pm\sqrt{\{2\mu[E - U(x)]\}}. \tag{96.2}$$

The sign must be chosen in accordance with the direction of motion of the particle. If the energy E of the particle exceeds the 'height' U_m of the barrier, the particle will pass the barrier without hindrance from left to right if the initial momentum $p > 0$ and in the opposite direction if the initial momentum $p < 0$.

Let us assume that the particle is moving from the left with a total energy E less than U_m. Then at some point x_1 where the potential energy $U(x_1) = E$, $p(x_1) = 0$ and the particle will come to rest. Its energy is completely converted into potential energy and motion in the opposite direction begins: x_1 is a turning point. For $E < U_m$, therefore, a particle moving from the left will not pass through the region of maximum potential energy ($x = x_0$) and will not penetrate into the other region $x > x_0$. Similarly, if a particle is moving from right to left with $E < U_m$ it will not penetrate into the region beyond the second turning point x_2, at which $U(x_2) = E$ (see Figure 71). Thus the potential barrier forms an 'opaque' partition for all particles of energy less than U_m (but is 'transparent' for particles having energy $E > U_m$). This explains the name 'potential barrier'.

The phenomena occurring near potential barriers are entirely different in the motions of microparticles in microfields, i.e. motions in which quantum effects are not negligible. In this case, as we shall now see, in contrast to the results of classical mechanics, particles of energy E greater than the height U_m of the barrier are partly reflected from the barrier, while those with energy E less than U_m partly penetrate the barrier.

In order to see this, let us consider the very simple barrier shown in Figure 72,

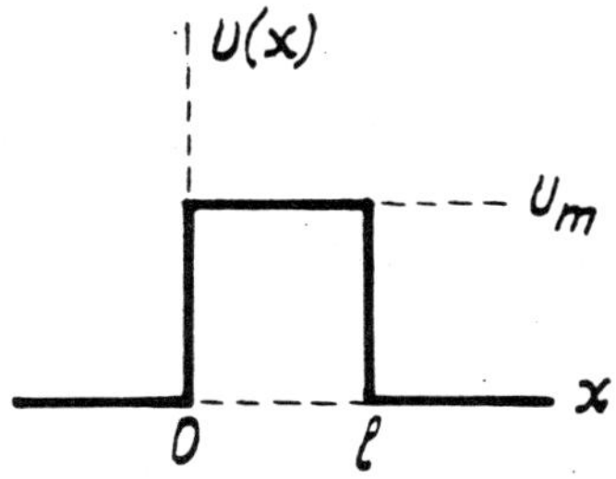

Fig. 72. Simplest potential barrier.

i.e. suppose that the potential energy $U(x)$ of the particle is zero everywhere except in the range $0 \leqslant x \leqslant l$, where it is constant and equal to U_m. Such a barrier is, of course, an idealisation, but it enables the problem under consideration to be examined with especial simplicity. We can imagine such a rectangular barrier to result from a continuous deformation of the smooth barrier shown in Figure 71.

We shall seek stationary states of a particle moving in the field of such a barrier. With the potential energy denoted by $U(x)$, Schrödinger's equation becomes

$$-\frac{\hbar^2}{2\mu}\frac{d^2\psi}{dx^2} + U(x)\psi = E\psi . \tag{96.3}$$

Denoting differentiation with respect to x by a prime, and using the notation familiar in optics

$$k_0^2 = 2\mu E/\hbar^2, \qquad (2\mu/\hbar^2)[E - U(x)] = k_0^2 n^2(x), \tag{96.4}$$

where $n(x)$ is the refractive index (see Section 36), we can write Equation (96.3) as

$$\psi'' + k_0^2 n^2(x)\psi = 0. \tag{96.5}$$

Equation (96.5) gives three equations for the three regions of space:

$$\psi'' + k_0^2\psi = 0, \qquad x < 0, \qquad U(x) = 0, \tag{96.5'}$$

$$\psi'' + k_0^2 n_m^2(x)\psi = 0, \qquad 0 \leqslant x \leqslant l, \qquad U(x) = U_m, \tag{96.5''}$$

$$\psi'' + k_0^2\psi = 0, \qquad x > l, \qquad U(x) = 0. \tag{96.5'''}$$

The solutions in these regions can be written down immediately:

$$\psi(x) = \psi_{\mathrm{I}}(x) = Ae^{ik_0x} + Be^{-ik_0x}, \tag{96.6}$$

$$\psi(x) = \psi_{\mathrm{II}}(x) = \alpha e^{ik_0n_mx} + \beta e^{-ik_0n_mx}, \tag{96.6'}$$

$$\psi(x) = \psi_{\mathrm{III}}(x) = ae^{ik_0x} + be^{-ik_0x}, \tag{96.6''}$$

where A, B, α, β, a and b are arbitrary constants. These are, however, the general solutions of the three independent Equations (96.5′), (96.5″) and (96.5‴), and in general do not represent a single wave function describing the state of the particle moving in the force field $U(x)$. In order that they should in fact form one function $\psi(x)$ we must satisfy certain boundary conditions which will now be established.

For this purpose we shall regard $U(x)$, and therefore $n(x)$, as smooth functions of x. Then, integrating Equation (96.5) near the point $x = 0$, we obtain

$$\int_{-\Delta}^{\Delta} \psi'' \mathrm{d}x + k_0^2 \int_{-\Delta}^{\Delta} n^2(x)\psi \,\mathrm{d}x = 0.$$

Hence

$$\psi'(\Delta) - \psi'(-\Delta) = -k_0^2 \int_{-\Delta}^{\Delta} n^2(x)\psi(x)\,\mathrm{d}x. \tag{96.7}$$

Taking the limit as $\Delta \to 0$, we obtain the boundary condition[1]

$$\psi'(0+) = \psi'(0-). \tag{96.7'}$$

According to the general requirement of continuity of the wave functions we have also the second condition

$$\psi(0+) = \psi(0-). \tag{96.7''}$$

The point $x = 0$ is in no way exceptional, and so the conditions (96.7′) and (96.7″) must be satisfied at any point, and in particular at $x = l$.

In order that the solutions (96.6), (96.6′), (96.6″) of the three equations (96.5) may

[1] See Appendix VIII.

be regarded as the limit of the solution of a single equation when $U(x)$ becomes discontinuous instead of smooth, it is necessary that these solutions should satisfy at $x = 0$ and $x = l$ the boundary conditions (96.7′) and (96.7″), i.e.

$$\left.\begin{aligned} \psi_{\mathrm{I}}(0) &= \psi_{\mathrm{II}}(0), & \psi'_{\mathrm{I}}(0) &= \psi'_{\mathrm{II}}(0), \\ \psi_{\mathrm{II}}(l) &= \psi_{\mathrm{III}}(l), & \psi'_{\mathrm{II}}(l) &= \psi'_{\mathrm{III}}(l). \end{aligned}\right\} \tag{96.8}$$

Substituting here the values of the functions from (96.6), (96.6′) and (96.6″), we have

$$\left.\begin{aligned} &A + B = \alpha + \beta, \\ &ik_0(A - B) = ik_0 n_m(\alpha - \beta), \\ &\alpha e^{ik_0 n_m l} + \beta e^{-ik_0 n_m l} = a e^{ik_0 l} + b e^{-ik_0 l}, \\ &ik_0 n_m(\alpha e^{ik_0 n_m l} - \beta e^{-ik_0 n_m l}) = ik_0(a e^{ik_0 l} - b e^{-ik_0 l}). \end{aligned}\right\} \tag{96.9}$$

We have here four equations for six constants. The arbitrariness in the choice of the constants is due to the fact that waves may be incident on the barrier from the left or from the right.

If we take, for example, A, $B \neq 0$ and $b = 0$, Ae^{ik_0x} can be regarded as the incident wave, Be^{-k_0x} as the reflected wave and ae^{ik_0x} as the transmitted wave. If we take $b \neq 0$, this means that there is also a wave incident from the other side of the barrier. These possibilities correspond to the cases in classical mechanics where particles move towards the barrier from the left and from the right.

For definiteness, let us take the case of particles incident from the left. Then we must put $b = 0$. Moreover, we can without loss of generality take the amplitude of the incident wave as unity ($A = 1$). The equations (96.9) then become

$$\left.\begin{aligned} &1 + B = \alpha + \beta, \\ &1 - B = n_m(\alpha - \beta), \\ &\alpha e^{ik_0 n_m l} + \beta e^{-ik_0 n_m l} = a e^{ik_0 l}, \\ &n_m(\alpha e^{ik_0 n_m l} - \beta e^{-ik_0 n_m l}) = a e^{ik_0 l}. \end{aligned}\right\} \tag{96.10}$$

From these algebraic equations we find

$$\alpha = \frac{2e^{-ik_0 n_m l}(1 + n_m)}{e^{-ik_0 n_m l}(1 + n_m)^2 - e^{ik_0 n_m l}(1 - n_m)^2}, \tag{96.11}$$

$$\beta = \frac{2e^{ik_0 n_m l}(n_m - 1)}{e^{-ik_0 n_m l}(1 + n_m)^2 - e^{ik_0 n_m l}(1 - n_m)^2}, \tag{96.12}$$

$$B = \frac{(e^{-ik_0 n_m l} - e^{ik_0 n_m l})(1 - n_m)^2}{e^{-ik_0 n_m l}(1 + n_m)^2 - e^{ik_0 n_m l}(1 - n_m)^2}, \tag{96.13}$$

$$a e^{ik_0 l} = \frac{4n_m}{e^{-ik_0 n_m l}(1 + n_m)^2 - e^{ik_0 n_m l}(1 - n_m)^2}. \tag{96.14}$$

If the energy E of the particle is greater than the height U_m of the barrier, the refractive index n_m is real, and the intensity of the reflected wave $|B|^2$ is

$$|B|^2 = \frac{4(1 - n_m^2)^2 \sin^2 k_0 n_m l}{(1 + n_m)^4 + (1 - n_m)^4 - 2(1 - n_m^2)^2 \cos 2k_0 n_m l}, \tag{96.15}$$

and that of the transmitted wave is

$$|a|^2 = \frac{16 n_m^2}{(1 + n_m)^4 + (1 - n_m)^4 - 2(1 - n_m^2)^2 \cos 2k_0 n_m l}. \tag{96.15'}$$

From the current-density formula we can calculate the particle fluxes in the incident wave (J_0), the reflected wave (J_r), and the transmitted wave (J_d). From (29.5) we have

$$\begin{aligned} J_0 &= (\hbar k_0/\mu)|A|^2 = \hbar k_0/\mu, \qquad J_r = -(\hbar k_0/\mu)|B|^2, \\ J_d &= (\hbar k_0/\mu)|a|^2. \end{aligned} \tag{96.16}$$

The ratio of the fluxes of reflected and incident particles is

$$|J_r|/J_0 = |B|^2/|A|^2 = |B|^2 = R, \tag{96.17}$$

and is called the *reflection coefficient*. The ratio of the fluxes of transmitted and incident particles is

$$J_d/J_0 = |a|^2/|A|^2 = |a|^2 = D, \tag{96.18}$$

and is called the *transmission coefficient* of the barrier.

From the law of conservation of particles (the equation of continuity for the current) it follows that

$$R + D = 1; \tag{96.19}$$

from the above expressions for R and D it can be directly verified that this equation is satisfied.

According to classical mechanics, if $E > U_m$ we must have $R = 0$ and $D = 1$, the barrier being completely transparent. From (96.15) it is seen that $|B|^2 \neq 0$, and therefore in quantum mechanics $R > 0$ and $D < 1$. Particles are partly reflected, in the same way as light waves at the interface between two media.

If the particle energy E is less than the height U_m of the barrier, then according to classical mechanics we have total reflection: $D = 0$, $R = 1$. The particles do not penetrate the barrier at all. In optics this corresponds to total internal reflection, and according to geometrical optics a ray of light does not penetrate into the second medium.

A more precise discussion using wave optics shows that in reality a light-wave field still penetrates into the reflecting medium even when there is total reflection, and if this medium is a very thin plate the light partly penetrates it. Quantum mechanics leads, in the case $E < U_m$ (the reflection case), to a conclusion similar to that of wave optics (see the discussion of analogies in Section 36). For, if $E < U_m$, the refractive index n_m is purely imaginary (see (96.4)). Hence we put

$$n_m = i|n_m| = i\sqrt{[(U_m - E)/E]}. \tag{96.20}$$

Substituting this expression for n_m in (96.14), we can calculate $|a|^2$, and if we assume that $e^{k_0|n_m|l} \gg 1$ the result is

$$D = |a|^2 = \frac{16|n_m|^2}{(1+|n_m|^2)^2} e^{-2k_0|n_m|l}. \tag{96.21}$$

Denoting the first factor by D_0 (which does not differ greatly from unity), and bearing in mind the value of k_0, we can write

$$D = D_0 e^{-2\sqrt{[2\mu(U_m - E)]}\, l/\hbar}. \tag{96.22}$$

Thus for $E < U_m$ particles penetrate the barrier, in contrast to the results of classical mechanics.

The phenomenon of penetration through a potential barrier is descriptively called the *tunnel effect*.[2] It is evident that this effect will be significant only when D is not very small, i.e. when

$$2\sqrt{[2\mu(U_m - E)]}\, l/\hbar \approx 1. \tag{96.23}$$

It is easy to see that the tunnel effect can occur only in the region of microphenomena. For example, when $U_m - E \sim 10^{-11}$ erg (about 10 eV), $\mu \sim 10^{-27}$ g (the mass of the electron) and $l \approx 10^{-8}$ cm, (96.22) shows that $D \approx 1/e$. But if we take, for example, $l = 1$ cm, the same formula gives $D \sim e^{-10^8}$. An increase in the particle mass and an increase of U_m relative to E will decrease D still further. It can similarly be shown that the reflection described above disappears as the particle energy increases; quantum mechanics and classical mechanics then coincide.

Formula (96.22) for the transmission coefficient D has been derived for a rectangular barrier, but can be generalised to a barrier of any shape. Here we shall do this by a simple though not entirely rigorous method.

Let the potential barrier $U(x)$ shown in Figure 71 be approximately represented as a series of rectangular barriers of width dx and height $U(x)$. These barriers are shown in Figure 71 by the shaded areas. A particle of energy E reaches the barrier at the point $x = x_1$ and leaves it at $x = x_2$. According to (96.22) the transmission coefficient for one of these elementary barriers is

$$D' = D_0' e^{-2\sqrt{[2\mu(U_m - E)]}\, dx/\hbar};$$

the potential energy $U(x)$ must be sufficiently smooth for dx to be taken fairly large. The transmission coefficient for the whole barrier must equal the product of the coefficients for the various elementary barriers. The exponents in the above formula for D are added, and the result is[3]

$$D = D_0 \exp\left[-\frac{2}{\hbar}\int_{x_1}^{x_2} \sqrt{\{2\mu[U(x) - E]\}}\, dx\right]. \tag{96.24}$$

[2] This phenomenon was first discussed by L. I. Mandel'shtam and M. A. Leontovich in connection with the quantum theory of the anharmonic oscillator (cf. the end of Section 67).

[3] This formula can be derived by the more rigorous method of the quasiclassical approximation (Section 37). See also [71], Section 12.

97. The apparent paradox of the 'tunnel effect'

The passage of particles through potential barriers appears paradoxical at first sight: a particle within a potential barrier and having a total energy E less than the height U_m of the barrier must have a negative kinetic energy $T = p^2/2\,\mu$, since, as in classical mechanics, the total energy is the sum of the kinetic and potential energies: $E = p^2/2\,\mu + U(x)$. In the range where $U(x) > E$, $p^2/2\,\mu < 0$, and this is absurd, since the momentum p is a real quantity. It is just these regions which are inaccessible to a particle in classical mechanics. According to quantum mechanics, however, a particle may also be found in this 'forbidden' region. Thus we find that quantum mechanics seems to lead to the conclusion that a particle can have a negative kinetic energy and an imaginary momentum. This is the paradox of the 'tunnel effect'.

There is, however, no paradox here, the conclusion itself being false. The reason is that, since the tunnel effect is a quantum phenomenon (the transmission coefficient D (96.24) tending to zero with $\hbar$), it can be discussed only in quantum-mechanical terms. The total energy of the particle can be regarded as the sum of the kinetic and potential energies only in classical mechanics. The formula $E = p^2/2\,\mu + U(x)$ assumes that we know simultaneously both the kinetic energy T and the potential energy $U(x)$. In other words we assign definite values simultaneously to the particle co-ordinate x and the particle momentum p, in contradiction with quantum mechanics. Dividing the total energy into potential and kinetic energy is meaningless in quantum mechanics, and so the paradox based on the possibility of representing the total energy E as the sum of the kinetic energy (a function of momentum) and the potential energy (a function of co-ordinates) is without foundation.

We need now only consider whether, nevertheless, by measuring the position of the particle we find it within the potential barrier, while at the same time its potential energy is less than the height of the barrier.

The particle may in fact be found within the barrier even if $E < U_m$, but as soon as the co-ordinate x of the particle is fixed this causes, in accordance with the uncertainty relation, a further dispersion $\overline{\Delta p^2}$ of the momentum, so that we cannot say that the energy of the particle is E after its position has been determined (cf. Sections 14, 15).

From the formula for the transmission coefficient it follows that particles penetrate appreciably only to a depth l given by Equation (96.23). In order to find the particle within the barrier, we must fix its co-ordinate with accuracy $\Delta x < l$, and this necessarily causes a dispersion of momentum $\overline{\Delta p^2} > \hbar^2/4\,\overline{\Delta x^2} = \hbar^2/4l^2$. Substituting l^2 from (96.23), we find

$$\overline{\Delta p^2}/2\mu > U_m - E\,, \tag{97.1}$$

i.e. the change in the kinetic energy of the particle due to the act of measurement must be greater than the difference between the energy of the particle and the height U_m of the barrier.

The following example illustrates this. Suppose that we wish to determine the co-ordinate of a particle within a potential barrier by sending a narrow beam of light in a direction perpendicular to the direction of motion of the particle. If the beam is scattered, the particle must have been in its path.

As shown above, the accuracy of the measurement must be such that $\Delta x < l$; but we cannot create a beam of light whose width is less than the wavelength λ of the light. Thus $\Delta x > \lambda$, and so the wavelength must be less than l, i.e.

$$\lambda < \hbar/\sqrt{[2\mu(U_m - E)]} \; ; \tag{97.2}$$

since $\lambda = c/\nu$, where ν is the frequency of the light vibrations and c the velocity of light, it follows that

$$\hbar^2\nu^2 > 2\mu c^2(U_m - E).$$

The energies occurring in non-relativistic mechanics must be less than the rest energy μc^2 of the particle, and so

$$\hbar\nu > U_m - E, \tag{97.3}$$

i.e. the energy of the quanta in the beam of light used must be greater than the difference between the height of the potential barrier and the energy of the particle.

Thus this example illustrates the unavoidability of using, for the measurement of the co-ordinate, apparatus of sufficiently great energy to localise the particle.

98. Cold emission of electrons from a metal

If a large electric field (of the order of 10^6 V/cm) is applied to a metal so as to make it the cathode, electrons are extracted by the field and an electric current results. This phenomenon is called *cold emission*. It can easily be interpreted on the basis of the quantum theory of the passage of particles through a potential barrier, and the interpretation is broadly in agreement with experiment.

In this section we shall consider the theory of this effect, which is one of the simplest applications of the theory of potential-barrier penetration, and first examine the motion of electrons in a metal in the absence of an external electric field.

In order to remove an electron from a metal, some work must be done. The potential energy of the electron in the metal is therefore less than outside the metal. This can be most simply expressed if we take the potential energy $U(x)$ of the electron inside the metal as zero, and outside as $U(x) = C > 0$, so that the potential energy has the form shown in Figure 73. By using this schematic representation of the true

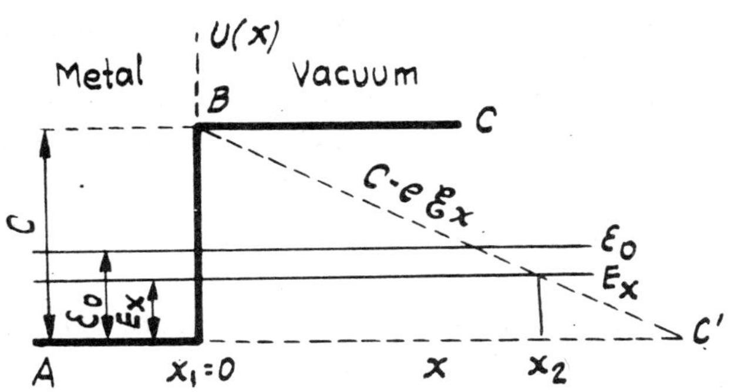

Fig. 73. The field at a metal boundary. Continuous line: in absence of external field; broken line: in presence of external field $\mathscr{E}$. When the field is present, a barrier OBC' is formed.

variation of the potential energy we are essentially taking the mean field in the metal, since the potential within the metal varies from point to point with a period equal to the lattice constant. Our approximation corresponds to the assumption of free electrons, since $U(x) = 0$ and so there is no force acting on an electron within the metal.

We cannot discuss here the extent to which this approximation is valid.[4] We shall merely mention that regarding the electrons in a metal as freely moving particles (an 'electron gas') makes it possible to account for many phenomena in metals and is therefore to some extent legitimate. The energy distribution of the electrons in this gas is such that the great majority of them have energies $E < C$; at absolute zero temperature the electrons occupy all energy levels from $E = 0$ to $E = \varepsilon_0 < C$, where ε_0 is the zero-point energy (see Section 120). Let the flux of electrons incident on the surface of a metal from its interior be J_0. Since the electrons have energy $E < C$, this flux is totally reflected from the potential discontinuity C which exists at the metal-vacuum boundary.

Let us now suppose that an electric field $\mathscr{E}$ is applied, directed towards the surface of the metal. Then the potential energy $-e\mathscr{E}x$ of the electron in the constant field $\mathscr{E}$ ($-e$ being the charge on the electron) is added to the potential energy $U(x)$ (Figure 73). The total potential energy of the electron is now

$$\left.\begin{aligned} U'(x) &= U(x) - e\mathscr{E}x = C - e\mathscr{E}x \quad (x > 0), \\ U'(x) &= 0 \qquad\qquad\qquad\qquad\qquad (x < 0). \end{aligned}\right\} \tag{98.1}$$

The potential energy curve has the modified form shown by the broken line in Figure 73. Within the metal a large field cannot be produced, and therefore the change in $U(x)$ occurs only outside the metal.

We see that a potential barrier is formed. According to classical mechanics an electron could penetrate the barrier only if its energy E were greater than C. Such electrons are very few in our case, and cause a small thermionic emission. There is therefore no electron current, according to classical mechanics, when the field is applied.[5] However, if the field $\mathscr{E}$ is sufficiently large, there is a rapid change in the potential energy, classical mechanics becomes inapplicable, and the electrons will penetrate the narrow potential barrier.

Let us calculate the transmission coefficient of this barrier for electrons whose energy of motion along the axis OX is E_x. According to (96.24) we have to calculate the integral

$$S = \int_{x_1}^{x_2} \sqrt{[2\mu[U'(x) - E_x]]}\, dx,$$

where x_1 and x_2 are the co-ordinates of the turning points. The first turning point

[4] See, for example, [79].

[5] If the field reduces the height of the barrier to a value less than ε_0, this will also occur in classical mechanics, but the current would be extremely large, with the electrons pouring through the barrier. In reality there is only a gradual increase in the current as the field increases.

is evidently $x_1 = 0$ (see Figure 73), since for any energy $E_x < C$ the horizontal line E_x representing the energy of motion along OX intersects the potential-energy curve at the point $x = 0$. The second turning point x_2 occurs, as we see from the diagram, when $E_x = C - e\mathscr{E}x$, whence $x_2 = (C - E_x)/e\mathscr{E}$; thus

$$S = \int_0^{(C-E_x)/e\mathscr{E}} \sqrt{\{2\mu[C - e\mathscr{E}x - E_x]\}}\,\mathrm{d}x. \tag{98.2}$$

Using a new variable of integration $\xi = e\mathscr{E}x/(C - E_x)$, we obtain

$$S = \sqrt{(2\mu)}\frac{(C - E_x)^{\frac{3}{2}}}{e\mathscr{E}}\int_0^1 \sqrt{(1 - \xi)}\,\mathrm{d}\xi = \tfrac{2}{3}\sqrt{(2\mu)}\frac{(C - E_x)^{\frac{3}{2}}}{e\mathscr{E}}. \tag{98.3}$$

Thus the transmission coefficient D, for electrons the energy of whose motion along the axis OX is E_x, is

$$D(E_x) = D_0 \exp\left[-\frac{4}{3}\frac{\sqrt{(2\mu)}}{\hbar}\frac{(C - E_x)^{\frac{3}{2}}}{e\mathscr{E}}\right]. \tag{98.4}$$

This coefficient varies slightly with E_x, but since $C > E_x$ the transmission coefficient averaged with respect to electron energy will have the form

$$\bar{D} = \bar{D}_0\, e^{-\mathscr{E}_0/\mathscr{E}}, \tag{98.5}$$

where $\bar{D}_0$ and $\mathscr{E}_0$ are constants for a given metal. The cold-emission current is

$$J(\mathscr{E}) = J_0 \bar{D} = Ae^{-\mathscr{E}_0/\mathscr{E}}. \tag{98.6}$$

This relation between the current and the field is fully confirmed by experiments.[6]

99. A three-dimensional potential barrier. Quasistationary states

The discussion in Sections 97 and 98 of the problem of penetration through a potential barrier related to a flux of particles coming from infinity and encountering a potential barrier. Later (in the theory of radioactive decay and self-ionisation of atoms) we shall find cases of a flux of particles leaving some bounded region of space (an atomic nucleus or an atom) which is surrounded by a potential barrier.

Let a sphere with centre O and radius r_0 (Figure 74a) be a surface on which the

Fig. 74. A potential barrier bounding a closed region $r < r_0$.

[6] Done by P. I. Lukirskiĭ.

potential energy $U(r)$ takes a maximum value, so that $U < U_m$ for $r < r_0$ and for $r > r_0$. Figure 74b shows a corresponding graph of $U(r)$. Let us assume that we wish to find the penetration through the barrier of particles initially within it. In accordance with the assumption that there are no particles incident from outside (no 'bombardment'), we must take only the outgoing waves

$$\psi = Ce^{ikr}/r, \qquad k > 0 \tag{99.1}$$

beyond the barrier. We call this the *condition of emission*. Evidently Schrödinger's equation

$$i\hbar\partial\psi/\partial t = -(\hbar^2/2\mu)\nabla^2\psi + U(r)\psi \tag{99.2}$$

can only have non-stationary solutions in this case. For, if we apply the law of conservation of number of particles to a sphere of radius r, we have

$$\frac{\mathrm{d}}{\mathrm{d}t}\int_V \psi^*\psi\,\mathrm{d}v = -\int_S J_r\,\mathrm{d}s = -\int J_r r^2\,\mathrm{d}\Omega\,, \tag{99.3}$$

and from (99.1)

$$J_r = \frac{i\hbar}{2\mu}\left(\psi\frac{\partial\psi^*}{\partial r} - \frac{\partial\psi}{\partial r}\psi^*\right) = \frac{\hbar k\,|C|^2}{\mu r^2}, \tag{99.4}$$

so that

$$\frac{\mathrm{d}}{\mathrm{d}t}\int_V \psi^*\psi\,\mathrm{d}v = -\frac{\hbar k}{\mu}\int |C|^2\,\mathrm{d}\Omega < 0, \tag{99.5}$$

i.e. the mean number in the sphere volume V decreases, and ψ cannot be a harmonic function of time.

The problem of outflow of particles through a barrier can be solved by using Equation (99.2) with the initial condition such that the function $\psi(r, 0)$ is non-zero only within the barrier (to express the fact that the particle was within the barrier at $t = 0$). We can also, however, use a different and to some extent contrary condition by assuming that the outflow of particles has continued for some time and a considerable fraction of them are already outside the barrier. This approach will be considered in more detail. It is convenient because it allows the separation of the variables r and t in Equation (99.2). We can put immediately

$$\psi(\mathbf{r}, t) = \psi(\mathbf{r})\,e^{-iEt/\hbar}. \tag{99.6}$$

The quantity E is complex, and cannot be regarded as the energy of the particles (see below). We write[7]

$$E = E_0 - \tfrac{1}{2}i\hbar\lambda. \tag{99.7}$$

Then the mean number of particles in the volume V_0 within the barrier is, from (99.6) and (99.7),

$$N(t) = \int_{V_0} \psi^*\psi\,\mathrm{d}v = e^{-\lambda t}\int_{V_0} \psi^*(r)\,\psi(r)\,\mathrm{d}v\,,$$

[7] It is seen from (99.6) and (99.7) that if we take $\lambda = 0$ we obtain stationary states, which contradicts the emission condition by (99.5).

i.e.

$$N(t) = e^{-\lambda t} N(0). \tag{99.8}$$

We shall call the quantity λ the *decay constant*. Substitution of (99.6) in (99.2) gives

$$-(\hbar^2/2\mu)\nabla^2\psi + U(r)\psi = (E_0 - \tfrac{1}{2}i\hbar\lambda)\psi. \tag{99.9}$$

In order to see the essential features of the problem, let us consider an idealised example, taking the barrier $U(r)$ to have the form shown in Figure 75, and for simplicity discuss states with zero orbital angular momentum ($l = 0$). Then, putting

$$\psi(r) = u(r)/r, \tag{99.10}$$

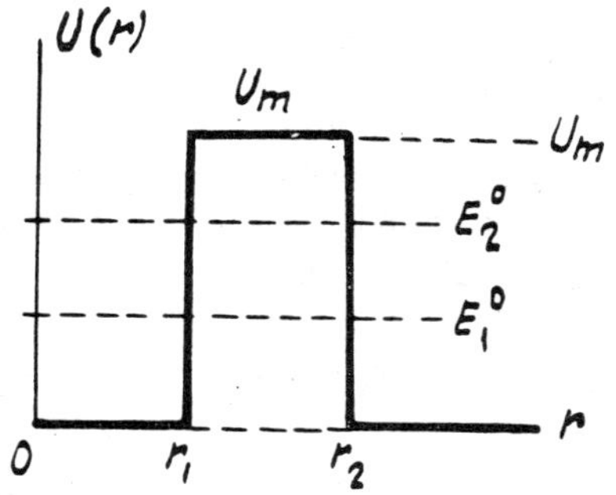

Fig. 75. A potential barrier bounding a closed region $r < r_1$ and having a simple rectangular shape.*

we obtain from (99.9)

$$-\frac{\hbar^2}{2\mu}\frac{d^2u}{dr^2} + U(r)u = (E_0 - \tfrac{1}{2}i\hbar\lambda)u. \tag{99.11}$$

In accordance with our assumption regarding the form of $U(r)$, Equation (99.11) separates into three:

$$u'' + k^2u = 0 \qquad (0 < r < r_1), \tag{99.12}$$

$$u'' - q^2u = 0 \qquad (r_1 < r < r_2), \tag{99.12'}$$

$$u'' + k^2u = 0 \qquad (r_2 < r), \tag{99.12''}$$

where

$$k^2 = (2\mu/\hbar^2)(E_0 - \tfrac{1}{2}i\hbar\lambda), \qquad q^2 = (2\mu/\hbar^2)(U_m - E_0 + \tfrac{1}{2}i\hbar\lambda). \tag{99.13}$$

The solutions of these equations are of the form

$$u_1 = A'e^{ikr} + Be^{-ikr}(0 < r < r_1), \tag{99.14}$$

* The potential curve Or_1U_m corresponds to the potential well obtained by moving r_2 to infinity, $E_1{}^0$, $E_2{}^0$ are the energy levels in such a well.

$$u_{\mathrm{II}} = \alpha e^{qr} + \beta e^{-qr} (r_1 < r < r_2), \tag{99.14'}$$

$$u_{\mathrm{III}} = a e^{ikr} + b e^{-ikr} (r_2 < r). \tag{99.14''}$$

The condition for ψ to be finite at $r = 0$ shows that

$$A' = B, \qquad u_{\mathrm{I}} = A \sin kr. \tag{99.15}$$

Moreover, the emission condition gives $b = 0$ (outgoing waves only). The boundary conditions at $r = r_1$ and $r = r_2$ reduce, as shown in Section 96, to the equality of the wave functions and their first derivatives:

$$A \sin kr_1 = \alpha e^{qr_1} + \beta e^{-qr_1}, \tag{99.16}$$

$$kA \cos kr_1 = q (\alpha e^{qr_1} - \beta e^{-qr_1}) \qquad \text{for} \qquad r = r_1, \tag{99.16'}$$

$$\alpha e^{qr_2} + \beta e^{-qr_2} = a e^{ikr_2}, \tag{99.17}$$

$$q (\alpha e^{qr_2} - \beta e^{-qr_2}) = ika e^{ikr_2} \qquad \text{for} \qquad r = r_2. \tag{99.17'}$$

This time we have four homogeneous equations for four coefficients A, α, β, a. It is therefore necessary for the determinant Δ of Equations (99.16)–(99.17′) to be zero. After a straightforward calculation we find

$$\Delta(k) = e^{-ql} \left(\frac{q}{k} \tan kr_1 - 1 \right) \frac{ik + q}{ik - q} + e^{ql} \left(\frac{q}{k} \tan kr_1 + 1 \right) = 0, \tag{99.18}$$

where l denotes the width $r_2 - r_1$ of the barrier. Equation (99.18) is a transcendental equation for k. We may determine its roots approximately if we assume that $ql \gg 1$. Then in the zero-order approximation the term in e^{-ql} may be omitted, leaving

$$(q/k) \tan kr_1 + 1 = 0. \tag{99.19}$$

This is the exact equation for the eigenvalues for the potential well Or_1U_m shown in Figure 75 and obtained from the potential barrier in that figure by making $r_2 \to \infty$. In such a potential well there are discrete energy levels (for $E < U_m$). If the roots of Equation (99.19) are denoted by $k_{01}, k_{02}, \ldots, k_{0n}, \ldots$, then the energies of these levels are, according to (99.13),

$$E_{0n} = \hbar^2 k_{0n}^2 / 2\mu + \tfrac{1}{2} i\hbar\lambda, \qquad n = 1, 2, 3, \ldots. \tag{99.20}$$

The roots are real[8] if $\lambda = 0$, and in order of magnitude are equal to $1/r_1$. In that case we have stationary solutions. When the width of the barrier is finite the asymptotic behaviour of the potential energy is such that $U(r) < E$ for $r \to \infty$, and instead of the discrete spectrum (99.20) we obtain a continuous spectrum. The emission condition, however, selects from the continuous spectrum levels close to E_{0n}, but these are no longer stationary ($\lambda_n \neq 0$). When λ_n is small the levels are almost stationary, and these are the *quasistationary levels* mentioned in Section 67. To find the value of λ_n, assuming it small, we expand the term in e^{ql} in (99.18) in powers of $\Delta k =$

[8] For a sufficiently deep well ($U_m \to \infty$), $q_m \to \infty$, and instead of (99.19) we have $\tan kr_1 = 0$, or $k_n r_1 = n\pi$ ($n = 1, 2, 3, \ldots$).

$k - k_0$, where k_0 is one root of Equation (99.19) for stationary states of the potential well, and in the term in e^{-ql} we put $k = k_0$; since $dq/dk = -k/q$, $\tan kr_1 = -k_0/q_0$, we obtain

$$2e^{-q_0 l}\frac{(q_0 + ik_0)^2}{q_0^2 + k_0^2} + e^{q_0 l}\frac{k_0^2 + q_0^2}{k_0 q_0^2}(1 + q_0 r_1)\Delta k + \ldots = 0.$$

This gives Δk.

The small correction to the real part of k_0 is of no interest and may also be omitted; the imaginary part is

$$Im(k - k_0) = Im\,\Delta k = k_0\frac{4e^{-2q_0 l}k_0 q_0^3}{(q_0^2 + k_0^2)^2(1 + q_0 r_1)}. \tag{99.21}$$

Neglecting also the small correction to the real part of k in (99.13), we can put $2\mu E_0/\hbar^2 = k_0^2$. From (99.13)

$$k = k_0 - i\lambda\mu/2k_0\hbar, \tag{99.22}$$

and comparison of this with the above expression for Δk gives

$$\lambda = \frac{\hbar}{\mu}\frac{8k_0^3 q_0^3}{(q_0^2 + k_0^2)^2(1 + q_0 r_1)}e^{-2q_0 l}. \tag{99.23}$$

Since $\hbar k_0/\mu$ is the velocity v_0 of a particle within the barrier and $k_0 \approx 1/r_1 = 1/r_0$ (r_0 being the radius of the well), we have from (99.23) and (99.13)

$$\lambda \approx \frac{v_0}{2r_0}e^{-2\sqrt{[2\mu(U_m - E)]}l/\hbar}. \tag{99.24}$$

This formula has a simple interpretation: $v_0/2r_0$ is the number of times a particle strikes the inner wall of the barrier per second, and the exponential factor is the transmission coefficient.

Some further properties of the problem under consideration may be noted. The imaginary value of the wave vector $\mathbf{k}$ has the result that the intensity $(a/r)e^{ikr}$ of the emitted wave increases without limit away from the potential barrier:

$$\psi_{\rm III} = \frac{ae^{ikr}}{r} = a\frac{e^{ik_0 r + \lambda\mu r/2k_0\hbar}}{r}.$$

The increase of $\psi_{\rm III}$ results from the requirement that there should be only emission, and corresponds to the fact that at large distances we find particles which emerged at an earlier time when the intensity $|\psi_{\rm I}|^2$ within the barrier was greater. In our method of solution, however, we have neglected the facts that the emission must actually have begun at some instant (and not have lasted for all time since $t = -\infty$) and that at the time when emission began $|\psi_{\rm I}|^2$ was finite. Hence our conclusion that $\psi_{\rm III} \to \infty$ when $r \to \infty$, which relates to particles which emerged a long time previously, is invalid, and the solution found is valid only for small r, viz. for $r \ll 2k_0\hbar/\lambda\mu$.

It may also be noted that the term 'imaginary energy' is often used in the literature in connection with Formula (99.7). It should be borne in mind that this terminology is purely formal. The state

$$\psi(\mathbf{r}, t) = \psi_0(\mathbf{r})\, e^{-iE_0 t/\hbar - \frac{1}{2}\lambda t} \tag{99.25}$$

which we have found is not a stationary state with a definite value of the energy; stationary states have a harmonic dependence on time.

In order to determine the probability of finding a given value of the energy E in this state, we must expand $\psi(\mathbf{r}, t)$ in terms of the eigenfunctions $\psi_E(\mathbf{r})$ of the operator $\boldsymbol{H}$. Since $U(r) > 0$, the eigenvalues of this operator form a continuous spectrum $0 \leqslant E < \infty$ (cf. Section 49). If we write

$$\psi(\mathbf{r}, t) = \int_0^\infty C(E) e^{-iEt/\hbar} \psi_E(\mathbf{r}) \, \mathrm{d}E, \tag{99.26}$$

then $w(E)\mathrm{d}E = |C(E)|^2 \, \mathrm{d}E$ gives the required probability. We cannot, however, use the function $\psi(\mathbf{r}, t)$ (99.25) to calculate $C(E)$, since it is valid only when r is not too large. We therefore proceed by supposing that $\psi(\mathbf{r}, t)$ has the correct behaviour at infinity, and that the initial function $\psi(\mathbf{r}, 0)$ is appreciably different from zero only within the barrier, so that the form of the function $\psi(\mathbf{r}, 0)$ corresponds to the fact that at $t = 0$ the particle is within the barrier. The amplitude $a(t)$ of the state $\psi(\mathbf{r}, 0)$ in the state $\psi(\mathbf{r}, t)$ is

$$a(t) = \int \psi(\mathbf{r}, t) \psi^*(\mathbf{r}, 0) \, \mathrm{d}v. \tag{99.27}$$

Substituting $\psi(\mathbf{r}, t)$ and $\psi^*(\mathbf{r}, 0)$ from (99.26) and using the orthogonality of the functions $\psi_E(\mathbf{r})$, we find

$$a(t) = \int_0^\infty e^{-iEt/\hbar} C(E) C^*(E) \, \mathrm{d}E = \int_0^\infty e^{-iEt/\hbar} w(E) \, \mathrm{d}E. \tag{99.28}$$

The quantity $P(t) = |a(t)|^2$ evidently gives the law of decay of the state $\psi(\mathbf{r}, 0)$. It is seen that the form of this law is determined by the distribution of energy $w(E)\mathrm{d}E$ in the initial state.[9]

Let us now return to the problem, and take $\psi(\mathbf{r}, 0) = \psi_0(\mathbf{r})$ within the barrier and $\psi(\mathbf{r}, 0) = 0$ outside it. Substituting $\psi(\mathbf{r}, t)$ from (99.25) in (99.27), we can then ignore the increase of $\psi_0(\mathbf{r})$ outside the barrier, since $\psi(\mathbf{r}, 0) = 0$ there. Since $\psi(\mathbf{r}, 0)$ and $\psi_0(\mathbf{r})$ are the same within the barrier, if we assume that $\psi(\mathbf{r}, 0)$ is normalised to unity we have

$$a(t) = e^{-iE_0 t/\hbar - \frac{1}{2}\lambda t}. \tag{99.29}$$

From (99.28) we can now see that $w(E)\mathrm{d}E$ must be[10]

$$w(E)\mathrm{d}E = \frac{\lambda \hbar}{2\pi} \frac{\mathrm{d}E}{(E - E_0)^2 + \frac{1}{4}\lambda^2 \hbar^2}, \tag{99.30}$$

i.e. we have a dispersion formula for the energy distribution. The quantity $\Delta E = \frac{1}{2}\lambda\hbar$ is called the *width* of the quasistationary level E_0. If $\tau = 1/\lambda$ denotes the mean lifetime of a particle in the state $\psi(\mathbf{r}, 0) = \psi_0(\mathbf{r})$, we obtain

$$\tau \cdot \Delta E = \tfrac{1}{2}\hbar, \tag{99.31}$$

a relation between the width of a quasistationary level and the lifetime of a particle in that level.

[9] This theorem is due to Krylov and Fok [56].
[10] The integral in (99.28) is here easily calculated by using residues in the complex plane.

100. The theory of α decay

It is well known that many radioactive elements decay by emitting α particles. After leaving the nucleus, the α particle, which has a double positive charge $(+ 2e)$, is accelerated by the Coulomb field of the nucleus, whose charge will be denoted by Ze (Z being the atomic number of the element *after* the emission of the α particle; $Z = Z' - 2$, where Z' is the atomic number before the radioactive decay).

On account of the great stability of the α particle we can suppose that it exists in the nucleus as an independent entity, being one of the basic units of which the atomic nucleus is built up.[11] Evidently the α particle can exist for long in the nucleus only if the region near the nucleus is a minimum for the potential energy of the α particle. The Coulomb potential energy of the α particle is $2Ze^2/r$, where r is the distance from the nucleus to the α particle, and this increases monotonically as the nucleus is approached, as shown by the broken line in Figure 76. A minimum of

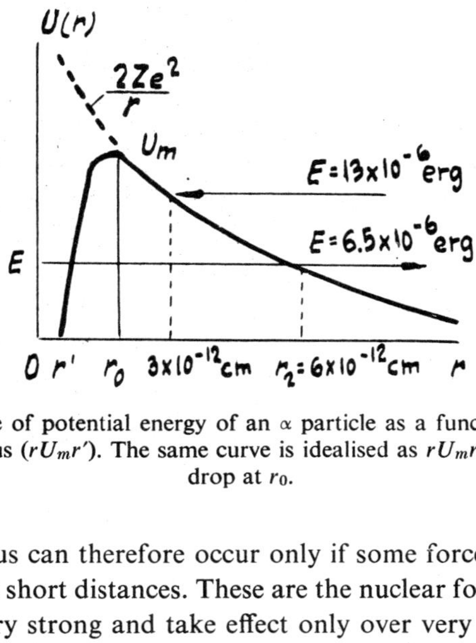

Fig. 76. Curve of potential energy of an α particle as a function of distance from the nucleus (rU_mr'). The same curve is idealised as rU_mr_0, with a sudden drop at r_0.

energy near the nucleus can therefore occur only if some forces other than electrical act on the α particle at short distances. These are the nuclear forces which act between nucleons; they are very strong and take effect only over very short distances. These forces bring about a change from the Coulomb repulsion to a strong attraction near the nucleus, as shown by the continuous curve in Figure 76. Such behaviour of the potential is described as forming a *potential well.* When such forces act, an α particle in the region $r < r_0$, i.e. in the field of attractive forces, will remain for a long time within the nucleus.

The mechanism of α decay has been a problem since the days of Kelvin, who supposed that the particles emitted by a radioactive element 'boil' within the po-

[11] This supposition is not obligatory. It is possible that, before its emergence from the nucleus, the α particle is formed from simpler particles (neutrons and protons). In what follows we shall assume that it exists permanently in the nucleus.

tential well. From time to time one of the particles acquires an amount of energy above the average, overcomes the barrier and is accelerated beyond it to higher energies by the repulsive field.

However, Rutherford showed that this simple picture is contradicted by an experiment which we shall now describe.

Rutherford bombarded atoms of radioactive uranium with α particles from thorium C'. The energy of thorium C' α particles is 13×10^{-6} erg. Such particles can overcome the Coulomb repulsion and come very close to the nucleus. To estimate the distance of closest approach r_1 we note that this is the distance at which the potential energy $2Z'e^2/r_1$ of the particle equals its original kinetic energy, i.e. $2Z'e^2/r_1 = 13 \times 10^{-6}$ erg. Z', the atomic number of uranium, is 92, and this gives $r_1 = 3 \times 10^{-12}$ cm.

Observation shows that the scattering of such particles is exactly as it should be if a Coulomb field is acting on the α particle. This means that nuclear forces begin to act on the α particle at distances less than 3×10^{-12} cm. Hence the α particles in the nucleus are in a region of radius less than this value.

Uranium itself, however, is a radioactive element and emits α particles. A measurement of their energy shows that it is 6.6×10^{-6} erg. These α particles leave the nucleus, i.e. escape from distances less than 3×10^{-12} cm. Then, undergoing acceleration in the Coulomb field, they should acquire an energy equal to the height of the potential barrier (see Figure 76) and certainly greater than 13×10^{-6} erg. The result is as if they left from a distance $r = 6 \times 10^{-12}$ cm. Thus experiment led, on the basis of classical physics, to a paradox: it was necessary to suppose that the Coulomb field of the nucleus acts on α particles incident from outside, but not on those leaving the nucleus, or else that the law of conservation of energy does not apply to radioactive decay.

The resolution of this paradox is provided by quantum mechanics, which leads to the possibility of the tunnel effect through a potential barrier separating the region of attraction ($r < r_0$) from that of repulsion ($r > r_0$).

This completely removes the paradox, since a particle within the nucleus may have an energy less than the height of the barrier and nevertheless penetrate through the barrier, but a particle incident from outside will be captured by the nucleus only very rarely (since the transparency of the barrier is small and the particle spends very little time near the nucleus). Hence the scattering of α particles incident from outside will be due to the Coulomb forces which act beyond the barrier. The assumed smallness of the transparency of the barrier is in accordance with the fact that the half-lives of α decay are very long.

By applying the theory of potential-barrier penetration it is easy to put the idea described above into mathematical form and derive an expression for the decay constant λ. This constant is defined as follows. If the number of atoms which have not yet decayed at time t is N, then during a time interval $\mathrm{d}t$ the mean number $\mathrm{d}N$ of atoms decaying is

$$\mathrm{d}N = -\lambda N \,\mathrm{d}t, \qquad \text{and} \qquad N(t) = N(0)\, e^{-\lambda t}. \tag{100.1}$$

To calculate the decay constant λ we can use the quantum theory of particle penetration through a barrier given in the preceding section. According to this theory, an α particle within the nucleus is to be regarded as in a 'quasistationary' state. Denoting the velocity of the particle in this state by v_i, the radius of the barrier by r_0 and the transmission coefficient by D, we obtain

$$\lambda = v_i D/2r_0 . \tag{100.2}$$

It remains to calculate D. On account of the greater complexity of the barrier we have instead of (99.24)

$$\lambda = \frac{v_i}{2r_0} D_0 \exp\left[-\frac{2}{\hbar}\int_{r_1}^{r_2} \sqrt{\{2\mu[U(r) - E]\}}\, dr\right]; \tag{100.3}$$

cf. (96.24). From Figure 76 it follows that the first turning point r_1 is r_0 (the radius of the nucleus), and the second turning point (r_2) is given by the condition

$$2Ze^2/r_2 = E, \qquad r_2 = 2Ze^2/E, \tag{100.4}$$

Thus

$$S = \int_{r_1}^{r_2} \sqrt{\{2\mu[U(r) - E]\}}\, dr = \sqrt{(2\mu)} \int_{r_0}^{2Ze^2/E} \sqrt{\left[\frac{2Ze^2}{r} - E\right]} dr. \tag{100.5}$$

Using a new variable $\xi = r/r_2$, we obtain

$$S = 2Ze^2 \sqrt{\frac{2\mu}{E}} \int_{r_0/r_2}^{1} \sqrt{\left(\frac{1}{\xi} - 1\right)} d\xi, \tag{100.5'}$$

and finally putting $\xi = \cos^2 u$, we easily find the value of the integral, so that

$$\left.\begin{aligned} S &= Ze^2 \sqrt{\frac{2\mu}{E}}(2u_0 - \sin 2u_0), \\ \cos^2 u_0 &= r_0/r_2 = r_0 E/2Ze^2 . \end{aligned}\right\} \tag{100.6}$$

Using the fact that the ratio r_0/r_2 is less than unity, we expand u_0 and $\sin 2u_0$ in powers of r_0/r_2, and need take only the first two terms. This gives

$$S = 2\pi e^2 Z/v - 2e\sqrt{\mu}\sqrt{(Zr_0)}, \tag{100.7}$$

where v is the velocity far from the nucleus, equal to $\sqrt{(2E/\mu)}$. Thus the expression for the decay constant (100.3) becomes

$$\lambda = \frac{\hbar D_0}{2\mu r_0^2} \exp\left[-\frac{4\pi e^2 Z}{\hbar v} + \frac{4e\sqrt{\mu}}{\hbar}\sqrt{(Zr_0)}\right], \tag{100.8}$$

or

$$\log \lambda = -\frac{4\pi e^2 Z}{\hbar v} + \frac{4e\sqrt{\mu}}{\hbar}\sqrt{(Zr_0)} + \log \frac{\hbar D_0}{2\mu r_0^2}. \tag{100.9}$$

The most noteworthy conclusion which follows from this equation is the form of the relation between λ and the velocity v of the α particle. Such a relation was established experimentally by Geiger and Nuttall long before the quantum theory of the phenomenon was developed.

We also see that $\log \lambda$ depends on the atomic number Z of the element ($Z = Z' - 2$) and on the radius of the nucleus.

It is known from experiment that the decay constant varies over a very wide range, from 10^6 sec^{-1} to 10^{-18} sec^{-1}. If the parameters which determine λ had to vary over a similar range, the theory would certainly be incorrect. A noteworthy consequence of Formula (100.9) is that if the radii of nuclei are determined from experimental values of λ they are all found to lie within a narrow range from about 5×10^{-12} to 9×10^{-12} cm. The considerable difference in the value of λ for different elements is due not to a difference in nuclear radii but to that in the energy of the emergent particles. The slight dependence of λ on r_0 and the marked dependence on v must be regarded as a confirmation of the theory.[12]

101. Ionisation of atoms in strong electric fields

Just as a strong electric field detaches electrons from metals (cold emission, Section 98), it also detaches them from individual atoms of a gas. This phenomenon is sometimes called 'self-ionisation' of atoms, and the reason for it is easily understood if we consider the form of the potential energy of an electron in an atom when there is an external electric field. Let the potential energy of the electron in the absence of the external field be $U(r)$, and let the external electric field $\mathscr{E}$ be in the direction of the axis OZ. Then the total potential energy of the electron is

$$U'(r) = U(r) + e\mathscr{E}z. \tag{101.1}$$

Let us consider the form of the potential curve along the axis OZ ($x = y = 0, r = |z|$). In the absence of the external field ($\mathscr{E} = 0$), $U' = U(r)$ and is as shown by the broken line in Figure 77. The additional potential energy $e\mathscr{E}z$ in the external field is represented by the straight line aa'. The curve of the total potential energy U', obtained by addition, is shown in Figure 77 by the continuous line $a'b'$ and ab. We see that near the point z_0 a potential barrier is formed which divides two regions of space, an inner region $z > z_0$ and an outer region $z < z_0$, in each of which the potential energy U' is less than $U'(z_0) = U_m$. The diagram also shows two energy levels E' and E''. If the energy $E = E'' > U_m$, the electron will not remain within the atom but will move away into the region of negative z; if its energy $E = E' < U_m$, then according to the laws of classical mechanics the electron will remain in the inner region. According to quantum mechanics, penetration through the barrier still occurs in this case. Thus the situation is entirely analogous to that which occurs in radioactive decay.

It is now easy to see the reason for ionisation of the atoms by the field. When the field is applied, a barrier is formed through which electrons penetrate into the outer

[12] Details of the theory of radioactive decay are given in [28]; [94], p. 279.

region. If the height U_m of the barrier is less than the energy of the electron, the particles will emerge even according to classical mechanics (passage 'above the barrier'). Thus classical mechanics also leads to the possibility of ionisation of an atom by an external electric field. The only difference is that, according to the laws of quantum mechanics, this ionisation should occur at lower values of the field than in classical mechanics, because in quantum mechanics the possibility of ionisation does not require the barrier to be lower than the energy of the electron. It is clear, however, that for weak fields the barrier will be very wide and its transparency will be very small.

The phenomenon of self-ionisation can be observed as follows. Let us consider some spectral line due to an electron transition from E' to E_0 (see Figure 77). As the electric field increases this line will be displaced by the Stark effect, and if the

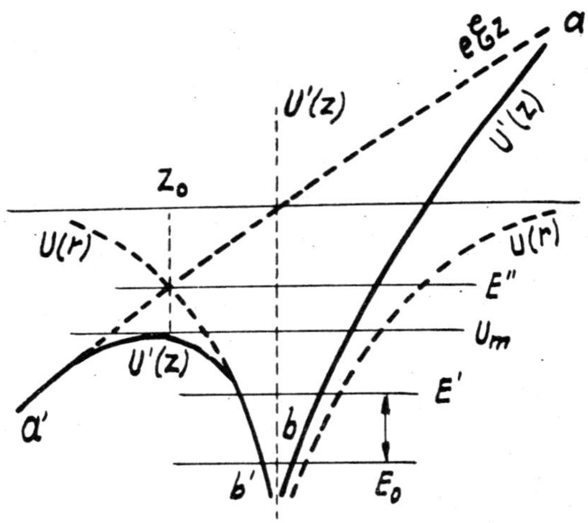

Fig. 77. Addition of atomic and external fields. A potential barrier is formed in the region of z_0.

field becomes so large that the transparency of the barrier is high, an electron in the state E' will more often leave the atom through the barrier (ionisation) than fall to the lower state (E_0) with emission of light. In consequence the spectral line will become less intense and finally disappear. This behaviour can be observed in the Balmer series of the hydrogen atom.[13]

In order to be able to examine the effect of electric fields of varying strength, the different parts of the spectral line are caused to be due to light from atoms in fields of different strengths. In a luminous gas the electric field increases in a direction parallel to the spectrograph slit (up to some limit, after which it drops again). Figure

[13] It may be noted that an observation of the number of electrons detached by the field is difficult in this case, since in gas-discharge conditions it is difficult to establish what causes are responsible for the increased electron current.

53 (Section 72) shows the results of such an experiment. The letters β, γ, δ, ε denote the lines of the Balmer series (H_β: $n = 4 \rightarrow n = 2$; H_γ: $n = 5 \rightarrow n = 2$; H_δ: $n = 6 \rightarrow n = 2$; H_ε: $n = 7 \rightarrow n = 2$). The applied electric field increases upwards. The white lines on the photograph are lines of equal field strength. The photograph shows that the lines are initially split. The splitting increases with the field; the position of the line of maximum field is easily seen from the splitting of the H_β line. At some value of the field the spectral line disappears.

A comparison of the lines β, γ, δ and ε shows that they disappear in the order ε, δ, γ (β does not disappear entirely at the fields used). This is the order of decreasing energy of the excited state. Figure 77 shows that, the higher the energy of the electron, the smaller the width and height of the barrier, i.e. the greater its transparency. Thus the observed order of disappearance of the spectral lines is in complete agreement with our interpretation of this phenomenon as a result of the tunnel effect. The fact that the red components of the split lines disappear before the violet components is also entirely explained if the wave functions of the electron are considered in more detail: the states which correspond to lines shifted to the red have the property that in them the electron cloud is denser in the barrier region than for states corresponding to the violet components. For this reason ionisation is favoured.

We may formulate somewhat more precisely the conditions under which a spectral line may be expected to disappear in an electric field. Let the probability of an optical transition of the electron to a lower state be $1/\tau$ (τ being the lifetime in the excited state, 10^{-8} sec). The probability of the tunnel effect (ionisation) is equal (as in the calculation of radioactive decay) to the number of collisions of the electron with the inner wall of the potential barrier per second, multiplied by the transmission coefficient D. The number of collisions is in order of magnitude $v/2r_0$, where v is the velocity of the electron and r_0 the radius of the barrier, which is approximately equal to the orbit radius a. The velocity is, again in order of magnitude, $v = \sqrt{(2|E|/\mu)}$, where $|E|$ is the electron energy and μ the electron mass.

Hence

$$v/2r_0 = \sqrt{(e^2/\mu a^3)} \approx 10^{16}\,\text{sec}^{-1}, \tag{101.2}$$

since $E = -e^2/2a$, $a = \hbar^2/\mu e^2$. The probability of self-ionisation is $10^{16}\ D\ \text{sec}^{-1}$. In order that this should predominate we must have $1/\tau < D \cdot 10^{16}$, or $D > 10^{-8}$. This is the condition for the spectral line to disappear.

The quantitative theory of self-ionisation is in good agreement with experiment.[14]

[14] See [5], pp. 412–415.

CHAPTER XVII

THE MANY-BODY PROBLEM

102. General remarks on the many-body problem

The quantum mechanics of a single particle in an external field can be generalised to the motion of several particles. To do so it is sufficient, as in classical mechanics, to consider a system of N particles as a single particle with $3N$ degrees of freedom (neglecting the spin of the particles, otherwise with $4N$ degrees of freedom). All the general results of quantum mechanics which are valid for systems with several degrees of freedom can be applied immediately to a system of N particles. There are, nevertheless, certain properties peculiar to many-particle systems which must receive special consideration. Among these there are some of particular importance for systems consisting of identical particles, and in what follows we shall be especially concerned with such systems, the properties of which form one of the most remarkable topics in quantum mechanics. For the present, however, we shall consider some problems which are common to all systems of particles.

Can an assembly of particles always be regarded as a mechanical system with an appropriately large number of degrees of freedom? The answer must be negative. The treatment of a system of particles with co-ordinates $x_1, y_1, z_1; x_2, y_2, z_2; \ldots; x_N, y_N, z_N$ as a mechanical system with $3N$ degrees of freedom is possible only if there are no retarded forces between the particles (or if an approximate treatment of such forces is used). In other words, all forces of interaction must depend only on the instantaneous values of the mechanical quantities belonging to the particles (e.g. on their co-ordinates and velocities at a given instant), and not on their previous values as they do when retarded forces are present. This condition is not peculiar to quantum mechanics; it is the same in classical mechanics.

The condition may be explained for the example of electromagnetic forces. Let the distance between the jth and kth particles be r_{jk}. Then the time needed for an electromagnetic perturbation to be propagated from one particle to the other is $\tau = r_{jk}/c$, where c is the velocity of light. If forces can be regarded as instantaneous, the distance between the particles must not vary greatly in the time τ. If the relative velocity of the particles along r_{jk} is v_{jk}, the change in r_{jk} in time τ is $\Delta r_{jk} = v_{jk}\tau = v_{jk}r_{jk}/c$. Thus the condition is

$$v_{jk}r_{jk}/c \ll r_{jk}, \qquad \text{i.e.} \qquad v_{jk} \ll c$$

The relative velocities of the particles must therefore be much less than the velocity of light c. This can be concisely expressed by saying that only non-relativistic velocities are considered.

If $v \sim c$, we must take into account both relativistic and quantum effects, and moreover we must use not only the mechanical equations for the particles but also the equations of the electromagnetic field, which govern the propagation of interactions from one particle to another. The resulting problems are beyond the scope of this book and have in fact not yet been completely resolved.[1]

If $v \ll c$, however, we can regard the quantum mechanics of a system of particles as the mechanics of a single particle with a large number of degrees of freedom.

If we have N particles with co-ordinates x_k, y_k, z_k $(k = 1, 2, 3, \ldots, N)$ and masses m_k, the wave function ψ will be, as usual, a function of the co-ordinates of all the degrees of freedom of the system and of the time t, i.e. a function of $3N + 1$ arguments[2]:

$$\psi = \psi(x_1, y_1, z_1, \ldots, x_k, y_k, z_k, \ldots, x_N, y_N, z_N, t). \tag{102.1}$$

It is thus defined in a space of $3N$ dimensions, called the *configuration space* of the system. The name of this fictitious space derives from the fact that specifying the co-ordinates of a point in this space is equivalent to specifying the three-dimensional co-ordinates (x_k, y_k, z_k) for all the particles $(k = 1, 2, 3, \ldots, N)$ in the system considered, and thus defines the position or configuration of all the particles in the system in three-dimensional space. The point in configuration space with the $3N$ co-ordinates $(x_1, y_1, z_1, \ldots, x_N, y_N, z_N)$ is therefore said to *represent* the system.

Let $\mathrm{d}\Omega$ denote an infinitesimal volume element in configuration space:

$$\mathrm{d}\Omega = \mathrm{d}x_1\,\mathrm{d}y_1\,\mathrm{d}z_1 \ldots \mathrm{d}x_k\,\mathrm{d}y_k\,\mathrm{d}z_k \ldots \mathrm{d}x_N\,\mathrm{d}y_N\,\mathrm{d}z_N. \tag{102.2}$$

Then the quantity

$$w(x_1, y_1, z_1, \ldots, x_k, y_k, z_k, \ldots, x_N, y_N, z_N, t)\,\mathrm{d}\Omega = \psi^*\psi\,\mathrm{d}\Omega \tag{102.3}$$

is the probability that the representative point lies in the volume element $\mathrm{d}\Omega$ in configuration space at time t, i.e. the probability of a configuration of the system in which at time t the co-ordinates of the first particle are between x_1 and $x_1 + \mathrm{d}x_1$, y_1 and $y_1 + \mathrm{d}y_1$, and z_1 and $z_1 + \mathrm{d}z_1$, those of the kth particle between x_k and $x_k + \mathrm{d}x_k$, y_k and $y_k + \mathrm{d}y_k$, and z_k and $z_k + \mathrm{d}z_k$, and so on.

Together with the volume element (102.2) we may consider volume elements in subspaces, of the type $\mathrm{d}\Omega_k$, $\mathrm{d}\Omega_{kj}$, ..., defined by

$$\mathrm{d}\Omega = \mathrm{d}x_k\,\mathrm{d}y_k\,\mathrm{d}z_k\,\mathrm{d}\Omega_k, \tag{102.4}$$

$$\mathrm{d}\Omega = \mathrm{d}x_k\,\mathrm{d}y_k\,\mathrm{d}z_k\,\mathrm{d}x_j\,\mathrm{d}y_j\,\mathrm{d}z_j\,\mathrm{d}\Omega_{kj}, \text{etc.} \tag{102.4'}$$

Integrating (102.3) over the co-ordinates of all particles except the kth, i.e. over $\mathrm{d}\Omega_k$, we find the probability that the co-ordinates of the kth particle lie between x_k and $x_k + \mathrm{d}x_k$, y_k and $y_k + \mathrm{d}y_k$, and z_k and $z_k + \mathrm{d}z_k$ irrespective of the positions of

[1] See [47, 29 (Chapter XII), 89] and especially [1].

[2] In order not to complicate the problem, we here ignore the spin of the particles.

the other particles, i.e. the probability that the kth particle is at a given point in space. Denoting this probability by $w(x_k, y_k, z_k, t)$, we have

$$w(x_k, y_k, z_k, t)\,dx_k\,dy_k\,dz_k = dx_k\,dy_k\,dz_k \int \psi^* \psi\, d\Omega_k. \tag{102.5}$$

Similarly, the quantity

$$\begin{aligned} &w(x_k, y_k, z_k, x_j, y_j, z_j, t)\,dx_k\,dy_k\,dz_k\,dx_j\,dy_j\,dz_j \\ &= dx_k\,dy_k\,dz_k\,dx_j\,dy_j\,dz_j \int \psi^* \psi\, d\Omega_{kj} \end{aligned} \tag{102.5'}$$

is the probability that the kth particle is near the point x_k, y_k, z_k and the jth particle is at the same time near the point x_j, y_j, z_j. Thus, if we know the wave function ψ in configuration space, we can find the probability of a given configuration of the system (102.3), the position probability of any one particle (102.5), that of a pair of particles (102.5′), and so on. Likewise, from the general formulae of quantum mechanics, by expanding ψ in terms of the eigenfunctions of any desired operator, we can calculate the probabilities of various values of any mechanical quantity.

We shall suppose that the wave function $\psi(x_1, \ldots, z_N, t)$, like that of a single particle, satisfies Schrödinger's equation:

$$i\hbar \partial\psi/\partial t = \boldsymbol{H}\psi\,; \tag{102.6}$$

$\boldsymbol{H}$ here denotes the Hamiltonian of the system. The latter can be written as

$$\begin{aligned} \boldsymbol{H} = &\sum_{k=1}^{N} \left\{ -\frac{\hbar^2}{2m_k} \nabla_k^2 + U_k(x_k, y_k, z_k, t) \right\} + \\ &+ \sum_{k \neq j = 1}^{N} U_{kj}(x_k, y_k, z_k, x_j, y_j, z_j), \end{aligned} \tag{102.6'}$$

where

$$\nabla_k^2 = \frac{\partial^2}{\partial x_k^2} + \frac{\partial^2}{\partial y_k^2} + \frac{\partial^2}{\partial z_k^2}$$

and $U_k(x_k, y_k, z_k, t)$ is the force function of the kth particle in the external field, and $U_{kj}(x_k, \ldots, z_j)$ the energy of interaction of the kth and jth particles. This is completely analogous to the classical Hamilton's function for a system of N particles with masses $m_1, \ldots, m_k, \ldots, m_N$, namely

$$H = \sum_{k=1}^{N} \left\{ \frac{p_k^2}{2m_k} + U_k(x_k, y_k, z_k, t) \right\} + \sum_{k \neq j = 1}^{N} U_{kj}(x_k, y_k, z_k, x_j, y_j, z_j).$$

This Hamiltonian is evidently a simple generalisation of the Hamiltonian for a single particle.[3]

[3] The Hamiltonian could also be written for the case where a magnetic field is present and spin is taken into account. It is equal to the sum of the Hamiltonians for the individual particles plus terms representing the interactions between the particles.

From Equation (102.6) we can derive the equation of continuity for the probability w in configuration space. To do so, we multiply (102.6) by ψ^* and subtract the complex conjugate. Using the Hamiltonian (102.6′), we obtain

$$i\hbar\frac{\partial}{\partial t}(\psi^*\psi) = -\tfrac{1}{2}\hbar^2\sum_{k=1}^{N}\frac{1}{m_k}\{\psi^*\nabla_k^2\psi - \psi\nabla_k^2\psi^*\}.$$

Putting

$$\mathbf{J}_k = \frac{i\hbar}{2m_k}\{\psi\nabla_k\psi^* - \psi^*\nabla_k\psi\}, \tag{102.7}$$

where ∇_k is the operator whose components are $\partial/\partial x_k$, $\partial/\partial y_k$, $\partial/\partial z_k$, we can write the above formula as

$$\frac{\partial w}{\partial t} + \sum_{k=1}^{N}\operatorname{div}_k\mathbf{J}_k = 0. \tag{102.8}$$

This equation shows that the change in the configuration probability w is due to the flux of this probability. Thus $\mathbf{J}_k$ is a function of the co-ordinates of all the particles (and of the time) and represents the current density due to the motion of the kth particle when the co-ordinates of the other $N-1$ particles are fixed. In order to find the current density $\mathbf{j}_k$ of the kth particle for any positions of the other particles, we must integrate (102.7) over all co-ordinates except those of the kth particle:

$$\mathbf{j}_k(x_k, y_k, z_k, t) = \int\mathbf{J}_k(x_1, \ldots, x_k, y_k, z_k, \ldots, z_N, t)\,\mathrm{d}\Omega_k. \tag{102.9}$$

This current also satisfies the equation of continuity, but in three-dimensional space: integrating (102.8) with respect to $\mathrm{d}\Omega_k$, we obtain

$$\int\frac{\partial}{\partial t}w(x_1, \ldots, z_N, t)\,\mathrm{d}\Omega_k$$
$$= \frac{\partial}{\partial t}\int w(x_1, \ldots, z_N, t)\,\mathrm{d}\Omega_k = \frac{\partial}{\partial t}w(x_k, y_k, z_k, t).$$

Also

$$\sum_{k'=1}^{N}\int\operatorname{div}_{k'}\mathbf{J}_{k'}\,\mathrm{d}\Omega_k = \int\operatorname{div}_k\mathbf{J}_k\,\mathrm{d}\Omega_k + \sum_{k'\neq k}^{N}\int\operatorname{div}_{k'}\mathbf{J}_{k'}\,\mathrm{d}\Omega_k.$$

Since $\mathrm{d}\Omega_k$ involves the co-ordinates of all particles except the kth (see (102.4)), the integrals of the form $\int\operatorname{div}_{k'}\mathbf{J}_{k'}\,\mathrm{d}\Omega_k$ can be transformed into surface integrals and are zero if ψ vanishes at infinity; in the integral $\int\operatorname{div}_k\mathbf{J}_k\mathrm{d}\Omega_k$ the differentiation and integration are with respect to different variables, we have

$$\int\operatorname{div}_k\mathbf{J}_k\,\mathrm{d}\Omega_k = \operatorname{div}_k\int\mathbf{J}_k\,\mathrm{d}\Omega_k = \operatorname{div}_k\mathbf{j}_k.$$

Thus the conservation law for each particle is obtained:

$$\partial w(x_k, y_k, z_k, t)/\partial t + \operatorname{div}_k\mathbf{j}_k(x_k, y_k, z_k, t) = 0, \tag{102.10}$$

formulated in three-dimensional space (x_k, y_k, z_k).

103. The law of conservation of the total momentum of a system of microparticles

In classical mechanics the total momentum of a system of particles subject only to internal forces remains constant, and the centre of mass of the system moves uniformly in a straight line. If there are external forces, however, the change in the total momentum per unit time is equal to the resultant of all the external forces acting on the particles of the system. We shall show that these results of classical mechanics remain valid in the quantum region. To do so, we define the total-momentum operator $\vec{\boldsymbol{P}}$ for all the particles in the system as the sum of the momentum operators $\vec{\boldsymbol{P}}_k$ for all the particles, $k = 1, 2, \ldots, N$:

$$\vec{\boldsymbol{P}} = \sum_{k=1}^{N} \vec{\boldsymbol{P}}_k = -i\hbar \sum_{k=1}^{N} \nabla_k . \tag{103.1}$$

The operator of the time derivative of the momentum $\vec{\boldsymbol{P}}$ is found as follows. According to the general formulae of quantum mechanics,

$$\mathrm{d}\vec{\boldsymbol{P}}/\mathrm{d}t = (i/\hbar)(\boldsymbol{H}\cdot\vec{\boldsymbol{P}} - \vec{\boldsymbol{P}}\cdot\boldsymbol{H}). \tag{103.2}$$

Substituting $\boldsymbol{H}$ from (102.6′) and noting that $\vec{\boldsymbol{P}}$ commutes with the operator of the kinetic energy of the particles,

$$\boldsymbol{T} = -\tfrac{1}{2}\hbar^2 \sum_{k=1}^{N} \frac{1}{m_k}\nabla_k^2 ,$$

we find that

$$\frac{\mathrm{d}\vec{\boldsymbol{P}}}{\mathrm{d}t} = \Big\{\Big(\sum_{k=1}^{N} U_k + \sum_{k\neq j=1}^{N} U_{kj}\Big)\Big(\sum_{k=1}^{N} \nabla_k\Big) - \Big(\sum_{k=1}^{N} \nabla_k\Big)\Big(\sum_{k=1}^{N} U_k + \\ + \sum_{k\neq j=1}^{N} U_{kj}\Big)\Big\}. \tag{103.2′}$$

Also,

$$U_k\Big(\sum_{k=1}^{N} \nabla_k\Big) - \Big(\sum_{k=1}^{N} \nabla_k\Big) U_k = -\nabla_k U_k . \tag{103.3}$$

Finally we calculate the commutator of the operator

$$\sum_{k=1}^{N} \nabla_k$$

and the mutual energy of the particles

$$\sum_{k\neq j} U_{kj} .$$

We assume that the forces between the particles depend only on the distances r_{kj} between the particles, i.e. $U_{kj} = U_{kj}(r_{kj})$. Then only those operators $\nabla_{k'}$ in the sum $\sum \nabla_{k'}$ act on U_{kj} for which $k' = k$ or $k' = j$, i.e. $\nabla_k + \nabla_j$ acts on U_{kj}; and

$$U_{kj}(\nabla_k + \nabla_j) - (\nabla_k + \nabla_j)\, U_{kj} = -\nabla_k U_{kj} - \nabla_j U_{kj}. \tag{103.4}$$

But

$$\nabla_k U_{kj} = \frac{\mathrm{d}U_{kj}}{\mathrm{d}r_{kj}} \nabla_k r_{kj} = \frac{\mathrm{d}U_{kj}\mathbf{r}_{kj}}{\mathrm{d}r_{kj}\mathbf{r}_{kj}},$$

$$\nabla_j U_{kj} = \frac{\mathrm{d}U_{kj}}{\mathrm{d}r_{kj}} \nabla_j r_{kj} = -\frac{\mathrm{d}U_{kj}\mathbf{r}_{kj}}{\mathrm{d}r_{kj}r_{kj}}.$$

Hence

$$\nabla_k U_{kj} + \nabla_j U_{kj} = 0. \tag{103.5}$$

This is the law of action and reaction, from which it follows that the commutator (103.4) is zero. Thus

$$\frac{\mathrm{d}\vec{\boldsymbol{P}}}{\mathrm{d}t} = -\sum_{k=1}^{N} \nabla_k U_k(x_k, y_k, z_k, t), \tag{103.6}$$

i.e. the operator of the time derivative of the total momentum is equal to the operator of the resultant force exerted on the system by external fields.

This result is entirely analogous to the classical theorem concerning the motion of the centre of mass of the system. The only difference is that in quantum mechanics it is formulated not for the mechanical quantities themselves but for operators representing these quantities, and therefore for mean values of quantities.

If there are no external forces ($U_k = C$), (103.6) shows that

$$\mathrm{d}\vec{\boldsymbol{P}}/\mathrm{d}t = 0, \tag{103.7}$$

i.e. the total momentum of a system of interacting particles is conserved if external forces are absent.

The operator equation (103.7), it may be recalled, signifies that (1) the mean value of the total momentum does not vary with time, (2) the probability $w(p')$ of a particular value p' also remains unchanged.

104. Motion of the centre of mass of a system of microparticles

We shall prove a theorem, important in applications, which states that the motion of the centre of mass of a system is independent of the relative motions of particles forming the system. To do so, we transform the Hamiltonian $\boldsymbol{H}$ of a system of particles subject only to internal forces:

$$H = -\tfrac{1}{2}\hbar^2 D + W, \tag{104.1}$$

$$D = \sum_{k=1}^{N} \frac{1}{m_k}\nabla_k^2 \quad \text{and} \quad W = \sum_{k \neq j = 1}^{N} U_{kj}(r_{kj}), \tag{104.2}$$

to new co-ordinates, viz. the co-ordinates X, Y, Z of the centre of mass of the system and $3N - 3$ relative co-ordinates. It is convenient to use what are called *Jacobi co-ordinates*, defined as follows:

$$\left.\begin{aligned} \xi_1 &= \frac{m_1 x_1}{m_1} - x_2 (= x_1 - x_2), \\ \xi_2 &= \frac{m_1 x_1 + m_2 x_2}{m_1 + m_2} - x_3, \\ \ldots & \qquad \ldots \\ \xi_j &= \frac{m_1 x_1 + \ldots + m_j x_j}{m_1 + m_2 + \ldots + m_j} - x_{j+1}, \\ \xi_N &= \frac{m_1 x_1 + \ldots + m_N x_N}{m_1 + \ldots + m_N} (= X). \end{aligned}\right\} \tag{104.3}$$

Similar formulae are valid for the axes OY and OZ:

$$\left.\begin{aligned} \eta_j &= \frac{m_1 y_1 + \ldots + m_j y_j}{m_1 + \ldots + m_j} - y_{j+1}, \qquad \eta_N = Y, \\ \zeta_j &= \frac{m_1 z_1 + \ldots + m_j z_j}{m_1 + \ldots + m_j} - z_{j+1}, \qquad \zeta_N = Z. \end{aligned}\right\} \tag{104.3'}$$

These formulae are a generalisation of the usual formulae for the co-ordinates of the centre of mass and relative co-ordinates of two particles. The Jacobi co-ordinates are orthogonal. By means of the usual rules for changing from differentiation with respect to one set of variables to that with respect to another set, it can be shown[4] that

$$D = \frac{1}{M}\nabla^2 + \sum_{j=1}^{N-1} \frac{1}{\mu_j}\nabla_j^2, \tag{104.4}$$

where

$$\nabla^2 = \frac{\partial^2}{\partial \xi_N^2} + \frac{\partial^2}{\partial \eta_N^2} + \frac{\partial^2}{\partial \zeta_N^2} = \frac{\partial^2}{\partial X^2} + \frac{\partial^2}{\partial Y^2} + \frac{\partial^2}{\partial Z^2}, \tag{104.5}$$

$$\nabla_j^2 = \frac{\partial^2}{\partial \xi_j^2} + \frac{\partial^2}{\partial \eta_j^2} + \frac{\partial^2}{\partial \zeta_j^2}, \tag{104.6}$$

[4] See Appendix XI.

M is the mass of the whole system, and μ_j the reduced mass of the $(j+1)$th particle and the centre of mass of the first j particles:

$$M = \sum_{k=1}^{N} m_k, \tag{104.7}$$

$$\frac{1}{\mu_j} = \frac{1}{\sum_{k=1}^{j} m_k} + \frac{1}{m_{j+1}}, \tag{104.8}$$

and

$$\sum_{k=1}^{N} \frac{\partial}{\partial x_k} = \frac{\partial}{\partial \xi_N} = \frac{\partial}{\partial X}. \tag{104.9}$$

These formulae show that the Hamiltonian (104.1) can be written as

$$\begin{aligned} \boldsymbol{H} = &-\frac{\hbar^2}{2M}\nabla^2 - \\ &-\sum_{j=1}^{N-1} \frac{\hbar^2}{2\mu_j}\nabla_j^2 + W(\xi_1, \dots, \xi_{N-1}, \eta_1, \dots, \eta_{N-1}, \zeta_1, \dots, \zeta_{N-1}), \end{aligned} \tag{104.10}$$

the operator

$$\boldsymbol{T} = -\frac{\hbar^2}{2M}\nabla^2 = -\frac{\hbar^2}{2M}\left(\frac{\partial^2}{\partial X^2} + \frac{\partial^2}{\partial Y^2} + \frac{\partial^2}{\partial Z^2}\right) \tag{104.11}$$

being the operator of the kinetic energy of the centre of mass of the whole system, and the operator

$$\boldsymbol{T}_i = -\sum_{j=1}^{N-1} \frac{\hbar^2}{2\mu_j}\nabla_j^2 \tag{104.12}$$

being the operator of the kinetic energy of the relative motion of the particles. It is important to note that the interaction energy W does not involve the co-ordinates of the centre of mass. By transforming $\xi_1, \dots, \xi_{N-1}, \eta_1, \dots, \eta_{N-1}, \zeta_1, \dots, \zeta_{N-1}$ to any new relative co-ordinates, $q_1, q_2, \dots, q_{3N-3}$, we do not alter $\boldsymbol{T}$. Hence (102.6′) can be replaced by

$$\boldsymbol{H} = -\frac{\hbar^2}{2M}\nabla^2 + \boldsymbol{H}_i(q_1, q_2, \dots, q_{3N-3}), \tag{104.13}$$

where $\boldsymbol{H}_i$ is the Hamiltonian for the relative motion and does not involve the co-ordinates of the centre of mass. From (104.9) and (103.1) we obtain a new expression for the total-momentum operator:

$$\boldsymbol{P}_x = -i\hbar\frac{\partial}{\partial X}, \qquad \boldsymbol{P}_y = -i\hbar\frac{\partial}{\partial Y}, \qquad \boldsymbol{P}_z = -i\hbar\frac{\partial}{\partial Z}. \tag{104.14}$$

The wave function Ψ will be regarded as a function of the co-ordinates X, Y, Z of the centre of mass and the relative co-ordinates $q_1, q_2, \ldots, q_{3N-3}$. Schrödinger's equation with the Hamiltonian (104.13) allows a separation of the variables if we put

$$\begin{aligned}&\Psi(X, Y, Z, q_1, q_2, \ldots, q_{3N-3}, t)\\&= \Phi(X, Y, Z, t)\,\psi(q_1, q_2, \ldots, q_{3N-3}, t).\end{aligned} \tag{104.15}$$

Substituting (104.15) in Schrödinger's equation, we obtain

$$i\hbar\frac{\partial\Phi}{\partial t}\psi + i\hbar\Phi\frac{\partial\psi}{\partial t} = -\psi\frac{\hbar^2}{2M}\nabla^2\Phi + \Phi\boldsymbol{H}_i\psi. \tag{104.16}$$

Dividing by $\Phi\psi$ and equating separately the terms which depend on X, Y, Z and $q_1, q_2, \ldots, q_{3N-3}$, we obtain two equations:

$$i\hbar\,\partial\Phi/\partial t = -(\hbar^2/2M)\nabla^2\Phi, \tag{104.17}$$

$$i\hbar\,\partial\psi/\partial t = \boldsymbol{H}_i\psi. \tag{104.18}$$

The first of these equations relates to the motion of the centre of mass, and the second to the relative motion. It is seen that the former is the equation of motion of a free particle with mass M: in the absence of external forces, the centre of mass moves like a free point mass. A simple particular solution of Equation (104.17) is the de Broglie wave

$$\Phi(X, Y, Z, t) = \frac{1}{(2\pi\hbar)^{\frac{3}{2}}}e^{(i/\hbar)(Et - P_xX - P_yY - P_zZ)}. \tag{104.19}$$

This is seen to be the eigenfunction of the total-momentum operator $\boldsymbol{P}_x$, $\boldsymbol{P}_y$, $\boldsymbol{P}_z$ which belongs to the eigenvalues P_x, P_y, P_z. E is the eigenvalue of the kinetic energy of the motion of the centre of mass of the system:

$$E = \frac{1}{2M}(P_x^2 + P_y^2 + P_z^2).$$

The wavelength λ of these waves is shown by (104.19) to be the same as for a particle:

$$\left.\begin{aligned}&\lambda = 2\pi\hbar/P = 2\pi\hbar/MV,\\&P = \sqrt{(P_x^2 + P_y^2 + P_z^2)},\end{aligned}\right\} \tag{104.20}$$

where V is the group velocity of the motion of the centre of mass.

This result is important, since it emphasises the fact that de Broglie waves are not a kind of oscillation related to the nature (e.g. the structure) of the particles, but express in the quantum region a general law of motion of free particles or law of motion of the centre of mass of a system not subject to external forces.

105. The law of conservation of the angular momentum of a system of microparticles

Suppose that we have a system of N particles. Let the operators of the components of the angular momentum of the kth particle along the co-ordinate axes be denoted by $\boldsymbol{M}_{kx}$, $\boldsymbol{M}_{ky}$, $\boldsymbol{M}_{kz}$:

$$\boldsymbol{M}_{kx} = -i\hbar\left(y_k\frac{\partial}{\partial z_k} - z_k\frac{\partial}{\partial y_k}\right), \tag{105.1}$$

$$\boldsymbol{M}_{ky} = -i\hbar\left(z_k\frac{\partial}{\partial x_k} - x_k\frac{\partial}{\partial z_k}\right), \tag{105.1'}$$

$$\boldsymbol{M}_{kz} = -i\hbar\left(x_k\frac{\partial}{\partial y_k} - y_k\frac{\partial}{\partial x_k}\right), \tag{105.1''}$$

where x_k, y_k, z_k are the co-ordinates of the kth particle.

The operators $\boldsymbol{M}_x$, $\boldsymbol{M}_y$, $\boldsymbol{M}_z$ of the components of the total angular momentum of the whole system of particles are accordingly given by

$$\boldsymbol{M}_x = \sum_{k=1}^{N} \boldsymbol{M}_{kx}, \tag{105.2}$$

$$\boldsymbol{M}_y = \sum_{k=1}^{N} \boldsymbol{M}_{ky}, \tag{105.2'}$$

$$\boldsymbol{M}_z = \sum_{k=1}^{N} \boldsymbol{M}_{kz}, \tag{105.2''}$$

We shall show that the operator of the time derivative of the angular momentum is equal to the moment of the forces acting on the system (more precisely, the operator of that moment). According to the general definition of the derivative of an operator we have

$$\frac{\mathrm{d}\boldsymbol{M}_x}{\mathrm{d}t} = \frac{i}{\hbar}(\boldsymbol{H}\boldsymbol{M}_x - \boldsymbol{M}_x\boldsymbol{H}). \tag{105.3}$$

The Hamiltonian $\boldsymbol{H}$ is, according to (102.6′),

$$\boldsymbol{H} = \sum_{k=1}^{N}\left\{-\frac{\hbar^2}{2m_k}\nabla_k^2 + U_k\right\} + \sum_{k\neq j=1}^{N} U_{kj}. \tag{105.4}$$

To calculate the commutator in (105.3) we must bear in mind that each component $\boldsymbol{M}_{kx}$ in the operator $\boldsymbol{M}_x$ acts only on the terms in $\boldsymbol{H}$ which contain the co-ordinates of the kth particle.

The operators ∇_k^2 commute with the operator $\boldsymbol{M}_{kx}$. For, as we know, the kinetic-energy operator can be written in the form

$$-\frac{\hbar^2}{2m_k}\nabla_k^2 = \boldsymbol{T}_{r_k} + \frac{(\boldsymbol{M}_k)^2}{2m_k r_k^2}, \tag{105.5}$$

where $\boldsymbol{T}_{r_k}$ is the operator of the part of the kinetic energy of the particle which corresponds to its motion along the radius vector $\mathbf{r}_k$, and $(\boldsymbol{M}_k)^2$ is the squared angular momentum of the kth particle. $\boldsymbol{M}_{kx}$ commutes with $\boldsymbol{T}_{r_k}$ and with $(\boldsymbol{M}_k)^2$, and therefore also with $-(\hbar^2/2m_k)\nabla_k^2$.

Let us now calculate the commutator of M_{kx} and U_k. We have

$$U_k M_{kx} - M_{kx} U_k$$
$$= -i\hbar \left\{ U_k \left(y_k \frac{\partial}{\partial z_k} - z_k \frac{\partial}{\partial y_k} \right) - \left(y_k \frac{\partial}{\partial z_k} - z_k \frac{\partial}{\partial y_k} \right) U_k \right\}$$
$$= i\hbar \left(y_k \frac{\partial U_k}{\partial z_k} - z_k \frac{\partial U_k}{\partial y_k} \right). \tag{105.6}$$

Finally, there is the commutator

$$U_{kj} M_{kx} - M_{kx} U_{kj} = i\hbar \left(y_k \frac{\partial U_{kj}}{\partial z_k} - z_k \frac{\partial U_{kj}}{\partial y_k} \right)$$
$$= i\hbar \frac{\partial U_{kj}}{\partial r_{kj}} \left(y_k \frac{z_k - z_j}{r_{kj}} - z_k \frac{y_k - y_j}{r_{kj}} \right) \tag{105.7}$$
$$= i\hbar (z_k y_j - z_j y_k) \frac{\partial U_{kj}}{\partial r_{kj}} \frac{1}{r_{kj}}.$$

Substituting (105.6) and (105.7) in (105.3), we find

$$\frac{dM_x}{dt} = -\sum_{k=1}^{N} \left(y_k \frac{\partial U_k}{\partial z_k} - z_k \frac{\partial U_k}{\partial y_k} \right) - \sum_{k \neq j = 1}^{N} (z_k y_j - y_k z_j) \frac{\partial U_{kj}}{\partial r_{kj}} \cdot \frac{1}{r_{kj}}.$$

The latter sum is zero, as we see immediately by interchanging the suffixes k and j. Hence we have

$$\frac{dM_x}{dt} = -\sum_{k=1}^{N} \left(y_k \frac{\partial U_k}{\partial z_k} - z_k \frac{\partial U_k}{\partial y_k} \right). \tag{105.8}$$

The expression on the right-hand side is just the operator of the component along the axis OX of the sum of the moments of the external forces acting on the system. Similarly

$$\frac{dM_y}{dt} = -\sum_{k=1}^{N} \left(z_k \frac{\partial U_k}{\partial x_k} - x_k \frac{\partial U_k}{\partial z_k} \right), \tag{105.8'}$$

$$\frac{dM_z}{dt} = -\sum_{k=1}^{N} \left(x_k \frac{\partial U_k}{\partial y_k} - y_k \frac{\partial U_k}{\partial x_k} \right). \tag{105.8''}$$

Thus we obtain the theorem of classical mechanics that the rate of change of the angular momentum is equal to the moment of the external forces acting on the system. In quantum mechanics this theorem, like that concerning the total momentum, is expressed in terms of operators.

If the moment of the external forces is zero, the total angular momentum of the system is conserved:

$$\mathrm{d}\boldsymbol{M}_x/\mathrm{d}t = \mathrm{d}\boldsymbol{M}_y/\mathrm{d}t = \mathrm{d}\boldsymbol{M}_z/\mathrm{d}t = 0\,. \tag{105.9}$$

Consequently, in the absence of external forces the mean values $\bar{M}_x$, $\bar{M}_y$, $\bar{M}_z$ of the angular momentum and the probabilities $w(M_x)$, $w(M_y)$, $w(M_z)$ of finding a particular value of any component of the angular momentum remain unchanged in the course of time.

If the spin of the particles is taken into account, the operator of the total angular momentum must be defined by

$$\boldsymbol{M}_x = \sum_{k=1}^{N} (\boldsymbol{M}_{kx} + \boldsymbol{s}_{kx}), \tag{105.10}$$

$$\boldsymbol{M}_y = \sum_{k=1}^{N} (\boldsymbol{M}_{ky} + \boldsymbol{s}_{ky}), \tag{105.10'}$$

$$\boldsymbol{M}_z = \sum_{k=1}^{N} (\boldsymbol{M}_{kz} + \boldsymbol{s}_{kz}), \tag{105.10''}$$

where $\boldsymbol{s}_{kx}$, $\boldsymbol{s}_{ky}$, $\boldsymbol{s}_{kz}$ are the operators (two-by-two matrices) of the components of the intrinsic angular momentum of the kth particle. The theorem of conservation of the total angular momentum remains valid in this case. If there are no forces acting on the spins, the proof is the same as above, since the Hamiltonian of the system then commutes with all the operators $\vec{\boldsymbol{s}}_k$.

Since the operators $\boldsymbol{M}_{kx}$, $\boldsymbol{M}_{ky}$, $\boldsymbol{M}_{kz}$, $\boldsymbol{s}_{kx}$, $\boldsymbol{s}_{ky}$, $\boldsymbol{s}_{kz}$ belonging to different particles (different k) commute, it is easy to obtain, from the known commutation rules for the components of orbital angular momentum (25.5) etc. and of spin angular momentum (59.1) for a single particle, the commutation rules for the total angular momentum of a system of particles:

$$\boldsymbol{M}_x\boldsymbol{M}_y - \boldsymbol{M}_y\boldsymbol{M}_x = i\hbar\boldsymbol{M}_z\,, \tag{105.11}$$

$$\boldsymbol{M}_y\boldsymbol{M}_z - \boldsymbol{M}_z\boldsymbol{M}_y = i\hbar\boldsymbol{M}_x\,, \tag{105.11'}$$

$$\boldsymbol{M}_z\boldsymbol{M}_x - \boldsymbol{M}_x\boldsymbol{M}_z = i\hbar\boldsymbol{M}_y\,; \tag{105.11''}$$

$$\boldsymbol{M}^2\boldsymbol{M}_x - \boldsymbol{M}_x\boldsymbol{M}^2 = 0\,, \tag{105.12}$$

$$\boldsymbol{M}^2\boldsymbol{M}_y - \boldsymbol{M}_y\boldsymbol{M}^2 = 0\,, \tag{105.12'}$$

$$\boldsymbol{M}^2\boldsymbol{M}_z - \boldsymbol{M}_z\boldsymbol{M}^2 = 0\,, \tag{105.12''}$$

where $\boldsymbol{M}^2$ is the operator of the squared total angular momentum:

$$\boldsymbol{M}^2 = \boldsymbol{M}_x^2 + \boldsymbol{M}_y^2 + \boldsymbol{M}_z^2\,. \tag{105.13}$$

It is shown below from these commutation rules that the total angular momentum for a system of particles is quantised according to the formulae

$$M^2 = \hbar^2 J(J+1)\,, \tag{105.14}$$

$$M_z = \hbar m\,, \quad |m| \leqslant J\,, \tag{105.15}$$

J being either an integer (0, 1, 2, 3, ...) or half an integer ($\frac{1}{2}$, $\frac{3}{2}$, $\frac{5}{2}$, ...), depending on the number of particles and on their spin. The inequality $|m| \leqslant J$ means that $m = J, J-1, J-2, ..., -J$. In other words, we always have $2J + 1$ quantum orientations of the total angular momentum about any direction OZ.

It may be noted that, since the spin of the electron is half-integral (viz. $\frac{1}{2}$), J is always integral for an even number of electrons and half-integral for an odd number.

The components (105.2)–(105.2″) of the total orbital angular momentum

$$\vec{M}_L = \sum_{k=1}^{N} \vec{M}_k \tag{105.16}$$

and of the total spin angular momentum

$$\vec{M}_s = \sum_{k=1}^{N} \vec{s}_k \tag{105.17}$$

obey the same commutation rules as the components of the total angular momentum. They are therefore quantised according to analogous formulae:

$$M_L^2 = \hbar^2 L(L+1), \quad L = 0, 1, 2, 3, ..., \tag{105.18}$$

$$M_{Lz} = \hbar m_L, \quad |m_L| \leqslant L; \tag{105.19}$$

$$M_s^2 = \hbar^2 S(S+1), \quad S = 0, 1, 2, 3, ... \quad \text{or} \quad S = \tfrac{1}{2}, \tfrac{3}{2}, \tfrac{5}{2}, ..., \tag{105.20}$$

$$M_{sz} = \hbar m_s, \quad |m_s| \leqslant S. \tag{105.21}$$

For given values of the total orbital angular momentum L and of the total spin angular momentum S, various values of J are possible, depending on the relative orientation of the vectors $\mathbf{M}_l$ and $\mathbf{M}_s$. Figure 48 (Section 65) illustrates the addition of these angular momenta.

Evidently J can take all values from $L + S$, corresponding to the parallel orientation of $\mathbf{M}_l$ and $\mathbf{M}_s$, to $|L - S|$, corresponding to the antiparallel orientation of these vectors, i.e.

$$J = L + S, \quad |L + S - 1|, \quad |L + S - 2|, \quad ..., \quad |L - S|, \tag{105.22}$$

a total of $2S + 1$ values. The states with the same values of L and S form a multiplet – a group of levels which are close together, owing to the weakness of the interaction between the spin and the orbital motion (cf. Section 65). The multiplicity (number of levels) is seen to be $2S + 1$.

The total angular momentum J of the system, its orbital angular momentum L, and the spin angular momentum S serve to label an atomic term as a whole. As for a single electron (cf. Section 65), the terms with $L = 0, 1, 2, 3, ...$ are denoted by $S, P, D, F, ...$ (capital letters in this case) respectively. The value of the total angular momentum J is shown at the bottom right, and the multiplicity of the multiplet containing the term at the top left. This also indicates the total spin. For example, ${}^4F_{\frac{3}{2}}$ denotes a term for which $L = 3$, $J = \frac{3}{2}$, $S = \frac{3}{2}$; ${}^6S_{\frac{5}{2}}$ denotes a term with $L = 0$, $J = \frac{5}{2}$, $S = \frac{5}{2}$.

Formula (105.15) is proved immediately if we note that the various terms in the sum (105.10″) commute and therefore can be brought simultaneously to diagonal form so that the eigenvalue M_z is equal to the sum of the eigenvalues $M_{kz} + s_{kz}$. The eigenvalues of the latter are $\hbar m_k$, where m_k is integral or half-integral, depending on the value of the spin of the particles. Thus

$$M_z = \sum_{k=1}^{N} \hbar m_k = \hbar m\,, \qquad m = \sum_{k=1}^{N} m_k\,. \tag{105.23}$$

To determine the eigenvalues M^2 we define the operators $\boldsymbol{A} = \boldsymbol{M}_x + i\boldsymbol{M}_y$, $\boldsymbol{B} = \boldsymbol{M}_x - i\boldsymbol{M}_y$. Using (105.12) etc. we obtain

$$\boldsymbol{A}\boldsymbol{M}_z - \boldsymbol{M}_z\boldsymbol{A} = -\hbar\boldsymbol{A}\,, \qquad \boldsymbol{B}\boldsymbol{M}_z - \boldsymbol{M}_z\boldsymbol{B} = \hbar\boldsymbol{B}\,. \tag{105.24}$$

These equations can be written as products of matrices, using a representation in which M_z is diagonal. Then we have

$$\left.\begin{aligned} A_{m'm''}\hbar m'' - \hbar m' A_{m'm''} &= -\hbar A_{m'm''}\,,\\ B_{m'm''}\hbar m'' - \hbar m' B_{m'm''} &= \hbar B_{m'm''}\,, \end{aligned}\right\} \tag{105.25}$$

or

$$A_{m'm''}(m'' - m' + 1) = 0\,, \qquad B_{m'm''}(m'' - m' - 1) = 0\,. \tag{105.26}$$

Hence it follows that the only non-zero elements of $\boldsymbol{A}$ and $\boldsymbol{B}$ are $A_{m,\,m-1}$ and $B_{m,m+1}$. The operator $\boldsymbol{M}^2$ of the squared total angular momentum can be expressed in terms of the operators $\boldsymbol{A}$ and $\boldsymbol{B}$ in two ways:

$$\boldsymbol{M}^2 = \boldsymbol{A}\boldsymbol{B} + \boldsymbol{M}_z^2 - \hbar\boldsymbol{M}_z\,, \tag{105.27}$$

and

$$\boldsymbol{M}^2 = \boldsymbol{B}\boldsymbol{A} + \boldsymbol{M}_z^2 + \hbar\boldsymbol{M}_z\,. \tag{105.27′}$$

Hence

$$\boldsymbol{A}\boldsymbol{B} = \boldsymbol{M}^2 + \tfrac{1}{4}\hbar^2 - (\boldsymbol{M}_z - \tfrac{1}{2}\hbar)^2\,, \tag{105.28}$$

$$\boldsymbol{B}\boldsymbol{A} = \boldsymbol{M}^2 + \tfrac{1}{4}\hbar^2 - (\boldsymbol{M}_z + \tfrac{1}{2}\hbar)^2\,. \tag{105.28′}$$

The diagonal element mm of these equations gives

$$(\boldsymbol{A}\boldsymbol{B})_{mm} = A_{m,\,m-1}\,B_{m-1,\,m} = M^2 + \tfrac{1}{4}\hbar^2 - \hbar^2(m - \tfrac{1}{2})^2\,. \tag{105.29}$$

$$(\boldsymbol{B}\boldsymbol{A})_{mm} = B_{m,\,m+1}\,A_{m+1,\,m} = M^2 + \tfrac{1}{4}\hbar^2 - \hbar^2(m + \tfrac{1}{2})^2\,. \tag{105.29′}$$

Let us now regard M^2 as given. Then the possible values of $|m|$ are necessarily restricted, since $M^2 = M_x^2 + M_y^2 + M_z^2$ and the eigenvalue $M_x^2 + M_y^2$ cannot be negative. Let the lowest value of m be m', and the highest m''. Then (105.29) and (105.29′) give

$$M^2 + \tfrac{1}{4}\hbar^2 = \hbar^2(m' - \tfrac{1}{2})^2\,,$$

$$M^2 + \tfrac{1}{4}\hbar^2 = \hbar^2(m'' + \tfrac{1}{2})^2\,,$$

since $A_{m',\,m'-1} = 0$, $B_{m'-1,\,m'} = 0$ and $A_{m''+1,\,m''} = 0$, $B_{m'',\,m''+1} = 0$ Hence

$$m' = \tfrac{1}{2} - \sqrt{(M^2/\hbar^2 + \tfrac{1}{4})}\,, \tag{105.30}$$

$$m'' = -\tfrac{1}{2} + \sqrt{(M^2/\hbar^2 + \tfrac{1}{4})}\,. \tag{105.30′}$$

Thus $m'' = -m'$. The difference $m'' - m' + 1$ is an integer equal to the number of different possible M_z for given M^2. Let $m'' - m' + 1 = 2J + 1$. Then from (105.30) and (105.30′) we have

$$2J + 1 = 2\sqrt{(M^2/\hbar^2 + \tfrac{1}{4})}\,,$$

or

$$M^2 = \hbar^2 J\,(J + 1)\,. \tag{105.31}$$

From (105.15),

$$|m| \leqslant J, \text{ where } m = 0, \pm 1, \pm 2, \ldots, \pm J$$

$$\text{or } m = \pm \tfrac{1}{2}, \pm \tfrac{3}{2}, \ldots, \pm J\,.$$

In this proof we have used only the commutation rules (105.11) etc. for the operators of the angular momentum components. Since the components of the operators of the total orbital angular momentum (105.16) and the total spin angular momentum (105.17) obey the same commutation rules, this proves also Formulae (105.18), (105.19) and (105.20), (105.21).

From these formulae and (105.14) it follows that the operator of the scalar product $2\mathbf{M}\cdot\mathbf{M}_s = M^2 - M_L{}^2 + M_s{}^2$ has eigenvalues

$$2\mathbf{M}\cdot\mathbf{M}_s = \hbar^2 \{J(J+1) - L(L+1) + S(S+1)\}, \tag{105.32}$$

so that Formula (64.14) for a single particle is a particular case of (105.32).

Repeating the arguments of Section 74, we can easily derive a formula for the energy of a system of particles in a magnetic field:

$$W = \hbar O_L m \left\{1 + \frac{J(J+1) - L(L+1) + S(S+1)}{J(J+1)}\right\}, \tag{105.33}$$

so that (74.23) for a single particle is a particular case of (105.33), which gives the splitting of the levels in a magnetic field for a system of electrons (a complex atom).

106. Eigenfunctions of the angular momentum operator of the system. Clebsch-Gordan coefficients

The eigenfunctions of the total angular momentum operator of a system are complicated functions of the angular and spin co-ordinates of the component parts of the system and of their quantum numbers, but in many cases frequently encountered they can be expressed in terms of functions of the angular momenta of the separate component parts.

Let us take the simplest case, that of a system consisting of two sub-systems: let $\vec{M}_1$ and $\vec{M}_2$ be the angular momentum operators of these sub-systems, and let them commute. $\mathbf{M}_1$ and $\mathbf{M}_2$ may be the orbital angular momenta of two particles, the orbital and spin angular momenta of a single particle, and so on.

We shall assume that the total angular momentum is an integral of the motion. The state of the system may be described either by the quantum numbers j_1, j_2, m_1, m_2 (where j_1, j_2 are the eigenvalues of the angular momenta of the sub-systems, and m_1, m_2 their components), or by the set of four numbers J, m, j_1, j_2 (where J, m are the eigenvalues of the total angular momentum of the system and its component, with $m = m_1 + m_2$ (105.23)).

Let us consider the problem of determining the wave functions of the system in terms of those of the sub-systems. Let $Y_{j_1m_1}$ be common eigenfunctions of the operators M_1^2 and M_{z1}, and $Y_{j_2m_2}$ those of M_2^2 and M_{z2}. Then the product $Y_{j_1m_1}Y_{j_2m_2}$ will be an eigenfunction of the operator

$$M_z = M_{z1} + M_{z2}$$

of the component of the total angular momentum, with eigenvalue $m = m_1 + m_2$.

Let $Y^m_{Jj_1j_2}$ be a common eigenfunction of the operators M^2 and M_z. This can be represented as a linear combination of the products $Y_{j_1m_1}Y_{j_2m_2}$:

$$Y^m_{Jj_1j_2} = \sum_{m_1=-j_1}^{j_1} \sum_{m_2=-j_2}^{j_2} (j_1j_2m_1m_2 \,|\, Jm)\, Y_{j_1m_1} Y_{j_2m_2}. \tag{106.1}$$

The coefficients $(j_1j_2m_1m_2|Jm)$ are real numbers, and are called the *Clebsch-Gordan coefficients*.[5] They are equal to zero when $m \neq m_1 + m_2$, so that the double sum in

[5] For details see [26], and concerning the notation see [28].

(106.1) actually reduces to a single sum. The functions $Y^m_{Jj_1j_2}$ depend on the same variables as the functions $Y_{j_1m_1}$, $Y_{j_2m_2}$. In particular, if one of these is a function of angle co-ordinates and the other of spin co-ordinates, the corresponding $Y^m_{Jj_1j_2}$ is called a spherical harmonic with spin. We have discussed this case in Section 64, where the eigenfunctions of the total angular momentum were found (spin and orbital for one particle). The coefficients of Y_{lm} and $Y_{l,m+1}$ in (64.28) and (64.28′) are just the Clebsch-Gordan coefficients for the case[6] $j_2 = \frac{1}{2}$. The spin wave functions in these formulae are replaced by their values

$$\begin{pmatrix}1\\0\end{pmatrix}$$

or

$$\begin{pmatrix}0\\1\end{pmatrix}.$$

The expression (106.1) can be inverted to give

$$Y_{j_1m_1}Y_{j_2m_2} = \sum_{J=|j_1-j_2|}^{j_1+j_2}\sum_{m=-J}^{J}(j_1j_2m_1m_2|Jm)\,Y^m_{Jj_1j_2}, \tag{106.2}$$

the sum over m in fact containing only the term with $m = m_1 + m_2$.

From the conditions of orthogonality of the functions Y_{jm} and $Y^m_{Jj_1j_2}$ we obtain the following orthogonality conditions for the Clebsch-Gordan coefficients:

$$\sum_{m_1=-j_1}^{j_1}\sum_{m_2=-j_2}^{j_2}(j_1j_2m_1m_2|Jm)(j_1j_2m_1m_2|J'm') = \delta_{JJ'}\delta_{mm'}, \tag{106.3}$$

$$\sum_{J=|j_1-j_2|}^{j_1+j_2}\sum_{m=-J}^{J}(j_1j_2m_1m_2|Jm)(j_1j_2m_1'm_2'|Jm) = \delta_{m_1m_1'}\delta_{m_2m_2'}, \tag{106.4}$$

$$\sum_{m_1,m}(j_1j_2m_1m_2|Jm)(j_1j_2'm_1m_2'|Jm) = \frac{2J+1}{2j_2+1}\delta_{j_2j_2'}\delta_{m_2m_2'}. \tag{106.5}$$

The Clebsch-Gordan coefficients also satisfy certain symmetry conditions:

$$(j_1j_2m_1m_2|Jm) = (-1)^{j_1+j_2-J}(j_1j_2,-m_1,-m_2|J,-m), \tag{106.6}$$

$$(j_1j_2m_1m_2|Jm) = (-1)^{j_1+j_2-J}(j_2j_1m_2m_1|Jm), \tag{106.7}$$

$$\sqrt{(2j_1+1)}(j_1j_2m_1m_2|Jm) = (-1)^{j_2+m_2}\sqrt{(2J+1)}(Jj_2,-m,m_2|j_1,-m_1), \tag{106.8}$$

$$\sqrt{(2j_2+1)}(j_1j_2m_1m_2|Jm) = (-1)^{j_1-m_1}\sqrt{(2J+1)}(j_1Jm_1,-m|j_2,-m_2), \tag{106.9}$$

$$\sqrt{(2j_1+1)}(j_1j_2m_1m_2|Jm) = (-1)^{j_1-J+m_2}\sqrt{(2J+1)}(j_2Jm_2,-m|j_1,-m_1). \tag{106.10}$$

The following tables show the Clebsch-Gordan coefficients for $j_2 = \frac{1}{2}$ and $j_2 = 1$:

[6] m in (64.28′) corresponds to m_1 in (106.1), l becomes j_1, and $m \pm \frac{1}{2}$ becomes n.

TABLE 1

CLEBSCH-GORDAN COEFFICIENTS $(j_1, \frac{1}{2}, m_1m_2 \mid Jm)$

J	$m_2 = \frac{1}{2}$	$m_2 = -\frac{1}{2}$
$j_1 + \frac{1}{2}$	$\left(\frac{j_1 + m + \frac{1}{2}}{2j_1 + 1}\right)^{\frac{1}{2}}$	$\left(\frac{j_1 - m + \frac{1}{2}}{2j_1 + 1}\right)^{\frac{1}{2}}$
$j_1 - \frac{1}{2}$	$-\left(\frac{j_1 - m + \frac{1}{2}}{2j_1 + 1}\right)^{\frac{1}{2}}$	$\left(\frac{j_1 + m + \frac{1}{2}}{2j_1 + 1}\right)^{\frac{1}{2}}$

TABLE 2

CLEBSCH-GORDAN COEFFICIENTS $(j_1, 1, m_1m_2 \mid Jm)$

J	$m_2 = 1$	$m_2 = 0$	$m_2 = -1$
j_1+1	$\left[\frac{(j_1+m)(j_1+m+1)}{(2j_1+1)(2j_1+2)}\right]^{\frac{1}{2}}$	$\left[\frac{(j_1-m+1)(j_1+m+1)}{(2j_1+1)(j_1+1)}\right]^{\frac{1}{2}}$	$\left[\frac{(j_1-m)(j_1-m+1)}{(2j_1+1)(2j_1+2)}\right]^{\frac{1}{2}}$
j_1	$-\left[\frac{(j_1+m)(j_1-m+1)}{2j_1(j_1+1)}\right]^{\frac{1}{2}}$	$\frac{m}{\sqrt{[j_1(j_1+1)]}}$	$\left[\frac{(j_1-m)(j_1+m+1)}{2j_1(j_1+1)}\right]^{\frac{1}{2}}$
j_1-1	$\left[\frac{(j_1-m)(j_1-m+1)}{2j_1(2j_1+1)}\right]$	$-\left[\frac{(j_1-m)(j_1+m)}{j_1(2j_1+1)}\right]^{\frac{1}{2}}$	$\left[\frac{(j_1+m+1)(j_1+m)}{2j_1(2j_1+1)}\right]^{\frac{1}{2}}$

Owing to the symmetry properties of the Clebsch-Gordan coefficients, these tables can be used whenever any of the quantum numbers j_1, j_2, J is $\frac{1}{2}$ or 1. Attention may be drawn to the values of some of the Clebsch-Gordan coefficients. If $J = j_1 + j_2$, then

$$(j_1 j_2, j_1 j_2 | JJ) = (j_1 j_2, -j_1, -j_2 | J, -J) = 1 \tag{106.11}$$

for all values of j_1 and j_2. When two antiparallel spins are added, we have

$$(\tfrac{1}{2}, -\tfrac{1}{2}, \tfrac{1}{2}, -\tfrac{1}{2} | 00) = -(-\tfrac{1}{2}, \tfrac{1}{2}, -\tfrac{1}{2}, \tfrac{1}{2} | 00) = 1/\sqrt{2}, \tag{106.12}$$

i.e. the wave function of the system of two antiparallel spins is

$$S(s_{z1}, s_{z2}) = \frac{1}{\sqrt{2}} [S_{+\frac{1}{2}}(s_{z1}) S_{-\frac{1}{2}}(s_{z2}) - S_{-\frac{1}{2}}(s_{z1}) S_{+\frac{1}{2}}(s_{z2})] \tag{106.13}$$

(see (121.13)).

A general formula for the Clebsch-Gordan coefficients is given by Wigner [91].

107. The relation of the conservation laws to the symmetry of space and time

Physical space has the properties of homogeneity and isotropy; time has the property of homogeneity. Moreover, for reversible processes there is equivalence as regards the sign of the time.

These properties of space and time correspond to the fundamental conservation laws for a closed system in quantum mechanics.

A. THE LAW OF CONSERVATION OF ENERGY

Let us first consider the consequences of the homogeneity of time. Under an infinitesimal time displacement Δt, the wave function ψ of the system becomes $\psi' = \psi(\mathbf{x}_1, \mathbf{x}_2, \ldots, \mathbf{x}_N, t + \Delta t)$. This change in the wave function can be regarded as the result of an infinitesimal unitary transformation S_t (see Section 28):

$$\psi' = S_t \psi , \tag{107.1}$$

where $S_t = 1 + i\boldsymbol{L}\Delta t$ and $\boldsymbol{L}$ is an Hermitian operator. Since also $\psi' - \psi = (\partial\psi/\partial t)\Delta t$, (107.1) shows that $\partial\psi/\partial t = i\boldsymbol{L}\psi$. This is the same as Schrödinger's equation, and $\boldsymbol{L} = -(1/\hbar)\boldsymbol{H}$. Owing to the homogeneity of time, however, $\boldsymbol{L}$, and therefore $\boldsymbol{H}$, must be independent of time, i.e. $\partial\boldsymbol{H}/\partial t = 0$, and therefore

$$\mathrm{d}\boldsymbol{H}/\mathrm{d}t = [\boldsymbol{H}, \boldsymbol{H}] = 0 ; \tag{107.2}$$

this is the law of conservation of energy for a closed system.

B. THE LAW OF CONSERVATION OF MOMENTUM

Let us consider a closed system of particles, and apply an infinitesimal displacement $\Delta\mathbf{x}$ to all the co-ordinates (radii vectores) $\mathbf{x}_k$. Then

$$\psi' = \psi(\mathbf{x}_1 + \Delta\mathbf{x}, \ldots, \mathbf{x}_N + \Delta\mathbf{x}, t) = \psi(\mathbf{x}_1, \ldots, \mathbf{x}_N, t) + \Delta\mathbf{x} \cdot \sum_{k=1}^{N} \nabla_k \psi , \tag{107.3}$$

where ∇_k is the gradient with respect to the co-ordinates of the kth particle.

Regarding this displacement as an infinitesimal unitary transformation $S_x = 1 + i\vec{\boldsymbol{g}} \cdot \Delta\mathbf{x}$, where $\vec{\boldsymbol{g}}$ is an Hermitian operator, we find

$$\vec{\boldsymbol{g}} = -i \sum_{k=1}^{N} \nabla_k . \tag{107.4}$$

The operator $\vec{\boldsymbol{g}}$ differs only by a factor $\hbar$ from the operator $\vec{\boldsymbol{P}}$ of the total momentum of the system (103.1). Since the space displacement operator S_x and the time displacement operator S_t can be applied in either order (in the absence of external forces), S_x and S_t commute, i.e. $[\boldsymbol{L}, \vec{\boldsymbol{g}}] = 0$, and therefore $[\vec{\boldsymbol{P}}, \boldsymbol{H}] = 0$. This implies that

$$\mathrm{d}\vec{\boldsymbol{P}}/\mathrm{d}t = 0 , \tag{107.5}$$

i.e. that the total momentum of a closed system is conserved.

C. THE LAW OF CONSERVATION OF ANGULAR MOMENTUM

Let us consider an infinitesimal rotation of the system in isotropic space through an angle $\Delta\phi_z$ around the axis OZ. This rotation leads to a change in the co-ordinates of the kth particle by

$$\Delta\mathbf{x}_k = \{y_k \Delta\phi_z, -x_k \Delta\phi_z, 0\} . \tag{107.6}$$

The new wave function $\psi' = \psi(\mathbf{x}_1 + \Delta\mathbf{x}_1, \ldots, \mathbf{x}_N + \Delta\mathbf{x}_N, t)$ can be obtained from the initial one by applying an infinitesimal unitary transformation

$$S_{\phi_z} = 1 + i m_z \Delta\phi_z\,; \qquad \psi' = S_{\phi_z}\psi\,. \tag{107.7}$$

Using (107.6), we have

$$\begin{aligned}\psi(\mathbf{x}_1 + \Delta\mathbf{x}_1, \ldots, \mathbf{x}_N + \Delta\mathbf{x}_N, t)\\ = \psi(\mathbf{x}_1, \ldots, \mathbf{x}_N, t) - \sum_{k=1}^{N}\left(x_k \frac{\partial\psi}{\partial y_k} - y_k \frac{\partial\psi}{\partial x_k}\right)\Delta\phi_z\,.\end{aligned} \tag{107.8}$$

Comparison of (107.7) and (107.8) gives

$$m_z = i\sum_{k=1}^{N}\left(x_k \frac{\partial}{\partial y_k} - y_k \frac{\partial}{\partial x_k}\right), \tag{107.9}$$

i.e. m_z differs only by a factor from the operator M_z of the component of the total angular momentum along OZ. Similar relations are obtained for rotation about the other two axes, so that

$$S_\phi = 1 - (i/\hbar)\vec{M}\cdot\vec{\Delta\phi}\,, \tag{107.10}$$

where $\vec{M}$ is the operator of the angular momentum of the system.

On account of the isotropy of space and homogeneity of time the operators S_ϕ and S_t, and therefore $\vec{M}$ and H, commute, i.e. $[\vec{M}, H] = 0$. Hence

$$\mathrm{d}\vec{M}/\mathrm{d}t = 0\,, \tag{107.11}$$

i.e. the angular momentum of the system is an integral of the motion.

D. THE LAW OF REVERSIBILITY OF PROCESSES IN QUANTUM MECHANICS

Let us consider the time-reversal transformation T, i.e. $t \to -t$. The equations of motion are invariant with respect to this transformation for reversible processes. In quantum mechanics, all processes are reversible.[7] Hence the operation T must correspond to some unitary transformation of the wave function and of the operators, which represents the property of reversibility.

Schrödinger's equation in the absence of electromagnetic fields is

$$i\hbar\,\partial\psi/\partial t = H\psi\,; \qquad H = (1/2m)(-i\hbar\nabla)^2 + U\,. \tag{107.12}$$

When t is replaced by $-t$, we obtain

$$-i\hbar\partial\psi'/\partial t = H\psi'\,, \tag{107.12'}$$

where $\psi' = \psi(\mathbf{x}_1, \ldots, \mathbf{x}_N, -t) = S_T\,\psi$.

[7] This statement does not apply to the process of measurement, which may be irreversible.

Comparison of (107.12′) with Schrödinger's equation for the complex conjugate wave function,

$$-i\hbar\,\partial\psi^*/\partial t = \boldsymbol{H}\psi^*, \tag{107.12″}$$

shows that

$$\psi' = S_T\psi = \psi^*, \tag{107.13}$$

i.e. the wave function which describes the reversed motion is the complex conjugate function.

For charged particles moving in an external electromagnetic field, the reversal of time must be accompanied by a change of sign of the magnetic field and of the spins:

$$S_T\mathbf{A} = -\mathbf{A}S_T, \tag{107.14}$$

$$S_T\boldsymbol{\sigma} = -\boldsymbol{\sigma}S_T. \tag{107.15}$$

For, when this transformation is applied, Pauli's equation (61.5)

$$i\hbar\frac{\partial\psi}{\partial t} = \frac{1}{2m}\left[\left(-i\hbar\nabla + \frac{e}{c}\mathbf{A}\right)^2 - eV + \frac{e\hbar}{2mc}\boldsymbol{\sigma}\cdot\mathbf{H}\right]\psi \tag{107.16}$$

(with $t \to -t$, $\mathbf{A} \to -\mathbf{A}$, $\boldsymbol{\sigma} = -\boldsymbol{\sigma}$, $\mathbf{H} = \mathbf{curl}\,\mathbf{A} \to -\mathbf{H}$) becomes the equation for the complex conjugate function ψ^*, i.e. Equation (107.13) remains valid.[8]

E. THE LAW OF CONSERVATION OF PARITY

Let us now consider the inversion transformation P, i.e. $x \to -x$, $y \to -y$, $z \to -z$. This corresponds to a change from a right-handed to a left-handed co-ordinate system.

In our space there is no difference between right-handed and left-handed screws if weak interactions[9] are not involved. The theory must then be invariant with respect to the inversion transformation $\boldsymbol{P}$. This requirement imposes a condition on the possible Hamiltonians, viz.

$$\boldsymbol{PH} = \boldsymbol{HP}. \tag{107.17}$$

The corresponding unitary transformation of the wave function is

$$\psi' \equiv \psi(-x, -y, -z, t) = \boldsymbol{P}\psi(x, y, z, t). \tag{107.18}$$

The equation (107.17) signifies that the inversion operator is an integral of the motion:

$$\mathrm{d}\boldsymbol{P}/\mathrm{d}t = 0. \tag{107.19}$$

It is also evident that $\boldsymbol{P}^2\psi = \psi$. Hence the eigenvalues of the inversion operator must be ± 1. The wave functions (or states) with $P = 1$ are said to be *even* (+), and those with $P = -1$ are said to be *odd* (−).

If a state has a given parity at some instant, then by (107.19) this parity cannot change. The parity is therefore one of the characteristic properties of a quantum system.

In particular, for a particle in a state with orbital angular momentum l, the parity

[8] See Section 44 and [26].
[9] See [102].

is $(-1)^l$ (Section 25). For a system of particles having angular momenta $l_1, \ldots, l_N$ the parity of the state is that of the product $Y_{l_1 m_1} \ldots Y_{l_N m_N}$, i.e. $(-1)^{l_1 + l_2 + \ldots + l_N}$.

In conclusion, it may be noted that, if a quantum system is not in empty space but in some medium, in an external field or in a crystal, the symmetry properties of the medium will also give rise to the existence of certain integrals of the motion.

For example, if an atom is situated within a crystal having an axis of symmetry of order n, the medium is unchanged by a rotation through an angle $2\pi/n$. The operation of rotation through this angle is an integral of the motion, and the wave function ψ of the atom undergoes a certain unitary transformation.

CHAPTER XVIII

SIMPLE APPLICATIONS OF THE THEORY OF MOTION OF MANY BODIES

108. Allowance for the motion of the nucleus in an atom

In discussing the motion of the optical electron in an atom we have assumed the nucleus of the atom to be at rest, regarding it as a source of central forces. This approximation is a good one when the mass m of the nucleus is large. By means of the theorem proved above concerning the centre of mass, we can easily calculate the corrections due to the finite mass of the nucleus. The equation for the energy E and the eigenfunctions Ψ_E, allowing for the motion of the nucleus, will be

$$\left\{-\frac{\hbar^2}{2m_1}\left(\frac{\partial^2}{\partial x_1^2}+\frac{\partial^2}{\partial y_1^2}+\frac{\partial^2}{\partial z_1^2}\right)-\frac{\hbar^2}{2m_2}\left(\frac{\partial^2}{\partial x_2^2}+\frac{\partial^2}{\partial y_2^2}+\frac{\partial^2}{\partial z_2^2}\right)\right\}\Psi+ \\ +U(r)\Psi=E\Psi, \tag{108.1}$$

where m_1 is the mass of the nucleus, x_1, y_1, z_1 its co-ordinates, m_2 the mass of the electron, x_2, y_2, z_2 its co-ordinates, and r the distance between the nucleus and the electron:

$$r^2=(x_1-x_2)^2+(y_1-y_2)^2+(z_1-z_2)^2\,. \tag{108.2}$$

In terms of the Jacobi co-ordinates (104.3) we have

$$\xi_1=x_1-x_2=x\,,\quad \xi_2=(m_1x_1+m_2x_2)/(m_1+m_2)=X\,, \tag{108.3}$$

$$\eta_1=y_1-y_2=y\,,\quad \eta_2=(m_1y_1+m_2y_2)/(m_1+m_2)=Y, \tag{108.3'}$$

$$\zeta_1=z_1-z_2=z,\quad \zeta_2=(m_1z_1+m_2z_2)/(m_1+m_2)=Z, \tag{108.3''}$$

so that ξ_1, η_1, ζ_1 are in this case simply the relative co-ordinates of the nucleus and electron, and X, Y, Z the co-ordinates of the centre of mass of the electron and the nucleus. In these co-ordinates the Hamiltonian of Equation (108.1) is transformed, similarly to (104.10), to give

$$-\frac{\hbar^2}{2M}\left(\frac{\partial^2\Psi}{\partial X^2}+\frac{\partial^2\Psi}{\partial Y^2}+\frac{\partial^2\Psi}{\partial Z^2}\right)-\frac{\hbar^2}{2\mu}\left(\frac{\partial^2\Psi}{\partial x^2}+\frac{\partial^2\Psi}{\partial y^2}+\frac{\partial^2\Psi}{\partial z^2}\right)+ \\ +U(r)\Psi=E\Psi, \tag{108.1'}$$

where

$$M=m_1+m_2\,,\qquad \frac{1}{\mu}=\frac{1}{m_1}+\frac{1}{m_2}. \tag{108.4}$$

Separating the variables X, Y, Z and x, y, z as in Section 104 (see (104.15)), we have

$$\Psi(X, Y, Z, x, y, z) = e^{-i(p_x X + p_y Y + p_z Z)/\hbar}\,\psi(x, y, z). \tag{108.5}$$

This solution represents a free motion of the centre of mass of the atom with momentum p_x, p_y, p_z. For the function $\psi(x, y, z)$ describing the relative motion we obtain

$$-\frac{\hbar^2}{2\mu}\left(\frac{\partial^2\psi}{\partial x^2} + \frac{\partial^2\psi}{\partial y^2} + \frac{\partial^2\psi}{\partial z^2}\right) + U(r)\psi = \varepsilon\psi, \tag{108.6}$$

where

$$\varepsilon = E - p^2/2M, \qquad E = \varepsilon + p^2/2M. \tag{108.7}$$

Equation (108.6) is exactly the same as that of the motion of a particle of mass μ in a given force field $U(r)$; ε signifies the internal energy of the atom (the energy of the relative motion), and the total energy E consists of the energy ε of the relative motion and that of the motion of the centre of mass of the atom ($p^2/2M$). A similar equation to (108.6), but with the mass of the electron in place of the reduced mass, occurred in the problem of the motion of an electron in an atom. We have therefore no need to solve afresh the problem of the motion of an electron in an atom when the motion of the nucleus is taken into account. In order to find ε and $\psi(x, y, z)$ we need only replace the electron mass in each formula by the reduced mass μ. Since the mass m_1 of the nucleus is much greater than the electron mass m_2, the corrections to ε and ψ will be small. If the mass of the nucleus is assumed to be infinite, then $\mu = m_2$, the electron mass. In Section 51 with this condition the value of Rydberg's constant R (which we now denote by R_∞) was found to be

$$R_\infty = e^4 m_2/4\pi\hbar^3 c. \tag{108.8}$$

We see that, in order to obtain the true value of Rydberg's constant, which determines the optical frequencies of the electron moving in a Coulomb field, m_2 must be replaced by the reduced mass μ. Since μ is different for different atoms, the mass of the electron can in this way be determined from spectroscopic observations. This was done by Houston by means of precise measurements of the H_α and H_β lines of hydrogen and a comparison of them with the corresponding lines of the helium ion He^+. For example, the frequency ν_H for H_α of hydrogen is

$$\nu_H = R_H\left(\frac{1}{2^2} - \frac{1}{3^2}\right) = \tfrac{5}{36} R_H,$$

where R_H is Rydberg's constant for hydrogen. For the same transition in the helium ion we have

$$\nu_{He} = 4R_{He}\left(\frac{1}{2^2} - \frac{1}{3^2}\right) = \tfrac{20}{36} R_{He},$$

where R_{He} is Rydberg's constant for He^+. The factor 4 appears because the atomic terms are proportional to the square of the nuclear charge, Z^2; the charge on the helium nucleus is twice that on the hydrogen nucleus. From the above formulae it follows that

$$\gamma = \frac{\frac{1}{4}\nu_{\mathrm{He}} - \nu_{\mathrm{H}}}{\nu_{\mathrm{H}}} = \frac{\mu_{\mathrm{He}} - \mu_{\mathrm{H}}}{\mu_{\mathrm{H}}}, \tag{108.9}$$

where μ_{He} and μ_{H} are the reduced masses of the helium ion and hydrogen. According to (108.4) we have

$$\frac{1}{\mu_{\mathrm{H}}} = \frac{1}{m_{\mathrm{H}}} + \frac{1}{m_2}, \qquad \frac{1}{\mu_{\mathrm{He}}} = \frac{1}{m_{\mathrm{He}}} + \frac{1}{m_2}, \tag{108.10}$$

where m_{H} is the mass of the hydrogen nucleus and m_{He} that of the helium nucleus. Substitution in the preceding formula gives

$$\gamma = \frac{m_{\mathrm{He}} - m_{\mathrm{H}}}{m_{\mathrm{He}} + m_{\mathrm{H}}} \cdot \frac{m_2}{m_{\mathrm{H}}}. \tag{108.9'}$$

Hence we see that, by determining γ spectroscopically and knowing the atomic weights of H and He, we can calculate the ratio m_2/m_{H}, i.e. the 'atomic weight' of the electron. In this way Houston found

$$m_2/m_{\mathrm{H}} = 0.000548\,, \qquad m_{\mathrm{H}}/m_2 = 1838.2 \pm 1.8\,. \tag{108.11}$$

The same effect can be used to determine isotope masses. The lines corresponding to the same quantum transitions are slightly different for different isotopes, because of the difference in the reduced masses. In this way the mass of heavy hydrogen (deuterium), $m_0 = 2m_{\mathrm{H}}$, has been found.

109. A system of microparticles executing small oscillations

Let us first consider a system of two identical particles executing small oscillations. Let the displacement of the first particle from the equilibrium position be x_1, and that of the second particle x_2. The potential energy $U(x_1, x_2)$ for small displacements can be expanded as a series:

$$U(x_1, x_2) = \tfrac{1}{2}\mu\omega_0^2 x_1^2 + \tfrac{1}{2}\mu\omega_0^2 x_2^2 + \lambda x_1 x_2 + \ldots, \tag{109.1}$$

where μ is the mass of each particle, ω_0 the frequency of oscillation of the particles when there is no interaction between them, and $\lambda x_1 x_2$ the interaction energy (for small x_1 and x_2).

The total-energy operator of particles having the potential energy (109.1) is

$$H = -\frac{\hbar^2}{2\mu}\frac{\partial^2}{\partial x_1^2} + \tfrac{1}{2}\mu\omega_0^2 x_1^2 - \frac{\hbar^2}{2\mu}\frac{\partial^2}{\partial x_2^2} + \tfrac{1}{2}\mu\omega_0^2 x_2^2 + \lambda x_1 x_2\,. \tag{109.2}$$

We know from classical mechanics that, for a system of particles executing small oscillations, we can define *normal co-ordinates* q_1, q_2, in which the potential energy U is given by the sum of the squares of q_1 and q_2, and the kinetic energy by the sum of the squares of the corresponding momenta, so that we have two independent normal oscillations. In the particular case considered here, the normal co-ordinates are related to x_1 and x_2 by the formulae

$$x_1 = \frac{1}{\sqrt{2}}(q_1 + q_2), \qquad x_2 = \frac{1}{\sqrt{2}}(q_1 - q_2). \tag{109.3}$$

This property of normal co-ordinates continues to hold in quantum mechanics. In (109.1) we express x_1 and x_2 in terms of the normal co-ordinates q_1 and q_2, noting that

$$\frac{\partial\psi}{\partial q_1} = \frac{\partial\psi}{\partial x_1}\frac{\partial x_1}{\partial q_1} + \frac{\partial\psi}{\partial x_2}\frac{\partial x_2}{\partial q_1} = \frac{1}{\sqrt{2}}\left(\frac{\partial\psi}{\partial x_1} + \frac{\partial\psi}{\partial x_2}\right),$$

$$\frac{\partial^2\psi}{\partial q_1^2} = \tfrac{1}{2}\left(\frac{\partial^2\psi}{\partial x_1^2} + \frac{\partial^2\psi}{\partial x_2^2} + 2\frac{\partial^2\psi}{\partial x_1\partial x_2}\right);$$

similarly

$$\frac{\partial^2\psi}{\partial q_2^2} = \tfrac{1}{2}\left(\frac{\partial^2\psi}{\partial x_1^2} + \frac{\partial^2\psi}{\partial x_2^2} - 2\frac{\partial^2\psi}{\partial x_1\partial x_2}\right).$$

Hence

$$\frac{\partial^2\psi}{\partial x_1^2} + \frac{\partial^2\psi}{\partial x_2^2} = \frac{\partial^2\psi}{\partial q_1^2} + \frac{\partial^2\psi}{\partial q_2^2},$$

and so

$$\boldsymbol{H} = -\frac{\hbar^2}{2\mu}\left(\frac{\partial^2}{\partial q_1^2} + \frac{\partial^2}{\partial q_2^2}\right) + \tfrac{1}{2}\mu\omega_1^2 q_1^2 + \tfrac{1}{2}\mu\omega_2^2 q_2^2, \tag{109.4}$$

where

$$\mu\omega_1^2 = \mu\omega_0^2 + \lambda, \qquad \mu\omega_2^2 = \mu\omega_0^2 - \lambda. \tag{109.5}$$

It follows from (109.4) that the Hamiltonian of two coupled oscillators in normal co-ordinates is the sum of the Hamiltonians of two independent oscillators, one with frequency ω_1 and the other with frequency ω_2 – the same result as in classical mechanics.

Let us now find the quantum levels and the corresponding eigenfunctions for a system of coupled oscillators. The operator involves the co-ordinates q_1 and q_2, and so the wave function ψ must be regarded as a function of q_1 and q_2. Schrödinger's equation for the stationary states of the system is

$$-\frac{\hbar^2}{2\mu}\frac{\partial^2\psi}{\partial q_1^2} + \tfrac{1}{2}\mu\omega_1^2 q_1^2\psi - \frac{\hbar^2}{2\mu}\frac{\partial^2\psi}{\partial q_2^2} + \tfrac{1}{2}\mu\omega_2^2 q_2^2\psi = E\psi. \tag{109.6}$$

This equation is easily solved by separating the variables, putting

$$\psi(q_1, q_2) = \psi_1(q_1)\cdot\psi_2(q_2) \tag{109.7}$$

and

$$E = E_1 + E_2. \tag{109.8}$$

Substituting (109.7) and (109.8) in (109.6), dividing by $\psi_1(q_1)\,\psi_2(q_2)$ and equating to the constants E_1 and E_2 separately the terms on the left-hand side which depend on

q_1 and q_2 respectively, we obtain

$$-\frac{\hbar^2}{2\mu}\frac{d^2\psi_1}{dq_1^2} + \tfrac{1}{2}\mu\omega_1^2 q_1^2 \psi_1 = E_1\psi_1\,, \tag{109.9}$$

$$-\frac{\hbar^2}{2\mu}\frac{d^2\psi_2}{dq_2^2} + \tfrac{1}{2}\mu\omega_2^2 q_2^2 \psi_2 = E_2\psi_2\,. \tag{109.9'}$$

The first of these equations is that of an oscillator of frequency ω_1, and the second is that of an oscillator of frequency ω_2. The eigenfunctions of Equation (109.9) are therefore

$$\psi_{n_1}(q_1) = \left(\frac{\mu\omega_1}{\hbar}\right)^{\frac{1}{4}} e^{-\frac{1}{2}\xi_1^2} H_{n_1}(\xi_1)\left(\xi_1 = \sqrt{\frac{\mu\omega_1}{\hbar}}\, q_1\right), \tag{109.10}$$

and the eigenvalues

$$E_{n_1} = \hbar\omega_1 (n_1 + \tfrac{1}{2}), \qquad n_1 = 0, 1, 2, \ldots. \tag{109.11}$$

Similarly for Equation (109.9′)

$$\psi_{n_2}(q_2) = \left(\frac{\mu\omega_2}{\hbar}\right)^{\frac{1}{4}} e^{-\frac{1}{2}\xi_2^2} H_{n_2}(\xi_2)\left(\xi_2 = \sqrt{\frac{\mu\omega_2}{\hbar}}\, q_2\right), \tag{109.10'}$$

$$E_{n_2} = \hbar\omega_2 (n_2 + \tfrac{1}{2}), \qquad n_2 = 0, 1, 2, \ldots. \tag{109.11'}$$

Hence it follows that the eigenfunctions of the original equation (109.6) are

$$\psi_{n_1 n_2}(q_1, q_2) = \psi_{n_1}(q_1)\,\psi_{n_2}(q_2)\,, \tag{109.12}$$

and the corresponding eigenvalues of the energy operator are

$$E_{n_1 n_2} = \hbar\omega_1 (n_1 + \tfrac{1}{2}) + \hbar\omega_2 (n_2 + \tfrac{1}{2})\,. \tag{109.13}$$

The zero-point energy of the system is

$$E_{00} = \tfrac{1}{2}\hbar\omega_1 + \tfrac{1}{2}\hbar\omega_2\,. \tag{109.14}$$

The probability of finding normal co-ordinates in the ranges q_1 to $q_1 + dq_1$ and q_2 to $q_2 + dq_2$ is

$$w(q_1, q_2)\, dq_1\, dq_2 = \psi_{n_1 n_2}^2 (q_1, q_2)\, dq_1\, dq_2\,. \tag{109.15}$$

If we wish to determine the probability that the co-ordinates of the particles will lie in the ranges x_1 to $x_1 + dx_1$ and x_2 to $x_2 + dx_2$, we need only use the fact that $dq_1\, dq_2 = dx_1\, dx_2$ and express q_1 and q_2 in (109.15) in terms of x_1 and x_2. This gives

$$w(x_1, x_2)\, dx_1\, dx_2 = \psi_{n_1 n_2}^2\left(\frac{x_1 + x_2}{\sqrt{2}}, \frac{x_1 - x_2}{\sqrt{2}}\right) dx_1\, dx_2\,. \tag{109.16}$$

Similar results are obtained for a system with any number of degrees of freedom. Suppose that we have N particles executing small oscillations about the equilibrium position. Let the displacements of the kth particle from this position be x_k, y_k, z_k.

Then the potential energy is

$$U = \tfrac{1}{2} \sum_{i,k=1}^{N} (A_{ik}x_i x_k + B_{ik}y_i y_k + C_{ik}z_i z_k + D_{ik}x_i y_k + E_{ik}x_i z_k + F_{ik}y_i z_k) + \dots, \tag{109.17}$$

where A_{ik}, B_{ik}, C_{ik}, D_{ik}, E_{ik}, F_{ik} are the second derivatives of the potential energy with respect to the displacements: for example, $A_{ik} = \partial^2 U/\partial x_i \partial x_k$ for $x_i, x_k = 0$. It is known from classical mechanics[1] that in this case we can define normal co-ordinates q_s $(s = 1, 2, \dots, 3N)$ such that the Hamiltonian function consists of a sum of Hamiltonian functions for harmonic oscillators.

The normal co-ordinates q_s and the Cartesian co-ordinates x_k, y_k, z_k are related by an orthogonal transformation:

$$q_s = \sum_k (\alpha_{sk}x_k + \beta_{sk}y_k + \gamma_{sk}z_k), \qquad s = 1, 2, \dots, 3N, \tag{109.18}$$

where α_{sk}, β_{sk}, γ_{sk} are coefficients of the transformation. In the normal co-ordinates q_s, the Hamiltonian

$$\boldsymbol{H} = \sum_{k=1}^{N} \left(-\frac{\hbar^2}{2\mu_k} \nabla_k^2 \right) + \tfrac{1}{2} \sum_{i,k=1}^{N} (A_{ik}x_i x_k + \dots + F_{ik}y_i z_k) \tag{109.19}$$

becomes

$$\boldsymbol{H} = \sum_{s=1}^{3N} \left(-\frac{\hbar^2}{2\mu} \frac{\partial^2}{\partial q_s^2} + \tfrac{1}{2}\mu\omega_s^2 q_s^2 \right), \tag{109.20}$$

where μ is some effective mass and ω_s are the frequencies of the normal oscillations. Schrödinger's equation for stationary states is

$$\left[\sum_{s=1}^{3N} \left(-\frac{\hbar^2}{2\mu} \frac{\partial^2}{\partial q_s^2} + \tfrac{1}{2}\mu\omega_s^2 q_s^2 \right) \right] \Psi(q_1, q_2, \dots, q_{3N}) = E\Psi(q_1, q_2, \dots, q_{3N}). \tag{109.21}$$

This equation can evidently be resolved into $3N$ equations for $3N$ independent oscillators if Ψ is represented as a product of functions of $q_1, q_2, \dots, q_{3N}$. The equation for the oscillator which executes the sth normal oscillation is

$$-\frac{\hbar^2}{2\mu} \frac{\mathrm{d}^2\psi(q_s)}{\mathrm{d}q_s^2} + \tfrac{1}{2}\mu\omega_s^2 q_s^2 \psi(q_s) = E_s \psi(q_s). \tag{109.22}$$

Hence

$$\psi_{n_s}(q_s) = \left(\frac{\mu\omega_s}{\hbar} \right)^{\frac{1}{4}} e^{-\frac{1}{2}\xi_s^2} H_{n_s}(\xi_s) \left(\xi_s = \sqrt{\frac{\mu\omega_s}{\hbar}}\, q_s \right), \tag{109.23}$$

$$E_{n_s} = \hbar\omega_s (n_s + \tfrac{1}{2}), \qquad n_s = 0, 1, 2, \dots. \tag{109.24}$$

[1] See, e.g., [21].

The eigenfunctions and eigenvalues of the whole system of oscillators are given by

$$\begin{aligned}&\psi_{n_1 n_2 \ldots n_s \ldots n_{3N}}(q_1, q_2, \ldots, q_s, \ldots, q_{3N}) = \\ &\psi_{n_1}(q_1)\psi_{n_2}(q_2)\ldots\psi_{n_s}(q_s)\ldots\psi_{n_{3N}}(q_{3N}),\end{aligned} \tag{109.25}$$

$$\begin{aligned}&E_{n_1 n_2 \ldots n_s \ldots n_{3N}} = \\ &\hbar\omega_1(n_1 + \tfrac{1}{2}) + \ldots + \hbar\omega_s(n_s + \tfrac{1}{2}) + \ldots + \hbar\omega_{3N}(n_{3N} + \tfrac{1}{2}),\end{aligned} \tag{109.26}$$

where $n_1, n_2, \ldots, n_s, \ldots, n_{3N}$ are positive integers or zero. The zero-point energy of the system is

$$E_0 = \tfrac{1}{2}\hbar(\omega_1 + \omega_2 + \ldots + \omega_s + \ldots + \omega_{3N}). \tag{109.27}$$

By taking all possible values of the numbers n_s in (109.26), we obtain all the quantum levels of the system of oscillating particles. It follows from (109.26) that a knowledge of the frequencies ω_s of the normal oscillations is sufficient to determine these levels.

Molecules and solids afford examples of systems which have quantum levels of the form (109.26). In both, the atoms execute small oscillations about their positions of equilibrium.[2]

It may be noted that, when the amplitude of the oscillations is large, higher terms must be taken into account in the expansion of the potential energy, i.e. terms of the form

$$\frac{1}{3!}\frac{\partial^3 U}{\partial x_i \partial y_k \partial z_l} x_i y_k z_l$$

etc. The oscillations are then non-linear, and the above results are only approximate. In particular, Formula (109.26) is valid only for small quantum numbers n_s.

110. Motion of an atom in an external field

Let us consider the motion of a system of particles (an atom or molecule) in an external field of force. For definiteness we shall take a system of two particles, of masses m_1 and m_2 and co-ordinates $x_1, y_1, z_1; x_2, y_2, z_2$. The generalisation to a larger number of particles is trivial.

Let the energy of interaction of the particles be $W(x_1 - x_2, y_1 - y_2, z_1 - z_2)$, that of the first particle in the external field be $U_1(x_1, y_1, z_1)$, and that of the second be $U_2(x_2, y_2, z_2)$. Schrödinger's equation for the wave function $\Psi(x_1, y_1, z_1, x_2, y_2, z_2, t)$ of the system is

$$i\hbar\frac{\partial\Psi}{\partial t} = -\frac{\hbar^2}{2m_1}\nabla_1^2\Psi - \frac{\hbar^2}{2m_2}\nabla_2^2\Psi + U_1\Psi + U_2\Psi + W\Psi. \tag{110.1}$$

In this equation we replace the co-ordinates $x_1, y_1, z_1; x_2, y_2, z_2$ of the particles by those of the centre of mass (X, Y, Z) and the relative co-ordinates (x, y, z) (see

[2] The quantisation of the energy of oscillations of atoms in a solid is shown by the quantum nature of the specific heat of a solid, which at sufficiently low temperatures is less than that given by the classical theory ($3k$, where k is Boltzmann's constant), decreasing as T^3. The calculation of the specific heat of a solid, using quantum theory, is described in most textbooks of statistical physics.

(108.3)–(108.3″)). In terms of these new co-ordinates, noting that

$$\left.\begin{aligned} x_1 &= X + \gamma_1 x, & x_2 &= X - \gamma_2 x, \\ y_1 &= Y + \gamma_1 y, & y_2 &= Y - \gamma_2 y, \\ z_1 &= Z + \gamma_1 z, & z_2 &= Z - \gamma_2 z, \end{aligned}\right\} \tag{110.2}$$

$$\gamma_1 = m_2/(m_1 + m_2), \qquad \gamma_2 = m_1/(m_1 + m_2), \tag{110.3}$$

we obtain from (110.1)

$$i\hbar \frac{\partial \Psi}{\partial t} = -\frac{\hbar^2}{2M}\nabla_X^2 \Psi - \frac{\hbar^2}{2\mu}\nabla_x^2 \Psi + U_1(X + \gamma_1 x, Y + \gamma_1 y, Z + \gamma_1 z)\Psi +$$
$$+ U_2(X - \gamma_2 x, Y - \gamma_2 y, Z - \gamma_2 z)\Psi + W(x, y, z)\Psi, \tag{110.1'}$$

where

$$\nabla_X^2 = \frac{\partial^2}{\partial X^2} + \frac{\partial^2}{\partial Y^2} + \frac{\partial^2}{\partial Z^2}, \qquad \nabla_x^2 = \frac{\partial^2}{\partial x^2} + \frac{\partial^2}{\partial y^2} + \frac{\partial^2}{\partial z^2}.$$

The variables X, Y, Z and x, y, z in this equation cannot be separated, owing to the presence of the field (U_1 and U_2). In general, therefore, the analysis of the equation is very difficult.

Let us assume, however, that the dimensions of the system are small. This means that we consider only systems and states such that the wave function Ψ decreases sufficiently rapidly with increasing relative distance $r = \sqrt{(x^2 + y^2 + z^2)}$ of the two particles. Let this decrease be such that the probability of finding the particles at a distance $r > a$ apart is practically zero. Then a may be regarded as the dimension of the system (the 'radius' of the atom, the 'length' of the molecule, and so on).

In this case only the region of x, y, z for which $r < a$ is of importance in Equation (110.1′). On this hypothesis we can expand U_1 and U_2 in powers of x, y, z (if U_1 and U_2 are sufficiently smooth functions). This expansion may be written

$$\begin{aligned} &U_1(X + \gamma_1 x, Y + \gamma_1 y, Z + \gamma_1 z) + U_2(X - \gamma_2 x, Y - \gamma_2 y, Z - \gamma_2 z) \\ &= U_1(X, Y, Z) + U_2(X, Y, Z) + \frac{\partial U_1}{\partial x} x + \ldots + \frac{\partial U_2}{\partial z} z + \ldots \\ &= V(X, Y, Z) + w(X, Y, Z, x, y, z) + \ldots, \end{aligned} \tag{110.4}$$

where $V(X, Y, Z)$ is the potential energy of the centre of mass of the system, and w denotes the terms involving x, y, z. This term relates the motion of the centre of mass to the relative motion. Schrödinger's equation (110.1′) may now be written

$$i\hbar \frac{\partial \Psi}{\partial t} =$$
$$\left[-\frac{\hbar^2}{2M}\nabla_X^2 + V(X, Y, Z)\right]\Psi + \left[-\frac{\hbar^2}{2\mu}\nabla_x^2 + W(x, y, z)\right]\Psi +$$
$$+ w(X, Y, Z, x, y, z)\Psi \tag{110.5}$$

Let the eigenfunctions of the internal motion, in the absence of the external field, be $\psi_n^0(x, y, z)$, and the energy eigenvalues E_n^0. Evidently ψ_n^0 satisfies the equation

$$-\frac{\hbar^2}{2\mu}\nabla_x^2\psi_n^0 + W(x, y, z)\psi_n^0 = E_n^0\psi_n^0. \tag{110.6}$$

If the effect of the external field is taken into consideration, the term $w(X, Y, Z, x, y, z)$ must be added, giving

$$-\frac{\hbar^2}{2\mu}\nabla_x^2\psi + W(x, y, z)\psi + w(X, Y, Z, x, y, z)\psi = E\psi. \tag{110.7}$$

In this equation the co-ordinates X, Y, Z of the centre of mass appear as parameters, and both the wave functions and the eigenvalues will depend on these co-ordinates.

In many cases $w(X, Y, Z, x, y, z)$ may be regarded as a perturbation, and so the equation can be solved if the solutions of Equation (110.6) are known. Let the eigenfunctions of Equation (110.7) and its eigenvalues be

$$\psi_n = \psi_n(x, y, z, X, Y, Z), \qquad E_n = E_n(X, Y, Z). \tag{110.8}$$

We now expand $\Psi(x, y, z, X, Y, Z, t)$ in terms of the eigenfunctions ψ_n. This gives

$$\Psi(x, y, z, X, Y, Z, t) = \sum_n \Phi_n(X, Y, Z, t)\psi_n(x, y, z, X, Y, Z). \tag{110.9}$$

Substituting this expansion in (110.5), multiplying by $\psi_m^*(x, y, z, X, Y, Z)$ and integrating with respect to x, y, z, we obtain (since the functions ψ_n are orthogonal) the equations for the functions Φ_n:

$$i\hbar\frac{\partial\Phi_m}{\partial t} = -\frac{\hbar^2}{2M}\nabla_X^2\Phi_m + $$
$$+ [V(X, Y, Z) + E_m(X, Y, Z)]\Phi_m - \frac{\hbar^2}{2M}\sum_n (2a_{mn}\nabla_X\Phi_n + b_{mn}\Phi_n), \tag{110.9'}$$

where

$$a_{mn} = \int \psi_m^* \cdot \nabla_X\psi_n \cdot dx\, dy\, dz, \tag{110.10}$$

$$b_{mn} = \int \psi_m^* \cdot \nabla_X^2\psi_n \cdot dx\, dy\, dz. \tag{110.10'}$$

These two terms are zero except when the functions ψ_n depend on the co-ordinates X, Y, Z of the centre of mass, and they allow the system to change from one state to another. For, even if when $t = 0$ all the Φ_n are zero except $\Phi_{n'}$, for $t \neq 0$ $\partial\Phi_n/\partial t \neq 0$, and in the course of time the superposition (110.9) will develop from the state $\Phi_{n'}$.

If the ψ_n are independent of X, Y, Z, then a_{mn} and b_{mn} are zero. If this independence is even approximately valid, we may neglect a_{mn} and b_{mn} in (110.9'), obtaining

$$i\hbar\frac{\partial\Phi_n}{\partial t} = -\frac{\hbar^2}{2M}\nabla_X^2\Phi_n + [V(X, Y, Z) + E_n(X, Y, Z)]\Phi_n. \tag{110.11}$$

This is the equation for the motion of the centre of mass of the system in a field with potential energy

$$U_n = V(X, Y, Z) + E_n(X, Y, Z), \tag{110.12}$$

with the condition that the internal state of the system is the nth quantum state. Equation (110.11) is the same as the equation of motion of a point mass.

111. Determination of the energy of stationary states of atoms from their deflection in an external field

In this section we shall consider the theory of experiments in which the energy of the stationary states of an atom is determined by deflecting a beam of atoms by means of an external field. The most important of these is the Stern-Gerlach experiment. This is usually regarded as an experiment to determine the magnetic moment of an atom, but on closer examination it is seen to be an experiment to determine the energy of an atom in an external magnetic field.

It follows from the theory of the motion of an atomic electron in the presence of a magnetic field (Section 62) that, if higher powers of the magnetic field are neglected, the effect of the field may be expressed in terms of the additional potential energy (62.7), which is equal to the energy of the (orbital and spin) magnetic dipole in the magnetic field. We can therefore apply the theory given in Section 110. It follows from the analysis in Section 62 that in this approximation the electron wave function ψ_{nlm} is independent of the magnetic field, and the eigenvalues of the energy are (62.13) and (62.13′):

$$E_{nlm} = E_{nl}^0 + \frac{e\hbar\mathscr{H}}{2\mu c}(m \pm 1). \tag{111.1}$$

Here we have assumed that the field $\mathscr{H}$ is uniform. If it varies sufficiently smoothly (as is always true for macroscopic fields) it may be regarded as a function of the coordinates X, Y, Z of the centre of the atom without rendering (111.1) invalid.[3]

Thus we can write

$$E_{nlm}(X, Y, Z) = E_{nl}^0 + \frac{e\hbar}{2\mu c}(m \pm 1)\,\mathscr{H}(X, Y, Z). \tag{111.2}$$

The wave functions ψ_{nlm} will not depend on X, Y, Z, since they are independent of the field $\mathscr{H}$. Thus we have a case where Equations (110.11) replace the general Equations (110.9′) for the wave functions Φ which describe the motion of the centre of mass. In the present problem Equations (110.11) become

$$i\hbar\frac{\partial\Phi_{nlm}}{\partial t} = -\frac{\hbar^2}{2M}\nabla_X^2\,\Phi_{nlm} + E_{nlm}(X, Y, Z)\,\Phi_{nlm}. \tag{111.3}$$

[3] For this to be true it is sufficient that the field $\mathscr{H}$ should vary only slightly over the dimension a of the atom, i.e. that

$$\mathscr{H} \gg \left|\frac{\partial\mathscr{H}}{\partial X}a\right|, \quad \left|\frac{\partial\mathscr{H}}{\partial Y}a\right|, \quad \left|\frac{\partial\mathscr{H}}{\partial Z}a\right|.$$

Since the mass M of the atom is large and the external field $\mathscr{H}$ always varies smoothly from point to point, we have just the conditions in which classical mechanics is valid. Putting

$$\Phi_{nlm}(X, Y, Z, t) = \sqrt{\rho_{nlm}(X, Y, Z, t)}\, e^{-iS_{nlm}(X,Y,Z,t)/\hbar}, \tag{111.4}$$

where S_{nlm} is the action function and ρ_{nlm} the spatial density of atoms, we obtain for S_{nlm} and ρ_{nlm} in the first approximation the classical equations (35.8) and (35.13):

$$\frac{\partial S_{nlm}}{\partial t} = \frac{1}{2M}(\nabla S_{nlm})^2 + E_{nlm}(X, Y, Z), \tag{111.5}$$

$$\frac{\partial \rho_{nlm}}{\partial t} - \frac{1}{M}\operatorname{div}(\rho_{nlm}\nabla S_{nlm}) = 0. \tag{111.6}$$

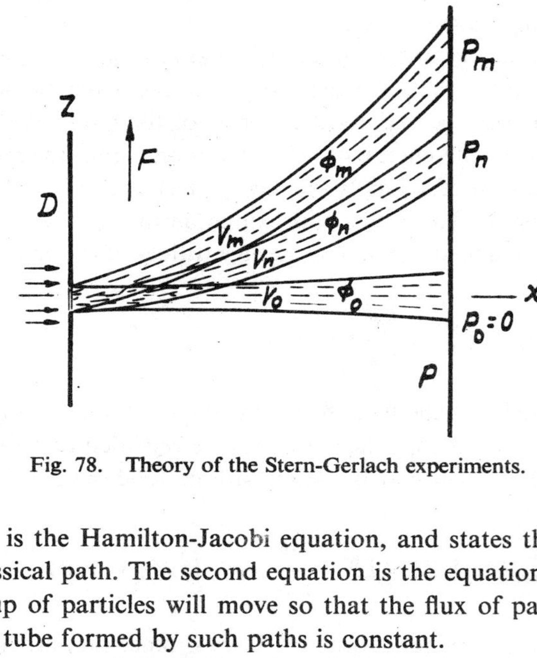

Fig. 78. Theory of the Stern-Gerlach experiments.

The first of these is the Hamilton-Jacobi equation, and states that the particle will move along a classical path. The second equation is the equation of continuity, and states that a group of particles will move so that the flux of particles through any cross-section of a tube formed by such paths is constant.

In Figure 78, let a magnetic field along the axis OZ act between D and P. Let atoms enter through a diaphragm at D, having a slit of width ΔZ_0. The beam of atoms entering at D will be split. The atoms which are in a state with magnetic moment $\mathfrak{M}_m = 0$ will move under no forces. From Equations (111.5) and (111.6) we obtain the flux which passes through without deflection. The atoms in all other states, with $\mathfrak{M}_m \neq 0$, form deflected beams (of which two are shown in Figure 78).

It is important to note that the magnetic moment $\mathfrak{M}_m$ varies discontinuously from one state to another. Consequently the beams will in general be separated so that, from the position at which atoms strike a screen (or a photographic plate) P, we can decide in which possible state ψ_m they are, i.e. we can determine their stationary states. The paths belonging to the beams are easily calculated from Equations (111.5) and (111.6), taking into account the position and shape of the diaphragm D and the initial velocity distribution of the atoms.

We can also make direct use of Newton's equations:

$$\left.\begin{aligned} M\,\mathrm{d}^2X/\mathrm{d}t^2 &= -\,\partial E_{nlm}/\partial X\,, \\ M\,\mathrm{d}^2Y/\mathrm{d}t^2 &= -\,\partial E_{nlm}/\partial Y\,, \\ M\,\mathrm{d}^2Z/\mathrm{d}t^2 &= -\,\partial E_{nlm}/\partial Z\,. \end{aligned}\right\} \tag{111.7}$$

We shall suppose that the magnetic field $\mathscr{H}$ depends only on z, at least over the greater part of the range DP. Then (111.7) gives

$$X = vt + X_0\,, \tag{111.8}$$

$$Y = Y_0\,, \tag{111.8'}$$

$$Z_m = -\frac{1}{2M}\frac{\partial E_{nlm}}{\partial Z}t^2 + Z_0\,, \tag{111.8''}$$

where v is the velocity of the atoms, which are assumed to be initially moving parallel to OX: we also assume that the field gradient $\mathrm{d}\mathscr{H}/\mathrm{d}Z$ is almost constant within the region traversed by the atoms. Denoting the length DP by l and using (111.2), we find the deflection

$$Z_m - Z_0 = -\frac{1}{2M}\frac{e\hbar}{2\mu c}(m \pm 1)\frac{\mathrm{d}\mathscr{H}}{\mathrm{d}Z}\frac{l^2}{v^2}. \tag{111.9}$$

This calculation is only approximate: in reality the atoms passing through the diaphragm will not move in classical paths, and the beam will spread out.

In order to take account of this, the approximate solution of Equation (111.3) must be taken a step further, and allowance made for the terms in S_{nlm} and ρ_{nlm} which contain the first power of $\hbar$ (see Section 35). We shall not do this here, but merely make some estimates.

Let the width of the beam in the direction of OZ be ΔZ_0. Then the velocities of the atoms in the direction of OZ cannot be zero (as was assumed in the classical treatment), because of the uncertainty relation

$$\Delta Z_0 \cdot \Delta p_z > \tfrac{1}{2}\hbar\,. \tag{111.10}$$

If the mean value $\bar{p}_z = 0$, it follows from (111.10) that $\Delta Z_0 \cdot Mv_z > \frac{1}{2}\hbar$, i.e.

$$v_z > \tfrac{1}{2}\hbar/M \cdot \Delta Z_0\,. \tag{111.10'}$$

During passage through the field for a time t, owing to the velocity spread v_z the beam width ΔZ increases, becoming

$$\Delta Z_t \approx v_z t > \tfrac{1}{2}\hbar t/M \cdot \Delta Z_0\,. \tag{111.11}$$

In order to be able to decide to which of the states E_{nlm} or $E_{nlm'}$ an atom striking the screen P belongs, we must have $|Z_{m'} - Z_m| \geqslant \Delta Z_t$, i.e. from (111.8″)

$$\frac{t^2}{2M}\left|\frac{\partial E_{nlm'}}{\partial Z} - \frac{\partial E_{nlm}}{\partial Z}\right| \geqslant \frac{\frac{1}{2}\hbar t}{M \cdot \Delta Z_0} \tag{111.12}$$

or

$$\left|\frac{\partial E_{nlm'}}{\partial Z}\Delta Z_0 - \frac{\partial E_{nlm}}{\partial Z}\Delta Z_0\right| t \geqslant \hbar . \tag{111.13}$$

Since the dependence of $E_{nlm'}$ and E_{nlm} on Z is only slight,

$$|E_{nlm'} - E_{nlm}| > \left|\frac{\partial E_{nlm'}}{\partial Z}\Delta Z_0 - \frac{\partial E_{nlm}}{\partial Z}\Delta Z_0\right|,$$

and the inequality (111.13) may be written

$$|E_{nlm'} - E_{nlm}|\, t > \hbar , \tag{111.14}$$

i.e. in order to distinguish the stationary states of the atom ($E_{nlm'}$ or E_{nlm}) the measurement must be made for a sufficient length of time t:

$$t > \hbar / |E_{nlm'} - E_{nlm}| . \tag{111.15}$$

This result will be further discussed in Section 112.

To conclude the theory of these experiments determining the stationary states of atoms by deflecting a beam of atoms in an external field, let us consider a more complex case, where the original wave function ψ represents a state with an indeterminate value of the energy.

According to the general theory, in such a state the probability of finding by measurement the value E_n for the energy is $|c_n|^2$, where c_n is the amplitude in an expansion of ψ in terms of the eigenfunctions of the energy operator.[4] We shall show how this general statement is related to the determination of the energy by the method of beam deflection. If the system is in an internal state $\psi_n(x, y, z)$, the total wave function, taking account of the motion of the centre of mass, is

$$\Psi(X, Y, Z, x, y, z) = \Phi_n(X, Y, Z)\,\psi_n(x, y, z), \tag{111.16}$$

where Φ_n is given by Equation (111.3) (or, in general, (110.11)). If the state ψ is a superposition of ψ_n, then, by the linearity of the equations of quantum mechanics, the total wave function is

$$\Psi(X, Y, Z, x, y, z) = \sum_n c_n \Phi_n(X, Y, Z)\,\psi_n(x, y, z). \tag{111.17}$$

In the experiment we measure directly not the required internal energy of the atom, but the position X, Y, Z of the atom. The probability that the atom is in the range X to $X + \mathrm{d}X$, Y to $Y + \mathrm{d}Y$, Z to $Z + \mathrm{d}Z$ is

$$\begin{aligned} w(X, Y, Z)\,\mathrm{d}X\,\mathrm{d}Y\,\mathrm{d}Z &= \mathrm{d}X\,\mathrm{d}Y\,\mathrm{d}Z \int |\Psi|^2\,\mathrm{d}x\,\mathrm{d}y\,\mathrm{d}z \\ &= \sum_n |c_n|^2\,|\Phi_n|^2\,\mathrm{d}X\,\mathrm{d}Y\,\mathrm{d}Z. \end{aligned} \tag{111.18}$$

The measurement of the energy E_n of the atom consists in deciding to which of the beams (Figure 78) the atom belongs. Each beam is described by a different wave

[4] For simplicity we denote all the quantum numbers (n, l, m) by the single letter n.

function $\Phi_n(X, Y, Z)$. In order that the experiment should in fact measure the energy of the atom, the various beams must be separated, i.e. the functions $\Phi_n(X, Y, Z)$ must be non-zero in different regions of space (and so the condition (111.15) must hold).

Let us now find the probability w_m that the atom belongs to beam m. For this purpose it is necessary to integrate (111.18) over the volume of that beam, denoted by V_m:

$$w_m = \int_{V_m} w(X, Y, Z) \mathrm{d}X \, \mathrm{d}Y \, \mathrm{d}Z = \sum_n |c_n|^2 \int_{V_m} |\Phi_n|^2 \mathrm{d}X \, \mathrm{d}Y \, \mathrm{d}Z . \quad (111.19)$$

If the beams are separated, all the integrals are zero except that of $|\Phi_m|^2$, which is equal to unity, since Φ_m is normalised. Thus

$$w_m = |c_m|^2 . \quad (111.20)$$

But w_m is just the probability that the energy of the atom is E_m (since atoms with different energies belong to different beams). Hence our determination of the energy of the atom is entirely in agreement with the interpretation of the quantities $|c_n|^2$ as the probabilities of finding a value E_n for the energy of the atom. The measuring apparatus here is the atom itself: the internal energy E_n is determined from the position of the centre of mass of the atom.

One other important point should be mentioned. It has been stated in Section 16 that measurement always changes a pure ensemble into a mixed one. It is easily seen that in the present case this change does in fact occur. The probability of finding an electron in the neighbourhood of the point x, y, z for a given position X, Y, Z of the centre of mass of the atom is

$$\begin{aligned} w(x, y, z, X, Y, Z) &= |\Psi|^2 \\ &= \sum_n \sum_m c_n c_m^* \psi_n(x, y, z) \psi_m^*(x, y, z) \Phi_n(X, Y, Z) \Phi_m^*(X, Y, Z) . \quad (111.21) \end{aligned}$$

In the region where Φ_n and Φ_m overlap there is interference of the states ψ_n, ψ_m, and the phases of c_n, c_m^* are important in the determination of w. In the range where Φ_n and Φ_m do not overlap and measurement is possible we have

$$w(x, y, z, X, Y, Z) = |\Psi|^2 = \sum |c_n|^2 |\psi_n(x, y, z)|^2 |\Phi_n(X, Y, Z)|^2 , \quad (111.21')$$

i.e. the phases of the c_n do not appear. The probability w is now formed non-coherently from the ψ_n, as is characteristic of a mixed ensemble (see Section 16).

112. Inelastic collisions between electrons and atoms. Determination of the energy of stationary states of atoms by the collision method

One simple application of the theory of the motion of many bodies is the calculation of inelastic collisions with atoms. Such collisions have been mentioned in connection with the experiments of Franck and Hertz (Section 3). We shall not, however, be able to apply our results directly to these experiments, since we shall suppose that the incident electron has an energy considerably exceeding that of an electron in an atom (and so perturbation theory can be applied). The total-energy operator of two electrons[5] is

$$\boldsymbol{H}(\mathbf{r}_1, \mathbf{r}_2) = \boldsymbol{H}(\mathbf{r}_1) + \boldsymbol{H}(\mathbf{r}_2) + W(\mathbf{r}_1, \mathbf{r}_2) = \boldsymbol{H}^0(\mathbf{r}_1, \mathbf{r}_2) + W(\mathbf{r}_1, \mathbf{r}_2), \tag{112.1}$$

$$\boldsymbol{H}(\mathbf{r}_1) = -(\hbar^2/2\mu)\nabla_1^2 + U(r_1), \tag{112.2}$$

$$\left.\begin{aligned} \boldsymbol{H}(\mathbf{r}_2) &= -(\hbar^2/2\mu)\nabla_2^2, \\ W(\mathbf{r}_1, \mathbf{r}_2) &= U(r_2) + e^2/|\mathbf{r}_1 - \mathbf{r}_2|. \end{aligned}\right\} \tag{112.3}$$

Here $U(r_1)$ denotes the potential energy of an atomic electron in the field of the nucleus and the remaining electrons; $e^2/|\mathbf{r}_1 - \mathbf{r}_2|$ is the Coulomb energy of interaction of the atomic electron with the incident electron; $U(r_2)$ is the energy of the latter in the field of the rest of the atom. The meaning of the remaining terms is clear.

The kinetic energy of the incident electron will be assumed so large that the whole of its interaction W with the atom may be regarded as a perturbation. Then Schrödinger's equation for the unperturbed motion is

$$\boldsymbol{H}^0(\mathbf{r}_1, \mathbf{r}_2)\psi^0(\mathbf{r}_1, \mathbf{r}_2) = E\psi^0(\mathbf{r}_1, \mathbf{r}_2), \tag{112.4}$$

and its solution is

$$\psi^0_{n,\mathbf{p}_0}(\mathbf{r}_1, \mathbf{r}_2) = \psi^0_n(\mathbf{r}_1)\psi_{\mathbf{p}_0}(\mathbf{r}_2), \tag{112.5}$$

$$E = E^0_n + p_0^2/2\mu, \tag{112.6}$$

where ψ^0_n is the wave function of the stationary state of the electron in the atom corresponding to the energy E^0_n, and $\psi_{\mathbf{p}_0}$ is the de Broglie wave which describes the free motion of the incident electron with momentum $\mathbf{p}_0$.

We wish to find the probability that the system of two electrons will undergo a transition to some other state

$$\psi^0_{m\mathbf{p}}(\mathbf{r}_1, \mathbf{r}_2) = \psi^0_m(\mathbf{r}_1)\psi_{\mathbf{p}}(\mathbf{r}_2). \tag{112.7}$$

To calculate this probability, we use the theory of quantum transitions under a perturbation independent of time (Section 85). The perturbation is the energy W (112.3). Let us first derive the matrix element of this perturbation for the transition $n, \mathbf{p}_0 \to m, \mathbf{p}$. We have

$$W_{m\mathbf{p},n\mathbf{p}_0} = \iint dv_1\, dv_2\, \psi^{0*}_{m\mathbf{p}}(\mathbf{r}_1, \mathbf{r}_2)\left\{U(r_2) + \frac{e^2}{|\mathbf{r}_1 - \mathbf{r}_2|}\right\}\psi^0_{n\mathbf{p}_0}(\mathbf{r}_1, \mathbf{r}_2), \tag{112.8}$$

where the integration is over the co-ordinates of the first and second electrons. First calculating the integral with respect to v_1, we define

$$\rho_{mn}(\mathbf{r}_1) = -e\psi^{0*}_m(\mathbf{r}_1)\psi^0_n(\mathbf{r}_1), \tag{112.9}$$

and call this the matrix element of charge density for the transition $n \to m$. Evidently

[5] The motion of the atom as a whole may be ignored owing to the large mass of the nucleus in comparison with that of the electron.

ρ_{nn} is the mean value of the density in the state ψ_n^0. Since the functions ψ_n^0 are orthogonal, we have

$$\int dv_1 \, \psi_m^{0*}(\mathbf{r}_1) \cdot \psi_n^0(\mathbf{r}_1) \{U(r_2) + e^2/|\mathbf{r}_1 - \mathbf{r}_2|\} = U(r_2)\delta_{mn} - e\int \rho_{mn}(\mathbf{r}_1) \, dv_1/|\mathbf{r}_1 - \mathbf{r}_2| = V_{mn}(\mathbf{r}_2). \tag{112.10}$$

This quantity may be regarded as the matrix element of the potential energy of the incident electron (2) in the field of the nucleus and the atomic electron (1).

If $m = n$, the collision is elastic. V_{nn} is the perturbation energy which appeared in Section 79 in the theory of elastic scattering of electrons. Substituting $V_{mn}(\mathbf{r}_2)$ in (112.8) and bearing in mind that

$$\psi_{\mathbf{p}_0}(\mathbf{r}_2) = \frac{e^{i\mathbf{p}_0 \cdot \mathbf{r}_2/\hbar}}{(2\pi\hbar)^{\frac{3}{2}}}, \qquad \psi_{\mathbf{p}}(\mathbf{r}_2) = \frac{e^{i\mathbf{p} \cdot \mathbf{r}_2/\hbar}}{(2\pi\hbar)^{\frac{3}{2}}}, \tag{112.11}$$

we obtain

$$W_{m\mathbf{p},n\mathbf{p}_0} = \frac{1}{(2\pi\hbar)^3} \int dv_2 \, V_{mn}(\mathbf{r}_2) \, e^{i(\mathbf{p}_0 - \mathbf{p}) \cdot \mathbf{r}_2/\hbar} = \frac{1}{(2\pi\hbar)^3} F_{mn}(\mathbf{K}), \tag{112.12}$$

where $\mathbf{K}$ denotes the vector

$$\mathbf{K} = (\mathbf{p}_0 - \mathbf{p})/\hbar = \mathbf{k}_0 - \mathbf{k}, \tag{112.13}$$

where $\mathbf{k}_0$ and $\mathbf{k}$ are the electron wave-vectors before and after the collision.

To calculate the probability per second of a transition from the initial state E_n, p_x^0, p_y^0, p_z^0 to the final state E_m, p, $d\Omega$ (where $d\Omega$ is a solid-angle element which contains the direction of the electron momentum $\mathbf{p}$ after the collision) we use Formula (85.3). The density of states in a range of total energy of the system, denoted in (85.3) by $\rho(E)$, will here be the same as for one particle (84.23), since $dE = d(E_m + p^2/2\mu) = d(p^2/2\mu)$. Thus $\rho(E) = \mu p$, and so (85.3) and (112.12) give

$$P_{n\mathbf{p}_0}(m, \mathbf{p}) \, d\Omega = \frac{2\pi}{\hbar} \frac{1}{(2\pi\hbar)^6} |F_{mn}(\mathbf{K})|^2 \, \mu p \, d\Omega. \tag{112.14}$$

To obtain from this the cross-section $q(\mathbf{p}_0, p, \theta, \phi)$ for the transition $\mathbf{p}_0 \to p$, $d\Omega$, we must normalise the wave functions of the incident electron differently, namely so that the flux per cm^2 per second is unity. Thus (112.11) must be replaced by

$$\psi'_{\mathbf{p}_0}(\mathbf{r}_2) = \frac{e^{i\mathbf{p}_0 \cdot \mathbf{r}_2/\hbar}}{\sqrt{v_0}}, \tag{112.11'}$$

where v_0 is the velocity of the incident electron:

$$v_0 = |\mathbf{p}_0|/\mu = p_0/\mu. \tag{112.15}$$

The functions (112.11) and (112.11′) differ by a factor:

$$\psi'_{\mathbf{p}_0} = \psi_{\mathbf{p}_0} (2\pi\hbar)^{\frac{3}{2}} \sqrt{(\mu/p_0)}.$$

Since the initial wave function appears in a square in the probability (112.14), the change from the normalisation (112.11) to (112.11′) (for incident particles) gives a factor $(2\pi\hbar)^3\mu/p_0$ in (112.14), and the probability $P_{n\mathbf{p}_0}(m, \mathbf{p})$ now has the dimensions of area. Since the normalisation for the incident wave is now precisely that used in calculating the cross-section (one particle per cm^2 per second), the resulting probability is equal to the cross-section, which is therefore

$$q_{n\mathbf{p}_0}(m, \mathbf{p})\,\mathrm{d}\Omega = \frac{p}{p_0}\left(\frac{\mu}{2\pi\hbar^2}\right)^2 |F_{mn}(\mathbf{K})|^2\,\mathrm{d}\Omega\,. \tag{112.16}$$

The resonance condition, which for a perturbation independent of time is the same as the law of conservation of energy, is

$$E_n + p_0^2/2\mu = E_m + p^2/2\mu\,. \tag{112.17}$$

For elastic scattering $m = n, p = p_0$, and Formula (112.16) coincides with that derived in Section 78 by the method of stationary states. For inelastic scattering the atomic factor F_{mn} is somewhat different in form (see (112.12)). Moreover, q contains the factor p/p_0, the significance of which is easily seen. $q\,\mathrm{d}\Omega$ is the ratio of the flux scattered into the angle $\mathrm{d}\Omega$ to the incident flux. This ratio of fluxes involves the ratio of velocities, which is just p/p_0 and is unity for elastic scattering. With p written as p_{mn}, and $\mathbf{K}$ as $\mathbf{K}_{mn} = (\mathbf{p}_0 - \mathbf{p}_{mn})/\hbar$, Equation (112.16) is often given in the form

$$q_{mn}(\theta, \phi)\,\mathrm{d}\Omega = \frac{p_{mn}}{p_0}\left(\frac{\mu}{2\pi\hbar^2}\right)^2 \cdot |F_{mn}(\mathbf{K}_{mn})|^2\,\mathrm{d}\Omega\,. \tag{112.16'}$$

Hence, integrating over all scattering angles, we obtain the cross-section for all collisions in which the electron energy changes by $E_m - E_n$ and the atom goes from E_n to E_m:

$$Q_{mn} = \int_{4\pi} q_{mn}(\theta, \phi)\,\mathrm{d}\Omega\,. \tag{112.18}$$

If E_n is the ground state of the atom, the incident electron can only excite the atom ($E_m > E_n$). Then Q_{mn} is called the *excitation cross-section* of the atom. Figure 79 shows a typical curve of this cross-section as a function of the electron energy. From the law of conservation of energy (112.17), if we measure the change in energy $(p^2/2\mu) - (p_0^2/2\mu)$ of the incident electrons, we can find the difference $E_m - E_n$ and so determine the

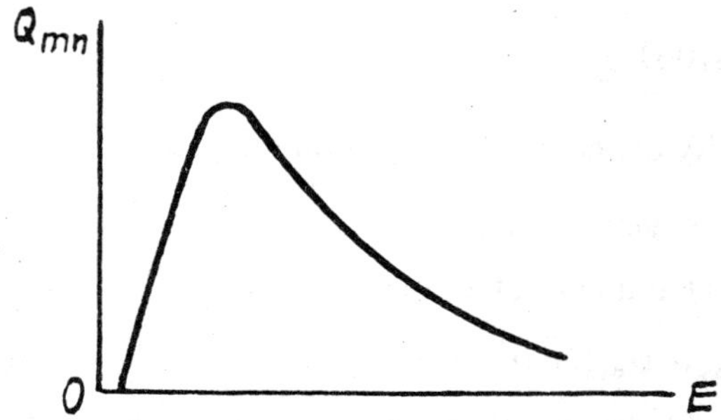

Fig. 79. The cross-section Q_{mn} for excitation by electron impact as a function of the electron energy E.

energy spectrum of the atom. This was first done in Franck and Hertz's experiments.

If, as is usually done, we take the boundary ($m = \infty$) between the discrete and continuous spectra of the atom as the origin of energy for the atomic electron, then by determining the energy loss $(p^2/2\mu) - (p_0^2/2\mu)$ of the primary electrons at which ionisation of the atom begins (i.e. secondary electrons appear), we can also measure the energy of the initial state of the atom before the collision. For in this case we have, from (112.17),

$$E_n = \frac{p^2}{2\mu} - \frac{p_0^2}{2\mu} + E_\infty = \frac{p^2}{2\mu} - \frac{p_0^2}{2\mu}. \tag{112.17'}$$

In this way we can therefore determine the stationary state of the atom. The difference between this measurement and those of the deflection of atoms by an external field is that the state of the atom is changed by the measurement (e.g. ionisation occurs), whereas in the deflection experiments it is not.

It should be pointed out that, in measuring the energy of the atom by the collision method, a certain minimum time is necessary, just as in the deflection method. The measurement is based on the law of conservation of energy (112.17), which is expressed by the appearance of a δ function in the transition probability; cf. (84.15)–(84.17), where we must put $\hbar\omega = 0$.

In reality we have not a δ function but the approximation thereto given by (84.14):

$$\delta'(E - E_0) = \frac{1}{\pi}\frac{\sin\left[(E - E_0)t/\hbar\right]}{E - E_0}, \tag{112.19}$$

which tends to $\delta(E - E_0)$ only as $t \to \infty$. The function $\delta'(E - E_0)$ is appreciably different from zero only in a range $\Delta(E - E_0)$ such that

$$\Delta(E - E_0)\cdot t \approx \hbar,$$

and is small when

$$\Delta(E - E_0)\cdot t \gg \hbar; \tag{112.20}$$

there is therefore an uncertainty in the difference of the initial and final energies E_0 and E related to the time interval t between the beginning of the measurement (when the incident electron begins to interact with the atom) and the end of the measurement (determination of the energy of the incident electron after the collision).

Let us now suppose that the energies of the incident electron both before and after the collision are precisely known. Then (112.20) gives a relation between the time of measurement t and the uncertainty $\Delta(E_n - E_{n'})$ in the difference between the initial and final energies of the measured system (the atom):

$$\Delta(E_n - E_{n'})\cdot t \geq \hbar. \tag{112.21}$$

In order to determine the levels of the system (as in Franck and Hertz's experiment) we have also to fix the final energy. To do this, we note cases where the atom is ionised by the collision ($E_{n'} > 0$) and measure the energy of the electron leaving the atom. Then the uncertainty is entirely transferred to the initial state, so that (112.21) gives

$$\Delta(E_n)\cdot t \geqslant \hbar . \tag{112.21'}$$

In order to know whether the energy of the atom before the collision was E_n or E_m, we must have $\Delta(E_n) < |E_n - E_m|$, i.e.

$$|E_n - E_m|\cdot t \geqslant \hbar . \tag{112.22}$$

Thus, in order to distinguish the state of the atom before the experiment, the measurement must continue for a sufficient time (the energy after the experiment being supposed known). If only the energy before the experiment (in the initial state) or that after the experiment is determined, the relation (112.21′) does not apply.

113. The law of conservation of energy and the special significance of time in quantum mechanics

In classical theory, the law of conservation of energy states that the energy of a closed system remains constant, and so, if the energy of such a system at time $t = 0$ is denoted by E_0, and that at time t by E_t, then

$$E_0 = E_t . \tag{113.1}$$

In quantum mechanics the law of conservation of energy is formulated similarly. According to Section 33 the energy is an integral of the motion, and the probability $w(E, t)$ of finding that the energy has the value E at time t is independent of time:

$$\mathrm{d}w(E,t)/\mathrm{d}t = 0 . \tag{113.2}$$

The law of conservation of energy in the form just stated presupposes the possibility of determining the energy at a given instant without causing it to change in an uncontrollable manner. In classical mechanics there is no doubt that such a measurement can be carried out. In quantum mechanics, on the contrary, the possibility is not obvious, because the interaction of the apparatus will in general alter the state of the system.

The measuring apparatus for determining the energy discussed in Sections 111 and 112 shows that the energy can be measured, without altering its value, only with accuracy

$$\Delta E \geqslant \hbar/\tau , \tag{113.3}$$

where τ is the duration of the measurement. This does not involve any difficulty as regards the law of conservation of energy, however, since the energy is an integral of the motion and any desired interval of time can be used to carry out the measurement. For example, if a measurement is made for a time τ, and the system is then left to itself for a time T, after which the energy is again determined, then the law of conservation of energy (113.2) states that the result of the second measurement will agree with that of the first to within $\Delta E \approx \hbar/\tau$. If we do not require that the energy should remain unchanged when it is measured, there is no restriction on the accuracy of a short (instantaneous) measurement of the energy, since the relation (113.3) involves only the uncertainty ΔE in the difference of the energies before and after the experiment (cf. (112.21)). We can therefore find a value of the energy at a given instant which is of any degree of accuracy, if only its value before or after the experiment is considered. For example, we can determine the energy at time $t = 0$ after the experiment, or $t = T$ before the experiment. The law of conservation of energy states that the two values will be equal.

Finally, it may be mentioned that a relation between the uncertainty ΔE in the energy E at a given time t and the accuracy Δt with which the instant t is determined:

$$E \cdot \Delta t \geqslant \tfrac{1}{2}\hbar, \tag{113.4}$$

similar to the relation for the momentum and the conjugate co-ordinate:

$$\Delta p_x \cdot \Delta x \geqslant \tfrac{1}{2}\hbar, \tag{113.5}$$

does *not* exist in quantum mechanics, just as there is no relation $t\boldsymbol{H} - \boldsymbol{H}t = i\hbar$ as distinct from the relation $x\boldsymbol{P}_x - \boldsymbol{P}_x x = i\hbar$.

We could expect such a relation only if the energy E could be correlated with an operator $i\hbar\partial/\partial t$ in the same way as p_x is correlated with the operator $-i\hbar\partial/\partial x$. For in quantum mechanics the energy operator $\boldsymbol{H}$ is a 'function' of the operators of the momentum and co-ordinates: $\boldsymbol{H} \equiv \boldsymbol{H}(\boldsymbol{P}_x, \boldsymbol{P}_y, \boldsymbol{P}_z, x, y, z)$. Thus from the point of view of quantum mechanics the energy is a quantity which at a given instant can have an entirely definite value, but the time t, unlike the co-ordinates x, y, z, is not an operator.

We can, however, obtain the relation (113.4) if the quantities ΔE and Δt are suitably interpreted. The following are some examples. Let a wave packet (Sections 7 and 14) be moving with group velocity v and have dimension (uncertainty of co-ordinate) Δx. Let $\Delta t = \Delta x/v$ be the time during which the wave packet passes through some fixed point x in space. Since

$$\Delta E = \Delta(p_x^2/2\mu) = v \cdot \Delta p_x, \tag{113.6}$$

we have from (113.5)

$$v\Delta p_x \cdot \Delta x/v = \Delta E \cdot \Delta t \geqslant \tfrac{1}{2}\hbar. \tag{113.7}$$

Here ΔE is the uncertainty in energy, and Δt the time for the wave packet to pass through a fixed point x in space, in other words, the time during which the mean value $\bar{x}$ changes by the uncertainty Δx of the co-ordinate.

Another example of a relation of the type (113.4) is given by the phenomenon of disintegration considered in Section 99 – the disappearance of some given initial state $\psi(x, 0)$. It has been shown in Section 99 that, if the energy uncertainty ΔE is identified with the width $\frac{1}{2}\hbar\lambda$ of the quasistationary level, and Δt is taken as the lifetime of the state $\tau = 1/\lambda = \Delta t$, then ΔE and Δt are related by (113.4) (cf. (99.31)).

It has been shown by Mandel'shtam and Tamm [63] that these examples are a particular case of a very general interpretation of the relation (113.4). Let L be any mechanical quantity which is not an integral of the motion. Then, if the state is non-stationary, the mean value $\bar{L}$ will vary with time. Let Δt be the time during which the mean value $\bar{L}$ varies by the uncertainty ΔL (the square root of the mean square deviation $\overline{\Delta L^2}$):

$$|\bar{L}(t + \Delta t) - \bar{L}(t)| = \Delta L .$$

Then Δt is related to the energy uncertainty $\Delta E = \sqrt{(\overline{\Delta E^2})}$ by (113.4).

CHAPTER XIX

SYSTEMS OF IDENTICAL MICROPARTICLES

114. The identity of microparticles

Let us now go on to consider systems consisting of identical particles, by which we mean particles having the same mass m, charge e, spin s etc., so that they behave in the same manner under the same conditions (external field, presence of other particles).

From the atomistic point of view it is reasonable, though not necessary, to suppose that all particles of a given species (electrons, protons, neutrons, etc.) are identical with one another, since the measurement of quantities pertaining to the particles (m, e, s), is, of course, possible only to within a certain accuracy (Δm, Δe, Δs), and it is always permissible to suppose that different particles of a given species may differ at least within the accuracy of measurement.

The question whether all particles of a given species are identical or not could be resolved only if the behaviour of a group of identical particles were different from that of a group of particles differing from one another to some extent, however slight. Quantum mechanics in fact leads to such a qualitative difference between the properties of a group of identical particles and those of a group of different particles. Thus quantum mechanics and experiment can be used to resolve the at first sight intractable problem whether all particles of a given species are or are not identical.

In order to see how the question is resolved, we must first discuss the simplest properties of groups of identical particles. Let such a group contain N particles, and let the co-ordinates of the kth particle be denoted by q_k, so that q_k must be taken to include the three co-ordinates (x_k, y_k, z_k) giving the position of the centre of mass of the particle, and also a fourth co-ordinate s_k giving the spin of the particle, if any.

Let the mass of the particles be m, the energy in the external field be $U(q_k, t)$, and the energy of interaction of the kth and jth particles be $W(q_k, q_j)$. Then the Hamiltonian of a system of such particles is

$$\begin{aligned} &\boldsymbol{H}(q_1, q_2, \ldots, q_k, \ldots, q_j, \ldots, q_N, t) \\ &= \sum_{k=1}^{N} \left[-\frac{\hbar^2}{2m} \nabla_k^2 + U(q_k, t) \right] + \sum_{k \neq j = 1}^{N} W(q_k, q_j). \end{aligned} \tag{114.1}$$

The assumption that the particles are identical is here expressed by the fact that the masses of the particles, the energy U in the external field and the interaction energy

W are taken to have the same form for all the particles. This property of the Hamiltonian holds in any external field, since any external field acts in the same way on identical particles.

In order to derive general conclusions it is not very convenient to start from the particular form (114.1) of the Hamiltonian.[1] We must therefore express the fact that the Hamiltonian describes a system of identical particles, without using its explicit form.

From (114.1) it is easy to see what is the necessary and most general property of the Hamiltonian of a system of identical particles. If the co-ordinates q_k of the kth particle and q_j of the jth particle are interchanged in (114.1), the Hamiltonian is unchanged, since this process simply rearranges the terms in the sums in the Hamiltonian:

$$\boldsymbol{H}(q_1, q_2, \ldots, q_k, \ldots, q_j, \ldots, q_N, t) = \boldsymbol{H}(q_1, q_2, \ldots, q_j, \ldots, q_k, \ldots, q_N, t) \tag{114.2}$$

for all pairs (j, k) of the N particles forming the system. If the N particles included even one which was different, the equality would not hold when this different particle was interchanged with any other. Thus Equation (114.2) expresses the most general property of a Hamiltonian belonging to a group of identical particles.

This property may be briefly stated as follows: the Hamiltonian of a system of identical particles is invariant (symmetrical) with respect to the interchange of the co-ordinates of any pair of particles.

Since we shall often have to mention interchanges, it is convenient to define a new operator, the particle interchange operator $\boldsymbol{P}_{kj}$, which signifies that the co-ordinates of the kth and jth particles are to be interchanged. For example, if we have a function $f(\ldots, q_k, \ldots, q_j, \ldots)$, then

$$\boldsymbol{P}_{kj} f(\ldots, q_k, \ldots, q_j, \ldots) = f(\ldots, q_j, \ldots, q_k, \ldots). \tag{114.3}$$

This operator is evidently linear, since in order to interchange the co-ordinates in a sum of two functions we must interchange them in each of the functions.

In terms of the operator $\boldsymbol{P}_{kj}$, Equation (114.2) may be written

$$\boldsymbol{P}_{kj} \boldsymbol{H}(q_1, \ldots, q_k, \ldots, q_j, \ldots, q_N, t) = \boldsymbol{H}(q_1, \ldots, q_k, \ldots, q_j, \ldots, q_N, t) \boldsymbol{P}_{kj} \tag{114.4}$$

for all pairs k, j. Thus the operator $\boldsymbol{P}_{kj}$ commutes with the Hamiltonian of a system of identical particles. For, if we apply the operator $\boldsymbol{PH}$ to some function ψ, (114.2) shows that the result is the same as that of applying the operator $\boldsymbol{HP}$, since the operator $\boldsymbol{P}$ does not change the Hamiltonian $\boldsymbol{H}$.

Using this property of the Hamiltonian, we shall prove an important lemma concerning wave functions which describe the state of systems of identical particles. Let the wave function of a system of N particles be $\Psi(q_1, \ldots, q_k, \ldots, q_j, \ldots, q_N, t)$; it must

[1] By writing the Hamiltonian $\boldsymbol{H}$ in the form (114.1) we have excluded non-potential fields (e.g. a magnetic field), and also interactions depending on the velocities of the particles (magnetic forces). These could be taken into account without affecting the following discussion.

satisfy Schrödinger's equation

$$i\hbar\frac{\partial}{\partial t}\Psi(q_1,\ldots,q_k,\ldots,q_j,\ldots,q_N,t)$$
$$=\boldsymbol{H}(q_1,\ldots,q_k,\ldots,q_j,\ldots,q_N,t)\,\Psi(q_1,\ldots,q_k,\ldots,q_j,\ldots,q_N,t). \tag{114.5}$$

In this equation we interchange the co-ordinates of the kth and jth particles by applying to both sides the operator $\boldsymbol{P}_{kj}$:

$$i\hbar\frac{\partial}{\partial t}(\boldsymbol{P}_{kj}\Psi)=\boldsymbol{P}_{kj}(\boldsymbol{H}\Psi). \tag{114.5'}$$

Since the Hamiltonian $\boldsymbol{H}$ for identical particles is symmetrical with respect to interchange of particles, we can, by (114.4), interchange the operators $\boldsymbol{P}_{kj}$ and $\boldsymbol{H}$ in (114.5'). This gives

$$i\hbar\frac{\partial}{\partial t}(\boldsymbol{P}_{kj}\Psi)=\boldsymbol{H}(\boldsymbol{P}_{kj}\Psi). \tag{114.6}$$

A comparison of (114.6) with the original equation (114.5) shows that, if $\Psi(q_1, \ldots, q_k, \ldots, q_j, \ldots, q_N, t)$ is a solution of Schrödinger's equation (114.5), then so is

$$\Psi'=\boldsymbol{P}_{kj}\Psi=\Psi(q_1,\ldots,q_j,\ldots,q_k,\ldots,q_N,t), \tag{114.7}$$

and so Ψ' as well as Ψ represents one of the possible states of the system. It differs from Ψ in that the kth particle is now in the state previously occupied by the jth particle and *vice versa*. Continuing with further interchanges, we can obtain new possible states Ψ'', Ψ''', ... of the system, differing in the distribution of particles among the states.

In asserting that the first particle is in state a (the first position in the wave function), the second particle in state b (the second position), etc., we meet with one characteristic difficulty. This arises because, if from the atomistic point of view we regard all particles of a given species as identical, we can distinguish particles only by their states (for example, by their distribution in space, the values of their momentum or energy, and so on). In the course of time the states of the particles will naturally vary, and they may exchange states. In classical mechanics, since it is in principle possible to follow the paths of particles, we can, for instance, label the particles by their position at time $t = 0$, and then say at any instant whether the particle called the first or that called the second is at a given point. In the quantum region this cannot be done. If we were to label the particles by their positions at time $t = 0$, the wave packets belonging to the different particles would rapidly spread and overlap, so that, on finding some particle at a given point at a time $t > 0$, we should have no way of saying whether this was the first or the second particle.

These considerations are illustrated by Figure 80. Figure 80a shows the positions of the particles x_1 and x_2 at time $t = 0$ and their further motion along the classical paths. Figure 80b shows the wave packets of the particles around x_1 and x_2 at time $t = 0$ (the hatched regions) and their subsequent spread. It should be noted that only those regions where $|\Psi|^2$ is large are hatched, so that in the remaining regions the wave packets still overlap but $|\Psi|^2$ is small. On finding a particle in the region where the wave packets overlap, we cannot decide which of the two particles it is.

Another example is the following. Let particles be in a box divided by a partition (Figure 81). The walls of the box are impenetrable, i.e. the potential energy of the particles increases as the walls are approached. In particular, the partition is a potential barrier, as shown in the lower part of Figure 81. If the particle energy is less than the height of the barrier, then according to classical mechanics the particles cannot penetrate the barrier. We can therefore distinguish particles by their position in the left or right half of the box.

According to quantum mechanics, for any barrier of finite height there is a certain probability that a particle will penetrate through the barrier owing to the tunnel effect.

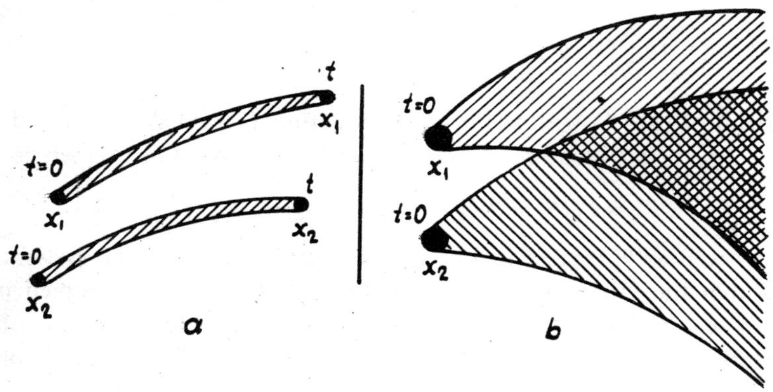

Fig. 80. Numbering of particles from their position in space: (a) in classical mechanics, (b) in quantum mechanics. The numbering becomes indefinite in the cross-hatched region.

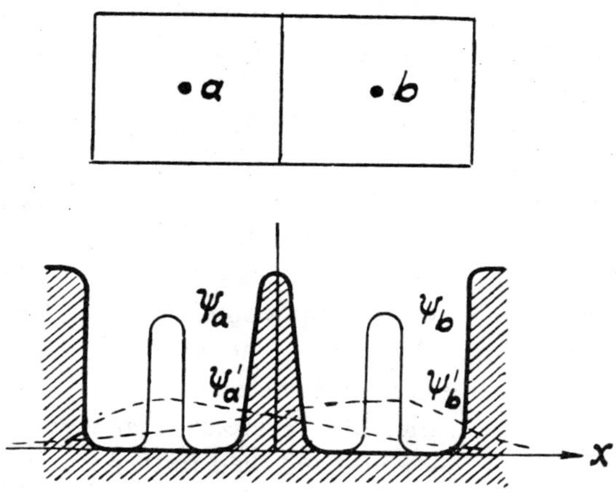

Fig. 81. Two particles in a box divided by a partition. The form of the potential near the walls and the wave functions of the particles are shown beneath.

If the wave functions of the particles are initially Ψ_a and Ψ_b (Figure 81), after a certain time they become Ψ_a' and Ψ_b' (the broken curves), so that the particle a may be found on the right and the particle b on the left. For $t \to \infty$ the wave functions Ψ_a' and Ψ_b' become equal and will have symmetrically placed maxima in the two halves of the box. The probability of finding particle a in one part of the box is equal to the corresponding probability for particle b, so that no trace of the original asymmetry remains.

Similar arguments can be given in cases where the particles are labelled not by their position in space, as in the above examples, but by some other properties describing their state. For example, let the particle a have momentum p_a at time $t = 0$, and particle b have momentum p_b. Since the states with given momentum occupy all space, there is always some probability that the particles will collide, exchanging momenta, so that particle a acquires momentum p_b, and particle b acquires momentum p_a.

Thus in the quantum region the only way in which identical particles might be distinguished, namely the distinction by means of their states, fails. In this connection it is conceivable that the systems found in Nature are so constructed that the problem of distinguishing identical particles is illusory: that is, the states of a group of identical particles are always such that we can speak only of the state of the group as a whole, and not of the distribution of the particles among states. This is in fact a correct hypothesis. It may be formulated in the form of the following principle: in a group of identical particles, only those states occur which are unaltered when identical particles are interchanged. This means that the probability of finding a value L', in measuring some mechanical quantity L pertaining to a system of identical particles or to part of it, is unaltered when the states of the particles are interchanged.

The above principle does not follow from the results of quantum mechanics previously given, but, as we shall see, it is entirely compatible with quantum mechanics and is necessary in order to derive results in agreement with experiment.

115. Symmetric and antisymmetric states

Let $\Psi(q_1, ..., q_k, ..., q_j, ..., q_N, t)$ be a wave function describing the state of a system of N identical particles. Then, if the states of the kth and jth particles, say, are interchanged, we obtain a new state of the system, which according to (114.7) is a possible state, described by the wave function $\Psi'(q_1, ..., q_j, ..., q_k, ..., q_N, t)$. The principle derived above asserts that this new state is indistinguishable from the previous one, i.e. that Ψ' and Ψ in fact describe the same state of the system.

Wave functions which describe the same physical state can differ only by a constant factor. It therefore follows from the principle of indistinguishability that

$$\Psi'(q_1, ..., q_j, ..., q_k, ..., q_N, t) = \lambda \Psi(q_1, ..., q_k, ..., q_j, ..., q_N, t),$$

where λ is some constant factor. This equation can be written in terms of the interchange operator as

$$P_{kj} \Psi = \lambda \Psi. \qquad (115.1)$$

On the left of this equation the operator P_{kj} acts on the function Ψ, and on the right the same function is multiplied by the number λ. Thus Equation (115.1) is an equation giving the eigenfunctions Ψ and eigenvalues λ of the interchange operator $\boldsymbol{P}_{kj}$. We can therefore say that the condition (115.1) imposed by the principle of indistinguishability on the possible states of the system is that the wave functions Ψ which describe the state of the system must be eigenfunctions of the operators $\boldsymbol{P}_{kj}$ for all k and j. It is easy to determine these eigenfunctions and eigenvalues λ. To do so, we again apply the interchange operator $\boldsymbol{P}_{kj}$ to (115.1):

$$\boldsymbol{P}_{kj}^2 \Psi = \lambda \boldsymbol{P}_{kj} \Psi. \tag{115.2}$$

When the interchange operator $\boldsymbol{P}_{kj}$ is applied twice, the function Ψ is unaffected. Thus the left-hand side of (115.2) is just $\Psi(..., q_k, ..., q_j, ...)$ and the right-hand side, by (115.1), is $\lambda^2 \Psi(..., q_k, ..., q_j, ...)$, so that (115.2) becomes $\Psi = \lambda^2 \Psi$, or

$$\lambda^2 = 1. \tag{115.3}$$

Hence we obtain the eigenvalues of the interchange operator $\boldsymbol{P}_{kj}$:

$$\lambda = \pm 1, \tag{115.4}$$

and from (115.1) the corresponding eigenfunctions have the following properties:

$$\boldsymbol{P}_{kj} \Psi = +\Psi \quad \text{for} \quad \lambda = +1 \tag{115.5}$$

and

$$\boldsymbol{P}_{kj} \Psi = -\Psi \quad \text{for} \quad \lambda = -1, \tag{115.6}$$

i.e. the eigenfunctions of the interchange operator $\boldsymbol{P}_{kj}$ are functions which either are unchanged (115.5) or change sign (115.6) when the co-ordinates q_k of the kth particle and q_j of the jth particle are interchanged. We call the functions satisfying (115.5) *symmetric* functions, and those satisfying (115.6) *antisymmetric* functions, with respect to the interchange of the kth and jth particles.

Thus the possible states of a system of N identical particles must be described by wave functions $\Psi(q_1, ..., q_k, ..., q_j, ..., q_N, t)$ which either change sign or remain unaltered when any pair of particles (k, j) are interchanged. From considerations of the equivalence of all particles it is easily seen that the possible functions Ψ are either symmetric with respect to all pairs of identical particles or antisymmetric with respect to all pairs of particles; there cannot be functions symmetric with respect to some particles and antisymmetric with respect to others.[2] The principle of indistinguishability of particles shows that only two kinds of state are possible for identical

[2] If interchanges of both kinds occur, then $\Psi = 0$. For, let Ψ be symmetric for the interchanges $k - j$ and $j - i$, but antisymmetric for the interchange $i - k$. Then

$$\begin{aligned} \Psi(\ldots, q_i, \ldots, q_k, \ldots, q_j, \ldots) &= -\Psi(\ldots, q_k, \ldots, q_i, \ldots, q_j, \ldots) \\ &= -\Psi(\ldots, q_k, \ldots, q_j, \ldots, q_i, \ldots) \\ &= -\Psi(\ldots, q_j, \ldots, q_k, \ldots, q_i, \ldots), \\ &= -\Psi(\ldots, q_i, \ldots, q_k, \ldots, q_j, \ldots) \end{aligned}$$

so that $\Psi(\ldots, q_i, \ldots, q_k, \ldots, q_j, \ldots) = 0$. The proof is similar if two interchanges are antisymmetric and one symmetric.

particles:

$$P_{kj}\,\Psi_s = \Psi_s \text{(for all } k,j), \tag{115.7}$$

which are symmetric with respect to all the particles, and

$$P_{kj}\,\Psi_a = -\,\Psi_a \text{(for all } k,j), \tag{115.8}$$

which are antisymmetric with respect to all the particles.

We shall now show that there can be no transitions between the two kinds of state: if at some instant the system is in a symmetric state (Ψ_s) or an antisymmetric state (Ψ_a), then it will always be in a symmetric or antisymmetric state respectively. To prove this important result, we need only make use of Schrödinger's equation and the fact that the Hamiltonian must be symmetrical with respect to identical particles. Schrödinger's equation

$$i\hbar\partial\Psi/\partial t = \boldsymbol{H}\Psi \tag{115.9}$$

may be conveniently written as

$$\mathrm{d}_t\,\Psi = (1/i\hbar)\,\boldsymbol{H}\Psi\,\mathrm{d}t, \tag{115.10}$$

where d_t denotes the increment in the wave function in time $\mathrm{d}t$. Let us assume that, at time $t = 0$, Ψ is a symmetric function of the co-ordinates of the particles ($\Psi = \Psi_s$). Then, owing to the symmetry of $\boldsymbol{H}$, the quantity $\boldsymbol{H}\Psi$ is also a symmetric function of the co-ordinates of the particles, and so the increment $\mathrm{d}_t\Psi$ is a symmetric function of them.

In terms of the interchange operator, this argument may be written in the form

$$\boldsymbol{P}_{kj}(\boldsymbol{H}\Psi_s) = \boldsymbol{H}(\boldsymbol{P}_{kj}\,\Psi_s) = \boldsymbol{H}\Psi_s,$$

whence, using (115.10), we have

$$\boldsymbol{P}_{kj}(\mathrm{d}_t\Psi_s) = \mathrm{d}_t\,\Psi_s \tag{115.11}$$

for all pairs k, j. Thus the proof shows that a function symmetric at any instant ($t = 0$) remains symmetric at neighbouring instants both earlier and later, since the proof is equally valid for $\mathrm{d}t > 0$ and $\mathrm{d}t < 0$. The symmetry of the function is therefore the same at every instant from $t = -\infty$ to $t = +\infty$. The corresponding proof for antisymmetric functions is exactly similar. Let the function Ψ describing the state of the system at time $t = 0$ be antisymmetric ($\Psi = \Psi_a$). Then

$$\boldsymbol{P}_{kj}\,\Psi_a = -\,\Psi_a,$$

and

$$\boldsymbol{P}_{kj}(\boldsymbol{H}\Psi_a) = \boldsymbol{H}(\boldsymbol{P}_{kj}\Psi_a) = -\,\boldsymbol{H}\Psi_a,$$

so that (115.10) gives

$$\boldsymbol{P}_{kj}(\mathrm{d}_t\Psi_a) = -\,\mathrm{d}_t\Psi_a, \tag{115.12}$$

i.e. the increment of an antisymmetric function Ψ_a is itself antisymmetric. If, therefore, the system is in a state described by an antisymmetric function Ψ_a, then it will always be in such a state. The theorem thus proved shows that the division of states into two classes is 'absolute': if a system is found to be in a state of one or the other class (Ψ_s or Ψ_a) at some instant, it will never change to the other class. Such a change is impossible, however the external field may vary, since any external field acts identically on identical particles, and so the Hamiltonian always remains symmetrical.

We now have to decide which of the two possibilities ($\Psi = \Psi_s$ or $\Psi = \Psi_a$) should be used to describe the state of a given system consisting of identical particles.

116. Bose particles and Fermi particles. The Pauli principle

We have seen that in quantum mechanics the principle of indistinguishability leads to the existence of two classes of states between which there is no mixing. The choice of one or the other class for a given system of particles can therefore be determined only by the nature of the particles, and not by that of the external field or any similar circumstance.

It has been established by experiment that there are in Nature particles of both classes, and it is observed that particles whose spin is an integral number of times Planck's constant,

$$s = \hbar m\,, \qquad m = 0, 1, 2, \tag{116.1}$$

are described by symmetric functions (Ψ_s). We shall call these *Bose particles* or *bosons*, and assemblies of them *Bose-Einstein ensembles*, after the physicists who derived the statistics of such particles. Particles whose spin is a half-integral number of times Planck's constant,

$$s = \hbar m\,, \qquad m = \tfrac{1}{2}, \tfrac{3}{2}, \tfrac{5}{2}, \tag{116.2}$$

are described by antisymmetric functions (Ψ_a). We shall call these *Fermi particles* or *fermions*, and assemblies of them *Fermi-Dirac ensembles*, after the physicists who set up the statistics of particles of this kind.[3]

All the simple 'elementary' particles have spin 0, $\frac{1}{2}$ or 1 (see Figure 93). Electrons, protons, neutrons, hyperons, the μ meson, the neutrino and their antiparticles have spin $\frac{1}{2}$, and are therefore fermions; π mesons and K mesons have spin 0 and are bosons. The only elementary particle with spin 1 is the photon, which also obeys Bose statistics.

Whether a complex system such as an atom or nucleus belongs to one or the other class is determined by the number and class of the simpler particles of which it consists. As an example, let us consider a hydrogen atom. This consists of two fermions, a proton and an electron. The total angular momentum of the hydrogen atom in the ground state is the sum of the spins of the proton and the electron. Since each of these has angular momentum $\pm \frac{1}{2}\hbar$, the total angular momentum of the hydrogen atom in the ground state may be 0 or $\pm \hbar$, i.e. it is an integral number of times Planck's constant.

Let us now consider an assembly of hydrogen atoms, denoting the co-ordinates of

[3] By means of the theory of relativity Pauli showed that this rule may be accounted for theoretically. The argument cannot be given here, and the reader is referred to the original work [72].

the proton in the kth atom by Q_k and those of the corresponding electron by ξ_k. Then the wave function describing the assembly of N hydrogen atoms is

$$\Psi = \Psi(Q_1, \xi_1, \ldots, Q_k, \xi_k, \ldots, Q_j, \xi_j, \ldots, Q_N, \xi_N, t). \tag{116.3}$$

We shall regard each hydrogen atom as one particle (as in possible for all phenomena where the possibility of excitation of the electron in the hydrogen atom may be neglected). Then an exchange of states between two hydrogen atoms, the kth and jth, signifies a simultaneous interchange in Ψ of the nucleus co-ordinates Q_k, Q_j and the electron co-ordinates ξ_k, ξ_j belonging to the kth and jth atoms. Since protons and electrons are regarded as fermions, the wave function Ψ must be antisymmetric with respect to interchange of any pair of nuclei Q_k and Q_j. It must likewise be antisymmetric with respect to interchange of any pair of electrons ξ_k and ξ_j. Thus Ψ changes sign when the kth and jth protons are interchanged, and again when the kth and jth electrons are interchanged. Hence Ψ is unaltered when atoms of hydrogen are interchanged by the simultaneous interchange of protons and electrons, i.e. Ψ is symmetric with respect to interchange of hydrogen atoms, and the latter, regarded as simple particles, are bosons.

A similar argument can be given for the α particle, which consists of two protons and two neutrons. Since the wave function for a system of α particles must be antisymmetric with respect to interchange of protons and with respect to interchange of neutrons, we easily conclude that it must be symmetric with respect to interchange of α particles, i.e. α particles must be bosons. This result corresponds to the fact that the total angular momentum of the α particle must be an integral multiple of $\hbar$, since it must consist of four spins each of $\frac{1}{2}\hbar$. In fact the angular momentum of the α particle is zero.

Let us now consider a fundamental property of fermions, namely that they obey what is called the *Pauli principle*, formulated by Pauli from an analysis of empirical data on atomic spectra before the detailed development of quantum mechanics.

This principle, in its simplest form, states that in a given system no more than one electron can be in any one quantum state.

As an example, let us consider the quantum state of an electron moving in a field of central forces, which is described by three quantum numbers n, l, m giving the energy of the electron (n), its orbital angular momentum (l) and the component of the orbital angular momentum in some one direction (m), and also by a fourth quantum number m_s $(= \pm \frac{1}{2})$ giving the component of the electron spin s in the same direction. Thus the quantum state is completely defined by the four numbers n, l, m, m_s. The Pauli principle asserts that there is either no electron or just one electron in such a state, but not more than one. In a state with the same quantum numbers n, l, m relating to the motion of the centre of mass of the electrons, there can be two electrons with oppositely directed spins $m_s = \pm \frac{1}{2}$.

The above formulation of the Pauli principle is simple, but suffers from the defect of being approximate, since when a second electron is placed in a state with given numbers n, l, m, the state itself is altered by the interaction between the first and the second electron. In the elementary formulation it is therefore not entirely clear which is the state in which not more than one electron can be placed. Nevertheless, since in many cases the state of the electrons is altered only slightly by their interaction, this formulation of the Pauli principle is already very useful.

We may formulate the Pauli principle so as to avoid this difficulty. To do so, we note that the electron (or any particle with spin $\frac{1}{2}\hbar$) has four degrees of freedom: three in the motion of its centre of mass, and one in its spin. To specify the state of an individual electron, either in a larger system or alone, it is therefore sufficient to measure four quantities L_1, L_2, L_3, s, which must have the following properties: (a) they can all be measured simultaneously, (b) the first three specify the motion of the centre of mass and are independent, (c) the fourth determines the state of the spin of the electron.

A set of four such quantities forms a complete set of mechanical quantities for the electron. A simultaneous measurement of them is a complete measurement, leading to a state $\psi_{L_1L_2L_3s}(q_k)$ in which the four quantities L_1, L_2, L_3, s have given values. For brevity we denote this set of values of the four quantities by the single letter n, so that

$$\psi_n(q_k) = \psi_{L_1L_2L_3s}(q_k). \tag{116.4}$$

Some examples of such sets of four quantities are as follows. Three of them may be the momentum components p_x, p_y, p_z, and the fourth, which determines the spin of the electron, may be, for example, the component of the spin in the direction of the electron momentum (s_p). Then $L_1 = p_x$, $L_2 = p_y$, $L_3 = p_z$, $s = s_p$. The necessary independence of the three quantities L_1, L_2, L_3 excludes, for example, a choice of $L_1 = p_x$, $L_2 = p_y$, $L_3 = p_x^2$, since then L_3 is a function of L_1. Another choice may be as follows: L_1 is the energy E_{nlm} of the motion of the electron in the field of the nucleus ($L_1 = E_{nlm}$), L_2 is the angular momentum of the electron ($L_2 = M$), L_3 is the component of the angular momentum in some direction ($L_3 = M_z$), and finally the spin state is defined by the component of the spin along the axis OZ ($s = s_z$). With the former choice of quantities L_1, L_2, L_3, s measurement yields the state

$$\psi_n(q_k) = \psi_{p_xp_yp_zs_p}(q_k), \tag{116.5}$$

and with the second choice

$$\psi_n(q_k) = \psi_{nlms_z}(q_k). \tag{116.5'}$$

These states resulting from the measurement will not be stationary states; this is evident merely from the fact that in a system of electrons neither the momentum nor the energy of an individual electron is an integral of the motion. In the present discussion another aspect of the matter is important. By considering states $\psi_n(q_k)$ of an individual electron which arise as a result of a measurement made on an electron of the system, we avoid the use of the imprecise term 'state of an electron in the system', since the state of the system is described by a single wave function $\psi(q_1, \ldots, q_k, \ldots, q_N, t)$, and the state of a single electron cannot be separated without altering the system. If we make a measurement of quantities pertaining to an individual electron (L_1, L_2, L_3, s), then, at least at the instant $t = 0$ when the measurement was made, the state of the electron will be $\psi_n(q_k)$. Thus, instead of the 'state of an individual electron in the system', we use the state of an individual electron occurring as a result of a complete measurement on it. In this way the Pauli principle can be formulated in the most general form without using the inexact expression 'quantum states of an individual electron'.

The general form of the Pauli principle thus runs: in a system of electrons, at any instant, on measuring any four quantities L_1, L_2, L_3, s which describe the state of an individual electron, each set of values of these four quantities can occur for only one electron in the system.

We shall now show that the principle empirically established by Pauli is a consequence of the identity of particles in quantum mechanics: particles described by antisymmetric wave functions (fermions) obey the Pauli principle.

First of all we shall prove this, for simplicity, for an assembly of only two particles; the generalisation to any number of particles is entirely straightforward. Let us assume that the state of the particles is described by an antisymmetric wave function $\Psi(q_1, q_2, t)$, where q_1, q_2 denote, as before, the set of all co-ordinates and the spin of the first and second particle respectively. Let us also assume that we measure for the first electron a set of four quantities which completely describe its state, and denote their values by the single letter n_1. The values of the same quantities for the second electron are denoted by n_2.

The state of the first electron when the measured quantities have the value n_1 is described by a wave function $\psi_{n_1}(q_1)$, say, and that of the second electron by $\psi_{n_2}(q_2)$. Since we are concerned with the measurement of mechanical quantities, the function $\psi_{n_1}(q_1)$ is an eigenfunction of the operators of those quantities, and so the functions for different values of n_1 form an orthogonal set of functions:

$$\int \psi^*_{n'_1}(q_1)\psi_{n_1}(q_1)\,\mathrm{d}q_1 = \delta_{n_1' n_1}. \tag{116.6}$$

The same is true, of course, of the function $\psi_{n_2}(q_2)$. Since n_2 represents the same mechanical quantities as n_1, the ψ_{n_2} are the same functions as the ψ_{n_1}, except that they refer to the second electron, and so their argument is q_2 instead of q_1.

We may expand the function $\Psi(q_1, q_2, t)$ which describes the state of the system in terms of the eigenfunctions of the quantities which are measured for the electrons, i.e. of $\psi_{n_1}(q_1)$ and $\psi_{n_2}(q_2)$. This gives

$$\Psi(q_1, q_2, t) = \sum_{n_1}\sum_{n_2} c(n_1, n_2, t)\psi_{n_1}(q_1)\psi_{n_2}(q_2), \tag{116.7}$$

where

$$c(n_1, n_2, t) = \int \Psi(q_1, q_2, t)\psi^*_{n_1}(q_1)\psi^*_{n_2}(q_2)\,\mathrm{d}q_1\,\mathrm{d}q_2\,; \tag{116.8}$$

in writing the sum over n_1 and n_2 in (116.7), we have assumed that the measured quantities take only discrete values. If they take continuous values, the sums must be replaced by integrals, but this does not affect the subsequent argument, and so for definiteness we use the sum notation. The sum over n_1 and n_2 is over all values of n_1 and n_2, which take the same range of values (since the mechanical quantities concerned are the same for the two electrons).

According to the general theory, the quantity

$$w(n_1, n_2, t) = |c(n_1, n_2, t)|^2 \tag{116.9}$$

is the probability that at time t the value n_1 will be measured for the first electron and n_2 for the same quantities for the second electron. In $\Psi(q_1, q_2, t)$ we interchange the two electrons. Since by hypothesis fermions are involved, the function Ψ thereby changes sign. Hence

$$\Psi(q_2, q_1, t) = \sum_{n_1}\sum_{n_2} c(n_1, n_2, t)\,\psi_{n_1}(q_2)\,\psi_{n_2}(q_1) = -\Psi(q_1, q_2, t), \tag{116.10}$$

i.e.

$$\sum_{n_1}\sum_{n_2} c(n_1, n_2, t)\,\psi_{n_1}(q_2)\,\psi_{n_2}(q_1) = -\sum_{n_1}\sum_{n_2} c(n_1, n_2, t)\,\psi_{n_1}(q_1)\,\psi_{n_2}(q_2). \tag{116.11}$$

If we now transpose the symbols n_1 and n_2 there is no change, since the sums are over all values of n_1 and n_2 and these take the same values. This shows that (116.11) can be written as

$$\sum_{n_2}\sum_{n_1} c(n_2, n_1, t)\,\psi_{n_2}(q_2)\,\psi_{n_1}(q_1) = -\sum_{n_1}\sum_{n_2} c(n_1, n_2, t)\,\psi_{n_1}(q_1)\,\psi_{n_2}(q_2). \tag{116.12}$$

These series of orthogonal functions can be equal only if the coefficients of each function are equal:

$$c(n_1, n_2, t) = -c(n_2, n_1, t). \tag{116.13}$$

For $n_1 = n_2$ this gives

$$c(n_1, n_1, t) = -c(n_1, n_1, t) = 0, \tag{116.13'}$$

i.e.

$$c(n, n, t) = 0. \tag{116.14}$$

Substitution in (116.9) shows that, if the values of n_1 and n_2 are the same, the probability $w(n_1, n_2, t)$ is zero:

$$w(n, n, t) = 0. \tag{116.15}$$

Thus we have shown that the probability of finding the same values for each of the two electrons in a simultaneous measurement of the same set of mechanical quantities describing the state of an electron is zero. Such a result of measurement is therefore impossible, and this is the Pauli principle.

The generalisation to N particles is effected without difficulty by arguments similar to those just given for two particles. The wave function $\Psi(q_1, \ldots, q_k, \ldots, q_j, \ldots, q_N, t)$

of the system is written as

$$\begin{aligned}&\Psi(q_1, \ldots, q_k, \ldots, q_j, \ldots, q_N, t)\\&= \sum_{n_1} \cdots \sum_{n_k} \cdots \sum_{n_j} \cdots \sum_{n_N} c(n_1, \ldots, n_k, \ldots, n_j, \ldots, n_N, t)\, \psi_{n_1}(q_1) \ldots \times\\&\times \psi_{n_k}(q_k) \ldots \psi_{n_j}(q_j) \ldots \psi_{n_N}(q_N),\end{aligned} \tag{116.7'}$$

where

$$\begin{aligned}&c(n_1, \ldots, n_k, \ldots, n_j, \ldots, n_N, t)\\&= \int \ldots \int \mathrm{d}q_1 \ldots \mathrm{d}q_N \Psi(q_1, \ldots, q_N, t)\, \psi^*_{n_1}(q_1) \ldots \psi^*_{n_N}(q_N).\end{aligned} \tag{116.8'}$$

The probability of finding by measurement the values n_1 for the first electron, n_k for the kth, n_j for the jth, n_N for the Nth, is

$$w(n_1, \ldots, n_k, \ldots, n_j, \ldots, n_N, t) = |c(n_1, \ldots, n_k, \ldots, n_j, \ldots, n_N, t)|^2. \tag{116.9'}$$

Interchanging the kth and jth particles in (116.7′) and also the summations over n_k and n_j, we have in exact analogy to (116.11) and (116.12)

$$c(n_1, \ldots, n_j, \ldots, n_k, \ldots, n_N, t) = -\, c(n_1, \ldots, n_k, \ldots, n_j, \ldots, n_N, t), \tag{116.13''}$$

whence

$$c(n_1, \ldots, n_j, \ldots, n_k, \ldots, n_N, t) = 0 \qquad \text{for} \qquad n_k = n_j. \tag{116.14'}$$

Thus

$$w(n_1, \ldots, n_k, \ldots, n_j, \ldots, n_N, t) = 0 \qquad \text{for} \qquad n_k = n_j. \tag{116.15'}$$

Since this proof applies to any pair of particles (k, j) among the N particles, $w = 0$ unless all the n_k are different. Hence the probability of finding in a system of fermions any two for which the results of measurement of all quantities describing the state of the particle (n_k) are the same, is zero.

For example, two electrons cannot have the same momentum and spins in the same direction (in which case $n_k = n_j$, where n denotes p_x, p_y, p_z, s). Similarly two electrons cannot be found at a given point in space with the same spin orientation (in which case $q_k = q_j$, where q denotes x, y, z, s); for $q_k = q_j$ the functions (116.7), (116.7′) have a node, so that $|\Psi|^2 = 0$. The same statements apply to all fermions, including positrons, protons and neutrons.

Finally it may be mentioned that, since electrons are a constituent of atoms, and protons and neutrons a constituent of the atomic nucleus, the Pauli principle is of cardinal importance in both the theory of the electron shells of atoms and that of the nucleus.

117. Wave functions for a system of fermions and bosons

Let us consider in more detail the wave functions having the properties of symmetry or antisymmetry in the particles, and take first of all the antisymmetric functions corresponding to fermions, for the case of two particles. The antisymmetric function $\Psi(q_1, q_2, t)$ can be expanded in terms of the eigenfunctions $\psi_{n_1}(q_1)$ and $\psi_{n_2}(q_2)$ corresponding to the individual particles:

$$\Psi(q_1, q_2, t) = \sum_{n_1}\sum_{n_2} c(n_1, n_2, t)\,\psi_{n_1}(q_1)\,\psi_{n_2}(q_2). \tag{117.1}$$

The expression (117.1) can be written in a different form by dividing the sum into two parts, one with $n_1 > n_2$ and the other $n_1 < n_2$ ($n_1 = n_2$ does not occur, since $c(n_1, n_1, t) = 0$):

$$\begin{aligned}\Psi(q_1, q_2, t) &= \sum_{n_1 > n_2}\sum_{n_2} c(n_1, n_2, t)\,\psi_{n_1}(q_1)\,\psi_{n_2}(q_2) + \\ &+ \sum_{n_1 < n_2}\sum_{n_2} c(n_1, n_2, t)\,\psi_{n_1}(q_1)\,\psi_{n_2}(q_2).\end{aligned} \tag{117.1'}$$

Interchanging n_1 and n_2 in the second sum gives

$$\begin{aligned}\Psi(q_1, q_2, t) &= \sum_{n_1 > n_2}\sum_{n_2} c(n_1, n_2, t)\,\psi_{n_1}(q_1)\,\psi_{n_2}(q_2) + \\ &+ \sum_{n_2 < n_1}\sum_{n_1} c(n_2, n_1, t)\,\psi_{n_2}(q_1)\,\psi_{n_1}(q_2),\end{aligned} \tag{117.1''}$$

and finally, interchanging n_1 and n_2 in $c(n_2, n_1, t)$, we have from (116.13)

$$\begin{aligned}&\Psi(q_1, q_2, t) \\ &= \sum_{n_1 > n_2}\sum_{n_2} c(n_1, n_2, t)\,\{\psi_{n_1}(q_1)\,\psi_{n_2}(q_2) - \psi_{n_1}(q_2)\,\psi_{n_2}(q_1)\}.\end{aligned} \tag{117.2}$$

The expression in braces can be written as a determinant, giving

$$\Psi(q_1, q_2, t) = \sum_{n_1 > n_2}\sum_{n_2} c(n_1, n_2, t)\begin{vmatrix}\psi_{n_1}(q_1) & \psi_{n_1}(q_2)\\ \psi_{n_2}(q_1) & \psi_{n_2}(q_2)\end{vmatrix}. \tag{117.3}$$

Thus the antisymmetric wave function is a sum (or integral) of determinants of the form

$$\Phi_{n_1 n_2}(q_1, q_2) = \begin{vmatrix}\psi_{n_1}(q_1) & \psi_{n_1}(q_2)\\ \psi_{n_2}(q_1) & \psi_{n_2}(q_2)\end{vmatrix}. \tag{117.4}$$

If we have N particles, a similar argument using (116.13) easily gives

$$\begin{aligned}&\Psi(q_1, \ldots, q_k, \ldots, q_j, \ldots, q_N, t) \\ &= \sum_{n_1 > n_2 > \ldots > n_N} \cdots \sum c(n_1, \ldots, n_k, \ldots, n_j, \ldots, n_N, t) \times \\ &\quad \times \Phi_{n_1, \ldots, n_k, \ldots, n_j, \ldots, n_N}(q_1, \ldots, q_k, \ldots, q_j, \ldots, q_N),\end{aligned} \tag{117.5}$$

where

$$\begin{aligned}&\Phi_{n_1, \ldots, n_k, \ldots, n_j, \ldots, n_N}(q_1, \ldots, q_k, \ldots, q_j, \ldots, q_N) \\ &= \begin{vmatrix}\psi_{n_1}(q_1) & \ldots & \psi_{n_1}(q_k) & \ldots & \psi_{n_1}(q_j) & \ldots & \psi_{n_1}(q_N)\\ \psi_{n_2}(q_1) & \ldots & \psi_{n_2}(q_k) & \ldots & \psi_{n_2}(q_j) & \ldots & \psi_{n_2}(q_N)\\ \ldots & \ldots & \ldots & \ldots & \ldots & \ldots & \ldots\\ \psi_{n_k}(q_1) & \ldots & \psi_{n_k}(q_k) & \ldots & \psi_{n_k}(q_j) & \ldots & \psi_{n_k}(q_N)\\ \ldots & \ldots & \ldots & \ldots & \ldots & \ldots & \ldots\\ \psi_{n_N}(q_1) & \ldots & \psi_{n_N}(q_k) & \ldots & \psi_{n_N}(q_j) & \ldots & \psi_{n_N}(q_N)\end{vmatrix}.\end{aligned} \tag{117.6}$$

Expanding the determinant, we can also write Φ in the form

$$\Phi_{n_1,\dots,n_k,\dots,n_j,\dots,n_N}(q_1,\dots,q_k,\dots,q_j,\dots,q_N) = \sum_P (\pm) P\psi_{n_1}(q_1)\dots\psi_{n_k}(q_k)\dots\psi_{n_j}(q_j)\dots\psi_{n_N}(q_N). \tag{117.6'}$$

Here the sum is taken over all $N!$ permutations of the particles $q_1, \dots, q_N$, and the + or − sign is taken before each term in (117.6′) according as the relevant permutation of the q is obtained from the arrangement in order of increasing suffix $q_1, q_2, \dots, q_k, q_{k+1}, \dots, q_N$ by means of an even or odd number of binary interchanges.

The above representation of antisymmetric wave functions as a sum of determinants is of great importance in practical applications of the theory for the approximate solution of the problem of the motion of many bodies. Let us suppose that we are concerned with the wave functions of stationary states of two electrons in an atom. In general it is fairly difficult to find such functions, but the functions for a single electron are considerably easier to derive. Let us assume that we know these functions – $\psi_{n_1}(q_1)$ and $\psi_{n_2}(q_2)$, say. If the interaction of the electrons is not strong, the wave function of the system of two electrons is such that the state of each electron will differ only slightly from the state of one electron in the absence of the other. However, if one electron occupies a quantum state described by the quantities (quantum numbers) n_1, the probability of finding any other value n_1' in this state is zero. Similarly, on placing the second electron in the state n_2, we must say that the probability of finding n_2' is zero. If both electrons are simultaneously present in the atom, and the interaction between the electrons is weak, the state will not be much changed when the second electron is added. This means that, although the probability of finding n_1' and n_2' is no longer zero, it is still small, and so all the $c(n_1', n_2', t)$ in (117.3) except $c(n_1, n_2, t)$ are small. Neglecting all c except the latter, we obtain from (117.3) the wave function Ψ^0 for two electrons in an atom in the zero-order approximation:

$$\Psi^0(q_1, q_2, t) = c(n_1, n_2, t)\begin{vmatrix} \psi_{n_1}(q_1) & \psi_{n_1}(q_2) \\ \psi_{n_2}(q_1) & \psi_{n_2}(q_2) \end{vmatrix}, \tag{117.7}$$

or, since the common factor $c(n_1, n_2, t)$ is unimportant,

$$\Psi^0 = \Phi_{n_1 n_2}(q_1, q_2). \tag{117.8}$$

Similarly in the case of many particles with weak interaction, the zero-order approximation Ψ^0 is $\Phi_{n_1,\dots,n_k,\dots,n_j,\dots,n_N}(q_1,\dots,q_k,\dots,q_j,\dots,q_N)$ (117.6), if $\psi_{n_1}(q_1)$, $\psi_{n_2}(q_2), \dots, \psi_{n_N}(q_N)$ are the electron functions when the interaction is neglected.

Thus the representation of the antisymmetric wave function in the form of the determinant (117.4) or (117.6) gives an approximate method of representing the wave functions of a system of weakly interacting particles in terms of the wave functions of the individual particles when there is no interaction between them.

For bosons we have a different expansion of the wave function Ψ of a system of particles in terms of products of the wave functions of the individual particles: $\psi_{n_1}(q_1)\,\psi_{n_2}(q_2)\dots\psi_{n_k}(q_k)\dots\psi_{n_j}(q_j)\dots\psi_{n_N}(q_N)$. Interchanging the co-ordinates of the

kth and jth particles in the expansion of the wave function of the system

$$\Psi(q_1, \dots, q_k, \dots, q_j, \dots, q_N, t) = \sum_{n_1} \dots \sum_{n_N} c(n_1, \dots, n_N, t) \times \times \psi_{n_1}(q_1) \dots \psi_{n_k}(q_k) \dots \psi_{n_j}(q_j) \dots \psi_{n_N}(q_N), \tag{117.9}$$

and noting that for bosons the function Ψ must be unchanged by this operation, we find by comparing the coefficients of the various products that

$$c(n_1, \dots, n_k, \dots, n_j, \dots, n_N, t) = c(n_1, \dots, n_j, \dots, n_k, \dots, n_N, t). \tag{117.10}$$

For two particles we therefore have

$$\Psi(q_1, q_2) = \sum_{n_1 > n_2} c(n_1, n_2) \{\psi_{n_1}(q_1)\psi_{n_2}(q_2) + + \psi_{n_2}(q_1)\psi_{n_1}(q_2)\}. \tag{117.11}$$

If the interaction between the particles is weak, the approximate expression for the wave function of the state of two particles close to the state of non-interacting particles in which one is in the state n_1 and the other in the state n_2 is

$$\Psi^0 = \psi_{n_1}(q_1)\psi_{n_2}(q_2) + \psi_{n_2}(q_1)\psi_{n_1}(q_2). \tag{117.12}$$

For N particles a similar argument gives

$$\Psi^0 = \sum_P \psi_{n_1}(q_1) \dots \psi_{n_k}(q_k) \dots \psi_{n_j}(q_j) \dots \psi_{n_N}(q_N), \tag{117.13}$$

where $\sum_P$ denotes the sum over all $N!$ permutations of the co-ordinates $q_1, q_2, \dots, q_N$ of the particles.

CHAPTER XX

SECOND QUANTISATION AND QUANTUM STATISTICS

118. Second quantisation

Assemblies of identical particles may be treated by a special method known as *second quantisation*. The essence of this method is that the independent variables that describe the assembly are taken to be not the set of mechanical quantities representing the individual states of the particles but the numbers of particles in those states. Each state can be described by three variables (L_1, L_2, L_3) relating to the motion of the centre of mass of the particle and the spin variable s (if the particle has spin). To simplify the mathematics we shall suppose that these variables have a discrete spectrum, so that all the states may be labelled by a number n as in Section 116 (n standing for the set of values of the four quantities L_1, L_2, L_3, s).

The Hamiltonian is usually given in the co-ordinate representation, and so we first carry out the transformation from the co-ordinate representation to the L representation, which is assumed discrete.[1]

If the wave function of a system of N identical particles in the co-ordinate representation is $\psi(q_1, q_2, \ldots, q_N, t)$, Schrödinger's equation for the system is

$$i\hbar\frac{\partial\psi}{\partial t} = \{\sum_{k=1}^{N} \boldsymbol{H}(q_k) + \sum_{k\neq j}^{N} W(q_k, q_j)\}\,\psi\,, \tag{118.1}$$

where $\boldsymbol{H}(q_k) = -(\hbar^2/2\mu)\,\nabla_k^2 + U(q_k)$ is the energy operator of the kth particle, $U(q_k)$ the potential energy of that particle in the external field, and $W(q_k, q_j)$ the energy of interaction of the kth and jth particles. We now expand the wave function ψ in terms of the eigenfunctions $\psi_{n_k}(q_k)$ of the operators L_1, L_2, L_3, s in exactly the same way as in Section 116, obtaining

$$\begin{aligned}\psi(q_1, q_2, \ldots, q_N, t) &= \sum_{n_1}\sum_{n_2}\cdots\sum_{n_N} c(n_1, n_2, \ldots, n_N, t)\,\times \\ &\times\,\psi_{n_1}(q_1)\,\psi_{n_2}(q_2)\ldots\psi_{n_N}(q_N)\,;\end{aligned} \tag{118.2}$$

[1] The momentum representation ($L_1 = p_x$, $L_2 = p_y$, $L_3 = p_z$) is often used in the theory of second quantisation. This representation, however, is continuous. The device is therefore used of putting $p_x = 2\pi\hbar n_x/l$, $p_y = 2\pi\hbar n_y/l$, $p_z = 2\pi\hbar n_z/l$, where n_x, n_y, n_z are integers and l some length of large magnitude (cf. Section 120). The momentum representation then becomes discrete. The limit $l \to \infty$ is finally taken, and the artificial assumption is thereby dispensed with. A detailed theory of second quantisation, applicable also to a continuous sequence of states, has been developed by V. A. Fok [38].

$c(n_1, n_2, \ldots, n_N, t)$ is evidently the wave function of the system in the L representation, and $|c(n_1, n_2, \ldots, n_N, t)|^2$ is the probability that the first particle is in the state n_1 (i.e. has the values of L_1, L_2, L_3, s denoted by the single letter n_1), the second particle is in the state n_2 (has L_1', L_2', L_3', s' denoted by n_2), and so on. Substituting (118.2) in (118.1), multiplying on the left by $\psi_{m_1}^*(q_1)\, \psi_{m_2}^*(q_2) \ldots \psi_{m_N}^*(q_N)$ and integrating with respect to $q_1, q_2, \ldots, q_N$, we have

$$\begin{aligned} i\hbar \frac{\mathrm{d}}{\mathrm{d}t} c(m_1, m_2, \ldots, m_k, \ldots, m_j, \ldots, m_N, t) \\ = \sum_{k=1}^{N} \sum_{n_k} H_{m_k;\, n_k}\, c(m_1, m_2, \ldots, n_k, \ldots, m_j, \ldots, m_N, t) + \\ + \sum_{k\neq j}^{N} \sum_{n_k} \sum_{n_j} W_{m_k m_j;\, n_k n_j}\, c(m_1, m_2, \ldots, n_k, \ldots n_j, \ldots m_N, t). \end{aligned} \tag{118.3}$$

Here $H_{m_k;n_k}$ and $W_{m_k m_j;n_k n_j}$ are the matrix elements

$$H_{m_k;\, n_k} = \int \psi_{m_k}^*(q_k)\, \boldsymbol{H}(q_k)\, \psi_{n_k}(q_k)\, \mathrm{d}q_k, \tag{118.4}$$

$$W_{m_k m_j;\, n_k n_j} = \int \psi_{m_k}^*(q_k)\, \psi_{m_j}^*(q_j)\, W(q_k, q_j)\, \psi_{n_k}(q_k)\, \psi_{n_j}(q_j)\, \mathrm{d}q_k\, \mathrm{d}q_j. \tag{118.5}$$

Equation (118.3) is Equation (118.1) in the L representation. Since the particles are identical, the matrix elements (118.4), (118.5) depend only on the values of the quantum numbers m_k, m_j, n_k, n_j and not on the numbers k, j of the particles. Denoting some value of m_k by m, of n_k by n, and similarly m_j by m', n_j by n', and the co-ordinates of the kth and jth particles by q and q' respectively, we can write (118.4) and (118.5) as

$$\begin{aligned} H_{m_k;\, n_k} &= -\frac{\hbar^2}{2\mu} \int \psi_m^*(q)\, \nabla^2 \psi_n(q)\, \mathrm{d}q + \int \psi_m^*(q)\, U(q)\, \psi_n(q)\, \mathrm{d}q \\ &= \frac{\hbar^2}{2\mu} \int \nabla \psi_m^*(q) \cdot \nabla \psi_n(q)\, \mathrm{d}q + \int \psi_m^*(q)\, U(q)\, \psi_n(q)\, \mathrm{d}q = H_{mn}, \end{aligned} \tag{118.6}$$

$$W_{m_k m_j;\, n_k n_j} = \int \psi_m^*(q)\, \psi_{m'}^*(q')\, W(q, q')\, \psi_n(q)\, \psi_{n'}(q')\, \mathrm{d}q\, \mathrm{d}q' = W_{mm',nn'}. \tag{118.7}$$

The amplitudes $c(m_1, m_2, \ldots, m_N, t)$ (the wave functions in the L representation) are symmetric functions of the quantum numbers $m_1, m_2, \ldots, m_N$ for bosons and antisymmetric functions for fermions (see Section 116). The values of these amplitudes therefore depend only on how many of the N arguments $m_1, m_2, \ldots, m_N$ are respectively equal to m, m', m'', etc., and not on precisely which of these arguments are equal to m, m', m'', etc.; that is, the amplitudes are functions of the number of particles in each state. Let these numbers be $N_1, N_2, \ldots, N_m, \ldots, N_{m'}, \ldots, N_{m''}, \ldots$ etc. Thus N_m, for example, is equal to the number of the arguments of $c(m_1, m_2, \ldots, m_N, t)$ whose value is m, $N_{m'}$ the number whose value is m', and so on.

For bosons, the numbers N_m can take any values, but for fermions, by the Pauli principle, the function $c(m_1, m_2, \ldots, m_N, t)$ is zero if two or more of the numbers m_k

are equal, so that N_m can take only the two values 0 and 1: a state can be occupied, if at all, by only one particle.

The following calculations are for bosons. The problem now is to write Schrödinger's equation (118.3) with the numbers $N_1, N_2, \ldots, N_m, \ldots$ of particles in the various states as variables instead of the quantum numbers $m_1, m_2, \ldots, m_N$. To do so, we must first of all change the normalisation of the amplitudes c. For, if c is regarded as a function of the numbers $N_1, N_2, \ldots, N_m, \ldots$, then $|c(N_1, N_2, \ldots, N_m, \ldots, t)|^2$ is the probability of finding N_1 particles in state 1, N_2 particles in state 2, ..., N_m particles in state m, and so on. This probability is expressed in terms of $c(m_1, m_2, \ldots, m_N, t)$ by

$$|c(N_1, N_2, \ldots, N_m, \ldots, t)|^2 = \sum |c(m_1, m_2, \ldots, m_N, t)|^2, \tag{118.8}$$

where the sum is over all $c(m_1, m_2, \ldots, m_N, t)$ which have N_1 numbers $m_k = 1$, N_2 numbers $m_k = 2$, and so on. By symmetry, all these c are equal. Hence

$$|c(N_1, N_2, \ldots, N_m, \ldots, t)|^2 = \frac{N!}{N_1! N_2! \ldots N_m! \ldots} |c(m_1, m_2, \ldots, m_N, t)|^2, \tag{118.8'}$$

whence

$$c(N_1, N_2, \ldots, N_m, \ldots, t) = \left(\frac{N!}{N_1! N_2! \ldots N_m! \ldots}\right)^{\frac{1}{2}} c(m_1, m_2, \ldots, m_N, t). \tag{118.9}$$

Substituting in (118.3) the amplitudes $c(N_1, N_2, \ldots, N_m, \ldots, t)$ in place of $c(m_1, m_2, \ldots, m_N, t)$, we can carry out the summation over the particle numbers k and j by using (118.6) and (118.7) and noting that $c(m_1, m_2, \ldots, m_k, \ldots, m_j, \ldots, m_N, t)$ differs from $c(m_1, m_2, \ldots, n_k, \ldots, m_j, \ldots, m_N, t)$ in that the number of particles in the state $m_k = m$ is decreased by 1 and the number in the state $n_k = n$ is increased by 1. Similarly $c(m_1, m_2, \ldots, n_k, \ldots, n_j, \ldots, m_N, t)$ differs from $c(m_1, m_2, \ldots, m_k, \ldots, m_j, \ldots, m_N, t)$ in that the number of particles in the states $m_k = m$, $m_j = m'$ is decreased by 1, and the number in the states $n_k = n$, $n_j = n'$ is increased by 1.

Hence we find

$$\begin{aligned}
& i\hbar \frac{\mathrm{d}}{\mathrm{d}t} \left\{ \left(\frac{N_1! \ldots N_m! \ldots N_{m'}! \ldots N_n! \ldots N_{n'}! \ldots}{N!} \right)^{\frac{1}{2}} \times \right. \\
& \left. \times\, c(N_1, \ldots, N_m, \ldots, N_{m'}, \ldots, N_n, \ldots, N_{n'}, \ldots, t) \right\} \\
& = \sum_{n,m} N_m H_{mn} \left(\frac{N_1! \ldots (N_m - 1)! \ldots N_{m'}! \ldots (N_n + 1)! \ldots N_{n'}! \ldots}{N!} \right)^{\frac{1}{2}} \times \\
& \quad \times\, c(N_1, \ldots, N_m - 1, \ldots, N_{m'}, \ldots, N_n + 1, \ldots, N_{n'}, \ldots, t) + \\
& \quad + \tfrac{1}{2} \sum_{m,m'} \sum_{n,n'} N_m N_{m'} W_{mm', nn'} \times \\
& \quad \times \left(\frac{N_1! \ldots (N_m - 1)! \ldots (N_{m'} - 1)! \ldots (N_n + 1)! \ldots (N_{n'} + 1)! \ldots}{N!} \right)^{\frac{1}{2}} \times \\
& \quad \times\, c(N_1, \ldots, N_m - 1, \ldots, N_{m'} - 1, \ldots, N_n + 1, \ldots, N_{n'} + 1, \ldots, t).
\end{aligned} \tag{118.10}$$

Division by $(N_1!\,N_2!\,\ldots/N!)^{\frac{1}{2}}$ gives

$$\begin{aligned} i\hbar\frac{\mathrm{d}}{\mathrm{d}t}c(N_1,\ldots,N_m,\ldots,N_{m'},\ldots,N_n,\ldots,N_{n'},\ldots,t) \\ = \sum_{n,m} N_m^{\frac{1}{2}}(N_n+1)^{\frac{1}{2}}H_{mn}\,c(N_1,\ldots,N_m-1,\ldots,N_{m'},\ldots,N_n+ \\ +1,\ldots,N_{n'},\ldots,t)+\tfrac{1}{2}\sum N_m^{\frac{1}{2}}N_{m'}^{\frac{1}{2}}(N_n+1)^{\frac{1}{2}}(N_{n'}+1)^{\frac{1}{2}}W_{mm',nn'}\times \\ \times\, c(N_1,\ldots,N_m-1,\ldots,N_{m'}-1,\ldots,N_n+1,\ldots,N_{n'}+1,\ldots,t). \end{aligned} \tag{118.11}$$

This is the required equation, in which the independent variables are the numbers of particles in the various states. It can be written in a very convenient form by using operators $\boldsymbol{a}_n$ and $\boldsymbol{a}_n^*$ which act on functions of the numbers N_n in the following manner:

$$\begin{aligned} \boldsymbol{a}_n^* f(N_1,N_2,\ldots,N_n,\ldots,N_m,\ldots) \\ = (N_n+1)^{\frac{1}{2}} f(N_1,N_2,\ldots,N_n+1,\ldots,N_m,\ldots), \end{aligned} \tag{118.12}$$

$$\begin{aligned} \boldsymbol{a}_n f(N_1,N_2,\ldots,N_n,\ldots,N_m,\ldots) \\ = N_n^{\frac{1}{2}} f(N_1,N_2,\ldots,N_n-1,\ldots,N_m,\ldots), \end{aligned} \tag{118.12$'$}$$

$$\boldsymbol{a}_n f(N_1,N_2,\ldots,0_n,\ldots,N_m,\ldots)=0. \tag{118.12$''$}$$

These operators clearly have the following properties:

$$\boldsymbol{a}_n^*\boldsymbol{a}_n = N_n, \qquad \boldsymbol{a}_n\boldsymbol{a}_n^* = N_n+1, \tag{118.13}$$

$$\boldsymbol{a}_m\boldsymbol{a}_n^* - \boldsymbol{a}_n^*\boldsymbol{a}_m = \delta_{mn}. \tag{118.14}$$

It is now easy to see that in terms of these operators Equation (118.11) can be written

$$i\hbar\frac{\mathrm{d}c(N_1,N_2,\ldots,t)}{\mathrm{d}t} = \boldsymbol{H}c(N_1,N_2,\ldots,t), \tag{118.15}$$

where

$$\boldsymbol{H} = \sum_{m,n}\boldsymbol{a}_m^* H_{mn}\boldsymbol{a}_n + \tfrac{1}{2}\sum_{m,m'}\sum_{n,n'}\boldsymbol{a}_m^*\boldsymbol{a}_{m'}^* W_{mm',nn'}\boldsymbol{a}_n\boldsymbol{a}_{n'}. \tag{118.16}$$

The operator $\boldsymbol{H}$ is the Hamiltonian of the system expressed in terms of the operators $\boldsymbol{a}_n$ and $\boldsymbol{a}_n^*$. It is usually said to have undergone second quantisation. This equation is exactly equivalent to the original Equation (118.1) for N particles in configuration space. Equation (118.15) is essentially Equation (118.1) in the N representation, i.e. that in which the variables are the numbers of particles $N_1, N_2, \ldots, N_m, \ldots$ in the various quantum states, $1, 2, \ldots, m, \ldots$.

However, in one respect Equation (118.15) is more general than Equation (118.1), in that the latter is written for a system of N particles, whereas in Equation (118.15) the total number of particles does not appear explicitly, but is a constant of integration: the operator $\boldsymbol{H}$ (118.16) in each term contains the same number of operators $\boldsymbol{a}$ and $\boldsymbol{a}^*$. Since the operators $\boldsymbol{a}^*$ increase by one the number of particles in some state, and the operators $\boldsymbol{a}$ decrease by one the number of particles in some state, the total number of particles $N = \sum N_m$ is not affected by the operator $\boldsymbol{H}$, so that

$$dN/dt = [\boldsymbol{H}, N] = 0. \tag{118.17}$$

Thus N = constant. Hence Equation (118.15) is valid for any number N of identical bosons.

The Hamiltonian (118.16) of second quantisation theory can be written in another form which corresponds to the energy of a certain wave field. Let the wave function of one particle be $\psi(q)$. We expand this function in terms of the eigenfunctions $\psi_n(q)$ of the operators L_1, L_2, L_3, s:

$$\psi(q) = \sum_n a_n \psi_n(q), \tag{118.18}$$

and now regard the amplitudes a_n not as numbers but as operators with the properties (118.14). Then the function ψ is itself an operator,

$$\boldsymbol{\psi}(q) = \sum_n \boldsymbol{a}_n \psi_n(q), \tag{118.19}$$

acting on the numbers $N_1, N_2, \ldots, N_m, \ldots$. The change from (118.18) to (118.19) is a change from numbers to operators, i.e. from classical to quantum theory, as it were. But since the description of the motion of one particle by means of the wave field $\psi(q)$ is already quantised, the replacement of the amplitudes a_n by the operators $\boldsymbol{a}_n$ is called *second quantisation*, and the wave function $\boldsymbol{\psi}$ is called a *quantised wave function.*[2]

It may be noted that the change from the non-quantised wave function (118.18) to the quantised wave function (118.19) can be formulated directly, without using the operators $\boldsymbol{a}_n$. From (118.14) and (118.19) we have

$$\begin{aligned}
&\boldsymbol{\psi}(q)\boldsymbol{\psi}^*(q') - \boldsymbol{\psi}^*(q')\boldsymbol{\psi}(q) \\
&= \sum_{m,n} (\boldsymbol{a}_n \boldsymbol{a}_m^* - \boldsymbol{a}_m^* \boldsymbol{a}_n)\psi_n(q)\psi_m^*(q') \\
&= \sum_{m,n} \delta_{mn} \psi_m^*(q')\psi_n(q) \\
&= \sum_n \psi_n^*(q')\psi_n(q),
\end{aligned}$$

where the sum is over all the eigenfunctions. This sum can be shown to equal $\delta(q' - q)$. Hence the quantisation of the wave function can be written

$$\boldsymbol{\psi}(q)\boldsymbol{\psi}^*(q') - \boldsymbol{\psi}^*(q')\boldsymbol{\psi}(q) = \delta(q' - q). \tag{118.20}$$

In terms of the quantised wave function $\boldsymbol{\psi}(q)$ (118.19), the Hamiltonian (118.16) can be written

$$\begin{aligned}
\boldsymbol{H} = &\frac{\hbar^2}{2\mu}\int \nabla\boldsymbol{\psi}^*(q)\cdot\nabla\boldsymbol{\psi}(q)\,dq + \int \boldsymbol{\psi}^*(q)\,U(q)\,\boldsymbol{\psi}(q)\,dq + \\
&+ \tfrac{1}{2}\iint \boldsymbol{\psi}^*(q)\boldsymbol{\psi}^*(q')\,W(q,q')\,\boldsymbol{\psi}(q)\boldsymbol{\psi}(q')\,dq\,dq'.
\end{aligned} \tag{118.21}$$

[2] It should not be forgotten that the wave function, in the usual sense of the term, in the theory of second quantisation is $c(N_1, N_2, \ldots, N_m, \ldots, t)$, not $\boldsymbol{\psi}$.

The equivalence of (118.21) and (118.16) is easily seen by using (118.19) and the expressions for the matrix elements (118.16) and (118.7).

In this form (118.21) the Hamiltonian $\boldsymbol{H}$ can be regarded as the energy of some wave field, which is 'quantised' in the sense that the classical field $\psi(q)$ is replaced by the operator $\boldsymbol{\psi}(q)$. We can regard $\psi(q)$ as a de Broglie-Schrödinger wave field and assume that the individual elements of this field interact in such a way that the interaction energy of two elements is proportional to the product of the densities $|\psi(q)|^2\,|\psi(q')|^2$. The 'classical' equation for such a field is[3]

$$i\hbar\frac{\partial\psi(q)}{\partial t} = -\frac{\hbar^2}{2\mu}\nabla^2\psi(q) + U(q)\psi(q) + \psi(q)\int W(q,q')\,|\psi(q')|^2\,\mathrm{d}q' . \tag{118.22}$$

The total energy of such a field is[4]

$$H = \frac{\hbar^2}{2\mu}\int|\nabla\psi(q)|^2\,\mathrm{d}q + \int|\psi(q)|^2\,U(q)\,\mathrm{d}q + \frac{1}{2}\int|\psi(q)|^2\,|\psi(q')|^2\,W(q,q')\,\mathrm{d}q\,\mathrm{d}q' . \tag{118.23}$$

If we now arrange ψ and ψ^* suitably and replace them by the operators $\boldsymbol{\psi}$ and $\boldsymbol{\psi}^*$ satisfying the commutation rule (118.20), we obtain precisely the second-quantisation Hamiltonian (118.21). Hence we see that second-quantisation theory permits the following treatment of the theory of systems of identical particles. If we consider some classical field ψ, find the expression for its energy H, and in that expression replace the classical field ψ by the operator $\boldsymbol{\psi}$, then we obtain the Hamiltonian $\boldsymbol{H}$ of second-quantisation theory and can say of the particles belonging to the field ψ that after quantisation the field has a discrete, corpuscular nature. This procedure is called *field quantisation*, and its importance is due to the fact that it is applicable to any classical field.[5]

In the above example we have discussed the quantisation of a de Broglie-Schrödinger field for the case of bosons. The quantisation can be effected in an exactly similar manner for fermions. The only difference is in the properties of the operators $\boldsymbol{a}$ and $\boldsymbol{a}^*$. In order to find these operators, we must again carry out the transformation of Equation (118.3) from the variables $m_1, m_2, \ldots, m_N$ to the variables $N_1, N_2, \ldots, N_m, \ldots$, which for fermions is a little lengthier because the functions $c(m_1, m_2, \ldots, m_N, t)$ change sign when particles are interchanged, and also, as already mentioned, the

[3] This equation differs from the regular Schrödinger's equation for a single particle by the last term, which expresses the self-interaction of ψ waves assumed here.

[4] Using Equation (118.22) we can see that $\mathrm{d}H/\mathrm{d}t = 0$, i.e. H is an integral of the motion. Since the second term in (118.22) corresponds to the potential energy in the external field, the whole expression must be regarded as the total energy of the field, since $H = $ constant.

[5] The general theory of this quantisation is given by Wentzel [89].

numbers N_m can take only two values, 1 and 0. By means of similar calculations[6] we again obtain from (118.3) Equation (118.15) with the Hamiltonian (118.16), but the operators $\boldsymbol{a}_n$, $\boldsymbol{a}_n^*$ are now defined as follows:

$$\boldsymbol{a}_n^* f(N_1, N_2, \ldots, 0_n, \ldots, N_m, \ldots) = \pm f(N_1, N_2, \ldots, 1_n, \ldots, N_m, \ldots), \tag{118.24}$$

$$\boldsymbol{a}_n^* f(N_1, N_2, \ldots, 1_n, \ldots, N_m, \ldots) = 0, \tag{118.24'}$$

$$\boldsymbol{a}_n f(N_1, N_2, \ldots, 0_n, \ldots, N_m, \ldots) = 0, \tag{118.24''}$$

$$\boldsymbol{a}_n f(N_1, N_2, \ldots, 1_n, \ldots, N_m, \ldots) = \pm f(N_1, N_2, \ldots, 0_n, \ldots, N_m, \ldots), \tag{118.24'''}$$

the + or − sign being taken according as the state n is preceded by an even or odd number of occupied ($N_m = 1$) states when the states are arranged in order of increasing n.[7]

From these rules we have

$$\boldsymbol{a}_n^* \boldsymbol{a}_n = N_n (0 \text{ or } 1), \qquad \boldsymbol{a}_n \boldsymbol{a}_n^* = 1 - N_n, \tag{118.25}$$

$$\boldsymbol{a}_n \boldsymbol{a}_m^* + \boldsymbol{a}_m^* \boldsymbol{a}_n = \delta_{mn}. \tag{118.26}$$

It is seen that the commutation rule for the operators $\boldsymbol{a}$ for fermions is different as regards a sign from that for bosons.

Using (118.18) and repeating the calculations leading to (118.20), we obtain

$$\boldsymbol{\psi}(q)\boldsymbol{\psi}^*(q') + \boldsymbol{\psi}^*(q')\boldsymbol{\psi}(q) = \delta(q' - q). \tag{118.27}$$

All other formulae, and in particular the expression (118.21) for $\boldsymbol{H}$, remain unchanged. Thus the Hamiltonian $\boldsymbol{H}$ together with the quantisation rule (118.27) may be regarded as a second-quantised Hamiltonian for fermion waves for which the 'classical' equation is (118.23). The quantisation rule for both cases can be written as the single formula

$$[\boldsymbol{\psi}(q), \boldsymbol{\psi}^*(q')]_\pm = \delta(q' - q), \tag{118.28}$$

where the + sign is taken for fermions and the − sign for bosons.

In modern physics phenomena involving the creation and annihilation of particles occur. These phenomena are, strictly speaking, outside the scope of quantum mechanics, but, since the second-quantisation method does not involve explicitly the total number of particles, it allows a simple generalisation to the case of a variable number of particles and thus is suitable for the description of the creation and annihilation of particles. If a term of the form

$$\boldsymbol{Q} = \sum_n Q_n \boldsymbol{a}_n + \sum_n Q_n^* \boldsymbol{a}_n^*, \tag{118.29}$$

[6] See, for instance, [29], Section 80, or the paper by Fok [36].

[7] We can define the Wigner function ν_n by

$$\nu_n = \prod_{m \leqslant n} (1 - 2N_m),$$

and replace the ± sign in (118.24), (118.24‴) by ν_n ($\nu_n = \pm 1$).

where $\boldsymbol{Q}_n$, $\boldsymbol{Q}_n^*$ are certain operators representing the interaction of the particles with other particles capable of absorbing or emitting them, is added to the Hamiltonian $\boldsymbol{H}$ (118.16), then the total number of particles N will not be an integral of the motion, since $[\boldsymbol{Q}, N] \neq 0$. The terms containing $\boldsymbol{a}^*$ represent the creation of particles, and those containing $\boldsymbol{a}$ represent their annihilation (see (118.12) and (118.12′)).

If light quanta (photons) are regarded as particles, then processes of emission and absorption of light may be regarded as those of creation and annihilation of photons. Dirac[8] has derived a quantum theory of radiation on this basis. A similar method can be used to study phenomena of creation and annihilation of electrons and positrons in $\beta^{\pm}$ decay, pair production and annihilation, meson formation and decay, etc. All these are treated in quantum field theory.[9]

In addition to quantum field theory, the theory of second quantisation is widely used in quantum statistics.

119. The theory of quantum transitions and the second-quantisation method

Let us now calculate the probability of transition from one quantum state to another under the action of a perturbation in an assembly of identical particles, using the second-quantisation method. In order to make the problem specific, we shall consider transitions under the effect of a weak interaction between the particles.

Here it is convenient to choose the variables L_1, L_2, L_3, s which describe the state of the particles in such a way that one of them (say L_1) is equal to the particle energy: $L_1(q_k) = E(q_k)$. Then the matrix H_{mn} is diagonal. If ε_m denotes the eigenvalues of the particle energy, then $H_{mn} = \varepsilon_m \delta_{mn}$. With this choice of variables, Equation (118.15) becomes

$$\begin{aligned} i\hbar \frac{\mathrm{d}}{\mathrm{d}t} c(N_1, N_2, \ldots, t) &= \sum_m \varepsilon_m N_m c(N_1, N_2, \ldots, t) + \\ &+ \tfrac{1}{2} \sum_{mm', nn'} \boldsymbol{a}_m^* \boldsymbol{a}_{m'}^* W_{mm', nn'} \boldsymbol{a}_n \boldsymbol{a}_{n'} c(N_1, N_2, \ldots, t). \end{aligned} \tag{119.1}$$

The sum $\sum \varepsilon_m N_m = E$ is the total energy of all the particles, neglecting their interaction. Using, instead of the functions $c(N_1, N_2, \ldots, t)$, the slowly varying amplitudes $b(N_1, N_2, \ldots, t) = c(N_1, N_2, \ldots, t) \exp(i\sum \varepsilon_m N_m t/\hbar)$, we have instead of (119.1) the following equation for the $b(N_1, N_2, \ldots, t)$:

$$\begin{aligned} & i\hbar \frac{\mathrm{d}}{\mathrm{d}t} b(N_1, N_2, \ldots, t) \\ &= \tfrac{1}{2} \sum_{mm', nn'} \exp[-i(\varepsilon_m + \varepsilon_{m'} - \varepsilon_n - \varepsilon_{n'}) t/\hbar] \boldsymbol{a}_m^* \boldsymbol{a}_{m'}^* W_{mm', nn'} \times \\ &\quad \times \boldsymbol{a}_n \boldsymbol{a}_{n'} b(N_1, N_2, \ldots, t). \end{aligned} \tag{119.2}$$

Let us assume that at the initial instant the various states have occupation numbers

[8] See [29], Sections 62–67; [47].
[9] See [17, 1].

$N_1^0, N_2^0, \ldots$, so that all the amplitudes b are zero at $t = 0$ except

$$b^0 = b(N_1^0, N_2^0, \ldots, N_m^0, \ldots, N_{m'}^0, \ldots, N_n^0, \ldots, N_{n'}^0, \ldots) = 1\,.$$

Using the ordinary method of perturbation theory, we substitute on the right-hand side of Equation (119.2) the initial value b^0. Then, using the properties of the operators $\boldsymbol{a}_m^*, \boldsymbol{a}_{m'}^*, \boldsymbol{a}_n, \boldsymbol{a}_{n'}$ (see (118.12) and (118.12′)), we obtain an equation for $b^{(1)}$ in the first approximation:

$$\begin{aligned} &i\hbar \frac{\mathrm{d}}{\mathrm{d}t} b^{(1)}(N_1^0; N_2^0, \ldots, N_m^0 + 1, \ldots, N_{m'}^0 + 1, \ldots, N_n^0 - 1, \ldots, \\ &N_{n'}^0 - 1, \ldots, t) \\ &= \exp[-i(\varepsilon_m + \varepsilon_{m'} - \varepsilon_n - \varepsilon_{n'})t/\hbar](N_m^0 + 1)^{\frac{1}{2}}(N_{m'}^0 + 1)^{\frac{1}{2}} \times \\ &\quad \times (N_n^0)^{\frac{1}{2}}(N_{n'}^0)^{\frac{1}{2}} W_{mm',nn'}\,. \end{aligned} \tag{119.3}$$

Integrating this equation with respect to time and calculating the transition probability per unit time $P_{mm',nn'} = (\mathrm{d}/\mathrm{d}t)|b^{(1)}|^2$ (cf. Section 84), we find

$$\begin{aligned} P_{mm',nn'} &= (N_m^0 + 1)(N_{m'}^0 + 1) N_n^0 N_{n'}^0 \frac{2\pi}{\hbar} |W_{mm',nn'}|^2 \times \\ &\times \delta(\varepsilon_m + \varepsilon_{m'} - \varepsilon_n - \varepsilon_{n'}), \end{aligned} \tag{119.4}$$

the delta function expressing the law of conservation of energy.

Similarly, taking $\boldsymbol{a}_m^*, \boldsymbol{a}_{m'}^*, \boldsymbol{a}_n, \boldsymbol{a}_{n'}$, in (119.2) to be the Fermi-Dirac operators (118.24), (118.24‴), we obtain for fermions

$$\begin{aligned} P_{mm',nn'} &= (1 - N_m^0)(1 - N_{m'}^0) N_n^0 N_{n'}^0 \frac{2\pi}{\hbar} |W_{mm',nn'}|^2 \times \\ &\times \delta(\varepsilon_m + \varepsilon_{m'} - \varepsilon_n - \varepsilon_{n'}). \end{aligned} \tag{119.5}$$

These formulae show that in a system of identical particles the probability of transition from the initial state (n, n') to the final state (m, m') depends not only on the number of particles in the initial state but also on the number in the final state. This is an entirely new result of quantum theory, and does not hold good in classical mechanics. For bosons, the transition probability increases with the number of particles already in the final state. Thus bosons tend to accumulate in one state. For fermions, on the other hand, the transition probability is zero if the final state is already occupied ($N_m^0 = 1$ or $N_{m'}^0 = 1$). This is another expression of the Pauli principle.

120. The collision hypothesis. A Fermi-Dirac gas and a Bose-Einstein gas

In classical kinetic theory it is assumed that the probability that particles in certain states n and n' (particle energies ε_n and $\varepsilon_{n'}$) undergo transition to other states m and m' (particle energies ε_m and $\varepsilon_{m'}$) as a result of a collision is proportional to the numbers N_n and $N_{n'}$ of particles in the initial states:

$$P_{mm',nn'} = A_{mm',nn'} N_n N_{n'}\,. \tag{120.1}$$

If $\overline{N_n}$ and $\overline{N_{n'}}$ are the mean numbers of particles in the states n and n', it is *assumed* in accordance with (120.1) the that mean number of transitions from n, n' to m, m' is

$$\overline{P_{mm', nn'}} = A_{mm', nn'} \overline{N_n}\, \overline{N_{n'}}, \tag{120.1'}$$

with $A_{mm', nn'} = A_{nn', mm'}$ (the *principle of detailed balancing*).[10]

In quantum mechanics, for a gas consisting of identical particles, we must make some other hypothesis concerning the mean number of transitions due to collisions. It has been shown in Section 119 that the transition probability depends not only on the number of particles in the initial states but also on the population of the final states: instead of (120.1) we have according to (119.5) for the collision probability of fermions

$$P_{mm', nn'} = A_{mm', nn'}(1 - N_m)(1 - N_{m'}) N_n N_{n'}, \tag{120.2}$$

where $N_m, N_{m'}, N_n, N_{n'} = 1$ or 0. The Pauli principle is clearly expressed by this formula: if one of the final states is occupied ($N_m = 1$ or $N_{m'} = 1$) no transition can occur. Similarly for bosons, from (119.4),

$$P_{mm', nn'} = A_{mm', nn'}(N_m + 1)(N_{m'} + 1) N_n N_{n'}. \tag{120.3}$$

Here the factors $(N_m + 1)$ and $(N_{m'} + 1)$ do not have such an evident significance as $(1 - N_m)$ and $(1 - N_{m'})$ for fermions. The necessity of these factors has, however, been proved in Section 119, and, as mentioned there, bosons tend to associate, making transitions to the states already most densely occupied.[11]

The equality of the quantities $A_{mm', nn'}$ and $A_{nn', mm'}$ (the reverse transition) follows in quantum mechanics from the fact that $A_{mm', nn'}$ is proportional to the squared modulus of the matrix element of the interaction energy $W_{mm', nn'}$, and $W_{mm', nn'} = W^*_{nn', mm'}$ (see footnote 10).

In accordance with (120.2) and (120.3), for a gas of identical particles in quantum mechanics, the mean number of transitions due to collisions is, instead of (120.1′),

$$\overline{P_{mm', nn'}} = A_{mm', nn'}(1 \pm \overline{N_m})(1 \pm \overline{N_{m'}}) \overline{N_n}\, \overline{N_{n'}}, \tag{120.4}$$

with the minus sign for fermions and the plus sign for bosons. Formula (120.4) will be regarded as a new hypothesis concerning the mean number of particle collisions, based on quantum mechanics.[12] It is evident that (120.4) becomes the classical expression (120.1) if the mean number of particles in each state is small compared with unity.

[10] This principle is not always valid, but it is certainly valid in the first approximation of the theory of quantum transitions (see Sections 84, 85) and is rigorously valid if the interaction forces between particles are central forces (see Section 44 and [10]).

[11] This leads to a remarkable property of a boson gas: at low temperatures it undergoes a peculiar type of condensation, even if it is assumed to be a completely ideal gas with vanishing forces of interaction; see [33]. The theory of an ideal Bose gas has been developed by Bogolyubov [16]. By means of this theory the superfluidity of helium can be explained.

[12] We call (120.4) a hypothesis because in the expression (120.2) for the transition probability the true values of the level populations $N_n, N_{n'}, N_m, N_{m'}$ appear, while in (120.4) we have the mean values. The equation $\overline{(1 \pm N_m)(1 \pm N_{m'}) N_n N_{n'}} = (1 \pm \overline{N_m})(1 \pm \overline{N_{m'}}) \overline{N_n}\, \overline{N_{n'}}$ is not obvious and is not always satisfied.

Let us now find the energy distribution in thermal equilibrium for a gas of bosons or fermions. In thermal equilibrium the number of transitions to states n and n' owing to collisions of particles in states m and m' must be equal to the number of reverse transitions. We then have from (120.4) (since $A_{mm',nn'} = A_{nn',mm'}$)

$$(1 \pm \overline{N_m})(1 \pm \overline{N_{m'}})\overline{N_n}\,\overline{N_{n'}} = (1 \pm \overline{N_n})(1 \pm \overline{N_{n'}})\overline{N_m}\,\overline{N_{m'}}. \tag{120.5}$$

In equilibrium the mean number in each of the states may be regarded as depending only on the energy of that state: $\overline{N_m} = \overline{N}(\varepsilon_m)$. The law of conservation of energy in collisions (cf. (119.4) and (119.5)) gives

$$\varepsilon_m + \varepsilon_{m'} = \varepsilon_n + \varepsilon_{n'}\,. \tag{120.6}$$

From (120.5)

$$\frac{\overline{N_m}}{1 \pm \overline{N_m}} \cdot \frac{\overline{N_{m'}}}{1 \pm \overline{N_{m'}}} = \frac{\overline{N_n}}{1 \pm \overline{N_n}} \frac{\overline{N_{n'}}}{1 \pm \overline{N_{n'}}} = C\,, \tag{120.5'}$$

where C is some constant which, according to the above hypothesis regarding $\bar{N}$ and the conservation law (120.6), can depend only on the sum $\varepsilon_m + \varepsilon_{m'}$ $(= \varepsilon_n + \varepsilon_{n'})$. Thus

$$\frac{\overline{N_m}}{1 \pm \overline{N_m}} \cdot \frac{\overline{N_{m'}}}{1 \pm \overline{N_{m'}}} = C(\varepsilon_m + \varepsilon_{m'})\,. \tag{120.5''}$$

Denoting $\overline{N_m}/(1 \pm \overline{N_m})$ by $\phi(\varepsilon_m)$, we can write instead of (120.5″)

$$\phi(\varepsilon_m)\,\phi(\varepsilon_{m'}) = C(\varepsilon_m + \varepsilon_{m'})\,. \tag{120.7}$$

Differentiating this equation with respect to ε_m and with respect to $\varepsilon_{m'}$ and dividing the results we find

$$\frac{\phi'(\varepsilon_m)}{\phi(\varepsilon_m)} = \frac{\phi'(\varepsilon_{m'})}{\phi(\varepsilon_{m'})} = -\frac{1}{\Theta}, \tag{120.8}$$

where Θ is some constant independent of ε. Integrating (120.8) with respect to ε_m gives

$$\phi(\varepsilon_m) = e^{-(\varepsilon_m/\Theta) + \alpha}, \tag{120.9}$$

where α is a constant of integration. Hence we find the mean number of particles in the state with energy ε_m:

$$\overline{N_m} = \overline{N}(\varepsilon_m) = \frac{1}{e^{(\varepsilon_m/\Theta) - \alpha} \pm 1}, \tag{120.10}$$

with the minus sign for bosons and the plus sign for fermions. For high energies ($\varepsilon \to \infty$) the energy distribution law must coincide with the classical Boltzmann formula,

$$\bar{N}(\varepsilon_m) = \text{constant} \times e^{-\varepsilon_m/kT}, \tag{120.11}$$

where k is Boltzmann's constant and T the absolute temperature. Taking the limit

$\varepsilon_m \to \infty$ in (120.10) and comparing with (120.11), we see that $\Theta = kT$. Thus finally

$$\overline{N_m} = \frac{1}{e^{(\varepsilon_m/kT)-\alpha} \mp 1}. \tag{120.12}$$

The constant of integration α is determined from the condition that the number of particles in all states is equal to the total number of particles in the gas considered:

$$\sum_m \overline{N_m} = N. \tag{120.13}$$

An assembly of particles satisfying the distribution law (120.12) with the plus sign is called a Fermi-Dirac gas, and with the minus sign a Bose-Einstein gas. The formula (120.12) is written explicitly for discrete states.

Let the number of states in an energy range $d\varepsilon$ be $V\rho(\varepsilon)\,d\varepsilon$, where V is the volume of the entire gas. Then, summing (120.12) over all quantum states whose energy lies in the range ε to $\varepsilon + d\varepsilon$, we obtain the mean number of gas particles with energy between ε and $\varepsilon + d\varepsilon$ (the energy distribution law):

$$F(\varepsilon)\,d\varepsilon = \frac{V\rho(\varepsilon)\,d\varepsilon}{e^{(\varepsilon/\Theta)-\alpha} \mp 1},$$

and, dividing by V, the mean number per unit volume of the gas

$$f(\varepsilon)\,d\varepsilon = \frac{\rho(\varepsilon)\,d\varepsilon}{e^{(\varepsilon/\Theta)-\alpha} \mp 1}. \tag{120.14}$$

Thus (120.13) can be written as

$$\int_0^\infty f(\varepsilon)\,d\varepsilon = \int_0^\infty \frac{\rho(\varepsilon)\,d\varepsilon}{e^{(\varepsilon/\Theta)-\alpha} \mp 1} = n, \tag{120.15}$$

where $n = N/V$ is the number density of particles.[13]

The distribution (120.14) with the plus sign is called a *Fermi-Dirac distribution*, and with the minus sign a *Bose-Einstein distribution*. The most important property of a Fermi-Dirac distribution is the existence of a zero-point energy of the gas. To see this, we put $\alpha = \varepsilon_0/\Theta$, obtaining

$$\int_0^\infty \frac{\rho(\varepsilon)\,d\varepsilon}{e^{(\varepsilon-\varepsilon_0)/\Theta} + 1} = n, \qquad f(\varepsilon) = \frac{\rho(\varepsilon)}{e^{(\varepsilon-\varepsilon_0)/\Theta} + 1}. \tag{120.16}$$

As $\Theta \to 0$ (low temperatures), ε_0 must be positive if the energy ε is measured so that $\varepsilon > 0$, since otherwise $f(\varepsilon) \to 0$ as $\Theta \to 0$ and the first Equation (120.16) cannot

[13] Evidently $\rho(\varepsilon)$ cannot depend on the volume of the gas, since otherwise the distribution function would do so too. The quantity $\rho(\varepsilon)$ is always independent of V if the gas volume V is much greater than λ^3, where λ is a wavelength characteristic of a predominant number of the occupied states.

be satisfied. We also see that, as $\Theta \to 0$, $f(\varepsilon) = \rho(\varepsilon)$ for $\varepsilon < \varepsilon_0$ and $f(\varepsilon) = 0$ for $\varepsilon > \varepsilon_0$, i.e. at absolute zero all states in a Fermi-Dirac gas are occupied up to the state $\varepsilon = \varepsilon_0$, while all other states are unoccupied. The energy of the particles occupying the states from $\varepsilon = 0$ to $\varepsilon = \varepsilon_0$ is the zero-point energy of the gas. A more detailed discussion shows that this distribution varies only very slightly with temperature in the range where $\Theta = kT \ll \varepsilon_0$. Clearly ε_0 is the maximum particle energy in a Fermi-Dirac gas at the absolute zero of temperature.

We have derived the Fermi-Dirac and Bose-Einstein distributions from the collision hypothesis (120.4). The same distributions can be found from general results of statistical thermodynamics (the Gibbs ensemble) without any assumption regarding the kinetics of the processes.[14]

The difference between calculations based on quantum mechanics and those based on classical mechanics consists in a different method of computing the number of possible states. In quantum mechanics a state is specified by means of a symmetric or antisymmetric wave function Ψ, and various permutations of particles among individual states do not give a new state (Ψ remains unaltered or changes sign). In classical mechanics any such permutation represents a new state of the particles. Classical statistics based on this method of counting the states is a limiting case of quantum statistics, in which the number of states is computed from the number of different wave functions; it can be shown that classical statistics is obtained from quantum statistics if the number of particles in the volume λ^3 given by the mean wavelength is much less than unity. In the quantum region two statistics are distinguished: Fermi-Dirac statistics for particles obeying the Pauli principle, with antisymmetric Ψ, and Bose-Einstein statistics for bosons, with symmetric Ψ. The two statistics do not, of course, differ fundamentally.

We may apply Fermi-Dirac statistics to the conduction electrons in a metal. These may be approximately regarded as free particles.[15] Let us calculate the number of states $\rho(\varepsilon)$ per unit energy range. In a volume of the metal $L^3 = V$ the states of free particles will be stationary waves. It is more convenient to consider travelling waves, assuming the metal to be of infinite extent, but we shall suppose that the state is exactly repeated in each volume $L^3 = V$ (the 'periodicity condition'). Such a treatment is entirely legitimate if $L \gg \lambda$, where λ is a wavelength characteristic of a predominant number of the occupied states. The wave functions will be travelling plane waves of the form

$$\psi = \frac{e^{i(k_x x + k_y y + k_z z)}}{(2\pi L)^{\frac{3}{2}}} \qquad (120.17)$$

(normalised to unity in L^3), with

$$k_x = 2\pi n_x/L, \qquad k_y = 2\pi n_y/L, \qquad k_z = 2\pi n_z/L. \qquad (120.18)$$

[14] See [61]; [40].

[15] No rigorous proof that this approximation is possible has been given, nor have its limits of validity been established.

By this choice of k_x, k_y, k_z the state is repeated in each volume L^3. The states are labelled by n_x, n_y, n_z, and this set of numbers must be identified with the single suffix m in (120.12).

We take the sum $\sum \Delta n_x \Delta n_y \Delta n_z (\Delta n = \pm 1)$ over states lying in the energy range ε to $\varepsilon + d\varepsilon$. From (120.18)

$$\Delta n_x \, \Delta n_y \, \Delta n_z = (L/2\pi)^3 \, \Delta k_x \, \Delta k_y \, \Delta k_z \, ,$$

and so

$$\sum_{\varepsilon,\, \varepsilon + d\varepsilon} \Delta n_x \, \Delta n_y \, \Delta n_z = \frac{V}{(2\pi)^3} \sum_{\varepsilon,\, \varepsilon + d\varepsilon} \Delta k_x \, \Delta k_y \, \Delta k_z = \frac{V}{(2\pi)^3} \int_{\varepsilon,\, \varepsilon + d\varepsilon} dk_x \, dk_y \, dk_z$$

$$= \frac{V}{(2\pi)^3} \iint_{\varepsilon,\, \varepsilon + d\varepsilon} k^2 \frac{dk}{d\varepsilon} \, d\varepsilon \, d\Omega = \frac{4\pi V}{(2\pi)^3} k^2 \frac{dk}{d\varepsilon} \, d\varepsilon \, . \qquad (120.19)$$

Since for free particles $\varepsilon = \hbar^2 k^2 / 2\mu$, and two states with different orientations of the electron spin correspond to each value of $\mathbf{k}$, we have

$$V \rho(\varepsilon) \, d\varepsilon = \frac{8\pi V}{(2\pi\hbar)^3} \frac{(2\mu)^{\frac{3}{2}}}{2} \varepsilon^{\frac{1}{2}} \, d\varepsilon \, . \qquad (120.20)$$

Substituting this value of $\rho(\varepsilon)$ in (120.14), we find the law governing the distribution of free electrons:

$$f(\varepsilon) \, d\varepsilon = \frac{8\pi}{(2\pi\hbar)^3} \frac{(2\mu)^{\frac{3}{2}}}{2} \frac{\varepsilon^{\frac{1}{2}} \, d\varepsilon}{e^{(\varepsilon - \varepsilon_0)/\Theta} + 1} \, . \qquad (120.21)$$

Let us calculate the maximum energy ε_0 for $\Theta = 0$. Since, for $\Theta = 0$, $f(\varepsilon) = 0$ for $\varepsilon > \varepsilon_0$, we have from (120.16) and (120.21)

$$n = \int_0^{\varepsilon_0} f(\varepsilon) \, d\varepsilon = \frac{8\pi}{(2\pi\hbar)^3} \frac{(2\mu)^{\frac{3}{2}}}{2} \int_0^{\varepsilon_0} \varepsilon^{\frac{1}{2}} \, d\varepsilon = \frac{8\pi}{(2\pi\hbar)^3} \frac{(2\mu)^{\frac{3}{2}}}{2} \cdot \tfrac{2}{3} \varepsilon_0^{\frac{3}{2}} \, . \qquad (120.22)$$

Hence

$$\varepsilon_0 = \frac{(2\pi\hbar)^2}{2\mu} \left(\frac{3n}{8\pi} \right)^{\frac{2}{3}} . \qquad (120.23)$$

The maximum electron energy ε_0 for metals ($n \sim 10^{22}$ cm^{-3}) is of the order of a few electron-volts. The mean zero-point energy $\bar{\varepsilon}(0)$ of the electrons is of the same order of magnitude (more precisely, $\bar{\varepsilon}(0) = \frac{3}{5}\varepsilon_0$). According to the classical theory, the mean electron energy should be much less ($\frac{3}{2}kT$). A more detailed study shows that ε_0 depends only very slightly on the temperature if the latter is much less than $T_0 = \varepsilon_0 / k$. This temperature is $\sim 10000°$ for an electron gas. At temperatures $T \gg T_0$ it can be shown that the Fermi-Dirac distribution tends to the Maxwellian distribution

$$f(\varepsilon) \, d\varepsilon = \text{constant} \times e^{-\varepsilon/\Theta} \varepsilon^{\frac{1}{2}} \, d\varepsilon \, . \qquad (120.24)$$

The temperature T_0 is called the *degeneracy temperature* or *effective Fermi temperature* of the gas. The use of Fermi-Dirac statistics for the electron gas makes it possible to overcome numerous fundamental difficulties in the classical electron theory of metals, and has formed the basis of the modern theory.[16]

As an example of a Bose-Einstein distribution, let us consider black-body radiation, regarding the light quanta (photons) as particles. The relation between the energy ε and the wave number k for these particles is $\varepsilon = \hbar\omega = \hbar ck$, i.e. $\mathrm{d}\varepsilon/\mathrm{d}k = \hbar c$. Since the states of the photon are represented by plane waves, the number of states in the energy range $\mathrm{d}\varepsilon$ is given by (120.19), multiplied by two because there are two independent polarisations possible for each value of $\mathbf{k}$. Hence (120.19) gives

$$\rho(\varepsilon)\,\mathrm{d}\varepsilon = \frac{8\pi}{(2\pi)^3}\left(\frac{\varepsilon}{\hbar c}\right)^2 \frac{1}{\hbar c}\,\mathrm{d}\varepsilon\,. \tag{120.25}$$

Thus the photon energy distribution law is

$$f(\varepsilon)\,\mathrm{d}\varepsilon = \frac{8\pi}{(2\pi\hbar c)^3}\,\frac{\varepsilon^2\,\mathrm{d}\varepsilon}{e^{(\varepsilon/\Theta)-\alpha}-1}\,. \tag{120.26}$$

The total number of photons is indefinite, and so the condition (120.15) cannot be used to determine α. The energy per unit volume and in the range $\mathrm{d}\varepsilon$ will be $\varepsilon f(\varepsilon)\,\mathrm{d}\varepsilon$. Since $\varepsilon = \hbar\omega$, we can use the radiation density $u(\omega)$ per unit frequency range: $u(\omega)\,\mathrm{d}\omega = \varepsilon f(\varepsilon)\hbar\,\mathrm{d}\omega$. Hence

$$u(\omega) = \frac{\hbar\omega^3}{\pi^2 c^3}\,\frac{1}{e^{(\hbar\omega/\Theta)-\alpha}-1}\,. \tag{120.26'}$$

For $\hbar\omega \ll \Theta$ the distribution law must tend to the classical Rayleigh-Jeans law (Section 6), and so we must put $\alpha = 0$, obtaining

$$u(\omega) = \frac{\hbar\omega^3}{\pi^2 c^3}\,\frac{1}{e^{\hbar\omega/\Theta}-1}\,, \tag{120.26''}$$

i.e. Planck's formula.[17]

[16] The literature on the quantum theory of metals is very extensive; see, for example, [79, 73, 43].
[17] By using Gibbs' method, Formula (120.26″) can be derived directly without bringing in the classical Rayleigh–Jeans law [61]; [40].

CHAPTER XXI

MULTI-ELECTRON ATOMS

121. The helium atom

The helium atom, the second in the periodic system, is the simplest multi-electron atom, but even here classical mechanics failed completely. Attempts at calculations using classical mechanics (with Bohr's quantum conditions) led to the conclusion that classical mechanics could not be applied to atomic systems with two or more electrons. It was supposed that there must exist some kind of 'non-mechanical effects'. Modern quantum mechanics meets with no fundamental difficulty in the problem of multi-electron systems, although the mathematical difficulties are considerable.

We shall begin with a qualitative analysis of the possible states of the helium atom, using the general theory of systems of identical particles given in Sections 114–117, and first of all determine the form of the Hamiltonian operator $\boldsymbol{H}$ for the electrons in the helium atom. The interactions in the helium atom may be divided into two groups, one including the strong Coulomb interactions between the nucleus and the electrons, and the other the weak magnetic interactions due to the interaction of the electron spins with each other and with the orbital motion.[1]

Let the co-ordinates of the electrons be x_1, y_1, z_1 ($\mathbf{r}_1$) and x_2, y_2, z_2 ($\mathbf{r}_2$), and their spins be $\mathbf{s}_1$ and $\mathbf{s}_2$. The Coulomb interaction operator is

$$U = -\frac{2e^2}{r_1} - \frac{2e^2}{r_2} + \frac{e^2}{r_{12}}, \tag{121.1}$$

where the first two terms represent the energies of interaction of the two electrons with the nucleus of the atom, whose charge is $+2e$, while the third term gives the energy of Coulomb interaction of the electrons (Figure 82).

The magnetic interaction operator will be denoted by $\boldsymbol{W}$; it depends on the spins, positions and velocities of the electrons:

$$\boldsymbol{W} = \boldsymbol{W}(\mathbf{s}_1, \mathbf{s}_2, \mathbf{r}_1, \mathbf{r}_2, -i\hbar\nabla_1, -i\hbar\nabla_2). \tag{121.2}$$

Taking into account also the kinetic energy of the two electrons, we can write the

[1] This group includes the corrections due to the velocity dependence of the electron mass (cf. Section 65).

complete Hamiltonian of the electrons in the helium atom as

$$H(\mathbf{r}_1, \mathbf{r}_2, \mathbf{s}_1, \mathbf{s}_2) = -\frac{\hbar^2}{2\mu}\nabla_1^2 - \frac{\hbar^2}{2\mu}\nabla_2^2 - \frac{2e^2}{r_1} - \frac{2e^2}{r_2} + \frac{e^2}{r_{12}} + W. \tag{121.3}$$

The last term is, as we know (cf. Section 74), very small and is responsible for the multiplet structure of spectra. In this qualitative analysis of the multiplet structure of the helium levels, we can omit this term and use the Hamiltonian

$$H(\mathbf{r}_1, \mathbf{r}_2) = -\frac{\hbar^2}{2\mu}\nabla_1^2 - \frac{\hbar^2}{2\mu}\nabla_2^2 - \frac{2e^2}{r_1} - \frac{2e^2}{r_2} + \frac{e^2}{r_{12}}. \tag{121.4}$$

In this approximation, ignoring the small spin interactions, the variables pertaining to the motion of the centres of mass of the electrons and to their spins can be separated. Taking as the spin variables the components s_{z1} and s_{z2} of the spins in any

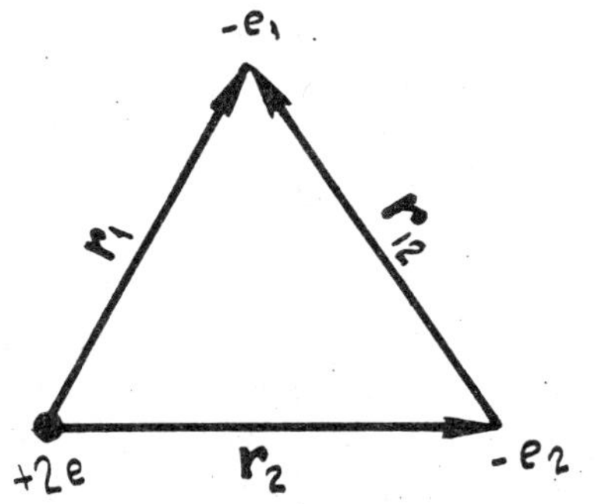

Fig. 82. Interactions in the helium atom.

one direction (for instance OZ), we can, as in Section 60, write the complete wave function of the two electrons of the helium atom as

$$\Psi(\mathbf{r}_1, \mathbf{r}_2, s_{z1}, s_{z2}) = \Phi(\mathbf{r}_1, \mathbf{r}_2) \cdot S(s_{z1}, s_{z2}), \tag{121.5}$$

where $S(s_{z1}, s_{z2})$ denotes the spin-dependent part of the wave function Ψ.

The Hamiltonian operator H (121.4) (and also the exact form (121.3)) is symmetrical with respect to the two electrons, because of their identity. In the present case, therefore, we can apply the general result (Section 115) whereby the wave function Ψ (121.5) must be antisymmetric or symmetric with respect to the particles, according as they obey the Pauli principle or not.

Experiment shows that electrons obey the Pauli principle (which, in fact, was first established for electrons). Consequently the wave function (121.5) must be antisymmetric with respect to interchange of electrons, i.e.

$$P_{12}\Psi(\mathbf{r}_1, \mathbf{r}_2, s_{z1}, s_{z2}) = -\Psi(\mathbf{r}_1, \mathbf{r}_2, s_{z1}, s_{z2}). \tag{121.6}$$

The transposition operator can be written as the product of two transposition operators P'_{12} and P''_{12}, of which the first transposes the co-ordinates $\mathbf{r}_1$ and $\mathbf{r}_2$ of the centres of mass of the electrons, and the second transposes their spins s_{z1} and s_{z2}. Then (121.6), using (121.5), can be written[2]

$$P'_{12}\,\Phi(\mathbf{r}_1,\mathbf{r}_2)\cdot P''_{12}\,S(s_{z1},s_{z2}) = -\,\Phi(\mathbf{r}_1,\mathbf{r}_2)\cdot S(s_{z1},s_{z2}). \qquad (121.7)$$

Thus there are two possibilities: either

$$P'_{12}\,\Phi(\mathbf{r}_1,\mathbf{r}_2) = +\,\Phi(\mathbf{r}_1,\mathbf{r}_2), \qquad (121.8)$$

and so

$$P''_{12}\,S(s_{z1},s_{z2}) = -\,S(s_{z1},s_{z2}), \qquad (121.9)$$

or else

$$P'_{12}\,\Phi(\mathbf{r}_1,\mathbf{r}_2) = -\,\Phi(\mathbf{r}_1,\mathbf{r}_2), \qquad (121.8')$$

and

$$P''_{12}\,S(s_{z1},s_{z2}) = +\,S(s_{z1},s_{z2}). \qquad (121.9')$$

The first possibility signifies that the co-ordinate function is symmetric and the spin function antisymmetric, while the second possibility represents the opposite situation. We have therefore two classes of wave functions for the possible states of the helium atom, namely

$$\Psi_{\mathrm{I}} = \Phi_s(\mathbf{r}_1,\mathbf{r}_2)\,S_a(s_{z1},s_{z2}), \qquad (121.10)$$

$$\Psi_{\mathrm{II}} = \Phi_a(\mathbf{r}_1,\mathbf{r}_2)\,S_s(s_{z1},s_{z2}), \qquad (121.10')$$

where the suffixes s and a denote symmetric and antisymmetric functions respectively.

Let us now consider further the spin functions S_a and S_s. Since we are neglecting the spin interaction, each function could be written as a product of the spin functions (60.6), (60.6′) for each electron separately, i.e. as

$$S(s_{z1},s_{z2}) = S_{\alpha_1}(s_{z1})\,S_{\alpha_2}(s_{z2}), \qquad (121.11)$$

where the suffixes α_1 and α_2 indicate the direction of the electron spin, parallel or antiparallel to the axis OZ. The function (121.11) is neither a symmetric nor an antisymmetric function of the electron spins, but it is easy to construct antisymmetric functions S_a and symmetric functions S_s from the functions (121.11).

Let us first take the case where the electron spins are in opposite directions. Then the wave function (121.11) is

$$S'(s_{z1},s_{z2}) = S_{+\frac{1}{2}}(s_{z1})\,S_{-\frac{1}{2}}(s_{z2}), \qquad (121.12)$$

but another state is also possible, where the spin of the first electron is antiparallel to the axis OZ and that of the second is parallel to that axis:

$$S''(s_{z1},s_{z2}) = S_{-\frac{1}{2}}(s_{z1})\,S_{+\frac{1}{2}}(s_{z2}). \qquad (121.12')$$

The two states both give a total spin of zero along the axis OZ and have the same energy E, and so any superposition of these states may be assigned to the energy E.

[2] Equation (121.6) is valid even if the spin interaction is not neglected, but the following argument is based on the approximation (121.5).

Among the superpositions the only one described by an antisymmetric function S_a is[3]

$$S_a(s_{z1}, s_{z2}) = \frac{1}{\sqrt{2}} \{S_{+\frac{1}{2}}(s_{z1}) S_{-\frac{1}{2}}(s_{z2}) - S_{-\frac{1}{2}}(s_{z1}) S_{+\frac{1}{2}}(s_{z2})\}. \quad (121.13)$$

Thus the form of the antisymmetric spin function is known. If the electron spins are parallel, antisymmetric states are evidently impossible. In this case we can have the following electron spin states:

$$S_s'(s_{z1}, s_{z2}) = S_{+\frac{1}{2}}(s_{z1}) S_{+\frac{1}{2}}(s_{z2}), \quad (121.14)$$

$$S_s''(s_{z1}, s_{z2}) = S_{-\frac{1}{2}}(s_{z1}) S_{-\frac{1}{2}}(s_{z2}). \quad (121.14')$$

These states are already symmetric with respect to the electron spins. In addition, the functions (121.12), (121.12′) yield a function symmetric with respect to the electron spins:

$$S_s'''(s_{z1}, s_{z2}) = \frac{1}{\sqrt{2}} \{S_{+\frac{1}{2}}(s_{z1}) S_{-\frac{1}{2}}(s_{z2}) + S_{-\frac{1}{2}}(s_{z1}) S_{+\frac{1}{2}}(s_{z2})\}. \quad (121.14'')$$

Thus we have altogether three functions S_s', S_s'' and S_s''' symmetric with respect to spin. The first two have total spin 1, but in the state S_s' the spin is parallel to the axis OZ, while in the state S_s'' it is antiparallel. It is somewhat less obvious that the state S_s''' also has total spin 1, but perpendicular to the axis OZ. This may be most easily seen as follows. We take as spin variables the components of the spin along the axis OZ. For a state in which the spin is perpendicular to the axis OZ these variables

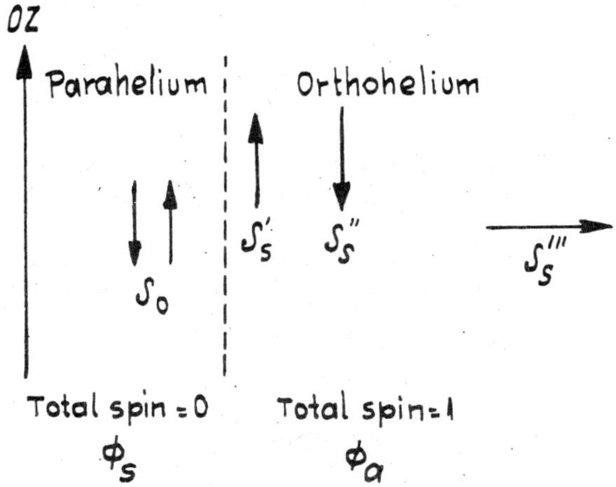

Fig. 83. Addition of spins of two electrons, with notation for the wave functions of the corresponding states as used in the text.

[3] The factor $1/\sqrt{2}$ is included in order to normalise S_a to unity. The functions $S_{\pm\frac{1}{2}}(s_z)$ are normalised to unity, by (60.7). If we take the product $S_a^*(s_{z1}, s_{z2})\, S_a(s_{z1}, s_{z2})$ and sum over both spins $s_{z1} = \pm \frac{1}{2}\hbar$, $s_{z2} = \pm \frac{1}{2}\hbar$, it is easily seen from (60.7) that the result is unity (see also Section 106).

s_{z1} and s_{z2} must have the indeterminate value $\pm \frac{1}{2}\hbar$, i.e. the state with spin perpendicular to OZ must be written in the variables s_{z1} and s_{z2} so that all possible values of s_{z1} and s_{z2} appear. The state must also be symmetric with respect to the spins. Then (121.14″) is the only possible wave function of this state.[4] Figure 83 shows diagrammatically the arrangement of the spins for the states which we have found.

Thus the states Φ_s symmetric in the co-ordinates of the centres of mass of the electrons are states with total electron spin zero; the states Φ_a antisymmetric in the co-ordinates of the centres of mass of the electrons are states with the electron spins parallel (total spin 1), and there are three such states, with three quantum orientations of the total spin. The levels of the helium atom therefore fall into two classes: those with antiparallel spins and those with parallel spins.

If we take into account the fact that the energy of a quantum level depends, though only very slightly, on the orientation of the spin with respect to the orbital motion, we must conclude that the levels with antiparallel spins occur singly (*singlet* levels), while those with parallel spins form sets of three close together corresponding to the three possible orientations of the total spin with respect to the magnetic field created by the orbital motion. Thus these levels are threefold (*triplet* levels).[5]

A notable property of these two classes of states in helium is that quantum transitions between them are (almost) impossible. The spin interactions are very slight, and if they are ignored the Hamiltonian of the helium atom is symmetrical with respect to the electron co-ordinates, even when an external field is present (e.g. on interaction with light), since the external field acts on both electrons in the same way. Thus

$$\boldsymbol{H}(\mathbf{r}_1, \mathbf{r}_2) = \boldsymbol{H}(\mathbf{r}_2, \mathbf{r}_1). \tag{121.15}$$

The change in the wave function $\Psi(\mathbf{r}_1, \mathbf{r}_2, s_{z1}, s_{z2}, t)$ in a time $\mathrm{d}t$ is given by Schrödinger's equation, which we write in the form

$$\mathrm{d}_t\Psi(\mathbf{r}_1, \mathbf{r}_2, s_{z1}, s_{z2}, t) = (1/i\hbar)\,\boldsymbol{H}(\mathbf{r}_1, \mathbf{r}_2)\,\Psi(\mathbf{r}_1, \mathbf{r}_2, s_{z1}, s_{z2}, t)\,\mathrm{d}t, \tag{121.16}$$

as in Section 115. If $\Psi(\mathbf{r}_1, \mathbf{r}_2, s_{z1}, s_{z2}, t)$ is at any instant a symmetric function of the electron co-ordinates $\mathbf{r}_1, \mathbf{r}_2$, the increment of this function $\mathrm{d}_t\Psi$ is, by (121.16)

[4] The statement that the states S_s', S_s'' and S_s''' have spin 1 (the electron spins are added) can be checked by direct calculation. If the electron spin operators defined by the matrices (59.12) are denoted by $\vec{s}_1$ and $\vec{s}_2$, the total-spin operator is given by the matrix

$$\vec{s}^2 = \vec{s}_1^2 + \vec{s}_2^2 + 2\vec{s}_1 \cdot \vec{s}_2.$$

The eigenfunction S of the operator $\vec{s}^2$ must satisfy the equation

$$\vec{s}^2 S = \hbar^2 l_s (l_s + 1) S,$$

where l_s is a number which determines the total spin. From this equation it can be seen that l_s has only two values: $l_s = 0$ (antiparallel spins) and $l_s = 1$ (parallel spins). By direct substitution of S_s', S_s'' and S_s''' in this equation we can show that these functions have $l_s = 1$. The straightforward calculations involved are left to the reader.

[5] The magnitude of the splitting is calculated in [5], Section 23.

and (121.15), also symmetric. Similarly, if $\Psi(\mathbf{r}_1, \mathbf{r}_2, s_{z1}, s_{z2}, t)$ is antisymmetric, the increment will be antisymmetric. Consequently, a state symmetric or antisymmetric in the co-ordinates will remain so under all possible changes. Thus transitions from states Ψ_{I} (121.10) to states Ψ_{II} (121.10′) or conversely are impossible.

It should be noted that there is a difference between the result just proved and the general theorem in Section 115. The functions Ψ_{I} and Ψ_{II} are antisymmetric in the particles, and so transitions between states with these functions are permitted by the theorem in Section 115. We have now proved that transitions between Ψ_{I} and Ψ_{II} are not possible if spin interactions are neglected. Since such interactions do in fact exist, transitions between Ψ_{I} and Ψ_{II} are indeed possible, but very improbable because the spin interaction is small. As an illustration, let us make an estimate for the interaction of a light wave. The energy of interaction of a light wave with an electron charge is, in order of magnitude, given by $W' = e\mathscr{E}a$, where a is the dimension of the atom, e the electron charge and $\mathscr{E}$ the electric field of the light wave, ea being the electric dipole moment of the atom. The interaction of the light wave with the magnetic moment of the electron is equal, in order of magnitude, to the product of this magnetic moment $(e\hbar/2\mu c)$ and the magnetic field $\mathscr{H}$ of the wave: $W'' = e\hbar\mathscr{H}/2\mu c$; since $\mathscr{E}$ and $\mathscr{H}$ are equal in the light wave, $W''/W \approx \hbar/2\mu ca$. $\hbar/a$ is equal in order of magnitude to the momentum of the electron in the atom, and $\hbar/\mu a$ to its velocity v; thus $W''/W' \approx v/c$. This ratio is less than $\frac{1}{100}$, and so it is very improbable that the light will cause a transition in which the direction of the electron spin is changed.[6] In other words, transitions without change of spin, i.e. transitions between states with like symmetry in the electron co-ordinates, will predominate. This agrees with the theorem proved above.

If helium is in a state with parallel spins (a state antisymmetric in the co-ordinates), it is therefore very unlikely that its state will change to one with antiparallel spins (symmetric in the co-ordinates), and conversely. The situation is as if there were two kinds of helium, with parallel and antiparallel spins. The first kind is called *orthohelium* and the second kind *parahelium* (Figure 84). In order to convert one kind of helium into the other, the direction of the spin of one electron must be changed. Owing to the smallness of the magnetic moment of the spin, this change is very

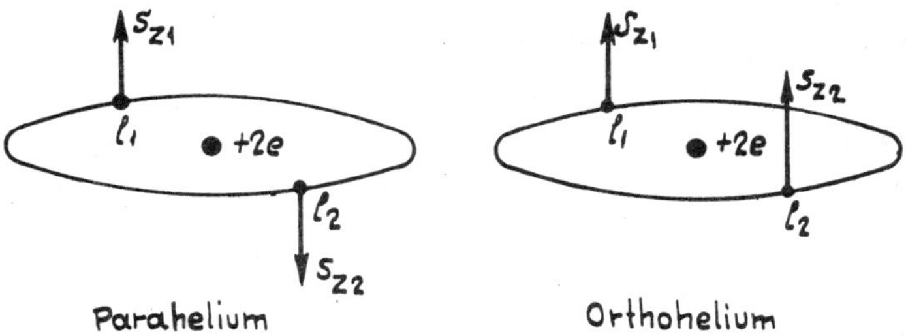

Fig. 84. Arrangement of spins in orthohelium and parahelium.

[6] It should be remembered that the transition probability is proportional to the square of the perturbation energy, and so the ratio of probabilities is 10^{-4}.

difficult to bring about. It is easily seen that the parahelium state must have the lower energy. As mentioned earlier, the lowest state has a wave function without nodes, while the antisymmetric function $\Phi_a(\mathbf{r}_1, \mathbf{r}_2)$ has a nodal surface at $\mathbf{r}_1 = \mathbf{r}_2$, since $\Phi_a(\mathbf{r}_1, \mathbf{r}_2) = -\Phi_a(\mathbf{r}_2, \mathbf{r}_1)$, and when $\mathbf{r}_1 = \mathbf{r}_2 = \mathbf{r}$ this gives $\Phi_a(\mathbf{r}, \mathbf{r}) = -\Phi_a(\mathbf{r}, \mathbf{r})$, i.e. $\Phi_a(\mathbf{r}, \mathbf{r}) = 0$. Hence the lowest state must belong to the symmetric function $\Phi_s(\mathbf{r}_1, \mathbf{r}_2)$, and is thus the state antisymmetric in the spins, i.e. the parahelium state. Hence parahelium is the ground state.

In this connection the problem arises of how orthohelium can be obtained. If irradiation with light is used, excited states will be produced but again with antiparallel spins, i.e. parahelium states. A different result is obtained, however, if helium is subjected to electron bombardment. Here three identical particles are involved: two electrons in the helium atom and one incident from outside. The above analysis of states for two identical particles will therefore be inapplicable. Physically the situation is that the incident electron can replace an atomic electron, the latter leaving the atom. Since the incident beam includes electrons with spins in all directions, this exchange may produce an atom whose electrons have spins in the same direction, and parahelium is thus converted to orthohelium.

The proof of the existence of two heliums (or, more precisely, of the existence of two classes of states of helium) has made possible a complete interpretation of the helium spectrum and its behaviour under various conditions. Figure 85 shows a diagram of levels of the helium atom. In parahelium the total spin is zero, and there is no multiplet structure; the lines are singlets. The corresponding terms are denoted by letters with the figure 1 at the top left (for instance ^1S, ^1P). The orthohelium terms, however, form close triplets. The spectral lines of orthohelium correspondingly consist of triplets. The orthohelium terms are denoted by adding the figure 3 (triplet) at the top left (for instance ^3S, ^3P). In Figure 85 the orthohelium state 2^3S is marked as being *metastable*. This is the lowest state of orthohelium, and the only transition to a lower state is that to the 1^1S state of parahelium, with change of spin direction. This has low probability, and a helium atom in such a state will remain there for a very long time, despite the extra energy of 19.77 eV.

This concludes our qualitative analysis of the states of the helium atom, and we shall now proceed to an approximate quantitative theory.

122. Approximate quantitative theory of the helium atom

To calculate the quantum levels of the helium atom we use a method which, though not giving the highest possible accuracy, is simple and clear. Schrödinger's equation to determine the quantum levels of the helium atom and the wave functions of the stationary states is

$$H(\mathbf{r}_1, \mathbf{r}_2, s_{z1}, s_{z2})\,\Psi(\mathbf{r}_1, \mathbf{r}_2, s_{z1}, s_{z2}) = E\Psi(\mathbf{r}_1, \mathbf{r}_2, s_{z1}, s_{z2}). \tag{122.1}$$

Since we are neglecting the spin interactions, (121.5) shows that $S(s_{z1}, s_{z2})$ can be cancelled from this equation, giving

$$H(\mathbf{r}_1, \mathbf{r}_2)\,\Phi(\mathbf{r}_1, \mathbf{r}_2) = E\Phi(\mathbf{r}_1, \mathbf{r}_2), \tag{122.2}$$

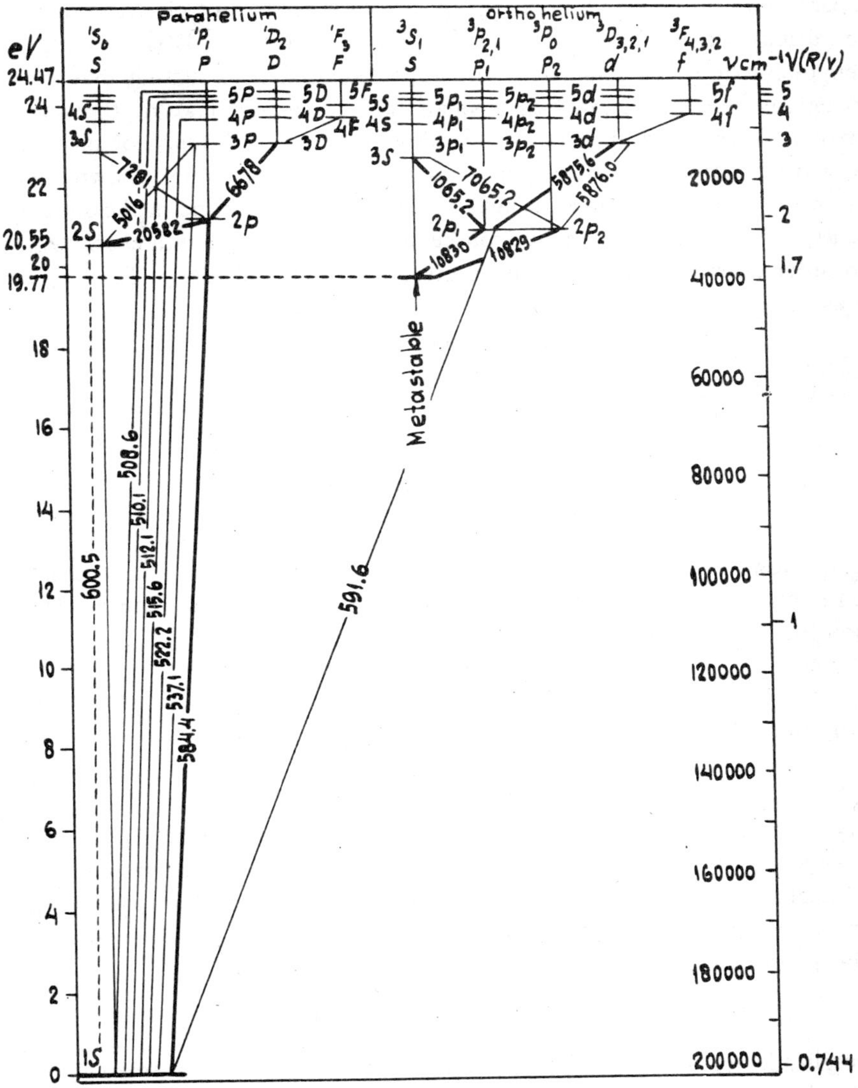

Fig. 85. Diagram of spectral terms of helium.

the total-energy operator being given by Formula (121.4). This operator may be written

$$H(\mathbf{r}_1, \mathbf{r}_2) = H_0(\mathbf{r}_1, \mathbf{r}_2) + W(\mathbf{r}_{12}), \tag{122.3}$$

where

$$H_0(\mathbf{r}_1, \mathbf{r}_2) = -\frac{\hbar^2}{2\mu}\nabla_1^2 - \frac{\hbar^2}{2\mu}\nabla_2^2 - \frac{2e^2}{r_1} - \frac{2e^2}{r_2} \tag{122.4}$$

$$= H_0(\mathbf{r}_1) + H_0(\mathbf{r}_2),$$

$$W(\mathbf{r}_{12}) = e^2/r_{12}. \tag{122.5}$$

The operator $\boldsymbol{H}_0(\mathbf{r}_1, \mathbf{r}_2)$ is the total-energy operator of the two electrons in the field of the nucleus without interaction between the electrons, and $\boldsymbol{W}(\mathbf{r}_{12})$ is the energy of their interaction. We shall make the approximation of assuming that the interaction energy can be regarded as a small correction, and take as the zero-order approximation the motion of non-interacting electrons in the field of the nucleus.[7]

The wave functions and quantum levels for such a motion are known, since the motion takes place in a Coulomb field. Let the first electron be in the state $\psi_n(\mathbf{r}_1)$, with energy E_n, and the second electron be in the state $\psi_m(\mathbf{r}_2)$, with energy E_m. Then a zero-order approximation function corresponding to the energy $E_n + E_m$ can be taken as

$$\psi_1(\mathbf{r}_1, \mathbf{r}_2) = \psi_n(\mathbf{r}_1)\psi_m(\mathbf{r}_2). \tag{122.6}$$

For

$$\begin{aligned} &\boldsymbol{H}_0(\mathbf{r}_1, \mathbf{r}_2)\psi_1(\mathbf{r}_1, \mathbf{r}_2) \\ &= \boldsymbol{H}_0(\mathbf{r}_1)\psi_n(\mathbf{r}_1)\psi_m(\mathbf{r}_2) + \boldsymbol{H}_0(\mathbf{r}_2)\psi_n(\mathbf{r}_1)\psi_m(\mathbf{r}_2) \\ &= E_n\psi_n(\mathbf{r}_1)\psi_m(\mathbf{r}_2) + E_m\psi_n(\mathbf{r}_1)\psi_m(\mathbf{r}_2) \\ &= (E_n + E_m)\psi_1(\mathbf{r}_1, \mathbf{r}_2). \end{aligned} \tag{122.7}$$

There is, however, another state belonging to the energy $E_n + E_m$, namely that where the first electron is in the state E_m and the second is in the state E_n. The wave function for this state is

$$\psi_2(\mathbf{r}_1, \mathbf{r}_2) = \psi_m(\mathbf{r}_1)\psi_n(\mathbf{r}_2). \tag{122.6'}$$

Similarly to (122.7) we find

$$\boldsymbol{H}_0(\mathbf{r}_1, \mathbf{r}_2)\psi_2(\mathbf{r}_1, \mathbf{r}_2) = (E_n + E_m)\psi_2(\mathbf{r}_1, \mathbf{r}_2). \tag{122.7'}$$

Thus two states ψ_1 and ψ_2, differing in that the states of the electrons 1 and 2 are interchanged, correspond to the level $E_n + E_m$ of the unperturbed system. This is a case of degeneracy, and is called *exchange degeneracy*. According to general perturbation theory (Section 69) the correct wave function of the zero-order approximation must be a superposition of degenerate states[8]:

$$\Phi(\mathbf{r}_1, \mathbf{r}_2) = c_1\psi_1(\mathbf{r}_1, \mathbf{r}_2) + c_2\psi_2(\mathbf{r}_1, \mathbf{r}_2). \tag{122.8}$$

The amplitudes c_1 and c_2 and the quantum levels E of the perturbed system are given by the fundamental equations of perturbation theory. Since we are considering

[7] It will appear finally that, although the interaction energy is not very small (and so the approximation is not especially good), it is less than the energy difference between the lowest levels by a factor of about three.

[8] Strictly speaking, the wave functions ψ_n should have three suffixes n, l, m, since, as we know, there are altogether n^2 different states belonging to the level E_n, by the degeneracy in the Coulomb field, and accordingly, for a correct calculation of the helium levels, the zero-order function should be taken as a superposition not only of states differing by the exchange of electrons, as above, but of all states belonging to the levels E_n and E_m and differing in their angular momenta and orientations thereof. We shall, however, calculate as if the levels E_n were not degenerate. This is done merely to exhibit the particular features of the problem which arise solely from the presence of two identical particles.

only twofold exchange degeneracy (the functions ψ_1 and ψ_2), we can apply directly the theory of twofold degeneracy given in Section 69. To determine the amplitudes c_1 and c_2 we then have Equations (69.5), which in the present case become

$$\left.\begin{aligned}(E^0_{nm} + W_{11} - E)\,c_1 + W_{12}\,c_2 = 0\,,\\ W_{21}\,c_1 + (E^0_{nm} + W_{22} - E)\,c_2 = 0\,,\end{aligned}\right\} \tag{122.9}$$

where E^0_{nm} is the energy of the unperturbed motion:

$$E^0_{nm} = E_n + E_m\,; \tag{122.10}$$

in Section 69 the suffixes n, m were denoted by the single letter k. The quantities W_{11}, W_{12}, W_{22}, W_{21} are the matrix elements of the perturbation W (see (69.6), (69.6′), (69.6″)). Since in (69.6) etc. the integration is over all the variables on which the wave functions depend, in our case these formulae become

$$W_{11} = \int \psi_1^*\, W \psi_1\, \mathrm{d}v_1\, \mathrm{d}v_2\,, \tag{122.11}$$

$$W_{12} = \int \psi_1^*\, W \psi_2\, \mathrm{d}v_1\, \mathrm{d}v_2\,, \tag{122.11′}$$

where $\mathrm{d}v_1 = \mathrm{d}x_1\,\mathrm{d}y_1\,\mathrm{d}z_1$, $\mathrm{d}v_2 = \mathrm{d}x_2\,\mathrm{d}y_2\,\mathrm{d}z_2$ and W is the perturbation energy (122.5).

The energy levels E of the perturbed system are determined from the secular Equation (69.7), which is unchanged in form:

$$\begin{vmatrix} W_{11} - \varepsilon & W_{12} \\ W_{21} & W_{22} - \varepsilon \end{vmatrix} = 0\,, \tag{122.12}$$

where, in the present notation, the correction to the energy is

$$\varepsilon = E - E^0_{nm} = E - (E_n + E_m)\,. \tag{122.13}$$

Before solving this equation we shall derive some properties of the matrix elements (122.11), (122.11′). Substituting for ψ_1 and ψ_2 their values from (122.6) and for W from (122.5), we obtain

$$W_{11} = e^2 \int \frac{|\psi_n(\mathbf{r}_1)|^2\, |\psi_m(\mathbf{r}_2)|^2}{r_{12}}\, \mathrm{d}v_1\, \mathrm{d}v_2 = W_{22}\,. \tag{122.14}$$

It is also evident that $W_{12} = W_{21}$, since

$$W_{12} = e^2 \int \frac{\psi_n^*(\mathbf{r}_1)\,\psi_m(\mathbf{r}_1)\,\psi_n(\mathbf{r}_2)\,\psi_m^*(\mathbf{r}_2)}{r_{12}}\, \mathrm{d}v_1\, \mathrm{d}v_2\,, \tag{122.15}$$

and

$$W_{21} = \int \psi_2^* \cdot W \psi_1 \cdot \mathrm{d}v_1\, \mathrm{d}v_2 = e^2 \int \frac{\psi_m^*(\mathbf{r}_1)\,\psi_n(\mathbf{r}_1)\,\psi_n^*(\mathbf{r}_2)\,\psi_m(\mathbf{r}_2)}{r_{12}}\, \mathrm{d}v_1\, \mathrm{d}v_2\,. \tag{122.16}$$

Since the variables of integration $\mathbf{r}_1(x_1, y_1, z_1)$ and $\mathbf{r}_2(x_2, y_2, z_2)$ take the same values, they can be interchanged (which amounts to a change of notation), and since $r_{12} =$

r_{21} we find that W_{12} becomes W_{21}. Hence

$$W_{12} = W_{21} = W_{12}^* = W_{21}^*, \tag{122.17}$$

i.e. the quantities W_{12} are real. We put

$$W_{11} = W_{22} = K, \qquad W_{12} = W_{21} = A, \tag{122.18}$$

where K and A are real. The secular equation (122.12) then becomes

$$\begin{vmatrix} K - \varepsilon & A \\ A & K - \varepsilon \end{vmatrix} = 0, \tag{122.19}$$

whence

$$(K - \varepsilon)^2 = A^2, \qquad \text{or} \qquad \varepsilon = K \pm A. \tag{122.20}$$

In this new notation, Equations (122.9) become

$$(K - \varepsilon)\, c_1 + A c_2 = 0, \qquad (K - \varepsilon)\, c_2 + A c_1 = 0. \tag{122.9'}$$

Substitution of the first root ε from (122.20) gives $c_1 = c_2$, and the second root gives $c_1 = - c_2$. Thus the solution (122.8) is

$$\Phi_s(\mathbf{r}_1, \mathbf{r}_2) = \frac{1}{\sqrt{2}}(\psi_1 + \psi_2), \qquad E_s = E_n + E_m + K + A, \tag{122.21}$$

$$\Phi_a(\mathbf{r}_1, \mathbf{r}_2) = \frac{1}{\sqrt{2}}(\psi_1 - \psi_2), \qquad E_a = E_n + E_m + K - A \tag{122.22}$$

(the factor $1/\sqrt{2}$ being introduced for purposes of normalisation).

Thus, owing to the exchange degeneracy, two kinds of state are obtained, symmetric Φ_s and antisymmetric Φ_a (according to (122.6) and (122.6'), when the electron coordinates are interchanged ψ_1 becomes ψ_2). The existence of these two kinds of state is in complete agreement with the general theory in Section 115. We know that the former are parahelium states and the latter are orthohelium states. Formulae (122.21) and (122.22) are therefore approximate expressions for the parahelium and orthohelium wave functions.

In discussing the qualitative theory of the helium atom we have pointed out that the ground state must be described by a symmetric function (parahelium). This result is also implied by the solutions (122.21) and (122.22). Only one wave function $\psi_{100}(\mathbf{r}_1)$ corresponds to the lowest level E_1. Hence, in order to obtain the lowest state of the helium atom, there is only one possibility: to place the second electron in the same state (when, from the elementary form of the Pauli principle, the second electron must have a spin in the direction opposite to that of the first). In the lowest state, therefore, $\psi_1 = \psi_2$ and $\Phi_a = 0$. Thus for this state we have the single solution

$$\Phi_s(\mathbf{r}_1, \mathbf{r}_2) = \psi_{100}(\mathbf{r}_1)\psi_{100}(\mathbf{r}_2), \tag{122.23}$$

$$E = 2E_1 + K + A. \tag{122.23'}$$

The difference in energy of the para and ortho states according to (122.21) and

(122.22) is $2A$. The levels of helium can therefore be divided into two groups, the parahelium and orthohelium levels, with different energies. To each level $E_n + E_m$ of the helium atom obtained when the interaction of the electrons is neglected, there correspond two levels when this interaction is taken into account: the parahelium level $E_n + E_m + K + A$ and the orthohelium level $E_n + E_m + K - A$. For example, if one electron is in the lowest state E_1 and the other in the next state E_2 (energy $E_1 + E_2$), when the exchange and interaction between the electrons is taken into account we have two levels: $E_1 + E_2 + K + A$ and $E_1 + E_2 + K - A$. This splitting and the level $2E_1 + K + A$ are shown in Figure 86. The diagram shows a smaller degree of splitting than that given in the full spectroscopic diagram in Figure 85 because we have ignored (for simplicity) the fact that the levels of the unperturbed system (E_2, for instance) are degenerate (except for the first level). A more complete

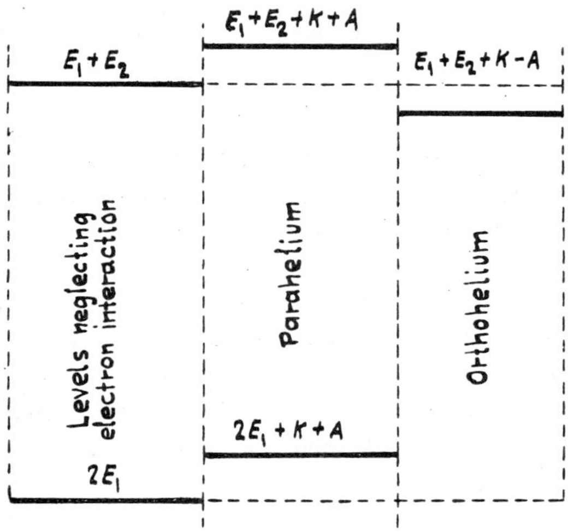

Fig. 86. Diagram of exchange splitting of helium levels.

calculation would show that the splitting of the levels is due not only to the exchange degeneracy but also to the removal of the l degeneracy. This is clear from the fact that the l degeneracy exists only in the Coulomb field of the nucleus, and must be removed by the presence of the second electron. When the removal of the l degeneracy is taken into account, the more complex pattern of level splitting shown in Figure 85 is obtained.

The reader who is interested in problems of computation is referred to the specialist literature[9]; here we shall merely indicate the present state of the theory of helium levels. Calculations by the method described above are by no means in complete agreement with experiment: the correction ε differs by 10 to 20% from that given by experimental measurement. The theoretical methods have been much improved, and

[9] See [5], Sections 11–24.

Hylleraas has obtained (in the eighth approximation) a value for the ground level of helium (the ionisation potential) of $I = 198{,}308\ \mathrm{cm}^{-1}$. (The energy is given in units of cm^{-1}, as is usual in spectroscopy.) The experimental value is $198{,}298 \pm 6\ \mathrm{cm}^{-1}$. The agreement is surprisingly good, especially if we bear in mind that the theory involves no arbitrary constants which might be fitted to the experimental data.

The calculation of the term values resulting from the l degeneracy is much more complicated, and the accuracy that has been achieved is considerably less than that of the ground state.

123. The exchange energy

Let us now consider in more detail the value of the correction $\varepsilon = K + A$ due to the Coulomb interaction of the electrons. Instead of the wave functions ψ_n and ψ_m, we use the new quantities

$$\rho_{nn}(\mathbf{r}_1) = -e\,|\psi_n(\mathbf{r}_1)|^2, \qquad \rho_{mm}(\mathbf{r}_2) = -e\,|\psi_m(\mathbf{r}_2)|^2, \tag{123.1}$$

$$\rho_{mn}(\mathbf{r}_1) = -e\psi_m^*(\mathbf{r}_1)\psi_n(\mathbf{r}_1), \qquad \rho_{mn}^*(\mathbf{r}_2) = -e\psi_m(\mathbf{r}_2)\psi_n^*(\mathbf{r}_2). \tag{123.2}$$

The first two of these have a simple physical significance: $\rho_{nn}(\mathbf{r}_1)$ is evidently the mean electric charge density at the point $\mathbf{r}_1$ due to an electron in the state $\psi_n(\mathbf{r}_1)$. Similarly, $\rho_{mm}(\mathbf{r}_2)$ is the mean electric charge density at the point $\mathbf{r}_2$ due to an electron in the state $\psi_m(\mathbf{r}_2)$.

The other two quantities, $\rho_{mn}(\mathbf{r}_1)$ and $\rho_{mn}^*(\mathbf{r}_2)$, have no such simple meaning. They are the charge densities due to the fact that each of the electrons may be partly in the state $\psi_n(\mathbf{r}_1)$ and partly in the state $\psi_m(\mathbf{r}_2)$, and we shall call them *exchange densities*. These quantities may be complex and so the term 'charge density' is mainly formal. Using the densities thus defined, the quantity K can be written, according to (122.18) and (122.14), as

$$K = \int \frac{\rho_{nn}(\mathbf{r}_1)\,\rho_{mm}(\mathbf{r}_2)}{r_{12}}\,dv_1\,dv_2\,, \tag{123.3}$$

and A, from (122.18) and (122.15), as

$$A = \int \frac{\rho_{mn}(\mathbf{r}_1)\,\rho_{mn}^*(\mathbf{r}_2)}{r_{12}}\,dv_1\,dv_2\,. \tag{123.4}$$

The significance of the quantity K is clear: the integral in (123.3) is just the mutual Coulomb energy of two charges, one distributed in space with density ρ_{nn} and the other with density ρ_{mm}. This energy may be descriptively interpreted as that of the Coulomb interaction of two electrons whose charges are spread out in space. This part of the interaction energy of the electrons is therefore called *the Coulomb part* (in the restricted sense of the word). The other part (A) has no such clear significance. Formally it may be regarded as the electrostatic energy of two charges distributed with densities ρ_{mn} and ρ_{mn}^*. This part of the electron interaction energy is called the *exchange energy*. In this sense we say that the energy of interaction of the two electrons consists of a Coulomb part K and an exchange part A.

It should be borne in mind that both K and A are due to the Coulomb interaction (since both are zero when $e = 0$). The difference between the Coulomb energy (in the restricted sense) and the exchange energy A is due to the approximate representation of the functions Φ_a and Φ_s in the form $(\psi_1 \pm \psi_2)/\sqrt{2}$. Nevertheless, this division of the interaction energy into the Coulomb and exchange parts is very useful and is therefore justified.

According to perturbation theory, the correction ε to the energy must be just the mean perturbation energy in the corresponding state. This statement is easily confirmed for the case under consideration, as follows. The perturbation energy is the Coulomb energy of interaction of the electrons, e^2/r_{12}. In order to calculate the mean value of this energy in some state $\Phi(\mathbf{r}_1, \mathbf{r}_2)$, e^2/r_{12} must be multiplied by the probability of the first electron's being in the range dv_1 and the second's being in the range dv_2, i.e. by $|\Phi|^2\, dv_1\, dv_2$, and the result integrated over all possible positions of the electrons:

$$\overline{e^2/r_{12}} = \int (e^2/r_{12})\,|\Phi|^2\, dv_1\, dv_2\,. \tag{123.5}$$

Substituting Φ_a or Φ_s from (122.21) and (122.22) for Φ, we find

$$\overline{e^2/r_{12}} = \tfrac{1}{2}\int (e^2/r_{12})\{|\psi_1|^2 + |\psi_2|^2 \pm \psi_1\psi_2^* \pm \psi_1^*\psi_2\}\, dv_1\, dv_2\,, \tag{123.6}$$

which, by (122.6) and (122.6′), gives

$$\overline{e^2/r_{12}} = K \pm A\,, \tag{123.6′}$$

i.e. the correction ε is just the mean energy of the Coulomb interaction of the electrons in the state Φ_s or Φ_a.

This calculation enables us to understand more thoroughly the origin of the exchange energy. The quantity $|\psi_1|^2\, dv_1\, dv_2$ is the probability that the first electron is in the range dv_1 and in the state n, and the second in the range dv_2 and in the state m. Similarly, $|\psi_2|^2\, dv_1\, dv_2$ is the probability that the first electron is in the range dv_1 and in the state m, and the second in the range dv_2 and in the state n. If the states ψ_1 and ψ_2 were independent, we should find that the probability of the first electron's being in the range dv_1 and the second's being in the range dv_2 is independent of the states of the electrons, and (assuming ψ_1 and ψ_2 equally probable) is

$$dP_{12} = \tfrac{1}{2}\{|\psi_1|^2 + |\psi_2|^2\}\, dv_1\, dv_2\,. \tag{123.7}$$

In reality the states ψ_1 and ψ_2 are not independent, and the states which actually occur are

$$\Phi = (\psi_1 \pm \psi_2)/\sqrt{2}\,. \tag{123.8}$$

The wave functions ψ_1 and ψ_2 are in definite phase relationships, and the expression for the probability of finding the particles in the ranges dv_1 and dv_2 contains the interference term

$$dP'_{12} = |\Phi|^2 \, dv_1 \, dv_2 = dP_{12} \pm (\psi_1 \psi_2^* + \psi_1^* \psi_2) \, dv_1 \, dv_2 , \tag{123.9}$$

which gives rise to the existence of the exchange energy.

It is easy to see that the exchange energy and the Coulomb interaction of the electrons are not related in any particular way. Let us assume any other interaction $W(r_{12})$ of the particles; then the mean energy $\overline{W(r_{12})}$ is again the sum of two parts: W, the energy in the restricted sense, obtained by substituting $W(r_{12})$ for e^2/r_{12} in (123.3), and the exchange energy A obtained by the same substitution in (123.4). Thus any classical interaction $W(r_{12})$ of two identical particles gives rise to an exchange energy. This energy has no analogue in classical mechanics, and its existence is one of the new fundamental results of the quantum theory.

The name 'exchange energy' is much more fully explained if we consider states Φ in which the distribution of particles between the states n and m is fixed. Since the states Φ_a and Φ_s are stationary states, their time dependence is given by

$$\left. \begin{aligned} \Phi_s &= \frac{1}{\sqrt{2}} (\psi_1 + \psi_2) \, e^{-i(E^0 + K + A)t/\hbar} , \\ \Phi_a &= \frac{1}{\sqrt{2}} (\psi_1 - \psi_2) \, e^{-i(E^0 + K - A)t/\hbar} . \end{aligned} \right\} \tag{123.10}$$

Let

$$\omega_0 = (E^0 + K)/\hbar , \qquad \delta = A/\hbar , \tag{123.11}$$

and let us consider instead of Φ_s and Φ_a a state which is a superposition of them (and will not be a stationary state):

$$\Phi = (\Phi_s + \Phi_a)/\sqrt{2} = \tfrac{1}{2} e^{-i\omega_0 t} \{ \psi_1 (e^{-i\delta t} + e^{i\delta t}) + \psi_2 (e^{-i\delta t} - e^{i\delta t}) \} , \tag{123.12}$$

or

$$\Phi = c_1(t) \psi_1 + c_2(t) \psi_2 , \tag{123.13}$$

where

$$c_1(t) = e^{-i\omega_0 t} \cos \delta t , \qquad c_2(t) = i e^{-i\omega_0 t} \sin \delta t . \tag{123.14}$$

According to the statistical significance of the amplitudes c_1 and c_2, the quantity $|c_1|^2$ is the probability of finding the system in the state ψ_1 (i.e. the first electron in the state n and the second in the state m), and $|c_2|^2$ is the probability that the system is in the state ψ_2 (i.e. the first electron in state m and the second in state n). We have

$$|c_1(t)|^2 = \cos^2 \delta t , \qquad |c_2(t)|^2 = \sin^2 \delta t ; \tag{123.15}$$

the state Φ (123.12) is therefore such that at $t = 0$ the first electron is in the state ψ_n and the second in the state ψ_m. After a time $\tau = \pi/2\delta$ we have $|c_1|^2 = 0$, $|c_2|^2 = 1$, i.e. the first electron has gone to the state ψ_m and the second to ψ_n; there is an exchange of states. From (123.11) we see that this exchange time may be expressed in terms of the exchange energy:

$$\tau = \pi\hbar/2A . \tag{123.16}$$

This leads to the important conclusion that the exchange time is inversely proportional to the exchange energy.

It is interesting to consider the conditions under which the exchange energy is so small, and the exchange time so long, that the exchange may be neglected. The exchange energy depends on the density $\rho_{mn}(\mathbf{r}) = \psi_m^*(\mathbf{r})\psi_n(\mathbf{r})$, and therefore depends on how much the functions ψ_m and ψ_n overlap. If $\psi_m = 0$ where $\psi_n \neq 0$, or $\psi_n = 0$ where $\psi_m \neq 0$, then $\rho_{mn} = 0$ and the exchange energy is zero. This limiting case is an idealisation, but we can nevertheless draw from it an important conclusion: if the states ψ_m and ψ_n are such that $|\psi_m|^2$ and $|\psi_n|^2$ are concentrated in different regions of space, then the exchange energy is small (tending to zero).

Let us now suppose that the states ψ_n are states of an electron in an atom, and that the energies E_n and E_m differ greatly, with $E_m \gg E_n$. Then the function ψ_n is concentrated in a region very close to the nucleus, while ψ_m extends to a considerable distance from the nucleus. Since both functions are normalised to unity, ψ_m must be small where ψ_n is fairly large. The density ρ_{mn} is therefore again small. Thus the exchange energy is small and the exchange can be neglected both for exchange of states concentrated in different regions of space and exchange of states of greatly differing energy.

This result, for example, in many cases justifies the neglect of the exchange between an optical electron and those of the inner shells.

124. Quantum mechanics of the atom and Mendeleev's periodic system of the elements

The periodic law discovered by Mendeleev is a most important law of Nature, and forms the basis not only of chemistry but also of the whole of modern atomic and nuclear physics.

The theory of the periodic law is still far from complete. The problem of nuclear structure is as yet in its initial stages, and it is the nucleus of the atom which entirely determines the structure of the electron shell, and therefore the chemical and physical properties of the atom as a whole. If, however, the properties of atomic nuclei are regarded as experimental data, quantum mechanics enables us to understand the periodicity in the structure of the electron shells of atoms on the basis of the theory of motion of a system of electrons in the electric field of the nucleus. Hence, to elucidate the nature of the periodicity, we need only consider the motion of electrons in atoms with a given nuclear mass and charge. The problem thus formulated is still extremely difficult mathematically, owing to the large number of electrons in atoms. It may be recalled that in classical mechanics even the problem of three bodies has not yet received a complete general solution. Fortunately the situation is better in atomic mechanics, and many results of practical importance can be derived by the use of approximate methods. This simplification is due to the discreteness of the states of electrons in atoms. Consequently, by means of the Pauli principle and the theory of the motion of an electron in a central field of force, it is possible to obtain important results concerning the distribution of electrons in atoms, and therefore the periodicity of the chemical properties of the elements.

The idea of the atomic number Z of an element in the periodic table is of great significance. This concept was used by Mendeleev himself, since in several places in the table he departed from his original principle of arranging the elements in order of increasing atomic weight, and gave precedence to the periodicity of chemical properties. It was later shown by the classical work of Rutherford and of Moseley that the atomic number has a profound physical significance: the atomic number Z is equal to the charge on the nucleus in units of the elementary charge ($+ e$), and is also equal to the number of electrons in the neutral atom. Thus, if the atomic number Z of an element is known, so are the nuclear charge and the number of electrons in the atom, and these are very important in atomic mechanics. It is now well known that atomic nuclei consist of uncharged particles: neutrons (charge 0, and mass 1.00898 if the mass of the oxygen atom is taken as 16), and protons (charge $+ e$ and mass 1.00759). The number of protons in the nucleus must be Z. Atoms with the same number of protons but different numbers of neutrons have the same Z but different atomic weights A. Such atoms are called *isotopes*. The chemical properties depend on the number of electrons in the neutral atom, i.e. on Z, and so isotopes are chemically equivalent[10], and the group of isotopes having a given Z forms a single chemical element. It is found that the atomic weight $A \approx 2Z$, so that the numbers of protons and neutrons in the nuclei are approximately equal. Consequently, the placing of the elements in order of increasing atomic weight leads (with a few exceptions) to the same arrangement as placing them in order of the nuclear charge $+ Ze$.

In order to analyse the distribution of electrons in the elements, we shall imagine each element to be formed from the preceding one by the addition of one proton (and the appropriate number of neutrons) to the nucleus, and the corresponding addition of one electron to the shell of the atom. We shall also neglect the interaction of the electrons, but make corrections for this interaction where necessary.[11]

The neutron may be regarded as an element of atomic number zero in the periodic system ($Z = 0$), forming the zeroth period. The first element is hydrogen ($Z = 1$), whose nucleus consists of one proton.[12]

The normal state of the single electron in the hydrogen atom is described by the quantum numbers $n = 1, l = 0, m = 0, m_s = \pm \frac{1}{2}$. The wave function is accordingly $\psi_{nlmm_s}(q)$, where q denotes the co-ordinates of the centre of mass of the electron and the spin co-ordinate.

When the nuclear charge is increased by $+ e$, the helium atom is obtained. A second electron may be placed in the state $n = 1, l = 0, m = 0$ if its spin is in the opposite direction to that of the first electron ($m_s = + \frac{1}{2}$ for one and $m_s = - \frac{1}{2}$ for the other). More precisely, we must form from the functions $\psi_{100\frac{1}{2}}(q_1)$ and

[10] This refers to valency properties. In reaction kinetics the mass of the atom, as well as the number of electrons, is important. We cannot say, therefore, that isotopes are chemically entirely identical, but the differences between isotopes are still very small, except perhaps for the isotopes of hydrogen, whose masses are very different (1, 2 and 3).

[11] This method of interpreting the periodic system on the basis of atomic mechanics was first suggested by Bohr [18].

[12] There are also the isotopes of hydrogen which occur naturally in small amounts: deuterium ($Z = 1, A = 2$) and tritium ($Z = 1, A = 3$). The former has been obtained in fairly large quantities as 'heavy water'.

$\psi_{100,-\frac{1}{2}}(q_2)$ an antisymmetric wave function, as in Section 117. The two helium electrons occupy all possible states belonging to $n = 1$. This group of states ($n = 1$, $l = 0$, $m = 0$, $m_s = \pm \frac{1}{2}$) is called the *K shell* (the nomenclature of the terms used in X-ray spectroscopy). Thus the K shell is filled, and this completes the first period of the system, consisting of only two elements, hydrogen and helium.

Again increasing the nuclear charge by $+e$ and adding one electron, we reach lithium. Here the approximate wave function must be the antisymmetric combination of $\psi_{n_1 l_1 m_1 m_{s1}}(q_1)$, $\psi_{n_2 l_2 m_2 m_{s2}}(q_2)$, $\psi_{n_3 l_3 m_3 m_{s3}}(q_3)$ which has the least energy (ground state of lithium). Continuing in this way, we can say that in our approximation the wave function of a multi-electron atom of atomic number Z is an antisymmetric combination of functions $\psi_{n_k l_k m_k m_{sk}}(q_k)$, each of which describes the motion of one electron in the Coulomb field of the nucleus of charge $+Ze$. From (117.6′) we can write this function as

$$\Phi(q_1, q_2, \ldots, q_z) = \sum_P (\pm 1) P \psi_{n_1 l_1 m_1 m_{s1}}(q_1) \ldots \psi_{n_z l_z m_z m_{sz}}(q_z). \quad (124.1)$$

This function is equal to zero if the numbers n, l, m, m_s are the same for any two electrons (the Pauli principle). Since we are concerned with the ground state of the atom, the numbers n_1, l_1, ..., n_z, l_z must be so chosen that the energy of the whole system of electrons,

$$E = \sum_{k=1}^{Z} E_{n_k l_k}, \quad (124.2)$$

is a minimum. If the functions $\psi_{n_k l_k m_k m_{sk}}$ are taken to be the wave functions for the motion in the Coulomb field of the nucleus (completely neglecting the interaction of the electrons), the energy E_{nl} of the various states depends only on n. In reality, E_{nl} depends also on l, since the electrons move not only in the field of the nucleus but also in that of the other electrons. This dependence is less strong, but nevertheless, for sufficiently large n, it may happen that states with larger n and smaller l will have a lower energy than those with smaller n and larger l. We shall see that this first occurs for potassium.

Thus for lithium the approximate wave function has the form (124.1) with $Z = 3$. Since the K shell is already filled, the third electron must enter the state $n = 2$, $l = 0$, $m = 0$, $m_s = \pm \frac{1}{2}$. The group of states with $n = 2$ is called the *L shell*, which therefore begins to be occupied in lithium. The L shell contains a total of $2n^2 = 2 \times 2^2 = 8$ states, of which two belong to the s term ($l = 0$, $m = 0$, $m_s = \pm \frac{1}{2}$), and six to the p term ($l = 1$, $m = 0$ or ± 1, $m_s = \pm \frac{1}{2}$).

On increasing further the nuclear charge and adding electrons, we go from lithium to beryllium, to boron, carbon, nitrogen, oxygen, fluorine to neon. In neon all the eight places in the L shell are occupied. We again have an inert gas, and this completes the second period of the system. Further electrons can be placed only in states with $n = 3$. This is called the *M shell*, and contains a total of $2 \times 3^2 = 18$ states ($l = 0$, 1 or 2). The group of states with $l = 0$ and $l = 1$ is entirely similar to the L shell and is filled between sodium and argon, forming the third period of the system. On increasing the charge of argon by $+e$ and adding an electron, we reach potassium. If the extra electron were put in the M shell, it would be in a state with $l = 2$ (a d term). However, the potassium atom is very similar, both optically and chemically, to those of lithium and sodium, which have the outermost valency electron in the s term. We must therefore place the potassium electron in the state $n = 4$, $l = 0$, beginning a new shell (the *N shell*) before completing the M shell. This means that the state $n = 4$, $l = 0$ has a lower energy (E_{40}) than the state $n = 3$, $l = 2$ (E_{32}), which is entirely possible if the interaction of the electrons is taken into account. Thus we have in potassium an electron distribution entirely similar to that in sodium (see Table 3).

The element following potassium is calcium ($Z = 20$). The spectroscopic data again indicate the necessity of placing an electron in the s term of the N shell. In the subsequent elements from scandium ($Z = 21$) to zinc ($Z = 30$) the M shell is then completed, followed by the filling of the N shell as far as krypton ($Z = 36$), an inert gas which ends the next period. Thus the inert gases (except helium) have a typical configuration of 8 electrons, 2 in the s state and six in the p state.

The element following krypton is rubidium ($Z = 37$), which is similar to sodium and potassium. Hence the outermost electron is not in the N shell but begins the *O shell* ($n = 5$). The electron in strontium (an alkaline earth) is again in the O shell, and strontium is similar to calcium. In the subsequent elements the O shell and the remainder of the N shell are filled (see Table 3). The *P shell* ($n = 6$) begins with caesium.

The rare-earth elements from lanthanum ($Z = 57$) to hafnium ($Z = 72$) have similar chemical properties, since all have the same arrangement of electrons in the O and P shells. They differ in the degree of occupation of the N shell, and in some cases in the occupation of the O shell (see Table 3), between cerium and lutetium. The rare-earth elements are often called *lanthanides*. Element 72 (hafnium) was for long regarded as a rare-earth element also, but, as we have seen, in lutetium the whole of the N shell is occupied, and the next electron must be placed in the $5d$ shell. This led Bohr to conclude that hafnium must be analogous to zirconium, and in fact it was soon found in zirconium ores.

Recently Mendeleev's table has been augmented by the newly discovered trans-uranic elements neptunium, plutonium, americium, curium and others. These form a group very similar to the rare-earth elements, with actinium in place of lanthanum, and so are called *actinides*. The elements in this group have similar outer shells and differ mainly in the occupation of the $5f$ shell.[13]

Table 3 could be replaced by symbolic formulae indicating the distribution of electrons among the various shells. For lithium, for example, the formula would be $(1s)^2\, 2s$, denoting that the atom has two electrons in the $1s$ state and one in the $2s$ state. The penultimate column in the table shows the ground terms of the atoms, denoted as usual by capital letters $S, P, D, F, \ldots$ for values of the quantum number $L = 0, 1, 2, 3, \ldots$ which gives the total orbital angular momentum (cf. Section 105).

[13] Details concerning electron shells and terms in the lanthanides and actinides are given in [48].

TABLE 3

ELECTRON DISTRIBUTIONS IN ATOMS

Element		*K*	*L*		*M*			*N*		*Ground term*	*Ionisation potential* (eV)
		1,0 1*s*	2,0 2*s*	2,1 2*p*	3,0 3*s*	3,1 3*p*	3,2 3*d*	4,0 4*s*	4,1 4*p*		
H	1	1	—	—	—	—	—	—	—	$^2S_{\frac{1}{2}}$	13.539
He	2	2	—	—	—	—	—	—	—	$^1S_0$	24.45
Li	3	2	1	—	—	—	—	—	—	$^2S_{\frac{1}{2}}$	5.37
Be	4	2	2	—	—	—	—	—	—	$^1S_0$	9.48
B	5	2	2	1	—	—	—	—	—	$^2P_{\frac{1}{2}}$	8.4
C	6	2	2	2	—	—	—	—	—	$^3P_0$	11.217
N	7	2	3	2	—	—	—	—	—	$^4S_{\frac{3}{2}}$	14.47
O	8	2	2	4	—	—	—	—	—	$^3P_2$	13.56
F	9	2	2	5	—	—	—	—	—	$^2P_{\frac{3}{2}}$	18.6
Ne	10	2	2	6	—	—	—	—	—	$^1S_0$	21.5
Na	11				1	—	—	—	—	$^2S_{\frac{1}{2}}$	5.12
Mg	12				2	—	—	—	—	$^1S_0$	7.61
Al	13				2	1	—	—	—	$^2P_{\frac{1}{2}}$	5.96
Si	14		As		2	2	—	—	—	$^3P_0$	7.39
P	15		neon		2	3	—	—	—	$^4S_{\frac{3}{2}}$	10.3
S	16				2	4	—	—	—	$^3P_2$	10.31
Cl	17				2	5	—	—	—	$^2P_{\frac{3}{2}}$	12.96
Ar	18				2	6	—	—	—	$^1S_0$	15.69
K	19						—	1	—	$^2S_{\frac{1}{2}}$	4.32
Ca	20						—	2	—	$^1S_0$	6.09
Sc	21						1	2	—	$^2D_{\frac{3}{2}}$	6.57
Ti	22						2	2	—	$^3F_2$	6.80
V	23						3	2	—	$^4F_{\frac{3}{2}}$	6.76
Cr	24						5	1	—	$^7F_3$	6.74
Mn	25						5	2	—	$^6S_{\frac{5}{2}}$	7.40
Fe	26						6	2	—	$^5D_4$	7.83
Co	27			As			7	2	—	$^4F_{\frac{9}{2}}$	7.81
Ni	28			argon			8	2	—	$^3F_4$	7.606
Cu	29						10	1	—	$^2S_{\frac{1}{2}}$	7.69
Zn	30						10	2	—	$^1S_0$	9.35
Ga	31						10	2	1	$^2P_{\frac{1}{2}}$	5.97
Ge	32						10	2	2	$^3P_0$	7.85
As	33						10	2	3	$^4S_{\frac{3}{2}}$	9.4
Se	34						10	2	4	$^3P_2$	
Br	35						10	2	5	$^2P_{\frac{3}{2}}$	11.80
Kr	36						10	2	6	$^1S_0$	13.940

TABLE 3 *(continued)*

Element		*Inner shells*	*N* 4,2 4d	*N* 4,3 4f	*O* 5,0 5s	*O* 5,1 5p	*O* 5,2 5d	*P* 6,0 6s	*Ground term*	*Ionisation potential (eV)*
Rb	37		—	—	1	—	—	—	$^2S_{\frac{1}{2}}$	4.16
Sr	38		—	—	2	—	—	—	$^1S_0$	5.67
Y	39		1	—	2	—	—	—	$^2D_{\frac{3}{2}}$	6.5
Zr	40	As	2	—	2	—	—	—	$^3F_2$	
Nb	41	krypton	4	—	1	—	—	—	$^6D_{\frac{1}{2}}$	
Mo	42		5	—	1	—	—	—	$^7S_3$	7.35
Tc	43		6	—	1	—	—	—	$^6S_{\frac{5}{2}}$	
Ru	44		7	—	1	—	—	—	$^5F_5$	7.7
Rh	45		8	—	1	—	—	—	$^4F_{\frac{9}{2}}$	7.7
Pd	46		10	—	—	—	—	—	$^1S_0$	8.5
Ag	47			—	1	—	—	—	$^2S_{\frac{1}{2}}$	7.54
Cd	48			—	2	—	—	—	$^1S_0$	8.95
In	49			—	2	1	—	—	$^2P_{\frac{1}{2}}$	5.76
Sn	50	As		—	2	2	—	—	$^3P_0$	7.37
Sb	51	palladium		—	2	3	—	—	$^4S_{\frac{3}{2}}$	8.5
Te	52			—	2	4	—	—	$^3P_2$	
I	53			—	2	5	—	—	$^2P_{\frac{3}{2}}$	10.44
Xe	54			—	2	6	—	—	$^1S_0$	12.078
Cs	55			—			—	1	$^2S_{\frac{1}{2}}$	3.88
Ba	56			—			—	2	$^1S_0$	5.19
La	57			—			1	2	$^2D_{\frac{3}{2}}$	
Ce	58			2			—	2	^3H	
Pr	59			3			—	2	^4I	
Nd	60	46		4	8		—	2	^5I	
Pm	61	in		5	in		—	2	^6H	
Sm	62	1s to 4d		6	5s & 5p		—	2	^7F	
Eu	63			7			—	2	^8S	
Gd	64			7			1	2	^9D	
Tb	65			9			—	2	^6H	
Dy	66			10			—	2	^5I	
Ho	67			11			—	2	^4I	
Er	68			12			—	2	^3H	
Tm	69			13			—	2	^2F	
Yb	70			14			—	2	^1S	
Lu	71			14			1	2	$^3D_{\frac{3}{2}}$	
Hf	72						2	2	$^3F_2$	
Ta	73						3	2	$^4F_{\frac{3}{2}}$	
W	74	68					4	2	$^5D_0$	
Re	75	in					5	2	$^6S_{\frac{5}{2}}$	
Os	76	1s to 5p					6	2	$^5D_4$	
Ir	77						7	2	$^4F_{\frac{9}{2}}$	
Pt	78						8	2	$^3D_3$	

TABLE 3 *(continued)*

Element		*Inner shells*	*O* 5,3 5*f*	*P* 6,0 6*s*	*P* 6,1 6*p*	*P* 6,2 6*d*	*Q* 7,0 7*s*	*Ground term*	*Ionisation potential (eV)*
Au	79		—	1	—	—	—	$^2S_{1/2}$	9.20
Hg	80		—	2	—	—	—	$^1S_0$	10.39
Tl	81	78	—	2	1	—	—	$^2P_{1/2}$	6.08
Pb	82	in	—	2	2	—	—	$^3P_0$	7.39
Bi	83	1*s* to 5*d*	—	2	3	—	—	$^4S_{3/2}$	8.0
Po	84		—	2	4	—	—	$^3P_2$	
At	85		—	2	5	—	—	$^2P_{3/2}$	
Rn	86		—	2	6	—	—	$^1S_0$	10.689
Fr	87		—	2	6	—	1	$^2S_{1/2}$	
Ra	88		—	2	6	—	2	$^1S_0$	
Ac	89		—	2	6	1	2	$^2D_{3/2}$	
Th	90		—	2	6	2	2	^3F	
Pa	91		2	2	6	1	2	^4K	
U	92		3	2	6	1	2	^5L	
Np	93		4	2	6	1	2	^6M	
Pu	94	78	6	2	6	—	2	^7F	
Am	95	in	7	2	6	—	2	^8S	
Cm	96	1*s* to 5*d*	7	2	6	1	2	^9D	
Bk	97		8	2	6	1	2		
Cf	98		10	2	6	—	2		
Es	99		11	2	6	—	2		
Fm	100		12	2	6	—	2		
Md	101		13	2	6	—	2		
No	102		14	2	6	—	2		
Lw	103		14	2	6	1	2		

The number J which gives the total angular momentum is shown at the bottom right, and the multiplicity $2S + 1$ of the term at the top left (S being the total spin number). For lithium the orbital angular momentum is zero, and the spins of the two inner electrons cancel. Thus the ground level of the lithium atom is the doublet $^2S_{1/2}$.

The corresponding formula for neon, for example, is $(1s)^2(2s)^2(2p)^6$. All the spins and orbital angular momenta cancel, and so the ground term of neon, and of all the other inert gases, is $^1S_0$. In aluminium we have one p electron ($3p$) whose orbital and spin angular momenta are not compensated, so that the ground term is $^2P_{1/2}$; the shell-structure formula is $(1s)^2(2s)^2(2p)^6(3s)^2 3p$. It is not difficult to derive the formulae for the other elements.

We see that the periodicity in the chemical properties of the elements, discovered by Mendeleev, represents in atomic mechanics a repetition in the structure of the outer electron shells. For example, the inert gases neon, argon, krypton, xenon and radon have identical outer shells of eight electrons. All the alkali metals have a single electron in the s term outside the inert-gas shell (term $^2S_{1/2}$). The alkaline-earth metals have two electrons outside the inert-gas shell (term $^1S_0$). The halogens fluorine,

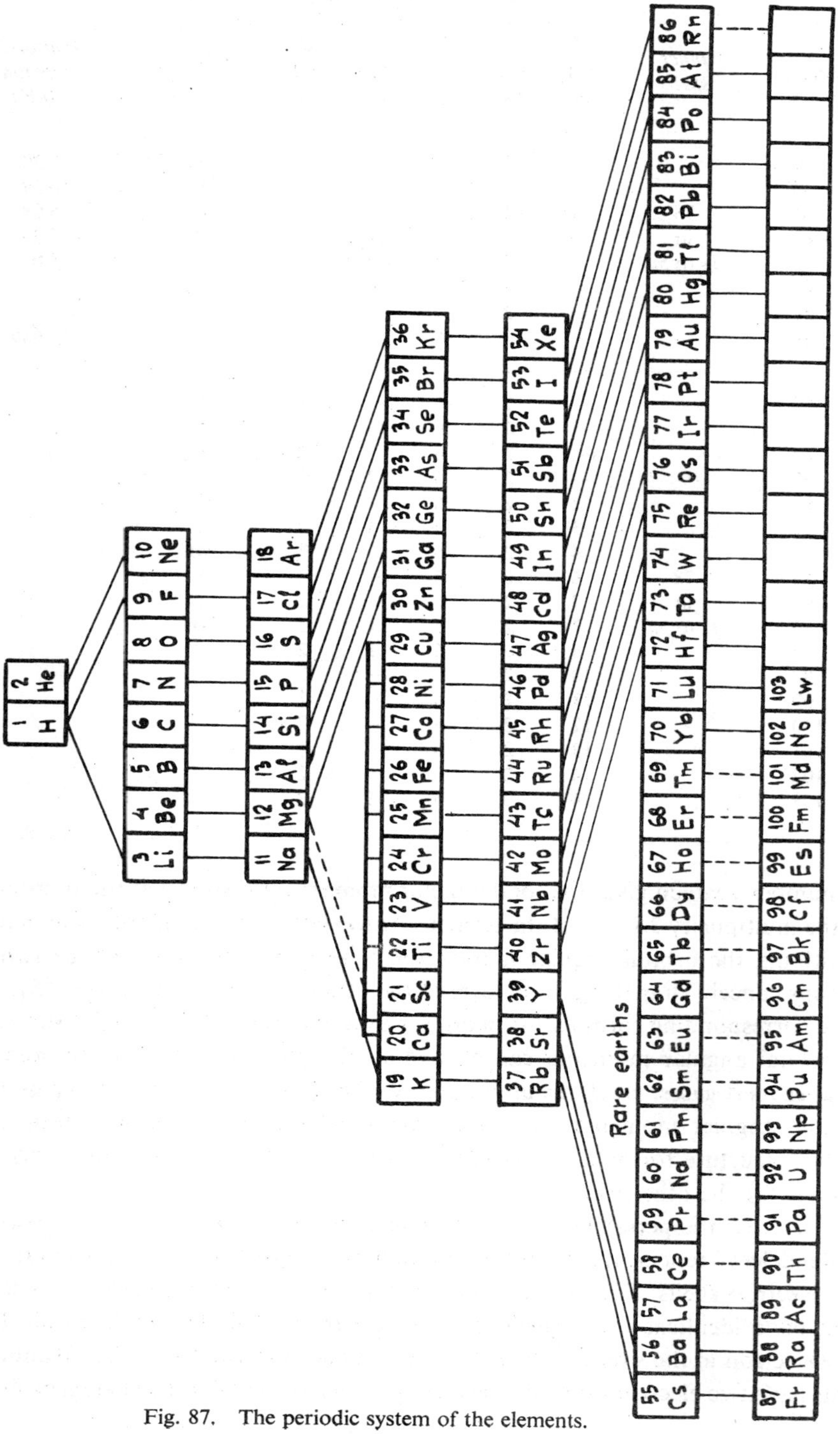

Fig. 87. The periodic system of the elements.

chlorine, bromine and iodine have shells lacking one electron from the inert-gas shell (term $^2P_{\frac{3}{2}}$). The length of the periods is determined essentially by the number of quantum states in each shell. This number is $2n^2$, according to (50.26), if we recall that two electrons with spins in opposite directions can be in each state; n is the principal quantum number of the shell. The lengths of the periods are therefore determined by the numbers 2, 8, 18, 32, ...

Thus modern atomic mechanics has made a very important contribution to the understanding of one of most remarkable laws of Nature, the periodicity of the chemical properties of the elements discovered by the great Russian chemist.

Figure 87 shows Mendeleev's table in a form due to Bohr.

As already mentioned, the representation of the wave function of a system of electrons as an antisymmetric combination of individual electron functions $\psi_{nlmm_s}(q)$ (124.1) is an approximation, which becomes very rough if these functions are taken as those for the motion of an electron in the Coulomb field of the nucleus and the interaction of the electrons is ignored.

The question may, however, be asked how to find functions $\psi_{nlmm_s}(q)$ such that the true wave function $\Phi(q_1, q_2, \ldots, q_N)$ of the electron system is best represented as the determinant (124.1). This is answered by Fok's method[14], which consists essentially in seeking $\psi_{nlmm_s}(q)$ so as to minimise the total energy of the system

$$E = \int \Phi^* \cdot \boldsymbol{H}\Phi \, \mathrm{d}q_1 \, \mathrm{d}q_2 \ldots \mathrm{d}q_N \,, \tag{124.3}$$

with the additional normalisation condition

$$\int \Phi^* \, \Phi \, dq_1 \, \mathrm{d}q_2 \ldots \mathrm{d}q_N = 1 \,. \tag{124.4}$$

Here $\boldsymbol{H}$ is the Hamiltonian of the whole system of electrons. This variational problem leads to a set of non-linear equations to determine the individual functions $\psi_{nlmm_s}(q)$. The resulting value of the energy E_0 of the lowest term is the most accurate compatible with the form (124.1).

The same variational problem can be solved directly by the methods of the calculus of variations (Ritz' method). In this method the zeroth-order approximation is taken as some class of functions Φ depending on parameters $a, b, \ldots$ (which may be, for example, the radii of the electron shells). Effecting the integration, we find E as a function of these parameters. From the minimum conditions

$$\partial E/\partial a = 0 \,, \qquad \partial E/\partial b = 0 \,, \qquad \ldots \tag{124.5}$$

and the condition (124.4) we find the values of the parameters which give the best approximation to E and Φ within the class of functions chosen. The accuracy of the approximation depends considerably on how well the type of function Φ is chosen for the selection of the first approximation. In practice the method is very effective (see [5], Sections 11–14).

14 See [5, 42].

CHAPTER XXII

FORMATION OF MOLECULES

125. The hydrogen molecule

Let us now consider the hydrogen molecule H_2 in terms of quantum mechanics. This molecule has a typical homopolar bond. Thus from this simple case of a homopolar molecule we can hope to elucidate the nature of the forces which are responsible for homopolar valency bonds. In order to calculate the force of interaction between two hydrogen atoms we must determine their potential energy $U(R)$ as a function of the distance R between the centres of the atoms (the nuclei). $U(R)$ consists of two parts: the energy $+ e^2/R$ of the Coulomb interaction of the nuclei and the energy E of the electrons, which depends on the distance between the nuclei and therefore appears in the potential energy of interaction of the two atoms. Thus we can write

$$U(R) = e^2/R + E(R), \tag{125.1}$$

and the problem reduces to the determination of the electron energy $E(R)$. When the distance R between the atoms is large, the effect of one atom on the motion of the electron in the other atom may clearly be neglected, and so for $R \to \infty$ the electron energy is just the sum of the energies of the electrons in the two atoms.

We shall be concerned mainly with the hydrogen molecule in the state of lowest energy, and accordingly when the atoms are moved to an infinite distance apart they will be in the ground state. Let the energy of the hydrogen atom in the ground state be $E_0(= -13.55 \text{ eV})$. Then, for the states of the molecule which are under consideration, the energy for large R is equal to $2E_0$. We put

$$E(R) = 2E_0 + \varepsilon(R), \tag{125.2}$$

where $\varepsilon(R)$ represents the change in the energy of the electrons as the hydrogen atoms approach and is to be determined.

The total energy $E(R)$ of the electrons is determined from Schrödinger's equation as the eigenvalue of the Hamiltonian operator for the system in question. The Hamiltonian is easily seen to be

$$\boldsymbol{H} = -\frac{\hbar^2}{2\mu}\nabla_1^2 - \frac{\hbar^2}{2\mu}\nabla_2^2 - \frac{e^2}{r_{a1}} - \frac{e^2}{r_{b2}} - \frac{e^2}{r_{b1}} - \frac{e^2}{r_{a2}} + \frac{e^2}{r_{12}}. \tag{125.3}$$

Here, in addition to the obvious kinetic-energy operators of the two electrons, we have

(a) the potential energy $-e^2/r_{a1}$ of the first electron (1) and the first nucleus, (b) the potential energy $-e^2/r_{b2}$ of the second electron (2) and the second nucleus, (c) the potential energy $-e^2/r_{b1}$ of the first electron and the second nucleus, (d) the potential energy $-e^2/r_{a2}$ of the second electron and the first nucleus, and finally (e) the energy e^2/r_{12} of the interaction of the two electrons. Figure 88 illustrates the notation used here for the distances $r_{a1}, r_{b1}, r_{b2}, r_{a2}, r_{12}$.

If the wave function for the system of electrons is denoted by $\Phi(\mathbf{r}_1, \mathbf{r}_2)$, then Schrödinger's equation determining Φ and E is

$$H\Phi(\mathbf{r}_1, \mathbf{r}_2) = E\Phi, \tag{125.4}$$

where H is given by (125.3).

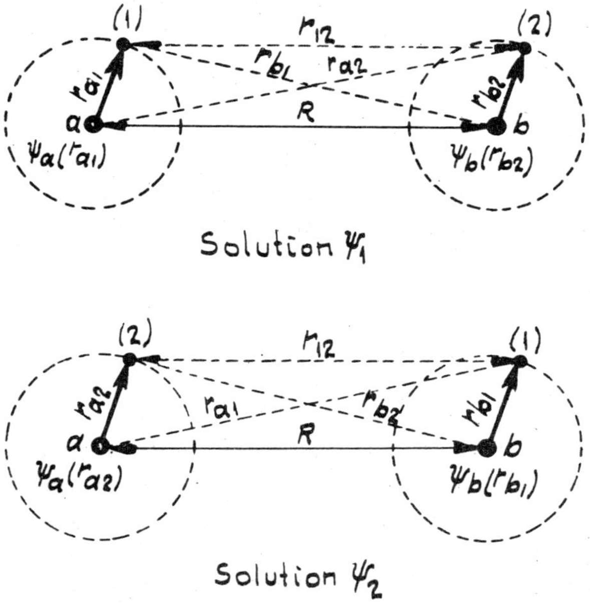

Fig. 88. Diagram of interactions in the hydrogen molecule. The continuous lines join particles whose interaction is taken into account in the solution ψ_1 or ψ_2. The broken lines join particles whose interaction is neglected in the zeroth-order approximation.

Equation (125.4) can be solved only approximately. Here we shall use a method which, though not the best as regards the accuracy attainable, is simple and clear; it is very similar to that used in Section 122 to solve the problem of the helium atom.

The wave functions of non-interacting hydrogen atoms are used as the initial approximation to the wave function in this method. Thus the zeroth-order approximation is the solution for two hydrogen atoms far apart ($R \to \infty$). The corresponding energy of the system is $2E_0$. We can regard the distance R as large if the change in the electron energy as the atoms approach is small in comparison with the difference between the lowest level $2E_0$ and the next higher level $E_0 + E_1$:

$$|\varepsilon(R)| \ll |E_1 - E_0|. \tag{125.5}$$

The latter difference is 10.15 eV. For such distances R the quantity $\varepsilon(R)$ may be regarded as a correction to the energy $2E_0$ of the non-interacting atoms, and the wave function Φ of the system of electrons as a function which is close to the wave function of the non-interacting hydrogen atoms.

In order to make the calculation in this way, i.e. starting from distant hydrogen atoms, we must consider further the Hamiltonian (125.3) of the system. Let $\boldsymbol{H}_a(1)$ be the part of the Hamiltonian $\boldsymbol{H}$ (125.3) given by

$$\boldsymbol{H}_a(1) = -\frac{\hbar^2}{2\mu}\nabla_1^2 - \frac{e^2}{r_{a1}}, \tag{125.6}$$

and $\boldsymbol{H}_b(2)$ be the part

$$\boldsymbol{H}_b(2) = -\frac{\hbar^2}{2\mu}\nabla_2^2 - \frac{e^2}{r_{b2}}. \tag{125.7}$$

The Hamiltonian $\boldsymbol{H}_a(1)$ is clearly that which corresponds to the motion of the first electron (1) round the nucleus (a), and $\boldsymbol{H}_b(2)$ is the Hamiltonian for the motion of the second electron (2) round the nucleus (b). The complete Hamiltonian may be written

$$\boldsymbol{H} = \boldsymbol{H}_a(1) + \boldsymbol{H}_b(2) + W(1,2), \tag{125.3'}$$

where

$$W(1,2) = -\frac{e^2}{r_{a2}} - \frac{e^2}{r_{b1}} + \frac{e^2}{r_{12}}. \tag{125.8}$$

Let us return now to the case where the distance R is large. Let the first electron be in atom a (near nucleus a) and the second electron in atom b (near nucleus b). Then the quantity $W(1,2)$ may be neglected, since this is the energy of the interaction of the second electron with nucleus a, plus that of the first electron with nucleus b, plus that of the two electrons. If the atoms are far apart, all three quantities are small. The quantity $W(1, 2)$ may therefore be neglected as an approximation in Equation (125.4), leaving

$$[\boldsymbol{H}_a(1) + \boldsymbol{H}_b(2)]\,\Phi = E\Phi. \tag{125.9}$$

This equation describes two non-interacting hydrogen atoms with the condition that the first electron is in atom a and the second in atom b. The solution can be written down immediately: it is simply the product of wave functions for the ground state of the hydrogen atom. For let $\psi_a(r_{a1})$ be the wave function of the ground state of hydrogen atom a for the first electron, and $\psi_b(r_{b2})$ that of the ground state of hydrogen atom b for the second electron. Then by (125.6) and (125.7)

$$\boldsymbol{H}_a(1)\,\psi_a(r_{a1}) = E_0\psi_a(r_{a1}), \tag{125.10}$$

$$\boldsymbol{H}_b(2)\,\psi_b(r_{b2}) = E_0\psi_b(r_{b2}). \tag{125.10'}$$

The solution of equation (125.9) can be taken as

$$\psi_1(\mathbf{r}_1, \mathbf{r}_2) = \psi_a(r_{a1})\psi_b(r_{b2}). \tag{125.11}$$

The corresponding value of the energy E is $2E_0$.

If there were no degeneracy, the solution (125.11) would be the zeroth-order approximation. In fact, however, the exchange degeneracy occurs in this problem. It is evident that, in addition to the solution ψ_1 (125.11), a solution is possible for which the second electron (2) is in the first atom (a), and the first electron (1) in the second atom (b). In order to examine this solution, we divide the Hamiltonian (125.3) into components as follows:

$$\boldsymbol{H} = \boldsymbol{H}_a(2) + \boldsymbol{H}_b(1) + W(2,1), \tag{125.3''}$$

where

$$\boldsymbol{H}_a(2) = -\frac{\hbar^2}{2\mu}\nabla_2^2 - \frac{e^2}{r_{a2}}, \tag{125.6'}$$

$$\boldsymbol{H}_b(1) = -\frac{\hbar^2}{2\mu}\nabla_1^2 - \frac{e^2}{r_{b1}} \tag{125.7'}$$

are the Hamiltonians for the hydrogen atoms a with the second electron and b with the first electron, and

$$W(2,1) = -\frac{e^2}{r_{a1}} - \frac{e^2}{r_{b2}} + \frac{e^2}{r_{12}} \tag{125.8'}$$

is the electron-electron and electron-nucleus interaction between different atoms. When the atoms a and b are sufficiently far apart this quantity may be neglected, and Equation (125.4) becomes simply

$$[\boldsymbol{H}_a(2) + \boldsymbol{H}_b(1)]\,\Phi = E\Phi. \tag{125.9'}$$

This equation, like (125.9), is that of two non-interacting hydrogen atoms, and its solution is

$$\psi_2(\mathbf{r}_1, \mathbf{r}_2) = \psi_a(r_{a2})\psi_b(r_{b1}), \tag{125.11'}$$

differing from (125.11) in that the electrons are interchanged. The corresponding value of the energy E is, of course, again $2E_0$. Thus for large R Equation (125.4) has two solutions (125.11) and (125.11′) corresponding to energy $2E_0$. These two solutions are illustrated in Figure 88. When the interaction $W(1,2)$ and $W(2,1)$ between the atoms is taken into account, the solution Φ will naturally not be either ψ_1 or ψ_2, but the zeroth-order approximation to Φ will be a linear combination of ψ_1 and ψ_2, as is always true when there is degeneracy. We can therefore put

$$\Phi = c_1\psi_1 + c_2\psi_2 + \phi, \tag{125.12}$$

where c_1 and c_2 are coefficients to be determined and ϕ is small (if the distance R is not very small), forming a correction to the zeroth-order approximation.

Regarding ϕ as a small correction, we may neglect the products $W(1,2)\,\phi$, $W(2,1)\phi$ and $\varepsilon\phi$, since W and ε are themselves considered to be small. Substituting (125.12) in (125.4) and using the notation (125.2), we obtain

$$\begin{aligned} &c_1 \boldsymbol{H}\psi_1 + c_2 \boldsymbol{H}\psi_2 + \boldsymbol{H}\phi \\ &= 2E_0(c_1\psi_1 + c_2\psi_2) + \varepsilon(c_1\psi_1 + c_2\psi_2) + (2E_0 + \varepsilon)\,\phi\,. \end{aligned} \tag{125.13}$$

This is separated into components in accordance with (125.3′) and (125.3″):

$$\begin{aligned} &c_1\left[\boldsymbol{H}_a(1) + \boldsymbol{H}_b(2) + W(1,2)\right]\psi_1 + c_2\left[\boldsymbol{H}_a(2) + \boldsymbol{H}_b(1) + \right. \\ &\left. + W(2,1)\right]\psi_2 + \left[\boldsymbol{H}_a(1) + \boldsymbol{H}_b(2)\right]\phi + W(1,2)\,\phi \\ &= 2E_0(c_1\psi_1 + c_2\psi_2) + \varepsilon(c_1\psi_1 + c_2\psi_2) + (2E_0 + \varepsilon)\,\phi\,. \end{aligned} \tag{125.14}$$

Since ψ_1 and ψ_2 are the solutions of Equations (125.9) and (125.9′) with $E = 2E_0$, we find, neglecting the products $W\phi$ and $\varepsilon\phi$,

$$\begin{aligned} \left[\boldsymbol{H}_a(1) + \boldsymbol{H}_b(2)\right]\phi - 2E_0\phi = &\left[\varepsilon - W(1,2)\right]c_1\psi_1 + \\ &+ \left[\varepsilon - W(2,1)\right]c_2\psi_2\,. \end{aligned} \tag{125.15}$$

This is an inhomogeneous equation to determine the corrections ϕ to the wave function and ε to the eigenvalue. We have not yet, however, determined the coefficients c_1 and c_2 which appear on the right-hand side of Equation (125.15). To determine these, we note that, if the right-hand side of (125.15) were zero, we should have a homogeneous equation for ϕ, the same as (125.9), whose solution is ψ_1. According to a well-known theorem, an inhomogeneous equation has a solution only if the right-hand side is orthogonal to the solution of the homogeneous equation. In other words, we must have

$$\int \{\left[\varepsilon - W(1,2)\right]c_1\psi_1 + \left[\varepsilon - W(2,1)\right]c_2\psi_2\}\,\psi_1\,\mathrm{d}v_1\,\mathrm{d}v_2 = 0\,, \tag{125.16}$$

where $\mathrm{d}v_1 = \mathrm{d}x_1\,\mathrm{d}y_1\,\mathrm{d}z_1$, $\mathrm{d}v_2 = \mathrm{d}x_2\,\mathrm{d}y_2\,\mathrm{d}z_2$. This gives one equation for the two coefficients c_1 and c_2; a second is easily found by substituting $\boldsymbol{H}\phi$ in (125.13) in another form:

$$\boldsymbol{H}\phi = \left[\boldsymbol{H}_a(2) + \boldsymbol{H}_b(1)\right]\phi + W(2,1)\,\phi\,;$$

again neglecting $W\phi$ as being of the second order of smallness, we have instead of (125.15)

$$\begin{aligned} \left[\boldsymbol{H}_a(2) + \boldsymbol{H}_b(1)\right]\phi - 2E_0\phi = &\left[\varepsilon - W(1,2)\right]c_1\psi_1 + \\ &+ \left[\varepsilon - W(2,1)\right]c_2\psi_2\,. \end{aligned} \tag{125.15′}$$

The left-hand side is the same as that of Equation (125.9′), whose solution is ψ_2. The right-hand side of the inhomogeneous equation for ϕ must again be orthogonal to the solution ψ_2 of the homogeneous equation. This gives the second equation

$$\int \{\left[\varepsilon - W(1,2)\right]c_1\psi_1 + \left[\varepsilon - W(2,1)\right]c_2\psi_2\}\,\psi_2\,\mathrm{d}v_1\,\mathrm{d}v_2 = 0\,. \tag{125.16′}$$

We shall use the notation

$$K = \int W(1,2)\psi_1\psi_1 \, dv_1 \, dv_2 = \int W(2,1)\psi_2\psi_2 \, dv_1 \, dv_2 , \qquad (125.17)$$

$$A = \int W(1,2)\psi_2\psi_1 \, dv_1 \, dv_2 = \int W(2,1)\psi_1\psi_2 \, dv_1 \, dv_2 . \qquad (125.18)$$

These equalities follow from the fact that $W(1, 2) = P_{12} W(2, 1)$ and $\psi_2 = P_{12}\psi_1$, so that the integrals differ only in the notation used for the variables of integration, and are therefore equal. The functions ψ_1 and ψ_2 are not orthogonal, and we therefore define a third integral[1]

$$S^2 = \int \psi_1\psi_2 \, dv_1 \, dv_2 . \qquad (125.19)$$

In this notation, (125.16) and (125.16′) become

$$(\varepsilon - K)c_1 + (\varepsilon S^2 - A)c_2 = 0 , \qquad (125.20)$$

$$(\varepsilon S^2 - A)c_1 + (\varepsilon - K)c_2 = 0 . \qquad (125.20')$$

Hence we find the equation for ε

$$(\varepsilon - K)^2 - (\varepsilon S^2 - A)^2 = 0 , \qquad (125.21)$$

which has the two roots

$$\varepsilon_1 = (K - A)/(1 - S^2) , \qquad (125.22)$$

$$\varepsilon_2 = (K + A)/(1 + S^2) . \qquad (125.22')$$

Substituting these values in (125.20), we obtain two solutions for c_1 and c_2: for $\varepsilon = \varepsilon_1$

$$c_1 = - c_2 \qquad (125.23)$$

and for $\varepsilon = \varepsilon_2$

$$c_1 = c_2 . \qquad (125.23')$$

The solutions may therefore be written

$$\left. \begin{aligned} E_a &= 2E_0 + (K - A)/(1 - S^2) , \\ \Phi_a &= \psi_1 - \psi_2 \end{aligned} \right\} \qquad (125.24)$$

(antisymmetric solution) and

$$\left. \begin{aligned} E_s &= 2E_0 + (K + A)/(1 + S^2) , \\ \Phi_s &= \psi_1 + \psi_2 \end{aligned} \right\} \qquad (125.24')$$

(symmetric solution).

Let us now consider further the values of the correction terms in the energy. To do so, we write out the precise form of the integrals (125.17) and (125.18). Substituting

[1] ψ_1 and ψ_2 are orthogonal only for $R \to \infty$; for $R = 0$, $S = 1$. The theory given here is therefore not a rigorous perturbation theory, which always presupposes orthogonality of the unperturbed solutions.

in (125.17) $W(1, 2)$ from (125.8) and ψ_1 from (125.11), we obtain[2]

$$K = \int\left\{-\frac{e^2}{r_{a2}} - \frac{e^2}{r_{b1}} + \frac{e^2}{r_{12}}\right\}\psi_a^2(r_{a1})\psi_b^2(r_{b2})\,\mathrm{d}v_1\,\mathrm{d}v_2\,;$$

since the term e^2/r_{b1} does not involve the co-ordinates of the second electron, nor e^2/r_{a2} those of the first, and since by normalisation $\int \psi_b^2(r_{b2})\,\mathrm{d}v_2 = 1$, $\int \psi_a^2(r_{a1})\,\mathrm{d}v_1 = 1$, we can write K in the form

$$K = \int\frac{e}{r_{a2}}\rho_b(2)\,\mathrm{d}v_2 + \int\frac{e}{r_{b1}}\rho_a(1)\,\mathrm{d}v_1 + \int\frac{\rho_a(1)\,\rho_b(2)}{r_{12}}\mathrm{d}v_1\,\mathrm{d}v_2\,, \tag{125.25}$$

where $\rho_b(2) = -e\psi_b^2(r_{b2})$ is the mean electric charge density in atom b due to electron 2, and $\rho_a(1) = -e\psi_a^2(r_{a1})$ the mean electric charge density in atom a due to electron 1. The first integral in (125.25) is the mean potential energy of electron 2 (atom b) in the field of nucleus a, and the second integral is that of electron 1 (atom a) in the field of nucleus b. The third integral is the mean potential energy of electrons 1 and 2 in the different atoms. Thus K is just the mean energy of electrostatic interaction of the atoms, except for the interaction of the nuclei, which is treated separately (see (125.1)).

The integral (125.18) represents the exchange energy. Substituting in (125.18) the values of $W(1, 2)$, ψ_1 and ψ_2, we have

$$A = \int\left\{-\frac{e^2}{r_{a2}} - \frac{e^2}{r_{b1}} + \frac{e^2}{r_{12}}\right\}\psi_a(r_{a1})\psi_b(r_{b2})\psi_a(r_{a2})\psi_b(r_{b1})\,\mathrm{d}v_1\,\mathrm{d}v_2\,.$$

Denoting the exchange density by a similar notation to that used for the helium atom:

$$\rho_{ab}(1) = -e\psi_a(r_{a1})\psi_b(r_{b1}),$$

$$\rho_{ab}(2) = -e\psi_a(r_{a2})\psi_b(r_{b2}),$$

we can write A in the form

$$A = S\int\frac{e}{r_{a2}}\rho_{ab}(2)\,\mathrm{d}v_2 + S\int\frac{e}{r_{b1}}\rho_{ab}(1)\,\mathrm{d}v_1 + \int\frac{\rho_{ab}(1)\,\rho_{ab}(2)}{r_{12}}\mathrm{d}v_1\,\mathrm{d}v_2\,; \tag{125.26}$$

the last term is the exchange energy of the electrons, and has exactly the same form as for the helium atom. The only difference is that in the latter case the exchange concerned electrons in states differing as regards the electron energy, whereas here the states ψ_a and ψ_b differ in the position of the electrons in atoms a and b. The exchange of electrons occurs between these atoms.

The first two terms are corrections to the exchange energy which arise from the non-orthogonality of the wave functions:

$$S = \int\psi_a(r_{a1})\psi_b(r_{b1})\,\mathrm{d}v_1 = \int\psi_a(r_{a2})\psi_b(r_{b2})\,\mathrm{d}v_2\,. \tag{125.19'}$$

[2] If we substitute $W(2, 1)$ from (125.8') and ψ_2 from (125.11'), this is easily seen to prove the equality of the two integrals in (125.17).

As $R \to \infty$ the wave functions ψ_a and ψ_b overlap to an extent that tends to zero, owing to their exponential decrease with increasing distance from the nuclei a and b; ψ_a is different from zero only near nucleus a, and ψ_b only near nucleus b. Thus S is very small and tends to zero. For $R = 0$, on the other hand, the nuclei a and b coincide, ψ_a and ψ_b are wave functions of the same hydrogen atom, and, by the normalisation of ψ_a and ψ_b, $S = 1$ for $R = 0$. Hence

$$0 \leqslant S \leqslant 1. \tag{125.27}$$

The quantity S^2 (125.19) likewise lies between the same limits. Thus Formulae (125.24) and (125.24′) for the energy of two hydrogen atoms have the same physical significance as Formulae (122.21) and (122.22) for the energy of the helium atom: in both cases the corrections consist of the Coulomb interaction energy K and the exchange energy A. The only difference is due to the non-orthogonality of the wave functions (the terms in S and S^2).

We can now write the energy $U(R)$ of two hydrogen atoms for the antisymmetric state Φ_a and the symmetric state Φ_s. From (125.1), (125.2), (125.24), (125.24′) we have

$$U_a = 2E_0 + \frac{e^2}{R} + \frac{K - A}{1 - S^2},$$

$$U_s = 2E_0 + \frac{e^2}{R} + \frac{K + A}{1 + S^2}.$$

These can also be written as

$$U_a = 2E_0 + \left(\frac{e^2}{R} + K\right) - A + S^2 \frac{K - A}{1 - S^2}, \tag{125.28}$$

$$U_s = 2E_0 + \left(\frac{e^2}{R} + K\right) + A - S^2 \frac{K + A}{1 + S^2}. \tag{125.28′}$$

The terms $(e^2/R) + K$ represent the mean Coulomb energy of the two hydrogen atoms at a distance R apart, and A the exchange energy. The last term, proportional to S^2, includes corrections for the non-orthogonality of the wave functions used as the zeroth-order approximation.

Both the Coulomb energy and the exchange energy can be calculated from Formulae (125.25) and (125.26). To do so, it is sufficient to substitute in these integrals the expression for the wave function of the ground state of hydrogen. This function is known, and is simply the exponential

$$\psi(r) = \frac{2}{\sqrt{(4\pi a^3)}} e^{-r/a}, \tag{125.29}$$

where r is the distance of the electron from the nucleus, and a the radius of the first Bohr orbit. In order to derive the functions $\psi_a(r_{a1})$, $\psi_b(r_{b2})$ etc. we must substitute r_{a1}, r_{b2} etc. for r, since these are just the distances of the electrons from the nuclei (see Figure 88).

We shall not give the calculation of these integrals, but merely note that both the integrals K and A involve wave functions belonging to different atoms (for example, $\psi_a(r_{a1})$ and $\psi_b(r_{b2})$; each of these functions decreases exponentially with increasing r_{a1} and r_{b2}). Hence the two integrals K and A differ from zero only to the extent that the wave functions, and therefore the electron shells of the atoms, overlap.

Consequently both integrals decrease as $e^{-2R/a}$ with increasing distance R between the atoms. Figure 89 shows the mutual energy $U_a(R)$ and $U_s(R)$ of the atoms as a function of the distance R, obtained by calculating the Coulomb energy K and the exchange energy A.[3] The quantity $2E_0$ is taken as the origin of energy. The distance R is measured in units of the Bohr radius, so that the abscissa is R/a. The diagram shows that for the antisymmetric state Φ_a the energy $U_a(R)$ corresponds to a repulsion between the two hydrogen atoms, so that the molecule H_2 cannot be formed. For

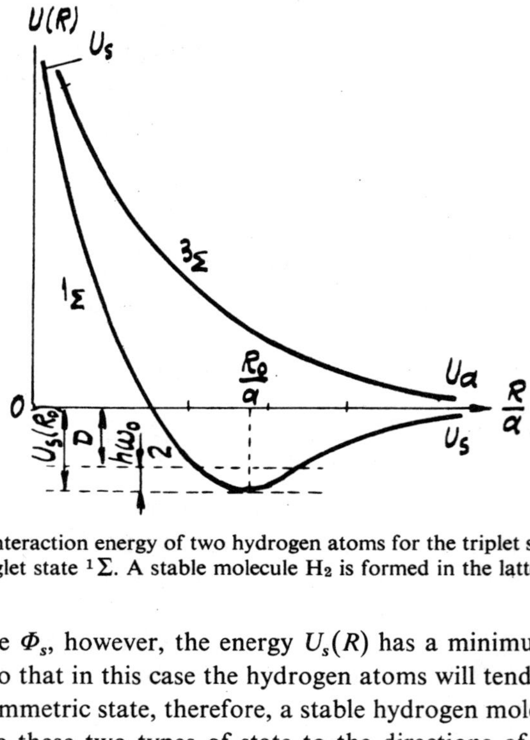

Fig. 89. Interaction energy of two hydrogen atoms for the triplet state $^3\Sigma$ and the singlet state $^1\Sigma$. A stable molecule H_2 is formed in the latter state.

the symmetric state Φ_s, however, the energy $U_s(R)$ has a minimum at $R_0 = 1.4a = 0.74 \times 10^{-8}$ cm, so that in this case the hydrogen atoms will tend to be at a distance R_0 apart. In the symmetric state, therefore, a stable hydrogen molecule H_2 is formed. We can now relate these two types of state to the directions of the electron spins. This is easily done by using the results obtained for the helium atom (Section 122). The wave functions derived above for the hydrogen molecule depend only on the co-ordinates $\mathbf{r}_1$ and $\mathbf{r}_2$ of the centres of mass of the electrons. The complete wave function Ψ must also depend on the electron spins s_{z1} and s_{z2}. Since the interaction of the spins with the orbital motion and with one another is neglected, the wave function Ψ is a product of a function Φ of the co-ordinates of the centres of mass of

[3] Concerning the calculation of the integrals K and A see [5], p. 535.

the electrons and a function S of the spins s_{z1} and s_{z2}. Since the electrons obey the Pauli principle, the wave function Ψ must be antisymmetric with respect to interchange of the electrons. As with the helium atom, we have two co-ordinate functions Φ, one symmetric (Φ_s) and the other antisymmetric (Φ_a).

In order that the complete function Ψ should be antisymmetric, it is necessary that for $\Phi = \Phi_s$ the spin function $S(s_{z1}, s_{z2})$ should be antisymmetric with respect to the spin ($S = S_a$). For the antisymmetric function $\Phi = \Phi_a$, the spin function must be symmetric ($S = S_s$). Evidently the spin functions S_a and S_s will be exactly the same as those derived in Section 122. S_a describes the state with antiparallel spins (see Section 122). Thus the state Φ_s with energy $U_s(R)$ is a singlet state (oppositely directed spins). In molecules, such a state is customarily denoted by the symbol $^1\Sigma$. The state Φ_a with energy $U_a(R)$, however, is a triplet state (parallel spins). This is denoted by $^3\Sigma$.

In terms of the curves for U_a and U_s in Figure 89 we can express the result as follows: two hydrogen atoms with their electrons having oppositely directed spins ($^1\Sigma$ state) attract each other and form a molecule; two hydrogen atoms with their electrons having parallel spins ($^3\Sigma$ state) repel each other.

The attraction or repulsion of hydrogen atoms depends on the sign of the exchange energy A (since the energies U_a and U_s differ in the sign of A). Thus the formation of a homopolar H_2 molecule is determined by exchange forces, and this explains why no theory of the homopolar bond was possible either in classical theories or in Bohr's old quantum theory.

Let us now consider some particular features of the potential energy $U_s(R)$ of the hydrogen molecule. In Figure 89 the curve $U_s(R)$ is shown separately from the triplet state curve $U_a(R)$. Knowing the analytical expression for $U_s(R)$, we can find the point of equilibrium ($R = R_0$) from the equation

$$\mathrm{d}U_s(R)/\mathrm{d}R = 0. \tag{125.30}$$

Expanding $U_s(R)$ in powers of the displacement from the equilibrium position, we have

$$\begin{aligned} U_s(R) = U_s(R_0) + \tfrac{1}{2}\left(\frac{\mathrm{d}^2 U_s}{\mathrm{d}R^2}\right)_{R_0} (R - R_0)^2 + \\ + \tfrac{1}{6}\left(\frac{\mathrm{d}^3 U_s}{\mathrm{d}R^3}\right)(R - R_0)^3 + \dots . \end{aligned} \tag{125.31}$$

This expansion is valid when $R - R_0$ is small. If sufficient accuracy were given by taking only the term in $(R - R_0)^2$ we should have a harmonic oscillator, whose frequency can be derived as follows. The potential energy of an oscillator of mass μ and frequency ω_0, executing oscillations about the equilibrium position R_0, is

$$U(R) = \text{constant} + \tfrac{1}{2}\mu\omega_0^2 (R - R_0)^2 .$$

Comparison with Formula (125.31) for $U_s(R)$ gives

$$\mu\omega_0^2 = (\mathrm{d}^2 U/\mathrm{d}R^2)_{R_0}, \tag{125.32}$$

whence

$$\omega_0 = \sqrt{\left[\frac{1}{\mu}\left(\frac{d^2U_s}{dR^2}\right)_{R_0}\right]}.$$

It may be noted that, since the relative motion of the nuclei is in question, μ should be taken as the reduced mass of the two hydrogen atoms, i.e.

$$1/\mu = 2/m_H, \tag{125.33}$$

where m_H is the mass of the hydrogen atom. Using Formula (125.32), we can find the frequency of the molecular vibrations ω_0 from the curvature d^2U_s/dR^2 of the potential curve $U_s(R)$ at the equilibrium point R_0. The third term in (125.31) gives a correction for deviation from harmonic motion.

For large energies of vibration this correction becomes increasingly important. If the energy of vibration E exceeds the potential energy $U_s(R)$ at infinity (in Figure 89 $U_s(\infty)$ is taken as zero, which means that this condition is $E > 0$), then the molecule will dissociate instead of vibrating. The energy D needed for dissociation is, according to classical mechanics, $-\ U_s(R_0)$.

In order to obtain the correct value for the dissociation energy of the molecule, it must be remembered that in the lowest state the molecule has a positive energy of vibration $\frac{1}{2}\hbar\omega_0$ (Figure 89), which must be subtracted. Thus

$$D = -\ U_s(R_0) - \tfrac{1}{2}\hbar\omega_0,$$

and this gives the dissociation energy.

The calculation has therefore given (1) the position of equilibrium R_0, (2) the frequency ω_0 of molecular vibrations, (3) the dissociation energy D of the hydrogen molecule. All these quantities have been determined experimentally. The quantity R_0 appears in the moment of inertia I of the molecule ($I = \mu R_0^2$), and can be determined from spectroscopic data by means of Deslandres' formula (54.20). The vibration frequency ω_0 is likewise found from spectroscopic data. The dissociation energy may be determined either optically or chemically. Table 4 gives the results of calculations by Hylleraas[4] for H_2, together with the experimental data. The agreement is excellent, particularly if we remember that R_0 and ω_0 depend very sensitively on the form of the curve $U_s(R)$. It may also be noted that the accuracy of Hylleraas' results is not complete. The success of quantum mechanics in the theory of the

TABLE 4

	Theoretical value	*Experimental value*
R_0	0.735×10^{-8} cm	0.753×10^{-8} cm
ω_0	4280 cm^{-1}	4390 cm^{-1}
D	4.37 eV	4.38 eV

[4] A summary of data for the hydrogen molecule is given in [5], p. 543.

hydrogen molecule, based on the simple fact that this molecule consists of two protons and two electrons and involving no arbitrary constants, is one of its greatest achievements.

126. The nature of chemical forces

Chemistry distinguishes between the two kinds of bond which bring about the formation of ionic (heteropolar) and homopolar molecules. The ionic bond occurs when the molecule can be regarded as consisting of positive and negative ions (for example, NaCl). The homopolar bond occurs when no such separation into ions is possible; a typical case is that of molecules consisting of like atoms, such as H_2.

The theory of ionic bonds was worked out with some success before the advent of quantum mechanics. The simplest view of the nature of the ionic bond (valency) is as follows: the heteropolar valency of an element is simply determined by the number of electrons which must be removed (from electropositive elements) or added (to electronegative elements) in order to obtain an ion having the electron shell structure of the nearest inert gas. For example, sodium must lose one electron to have the neon shell; chlorine must gain one electron to have the argon shell. Thus Na^+ and Cl^- are like charged inert-gas atoms.

Under these conditions the Coulomb attraction of oppositely charged ions should play the main part in the ionic bond, since the electron shells of the inert gases are chemically inactive. It is known, however, that electrostatic forces alone cannot bring about stable equilibrium. In addition to the Coulomb attraction $-e^2/r^2$ of the ion charges, therefore, a repulsion at small distances must also be postulated. Such repulsive forces were not predicted by the classical theory, but they appeared to be empirically justified, since the atoms of the inert gases repel one another at small distances. The repulsive forces were taken in the form α/r^{m+1}, where α and m are empirically determined constants. The total potential energy of the two ions is therefore[5]

$$U(r) = -\frac{e^2}{r} + \frac{\alpha}{r^m}. \tag{126.1}$$

Whereas this procedure threw some light on the problem of the heteropolar bond, that of the homopolar bond remained completely obscure. Attempts at a theory of the hydrogen molecule led to unsatisfactory results. The reason for these failures is evident from the preceding quantum theory of the hydrogen molecule. Exchange forces play a leading part in the formation of this molecule, and the existence of such forces is peculiar to quantum mechanics. The exchange forces themselves involve no new interaction of particles; they arise from the same Coulomb interaction of electrons in the molecule. We have also seen that the Pauli principle, i.e. the principle of indistinguishability of particles, has to be used in deriving a correct theory of the hydrogen molecule. Lack of knowledge of these facts made it impossible to solve the

[5] It may be noted that quantum mechanics gives a different form of the repulsion term, and one in better agreement with experiment.

problem of the structure of even the simplest molecule before the discovery of quantum mechanics.

The successful resolution of the problem of the hydrogen molecule by the methods of quantum mechanics was the starting point of the quantum theory of homopolar valency. This topic cannot be discussed in detail here, and we shall make only a few comments. For H_2 we have derived two states, with parallel and antiparallel spins. Figure 90 shows the distribution of electric charge density ρ of the electrons for the two states. The electric charge density at the point $\mathbf{r}$ is calculated from the wave function $\Phi(\mathbf{r}_1, \mathbf{r}_2)$ by means of the formula

$$\rho(\mathbf{r}) = -2e\int|\Phi(\mathbf{r}, \mathbf{r}')|^2 dv'. \tag{126.2}$$

If the spins of the atoms are parallel, $\Phi = \Phi_a$. At the point $\mathbf{r} = \mathbf{r}'$ the function $\Phi_a = 0$ (a nodal plane), and so the density ρ has a minimum in the region between

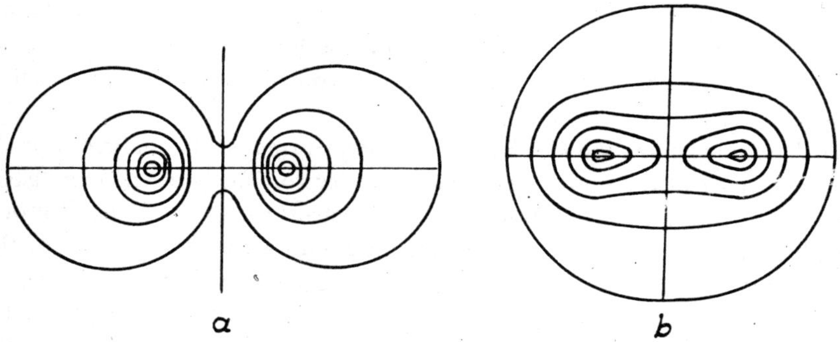

Fig. 90. Charge density distribution (a) in two mutually repelling hydrogen atoms ($^3\Sigma$), (b) in the hydrogen molecule ($^1\Sigma$).

the atoms (Figure 90a). In the state with antiparallel spins, however, $\Phi = \Phi_s$ and there is no nodal plane; the charge densities of the two atoms 'coalesce' (Figure 90b). The coalescence of the densities (formation of a homopolar bond) is represented by the valency dash: H – H. The presence of the minimum density corresponds to the absence of a valency bond.

It can be shown that homopolar bond force have the property of saturation – a feature characteristic of valency forces. It is easy to see that the attachment of a third hydrogen atom to the H_2 molecule does not lead to the occurrence of exchange forces between the electrons in the molecule and those in the third atom. Let the wave function of the electrons in the molecule (atoms a and b) be $\Phi_s(\mathbf{r}_1, \mathbf{r}_2)\, S_a(s_1, s_2)$ and that of the electron in the third atom c be $\psi_c(\mathbf{r}_3)\, S_{\frac{1}{2}}(s_3)$. The spin of the third electron is taken to be along the axis OZ. It could have been taken in the opposite direction; the only important point is that this spin must be in the direction opposite to that of one of the electrons in the molecule. In order to obtain the wave function of the entire system, a function antisymmetric in the particles has to be formed from

$\Phi_s S_a$ and $\psi_c S_{\frac{1}{2}}$ (the Pauli principle). The only antisymmetric function which can be formed from $\Phi_s S_a$ and $\psi_c S_{\frac{1}{2}}$ is

$$\Phi(\mathbf{r}_1, \mathbf{r}_2, \mathbf{r}_3, s_1, s_2, s_3) = \frac{1}{\sqrt{3}}\{\Phi_s(\mathbf{r}_1, \mathbf{r}_2)\psi_c(\mathbf{r}_3)S_a(s_1, s_2)S_{\frac{1}{2}}(s_3) + \Phi_s(\mathbf{r}_1, \mathbf{r}_3)\psi_c(\mathbf{r}_2)S_a(s_3, s_1)S_{\frac{1}{2}}(s_2) + \Phi_s(\mathbf{r}_2, \mathbf{r}_3)\psi_c(\mathbf{r}_1)S_a(s_2, s_3)S_{\frac{1}{2}}(s_1)\}. \quad (126.3)$$

According to (121.13),

$$S_a(s_1, s_2) = \frac{1}{\sqrt{2}}\{S_{\frac{1}{2}}(s_1)S_{-\frac{1}{2}}(s_2) - S_{\frac{1}{2}}(s_2)S_{-\frac{1}{2}}(s_1)\}.$$

From the orthogonality and normalisation of the spin functions $S_\alpha(s)$ ($\alpha = \pm\frac{1}{2}$) (see (60.7)) it is easily seen that all three spin functions in the superposition (126.3) are orthogonal to one another. If we take $|\Phi|^2$ and sum over both values ($\pm\frac{1}{2}\hbar$) of all the spins to get the probability $w(\mathbf{r}_1, \mathbf{r}_2, \mathbf{r}_3)$ that the electrons are near the points $\mathbf{r}_1, \mathbf{r}_2, \mathbf{r}_3$, the result is therefore

$$w(\mathbf{r}_1, \mathbf{r}_2, \mathbf{r}_3) = \sum_{s_1, s_2, s_3} |\Phi|^2 = \tfrac{1}{3}\{|\Phi_s(\mathbf{r}_1, \mathbf{r}_2)|^2 |\psi_c(\mathbf{r}_3)|^2 + |\Phi_s(\mathbf{r}_1, \mathbf{r}_3)|^2 |\psi_c(\mathbf{r}_2)|^2 + |\Phi_s(\mathbf{r}_2, \mathbf{r}_3)|^2 |\psi_c(\mathbf{r}_1)|^2\}. \quad (126.4)$$

Again denoting (as in Section 125) the electron charge density in atom a by ρ_a, that in atom b by ρ_b, and that in atom c by ρ_c, and the exchange density by ρ_{ab}, we can use the value of Φ_s from (125.24′), (125.11) and (125.11′) to write the resulting probability of the electron configuration as

$$\begin{aligned} w(\mathbf{r}_1, \mathbf{r}_2, \mathbf{r}_3) = \tfrac{1}{3}\{&[\rho_a(r_1)\rho_b(r_2) + 2\rho_{ab}(r_1)\rho_{ab}(r_2) + \\ &+ \rho_a(r_2)\rho_b(r_1)]\rho_c(r_3) + [\rho_a(r_1)\rho_b(r_3) + 2\rho_{ab}(r_1)\rho_{ab}(r_3) + \\ &+ \rho_a(r_3)\rho_b(r_1)]\rho_c(r_2) + [\rho_a(r_2)\rho_b(r_3) + 2\rho_{ab}(r_2)\rho_{ab}(r_3) + \\ &+ \rho_a(r_3)\rho_b(r_2)]\rho_c(r_1)\}, \end{aligned} \quad (126.5)$$

where

$$\rho_a(r_1) = -e\psi_a^2(r_{a1}), \qquad \rho_b(r_1) = -e\psi_b^2(r_{b1}), \qquad \rho_c(r_1) = -e\psi_c^2(r_{c1}), \quad (126.6)$$

and

$$\rho_{ab}(r_1) = -e\psi_a(r_{a1})\psi_b(r_{b1}). \quad (126.6')$$

From this expression we see that there is no exchange density of the type ρ_{ac}, ρ_{bc} for the third atom (c), and therefore no exchange forces with the atoms a and b which form the molecule. The Coulomb interaction remains, and the third atom is therefore repelled. This proves that exchange forces are capable of saturation and the validity of representing the coalescence of the electric charge density of two atoms by the valency dash.

It should be noted that there can be no precise demarcation between homopolar and ionic bonds. They are merely the two limiting cases. In a typical homopolar bond the charge is distributed symmetrically between the two atoms. If the atoms are unlike, the symmetry is destroyed, and if there is a very marked departure from symmetry so that the electron charge is concentrated mainly in the neighbourhood of one atom only, we have an ionic bond.

127. Dispersion forces between molecules

The valency forces discussed above are connected with the orientation of electron spins and have the property of saturation. Moreover, they act at short distances. They are determined by the degree of overlapping of the electron densities of the interacting atoms. Since the electron density decreases exponentially with increasing distance from the centre of the atom, the valency forces decrease exponentially with increasing distance between the atoms.

In addition to the valency forces there are other forces, which are always attractive, between atoms and molecules. These are the *intermolecular dispersion forces* or *van der Waals forces*. A notable feature of these forces is that they act between electrically neutral systems and between systems having no electric moment, for example between helium atoms, whose charge distribution has spherical symmetry, so that these atoms have no dipole, quadrupole or higher electric moment. A second important property of these forces is that they are independent of temperature. These forces are also a quantum phenomenon.

The van der Waals forces may be calculated by considering the interaction of atoms at a sufficient distance apart. When the distance between the atoms is large, the valency forces calculated from the first approximation of perturbation theory are very small, but at such distances it is no longer possible to disregard the second approximation, in which the deformation of the electron shells of the atoms is taken into account. This is explained by the fact that the additional corrections to the energy of interaction of the atoms in the second approximation decrease as $1/R^6$ with increasing distance R between the atoms, whereas the energy of the valency bonds decreases as $e^{-2R/a}$. For large R the second approximation term is larger than the first.

The van der Waals forces may be found by calculating the energy in the second approximation for large distances. Without going through these calculations, we may explain the basic idea of the quantum theory of van der Waals forces by means of a simple example which allows an exact solution of the problem. Instead of actual atoms, let us consider two one-dimensional oscillators of eigenfrequency ω_0 (a model of the atom which is used in classical dispersion theory). Let the co-ordinate of the electron in the first atom be denoted by x_1 and its momentum by p_1, and those of the electron in the second atom by x_2 and p_2; let the distance between the 'atoms' be R. The electric moments of the two atoms are ex_1 and ex_2. If the distance R between these atoms is sufficiently large, their interaction energy can be represented as the potential energy of interaction of two dipoles with moments ex_1 and ex_2. This energy is

$$W = ex_1 \cdot ex_2/R^3 = e^2 x_1 x_2/R^3 . \tag{127.1}$$

If the oscillators are at rest, then $x_1 = x_2 = 0$ and their dipole moments are zero. Since the two 'atoms' are also electrically neutral, there is no interaction between them.

According to classical theory, interaction occurs only between vibrating oscillators. Without giving the detailed calculation of this interaction, we can predict that it will depend on the temperature T, since at absolute zero there is no vibration and $x_1 = x_2 = 0$. The result given by quantum mechanics is different: even at absolute zero there are zero-point vibrations which result in a non-zero mean energy of interaction of the oscillators.

In order to calculate this energy, we return to the results of Section 109, which dealt with just this case of the interaction of two one-dimensional oscillators of frequency ω_0 and mass μ. The interaction energy of the oscillators was assumed to be of the form $\lambda x_1 x_2$ (see (109.1)). In our case the interaction energy is given by Formula (127.1). Thus, if we put

$$\lambda = e^2/R^3\,, \tag{127.2}$$

we can use all the results of Section 109. Here we are interested in the lowest zero-point energy of the oscillators, which is

$$E_0 = \tfrac{1}{2}\hbar\omega_1 + \tfrac{1}{2}\hbar\omega_2 = \tfrac{1}{2}\hbar(\omega_1 + \omega_2)\,, \tag{127.3}$$

where ω_1 and ω_2 are determined from (109.5):

$$\omega_1^2 = \omega_0^2 + \lambda/\mu\,, \qquad \omega_2^2 = \omega_0^2 - \lambda/\mu\,.$$

Hence, assuming that $\omega_0^2 \gg \lambda/\mu$, we find

$$\omega_1 = \omega_0\left[1 + \frac{\lambda}{2\mu\omega_0^2} - \frac{1}{8}\left(\frac{\lambda}{\mu\omega_0^2}\right)^2 + \ldots\right],$$

$$\omega_2 = \omega_0\left[1 - \frac{\lambda}{2\mu\omega_0^2} - \frac{1}{8}\left(\frac{\lambda}{\mu\omega_0^2}\right)^2 + \ldots\right]$$

and so

$$\omega_1 + \omega_2 = 2\omega_0 - \frac{1}{4}\frac{\lambda^2}{\mu^2\omega_0^3} + \ldots\,. \tag{127.4}$$

Using the value of λ in the present case (127.2), we find from (127.3) and (127.4) the zero-point energy of two oscillators interacting as dipoles:

$$E_0(R) = \hbar\omega_0 - \tfrac{1}{8}\hbar\,\frac{e^4}{\mu^2\omega_0^3}\,\frac{1}{R^6} + \ldots\,. \tag{127.5}$$

We see that the zero-point energy depends on the distance R between the 'atoms', and therefore corresponds to a potential energy of their interaction.

Omitting the unimportant additive constant $\hbar\omega_0$, we find for this potential energy the expression

$$U(R) = -\frac{\hbar}{8}\frac{e^4}{\mu^2\omega_0^3}\frac{1}{R^6}\,. \tag{127.6}$$

We see from the minus sign that this energy corresponds to forces of attraction, which may be regarded as the van der Waals forces for the idealised atoms considered. The quantum nature of these forces is evident from the fact that $U = 0$ for $\hbar = 0$, so that in the limit of classical mechanics the forces are zero.

Thus the van der Waals attraction results from a diminution in the zero-point energy when the oscillators approach. Formula (127.6) can be transformed by using the atomic polarisability coefficient β for a constant field. We know from dispersion theory that this coefficient for an oscillator of mass μ and frequency ω_0 is[6] (Section 92)

$$\beta = \frac{e^2}{\mu} \frac{1}{\omega_0^2 - \omega^2};$$

putting $\omega = 0$, we obtain the polarisability coefficient for a constant field:

$$\beta = e^2/\mu\omega_0^2. \tag{127.7}$$

Substituting this in Formula (127.6) for the potential energy of the van der Waals forces, we find

$$U(R) = -\tfrac{1}{8}\varepsilon\beta^2/R^6, \tag{127.8}$$

where

$$\varepsilon = \hbar\omega_0, \tag{127.9}$$

that is, the difference between the quantum levels of the oscillator. Since the polarisability coefficient given by dispersion theory appears in the formula for the van der Waals forces, these forces have recently been called *dispersion forces*.

A calculation using the second approximation for actual atoms gives essentially the same result as (127.8) for the linear-oscillator model of the atom. The quantum formula for the potential energy of the van der Waals forces for actual atoms is

$$U(R) = -kI\beta^2/R^6, \tag{127.10}$$

where β is the atomic polarisability in a constant field, I the ionisation potential of the atom, and k some numerical coefficient of the order of unity. This expression for the van der Waals interaction is in good agreement with the experimental data derived from a study of deviations from Clapeyron's law in gases.[7]

128. Nuclear spin in diatomic molecules

Atomic nuclei have spins and magnetic moments.[8] Hence the wave function of nuclei depends not only on their co-ordinates but also on their spins s_{z1}, s_{z2}. Using the relative co-ordinates $\mathbf{r}_{12}$ of the nuclei (in spherical polar co-ordinates r, θ, ϕ) and neglecting the interaction of the magnetic moment of the nuclei with their motion,

[6] This is the classical formula. The quantum Formula (92.5′) for an oscillator leads to the same result, as the reader may confirm by using the matrix x_{nk} for the oscillator co-ordinate and Formula (92.25).

[7] See [5], Section 63, where a review of the literature is also given.

[8] The magnetic moments of nuclei are of the order of magnitude of the nuclear Bohr magneton (cf. Section 63), which is less than the ordinary Bohr magneton by a factor 1842.

we can write the wave function of the nuclei as

$$\Psi(\mathbf{r}_{12}, s_{z1}, s_{z2}) = R_{nl}(r) P_l(\cos\theta) S(s_{z1}, s_{z2}). \quad (128.1)$$

The function $R_{nl}(r)$ describes the vibration of the nuclei, the function P_l their rotation (we take $m = 0$, since here we are not interested in the spatial orientation of the molecule), and the function S the spin state of the nuclei. According to the principle of indistinguishability, for identical nuclei (the same isotopes) the function Ψ must be symmetric or antisymmetric, according as the nuclei have integral or half-integral spin.

For definiteness, let us consider the latter case; this occurs in the H_2 molecule, where both nuclei are protons. The function Ψ must then be antisymmetric with respect to interchange of the protons.

Interchange of the protons corresponds to an inversion of the relative co-ordinates $\mathbf{r} = \mathbf{r}_1 - \mathbf{r}_2$. The function $R_{nl}(r)$ does not change sign. The parity of a state with respect to the co-ordinates of the particles is determined by the orbital quantum number (Sections 25, 107). The energy levels of a molecule with even l are called even terms, and those with odd l odd terms.

Since the complete function Ψ is antisymmetric, the parity of the terms depends on the relative orientation of the spins in the molecule. The two possibilities are as follows.

(1) Nuclear spins parallel. Then $S = S_s$ is a symmetric function, and P_l must therefore be odd. The H_2 molecule with parallel nuclear spins (*orthohydrogen*) can therefore have only odd orbital quantum numbers l. In particular, its lowest state corresponds to a state of rotation with $l = 1$.

(2) Nuclear spins antiparallel. Then $S = S_a$ is an antisymmetric function of the spins, and P_l must therefore be even. The H_2 molecule with antiparallel nuclear spins (*parahydrogen*) can therefore have only even orbital quantum numbers l. The lowest state is $l = 0$.

Thus, on account of the Pauli principle, the nuclear spin has a considerable indirect influence on the orbital motion of the nuclei in the molecule. This influence is seen particularly in the alternation of intensities in molecular vibrational spectra[9] and in the specific heats of molecules. To examine the latter phenomenon, let us suppose that thermal equilibrium is established at so low a temperature that the rotation is 'frozen' (cf. Section 54). Then hydrogen will be in the parahydrogen state ($l = 0$). If now the hydrogen in this state is heated, the probability of a change in direction of the nuclear spins when molecules collide is very small, because the interaction with the small magnetic moment of the nuclei is small. Hence, despite collisions, the hydrogen will remain in the parahydrogen state, and the specific heat on account of rotation will be determined by transitions $l = 0 \rightarrow l = 2 \rightarrow l = 4 \ldots$.

If, however, the hydrogen is left for some time (many days being needed) at the higher temperature, the nuclear spins will be redistributed, and some orthohydrogen will be formed. Then transitions of the type $l = 1 \rightarrow l = 3 \rightarrow l = 5 \ldots$ are also

[9] Concerning this phenomenon see [53, 55, 69]; [26, 97].

possible. Since the change in the rotational energy, $\Delta E = (\hbar^2/2I)\{(l+\frac{1}{2})^2 - (l'+\frac{1}{2})^2\}$, is different for even and odd l, the specific heats of parahydrogen and orthohydrogen are different. In consequence, the slow process of establishment of equilibrium between parahydrogen and orthohydrogen will be accompanied by a change in specific heat.

In equilibrium, the number of orthohydrogen molecules is three times that of parahydrogen molecules (since for parallel spins there are three symmetric functions S_s, but for antiparallel spins there is only one antisymmetric function S_a; cf. Section 121). Thus hydrogen is normally a mixture of orthohydrogen and parahydrogen in the ratio 3 : 1.

This remarkable phenomenon of the change in the specific heat of hydrogen is not only qualitatively explained by quantum mechanics as indicated above, but also capable of quantitative analysis in entire agreement with experiment.[10]

10 See [40, 55].

CHAPTER XXIII

MAGNETIC PHENOMENA

129. Paramagnetism and diamagnetism of atoms

A simple and fundamental problem of atomic mechanics in the field of magnetic phenomena is to calculate the magnetic moments of atoms placed in an external magnetic field. We have already given in Section 53 an elementary determination of the magnetic moments of the orbital currents in an atom; here we shall consider general methods.

The operators of the components of the magnetic moment can be most generally defined as minus the derivatives of the total-energy operator (more precisely, the Hamiltonian) with respect to the components of the magnetic field:

$$\mathfrak{M}_x = -\,\partial \boldsymbol{H}/\partial \mathscr{H}_x, \qquad \mathfrak{M}_y = -\,\partial \boldsymbol{H}/\partial \mathscr{H}_y, \qquad \mathfrak{M}_z = -\,\partial \boldsymbol{H}/\partial \mathscr{H}_z. \tag{129.1}$$

In particular, for a single electron the Hamiltonian $\boldsymbol{H}$ which describes the motion of the electron in a magnetic field is

$$\boldsymbol{H} = \frac{1}{2\mu}\left(\vec{\boldsymbol{P}} + \frac{e}{c}\mathbf{A}\right)^2 + U(r) + \frac{e}{\mu c}(\vec{s} \cdot \mathscr{H}) \tag{129.2}$$

(the plus sign being taken in front of the vector potential $\mathbf{A}$ because the electron charge is taken as $-e$). We take the axis OZ to be in the direction of the magnetic field, and the vector potential in the form

$$A_x = -\tfrac{1}{2}\mathscr{H}y, \qquad A_y = \tfrac{1}{2}\mathscr{H}x, \qquad A_z = 0. \tag{129.3}$$

Differentiation of $\boldsymbol{H}$ with respect to $\mathscr{H}_z$ gives

$$\mathfrak{M}_z = -\frac{e}{2\mu c}\left[\left(\boldsymbol{P}_y + \frac{e}{c}A_y\right)x - \left(\boldsymbol{P}_x + \frac{e}{c}A_x\right)y\right] - \frac{e}{\mu c}s_z. \tag{129.4}$$

The operator in the square brackets is the operator of the component of the true momentum along OZ,[1] and $\boldsymbol{P}_y x - \boldsymbol{P}_x y$ is the operator $\boldsymbol{M}_z$ of the component of the

[1] It will be recalled that in a magnetic field the velocity operator is not $(1/\mu)\vec{\boldsymbol{P}}$ but $(1/\mu)(\vec{\boldsymbol{P}} + e\mathbf{A}/c)$.

generalised momentum along OZ. Using (129.3), we can write (129.4) as

$$\begin{aligned}\mathfrak{M}_z &= -\frac{e}{2\mu c}(M_z + 2s_z) - \frac{e^2\mathscr{H}}{4\mu c^2}(x^2 + y^2)\\ &= -\frac{e}{2\mu c}(J_z + s_z) - \frac{e^2\mathscr{H}}{4\mu c^2}(x^2 + y^2).\end{aligned} \tag{129.5}$$

It is seen that this operator consists of two parts, one dependent on the magnetic field and the other not. The latter part,

$$\mathfrak{M}_z' = -(e/2\mu c)(J_z + s_z), \tag{129.6}$$

has the eigenvalues which we have already determined in the theory of the Zeeman effect. The perturbation energy in the magnetic field is $W = -\mathscr{H}\cdot\mathfrak{M}_z'$. The eigenvalues of the operator W are different for strong magnetic fields (the normal Zeeman effect) and weak magnetic fields (the anomalous Zeeman effect). In the latter case the eigenvalues of W are given by Formula (74.23). These differ from the eigenvalues of $\mathfrak{M}_z$ by the factor $-\mathscr{H}$. Hence (74.23) gives

$$\mathfrak{M}_z' = -\frac{e\hbar}{2\mu c} m_j \left\{1 + \frac{j(j+1) - l(l+1) + l_s(l_s+1)}{2j(j+1)}\right\}, \tag{129.7}$$

where m_j is the magnetic number, j a number giving the total angular momentum, l the orbital quantum number and l_s the spin quantum number. The potential energy of this angular momentum in the external magnetic field is just W. It can take both positive and negative values, depending on the value of $m_j = \pm\frac{1}{2}, \pm\frac{3}{2}, \ldots, \pm j$.

In thermodynamic equilibrium negative values of W, and therefore positive values of $\mathfrak{M}_z'$, will occur preferentially. The mean moment is therefore in the direction of the field; this is *paramagnetism*. It is important to note that $\mathfrak{M}_z'$ cannot be zero, and so atoms with a single electron are always paramagnetic.

The second term in (129.5),

$$\mathfrak{M}_z'' = -e^2\mathscr{H}(x^2 + y^2)/4\mu c^2, \tag{129.8}$$

is a magnetic moment which is obviously always in the opposite direction to the field. Thus this moment causes *diamagnetism*. It can never be zero, since $x^2 + y^2 > 0$, and the diamagnetic effect therefore occurs for all atoms. It is easily seen, however, that the moment $\mathfrak{M}_z''$ is considerably less than $\mathfrak{M}_z'$, and can be neglected in comparison with the latter: $\mathfrak{M}_z'$ is equal, in order of magnitude, to the magneton $e\hbar/2\mu c$, while $\mathfrak{M}_z'' \approx (e^2\mathscr{H}/2\mu c^2)\, a^2$, where a is the dimension of the atom, and so $\mathfrak{M}_z' \gg \mathfrak{M}_z''$ for all fields $\mathscr{H}$ such that

$$\mathscr{H} \ll \frac{e}{a^2}\frac{\hbar c}{e^2} = 137\frac{e}{a^2}. \tag{129.9}$$

This condition is satisfied by all fields attainable in practice.

If the number of electrons in the atom is even, the total angular momentum may be zero, and then so will be the magnetic moment $\mathfrak{M}_z'$ which causes paramagnetism.

Such an atom will be diamagnetic. For example, in the helium atom in the ground state, the orbital angular momentum is, as we know, zero, and the spin is zero because the spins of the electrons are in opposite directions. Hence $\mathfrak{M}_z' = 0$. Helium should be diamagnetic, and is in fact observed to be so. The diamagnetic susceptibility of the helium atom may be calculated by using the fact that for two electrons

$$\mathfrak{M}_z'' = -\frac{e^2\mathscr{H}}{4\mu c^2}\{\overline{x_1^2 + y_1^2} + \overline{x_2^2 + y_2^2}\}. \tag{129.10}$$

The mean values $\overline{x_1^2}, \overline{y_1^2}, \overline{x_2^2}, \overline{y_2^2}$ are all equal to $\frac{1}{3}\overline{r^2}$ (where $\overline{r^2}$ is the mean square of the radius vector), owing to the spherical symmetry of the ground state of helium and the symmetry of the electrons in that state. Thus

$$\mathfrak{M}_z'' = -\frac{e^2\mathscr{H}}{4\mu c^2}\cdot\frac{4}{3}\overline{r^2}.$$

The diamagnetic susceptibility per atom is

$$\chi = \partial\mathfrak{M}_z''/\partial\mathscr{H} = -(e^2/3\mu c^2)\overline{r^2}. \tag{129.11}$$

Using the wave functions (122.23) for the electrons in the helium atom, we can calculate the mean value $\overline{r^2}$ and derive the numerical value of the magnetic susceptibility. The result is $\chi = -1.87 \times 10^{-6}$, the experimental value being $\chi = -1.88 \times 10^{-6}$.

The expression (129.8) for the diamagnetic moment is the same as that given by the classical electron theory.[2] However, the quantity $\overline{x^2 + y^2}$ cannot be calculated from the atomic constants except by the use of quantum mechanics.

For an atom with more than one electron, we have instead of (129.7)

$$\mathfrak{M}_z' = -\frac{e\hbar}{2\mu c} m_J\left\{1 + \frac{J(J+1) - L(L+1) + S(S+1)}{J(J+1)}\right\} \tag{129.12}$$

(see (105.33)), where J is a number giving the total angular momentum of all the electrons, L a number giving the total orbital angular momentum, S a number giving the total spin angular momentum, and $|m_J| \leqslant J$ and gives the component of the total angular momentum along the magnetic field. If $J = 0$, which can occur only for an atom with an even number of electrons, then $\mathfrak{M}_z' = 0$ and the atom will be diamagnetic, with

$$\mathfrak{M}_z'' = -\frac{e^2\mathscr{H}}{4\mu c^2}\sum_{k=1}^{N}\overline{(x_k^2 + y_k^2)}, \tag{129.13}$$

where N is the number of electrons. If $J \neq 0$, the quantity $\mathfrak{M}_z''$ can be neglected in comparison with $\mathfrak{M}_z'$, and such atoms are paramagnetic.

130. Ferromagnetism

The cause of the permanent magnetism of ferromagnetic substances remained for long a complete mystery. The phenomenon is essentially that ferromagnetic bodies can remain magnetised even when no external magnetic field $\mathscr{H}$ is present. To account for the properties of ferromagnetic materials, Weiss proposed a theory which explained the permanent magnetism by the presence of an internal magnetic field $\mathscr{H}_i$ which

[2] Cf. [4], Section 29.

causes an orientation of the elementary magnets even if the external field is zero. Weiss' theory made possible an explanation of many properties of ferromagnetic substances, but the origin of the internal field $\mathscr{H}_i$ remained obscure.

In order to bring Weiss' theory into agreement with experiment, it is necessary to assume that the field $\mathscr{H}_i$ is extremely large, 10^6 Oe. Direct experiments[3] show that there is in fact no such magnetic field inside a ferromagnetic substance. Heisenberg was able to prove that the forces which orient the elementary magnets are exchange forces. This explained the nature of the Weiss field. Heisenberg supposed, in accordance with the experimental data of Einstein and de Haas (Section 58), that the magnetisation of ferromagnetic bodies is due not to the orbital motion of the electrons but to the spin magnetic moment. Moreover, ferromagnetism is apparently due not to the valency electrons (conduction electrons) but to the electrons in incomplete inner shells of ferromagnetic atoms (see the electron distribution in the atoms of iron, nickel and cobalt shown in Table 3 (Section 124)).

For simplicity we shall assume that each atom in a crystal has only one such electron. The interaction of this electron with neighbouring atoms may be considered small, and so we may take the wave function for all the electrons (N in number) which cause ferromagnetism as that corresponding to a system of non-interacting electrons.

To label the states we note that the positions of the centres of the atoms in the crystal (the lattice points) are given by

$$\mathbf{r} = n_1\mathbf{a}_1 + n_2\mathbf{a}_2 + n_3\mathbf{a}_3\,, \tag{130.1}$$

where n_1, n_2, n_3 are integers, and $\mathbf{a}_1$, $\mathbf{a}_2$, $\mathbf{a}_3$ the basis vectors of the lattice. Thus the position of any atom is specified by a set of three numbers n_1, n_2, n_3. For brevity we shall denote these by the single letter n, and call this the number of the atom. Let the wave function of the kth electron in the nth atom be

$$\Phi_n(\mathbf{r}_k, s_{zk}) = \psi_n(\mathbf{r}_k)\, S_\alpha(s_{zk})\,,$$

where S_α is the spin function.

Since we neglect the interaction with neighbouring atoms, the wave function of the whole crystal will be an antisymmetric combination, of the form (117.6′), of the products of the functions Φ_n for the individual electrons. The choice of the suffix α $(+\tfrac{1}{2}$ or $-\tfrac{1}{2})$ for each of the functions S_α will signify the choice of a definite distribution of the spins (parallel or antiparallel to the axis OZ) among the atoms of the crystals. If the spins of all the electrons are in the same direction, say along OZ, we have saturation (the maximum magnetisation). Let us consider a state where all the spins are along OZ except one which is in the opposite direction, in the lth atom. Then the above discussion shows that the wave function Ψ of all the N electrons is

$$\begin{aligned}\Psi_l = \sum_P (\pm 1) P \psi_1(\mathbf{r}_1)\, S_{+\frac{1}{2}}(s_{z1})\, \psi_2(\mathbf{r}_2)\, S_{+\frac{1}{2}}(s_{z2}) \ldots \times \\ \times\, \psi_l(\mathbf{r}_l)\, S_{-\frac{1}{2}}(s_{zl}) \ldots \psi_N(\mathbf{r}_N)\, S_{+\frac{1}{2}}(s_{zN})\,.\end{aligned} \tag{130.2}$$

[3] Ya. G. Dorfman has passed a beam of fast electrons through a magnetised ferromagnetic foil. A field of 10^6 Oe should have deflected the electrons, but no deflection was observed.

Let us now take into account the interaction of the electrons with the neighbouring atoms. To do so, we apply perturbation theory. Degeneracy is involved, since the electron whose spin is in the opposite direction to the axis OZ may evidently be in any atom. The correct zeroth-order approximation function will therefore be a linear superposition of the Ψ_l:

$$\Psi = \sum_{l'=1}^{N} a_{l'} \Psi_{l'}, \tag{130.3}$$

where the amplitudes have to be determined. To do this, we note that the total-energy operator $\boldsymbol{H}$ of the electrons is

$$\boldsymbol{H} = \boldsymbol{H}^0 + \sum_{n>m=1}^{N} \frac{e^2}{r_{mn}} + \sum_{n>m=1}^{N} U_n(r_m), \tag{130.4}$$

$$\boldsymbol{H}^0 = \sum_{n=1}^{N} \boldsymbol{H}_n(r_n), \qquad \boldsymbol{H}_n(r_n) = -\frac{\hbar^2}{2\mu}\nabla_n^2 + U_n(r_n), \tag{130.5}$$

where $\boldsymbol{H}_n$ is the total-energy operator of the nth electron in the nth atom, e^2/r_{mn} the energy of interaction of the nth and mth electrons, and $U_n(r_m)$ the energy of interaction of the mth electron with the nth ion ($n \neq m$). All the terms in $\boldsymbol{H}$ except $\boldsymbol{H}^0$ will be regarded as a perturbation. Substituting in Schrödinger's equation the approximate function (130.3) for Ψ and bearing in mind that

$$\boldsymbol{H}_n(\mathbf{r}_n)\psi_n(\mathbf{r}_n) = E_0\psi_n(\mathbf{r}_n), \tag{130.6}$$

where E_0 is the energy of the electron in the atom, we obtain

$$NE_0 \sum_{l'} a_{l'}\Psi_{l'} + \left[\sum_{n>m=1}^{N}\left(\frac{e^2}{r_{mn}} + U_n(r_m)\right)\right]\sum_{l'} a_{l'}\Psi_{l'} = E\sum_{l'} a_{l'}\Psi_{l'}. \tag{130.7}$$

Multiplying this equation by Ψ_l^*, integrating over the co-ordinates of all the electrons and summing over both values $s_z = \pm\frac{1}{2}\hbar$ of the spin of each electron, assuming that the functions $\psi_n(\mathbf{r})$ and $\psi_m(\mathbf{r})$ which belong to different atoms are orthogonal[4], and using in the summation over the spins the orthogonality of the functions $S_\alpha(s_z)$ (Section 60), we have from (130.7)

$$NE_0 a_l + \sum_{l'} I_{ll'}\left[a_{l'} - a_l\right] = Ea_l, \tag{130.8}$$

where $I_{ll'}$ is the exchange integral (the matrix element of the perturbation energy)

$$I_{ll'} = \tfrac{1}{2}\int \psi_l(\mathbf{r}_1)\psi_{l'}(\mathbf{r}_2)\psi_l^*(\mathbf{r}_2)\psi_{l'}^*(\mathbf{r}_1) \times \\ \times \left\{\frac{2e^2}{r_{12}} + U_l(\mathbf{r}_1) + U_{l'}(\mathbf{r}_2) + U_{l'}(\mathbf{r}_1) + U_l(\mathbf{r}_2)\right\} dv_1\, dv_2. \tag{130.9}$$

The wave functions $\psi_l(\mathbf{r})$ decrease rapidly with increasing distance r from the centre

[4] In reality this is only approximately true.

of the atom. Hence the exchange integral $I_{ll'}$ decreases rapidly as the distance between the atoms l and l' increases. In consequence only the matrix elements $I_{ll'}$ belonging to nearest neighbours need be considered in solving Equations (130.8). Since all nearest neighbour atoms in the crystal are equivalent the exchange integral has the same value I for each. Thus Equations (130.8) can be written

$$(E - NE_0)\,a_l + I\sum_{l'}\left[a_l - a_{l'}\right] = 0\,, \tag{130.8'}$$

where the sum is taken over the atoms l' adjoining the atom l. The number and arrangement of the nearest neighbours depend on the type of crystal lattice concerned. For a simple cubic lattice the neighbours of an atom l (l_1, l_2, l_3) are those with l' equal to $(l_1 \pm 1, l_2, l_3)$, $(l_1, l_2 \pm 1, l_3)$, $(l_1, l_2, l_3 \pm 1)$.

It is easily seen that equations (130.8′) then have the solutions

$$a_l = a_{l_1 l_2 l_3} = \text{constant} \times e^{i(q_1 l_1 + q_2 l_2 + q_3 l_3)}\,, \tag{130.10}$$

where q_1, q_2, q_3 are certain dimensionless quantities. Substitution of (130.10) in (130.8′) gives

$$E - NE_0 = 2I\left[3 - \cos q_1 - \cos q_2 - \cos q_3\right], \tag{130.11}$$

whence

$$E(q_1, q_2, q_3) = NE_0 + 2I\left[3 - \cos q_1 - \cos q_2 - \cos q_3\right]. \tag{130.12}$$

Noting that $l_1 a$, $l_2 a$, $l_3 a$ (where a is the lattice constant) are the co-ordinates of a lattice point, we see that (130.10) can be regarded as a plane wave with wave vector $\mathbf{k} = \mathbf{q}/a$ $(q_1/a, q_2/a, q_3/a)$. The probability of finding the spin opposite to OZ is $|a_l|^2 =$ constant, i.e. all positions of the spins are equally probable. Thus the amplitudes a_i which determine the spin state are very similar to the wave function of a freely moving particle having a given momentum. This analogy is reinforced by the fact that, at least for small k, the energy (130.12) can be written in the form

$$E = \text{constant} + (\hbar^2/2\mu^*)\,k^2 + \ldots, \tag{130.13}$$

where $\hbar^2/2\mu^* = Ia^2$, i.e. in the same form as the energy of a free particle. The quantity μ^* may be regarded as an effective mass. On account of this analogy between the propagation of a given spin orientation in a crystal and the motion of a free particle, the state (130.10) is called a *spin wave*.

If the crystal contains not one but r spins in the direction opposite to OZ the calculation is similar, but is complicated by the fact that in this case pairs of adjoining atoms may have such spins, and for these the exchange integrals are not zero. When r is small, however, such pairs are rare, and the complete solution can be regarded as an assembly of non-interacting spin waves of the form (130.10) (called, from the corpuscular viewpoint, a 'spin gas'). The energy will be the sum of the energies of the individual spin waves. If we denote by $\mathbf{q}_k$ the vector $\mathbf{q}$ for the kth spin wave, the total energy of the spin gas is

$$E = NE_0 + 2I\sum_{k=1}^{r}\left[3 - \cos q_{1k} - \cos q_{2k} - \cos q_{3k}\right]. \tag{130.14}$$

It follows from this formula that for negative I ferromagnetism cannot occur, since when $I < 0$ the energy has a minimum when r is greatest. In thermal equilibrium, therefore, the original orientation of all the spins along the axis will tend to be destroyed. When the exchange integral is positive, however, the minimum energy will be reached when r is least, so that, if some of the spins are opposite to OZ, they will tend to become oriented along OZ (thus decreasing r). A positive value of the exchange integral is therefore a necessary condition for ferromagnetism, since only then is the state of least energy that in which all the electron spins are in the same direction. The exchange forces, therefore, and not the fictitious magnetic field proposed by Weiss, are responsible for the uniform orientation of the spins, and ferromagnetism is a quantum phenomenon. We also see that ferromagnetism is not a property of individual atoms, but of the crystal. This is in accordance with the fact that ferromagnetic gases do not exist.

To calculate the magnetisation of a ferromagnetic substance at a temperature T, the mean value $\bar{r}$ must be found by statistical methods. Then the magnetic moment of a piece of the substance containing N electrons is clearly

$$\mathfrak{M} = \mathfrak{M}_B(N - 2\bar{r}), \tag{130.15}$$

where $\mathfrak{M}_B$ is the magnetic moment of one electron (the Bohr magneton). For the calculations and other details the reader is referred to the specialist literature.[5]

[5] See [86].

CHAPTER XXIV

THE ATOMIC NUCLEUS

131. Nuclear forces. Isotopic spin

The interaction of the nucleons in a nucleus is a problem which is as yet far from being solved. The principles of quantum mechanics can, however, be applied both to the motion of the nucleons in the nucleus and to the interaction of nucleons with the nucleus. Considerable success has been achieved in this way in recent years, and quantum mechanics has proved to be a sound guide to the physicist's understanding of the complex pattern of nuclear interactions.

The reader is referred for details to the textbooks on the subject[1]; here we shall discuss only the simplest and most important results.

No exact expressions have yet been derived for the potential of the nucleons (protons and neutrons) in an atomic nucleus. This is no doubt a very complicated function of the positions, velocities and spins of the nucleons. It is very probably not possible to represent it as a sum of interactions between pairs of individual nucleons.

The 'potential' has not been ascertained, however, even for a pair of nucleons. A simple representation of the forces exists in this case only when the nucleons are at a large distance apart. Nevertheless, some fairly far-reaching conclusions can be drawn concerning the nature of the nuclear interactions, which make possible an analysis of the extensive range of experimental results.

The interaction of two nucleons depends on the distance $\mathbf{r}_{12}$ between them, their relative velocity $\mathbf{v}_{12}$, and their spins $\mathbf{s}_1$ and $\mathbf{s}_2$, and is also shown by experiment to depend essentially on the types of nucleons concerned, whether two protons, two neutrons or one proton and one neutron. There may also occur what is known as *charge exchange* during the interaction, whereby a proton is converted into a neutron and *vice versa*.[2]

It is found that, if the proton and the neutron are regarded as two states of one particle, the nucleon, the principal properties of nucleon interactions can be expressed in the form of very simple relations in terms of what is called *charge spin* or more usually *isotopic spin*.

Since there are only two charge states of the nucleon, it is reasonable to define a new dynamical variable t_3 which takes only two values, so that the wave function of

[1] See [28]; [95], Part 9.
[2] See, for example, [12].

the nucleon (omitting at present the dependence on the ordinary spin s) can be written as a single-column matrix:

$$\Psi(x, t) = \begin{vmatrix} \psi_1(x) \\ \psi_2(x) \end{vmatrix} \quad \begin{matrix} \text{'proton' state} \\ \text{'neutron' state} \end{matrix}\Bigg\} \tag{131.1}$$

as in the theory of the ordinary spin (cf. (60.3) and (60.3′)). In accordance with the optical terminology of multiplets for states differing only in the spin component, the proton and neutron states are called an *isotopic* (or *charge*) *doublet*.

All operators which alter the charge states of nucleons can be expressed, as for the ordinary spin, in terms of two-by-two Pauli matrices like $\boldsymbol{\sigma}_x$, $\boldsymbol{\sigma}_y$, $\boldsymbol{\sigma}_z$ (cf. Section 59). We shall denote these matrices, which now act on the charge suffixes 1, 2, by

$$\boldsymbol{\tau}_1 = \begin{vmatrix} 0 & 1 \\ 1 & 0 \end{vmatrix}, \quad \boldsymbol{\tau}_2 = \begin{vmatrix} 0 & -i \\ i & 0 \end{vmatrix}, \quad \boldsymbol{\tau}_3 = \begin{vmatrix} 1 & 0 \\ 0 & -1 \end{vmatrix}. \tag{131.2}$$

Any operator acting on the pair of functions (ψ_1, ψ_2) can be expressed in terms of a linear combination of the matrices $\boldsymbol{\tau}_1$, $\boldsymbol{\tau}_2$, $\boldsymbol{\tau}_3$. We define the isotopic-spin vector $\mathbf{t}$ which is analogous to the ordinary spin vector $\mathbf{s}$:

$$\mathbf{t} = \tfrac{1}{2}\vec{\boldsymbol{\tau}}, \tag{131.3}$$

where $\vec{\boldsymbol{\tau}}$ is a vector with the three components $\boldsymbol{\tau}_1$, $\boldsymbol{\tau}_2$, $\boldsymbol{\tau}_3$. It is clear that this is not a vector in ordinary space; it is defined in an abstract charge space or isotopic-spin space.

'Rotations' in this space denote linear transformations of ψ_1 and ψ_2 such that the base functions are taken to be different linear combinations of the proton and neutron states of the nucleons. For example, instead of ψ_1 and ψ_2 we can take new base functions $\phi_1 = (\psi_1 + \psi_2)/\sqrt{2}$ and $\phi_2 = (\psi_1 - \psi_2)/\sqrt{2}$, which are respectively symmetric and antisymmetric. The change from ψ_1, ψ_2 to ϕ_1, ϕ_2 is a rotation in isotopic space.

By using the operator of the isotopic spin $\mathbf{t}$ of the nucleon we can apply the theory of the ordinary spin. In particular, it is clear that the operators of t^2 and t_3 can be simultaneously brought to diagonal form, and have the eigenvalues

$$t^2 = \tfrac{1}{2}(\tfrac{1}{2} + 1) = \tfrac{3}{4}, \qquad t_3 = \pm\tfrac{1}{2}; \tag{131.4}$$

cf. (59.14) and (59.15).

It may be noted that t^2 is an invariant with respect to rotation in isotopic space. It is also evident that the rules for addition of isotopic-spin vectors for a system of nucleons are the same as for the ordinary spin. In particular, we have for the total isotopic-spin vector of a system of N nucleons

$$\mathbf{I} = \sum_{k=1}^{N} \mathbf{t}_k \tag{131.5}$$

(where k is the number of the nucleon) and Formulae (105.20) and (105.21):

$$I^2 = T(T+1), \qquad T = 0, 1, 2, 3, \ldots, \tag{131.6}$$

$$\text{or} \qquad T = \tfrac{1}{2}, \tfrac{3}{2}, \tfrac{5}{2}, \ldots,$$

$$I_3 = T_3, \qquad |T_3| \leqslant T. \tag{131.7}$$

It is also clear that the scalar products of isotopic spins, of the form

$$\mathbf{t}' \cdot \mathbf{t}'' = t_1' t_1'' + t_2' t_2'' + t_3' t_3'' \tag{131.8}$$

(where t_s', $s = 1, 2, 3$, are the components of the vector $\mathbf{t}'$, and t_s'' those of the vector $\mathbf{t}''$ of the second nucleon), are invariants in isotopic space, like $t^2 = \mathbf{t} \cdot \mathbf{t}$.

We may also give a formula for the charge Q of a system of N nucleons in terms of the isotopic spin:

$$Q = e(\tfrac{1}{2}N + T_3); \tag{131.9}$$

in particular, for a single nucleon

$$Q = \tfrac{1}{2}e(1 + \tau_3). \tag{131.9'}$$

An important physical fact is that the interaction of two nucleons is isotopically invariant (i.e. does not depend on possible rotations in isotopic space) and the total isotopic spin is conserved in the interaction.[3] This justifies the introduction of the isotopic spin of the nucleon as a new dynamical variable.

The strong interactions of nucleons will be invariant with respect to rotations, reflections and inversions of co-ordinates in ordinary space. If we consider only small nucleon velocities and take into account only their relative distance r, their ordinary spins $\mathbf{s}_1$, $\mathbf{s}_2$ and their isotopic spins $\mathbf{t}_1$, $\mathbf{t}_2$, the following invariants can be constructed: r, $\mathbf{s}_1 \cdot \mathbf{s}_2$, $\mathbf{t}_1 \cdot \mathbf{t}_2$, $(\mathbf{s}_1 \cdot \mathbf{r})(\mathbf{s}_2 \cdot \mathbf{r})$. These in turn can be expressed in terms of the total spin $\mathbf{S} = \mathbf{s}_1 + \mathbf{s}_2$ and the total isotopic spin $\mathbf{I} = \mathbf{t}_1 + \mathbf{t}_2$, by using the following invariants in place of those given above:

$$\mathbf{s}_1 \cdot \mathbf{s}_2 \rightarrow S^2, \tag{131.10}$$

$$\mathbf{t}_1 \cdot \mathbf{t}_2 \rightarrow I^2, \tag{131.10'}$$

$$(\mathbf{s}_1 \cdot \mathbf{r})(\mathbf{s}_2 \cdot \mathbf{r}) \rightarrow S_{12} = 6\frac{(\mathbf{S} \cdot \mathbf{r})^2}{r^2} - 2S^2. \tag{131.10''}$$

The last invariant is such that its mean value with respect to angle is zero. This is the choice usually made. The interaction represented by this term is evidently non-central and is called a *tensor interaction*. If the velocity dependence is taken into account, many other invariants can be constructed. Experiment shows, however, that, if the velocities are small compared with that of light, only the spin-orbit interaction invariant $\mathbf{L} \cdot \mathbf{S}$ is important, where $\mathbf{L}$ denotes the vector of the total orbital angular

[3] This has been confirmed by very precise and complete experimental work at the Joint Institute for Nuclear Research, Dubna; see [31].

momentum of the nucleons. Instead of **L** we can use the vector $\mathbf{J} = \mathbf{L} + \mathbf{S}$ giving the total angular momentum of the nucleons, and the corresponding invariant $\mathbf{J} \cdot \mathbf{S}$.

Thus we can write the interaction energy of two nucleons as

$$U(1,2) = A(r, S^2, I^2) + B(r, S^2, I^2) \cdot S_{12}(\mathbf{r} \cdot \mathbf{S}) + C(r, S^2, I^2)\, \mathbf{J} \cdot \mathbf{S}. \tag{131.11}$$

Very little is known concerning the functions A, B and C. In the meson theory of nuclear forces these functions must depend on distance as $(1/r)e^{-r/a}$ for $r > a = \hbar/m_\pi c = 1.4 \times 10^{-13}$ cm, the Compton wavelength of the π meson.

The above form of possible interaction of nucleons (131.11) is therefore more useful for the systematics of possible states of nucleons than for quantitative calculations of levels or the scattering matrix.

132. Systematics of states of a system of nucleons

The Hamiltonian of a system of strongly interacting nucleons is invariant not only with respect to rotation, reflection and inversion transformations but also with respect to interchange of nucleons. It therefore follows, exactly as in Sections 115 and 116, that the wave function must be either symmetric or antisymmetric with respect to interchange of any pair of nucleons. Since the nucleons have spin $\frac{1}{2}$, we must use antisymmetric functions, the Pauli principle and Fermi statistics.

Let us now consider states of two nucleons, and first examine the isotopic spin. It is evident that only four states are possible: $T = 0$, and $T = 1$ with $T_3 = 0$ or ± 1. The first state is antisymmetric in the isotopic variables, the remaining three are symmetric (as for the ordinary spin; see the theory of the helium atom, Section 121). When $T = 1$, since the Hamiltonian is independent of T_3, the energy of the three states with $T_3 = 0$ and ± 1 is the same.

This equality of energy, however, ceases to be true when the relatively weak electromagnetic interactions are taken into account. Owing to the different charges and magnetic moments of the proton and neutron, the coincident levels $T_3 = 0, \pm 1$ are in general split. These three states are therefore called a *charge triplet*, and $T = 1$ a triplet state. The state $T = 0$ is a *charge singlet*.

States are further distinguished by the value of the total spin S. Four states are again possible: the triplet $S = 1$, $S_z = 0, \pm 1$ and the singlet $S = 0$. The symmetry of the wave function in the space co-ordinates is determined by the symmetry in the charge and spin variables. Table 5 shows all possible symmetries for two nucleons.

TABLE 5

SYMMETRY OF WAVE FUNCTIONS IN A SYSTEM OF TWO NUCLEONS

$T = 0$ a		$T = 1$ s	
$S = 0$	$S = 1$	$S = 0$	$S = 1$
a	s	a	s
L odd	L even	L even	L odd
a	s	s	a

Here the symbol a denotes an antisymmetric function and s a symmetric function. It will be recalled (cf. Section 114) that for two particles the transposition P_{12} is equivalent to the inversion operation I_{12}, i.e. replacing the relative co-ordinates $\mathbf{x}$ by $-\mathbf{x}$. The parity of the state is in this case the same as that of the orbital quantum number L.

If the notation $S, P, D, F, \ldots$ for $L = 0, 1, 2, 3, \ldots$ is retained to denote nucleon states, together with the usual notation for the total angular momentum J and the multiplicity, the complete symbol for a state will be

$$^{2T+1,\,2S+1}L_J^{\pm}.$$

The first index shows the isotopic spin multiplicity $2T + 1$, the second the spin multiplicity $2S + 1$, the index $\pm$ the parity of the term, the suffix J its total angular momentum, and L (= $S, P, D, F, \ldots$) the orbital angular momentum. For a two-nucleon system the symbol $\pm$ is omitted, since it is determined by the parity of L; the isotopic-spin index T is also often omitted.

For two nucleons we now have the following system of possible states for $J = 0, 1, 2, \ldots$:

TABLE 6

STATES OF TWO NUCLEONS

J	$T=0$		$T=1$	
	$S=0$	$S=1$	$S=0$	$S=1$
0	—	—	$^1S_0$	$^3P_0$
1	$^1P_1$	$^3S_1$, $^3D_1$	—	$^3P_1$
2	—	$^3D_2$	—	$^3P_2$, $^3F_2$

133. Theory of the deuteron

Deuterium is an isotope of hydrogen, and its nucleus consists of a proton and a neutron. Its spin is known to be $S = 1$, and there is only one charge state, so that $T = 0$. Table 6 shows that the ground state of the deuteron must be $T = 0$, $^3S_1$ or $^3D_1$. We know that in the ground state the wave function must have the smallest possible number of nodes, and so the ground term of the deuteron must be taken as $^3S_1$. Owing to the presence of tensor forces, the orbital angular momentum is not conserved in the deuteron, and so an admixture of the state $^3D_1$ is possible. This in fact occurs and causes the existence of a quadrupole electric moment of the deuteron. The magnitude of this moment shows that the admixture of $^3D_1$ state is small (about 5%).

Thus experiment shows that the state with $T = 0$, $S = 1$ is the lowest state. No other bound states are known in the two-nucleon system.

Since the functions $A(r)$, $B(r)$ and $C(r)$ in the nucleon interaction energy (131.11) are unknown, we shall determine the wave function of the deuteron in the ground state in an indirect manner, using the experimental fact that the binding energy of the nucleons in the deuteron, $E_0 = -2.1 \times 10^6$ eV, is small compared with the rest energy of the π meson, $m_\pi c^2 = 140 \times 10^6$ eV.

For given T, S and I (or L) the interaction energy $U(r)$ of the nucleons (131.11) is simply a function of their relative distance r. (We neglect the tensor and spin-orbit interactions, which in the deuteron give only small corrections in the form of an admixture of the state $^3D_1$.) Then the equation for the radial function of the deuteron, $\psi(r) = u(r)/r$, is

$$-\frac{\hbar^2}{2\mu}\frac{d^2u}{dr^2} + U(r)u = E_0 u, \tag{133.1}$$

where $1/\mu = 1/m_p + 1/m_n$ and μ is the reduced mass of the proton and the neutron (cf. (108.4)); m_p is the mass of the proton and m_n that of the neutron. Since these are almost equal, $\mu = \frac{1}{2}m_p$.

Equation (133.1) can be written in the form

$$\frac{d^2u}{dr^2} - \kappa^2 u = \frac{2\mu}{\hbar^2} Uu, \tag{133.2}$$

where $\kappa^2 = -2\mu E_0/\hbar^2$, $1/\kappa = 4.31 \times 10^{-13}$ cm. This length determines the asymptotic behaviour of the deuteron function $\psi(r)$, since for $r \to \infty$ $(U \to 0)$ (133.2) gives $u \cong e^{\pm\kappa r}$, i.e. $\psi(r) = Ce^{-\kappa r}/r$. The function $U(r)$ decreases as $(1/r)\,e^{-r/a}$, where $a = \hbar/m_\pi c = 1.4 \times 10^{-13}$ cm, i.e. much more rapidly than $\psi(r)$. We can therefore assume that nuclear forces act only over a very short distance, and neglect them entirely for $r > a$. This is illustrated in Figure 91, which shows the curve of the potential energy $U(r)$ for a proton-neutron system.

We may now find the constant C by normalising $\psi(r)$ to unity:

$$4\pi \int_0^\infty \psi^2(r)\,r^2\,dr = 1, \tag{133.3}$$

which gives $C = \sqrt{(\kappa/2\pi)}$. Thus we have

$$\psi_0(r) = \sqrt{\frac{\kappa}{2\pi}}\frac{e^{-\kappa r}}{r}. \tag{133.4}$$

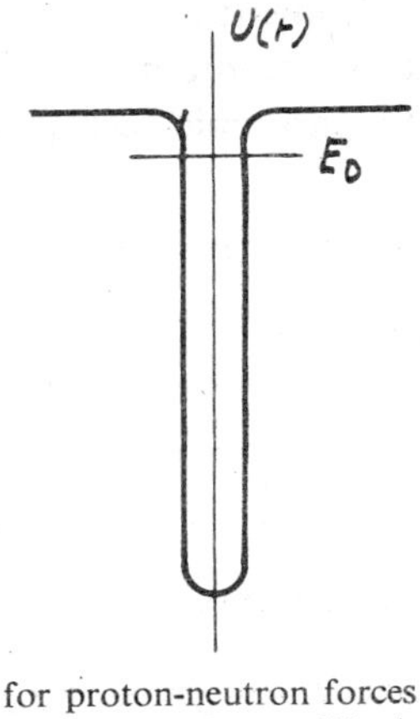

Fig. 91. Potential curve for proton-neutron forces in the deuteron. The level E_0 is at a depth of 2 MeV. The depth of the well is about 25 MeV, and its radius $a = 1.4 \times 10^{-13}$ cm.

This function can be used for calculations of the photodisintegration of the deuteron, and of certain nuclear reactions involving the deuteron in which large impact parameters are important.

The derivation of this function shows that it is not valid for distances r less than $a = 1.4 \times 10^{-13}$ cm.[4]

134. Scattering of nucleons

The problem of nucleon scattering is very extensive and includes such various phenomena as the scattering of slow thermal neutrons in hydrogen and collisions of fast nucleons up to very high energies, when in addition to elastic scattering there occur strong inelastic processes in which π mesons or other new particles are produced. Here we shall discuss two important cases.

A. SCATTERING OF SLOW NEUTRONS BY PROTONS

In this case only the S state is significant, since the wavelength $\lambda/2\pi$ is assumed to be much greater than the range of interaction a of nuclear forces. It will be recalled that the higher states are at distances greater than $\lambda/2\pi$ (cf. Figure 65). The table of possible states of two nucleons shows that both isotopic states ($T = 0$ and $T = 1$) are involved in *pn* scattering, and the possible S states differ in their total spin, being ${}^3S_1$ and ${}^1S_0$ respectively (triplet and singlet states). Thus we have to calculate the two phase shifts ${}^3\eta_1$ and ${}^1\eta_0$.

Let us first consider the triplet state. In this case the equation for the wave function $u(r)$ is the same as (133.2), but we now take $E > 0$ and put $2\mu E/\hbar^2 = k^2$. The asymptotic form of $u(r)$ for $r \gg a$ is

$$u(r) = C \sin(kr + {}^3\eta_1). \tag{134.1}$$

Assuming that the neutron energy E is small in comparison with the nucleon interaction energy $U(r)$, we can solve Equation (133.2) by neglecting the term E in comparison with U. This means that the logarithmic derivative u'/u for $r < a$ is almost independent of E when E is small; we denote it by α.

Since at the boundary $r = a$ the logarithmic derivatives must be equal, the solution (134.1) gives

$$(u'/u)_{r=a} = k \cot(ka + {}^3\eta_1) = \alpha. \tag{134.2}$$

Neglecting the small quantity ka, we have

$$\sin {}^3\eta_1 = -\frac{k}{\sqrt{(k^2 + \alpha^2)}}. \tag{134.3}$$

Hence, according to the general formula (80.16), the differential cross-section is

[4] The experiments of M. G. Meshcheryakov have shown that in collisions between fast nucleons and nuclei large numbers of deuterons are emitted from the nuclei. This indicates a very strong binding in the deuteron at short distances; see also [14].

$$\mathrm{d}q(\theta) = \frac{1}{k^2}\sin^2({}^3\eta_1)\,\mathrm{d}\Omega = \frac{2\pi}{k^2 + \alpha^2}\sin\theta\,\mathrm{d}\theta\,. \tag{134.4}$$

Let us now establish the relation between α and κ. According to Section 80 the phase shift η is $+ i\infty$ for a bound state. With ${}^3\eta_1 = + i\infty$ in (134.4) we find $k = + i\alpha$, and so the wave function $u(r)$ will behave as $e^{-\alpha r}$ for a bound state. Comparison with (133.4) shows that $\alpha = \kappa$. Thus Formula (134.4) can be written

$$\mathrm{d}^3q(\theta) = \frac{2\pi}{k^2 + \kappa^2}\sin\theta\,\mathrm{d}\theta\,, \tag{134.5}$$

the quantity κ being known from the binding energy of the deuteron. The total cross-section in the triplet state ($S = 1$) is

$${}^3q = 4\pi/(k^2 + \kappa^2)\,. \tag{134.6}$$

Similarly we obtain for the singlet state ($S = 0$)

$${}^1q = 4\pi/(k^2 + \kappa_1^2)\,, \tag{134.7}$$

where $1/\kappa_1$ is a new length determined by the interaction potential in the singlet state. Since this appears in the formula for the cross-section in exactly the same way as $\kappa_3 = \kappa$, the corresponding energy $E_1 = \hbar^2\kappa_1^2/2\mu > 0$ is called the energy of the *virtual level* of the deuteron.

B. ELASTIC SCATTERING OF NUCLEONS

In this section we shall consider the elastic scattering of nucleons by nucleons. It should be noted that mesons may be formed at nucleon energies $E_0 > 292$ MeV, but the contribution of this inelastic process is not large even for energies $E_0 \sim 400$ MeV.

Let us first consider an initial wave function Ψ^0 which represents the motion of the two nucleons before they are scattered. We shall examine only the relative motion, so that Ψ^0 depends only on the difference of the nucleon co-ordinates, $\mathbf{r} = \mathbf{r}_1 - \mathbf{r}_2$. Then

$$\Psi^0 = \psi^0(\mathbf{r})\,S^0(s_{z1}, s_{z2})\,T^0(t_{31}, t_{32})\,, \tag{134.8}$$

where S^0 is the spin function (Section 121) and T^0 the isotopic-spin function; s_{z1}, s_{z2} are the components of the nucleon spins along the axis OZ, and t_{31}, t_{32} the third components of the isotopic spins of the nucleons. According to (131.4) $t_3 = + \frac{1}{2}$ for the proton and $- \frac{1}{2}$ for the neutron. The structure of the function $T(t_{31}, t_{32})$ is exactly the same as that of the function $S(s_{z1}, s_{z2})$. We now regard the two nucleons as identical particles obeying the Pauli principle; the function Ψ^0 must therefore be antisymmetric with respect to interchange of the nucleons. This interchange converts $\mathbf{r}$ to $-\mathbf{r}$, so that the symmetry of $\psi^0(\mathbf{r})$ is the same as its parity. The symmetry of the functions $\psi^0(\mathbf{r})$, S^0 and T^0 must be so chosen that the whole function Ψ^0 is antisymmetric. If the co-ordinate function $\psi^0(\mathbf{r})$ represents an initial plane wave with momentum $\mathbf{p} = \hbar\mathbf{k}$, the function $e^{i\mathbf{k}\cdot\mathbf{r}}$ (cf. (80.5)) must be replaced by the symmetrised function

$$\psi^0_{a,s}(\mathbf{r}) = e^{i\mathbf{k}\cdot\mathbf{r}} \pm e^{-i\mathbf{k}\cdot\mathbf{r}}. \tag{134.9}$$

This symmetrisation expresses the fact that we no longer distinguish which of the nucleons 1 and 2 is the target and which is being scattered.

If now we denote by $A(\theta)$ the amplitude of the wave scattered at an angle θ from the direction of the original wave $e^{i\mathbf{k}\cdot\mathbf{r}}$, it is evident that the wave scattered from $e^{-i\mathbf{k}\cdot\mathbf{r}}$ will be $A(\pi - \theta)$: replacement of $\mathbf{r}$ by $-\mathbf{r}$ changes θ to $\pi - \theta$. Thus for identical particles, unlike (80.5), the whole wave (incident plus scattered) for large r has the form

$$\psi_{a,s}(\mathbf{r}) = e^{i\mathbf{k}\cdot\mathbf{r}} \pm e^{-i\mathbf{k}\cdot\mathbf{r}} + \frac{e^{i\mathbf{k}\cdot\mathbf{r}}}{r}\left[A(\theta) \pm A(\pi - \theta)\right]. \tag{134.10}$$

The corresponding differential cross-section $q(\theta)$ is

$$q(\theta) = \left[A(\theta) \pm A(\pi - \theta)\right]^2. \tag{134.11}$$

In (134.10) we have not written out the spin dependence of the function $\psi_{a,s}$ and the amplitudes A. If this is included, we have

$$\begin{aligned}\Psi(\mathbf{r}, s_{z1}, s_{z2}, t_{31}, t_{32}) &= \psi^0_{a,s}(\mathbf{r})\, S^0(s_{z1}, s_{z2})\, T^0(t_{31}, t_{32}) + \\ &+ \frac{e^{ikr}}{r}\left[A(\theta, s_{z1}, s_{z2}, t_{31}, t_{32}) \pm A(\pi - \theta, s_{z1}, s_{z2}, t_{31}, t_{32})\right].\end{aligned} \tag{134.12}$$

Let us now consider some particular cases, and take first the scattering of a proton by a proton (pp scattering). In this case $T = 1$, $T_3 = +1$, $S = 0$ or 1. The spin function $S^0(s_{z1}, s_{z2})$ coincides with one of the functions $S(s_{z1}, s_{z2})$ (121.13)–(121.14″), depending on the value of the spin S and its component along the axis OZ. The function T^0 for $T = 1$ and $T_3 = +1$ is

$$T^0(t_{31}, t_{32}) = S'_s(t_{31}, t_{32}), \tag{134.13}$$

where S'_s is the function (121.14) with s_{z1} replaced by t_{31} and s_{z2} by t_{32}.

The total proton scattering cross-section is given by the squared modulus of the amplitude of the outgoing wave e^{ikr}/r in (134.12). Let this cross-section for the triplet state $S = 1$ be denoted by

$$^3q(\theta) = |^3A\theta) - {}^3A(\pi - \theta)|^2, \tag{134.14}$$

where the spin variables are again omitted. The cross-sections for all three spin orientations $S_z = 0, \pm 1$ are clearly equal. The cross-section in the singlet state is

$$^1q(\theta) = |^1A(\theta) + {}^1A(\pi - \theta)|^2. \tag{134.15}$$

If all spin orientations in the original beam are equally probable (an unpolarised beam), each spin state will have a probability $\frac{1}{4}$, and the differential scattering cross-section for unpolarised protons is therefore

$$q_{pp}(\theta) = \tfrac{3}{4}\,{}^3q(\theta) + \tfrac{1}{4}\,{}^1q(\theta). \tag{134.16}$$

When electromagnetic interactions (interaction of charges and magnetic moments) are neglected, the operator T_3 does not appear in the Hamiltonian. In this approximation, therefore, the nucleon interaction must be *isotopically invariant*: that is, it can depend only on the value of the total isotopic spin, but not on its components.

For collisions of two neutrons (*nn* scattering) we have $T = 1$, $T_3 = -1$. Hence it follows that the scattering cross-section for two neutrons is equal to that for two protons:

$$q_{nn}(\theta) = q_{pp}(\theta). \tag{134.17}$$

The situation is somewhat more complex for a collision between a proton and a neutron (*pn* scattering). Here we have a superposition of two states: $T = 1$; $T_3 = 0$ and $T = 0$, $T_3 = 0$. For, if we consider the original wave in (134.8), we see that $T^0(t_{31}, t_{32})$ can be equal either to $S_s'''(t_{31}, t_{32})$ for $T = 1$, $T_3 = 0$ (cf. (121.14″)) or to $S_a(t_{31}, t_{32})$ for $T = 0$, $T_3 = 0$ (cf. (121.13)):

$$T^0(t_{31}, t_{32}) = \frac{1}{\sqrt{2}}\{S_{+\frac{1}{2}}(t_{31})S_{-\frac{1}{2}}(t_{32}) \pm S_{+\frac{1}{2}}(t_{32})S_{-\frac{1}{2}}(t_{31})\}, \tag{134.18}$$

the suffix $+\frac{1}{2}$ denoting a proton and $-\frac{1}{2}$ a neutron. The two possible states are both superpositions of proton and neutron states.

In order to obtain proton and neutron states, a superposition of states with $T = 1$ and $T = 0$ must be taken. For example, for the singlet state $S = 0$, the initial wave must be written as

$$\begin{aligned}\Psi^0 &= \frac{1}{\sqrt{2}}\psi_a^0(\mathbf{r})\overbrace{S_a(s_{z1}, s_{z2})S_a(t_{31}, t_{32})}^{T=0} + \\ &+ \frac{1}{\sqrt{2}}\psi_s^0(\mathbf{r})\overbrace{S_a(s_{z1}, s_{z2})S_s'''(t_{31}, t_{32})}^{T=1} \\ &= e^{i\mathbf{k}\cdot(\mathbf{r}_1-\mathbf{r}_2)}S_{+\frac{1}{2}}(t_{31})S_{-\frac{1}{2}}(t_{32})S_a(s_{z1}, s_{z2}) + \\ &+ e^{i\mathbf{k}\cdot(\mathbf{r}_2-\mathbf{r}_1)}S_{+\frac{1}{2}}(t_{32})S_{-\frac{1}{2}}(t_{31})S_a(s_{z1}, s_{z2}),\end{aligned} \tag{134.19}$$

since this superposition represents a wave such that a particle with momentum $+\mathbf{k}$ has isotopic spin $t_3 = +\frac{1}{2}$ (i.e. is a proton), while a particle with momentum $-\mathbf{k}$ has isotopic spin $t_3 = -\frac{1}{2}$ (i.e. is a neutron). This is the correct choice of an initial wave representing a proton with momentum $+\mathbf{k}$ or a neutron with momentum $-\mathbf{k}$. The numbering of particles 1 and 2 is without significance.

Owing to the linearity of the equations, the amplitude $F_{pn}(\theta)$ of the scattered *pn* wave will also be a superposition of amplitudes $F_1(\theta) = A_1(\theta) + A_1(\pi - \theta)$ and $F_0(\theta) = A_0(\theta) - A_0(\pi - \theta)$ for states $T = 1$ and $T = 0$ respectively, with the same coefficients as in the superposition of the original waves $(+1/\sqrt{2})$, i.e.

$$F_{pn}(\theta) = \frac{1}{\sqrt{2}}F_1(\theta) + \frac{1}{\sqrt{2}}F_0(\theta). \tag{134.20}$$

The differential cross-section for *pn* scattering is therefore

$$q_{pn}(\theta) = \tfrac{1}{2}q_1(\theta) + \tfrac{1}{2}q_0(\theta) + Re\left[F_0(\theta)F_1^*(\theta)\right]. \tag{134.21}$$

Let us now consider the sum $q_{pn}(\theta) + q_{pn}(\pi - \theta)$. It is evident that this sum gives the cross-section for some scattered particle (p or n) to be observed, since if the proton is scattered at an angle θ, the neutron is scattered at an angle $\pi - \theta$.

On replacing θ by $\pi - \theta$, however, we have $F_1(\pi - \theta) = F_1(\theta)$, since for $T = 1$ the co-ordinate function is symmetric, and $F_0(\pi - \theta) = - F_0(\theta)$, since for $T = 0$ it is antisymmetric. Hence

$$q_{pn}(\theta) + q_{pn}(\pi - \theta) = q_1(\theta) + q_0(\theta). \tag{134.22}$$

But $q_1(\theta) = q_{pp}(\theta) = q_{nn}(\theta)$. Thus, by measuring $q_{pn}(\theta)$ and $q_{pp}(\theta)$, we can calculate the scattering cross-section $q_0(\theta)$ in the isotopic state $T = 0$.

Figure 92 shows the dependence of $q_0(\theta)$ and $q_1(\theta)$ on the angle θ for energies 380–400 MeV [30]. It is seen that the interaction is entirely different in the states $T = 0$ and $T = 1$. The total cross-sections q_0 and q_1 are also quite different: q_1 is practically constant at high energies, but q_0 decreases with increasing energy.

135. Polarisation in the scattering of particles which have spin

We have seen that nuclear interactions depend on the spin of the particles. This has the consequence that in nucleon-nucleon or nucleon-nucleus collisions the amplitude

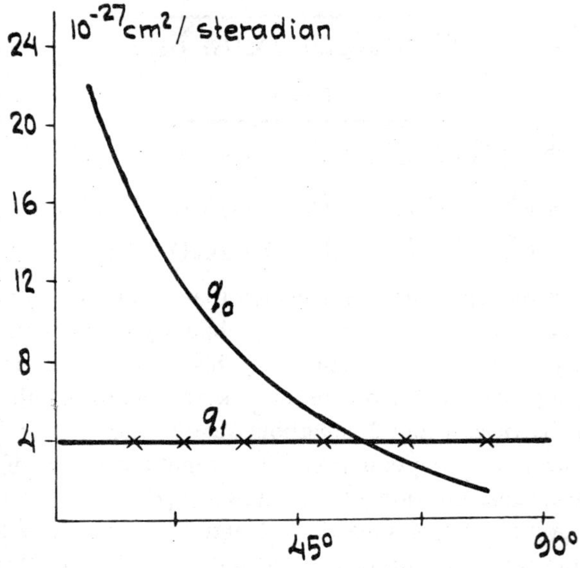

Fig. 92. Angle dependence of elastic scattering of nucleons in different isotopic states: $T = 0$ ($q_0(\theta)$) and $T =$ ($q_1(\theta)$) for nucleon energy 400 MeV. For $T = 1$ the scattering is isotropic.

of the scattered wave is different for different orientations of the spin of the scattered particles, i.e. there is spin polarisation. The original particles are usually unpolarised. The initial state is therefore usually not pure but mixed; it is an assembly of states with different spin orientations each having probability P_α. Such a beam may be more conveniently described by a density matrix $\boldsymbol{\rho}$ (see Section 45) than by a wave function.

Let us consider the polarisation of a particle with spin $\frac{1}{2}$, taking ϕ_1 and ϕ_2 as the basic spin functions. Let two spin states ψ_1 and ψ_2 be mixed in the original beam with probabilities P_1 and P_2. These states can be represented as a linear combination of the basic states ϕ_1 and ϕ_2:

$$\psi_i = \sum_{k=1}^{2} c_{ik}\phi_k, \qquad i = 1, 2. \tag{135.1}$$

According to (45.7) the density matrix elements are given by

$$\rho_{ik} = \sum_{n=1}^{2} P_n c_{ni} c_{nk}^{*}. \tag{135.2}$$

The mean value of any spin operator $\boldsymbol{O}$, according to the general formula (45.5), is

$$\bar{O} = \mathrm{Tr}(\boldsymbol{\rho}\boldsymbol{O}). \tag{135.3}$$

Since $\boldsymbol{\rho}$ is a two-by-two matrix, it can be represented as a linear combination of Pauli matrices:

$$\boldsymbol{\rho} = A\boldsymbol{\delta} + \mathbf{B}\cdot\vec{\boldsymbol{\sigma}}. \tag{135.4}$$

The coefficients A and $\mathbf{B}$ can now be expressed in terms of the mean value of the spin of the particle $\mathbf{S} = \frac{1}{2}\hbar\boldsymbol{\sigma}$, or more conveniently in terms of the mean value of $\boldsymbol{\sigma}$. To do this, we note that

$$\mathrm{Tr}\,\vec{\boldsymbol{\sigma}} = 0, \qquad \mathrm{Tr}\,\boldsymbol{\sigma}_x^2 = 2. \tag{135.5}$$

Hence

$$\bar{\boldsymbol{\sigma}}_x = \mathrm{Tr}(\boldsymbol{\rho}\boldsymbol{\sigma}_x) = A\,\mathrm{Tr}\,\boldsymbol{\sigma}_x + \mathrm{Tr}\,\boldsymbol{\sigma}_x(\mathbf{B}\cdot\vec{\boldsymbol{\sigma}}) = 2B_x,$$

i.e. $\bar{\boldsymbol{\sigma}} = 2\mathbf{B}$. The normalisation condition gives $\mathrm{Tr}\,\boldsymbol{\rho} = 2A = 1$, i.e. $A = \frac{1}{2}$. Thus the matrix

$$\boldsymbol{\rho} = \tfrac{1}{2}(\boldsymbol{\delta} + \bar{\boldsymbol{\sigma}}\cdot\vec{\boldsymbol{\sigma}}) \tag{135.6}$$

describes the state of polarisation in the original beam. It is seen to be expressed directly in terms of the spin vector $\vec{\boldsymbol{\sigma}}$ and its mean value $\bar{\boldsymbol{\sigma}}$. For an unpolarised beam $\boldsymbol{\rho} = \frac{1}{2}\boldsymbol{\delta}$. After scattering, the spin states are altered, and instead of a mixture of the states ψ_1 and ψ_2 we have a mixture of some new states ψ_1' and ψ_2', which can be expressed in terms of the former states by means of the scattering matrix $S_{\alpha\beta}(\theta)$:

$$\psi_\alpha' = S_{\alpha\beta}(\theta)\,\psi_\beta. \tag{135.7}$$

The elements of this matrix depend on the angle θ and the particle momentum $\mathbf{k}$. For $\theta \neq 0$ the scattering matrix $S(\theta)$ is proportional to the scattering amplitude $A(\theta)$.

According to the rules of matrix transformation, the new density matrix $\boldsymbol{\rho}'$ is

$$\boldsymbol{\rho}' = \boldsymbol{S}^{+}\boldsymbol{\rho}\boldsymbol{S}, \tag{135.8}$$

where $\boldsymbol{S}^{+}$ is the adjoint matrix to $\boldsymbol{S}$ (see Section 80). If the original beam was unpolarised, then $\boldsymbol{\rho} = \frac{1}{2}\boldsymbol{\delta}$ and

$$\boldsymbol{\rho}' = \tfrac{1}{2}\boldsymbol{S}^{+}\boldsymbol{S}. \tag{135.9}$$

This quantity is not normalised to unity, since $\boldsymbol{S}$ contains other variables (θ, k, . . .) besides the spin variables. The mean value after scattering must therefore be calculated from the formula

$$\bar{\boldsymbol{\sigma}}' = \mathrm{Tr}\,\boldsymbol{\rho}'\vec{\boldsymbol{\sigma}}/\mathrm{Tr}\,\boldsymbol{\rho}'. \tag{135.10}$$

This quantity is called the polarisation **P**:

$$\mathbf{P} = \bar{\boldsymbol{\sigma}}'. \tag{135.11}$$

The actual value of **P** depends on the scattering matrix $\boldsymbol{S}$, or equivalently, on the scattering amplitude A. It can be shown, however, that the polarisation vector **P** is perpendicular to the scattering plane formed by the wave vectors **k** and **k**′ before and after scattering. **P** is the mean value of $\vec{\boldsymbol{\sigma}}'$, and is therefore a pseudovector. The right-hand side of (135.10) is thus also a pseudovector. The only pseudovector which we can construct from quantities which appear in the scattering amplitude is the vector product $\mathbf{k} \times \mathbf{k}'$. We can therefore assert that

$$\mathbf{P} = \alpha\mathbf{k} \times \mathbf{k}', \tag{135.12}$$

where α is some proportionality factor which depends on angles and energy. Hence we see that for small angles the polarisation is zero. If **k** is taken along the axis OZ, the polarisation changes sign when the azimuthal angle of scattering ϕ is replaced by $\pi - \phi$ (in particular, scattering to the right or to the left).

The existence of polarisation is confirmed by experiment.[5] In proton-proton scattering at an energy of 600 MeV the polarisation reaches 40 %.

136. The application of quantum mechanics to the systematics of elementary particles

The table in Section 3 lists the considerable number of elementary particles now known.

An important property of the majority of elementary particles is that they are unstable, with short lifetimes, as shown in the last column of the table and decay into other elementary particles.

Among the transformations of these particles the interaction of particles with their antiparticles (electron-positron, proton-antiproton, etc.) has a special place. This process is called *annihilation*. In it, the particle and antiparticle as such disappear, and are converted into mesons, photons, electrons and neutrinos. These interaction processes cannot be treated by non-relativistic quantum mechanics, in which, as in classical mechanics, the number of particles is conserved. The theory of elementary particles therefore involves quantum field theory and relativistic quantum mechanics.

[5] See [31].

Nevertheless, the basic principles of quantum mechanics are sufficient to derive the systematics of elementary particles.

The elementary particles may first of all be divided by mass into heavy particles (baryons), medium particles (mesons) and light particles (leptons). The baryons include the nucleons (proton and neutron) and the even heavier hyperons: Λ_0 (lambda particle), Σ (sigma particle) and the cascade hyperon Ξ (xi particle). All the hyperons have a spin of $\frac{1}{2}$ and are therefore fermions (Section 116). The decay of hyperons leads finally to nucleons. The hyperons may therefore be regarded as excited states of the nucleon, the mass being a measure of the excitation. Accordingly, the hyperons are shown in Figure 93 as horizontal levels at the corresponding mass (in units of the

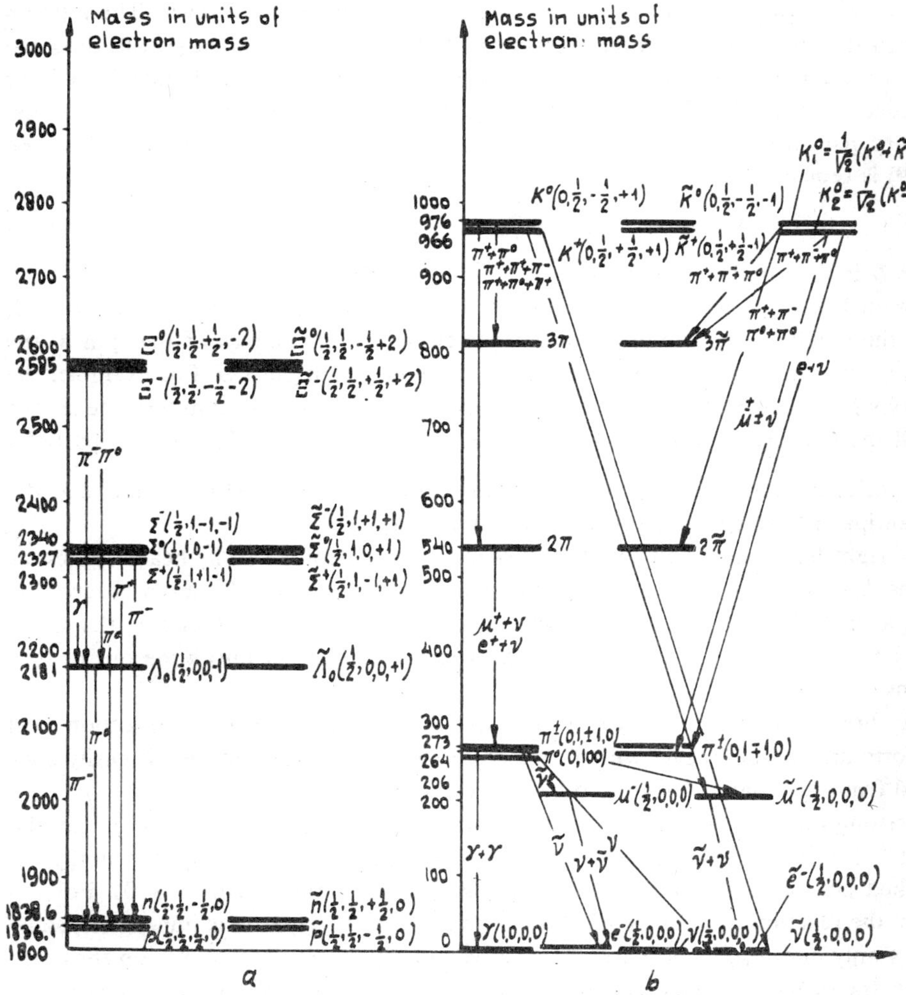

Fig. 93. Diagram of elementary particles and their decays: (a) baryons (nucleon levels), (b) mesons and leptons.

electron mass). The vertical lines indicate quantum transitions, accompanied by the emission of π mesons or photons, to lower levels of excitation (lighter hyperons).

The table shows that the nucleon levels consist of a group of lines representing particles with different charges and almost the same mass. A common value of the isotopic spin and different values of its components can be assigned to each group of particles, i.e. each such group is an isotopic multiplet (Section 131). The proton and neutron (the ground state) form the doublet $T = \frac{1}{2}$, $T_3 = \pm \frac{1}{2}$. The Λ_0 hyperon, a neutral particle, has no close neighbours, and has isotopic spin $T = 0$, $T_3 = 0$. The Σ hyperon has three charge states (0, $\pm e$), and accordingly its isotopic spin is $T = 1$, $T_3 = 0, \pm 1$. Finally the Ξ hyperon is a doublet (charge 0, $-e$), corresponding to isotopic spin $T = \frac{1}{2}$, $T_3 = \pm \frac{1}{2}$.

This simple picture of the hyperons encounters difficulties, however. The relation between the charge and isotopic spin of particles given by Formula (131.9) does not hold for excited states. To resolve this problem, Gell-Mann and Nishijima have proposed to generalise Formula (131.9) by introducing a new characteristic of elementary particles, the *strangeness*, expressed by a new quantum number S. Then (131.9) is replaced by

$$Q = e(\tfrac{1}{2}N + T_3 + \tfrac{1}{2}S), \tag{136.1}$$

where N is the number of baryons.

For nucleons $S = 0$, for Λ_0 and Σ hyperons $S = -1$, and for Ξ hyperons $S = -2$. Thus the complete description of a particle comprises the values of the spin σ, the isotopic spin and its component (T and T_3) and the strangeness S. For example, the Σ^- hyperon has $\sigma = \frac{1}{2}$, $T = 1$, $T_3 = -1$, $S = -1$. These four numbers are shown for all the particles in Figure 93 and in the table in Section 3. For the antiparticles the signs of T_3 and S are reversed. The antiparticles are denoted by a tilde, e.g. $\tilde{\Lambda}_0$. The antiparticle states are shown separately in Figure 93.

The right-hand side of Figure 93 shows mesons and leptons.[6] The heaviest mesons are the K mesons. According to existing data the spin σ of the K meson is zero, the isotopic spin $T = \frac{1}{2}$, $T_3 = \pm \frac{1}{2}$ (an isotopic doublet with charges 0 and e), the strangeness $S = 1$. The diagram shows that K mesons decay either into π mesons or into leptons.

The three π mesons (π^0 and $\pi^{\pm}$) have spin $\sigma = 0$. They are bosons (Section 116) and form an isotopic triplet with $T = 1$, $T_3 = 0, \pm 1$. The strangeness is easily confirmed from the charge formula to be $S = 0$.

In strong interactions of mesons and baryons the strangeness is conserved, i.e. the change of S in such reactions is zero. This is expressed by the experimentally established law of pair production of 'strange' particles (those with $S \neq 0$). For example, the reaction $\pi^- + p \rightarrow \Lambda_0 + \pi^0$ to produce a Λ_0 hyperon is impossible, since it gives only one strange particle, Λ_0. The reaction $\pi^- + p \rightarrow \Lambda_0 + K^0$, on the other hand is the ordinary reaction for producing Λ_0 hyperons and K^0 mesons. In decays of

[6] The $K_1{}^0$ and $K_2{}^0$ particles shown are superpositions of the states K^0 and $\tilde{K}^0$.

strange particles (called weak interactions) strangeness is not conserved, e.g. $\Delta S \neq 0$ in the decay $\Lambda_0 \rightarrow p + \pi^-$.

Finally, the leptons include the electron e (and its antiparticle the positron $\tilde{e}$), the neutrino v (and the antineutrino $\tilde{v}$) and the μ^- meson (and $\tilde{\mu}^-$). There is as yet no final systematics of leptons and it is not clear how the concepts of isotopic spin and strangeness can be applied to them.

A special place is occupied by the photon γ, whose spin $\sigma = 1$. It is possible that leptons and photons are of especially fundamental significance among the elementary particles, since all unstable particles ultimately decay either into electrons and neutrinos or into photons.

CHAPTER XXV

CONCLUSION

137. The formalism of quantum mechanics

In this exposition of the fundamental ideas of quantum mechanics no attempt has been made to preserve a strict sequence of deduction. The orderly logic of a deductive account would inevitably involve some degree of abstractness which would obscure the experimental foundation of any given general result. However, to conclude the book it is appropriate to summarise briefly the fundamental ideas and problems of quantum mechanics.

Quantum mechanics deals with statistical ensembles of microparticles, and solves three main problems: (1) to determine the possible values (spectrum of values) of physical quantities; (2) to calculate the probability of any particular value of these quantities in the ensemble of microparticles; (3) to examine the variation of an ensemble with time (the motion of microparticles).

In quantum mechanics the *wave function* ψ represents the fact that a microparticle belongs to a particular ensemble. It is a function of a *complete set* of quantities, which we denote[1] by x. The number of quantities in a complete set is determined by the nature of the system and is equal to the number of its degrees of freedom. The choice of the set of quantities which appear as arguments of the wave function is said to determine a particular *representation*.

The wave function also has a suffix (often omitted), such as n in $\psi_n(x)$, indicating another set which determines the wave function itself.

A statistical ensemble described by a particular wave function is called a *pure ensemble*; one which does not have a particular wave function is called a *mixed ensemble*, and is described by a *density matrix*.

The fundamental property of pure quantum ensembles is given by the principle of superposition: if two possible states are described by wave functions ψ_1 and ψ_2, there exists a third state described by the wave function

$$\psi = c_1\psi_1 + c_2\psi_2, \tag{I}$$

where c_1 and c_2 are arbitrary amplitudes.

All relations between physical quantities are expressed in quantum mechanics in

[1] Here x does not necessarily denote one or more co-ordinates. We use it to signify any group of variables, discrete or continuous, which form a complete set.

terms of linear self-adjoint operators so that to every real physical quantity L there corresponds a linear self-adjoint operator $\boldsymbol{L}$. The representation of quantities by means of operators is related to measurable quantities by a formula giving the mean value $\bar{L}$ of a quantity in the state ψ. This formula is

$$\bar{L} = (\psi, \boldsymbol{L}\psi) \tag{II}$$

with the normalisation condition[2]

$$1 = (\psi, \psi).$$

This definition of the mean value enables us to find the spectrum of the quantity L, i.e. its possible values. For this purpose we seek states in which the quantity L has only a single definite value, i.e. states in which $\overline{\Delta L^2} = 0$. This requirement leads to an equation for the eigenfunction f the operator $\boldsymbol{L}$ (cf. Section 20):

$$\boldsymbol{L}\psi_L(x) = L\psi_L(x). \tag{III}$$

Hence we find the spectrum (continuous or discrete) of L and the corresponding eigenstates $\psi_L(x)$. It is assumed that the eigenvalues of the operator $\boldsymbol{L}$ are those values of the quantity L which are experimentally observed.

Since the eigenfunctions form an orthogonal set, any wave function $\psi(x)$ can be expanded as a series of eigenfunctions $\psi_L(x)$:

$$\psi(x) = \sum_L c(L)\psi_L(x), \tag{137.1}$$

where

$$c(L) = (\psi_L, \psi), \tag{137.2}$$

and the sum is to be regarded as an integral $\int \mathrm{d}L \ldots$ if the spectrum of L is continuous.

This spectral resolution is in fact performed in an apparatus which resolves the ensemble $\psi(x)$ into sub-ensembles $\psi_L(x)$, and in particular in a measuring apparatus which measures the quantity L.

The probability of finding a value L in an ensemble described by a wave function $\psi(x)$ is $|c(L)|^2$ (for a continuous spectrum, $|c(L)|^2$ is the probability density); $c(L)$ is also the wave function of the ensemble in the L representation. That is, $c(L)$ and $\psi(x)$ represent the same ensemble.

A fourth fundamental point in quantum mechanics relates to the variation of ensembles with time. The variation with time of the wave function describing an ensemble is given by Schrödinger's equation

$$i\hbar\partial\psi/\partial t = \boldsymbol{H}\psi, \tag{IV}$$

[2] The symbol $(u, \boldsymbol{L}v)$ denotes the 'scalar product' of u and $\boldsymbol{L}v$, which for continuous variables is the integral

$$(u, \boldsymbol{L}v) = \int u^* \cdot \boldsymbol{L}v \cdot \mathrm{d}x,$$

and for discrete variables is the sum

$$(u, \boldsymbol{L}v) = \sum_n \sum_m u_n^* L_{nm} v_m.$$

where the operator $\boldsymbol{H}$ is the Hamiltonian of the system and depends only on the nature of the system and the kinds of external field acting on it. The operator $\boldsymbol{H}$ is the total-energy operator if the external fields are independent of time. Usually

$$\boldsymbol{H} = \boldsymbol{T} + U, \tag{137.3}$$

where $\boldsymbol{T}$ is the kinetic-energy operator and U an operator representing the potential energy or force function.

The operator $\boldsymbol{T}$ is a function of the momentum operator $\boldsymbol{P}$. Experiment shows that, in the absence of magnetic forces,

$$\boldsymbol{T} = \sum_k \frac{\boldsymbol{P}_k^2}{2m_k}, \tag{137.4}$$

where $\boldsymbol{P}_k$ is the momentum of the kth particle and m_k its mass. When a magnetic field is present, $\boldsymbol{P}_k$ must be replaced by

$$\boldsymbol{\Pi}_k = \boldsymbol{P}_k - \frac{e}{c}\mathbf{A}_k, \tag{137.5}$$

where $\mathbf{A}_k$ is the vector potential at the position of the kth particle.

From Schrödinger's equation (IV) and the definition of the mean value (II) it follows that

$$\frac{\mathrm{d}\overline{L}}{\mathrm{d}t} = \left(\psi, \frac{\partial \boldsymbol{L}}{\partial t}\psi\right) + (\psi, [\boldsymbol{H}, \boldsymbol{L}]\psi). \tag{137.6}$$

The operator $\mathrm{d}\boldsymbol{L}/\mathrm{d}t$ which represents the time derivative of the quantity L is therefore

$$\frac{\mathrm{d}\boldsymbol{L}}{\mathrm{d}t} = \frac{\partial \boldsymbol{L}}{\partial t} + [\boldsymbol{H}, \boldsymbol{L}], \tag{137.7}$$

where $[\boldsymbol{H}, \boldsymbol{L}] = (i/\hbar)(\boldsymbol{HL} - \boldsymbol{LH})$ is the quantum Poisson bracket.

The integrals of the motion are such that

$$\mathrm{d}\boldsymbol{L}/\mathrm{d}t = 0. \tag{137.8}$$

In the absence of external forces the most important integrals of the motion are the energy, the total momentum of the system

$$\boldsymbol{P} = \sum_k \boldsymbol{P}_k = -i\hbar \sum_k \nabla_k \tag{137.9}$$

and the angular momentum

$$\boldsymbol{M} = \sum_k \mathbf{r}_k \times \boldsymbol{P}_k + \sum_k \boldsymbol{S}_k, \tag{137.10}$$

where $\boldsymbol{S}_k$ is the spin angular momentum of the kth particle.

The form of the operator $\boldsymbol{P}$ can be determined from the very fact that it represents a quantity which is an integral of the motion, i.e. commutes with the operator $\boldsymbol{H}$ in the absence of external forces. Other more complex operators, whose physical significance may be highly specialised, can be constructed from the operators $\boldsymbol{P}_k$ and $\mathbf{r}_k$. Thus the form of the principal operators is automatically determined if the form of the Hamiltonian (or Schrödinger's equation) is postulated.

The last of the fundamental ideas of quantum mechanics relates to systems of identical particles, and is the *principle of indistinguishability*, according to which the interchange of any pair of identical particles (k, j) does not lead to a physically different state. Mathematically this is expressed as a condition on the wave functions:

$$\boldsymbol{P}_{kj}\Psi = \lambda\Psi, \tag{V}$$

where $\lambda = \pm 1$ is an eigenvalue of the interchange operator $\boldsymbol{P}_{kj}$. This condition leads to a division of states into two classes:

$$\Psi = \Psi_s \,(\text{symmetric}), \tag{137.11}$$

$$\Psi = \Psi_a \,(\text{antisymmetric}). \tag{137.11'}$$

It also follows from Schrödinger's equation that the symmetry cannot alter in the course of time. Hence the nature of particles alone determines whether they belong to the s type or the a type. Particles whose states are described by antisymmetric wave functions Ψ_a are fermions, and obey the Pauli principle, which is a consequence of the properties of an ensemble described by antisymmetric wave functions. Particles whose states are described by symmetric wave functions Ψ_s are called bosons.

Thus we see that quantum mechanics is based on five fundamental ideas: the principle of superposition of states (I), the definition of the mean value (II), the interpretation of eigenvalues as the only possible values (III), Schrödinger's equation (IV), and the principle of indistinguishability of identical particles (V). The physical foundations of these ideas have been discussed in detail in the relevant chapters.

138. The limits of applicability of quantum mechanics

The limits of applicability of a physical theory can be stated with complete rigour and precision only in terms of a more general theory which comprises the other as a particular or limiting case. At the present time there is no theory of microphenomena of greater breadth or depth than quantum mechanics. Thus the limits of quantum mechanics can be outlined only very provisionally. We can say for certain only that quantum mechanics is not applicable to systems consisting of particles moving with velocities comparable with that of light c, i.e. in the relativistic range.

Quantum mechanics is the mechanics of systems with a finite number of degrees of freedom. In this respect it is the analogue of the classical mechanics of systems of point masses. If the velocities of the particles become comparable with that of light, we can no longer say that the system has a finite number of degrees of freedom. For in this case the finite velocity of propagation of electromagnetic fields must be taken into account. If in a time Δt the distance r_{jk} between two particles varies by Δr_{jk}, and the relative velocity $\Delta r_{jk}/\Delta t$ of the particles is close to that of light, approximately the same time is needed for the propagation of the electromagnetic field through a distance Δr_{jk}. Hence we must consider not only the particles but also the electromagnetic field which they produce and which in turn acts on them. In other words, the system includes not only all the particles ($3N$ degrees of freedom for N spinless particles and $4N$ for N particles with spin), but also the electromagnetic field, the state of which has an infinite number of degrees of freedom.

This electromagnetic field, in a consistent theory, should also receive a quantum treatment, since it is known that momentum and energy of the field are transferred by photons.

When the photon or particle energy exceeds the rest energy m_0c^2 of the particles, production or annihilation of particles is possible. For example, a photon γ with energy $\hbar\omega \geqslant 2m_0c^2$ can be converted into a pair of particles: an electron (e^-, m_0) and a positron (e^+, m_0). Conversely, a positron and electron can be converted into a photon.[3]

These conversion processes can be expressed by

$$\gamma \rightleftarrows e^+ + e^- . \tag{138.1}$$

In this example particles are created and destroyed by *electromagnetic interaction.*

Another type of process whereby particles are created is *strong interaction.* The reaction

$$\pi^- + p \to p + K + \tilde{K} \tag{138.2}$$

is an example. Here a π^- meson collides with a proton and a pair of K mesons are produced.

Elementary particles are also converted into one another in *weak interactions,* which cause radioactive decay. For example, the neutron is spontaneously transformed into a proton, emitting an electron and a neutrino:

$$n \to p + e^- + \nu . \tag{138.3}$$

In positron decay of nuclei the opposite reaction can occur:

$$p \to n + e^+ + \tilde{\nu} . \tag{138.3'}$$

The following are examples of meson decay:

$$\pi^+ \to \mu^+ + \nu , \tag{138.4}$$

$$\mu^+ \to e^+ + \nu + \tilde{\nu} . \tag{138.4'}$$

A comparison of these formulae shows that a neutron can not be regarded as a complex particle consisting of a proton and an electron, nor can the proton be regarded as consisting of a neutron and a positron. The phenomenon is not an emission of already existing particles but the production of new particles (e^+, e^-, ν) in the transformation $n \rightleftarrows p$ (just as a light quantum emitted by an atom is not already present within the atom, but is created by the transformation of the energy of an

[3] The laws of conservation of momentum and energy require that a third body (for example, an atomic nucleus or a second photon) should participate in this process.

excited electron into radiation energy). Again, in reactions (138.4) and (138.4′) we do not have mesons decaying into already existing particles of which they are composed, but being transformed with the production of new particles.

These phenomena have nothing in common with those in a mechanical system of particles: even the number and nature of the particles undergo changes. The systems concerned have an indeterminate and infinitely large number of degrees of freedom. Such systems are akin to fields rather than to mechanical systems of material particles. In particular, at high energies we can no longer perceive the boundary between 'true' particles (electrons, protons, neutrons, atomic nuclei, atoms, etc.) and the 'transient' photons. The laws governing particles of the first type are essentially the subject of quantum mechanics, but the photons have been regarded as belonging to electromagnetic field theory.[4] The demarcation was based on the fact that the particles have a rest mass m_0, so that they remain unaltered and cannot be created at non-relativistic energies $E \ll m_0c^2$.

The rest mass of the photon, however, is zero, so that it is always a relativistic particle capable of being created and destroyed at arbitrarily low energies.

If the energies become comparable with the rest energies of the particles, all particles resemble photons in being created, destroyed and transformed into one another. At such high energies, therefore, it is more appropriate to speak of an electron-positron field, a meson field, a proton or neutron (nucleon) field, than of a system of given particles.[5]

In recent years the quantum theory of fields has been considerably developed, but it has not yet been completed. Already in the quantum theory of the electromagnetic field it has been found that fundamental difficulties arise in extending field theory beyond the simple processes of absorption, emission and scattering of photons to take account of all electromagnetic processes, including the interaction of particles. In such cases an infinite number of photons has to be considered, and it is found, as in the classical theory of the electron, that the electromagnetic mass of charged particles is infinite.

The same result occurs in other field theories. The problem of the mass of a particle appears to be one of particle structure, and this very difficult problem has not yet been solved.

The relativistic theory of the electron due to Dirac is of particular importance in the modern theory. Dirac's theory is a generalisation of the non-relativistic quantum mechanics of the electron to the high-velocity case.

This theory, in combination with quantum field theory, allows a treatment of many relativistic phenomena, such as the conversion of a light quantum into electrons and positrons, the converse process, and the scattering of light by electrons. It gives a complete theory of the motion of a fast electron in an external field, such as the Coulomb field of an atomic nucleus. The corrections to this motion due to zero-point oscillations of the electromagnetic field and vacuum polarisation are of par-

[4] Cf. Section 118.
[5] Similarly, the term 'photons' implies the use of the quantum theory of the electromagnetic field.

ticular importance. These effects have now been experimentally confirmed, and demonstrate the remarkable fact that in a vacuum there are continual zero-point oscillations like those in a solid, and moreover there is a polarisation of the vacuum owing to the production and subsequent annihilation of positron-electron pairs. All these effects can be calculated by the use of perturbation theory based on the smallness of the electron charge. To avoid infinities, special techniques of 'renormalisation' are used, which allow the infinity to be successively removed in each approximation.[6]

It has not yet been possible to apply these methods to strong interactions such as the interaction of a meson field with nucleons. The reason is that the 'renormalisation' methods themselves do not resolve the problem of the rest mass and structure of particles, but are merely an artifice for avoiding an explicit treatment of physical processes over very small distances.

139. Some epistemological problems

With the development of quantum mechanics, physical theory reached a stage of demolishing fundamental concepts which had seemed obvious and inviolable. This radical reformation of the basic ideas of physics was concerned principally with the concept of a particle and its motion, and is now entering an even more advanced phase.

Idealist philosophers have tended to represent this process as a critical threat to materialism.

At the time when Lenin's *Materialism and empiriocriticism* was published, reactionary philosophers likewise attempted to refute materialism with the aid of the 'latest' results of physics at that time. In his profound and acute analysis Lenin demonstrated the invalidity of these attempts and explained that the scientific foundations of dialectical materialism would not be shaken by the discovery that the nature of matter is 'electromagnetic' or anything else. From the point of view of materialistic epistemology, the upheaval of physical ideas is a necessary stage in the development of knowledge. Lenin showed what philosophical entanglements await the investigator who confuses the reformation of specific physical ideas concerning matter with the 'critical threat to materialism' preached by reactionary thinkers.

The present position in the methodology of theoretical physics in the capitalist countries is essentially no different from that dealt with by Lenin in his book. The bourgeois philosophy, through its social nature, is based from the start on idealism, and is even now attempting to turn the development of natural science to reactionary ends.

The idealistic philosophy has influenced the interpretation of the essence and significance of quantum mechanics in many Western scientific groups. The Copenhagen school from its beginning adhered to positivism and later did much to foster the development of subjective views of the essential nature of quantum mechanics.

Bohr, in his interpretation of quantum mechanics[7], starts from the *principle of*

[6] See [1, 17].
[7] See [20].

complementarity, according to which there are two classes of possible experimental systems. One allows the determination of momentum-energy relations and the other that of space-time relations. The simultaneous use of both is in principle impossible. Thus the 'quantum description' of phenomena falls into two separate categories, which complement each other in the sense that by combining them in classical physics a complete description is obtained.

It is evident from this account of the complementarity principle that it emphasises not the existence of new kinds of object but the possibility of macroscopic measuring apparatus. In other words, it relates principally not to objective properties of the microuniverse, which result in the inapplicability of the methods of classical physics thereto, but to the possibility of an observer employing macroscopic quantities and concepts.

This orientation of Bohr's complementarity principle leads to a twofold consequence.

Firstly, Bohr, and his successors even more, have elevated this principle into a particular philosophical conception of complementarity, which leads them to deny causality and objectivity of microphenomena.[8] On this basis Bohr speaks of the 'inadequacy of the customary viewpoint of natural philosophy for an account of physical phenomena of the type with which we are concerned in quantum mechanics'[9], while Jordan [52] declares outright the 'liquidation of materialism'.

Secondly, the use of this principle in physics involves a subjective interpretation of the wave function and the concept of states in quantum mechanics. The wave function is regarded not as an objective characteristic of a quantum ensemble but as an expression of information gained by the observer as a result of measurement. The reality of any particular state of microsystems then becomes equivalent to information gained by the observer concerning such a system, i.e. becomes subjective instead of objective.

In discussing these views it must be remembered that the Copenhagen school starts from the tenets of positivism, denying immediately the objective existence of matter and 'merely' analysing the 'results of observation'. In positivist thought neither classical nor quantum physics is a reflection of an objective universe; both are mathematical constructs. In the former it is possible to separate subjective and objective ideas, but in the latter it is not, since the subject 'leads to physical reality' through measurement. Thus it is a question not of analysis of relations between the percipient subject and the object as parts of an objective universe but of analysis of these constructs, that is, analysis in the realm of ideas.

On this basis the positivist tries to refute materialism by first linking it with certain restricted physical and philosophical notions and then demonstrating that these are invalid.

From the materialist viewpoint the possibility of perception is itself a consequence of the existence of material links between the percipient subject and the object of examination. In physics this connection is established by means of various types of

[8] See [19].
[9] See [20].

apparatus. The apparatus always acts on the object, and the object in turn acts on the apparatus. In classical physics it was supposed that this mutual interaction could be made arbitrarily small; in the quantum region it is found that this is not possible. We have seen that measuring apparatus in fact changes the states of systems and so transforms one ensemble into another. It would, however, be ridiculous to suppose that materialism implied that the interaction could be arbitrarily small. The finiteness of the interaction which is discovered in the quantum region does not undermine materialism and does not impose any limitations on perception.

For example, counters and other apparatus are used in the study of cosmic rays. The apparatus changes the state of the individual particles detected, transferring them to a new ensemble, but does not change the entire quantum ensemble which may be called the ensemble of cosmic rays. The effect of the apparatus on the phenomenon of cosmic rays as a whole is, of course, negligible, and so does not interfere with the derivation of the objective laws governing cosmic rays.

Thus quantum mechanics in fact deals with an objective nature of a quantum ensemble, existing independently of the observer. The properties of a single micro-phenomenon are examined through statistical laws, which are entirely objective. For instance, the disintegration of radioactive atoms follows a statistical law, and occurs spontaneously, without any interference through measurement.

It is therefore incorrect to assert that a phenomenon is statistical because of measurement. The statistical laws exist as objective natural laws, independently of measurements.

In classical physics a law may also be formulated in a non-statistical form. The state of an isolated system at any instant is uniquely determined by its state at some initial instant. This expression of laws governing physical phenomena is in reality an approximation. Isolation of a system can occur in Nature only to a certain degree of exactness, never completely, and even in classical physics it is possible only in relation to the simplest laws.

Such causality applied to all phenomena leads to a Calvinist predeterminism, where, in Engels' phrase [35], 'necessity is degraded into chance'. It is this narrow view of causality which the positivists would allot to materialism, and when it proves to be less than universally valid they announce the failure of materialism.[10]

In the quantum region an ensemble can be isolated only as a whole. The wave function which describes the state of the ensemble is uniquely determined by Schrödinger's equation:

$$i\hbar\partial\psi/\partial t = \boldsymbol{H}\psi \tag{139.1}$$

at any instant if it is known at the initial instant. Thus the simplest form of causal connection is maintained for the ensemble.

Individual events are governed by statistical laws, which are not due to the absence of relations within the universe of individual phenomena, as the positivists assert; on the contrary, they express precisely the general law governing individual phenomena.

[10] See [52].

It would not be correct to suppose that it is possible (or will in future be possible) to apply to individual microphenomena the concept of classical causality valid for isolated systems. Rather is it that such isolation does not exist as regards atomic phenomena.

If this is so, then the future development of atomic physics will involve the extension and perfection of the statistical method, through which will be perceived further laws of the structure of atoms, atomic nuclei and elementary particles.[11] The discovery of quantum levels of atoms, then their fine structure and finally their hyperfine structure (with energy differences 10^{-12}, 10^{-15} and 10^{-18} erg respectively), achieved by the development of statistical design of experiments, is an excellent illustration of the power of this method and a noteworthy confirmation of Lenin's doctrine of the knowability of matter and the inexhaustibility of its properties.

Let us now consider the position of the Copenhagen school in the understanding of the wave function, which can be very clearly seen from the discussion between Einstein and Bohr.[12] This discussion dealt with the following example. Let two particles 1 and 2 undergo a collision; let their state before the collision, at the initial instant, be described by a wave function

$$\Psi^0(x_1, x_2) = \psi^0(x_1)\,\phi^0(x_2). \tag{139.2}$$

The wave function of the particles after the collision, when a sufficiently long time has elapsed, is denoted by $\Psi(x_1, x_2)$; it is not a product of functions of x_1 and x_2 separately.

Let us now measure some quantity pertaining only to the first particle, say its momentum p_1. After this measurement the wave function of the first particle will be $\psi_{p_1}(x_1)$. Let us expand $\Psi(x_1, x_2)$ in terms of the functions $\psi_p(x_1)$:

$$\Psi(x_1, x_2) = \int \phi_p(x_2)\,\psi_p(x_1)\,\mathrm{d}p, \tag{139.3}$$

where $\phi_p(x_2)$ are the amplitudes in this expansion. If the measurement of the momentum of the first particle gives a value p_1, the wave function is reduced to a single term of the superposition (139.3):

$$\Psi(x_1, x_2) \to \phi_{p_1}(x_2)\,\psi_{p_1}(x_1). \tag{139.4}$$

Thus the state of the second particle is also changed, although no measurement has been made on it and it has long since ceased to interact with the first particle. Thus it is said that the 'information' about this particle has changed, and therefore so has its state, i.e. the concept of a state in this treatment is equivalent to that of information concerning a state.

[11] New physical phenomena may also be discovered which we do not yet suspect and which may permit the establishment of a non-statistical theory of microphenomena. Various statements which at present appear contradictory may be reconciled. For example, in the heyday of classical thermodynamics any statement that heat may of itself pass from a cold body to a hotter one would have been regarded as obviously unscientific and in contradiction to the 'second law' of thermodynamics. Yet we know that later developments made this possibility compatible with the classical formulation of the second law, in the new kinetic theory of matter.

[12] See [20, 34].

This is the subjective treatment of the wave function, which is due to the fact that the Copenhagen school does not lay stress on the statistical nature of quantum mechanics.

In quantum mechanics the state of a particle is not something in itself, but depends on the particle's belonging to a particular ensemble (pure or mixed). This is entirely objective and does not depend on observed information. If this information does not correspond to the nature of the ensemble, it can reveal nothing new except perhaps some absurd result.

Measuring apparatus, as has been explained in earlier chapters, performs spectral analysis, dividing the original ensemble into sub-ensembles whose nature depends not only on the nature of the ensemble but also essentially on the nature of the analysing apparatus.

In the above example the analysis is based on a characteristic of the first particle. But since in the initial ensemble $\Psi(x_1, x_2)$ there existed a correlation between the two particles owing to their interaction, the analysis by the characteristic p_1 at the same time separates a sub-ensemble for the second particle, i.e. after the measurement it belongs to a different sub-ensemble with wave function $\phi_{p_1}(x_2)$.

Thus the change in state of the second particle is due not to a change in 'information' about it but to the interaction between the two particles before measurement. If this interaction did not exist, changes in the state of the first particle would not affect the state of the second particle; the wave function $\Psi(x_1, x_2)$ would remain a product of functions of x_1 and x_2 separately. In our example the nature of the correlation due to the interaction is particularly clear. Let the momenta of the first and second particles before the collision be p_1^0 and p_2^0 respectively. Then, if the momentum of the first particle after the collision is p_1, that of the second particle must, by the law of conservation of momentum, be $p_2 = p_1^0 + p_2^0 - p_1$ (that is, $\phi_{p_1}(x_2)$ is a de Broglie wave with momentum $p_2 = p_1^0 + p_2^0 - p_1$). The classification of particle 1 according to momentum is therefore necessarily also a classification of particle 2 according to momentum.

Thus we see that the subjective treatment of the wave function arises from overlooking its statistical significance.

The wave function, or more generally the density matrix, in fact gives an entirely objective description of a quantum ensemble.[13] If Ψ (or the density matrix) is given, all possible resolutions of the ensemble with respect to any analysing/measuring apparatus are determined.

The process of analysis of initial ensembles with respect to any characteristic into sub-ensembles occurs not only in the laboratory but also in Nature. In every case where the phase relations between various states $\psi_n(x)$ in the superposition

$$\psi(x) = \sum_n c_n \psi_n(x) \tag{139.5}$$

[13] The significance of the wave function as a characteristic of a quantum ensemble is very fully discussed by K. V. Nikol'skiĭ [70].

become unimportant, we can speak of a 'measurement' of the quantity n, since then the pure state $\psi(x)$ can be replaced by a mixture of states $\psi_n(x)$ with probabilities $\sim |c_n|^2$.[14]

We can now return to the discussion between Einstein and Bohr. Einstein and his co-authors[15] asserted the incompleteness of quantum mechanics.[16] They pointed out that it is impossible to measure simultaneously p and x for a particle, despite the fact that each quantity separately can be measured without directly influencing the particle. In their example, a particular function is taken,

$$\Psi(x_1, x_2) = \int_{-\infty}^{\infty} e^{(i/\hbar)(x_1 - x_2 - x_0)p} \, dp = 2\pi \cdot \delta(x_1 - x_2 - x_0), \tag{139.6}$$

where x_0 is some constant. Suppose that we measure the momentum of the first particle and obtain the value p. Then (139.6) shows immediately that the momentum of the second particle is $-p$, but its co-ordinate x_2 is entirely indeterminate. Instead of the momentum we could measure the co-ordinate of the first particle, $x_1 = x$. Then from (139.6) the wave function of the second particle is $\sim \delta(x - x_2 - x_0)$, which gives $x_2 = x - x_0$. Thus we have determined the co-ordinate x_2 of the second particle, but its momentum p_2 in this state is entirely indeterminate. Einstein *et al.* deduced that quantum mechanics is incomplete, since it does not afford the possibility of simultaneously determining p_2 and x_2, even though the measuring apparatus does not act on the second particle.

Bohr, in his reply to Einstein *et al.*, repudiated this view on the basis of the complementarity principle. He asserted that measuring apparatus is necessarily always such that either only p or only x can be determined. Hence quantum mechanics is complete, as it corresponds fully to the possibilities of macroscopic apparatus.

Bohr's reply was only partly correct. By basing his answer on the complementarity principle, he laid most emphasis on the capabilities of measuring apparatus, whereas the essence of the matter lies in the novel nature of microparticles as objects of measurement to which the classical concept of motion in paths is inapplicable. Bohr also passed over the statistical interpretation of the wave function, from which it follows (as first clearly shown by L. I. Mandel'shtam[17]) that in Einstein's example we have a resolution of the initial ensemble $\Psi(x_1, x_2)$ into different mutually exclusive ensembles (with respect to p_1, and with respect to x_1). The change in the 'state' of a second particle, as explained above, is due not to the action of the apparatus but to the correlation of the states of the two particles caused by the interaction of these particles that occurred before the measurement.

14 For instance, this happens if beams belonging to different $\psi_n(x)$ diverge in space. Moreover, if such analysis of an ensemble did not occur objectively in Nature, there would be no sense in using the probabilities calculated by quantum mechanics in the kinetic equations; but this is, perhaps, one of the most important practical applications of quantum mechanics.

15 See [34].

16 This refers to the completeness of quantum mechanics within its range of application. There is no suggestion that quantum mechanics brings to an end the evolution of physical theories of the micro-universe.

17 In lectures on quantum mechanics given at Moscow University in 1939.

Thus Einstein *et al.*, in criticising quantum mechanics for the impossibility of simultaneously measuring p and x, even when there is no direct interaction of the apparatus, overlooked the fact that the nature of microparticles is fundamentally different. They implicitly assumed that microparticles do not differ from classical particles and that the uncertainty relation is due only to the action of the apparatus. But measuring apparatus is classical both in macroscopic physics and in quantum physics. Thus the apparatus is immaterial, and the impossibility of simultaneously measuring p and x results not from shortcomings of our present measuring apparatus but from the different nature of microparticles as compared with classical particles.

It is incorrect to suppose that a modern physical experiment is insufficiently accurate to measure the 'true' simultaneous values of the momenta and co-ordinates of microparticles. It is, on the contrary, accurate enough to demonstrate that for microparticles these two quantities cannot simultaneously exist. The following example illustrates this.[18] From the scattering of X-rays or electrons by atoms we can find the distribution of electrons within atoms (i.e. $|\psi(r)|^2$; cf. Section 79). This signifies a determination of the co-ordinates of electrons within the atom. Figure 64 shows the result of such an experiment for helium atoms, and indicates that a considerable fraction of the electrons are found so far from the centre of the atom that the total energy E_0 of the initial state is less than the potential energy $U(r)$. If we suppose that an electron in an atom has both a certain momentum p and a certain co-ordinate r, the total energy is $E_0 = p^2/2m + U(r)$ and for all electrons at a distance exceeding 0.6 Å we find that $p^2/2m < 0$, i.e. the momentum is imaginary. This is patently absurd no matter how we look at it.

Another possibility would be to suppose that the true energy of the electron in the atom is $E = p^2/2m + U(r)$ $(p^2/2m > 0)$, and that the energy E_0 considered in quantum theory is only some mean value of these 'true' energies: $E_0 = \bar{E}$. This hypothesis signifies that the ionisation energy of various atoms of the same substance, in the same state E_0 so far as quantum mechanics is concerned, is different. The variation $\Delta E = E - E_0$ of this energy is equal in order of magnitude to E_0, which for the helium atom is about 20 eV; the number of electrons involved (i.e. with $U(r) > E$) is 20%. This conclusion is completely contradictory to any experiment determining the ionisation energy, such as Franck and Hertz's experiment or those which determine the limit of the photoelectric effect. No such variation in the value of the ionisation energy is in fact observed.

Thus the hypothesis that an electron in an atom, in a state of given energy, has some concealed values of the two quantities p, x is disproved by experiment.

From the fact that the classical pair of quantities p, x is not a characteristic of the motion of microparticles the positivists conclude that particles exist outside time and space, and regard particles not as objectively real but only as a concept serving to bring the results of observation into an orderly mathematical system. In quantum mechanics, however, there is a simple expression of the fact that a particle exists in

[18] As would essentially any quantum experiment.

space and time independently of the observer, namely the normalisation condition

$$\int |\psi(x,t)|^2 \, dx = 1 . \tag{139.7}$$

This condition means that at any instant t the particle can be localised at one point of space x. That is, the particle can always be caused to interact so as to reveal its corpuscular nature. It then enters a state in which the momentum, as a physical characterististic of the state of the particle, ceases to have any significance. If we have a state in which the particle has a definite momentum (a very broad de Broglie wave packet), then the particle can be localised at any point in a region of space large compared with the wavelength, at any instant.

There is therefore no case where quantum mechanics makes use of objects outside time and space; an actual quantum ensemble is always realised in a finite region of space, and exists as such for a finite time. Quantum mechanics shows, however, that the motion of microparticles in space and time cannot be identified with the motion of material particles in paths. According to quantum mechanics, motion along paths is a particular type of motion which occurs only approximately under certain conditions (cf. Section 34).

The seeming paradox of quantum mechanics arises only where attempts are made to comprehend the new laws which it establishes, using the viewpoint of classical mechanics. But quantum mechanics gives a generalisation, extension and refinement of the concept of motion which goes beyond the narrow bounds of classical atomism. It would therefore be wrong to think of the idea of motion of particles in paths as being 'true in the last resort'. Lenin emphasised in his book [60] that the fundamental point of materialistic epistemology is the recognition of the objectivity of Nature and Nature's laws: 'But dialectical materialism insists on the approximate, relative character of every scientific theory of the structure of matter and its properties; it insists on the absence of absolute boundaries in nature, on the transformation of moving matter from one state into another, which is to us apparently irreconcilable with it, and so forth. However bizarre from the standpoint of "common sense" the transformation of imponderable ether into ponderable matter and *vice versa* may appear, however "strange" may seem the absence of any other kind of mass in the electron save electromagnetic mass, however extraordinary may be the fact that the mechanical laws of motion are confined only to a single sphere of natural phenomena and are subordinated to the more profound laws of electromagnetic phenomena, and so forth – all this is but another *corroboration* of dialectical materialism.'[19]

Quantum mechanics has demonstrated the limitations of classical atomism and has revealed entirely new properties of the microuniverse, which have been fully confirmed by practical physical experiment. And so, to dialectical materialism, quantum mechanics must be a major advance in the atomistics of the twentieth century. That advance bears witness to the extreme power of human thought, discerning in the seeming chaos of microphenomena laws that are remarkable both in their scope and in their exactitude.

[19] See [60], p. 268.

APPENDICES

I. The Fourier transformation

First of all we may recall the Dirichlet integral, which appears in the theory of Fourier integrals:

$$\lim_{m \to \infty} \frac{1}{\pi} \int_a^b \phi(z) \frac{\sin mz}{z} \mathrm{d}z, \tag{1}$$

where $\phi(z)$ is an arbitrary function. This integral has the following properties: (1) if $a, b > 0$ or $a, b < 0$ it is zero; (2) if $a < 0, b > 0$ it is equal to $\phi(0)$ (for continuous functions).[1] The presence of the function $(1/\pi z) \sin mz$ in the integrand and the taking of the limit $(m \to \infty)$ can both be denoted by the symbol $\delta(z)$, so that the integral becomes

$$\int_a^b \phi(z) \delta(z) \mathrm{d}z \begin{cases} = 0 & \text{if } a, b > 0 \text{ or } a, b < 0, \\ = \phi(0) & \text{if } a < 0, b > 0. \end{cases} \tag{2}$$

This $\delta(z)$ is often called a *delta function.* The general definition is given in Appendix III. Proceeding now to prove the equivalence of Formulae (13.1), (13.3) and (13.5), (13.6), we shall consider for simplicity the one-dimensional case and shall prove that

$$\overline{p_x^n} = \int_{-\infty}^{\infty} \phi^*(p_x) p_x^n \phi(p_x) \mathrm{d}p_x = \int_{-\infty}^{\infty} \psi^*(x) \left(-i\hbar \frac{\partial}{\partial x} \right)^n \psi(x) \mathrm{d}x, \tag{3}$$

where $\phi(p_x)$ is the Fourier transform of $\psi(x)$:

$$\phi(p_x) = \int_{-\infty}^{\infty} \psi(x) \frac{e^{-ip_x x/\hbar}}{\sqrt{(2\pi\hbar)}} \mathrm{d}x, \tag{4}$$

and n is a positive integer. To prove Equation (3) we substitute for $\phi(p_x)$ and $\phi^*(p_x)$ from (4), obtaining

$$\overline{p_x^n} = \int_{-\infty}^{\infty} \mathrm{d}p_x \int_{-\infty}^{\infty} \psi^*)x') \frac{e^{ip_x x'/\hbar}}{\sqrt{(2\pi\hbar)}} \mathrm{d}x' \cdot p_x^n \int_{-\infty}^{\infty} \psi(x) \frac{e^{-ip_x x/\hbar}}{\sqrt{(2\pi\hbar)}} \mathrm{d}x. \tag{5}$$

[1] See, for instance, [78], Vol. II, p.369; [105], p. 3.

Instead of the product $p_x^n e^{-ip_x x/\hbar}$ we can write $(i\hbar\,\partial/\partial x)^n e^{-ip_x x/\hbar}$. This gives

$$\overline{p_x^n} = \int_{-\infty}^{\infty} \frac{\mathrm{d}p_x}{2\pi\hbar} \int_{-\infty}^{\infty} \psi^*(x')\, e^{ip_x x'/\hbar}\,\mathrm{d}x' \int_{-\infty}^{\infty} \psi(x)\left(i\hbar\frac{\partial}{\partial x}\right)^n e^{-ip_x x/\hbar}\,\mathrm{d}x\,. \qquad (6)$$

Integrating n times by parts in the last integral, and assuming that $\psi(x)$ and its derivatives vanish at the limits of integration ($x = \pm\,\infty$), we find

$$\overline{p_x^n} = \int_{-\infty}^{\infty} \frac{\mathrm{d}p_x}{2\pi\hbar} \int_{-\infty}^{\infty} \psi^*(x')\, e^{ip_x x'/\hbar}\,\mathrm{d}x' \int_{-\infty}^{\infty} e^{-ip_x x/\hbar}\left(-\,i\hbar\frac{\partial}{\partial x}\right)^n \psi(x)\,\mathrm{d}x\,; \qquad (7)$$

we now interchange the order of integration, integrating first with respect to p_x:

$$\overline{p_x^n} = \int_{-\infty}^{\infty} \psi^*(x')\,\mathrm{d}x' \int_{-\infty}^{\infty} \left(-\,i\hbar\frac{\partial}{\partial x}\right)^n \psi(x)\,\mathrm{d}x \int_{-\infty}^{\infty} e^{ip_x(x'-x)/\hbar}\frac{\mathrm{d}p_x}{2\pi\hbar}. \qquad (8)$$

We next define variables $\zeta = p_x/\hbar$, $z = x' - x$. Performing the integration with respect to ζ in the last integral in (8) over a finite range from $-\,m$ to $+\,m$ and then letting $m \to \infty$, we can write

$$\begin{aligned}\overline{p_x^n} &= \int_{-\infty}^{\infty} \left[\left(-\,i\hbar\frac{\partial}{\partial x}\right)^n \psi(x)\right]\mathrm{d}x \cdot \lim_{m\to\infty} \int_{-\infty}^{\infty} \psi^*(x+z)\,\mathrm{d}z\,\frac{\sin mz}{\pi z} \\ &= \int_{-\infty}^{\infty} \left[\left(-\,i\hbar\frac{\partial}{\partial x}\right)^n \psi(x)\right]\mathrm{d}x \int_{-\infty}^{\infty} \psi^*(x+z)\,\delta(z)\,\mathrm{d}z\,.\end{aligned} \qquad (8')$$

From (2) with $a = -\,\infty$, $b = \infty$ and $\phi(z) = \psi^*(x+z)$ we have

$$\overline{p_x^n} = \int_{-\infty}^{\infty} \left[\left(-\,i\hbar\frac{\partial}{\partial x}\right)^n \psi(x)\right]\psi^*(x)\,\mathrm{d}x = \int \psi^*(x)\left(-\,i\hbar\frac{\partial}{\partial x}\right)^n \psi(x)\,\mathrm{d}x\,. \qquad (9)$$

This proves (3). An integral rational function of p_x has the form

$$F(p_x) = \sum_n a_n\, p_x^n\,.$$

Then

$$\begin{aligned}\overline{F(p_x)} &= \sum_n a_n\, \overline{p_x^n} \\ &= \sum_n a_n \int \psi^*(x)\left(-\,i\hbar\frac{\partial}{\partial x}\right)^n \psi(x)\,\mathrm{d}x \\ &= \int \psi^*(x)\, F\left(-\,i\hbar\frac{\partial}{\partial x}\right)\psi(x)\,\mathrm{d}x\,.\end{aligned} \qquad (10)$$

Thus the equivalence of (13.3) and (13.6) for the one-dimensional case is proved.

The generalisation to three dimensions simply involves increasing the number of integrations and is therefore trivial; it is sufficient to show the equivalence of (13.3) and (13.6) for the mean value of $p_x^n \cdot p_y^m \cdot p_z^l$, where m, n and l are positive integers.

The validity of the equation

$$\overline{x^n} = \int_{-\infty}^{\infty} \psi^*(x) x^n \psi(x) \mathrm{d}x = \int_{-\infty}^{\infty} \phi^*(p_x)\left(i\hbar \frac{\partial}{\partial p_x}\right)^n \phi(p_x) \mathrm{d}p_x \qquad (11)$$

follows from (3) if we note that by Fourier's inversion theorem

$$\psi(x) = \int_{-\infty}^{\infty} \phi(p_x) \frac{e^{ip_x x/\hbar}}{\sqrt{(2\pi\hbar)}} \mathrm{d}p_x. \qquad (4')$$

Replacing ψ by ϕ and p_x by x in (3) and changing the sign of i in the exponent in (4), we obtain Formulae (11) and (4′) from (3) and (4). From (11) we also have

$$\overline{F(x)} = \sum_n \overline{a_n x^n} = \int_{-\infty}^{\infty} \phi^*(p_x)\, F\left(i\hbar \frac{\partial}{\partial p_x}\right) \phi(p_x) \mathrm{d}p_x. \qquad (12)$$

This is the one-dimensional case of (13.5). The generalisation to three dimensions is again trivial.

II. Eigenfunctions when there is degeneracy

The eigenfunctions $\psi_{nk}(k = 1, 2, \ldots, f)$ corresponding to the eigenvalue L_n are linearly independent, i.e. they do not satisfy a relation of the form

$$\sum_{k=1}^{f} a_k \psi_{nk} = 0, \qquad (1)$$

where a_k are some constants. If such relations existed, they would imply that one or more of the functions could be expressed in terms of the others, i.e. the actual number of different eigenfunctions corresponding to L_n would be less than f. If the functions ψ_{nk} are not mutually orthogonal, we can define new functions derived from the ψ_{nk} by a linear transformation:

$$\phi_{n\alpha} = \sum_{k=1}^{f} a_{\alpha k} \psi_{nk}, \qquad \alpha = 1, 2, \ldots, f. \qquad (2)$$

Since the equation for the eigenfunctions is linear, the functions $\phi_{n\alpha}$ will again be eigenfunctions of the operator $\boldsymbol{L}$ and will belong to the eigenvalue L_n.

From the orthogonality condition on the functions $\phi_{n\alpha}$:

$$\int \phi_{n\alpha}^* \phi_{n\beta} \mathrm{d}x = \delta_{\alpha\beta} \qquad (3)$$

we have the conditions determining the coefficients $a_{\alpha k}$:

$$\sum_{k=1}^{f} \sum_{k'=1}^{f} a_{\alpha k}^* a_{\beta k'} s_{kk'} = \delta_{\alpha\beta}, \qquad (4)$$

where

$$s_{kk'} = \int \psi^*_{nk} \psi_{nk'} \, \mathrm{d}x \,. \tag{5}$$

The possibility of finding coefficients $a_{\alpha k}$ which satisfy the conditions (4) follows from a geometrical analogy. We regard the functions ψ_{nk} as unit vectors $\mathbf{j}_k$ in a space of f dimensions, and $s_{kk'}$ as scalar products $\mathbf{j}_k \cdot \mathbf{j}_{k'}$. Then Equation (2) can be regarded as a transformation in f-dimensional space from oblique to rectangular co-ordinates.[2] Hence it is clear that the transformation (2) is not unique; having obtained an orthogonal co-ordinate system, we can then rotate it in any manner.

For example, if the functions ψ_{nk} are already orthogonal, then $s_{kk'} = \delta_{kk'}$, and (4) gives

$$\sum_{k=1}^{f} a^*_{\alpha k} a_{\beta k} = \delta_{\alpha\beta} \,. \tag{6}$$

These are the conditions for the coefficients in an orthogonal transformation of a set of orthogonal functions ψ_{nk} into a new set of orthogonal functions $\phi_{n\alpha}$. Thus the eigenfunctions belonging to a single eigenvalue L_n are defined 'to within' an orthogonal transformation of the type (2) with coefficients which satisfy the condition (6).

III. Orthogonality and normalisation of eigenfunctions of the continuous spectrum. The δ function

If we integrate the equation for the eigenfunctions,

$$\boldsymbol{L}\psi(x, L) = L\psi(x, L) \,, \tag{1}$$

with respect to L over a small range ΔL, we obtain

$$\boldsymbol{L}\Delta\psi(x, L) = \int_{L}^{L+\Delta L} L\psi(x, L)\,\mathrm{d}L \,, \tag{2}$$

where

$$\Delta\psi(x, L) = \int_{L}^{L+\Delta L} \psi(x, L)\,\mathrm{d}L \,. \tag{3}$$

This quantity is called the *eigendifferential* of the operator $\boldsymbol{L}$. An example is the wave packet discussed in Section 7. We shall show that not the functions themselves but the eigendifferentials are orthogonal and can be normalised. To do so, we integrate similarly the complex conjugate equation

$$\boldsymbol{L}^*\psi(x, L') = L'\psi^*(x, L') \tag{4}$$

with respect to L', obtaining

$$\boldsymbol{L}^*\Delta\psi^*(x, L') = \int_{L'}^{L'+\Delta L'} L'\psi^*(x, L')\,\mathrm{d}L' \,. \tag{5}$$

[2] Details concerning the orthogonalisation of functions are given in [27], Chapter II, Section 1.

We multiply (2) by $\Delta\psi^*(x, L')$ and (5) by $\Delta\psi(x, L)$, subtract, and integrate with respect to x. This gives

$$\begin{aligned}&\int \mathrm{d}x \{\Delta\psi^*(x, L')\cdot \boldsymbol{L}\,\Delta\psi(x, L) - \Delta\psi(x, L)\cdot \boldsymbol{L}^*\Delta\psi^*(x, L)\}\\ &= \int \mathrm{d}x \int_L^{L+\Delta L} \mathrm{d}L \int_{L'}^{L'+\Delta L'} \mathrm{d}L' (L - L')\psi^*(x, L')\psi(x, L).\end{aligned} \tag{6}$$

The left-hand side is zero because of the self-adjointness of the operator $\boldsymbol{L}$, and on the right $L - L'$ may be taken outside the integral if ΔL and $\Delta L'$ are small. We then have

$$(L - L')\int \mathrm{d}x \cdot \Delta\psi^*(x, L')\Delta\psi(x, L) = 0. \tag{7}$$

If the intervals ΔL and $\Delta L'$ do not overlap, $L \neq L'$, and hence

$$\int \mathrm{d}x \cdot \Delta\psi^*(x, L')\Delta\psi(x, L) = 0, \tag{8}$$

i.e. the eigendifferentials are orthogonal. If ΔL and $\Delta L'$ coincide, the integral (8) need not be zero. It is easily shown that this integral is of the first order of smallness relative to ΔL. The integral

$$I = \int \mathrm{d}x \cdot \Delta\psi^*(x, L)\Delta\psi(x, L) \tag{9}$$

can be replaced by

$$I' = \int \mathrm{d}x \cdot \Delta\psi^*(x, L)\int_{L_1}^{L_2} \psi(x, L)\,\mathrm{d}L, \tag{10}$$

where L_1 and L_2 are so chosen that the range $(L, L + \Delta L)$ lies within the range (L_1, L_2). By the orthogonality of the eigendifferentials, the integrals over (L_1, L) and $(L + \Delta L, L_2)$ do not contribute to the value of the integral (9). Hence (9) and (10) are equal. But as $\Delta L \to 0$ the integral (10) tends to zero as ΔL. Hence, by taking a suitable normalisation factor, we can always ensure that

$$\lim_{\Delta L \to 0} (I/\Delta L) = 1,$$

i.e.

$$\int \mathrm{d}x \cdot \Delta\psi^*(x, L)\Delta\psi(x, L) = \Delta L \tag{11}$$

as $\Delta L \to 0$.

Formulae (8) and (11) may be combined in a form which expresses the normalisation and orthogonality of the eigendifferentials:

$$\int \mathrm{d}x \cdot \Delta\psi^*(x, L')\Delta\psi(x, L) = \Delta L \text{ or } 0 \tag{12}$$

according as the intervals L, $L + \Delta L$ and L', $L' + \Delta L$ do or do not coincide. Removing one integration (that with respect to L) in (12), we can write

$$\int \mathrm{d}x \cdot \Delta\psi^*(x, L')\psi(x, L) = 1 \text{ or } 0 \tag{12'}$$

according as the point $L' = L$ lies in the range L' to $L' + \Delta L$ or not. The orthogonality and normalisation condition (12) or (12′) can be formulated for the functions themselves by means of a special symbolism. To do this, we interchange the integrations with respect to x and L' in (12'):

$$\int_{L'}^{L'+\Delta L} \mathrm{d}L' \int \psi(x, L)\psi^*(x, L')\,\mathrm{d}x = 1 \text{ or } 0. \tag{13}$$

With the notation

$$\int \psi^*(x, L')\psi(x, L)\,\mathrm{d}x = \delta(L' - L), \tag{14}$$

(13) gives

$$\int_{L'}^{L'+\Delta L} \mathrm{d}L' \cdot \delta(L' - L) = 1 \text{ or } 0, \tag{15}$$

according as the point $L' = L$ lies in the range L' to $L' + \Delta L$ or not. The last equation may be regarded as a definition of the symbol $\delta(L' - L)$, which is called the *delta function* or *Dirac function* (it is in fact not a function but a symbol).

From (15) we have (see (21.11))

$$\int_a^b f(L')\delta(L' - L)\,\mathrm{d}L' = f(L) \text{ or } 0, \tag{16}$$

according as the point $L' = L$ lies in the range (a, b) or not. To prove (16) we need only divide this range into such small intervals that the function $f(L')$ may be taken outside the integral in each interval; for this to be possible, the function must be smooth. In each interval the integral is zero by (15), except that infinitesimal one which contains the point $L' = L$. In this interval the integral of δ is unity, by (15).

Instead of saying that the eigendifferentials are normalised and orthogonal (12) we shall say that the eigenfunctions are normalised by the δ function (14).

As an example we may take the normalisation of the eigenfunctions of the momentum operator $\boldsymbol{P}_x$. These functions are

$$\psi_{p_x}(x) = N_{p_x} e^{-ip_x x/\hbar}, \tag{17}$$

where N_{p_x} is the required normalising factor, which might *a priori* depend on p_x. We form the integral (14):

$$\begin{aligned} \int \psi^*_{p_x'}(x)\psi_{p_x}(x)\,\mathrm{d}x &= N^*_{p_x'} N_{p_x} \int_{-\infty}^{\infty} e^{i(p_x' - p_x)x/\hbar}\,\mathrm{d}x \\ &= N^*_{p_x'} N_{p_x} \hbar \lim_{m\to\infty} \int_{-m}^{m} e^{i(p_x' - p_x)x/\hbar} \frac{\mathrm{d}x}{\hbar} \\ &= N^*_{p_x'} N_{p_x} \hbar \lim_{m\to\infty} \frac{2\sin\left[(p_x' - p_x)m/\hbar\right]}{p_x' - p_x}. \end{aligned} \tag{18}$$

Comparing this with the Dirichlet factor

$$\lim_{m\to\infty} \frac{1}{\pi} \frac{\sin mz}{z},$$

which has the properties of $\delta(z)$ (see Appendix I, Formula (1)), we find that

$$\int \psi^*_{p_x'}(x)\psi_{p_x}(x)\,\mathrm{d}x = N^*_{p_x'} N_{p_x} \cdot 2 \quad \cdot \; (p_x' - p_x). \tag{19}$$

Hence we find the normalising factor:

$$|N_{p_x}|^2 \cdot 2\pi\hbar = 1\,, \qquad N_{p_x} = 1/\sqrt{(2\pi\hbar)}\,. \tag{20}$$

Of course a phase factor $e^{i\phi(p_x)}$, where ϕ is a real function, could be included, but there is no need to do so.

IV. The significance of commutability of operators

We shall show that, if two operators $\boldsymbol{L}$ and $\boldsymbol{M}$ have in common a complete set of eigenfunctions, then they commute. Let the common eigenfunctions be $\psi_n(x)$. Then

$$\boldsymbol{L}\psi_n = L_n\psi_n\,, \qquad \boldsymbol{M}\psi_n = M_n\psi_n\,. \tag{1}$$

By applying the operator $\boldsymbol{M}$ to the first equation and $\boldsymbol{L}$ to the second and subtracting, we obtain

$$\begin{aligned} &\boldsymbol{ML}\psi_n = L_n M_n \psi_n\,, \qquad \boldsymbol{LM}\psi_n = L_n M_n \psi_n\,, \\ &(\boldsymbol{ML} - \boldsymbol{LM})\,\psi_n = 0\,. \end{aligned} \tag{2}$$

Since any function can be expanded in terms of the functions ψ_n, we have

$$(\boldsymbol{ML} - \boldsymbol{LM})\,\phi = \sum_n c_n (\boldsymbol{LM} - \boldsymbol{LM})\,\psi_n = 0\,, \tag{3}$$

i.e. the application of the operator $\boldsymbol{ML} - \boldsymbol{LM}$ to any function gives zero. This means that these operators commute:

$$\boldsymbol{ML} - \boldsymbol{LM} = 0\,. \tag{4}$$

We shall now show that, if the operators $\boldsymbol{L}$ and $\boldsymbol{M}$ commute, they have common eigenfunctions. The equation for the eigenfunctions of the operator $\boldsymbol{L}$ is

$$\boldsymbol{L}\psi = L\psi\,. \tag{5}$$

Applying the operator $\boldsymbol{M}$ and replacing $\boldsymbol{ML}$ by $\boldsymbol{LM}$, we have

$$\boldsymbol{L}(\boldsymbol{M}\psi) = L(\boldsymbol{M}\psi)\,. \tag{6}$$

Hence it follows that $\psi' = \boldsymbol{M}\psi$ is also an eigenfunction of the operator $\boldsymbol{L}$, with eigenvalue L. If there is no degeneracy, only one function corresponds to the value L. Thus ψ' can differ from ψ only by a constant factor, i.e. $\psi' = M\psi$. Consequently

$$\boldsymbol{M}\psi = M\psi\,, \tag{7}$$

whence it follows that ψ is also an eigenfunction of the operator $\boldsymbol{M}$. When degeneracy is present, ψ' may be a linear combination of functions ψ_k ($k = 1, 2, \ldots, f$) belonging to the eigenvalue L:

$$\psi' = \boldsymbol{M}\psi_k = \sum_{k'=1}^{f} M_{kk'}\psi_{k'}\,, \qquad k = 1, 2, \ldots, f\,. \tag{8}$$

However, instead of the functions ψ_k we can take linear combinations of them (see

Appendix II):

$$\phi = \sum_{k=1}^{f} a_k \psi_k , \tag{9}$$

where the a_k may be so chosen that the new functions ϕ are eigenfunctions of the operator $\boldsymbol{M}$:

$$\boldsymbol{M}\phi = M\phi . \tag{10}$$

Substituting ϕ from (9) and using (8) we find on comparing coefficients of ψ_k

$$\sum_{k'=1}^{f} M_{kk'} a_{k'} = M a_k , \qquad k = 1, 2, \ldots, f . \tag{11}$$

This is a set of homogeneous algebraic equations to determine the coefficients a_k. It has a solution only if its determinant is zero:

$$\begin{vmatrix} M_{11} - M & M_{12} & \cdots & M_{1f} \\ M_{21} & M_{22} - M & \cdots & M_{2f} \\ \cdots & \cdots & \cdots & \cdots \\ M_{f1} & \cdots & \cdots & M_{ff} - M \end{vmatrix} = 0 . \tag{12}$$

From this equation we find the roots $M_1, M_2, \ldots, M_f$. For each root M_α we have a solution $a_{\alpha 1}, a_{\alpha 2}, \ldots, a_{\alpha f}$, and therefore a function ϕ from (9):

$$\phi_\alpha = \sum_{k=1}^{f} a_{\alpha k} \psi_k . \tag{13}$$

The new functions ϕ_α ($\alpha = 1, 2, \ldots, f$), being linear combinations of the ψ_k, are eigenfunctions of the operator $\boldsymbol{L}$ belonging to the eigenvalue L, and are also eigenfunctions of the operator $\boldsymbol{M}$ belonging to the eigenvalues $M = M_1, M_2, \ldots, M_\alpha, \ldots, M_f$ respectively.

V. The spherical harmonic functions $Y_{lm}(\theta, \phi)$

In the problem of finding the eigenfunctions of the angular momentum operator $\boldsymbol{M}^2$ we encounter Equation (25.14) for the spherical harmonic functions:

$$\frac{1}{\sin\theta}\frac{\partial}{\partial\theta}\left(\sin\theta\frac{\partial\psi}{\partial\theta}\right) + \frac{1}{\sin^2\theta}\frac{\partial^2\psi}{\partial\phi^2} + \lambda\psi = 0 . \tag{1}$$

We have to find the eigenfunctions of this equation (i.e. solutions continuous, single-valued and finite throughout the ranges $0 \leqslant \theta \leqslant \pi$, $0 \leqslant \phi \leqslant 2\pi$).

We first separate the variables θ and ϕ, putting

$$\psi = \Theta(\theta)\,\Phi(\phi) , \tag{2}$$

substituting (2) in (1), and taking

$$\mathrm{d}^2\Phi/\mathrm{d}\phi^2 = -\, m^2 \Phi , \tag{3}$$

whence

$$\Phi_m(\phi) = e^{im\phi}. \tag{4}$$

If Φ_m is to be a single-valued function of ϕ, it is necessary that m should be an integer:

$$m = 0, \qquad \pm 1, \qquad \pm 2, \ldots. \tag{5}$$

Substituting (4) in (1) and dividing by Φ_m, we obtain the following equation for Θ:

$$\frac{1}{\sin\theta}\frac{\partial}{\partial\theta}\left(\sin\theta\frac{\partial\Theta}{\partial\theta}\right) - \frac{m^2}{\sin^2\theta}\Theta + \lambda\Theta = 0. \tag{6}$$

Instead of θ we define the variable

$$\xi = \cos\theta, \qquad -1 \leqslant \xi \leqslant 1, \qquad d\xi = -\sin\theta\, d\theta \tag{7}$$

and regard Θ as a function of ξ. Then (6) gives

$$(1-\xi^2)\Theta'' - 2\xi\Theta' + \left(\lambda - \frac{m^2}{1-\xi^2}\right)\Theta = 0. \tag{8}$$

We consider the behaviour of the solution Θ near the singular points of the equation, $\xi = \pm 1$, and take first the point $\xi = +1$, using the variable $z = \xi - 1$. Then we obtain from (8)

$$\Theta'' + \frac{2z+1}{zz+2}\Theta' - \left[\frac{\lambda}{z(z+2)} + \frac{m^2}{z^2(z+2)^2}\right]\Theta = 0. \tag{9}$$

We seek Θ as a power series in z:

$$\Theta = z^{\gamma} v, \qquad v = a_0 + a_1 z + a_2 z^2 + \ldots + a_\nu z^\nu + \ldots. \tag{10}$$

We must first determine the power γ with which the series begins. As $z \to 0$, $\Theta \to a_0 z^{\gamma}$. Substituting this solution in (9) and neglecting small quantities of order higher than $z^{\gamma-2}$, we have from (9)

$$[\gamma(\gamma-1) + \gamma - \tfrac{1}{4}m^2]\, a_0 z^{\gamma-2} = 0,$$

whence

$$\gamma = \pm\tfrac{1}{2}m. \tag{11}$$

The same value of γ is obtained from an expansion near the singular point $\xi = -1$. If the solution remains finite at $\xi = \pm 1$, we must take in (10)

$$\gamma = \tfrac{1}{2}|m|, \tag{12}$$

i.e. $\gamma = \frac{1}{2}m$ for $m > 0$ and $\gamma = -\frac{1}{2}m$ for $m < 0$. The other solution from (11) becomes infinite. Thus we can take Θ as

$$\Theta = (1-\xi^2)^{\frac{1}{2}|m|}\, v, \tag{13}$$

where v is a power series in z. We may now write v as a power series in ξ:

$$v = \sum_{\nu=0}^{\infty} b_\nu \xi^\nu. \tag{14}$$

Substitution of (13) in (8) gives

$$(1-\xi^2)v'' - 2(|m|+1)\xi v' + (\lambda - |m| - m^2)v = 0. \tag{15}$$

With the series (14), comparison of coefficients of powers of ξ leads to a recurrence formula to determine the coefficients b_ν:

$$(\nu+2)(\nu+1)b_{\nu+2} = [\nu(\nu-1) + 2(|m|+1)\nu - \lambda + |m| + m^2]\,b_\nu. \tag{16}$$

If the series (14) terminates at some term $\nu = k$, v is a polynomial of degree k, and therefore (13) is a finite, continuous and single-valued solution, i.e. an eigenfunction of Equation (1). It follows from (16) that the series can terminate only if

$$k(k-1) + 2(|m|+1)k - \lambda + |m| + m^2 = 0,$$

i.e.

$$\lambda = (k+|m|)(k+|m|+1). \tag{17}$$

Putting

$$k + |m| = l, \tag{18}$$

we have

$$\lambda = l(l+1), \qquad l = 0, 1, 2, 3, \ldots, \tag{19}$$

$$|m| = 0, 1, 2, \ldots, l. \tag{20}$$

It can be shown that there are no other eigenfunctions of Equation (1).[3]

The solution Θ belonging to the numbers l and m is denoted by

$$\Theta(\xi) = P_l^{|m|}(\xi), \qquad \xi = \cos\theta. \tag{21}$$

If Equation (15) is differentiated with respect to ξ, we obtain an equation in which $|m|$ is replaced by $|m| + 1$. Hence, if the solution for $m = 0$ is denoted by $P_l(\xi)$, then

$$P_l^{|m|}(\xi) = (1-\xi^2)^{\frac{1}{2}|m|} \frac{\mathrm{d}^{|m|}}{\mathrm{d}\xi^{|m|}} P_l(\xi). \tag{22}$$

$P_l(\xi)$ is a polynomial of degree l and is called a *Legendre polynomial.* The coefficient is usually so normalised that

$$P_l(1) = 1. \tag{23}$$

From (16) with $|m| = 0$ we have

$$b_{\nu+2} = \frac{\nu(\nu+1) - l(l+1)}{(\nu+2)(\nu+1)} b_\nu. \tag{24}$$

Hence we see that, if we take $b_0 \neq 0$, $b_1 = 0$, the polynomial P_l will contain only even powers of ξ; if $b_0 = 0$, $b_1 \neq 0$, it will contain only odd powers. Choosing b_0 (for even l) or b_1 (for odd l) so as to satisfy (23), we can calculate all the coefficients

[3] See [3].

in the polynomial P_l. It can be verified that the resulting polynomial can be written as

$$P_l^0(\xi) = P_l(\xi) = \frac{1}{2^l l!} \frac{\mathrm{d}^l}{\mathrm{d}\xi^l}(\xi^2 - 1)^l. \tag{25}$$

Using (2), (4) and (21), we have the eigenfunction of Equation (1) in the form

$$Y_{lm}(\theta, \phi) = N_{lm} P_l^{|m|}(\cos\theta)\, e^{im\phi}, \tag{26}$$

where N_{lm} is a normalising factor. The calculation of this factor, which we omit[4], gives

$$N_{lm} = \sqrt{\frac{(l - |m|)!\,(2l + 1)}{(l + |m|)!\quad 4\pi}}. \tag{27}$$

The functions (26) form a complete set of orthogonal functions on the surface of a sphere θ, ϕ. Hence any single-valued function $\psi(\theta, \phi)$ of integrable square can be represented as a series

$$\psi(\theta, \phi) = \sum_{l=0}^{\infty} \sum_{m=-l}^{l} c_{lm} Y_{lm}(\theta, \phi), \tag{28}$$

where

$$c_{lm} = \int_0^{\pi} \int_0^{2\pi} \psi(\theta, \phi)\, Y_{lm}^*(\theta, \phi) \sin\theta \,\mathrm{d}\theta\, \mathrm{d}\phi. \tag{29}$$

Finally, we give the results of applying certain operators to the spherical harmonic functions, which will be useful in certain calculations:

(a) multiplications by $\cos\theta = \xi$ or $\sin\theta = \sqrt{(1 - \xi^2)}$:

$$\xi Y_{lm} = \sqrt{\frac{(l + m + 1)(l - m + 1)}{(2l + 1)(2l + 3)}}\, Y_{l+1,m} + \sqrt{\frac{(l + m)(l - m)}{(2l + 1)(2l - 1)}}\, Y_{l-1,m}, \tag{30}$$

$$\sqrt{(1 - \xi^2)}\, Y_{lm} = \left\{ -\sqrt{\frac{(l - m + 1)(l - m + 2)}{(2l + 1)(2l + 3)}}\, Y_{l+1,m-1} + \sqrt{\frac{(l + m)(l + m - 1)}{(2l + 1)(2l - 1)}}\, Y_{l-1,m-1} \right\} e^{i\phi}; \tag{31}$$

(b) action of the angular momentum component operators $\boldsymbol{M}_x$, $\boldsymbol{M}_y$, $\boldsymbol{M}_z$:

$$\boldsymbol{M}_z Y_{lm} = \hbar m Y_{lm}, \tag{32}$$

$$(\boldsymbol{M}_x + i\boldsymbol{M}_y) Y_{lm} = -\hbar\sqrt{[(l - m)(l + m + 1)]}\, Y_{l,m+1}, \tag{33}$$

$$(\boldsymbol{M}_x - i\boldsymbol{M}_y) Y_{lm} = -\hbar\sqrt{[(l + m)(l - m + 1)]}\, Y_{l,m-1}. \tag{34}$$

These are proved in textbooks on spherical harmonics; see also [5], Section 65.

VI. Hamilton's equations

Let $q_1, q_2, \ldots, q_s, \ldots, q_f$ be generalised co-ordinates defining the configuration of a system, and $p_1, p_2, \ldots, p_s, \ldots, p_f$ the corresponding generalised conjugate momenta.

[4] See, for example, [37], pp. 124–125; [100], App. VI.

The Hamiltonian H is a function of these co-ordinates and momenta and, in general, of the time t. Hamilton's equations are

$$\mathrm{d}p_s/\mathrm{d}t = -\,\partial H/\partial q_s, \qquad \mathrm{d}q_s/\mathrm{d}t = \partial H/\partial p_s. \tag{1}$$

The time derivative of any function F of generalised co-ordinates, momenta and time is

$$\frac{\mathrm{d}F}{\mathrm{d}t} = \frac{\partial F}{\partial t} + \sum_{s=1}^{f} \frac{\partial F}{\partial q_s}\frac{\mathrm{d}q_s}{\mathrm{d}t} + \sum_{s=1}^{f} \frac{\partial F}{\partial p_s}\frac{\mathrm{d}p_s}{\mathrm{d}t}. \tag{2}$$

Using Hamilton's equations (1), we can write (2) in the form

$$\mathrm{d}F/\mathrm{d}t = \partial F/\partial t + [H, F], \tag{3}$$

where

$$[H, F] = \sum_{s=1}^{f} \left\{ \frac{\partial F}{\partial q_s}\frac{\partial H}{\partial p_s} - \frac{\partial H}{\partial q_s}\frac{\partial F}{\partial p_s} \right\} \tag{4}$$

is called a *Poisson bracket*.

It is evident that Hamilton's equations (1) can also be written in terms of Poisson brackets:

$$\mathrm{d}p_s/\mathrm{d}t = [H, p_s], \qquad \mathrm{d}q_s/\mathrm{d}t = [H, q_s], \qquad s = 1, 2, \ldots, f, \tag{5}$$

putting in (3) $F = p_s$ and $F = q_s$. As we have seen in Section 31, the equations of motion can be written in an entirely analogous form in quantum mechanics. In the particular case of Cartesian co-ordinates and a single particle moving in a field of force derived from a force function $U(x, y, z, t)$ we have

$$H = \frac{p_x^2 + p_y^2 + p_z^2}{2\mu} + U(x, y, z, t) \tag{6}$$

($q_1 = x$, $q_2 = y$, $q_3 = z$, $p_1 = p_x$, $p_2 = p_y$, $p_3 = p_z$). From (5) we therefore have

$$\left.\begin{aligned} \mathrm{d}p_x/\mathrm{d}t &= [H, p_x] = -\,\partial H/\partial x = -\,\partial U/\partial x, \\ \mathrm{d}x/\mathrm{d}t &= [H, x] = \partial H/\partial p_x = p_x/\mu, \end{aligned}\right\} \tag{7}$$

with corresponding equations for the other two co-ordinates and momenta. From (7)

$$\mu\,\mathrm{d}^2x/\mathrm{d}t^2 = -\,\partial U/\partial x, \tag{8}$$

which is Newton's equation.

In the motion of a charged particle with charge e and mass μ in an electromagnetic field described by a scalar potential V and a vector potential $\mathbf{A}$, so that

$$\mathscr{E} = -\,\nabla V - \frac{1}{c}\frac{\partial \mathbf{A}}{\partial t}, \tag{9}$$

$$\mathscr{H} = \mathbf{curl}\,\mathbf{A}, \tag{10}$$

where $\mathscr{E}$ is the electric field and $\mathscr{H}$ the magnetic field, the Hamiltonian becomes

$$H = \frac{1}{2\mu}\left(\mathbf{p} - \frac{e}{c}\mathbf{A}\right)^2 + eV. \tag{6'}$$

We shall show that the resulting Hamilton's equations

$$\mathrm{d}p_x/\mathrm{d}t = -\,\partial H/\partial x\,,\ \mathrm{d}p_y/\mathrm{d}t = -\,\partial H/\partial y\,,\ \mathrm{d}p_z/\mathrm{d}t = -\,\partial H/\partial z\,, \tag{7'}$$

$$\mathrm{d}x/\mathrm{d}t = \partial H/\partial p_x,\ \mathrm{d}y/\mathrm{d}t = \partial H/\partial p_y,\ \mathrm{d}z/\mathrm{d}t = \partial H/\partial p_z \tag{7''}$$

are equivalent to Newton's equations for the same particle moving under the Lorentz force:

$$\mu\frac{\mathrm{d}^2x}{\mathrm{d}t^2} = e\left[\mathscr{E}_x + \frac{1}{c}\left(\frac{\mathrm{d}y}{\mathrm{d}t}\mathscr{H}_z - \frac{\mathrm{d}z}{\mathrm{d}t}\mathscr{H}_y\right)\right], \tag{8'}$$

$$\mu\frac{\mathrm{d}^2y}{\mathrm{d}t^2} = e\left[\mathscr{E}_y + \frac{1}{c}\left(\frac{\mathrm{d}z}{\mathrm{d}t}\mathscr{H}_x - \frac{\mathrm{d}x}{\mathrm{d}t}\mathscr{H}_z\right)\right], \tag{8''}$$

$$\mu\frac{\mathrm{d}^2z}{\mathrm{d}t^2} = e\left[\mathscr{E}_z + \frac{1}{c}\left(\frac{\mathrm{d}x}{\mathrm{d}t}\mathscr{H}_y - \frac{\mathrm{d}y}{\mathrm{d}t}\mathscr{H}_x\right)\right]. \tag{8'''}$$

Substituting H from (6′) in (7′) and (7″) and carrying out the differentiation, we obtain

$$\begin{aligned}\frac{\mathrm{d}p_x}{\mathrm{d}t} = \frac{e}{\mu c}\Bigg[&\left(p_x - \frac{e}{c}A_x\right)\frac{\partial A_x}{\partial x} + \left(p_y - \frac{e}{c}A_y\right)\frac{\partial A_y}{\partial x} + \\ &+ \left(p_z - \frac{e}{c}A_z\right)\frac{\partial A_z}{\partial x} - e\frac{\partial V}{\partial x}\Bigg].\end{aligned} \tag{9'}$$

From (7″)

$$\frac{\mathrm{d}x}{\mathrm{d}t} = \frac{1}{\mu}\left(p_x - \frac{e}{c}A_x\right),\ \frac{\mathrm{d}y}{\mathrm{d}t} = \frac{1}{\mu}\left(p_y - \frac{e}{c}A_y\right),\ \frac{\mathrm{d}z}{\mathrm{d}t} = \frac{1}{\mu}\left(p_z - \frac{e}{c}A_z\right). \tag{10'}$$

It follows from (10′) that

$$\frac{\mathrm{d}p_x}{\mathrm{d}t} = \mu\frac{\mathrm{d}^2x}{\mathrm{d}t^2} + \frac{e}{c}\frac{\mathrm{d}A_x}{\mathrm{d}t}. \tag{11}$$

Since the value of the vector potential A_x is taken at the point where the charge e is situated, the total time derivative of A_x is

$$\frac{\mathrm{d}A_x}{\mathrm{d}t} = \frac{\partial A_x}{\partial t} + \frac{\partial A_x}{\partial x}\frac{\mathrm{d}x}{\mathrm{d}t} + \frac{\partial A_y}{\partial y}\frac{\mathrm{d}y}{\mathrm{d}t} + \frac{\partial A_z}{\partial z}\frac{\mathrm{d}z}{\mathrm{d}t}. \tag{12}$$

Substituting in (9′) the values $(p_x - eA_x/c)$, $(p_y - eA_y/c)$, $(p_z - eA_z/c)$ from (10′) and $\mathrm{d}p_x/\mathrm{d}t$ from (11), and using (12), we find

$$\mu\frac{\mathrm{d}^2x}{\mathrm{d}t^2} = -\frac{e}{c}\frac{\partial A_x}{\partial t} - e\frac{\partial V}{\partial x} + \frac{e}{c}\left[\frac{\mathrm{d}y}{\mathrm{d}t}\left(\frac{\partial A_y}{\partial x} - \frac{\partial A_x}{\partial y}\right) + \frac{\mathrm{d}z}{\mathrm{d}t}\left(\frac{\partial A_z}{\partial x} - \frac{\partial A_x}{\partial z}\right)\right]. \tag{13}$$

Hence, from Formulae (9) and (10) which relate the fields and the potentials, we have

$$\mu \frac{d^2 x}{dt^2} = e\mathscr{E}_x + \frac{e}{c}\left(\frac{dy}{dt}\mathscr{H}_z - \frac{dz}{dt}\mathscr{H}_y\right), \tag{8''''}$$

i.e. Equation (8′). Equations (8″) and (8‴) are derived similarly.

Thus Hamilton's equations (7′) and (7″) derived from the Hamiltonian (6′) are equivalent to Newton's equations (8′), (8″), (8‴).

The potentials **A** and V can be chosen arbitrarily so long as the required electromagnetic field is obtained from (9) and (10). If we take instead of **A** and V

$$\mathbf{A}' = \mathbf{A} + \nabla f, \qquad V' = V - \frac{1}{c}\frac{\partial f}{\partial t}, \tag{14}$$

where f is an arbitrary function of co-ordinates and time, then $\mathscr{E}' = \mathscr{E}$, $\mathscr{H}' = \mathscr{H}$. Substituting $\mathbf{A}'$ and V' in the Hamiltonian (6′) instead of **A** and V we obtain the equation of motion (13) with $\mathbf{A}'$ and V' in place of **A** and V. By means of (14) it can be shown that the different choice of potentials does not affect Equations (8′)–(8‴). This property of Hamilton's equations is called *gauge invariance*.

It may be noted that, unlike the equations of motion (8′)–(8‴), the Hamiltonian H is changed by the transformation (14). For example, motion in a uniform constant electric field $\mathscr{E}$ along the axis OX can be described by potentials $\mathbf{A} = 0$, $V = -\mathscr{E}x$. Instead of these potentials we can take others in accordance with (14), such as $A'_x = -c\mathscr{E}t$, $A'_y = A'_z = 0$, $V' = 0$. The reader may verify that in both cases Newton's equation for uniformly accelerated motion is obtained, but with the first choice of potentials the Hamiltonian represents the total energy of the particle, while with the second choice it is the kinetic energy of the particle.

VII. Schrödinger's equation and the equations of motion in curvilinear co-ordinates

In Section 27 we have explained why the Cartesian co-ordinate system occupies a special place in quantum mechanics: in this system the measurement of the momentum components p_x, p_y, p_z also gives the value of the kinetic energy. The initial equations of quantum mechanics are therefore usually written in Cartesian co-ordinates. Schrödinger's equation can easily be written in any curvilinear co-ordinate system q_1, q_2, q_3, being known in the Cartesian system. In the latter it is

$$i\hbar \frac{\partial \psi(x,y,z,t)}{\partial t} = -\frac{\hbar^2}{2\mu}\nabla^2 \psi(x,y,z,t) + U(x,y,z,t)\,\psi(x,y,z,t). \tag{1}$$

For simplicity we give only the equation for a single particle in the absence of a magnetic field.[5] When we change from Cartesian to curvilinear co-ordinates, ψ and U will be functions of q_1, q_2, q_3, and the problem is simply that of transforming the Laplacian operator ∇^2. Let the square of the line element ds in curvilinear co-

[5] The general case is discussed in [71].

ordinates q be

$$ds^2 = dx^2 + dy^2 + dz^2 = \sum_{s,k=1}^{3} g_{sk}\, dq_s\, dq_k, \tag{2}$$

where g_{sk} are the components of the metric tensor, and let $D^2 = |g_{sk}|$ be the determinant of the matrix g_{sk}. We also define the elements of the inverse matrix g^{sk} so that

$$g_{s\alpha} g^{\alpha k} = \delta_s^k, \qquad \delta_s^k = 1 \text{ for } k = s, \qquad 0 \text{ for } k \neq s. \tag{3}$$

In (3) we sum over α from 1 to 3.

Then the operator ∇^2 in this notation is[6]

$$\nabla^2 = \frac{1}{D}\left(\frac{\partial}{\partial q_s}\left(D g^{sk} \frac{\partial}{\partial q_k}\right)\right) \tag{4}$$

(with summation over s and k), and accordingly Schrödinger's equation becomes

$$i\hbar \frac{\partial \psi(q_1, q_2, q_3, t)}{\partial t} = -\frac{\hbar^2}{2\mu}\frac{1}{D}\left(\frac{\partial}{\partial q_s}\left(D g^{sk} \frac{\partial \psi(q_1, q_2, q_3, t)}{\partial q_k}\right)\right) + U(q_1, q_2, q_3, t)\,\psi(q_1, q_2, q_3, t). \tag{5}$$

The Hamiltonian operator is

$$\boldsymbol{H} = -\frac{\hbar^2}{2\mu}\frac{1}{D}\left(\frac{\partial}{\partial q_s}\left(D g^{sk} \frac{\partial}{\partial q_k}\cdots\right)\right) + U(q_1, q_2, q_3, t). \tag{6}$$

By taking the Poisson bracket

$$dq^{(s)}/dt = [\boldsymbol{H}, q_s] \tag{7}$$

we obtain the contravariant velocity component $dq^{(s)}/dt$. Multiplying by the mass μ, we have the corresponding momentum component $P^{(s)}$. In order to obtain the covariant momentum component P_s, we transform $P^{(s)}$ by the formula relating contravariant and covariant components:

$$P_s = g_{sk} P^{(k)}. \tag{8}$$

As an example, let us consider spherical polar co-ordinates r, θ, ϕ. In this case

$$ds^2 = dr^2 + r^2 d\theta^2 + r^2 \sin^2\theta\, d\phi^2, \quad g_{11} = 1, \quad g_{22} = r^2, \quad g_{33} = r^2 \sin^2\theta; \tag{9}$$

$$g^{11} = 1, \quad g^{22} = 1/r^2, \quad g^{33} = 1/r^2 \sin^2\theta, \quad D = r^2 \sin\theta. \tag{9'}$$

The Hamiltonian is

$$\boldsymbol{H} = -\frac{\hbar^2}{2\mu}\left[\frac{\partial^2}{\partial r^2} + \frac{2}{r}\frac{\partial}{\partial r} + \frac{1}{r^2 \sin\theta}\frac{\partial}{\partial\theta}\left(\sin\theta \frac{\partial}{\partial\theta}\right) + \frac{1}{r^2 \sin^2\theta}\frac{\partial^2}{\partial\phi^2}\right] + U. \tag{10}$$

[6] See, for example, [32]; [103], p. 47.

The first group of equations (velocity operators) is found as follows. According to (7)

$$\mathrm{d}r/\mathrm{d}t = [\boldsymbol{H}, r], \qquad d\theta/\mathrm{d}t = [\boldsymbol{H}, \theta], \qquad \mathrm{d}\phi/\mathrm{d}t = [\boldsymbol{H}, \phi]. \tag{11}$$

We calculate the first Poisson bracket by noting that

$$r\left(\frac{\partial^2}{\partial r^2} + \frac{2}{r}\frac{\partial}{\partial r}\right) - \left(\frac{\partial}{\partial r^2} + \frac{2}{r}\frac{\partial}{\partial r}\right) r = -\frac{2}{r}\left(\frac{\partial}{\partial r} r \ldots\right),$$

and so the first Poisson bracket (11) gives

$$\mu\, \mathrm{d}r/\mathrm{d}t = -\, i\hbar \frac{1}{r}\left(\frac{\partial}{\partial r} r \ldots\right) = \boldsymbol{P}^{(r)}. \tag{12}$$

For the second Poisson bracket, since

$$\theta \frac{1}{\sin\theta}\frac{\partial}{\partial\theta}\left(\sin\theta \frac{\partial}{\partial\theta}\ldots\right) - \frac{1}{\sin\theta}\frac{\partial}{\partial\theta}\left(\sin\theta\frac{\partial}{\partial\theta}\ldots\right)\theta$$
$$= -\frac{2}{\sqrt{\sin\theta}}\frac{\partial}{\partial\theta}(\sqrt{\sin\theta}\ldots),$$

we have

$$\mu\, \mathrm{d}\theta/\mathrm{d}t = -\frac{1}{r^2}\frac{i\hbar}{\sqrt{\sin\theta}}\frac{\partial}{\partial\theta}(\sqrt{\sin\theta}\ldots) = \boldsymbol{P}^{(\theta)}. \tag{13}$$

Finally, for the third bracket we have simply

$$\mu\, \mathrm{d}\phi/\mathrm{d}t = -\frac{i\hbar}{r^2\sin^2\theta}\frac{\partial}{\partial\phi} = \boldsymbol{P}^{(\phi)}. \tag{14}$$

Changing by means of formula (8) to the covariant components $\boldsymbol{P}_r, \boldsymbol{P}_\theta, \boldsymbol{P}_\phi$, we obtain from (9), (12), (13) and (14)

$$\left.\begin{aligned} \boldsymbol{P}_r = -\, i\hbar\frac{\partial}{\partial r}(r\ldots), \qquad \boldsymbol{P}_\theta &= -\frac{i\hbar}{\sqrt{\sin\theta}}\frac{\partial}{\partial\theta}(\sqrt{\sin\theta}\ldots), \\ \boldsymbol{P}_\phi &= -\, i\hbar\,\partial/\partial\phi. \end{aligned}\right\} \tag{15}$$

We now calculate the second group of quantum Hamilton's equations:

$$\left.\begin{aligned} \mathrm{d}\boldsymbol{P}_r/\mathrm{d}t = [\boldsymbol{H}, \boldsymbol{P}_r], \qquad \mathrm{d}\boldsymbol{P}_\theta/\mathrm{d}t &= [\boldsymbol{H}, \boldsymbol{P}_\theta], \\ \mathrm{d}\boldsymbol{P}_\phi/\mathrm{d}t &= [\boldsymbol{H}, \boldsymbol{P}_\phi]. \end{aligned}\right\} \tag{16}$$

For this it is convenient to put (10) in the form

$$\boldsymbol{H} = \frac{\boldsymbol{P}_r^2}{2\mu} + \frac{\boldsymbol{M}^2}{2\mu r^2} + U(r, \theta, \phi), \tag{17}$$

where $\boldsymbol{M}^2$ is the squared angular momentum operator and $\boldsymbol{P}_r$ the first of the operators

(15). An easy calculation of the Poisson brackets (16) by means of (17) gives

$$\left.\begin{aligned} \mathrm{d}\boldsymbol{P}_r/\mathrm{d}t &= -\boldsymbol{M}^2/2\mu r^3 - \partial U/\partial r\,, \\ \mathrm{d}\boldsymbol{P}_\theta/\mathrm{d}t &= \frac{\cot\theta}{\mu r^2 \sin\theta}\{\boldsymbol{P}_\phi^2 - \tfrac{1}{4}\hbar^2\} - \partial U/\partial\theta\,, \\ \mathrm{d}\boldsymbol{P}_\phi/\mathrm{d}t &= -\partial U/\partial\phi\,. \end{aligned}\right\} \tag{18}$$

Of these three equations, the first and last are the same in form as the corresponding classical Hamilton's equations. The equation for $\boldsymbol{P}_\theta$ contains $\boldsymbol{P}_\phi^2 - \frac{1}{4}\hbar^2$ instead of $\boldsymbol{P}_\phi^2$. The appearance of the term $-\frac{1}{4}\hbar^2$ is due to the existence in quantum mechanics of stable states with $\boldsymbol{M}^2 = 0$, i.e. ultimately to the zero-point energy of quantum systems.

VIII. Conditions on the wave function

In formulating the conditions imposed on the wave function ψ it is most reasonable to begin from the properties of the Hamiltonian $\boldsymbol{H}$, since it is this operator which determines the physical nature of the system. From Schrödinger's equation for ψ and ψ^* we easily derive the equation

$$\int \frac{\partial(\psi^*\psi)}{\partial t}\mathrm{d}v = \frac{1}{i\hbar}\int \psi^*\cdot\boldsymbol{H}\psi\cdot\mathrm{d}v - \frac{1}{i\hbar}\int \psi\cdot\boldsymbol{H}^*\psi^*\cdot\mathrm{d}v = -\int \operatorname{div}\mathbf{J}\,\mathrm{d}v\,, \tag{1}$$

where the expression for the current density $\mathbf{J}$ is the same as that given in Section 29. The condition for the operator $\boldsymbol{H}$ to be self-adjoint is

$$\int \psi^*\cdot\boldsymbol{H}\psi\cdot\mathrm{d}v = \int \psi\cdot\boldsymbol{H}^*\psi^*\cdot\mathrm{d}v\,, \tag{2}$$

and so, for the class of wave functions for which this condition is satisfied, we must have

$$\frac{\mathrm{d}}{\mathrm{d}t}\int \psi^*\psi\,\mathrm{d}v = -\int \operatorname{div}\mathbf{J}\,\mathrm{d}v = -\int J_N\,\mathrm{d}s = 0\,. \tag{3}$$

Let us first take the one-dimensional case, with $-\infty < x < \infty$. We have $\mathrm{d}v = \mathrm{d}x$, $\operatorname{div}\mathbf{J} = \mathrm{d}J_x/\mathrm{d}x$. If at some point $x = x_1$ the potential energy $U(x)$ is discontinuous, this point must be excluded from the integration in (3). Integrating, we have

$$J_x(+\infty) - J_x(x_1 + 0) + J_x(x_1 - 0) - J_x(-\infty) = 0\,. \tag{4}$$

The current density $J_x(\pm\infty)$ must be zero, since otherwise the wave functions would not vanish at infinity and all the integrals would diverge. In discussing self-adjointness the eigenfunctions ψ_L of operators with a continuous spectrum L which do not vanish at infinity must be replaced by the eigendifferentials, which do (cf. Appendix III).

Thus from (4) we have the continuity of current density:

$$J_x(x_1 + 0) = J_x(x_1 - 0)\,. \tag{5}$$

Substituting here J_x from (29.5), we obtain

$$(\mathrm{d}\psi/\mathrm{d}x)_{x_1+0} = (\mathrm{d}\psi/\mathrm{d}x)_{x_1-0}, \tag{6}$$

$$(\psi)_{x_1+0} = (\psi)_{x_1-0}, \tag{6'}$$

i.e. the wave function and its first derivative are continuous.

Let us now consider a three-dimensional problem, with $r = 0$ a singular point of the Hamiltonian operator. At this point Gauss' theorem (3) is again inapplicable, and we must exclude it from the volume of integration by a sphere of small radius R. Then the surface integral in (3) has two parts: one over an infinitely remote surface, in the limit enclosing the whole volume of space, and one over the surface of a sphere whose radius R tends to zero:

$$\lim_{R\to 0} R^2 \int J_R \, \mathrm{d}\Omega + \int_\infty J_N \, \mathrm{d}s = 0, \tag{7}$$

where in the first integral the surface element of the sphere has been put in the form $\mathrm{d}s = R^2 \, \mathrm{d}\Omega$, $\mathrm{d}\Omega$ being an element of solid angle. Since the wave functions (or their eigendifferentials) vanish at infinity, the second integral is zero. Substituting in the first integral $J_R = (i\hbar/2\mu)(\psi \, \partial\psi^*/\partial R - \psi^* \, \partial\psi/\partial R)$ and putting $\psi = u/r^\alpha$, where u is regular as $r \to 0$, we have

$$\lim_{R\to 0} \frac{R^2}{R^{2\alpha}} \int \left(u \frac{\partial u^*}{\partial r} - u^* \frac{\partial u}{\partial r} \right)_{r=R} \mathrm{d}\Omega = 0, \tag{8}$$

which can occur only if $\alpha < 1$. Hence we see that the wave functions certainly cannot become infinite more rapidly than $1/r^\alpha$ with $\alpha < 1$.

A many-valuedness of the wave function can occur if we have cyclic co-ordinates, such as the angle ϕ measured round some axis. Then the angles ϕ and $\phi + 2\pi$ represent the same position in space, and so the probability $\psi^*\psi$, which is an observed quantity, must be a single-valued function of the angle ϕ. This cannot be said *a priori* of the function ψ itself. However, from the properties of the spherical harmonic functions and the equation of continuity (1) it can be shown, as in the above arguments, that the function ψ must be single-valued, since otherwise the operator $\boldsymbol{H}$ may not be self-adjoint.[7] Thus the natural conditions imposed on the wave function as a result of the conservation of particle number (3) reduce finally to the condition that the operator (2) should be self-adjoint.

The self-adjointness of other operators $\boldsymbol{L}$ will depend on their nature, since the class of permissible wave functions is determined by the operator $\boldsymbol{H}$ and the discontinuities which it is allowed to possess.

IX. The solution of the oscillator equation

Problems concerning the determination of the quantum levels of an oscillator lead

[7] See [71], Section 6.

to the equation

$$\psi'' + (\lambda - \xi^2)\psi = 0. \tag{1}$$

We desire to find the finite continuous solutions of this equation.

Let us first consider the asymptotic solution of (1), i.e. that for $\xi = \pm\infty$. These points are singular points of the equation. We put

$$\psi(\xi) = e^{f(\xi)}v(\xi). \tag{2}$$

Substitution in (1) gives

$$v'' + 2f'v' + [f'' + f'^2 + \lambda - \xi^2]v = 0. \tag{3}$$

If the function $e^{f(\xi)}$ is to determine the asymptotic behaviour of $\psi(\xi)$, f must be so chosen that the coefficient $f'' + f'^2 - \xi^2$ is regular at the singular points $\xi = \pm\infty$, i.e. so that the term in ξ^2 vanishes. This gives

$$f(\xi) = \pm\tfrac{1}{2}\xi^2. \tag{4}$$

Thus the solution of Equation (1) can be written in the form

$$\psi(\xi) = c_1 e^{-\frac{1}{2}\xi^2} v_1(\xi) + c_2 e^{\frac{1}{2}\xi^2} v_2(\xi). \tag{5}$$

We are concerned with finite solutions ψ, and therefore take the particular solution $c_2 = 0$:

$$\psi(\xi) = e^{-\frac{1}{2}\xi^2} v(\xi). \tag{6}$$

For the function v we now have the equation

$$v'' - 2\xi v' + (\lambda - 1)v = 0. \tag{7}$$

The point $\xi = 0$ is regular, and v may therefore be sought as a Taylor series

$$v = \sum_{k=0}^{\infty} a_k \xi^k. \tag{8}$$

Substituting (8) in (7) and collecting powers of ξ, we obtain a recurrence formula to determine the coefficients a_k:

$$(k+2)(k+1)a_{k+2} - 2ka_k + (\lambda - 1)a_k = 0, \tag{9}$$

whence

$$a_{k+2} = \frac{2k - (\lambda - 1)}{(k+2)(k+1)} a_k. \tag{10}$$

If the series (8) terminates at the nth term, v will be a polynomial of degree n. Then the solution (6) will be finite, continuous and single-valued throughout the range $-\infty < \xi < \infty$. Such solutions are eigenfunctions of Equation (1). It follows from (10) that the series can terminate only for values of λ such that

$$\lambda = 2n + 1, \qquad n = 0, 1, 2, \ldots. \tag{11}$$

This is Formula (47.6).

The polynomial $v(\xi)$ with coefficients given by Formula (10) with $\lambda = 2n + 1$ is called an *Hermite polynomial*. It is usually denoted by $H_n(\xi)$, and satisfies Equation (7) with $\lambda = 2n + 1$, i.e.

$$H_n'' - 2\xi H_n' + 2nH_n = 0. \tag{12}$$

It is easy to verify that this equation is satisfied by the polynomial

$$e^{\xi^2} \frac{d^n}{d\xi^n} (e^{-\xi^2}).$$

Hence H_n can differ from this polynomial only by a constant factor. The usual definition is

$$H_n(\xi) = (-1)^n e^{\xi^2} \frac{d^n}{d\xi^n} (e^{-\xi^2}). \tag{13}$$

It is not difficult to see that the polynomial (13) has coefficients which satisfy the recurrence formula (10) when $\lambda = 2n + 1$.

The polynomial H_n given in (47.8) differs from (13) by a factor $\sqrt{(2^n n!\ \sqrt{\pi})}$, which is chosen so that the function $\psi_n(\xi)$ is normalised to unity; thus that polynomial is the normalised Hermite polynomial

$$H_n(\xi) = \frac{(-1)^n}{\sqrt{(2^n n!\ \sqrt{\pi})}} e^{\xi^2} \frac{d^n}{d\xi^n} (e^{-\xi^2}). \tag{14}$$

The eigensolution of Equation (1) which corresponds to the eigenvalue $\lambda = 2n + 1$ can now be written

$$\psi_n(\xi) = e^{-\frac{1}{2}\xi^2} H_n(\xi), \tag{15}$$

where $H_n(\xi)$ is the normalised Hermite polynomial (14).

Since the operator which determines Equation (1) is self-adjoint, the functions $\psi_n(\xi)$ must be orthogonal. This is easily confirmed. For the two functions ψ_n and $\psi_{n'}$ we have

$$\frac{d^2\psi_n}{d\xi^2} + (2n + 1 - \xi^2)\psi_n = 0,$$

$$\frac{d^2\psi_{n'}}{d\xi^2} + (2n' + 1 - \xi^2)\psi_{n'} = 0.$$

Multiplying the first equation by $\psi_{n'}$ and the second by ψ_n, subtracting and integrating with respect to ξ, we obtain

$$\int_{-\infty}^{\infty} \left(\psi_{n'} \frac{d^2\psi_n}{d\xi^2} - \psi_n \frac{d^2\psi_{n'}}{d\xi^2}\right) d\xi = -2(n - n') \int_{-\infty}^{\infty} \psi_n \psi_{n'}\, d\xi.$$

The left-hand side is

$$\int_{-\infty}^{\infty} \frac{d}{d\xi}\left(\psi_{n'}\frac{d\psi_n}{d\xi} - \psi_n\frac{d\psi_{n'}}{d\xi}\right)d\xi = \left[\psi_{n'}\frac{d\psi_n}{d\xi} - \psi_n\frac{d\psi_{n'}}{d\xi}\right]_{-\infty}^{\infty} = 0,$$

i.e.

$$\int_{-\infty}^{\infty} \psi_n\psi_{n'}\,d\xi = 0.$$

Integration by parts also shows that

$$\int_{-\infty}^{\infty} \psi_n\psi_n\,d\xi = 1,$$

so that

$$\int_{-\infty}^{\infty} \psi_n\psi_{n'}\,d\xi = \delta_{nn'}, \tag{16}$$

i.e. the functions ψ_n form a set of orthogonal and normalised functions. Any function $\psi(\xi)$ (with restrictions which are here unimportant) can be represented as a series

$$\psi(\xi) = \sum_{n=0}^{\infty} c_n\psi_n(\xi), \tag{17}$$

where

$$c_n = \int_{-\infty}^{\infty} \psi(\xi)\psi_n(\xi)\,d\xi. \tag{18}$$

Let us now consider the properties of the non-normalised Hermite polynomials (13). By means of Cauchy's formula the derivative $(d^n/d\xi^n)\,e^{-\xi^2}$ can be written as an integral along a closed contour:

$$\frac{d^n}{d\xi^n}e^{-\xi^2} = \frac{n!}{2\pi i}\int\frac{e^{-z^2}}{(z-\xi)^{n+1}}dz, \tag{19}$$

where the contour encloses the point ξ. Hence (13) gives

$$e^{-\xi^2}H_n(\xi) = (-1)^n\frac{n!}{2\pi i}\int\frac{e^{-z^2}}{(z-\xi)^{n+1}}dz.$$

Putting $z = \xi - t$, we have

$$\frac{1}{n!}H_n(\xi) = \frac{1}{2\pi i}\int\frac{e^{-t^2+2t\xi}}{t^{n+1}}dt, \tag{20}$$

where the contour encloses the point $t = 0$. Hence

$$e^{-t^2+2t\xi} = \sum_{n=0}^{\infty}\frac{1}{n!}H_n(\xi)\,t^n \tag{21}$$

so that $e^{-t^2+2t\xi}$ is the generating function of $H_n(\xi)$.

This function leads to an important recurrence relation between Hermite polynomials. To obtain this, we differentiate (21) with respect to t:

$$e^{-t^2+2t\xi}(2\xi - 2t) = \sum_{n=1}^{\infty} \frac{1}{(n-1)!} H_n(\xi)\, t^{n-1},$$

i.e.

$$\sum_{n=0}^{\infty} \frac{2\xi}{n!} H_n(\xi)\, t^n - \sum_{n=0}^{\infty} \frac{2}{n!} H_n(\xi)\, t^{n+1} = \sum_{n=1}^{\infty} \frac{1}{(n-1)!} H_n(\xi)\, t^{n-1}. \quad (22)$$

Collecting coefficients of powers of t, we have

$$2\xi H_n(\xi) = H_{n+1}(\xi) + 2n\, H_{n-1}(\xi). \quad (23)$$

Multiplication by ξ, again using (23), gives

$$2\xi^2 H_n(\xi) = (2n+1) H_n(\xi) + \tfrac{1}{2} H_{n+2}(\xi) + 2n(n-1) H_{n-2}(\xi). \quad (24)$$

We multiply this equation by $e^{-\xi^2}$ and change to normalised Hermite polynomials (by multiplying and dividing each polynomial H_m in (23) and (24) by $\sqrt{(2^m m!\, \sqrt{\pi})}$). After cancelling we obtain the following recurrence formula for the wave functions (15):

$$\xi\psi_n(\xi) = \sqrt{(\tfrac{1}{2}n + \tfrac{1}{2})}\, \psi_{n+1}(\xi) + \sqrt{(\tfrac{1}{2}n)}\, \psi_{n-1}(\xi). \quad (25)$$

This gives the integral in Sections 47 and 48: multiplying (25) by $\psi_m(\xi)$, integrating with respect to ξ and using the orthogonality and normalisation of the functions ψ_n (16), we have

$$\int \psi_m \xi \psi_n \, \mathrm{d}\xi = \sqrt{(\tfrac{1}{2}n + \tfrac{1}{2})}\, \delta_{m,\, n+1} + \sqrt{(\tfrac{1}{2}n)}\, \delta_{m,\, n-1}, \quad (26)$$

which corresponds to (48.7).

In the same way, using (25) and the orthogonality relation, we can calculate the integrals of any positive integral power of ξ.

X. An electron in a uniform magnetic field

The Hamiltonian (see Appendix VI, Formula (6)) with our choice of the vector potential **A** (57.1) is

$$H = \frac{1}{2\mu}\left(p_x + \frac{e}{c}\mathscr{H} y\right)^2 + \frac{p_y^2}{2\mu} + \frac{p_z^2}{2\mu}. \quad (1)$$

Hence

$$\left.\begin{aligned} \frac{\mathrm{d}p_x}{\mathrm{d}t} &= -\frac{\partial H}{\partial x} = 0, \qquad \frac{\mathrm{d}p_y}{\mathrm{d}t} = -\frac{\partial H}{\partial y} = -\frac{e\mathscr{H}}{\mu c}\left(p_x + \frac{e}{c}\mathscr{H} y\right), \\ \frac{\mathrm{d}p_z}{\mathrm{d}t} &= -\frac{\partial H}{\partial z} = 0, \end{aligned}\right\} \quad (2)$$

$$\frac{dx}{dt} = \frac{\partial H}{\partial p_x} = \frac{1}{\mu}\left(p_x + \frac{e}{c}\mathscr{H} y\right), \qquad \frac{dy}{dt} = \frac{\partial H}{\partial p_y} = \frac{p_y}{\mu}, \qquad \frac{dz}{dt} = \frac{\partial H}{\partial p_z} = \frac{p_z}{\mu}. \tag{3}$$

Thus

$$p_x = \text{constant} = p_x^0, \qquad p_z = \text{constant} = p_z^0, \tag{4}$$

$$\mu \frac{d^2 y}{dt^2} = -\frac{e\mathscr{H}}{\mu c}\left(p_x^0 + \frac{e}{c}\mathscr{H} y\right). \tag{5}$$

Putting

$$y = Y - cp_x^0/e\mathscr{H}, \qquad \omega_0 = e\mathscr{H}/\mu c, \tag{6}$$

we obtain

$$d^2 Y/dt^2 = -\omega_0^2 Y, \qquad Y = a \sin \omega_0 t + b \cos \omega_0 t, \tag{7}$$

and so

$$y = a \sin \omega_0 t + b \cos \omega_0 t - cp_x^0/e\mathscr{H}. \tag{8}$$

Also

$$\mu \frac{dx}{dt} = p_x^0 + \frac{e}{c}\mathscr{H} y = p_x^0 + \frac{e}{c}\mathscr{H}(a \sin \omega_0 t + b \cos \omega_0 t - cp_x^0/e\mathscr{H}), \tag{9}$$

i.e.

$$x = -a \cos \omega_0 t + b \sin \omega_0 t + x_0. \tag{10}$$

Thus the motion is along the circle

$$(x - x_0)^2 + (y + cp_x^0/e\mathscr{H})^2 = a^2 + b^2,$$

whose centre is $x = x_0$, $y = -cp_x^0/e\mathscr{H}$ and radius $R = \sqrt{(a^2 + b^2)}$. The energy of the motion is independent of p_x^0, which determines only the position of the centre of the circle.

It is evident that this classical derivation is an exact parallel to the quantum derivation given in Section 57.

XI. Jacobi co-ordinates

According to the transformation Formulae (104.3) we have

$$\left.\begin{aligned} \partial\xi_j/\partial x_k &= m_k/M_j, & k &\leqslant j; \\ \partial\xi_j/\partial x_k &= -1, & k &= j+1; \\ \partial\xi_j/\partial x_k &= 0, & k &> j+1, \end{aligned}\right\} \tag{1}$$

with

$$M_j = \sum_{k=1}^{j} m_k \tag{2}$$

the mass of the first j particles. From (1) and (2) we find

$$\sum_{k=1}^{N} \frac{\partial\psi}{\partial x_k} = \sum_{k=1}^{N}\sum_{j=1}^{N} \frac{\partial\psi}{\partial\xi_j}\frac{\partial\xi_j}{\partial x_k} = \sum_{j=1}^{N} \frac{\partial\psi}{\partial\xi_j}\sum_{k=1}^{N}\frac{\partial\xi_j}{\partial x_k}$$

$$= \sum_{j=1}^{N} \frac{\partial\psi}{\partial\xi_j}\left\{\sum_{k=1}^{j}\frac{m_k}{M_j} + \frac{\partial\xi_j}{\partial x_{j+1}}\right\} = \frac{\partial\psi}{\partial\xi_N} = \frac{\partial\psi}{\partial X}, \tag{3}$$

i.e. Formula (104.9). The kinetic-energy operator is calculated similarly:

$$D_x\psi = \sum_{k=1}^{N} \frac{1}{m_k}\frac{\partial^2\psi}{\partial x_k^2} = \sum_{k=1}^{N}\frac{1}{m_k}\sum_{j=1}^{N}\sum_{j'=1}^{N}\frac{\partial^2\psi}{\partial\xi_j\partial\xi_{j'}}\frac{\partial\xi_j}{\partial x_k}\frac{\partial\xi_{j'}}{\partial x_k}. \tag{4}$$

From (1) and (2) we have

$$D_x\psi = \sum_{k=1}^{N}\frac{1}{m_k}\left\{\sum_{j,j'=k}^{N}\sum^{N}\frac{m_k^2}{M_j M_{j'}}\frac{\partial^2\psi}{\partial\xi_j\partial\xi_{j'}} - 2\sum_{j=k}^{N}\frac{m_k}{M_j}\frac{\partial^2\psi}{\partial\xi_j\partial\xi_{k-1}} + \frac{\partial^2\psi}{\partial\xi_{k-1}^2}\right\}$$

$$= \sum_{k=1}^{N}\frac{1}{m_k}\left\{2\sum_{j>j'=k}^{N}\sum^{N}\frac{m_k^2}{M_j M_{j'}}\frac{\partial^2\psi}{\partial\xi_j\partial\xi_{j'}} - 2\sum_{j=k}^{N}\frac{m_k}{M_j}\frac{\partial^2\psi}{\partial\xi_j\partial\xi_{k-1}}\right. + \tag{5}$$

$$+ \sum_{k=1}^{N}\frac{1}{m_k}\left\{\sum_{j=k}^{N}\frac{m_k^2}{M_j^2}\frac{\partial^2\psi}{\partial\xi_j^2} + \frac{\partial^2\psi}{\partial\xi_{k-1}^2}\right\}.$$

The first sum over k in (5) is easily seen to be zero by interchanging the order of summation over k, j and j'. The second sum is

$$\sum_{k=1}^{N}\frac{1}{m_k}\left\{\sum_{j=k}^{N}\frac{m_k^2}{M_j^2}\frac{\partial^2\psi}{\partial\xi_j^2} + \frac{\partial^2\psi}{\partial\xi_{k-1}^2}\right\} = \sum_{j=1}^{N}\sum_{k=1}^{j}\frac{m_k}{M_j^2}\frac{\partial^2\psi}{\partial\xi_j^2} + \sum_{k=1}^{N-1}\frac{1}{m_{k+1}}\frac{\partial^2\psi}{\partial\xi_k^2}$$

$$= \sum_{j=1}^{N}\frac{1}{M_j}\frac{\partial^2\psi}{\partial\xi_j^2} + \sum_{j=1}^{N-1}\frac{1}{m_{j+1}}\frac{\partial^2\psi}{\partial\xi_j^2} \tag{6}$$

$$= \frac{1}{M}\frac{\partial^2\psi}{\partial\xi_N^2} + \sum_{j=1}^{N-1}\left(\frac{1}{M_j} + \frac{1}{m_{j+1}}\right)\frac{\partial^2\psi}{\partial\xi_j^2},$$

i.e.

$$D_x\psi = \frac{1}{M}\frac{\partial^2\psi}{\partial\xi_N^2} + \sum_{j=1}^{N-1}\frac{1}{\mu_j}\frac{\partial^2\psi}{\partial\xi_j^2}, \tag{7}$$

where μ_j is the reduced mass of the $(j + 1)$th particle and the centre of mass of the first j particles:

$$1/\mu_j = 1/M_j + 1/m_{j+1}\,. \tag{8}$$

Since

$$D\psi = (D_x + D_y + D_z)\psi\,, \tag{9}$$

Equation (7) gives (104.4):

$$D\psi = \frac{1}{M}\nabla_N^2\psi + \sum_{j=1}^{N-1}\frac{1}{\mu_j}\nabla_j^2\psi\,. \tag{10}$$

REFERENCES

[1] A. I. Akhiezer and V. B. Berestetskiĭ, *Elements of quantum electrodynamics*, Israel Program for Scientific Translations, Jerusalem, 1963 (*Kvantovaya élektrodinamika*, 2nd edition, Fizmatgiz, Moscow, 1959).

[2] A. I. Akhiezer and I. Ya. Pomeranchuk, *Nekotorye voprosy teorii yadra* (Some problems of nuclear theory), Gostekhizdat, Moscow, 1948.

[3] K. Bechert, *Annalen der Physik* [4] **83** (1927) 905–912.

[4] R. Becker, *Theorie der Elektrizität* (Theory of electricity), vol. II, *Elektronentheorie* (Electron theory), 6th edition, Teubner, Leipzig, 1933.

[5] H. A. Bethe, 'Quantenmechanik der Ein- und Zwei-Elektronenprobleme' (Quantum mechanics of problems involving one and two electrons), *Handbuch der Physik* **24**, 1, 273–560, Springer, Berlin, 1933.

[6] G. A. Biberman, N. Sushkin and V. Fabrikant, *Doklady Akademii nauk SSSR* **66** (1949) 185–186.

[7] F. Bloch, 'Molekulartheorie des Magnetismus' (The molecular theory of magnetism), *Handbuch der Radiologie*, 2nd edition, **6** (II), 375–484, Akademische Verlagsgesellschaft, Leipzig, 1934.

[8] D. I. Blokhintsev, *Journal of physics* **2** (1940) 71–74.

[9] D. I. Blokhintsev, *Journal of physics* **10** (1946) 196.

[10] D. I. Blokhintsev, *Zhurnal éksperimental'noĭ i teoreticheskoĭ fiziki* **17** (1947) 924–929.

[11] D. I. Blokhintsev, *Vestnik Moskovskogo gosudarstvennogo universiteta* (1) (1948) 83–91.

[12] D. I. Blokhintsev, *Soviet Physics: JETP* **2** (1956) 23–25 (*Zhurnal éksperimental'noĭ i teoreticheskoĭ fiziki* **29** (1955) 33–36).

[13] D. I. Blokhintsev, *Mirnoe primenenie atomnoĭ énergii* (Peaceful uses of atomic energy), report at jubilee meeting of the Prague Institute of Technology, 1957.

[14] D. I. Blokhintsev, *Soviet Physics: JETP* **6** (1958) 995–999 (*Zhurnal éksperimental'noĭ i teoreticheskoĭ fiziki* **33** (1957) 1295–1299).

[15] D. I. Blokhintsev, V. S. Barashenkov and B. M. Barabashov, *Soviet Physics: Uspekhi* **2** (1959) 505–525 (*Uspekhi fizicheskikh nauk* **68** (1959) 417–447).

[16] N. N. Bogolyubov (Bogolubov), *Journal of physics* **11** (1947) 23–32.

[17] N. N. Bogoliubov and D. V. Shirkov, *Introduction to the theory of quantized fields*, Interscience, New York, 1959 (*Vvedenie v teoriyu kvantovykh poleĭ*, Gittl, Moscow, 1957).

[18] N. Bohr, *Nature* **107** (1921) 104–107; **108** (1921) 208–209.

[19] N. Bohr, *Atomic theory and the description of Nature*, University Press, Cambridge, 1934.

[20] N. Bohr, *Physical Review* **48** (1935) 696–702.

[21] M. Born and M. Göppert-Mayer, 'Dynamische Gittertheorie der Kristalle' (Dynamical theory of crystal lattices), *Handbuch der Physik* **24**, 2, 623–794, Springer, Berlin, 1933.

[22] L. N. Brillouin, *Les statistiques quantiques et leurs applications* (Quantum statistics and applications thereof), Presses Universitaires, Paris, 1930.

[23] L. N. Brillouin, *L'atome de Bohr* (The Bohr atom), 2nd edition, Presses Universitaires, Paris, 1931.

[24] L. de Broglie, *An introduction to the study of wave mechanics*, Methuen, London, 1930.

[25] A. Carst and R. Ladenburg, *Zeitschrift für Physik* **48** (1928) 192–204.

[26] E. U. Condon and G. H. Shortley, *The theory of atomic spectra*, University Press, Cambridge, 1935.

[27] R. Courant and D. Hilbert, *Methods of mathematical physics*, Vol. 1, Interscience, New York, 1953 (*Methoden der mathematischen Physik*, 2nd edition, Springer, Berlin, 1934).

[28] A. S. Davydov, *Teoriya atomnogo yadra* (Theory of the atomic nucleus), Fizmatgiz, Moscow, 1958.

[29] P. A. M. Dirac, *The principles of quantum mechanics*, 2nd edition, Clarendon Press, Oxford, 1935.

[30] V. P. Dzhelepov, B. M. Golovin, Yu. M. Kazarinov and N. N. Semenov, *Proceedings of the CERN symposium on high-energy accelerators and pion physics* **2** (1956) 115–124.

[31] V. P. Dzhelepov and B. M. Pontekorvo, *Uspekhi fizicheskikh nauk* **64** (1958) 15–54.

[32] A. Éĭkhenval'd, *Teoreticheskaya fizika* (Theoretical physics), Part I, Giz, Moscow, 157–166.

[33] A. Einstein, *Sitzungsberichte der preußischen Akademie der Wissenschaften, Physikalisch-mathematische Klasse* (1925) 3–14.

[34] A. Einstein, B. Podolsky and N. Rosen, *Physical Review* **47** (1935) 777–780.

[35] F. Engels, *Dialectics of nature*, Foreign Languages Publishing House, Moscow, 1954.

[36] V. A. Fok (Fock), *Zeitschrift für Physik* **75** (1932) 622–647.
[37] V. A. Fok, *Nachala kvantovoĭ mekhaniki* (Introduction to quantum mechanics), Kubuch, Kiev, 1932.
[38] V. A. Fok (Fock), *Physikalische Zeitschrift der Sowjetunion* **6** (1934) 425–469.
[39] V. A. Fok, *Izvestiya Akademii nauk SSSR, Seriya fizicheskaya* (1935) 169–187.
[40] R. H. Fowler, *Statistical mechanics,* 2nd edition, University Press, Cambridge, 1936.
[41] J. Frenkel, *Lehrbuch der Elektrodynamik* (Textbook of electrodynamics), Springer, Berlin, 1926.
[42] J. Frenkel, *Wave mechanics: advanced general theory,* Clarendon Press, Oxford, 1934.
[43] Ya. I. Frenkel', *Vvedenie v teoriyu metallov* (Introduction to the theory of metals), Gostekhizdat, Moscow, 1948.
[44] S. É. Frish, *Atomnye spektry* (Atomic spectra), Gtti, Moscow, 1935.
[45] W. Heisenberg, *The physical principles of the quantum theory,* University of Chicago Press, Chicago, 1930.
[46] W. Heisenberg, *Zeitschrift für Physik* **120** (1943) 513–538.
[47] W. Heitler, *The quantum theory of radiation,* 1st edition, Clarendon Press, Oxford, 1936.
[48] E. K. Hyde and G. T. Seaborg, 'The transuranic elements', *Handbuch der Physik* **42**, 205–308, Springer, Berlin, 1957.
[49] A. F. Ioffe (ed.), *Osnovanie novoĭ kvantovoĭ mekhaniki* (The foundation of the new quantum mechanics), Gosizdat, Moscow, 1927.
[50] R. W. James, G. W. Brindley and R. G. Wood, *Proceedings of the Royal Society* A **125** (1929) 401–419.
[51] L. Jánossy and Zs. Náray, *Nuovo cimento* [10] **9**, Supplement (1958) 588–598.
[52] P. Jordan, *Physics of the twentieth century,* Philosophical Library, New York, 1944.
[53] V. N. Kondrat'ev, *Struktura atomov i molekul* (Atomic and molecular structure), USSR Academy of Sciences, Moscow, 1946.
[54] H. Kopfermann and R. Ladenburg, *Zeitschrift für Physik* **65** (1930) 167–188.
[55] R. de L. Kronig, *Band spectra and molecular structure,* University Press, Cambridge, 1930.
[56] N. S. Krylov and V. A. Fok, *Zhurnal éksperimental'noĭ i teoreticheskoĭ fiziki* **17** (1947) 93–107; *Journal of physics* **11** (1947) 112–120.
[57] I. V. Kuznetsov, *Printsip sootvetstviya v sovremennoĭ fiziki i ego filosofskoe znachenie* (The correspondence principle in modern physics and its philosophical significance), Gostekhizdat, Moscow, 1948.
[58] L. D. Landau and E. M. Lifshitz, *Quantum mechanics*, Pergamon, London, 1958 (*Kvantovaya mekhanika,* Gittl, Moscow, 1948).
[59] G. S. Landsberg, *Optika* (Optics), Gittl, Moscow, 1947.
[60] V. I. Lenin, *Materialism and empiriocriticism,* International Publishers, New York, 1927.
[61] M. A. Leontovich, *Statisticheskaya fizika* (Statistical physics), Gostekhizdat, Moscow, 1944.
[62] H. A. Lorentz, *The theory of electrons,* 2nd edition, Teubner, Leipzig, 1916.
[63] L. I. Mandel'shtam and I. E. Tamm, *Izvestiya Akademii nauk SSSR, Seriya fizicheskaya* **9** (1945) 122–128.
[64] M. A. Markov, *Giperony i K-mezony* (Hyperons and K mesons), Fizmatgiz, Moscow, 1958.
[65] J. Howard McMillen, *Reviews of Modern Physics* **11** (1939) 84–110.
[66] R. A. Millikan, *The electron,* 1st edition, University of Chicago Press, Chicago, 1922.
[67] N. F. Mott and H. S. W. Massey, *The theory of atomic collisions,* 2nd edition, Clarendon Press, Oxford, 1950.
[68] J. von Neumann, *Nachrichten von der Gesellschaft der Wissenschaften zu Göttingen, Mathematisch-physikalische Klasse* (1927) 245–272.
[69] K. V. Nikol'skiĭ, *Kvantovaya mekhanika molekuly* (Quantum mechanics of molecules), Gtti, Moscow, 1934.
[70] K. V. Nikol'skiĭ, *Kvantovye protsessy* (Quantum processes), Gittl, Moscow, 1940.
[71] W. Pauli, 'Die allgemeinen Prinzipien der Wellenmechanik' (The general principles of wave mechanics), *Handbuch der Physik* **24**, 1, 83–272, Springer, Berlin, 1933.
[72] W. Pauli, *Reviews of Modern Physics* **13** (1941) 203–232.
[73] R. Peierls, 'Elektronentheorie der Metalle' (Electron theory of metals), *Ergebnisse der exakten Naturwissenschaften* **11**, 264–322, Springer, Berlin, 1932.
[74] O. Piccioni, *Proceedings of the annual international conference on high energy physics at CERN* 1958, 65–89.

[75] G. Placzek, 'Rayleigh-Streuung und Raman-Effekt' (Rayleigh scattering and the Raman effect), *Handbuch der Radiologie,* 2nd edition, **6**(II), 205–374, Akademische Verlagsgesellschaft, Leipzig, 1934.
[76] D. S. Rozhdestvenskiĭ, *Zhurnal" russkago fiziko-khimicheskagoob shchestva, Fizicheskiĭ otdêl"* **42** (1910) 87–97.
[77] É. V. Shpol'skiĭ, *Atomnaya fizika* (Atomic physics), 3rd edition, Gittl, Moscow, 1951.
[78] V. I. Smirnov, *Kurs vyssheĭ matematiki* (Textbook of higher mathematics), Gostekhizdat, Moscow, 1939.
[79] A. Sommerfeld and H. Bethe, 'Elektronentheorie der Metalle' (Electron theory of metals), *Handbuch der Physik* **24**, 2, 333–622, Springer, Berlin, 1933.
[80] A. Sommerfeld and G. Schur, *Annalen der Physik* [5] **4** (1930) 409–432.
[81] M. Stobbe, *Annalen der Physik* [5] **7** (1930) 661–715.
[82] I. E. Tamm, *Osnovy teorii élektrichestva* (Principles of the theory of electricity), Gostekhizdat, Moscow, 1949.
[83] P. S. Tartakovskiĭ, *Kvanty sveta* (Light quanta), Giz, Moscow, 1928.
[84] P. S. Tartakovskiĭ, *Éksperimental'nye osnovaniya volnovoĭ teorii materii* (Experimental foundations of the wave theory of matter), Gtti, Moscow, 1932.
[85] Ya. P. Terletskiĭ, *Zhurnal éksperimental'noĭ i teoreticheskoĭ fiziki* **7** (1937) 1290–1298.
[86] S. V. Vonsovskiĭ, *Uspekhi fizicheskikh nauk* **35** (1948) 514–555, **36** (1948) 30–82, **37** (1949) 1–64, 137–182.
[87] B. A. Vvedenskiĭ and G. S. Landsberg, *Vvedenie v sovremennuyu teoriyu magnetizma* (Introduction to the modern theory of magnetism), Giz, Moscow, 1929.
[88] Wang Kang-Ch'ang, Wang Ts'u-Chien, Ting Ta-Ts'ao, V. G. Ivanov, Yu. V. Katyshev, E. N. Kladnitskaya, L. A. Kulyukina, Nguyen Dinh Tu, A. V. Nikitin, S. Z. Otvinovskiĭ, M. I. Solov'ev, R. Sosnowski and M. D. Shafranov, *Soviet Physics: JETP* **11** (1960) 313-316 (*Zhurnal éksperimental'noĭ i teoreticheskoĭ fiziki* **38** (1960) 426–431).
[89] G. Wentzel, *Quantum theory of fields,* Interscience, New York, 1949 (*Einführung in die Quantentheorie der Wellenfelder,* Deuticke, Vienna, 1943).
[90] E. T. Whittaker and G. N. Watson, *A course of modern analysis,* 4th edition, University Press, Cambridge, 1927.
[91] E. P. Wigner, *Group theory and its application to the quantum mechanics of atomic spectra,* Academic Press, New York, 1959 (*Gruppentheorie und ihre Anwendung auf die Quantenmechanik der Atomspektren,* Vieweg, Brunswick, 1931).
[92] A. P. Zhdanov, *General meeting of the USSR Academy of Sciences,* 1–4 July 1946.

References added in the English translation

[93] J. M. Blatt and V. F. Weisskopf, *Theoretical nuclear physics,* Wiley, New York, 1952.
[94] D. Bohm, *Quantum theory,* Prentice-Hall, Englewood Cliffs, 1951.
[95] E. U. Condon and H. Odishaw (ed.), *Handbook of physics,* McGraw-Hill, New York, 1958.
[96] J. A. Crowther, *Ions, electrons, and ionizing radiations,* 7th edition, Arnold, London, 1938.
[97] G. Herzberg, *Molecular spectra and molecular structure,* Van Nostrand, New York, 1945 and 1950.
[98] J. H. Jeans, *The mathematical theory of electricity and magnetism,* 5th edition, University Press, Cambridge, 1925.
[99] L. Pauling and S. Goudsmit, *The structure of line spectra,* McGraw-Hill, New York, 1930.
[100] L. Pauling and E. B. Wilson, Jr., *Introduction to quantum mechanics,* McGraw-Hill, New York, 1935.
[101] A. S. Ramsey, *Electricity and magnetism,* 2nd edition, University Press, Cambridge, 1945.
[102] P. Roman, *Theory of elementary particles,* 2nd edition, North Holland, Amsterdam, 1961.
[103] J. A. Stratton, *Electromagnetic theory,* McGraw-Hill, New York, 1941.
[104] G. P. Thomson, *The wave mechanics of free electrons,* McGraw-Hill, New York, 1930.
[105] E. C. Titchmarsh, *Introduction to the theory of Fourier integrals,* Clarendon Press, Oxford, 1937.
[106] G. N. Watson, *A treatise on the theory of Bessel functions,* 2nd edition, University Press, Cambridge, 1944.

INDEX